Conservation of Modern Oil Paintings

Klaas Jan van den Berg • Ilaria Bonaduce
Aviva Burnstock • Bronwyn Ormsby
Mikkel Scharff • Leslie Carlyle
Gunnar Heydenreich • Katrien Keune
Editors

Conservation of Modern Oil Paintings

Springer

Editors
Klaas Jan van den Berg
Cultural Heritage Laboratory
Cultural Heritage Agency of the Netherlands
Amsterdam, The Netherlands

Ilaria Bonaduce
Department of Chemistry and Industrial Chemistry
University of Pisa
Pisa, Italy

Aviva Burnstock
Department of Conservation & Technology
Courtauld Institute of Art
London, UK

Bronwyn Ormsby
Conservation Science & Preventive Conservation
Tate Britain
London, UK

Mikkel Scharff
School of Conservation
KADK School of Conservation
Copenhagen, Denmark

Leslie Carlyle
Faculty of Sciences and Technology
Universidade Nova de Lisboa (FCT/UNL)
Caparica, Portugal

Gunnar Heydenreich
TH Köln
Cologne Institute of Conservation Sciences
Köln, Germany

Katrien Keune
Conservation & Science Department, Rijksmuseum
Amsterdam, The Netherlands

ISBN 978-3-030-19253-2 ISBN 978-3-030-19254-9 (eBook)
https://doi.org/10.1007/978-3-030-19254-9

Cover Photograph: Composition (detail), 1956. Jean-Paul Riopelle. Photo: Ida Antonia Tank Bronken/ Nasjonalmuseet, Børre Høstland.

This Springer imprint is published by the registered company Springer Nature Switzerland AG.
The registered company address is: Gewerbestrasse 11, 6330 Cham, Switzerland

Preface

Modern Oil Paints

Oil paints continue to be used by contemporary artists, despite the introduction of a wide range of synthetic media. Although modern oil paints essentially consist of the same medium as paints made before the nineteenth century, the scale of production and formulation has changed significantly. A large number of additives to artists' paints may be present in modern oil paints, leading to a wide variety in their composition. These developments have in some instances led to compromises in the stability of the modern oil paints. The deterioration of works made using modern oil pains may also be due to the way the materials have been used by artists. This combination of material formulation and use and the impact on the condition of works of art is discussed by many authors in this volume.

The ageing process of oil paint has been linked to observed defects in both paintings and paint samples including water sensitivity, efflorescence, yellowing, dripping, crack formation, wrinkling, blooming, blistering, flaking and peeling.[1] Many twentieth-century paintings are particularly vulnerable since they were deliberately not varnished, and as a result, the painting surfaces may have attracted dirt preferentially and be degraded and sensitive to common methods used for cleaning.

These issues are likely to pose problems for conservators and collection keepers in the future. A deeper knowledge of these processes is important to conservation, which must consider both material preservation and aesthetics.

[1]A modern paint damage atlas and assessment tool has been developed partly within the framework of CMOP: the URL for the website will be made public in 2020 through https://www.cultureelerfgoed.nl.

CMOP and ICOP

This volume attempts to address the relationship between the formulation of modern oil paints and the deterioration of works made using oil media. These issues and the implication for conservation of works in modern oil media are discussed in 46 peer-reviewed chapters that were presented at the Conference on Modern Oil Paints, held in Amsterdam in 2018. This conference attracted 325 people from more than 30 countries, with a variety of backgrounds in art history, science, conservation and paint manufacture.

The conference marked the end of the Cleaning Modern Oil Paints (CMOP), a collaborative European research project, which explored the challenges associated with the conservation of the twentieth- and twenty-first-century oil paintings, paint formulation and deterioration. CMOP explored a range of topics associated with modern oil paints, including inventories of degradation phenomena and paint manufacture as well as selection and preparation of model paint samples. Also included were studies of the chemical and physical analysis of degradation of modern oil paints and paintings. Important for conservation practice were the development, testing and evaluation of cleaning systems and case study conservation treatments. CMOP was a Joint Programme Initiative Heritage Plus EU project partly funded by EU/H2020, NWO, AHRC and MIUR.[2]

Both the CMOP Conference and its Proceedings can be seen as a follow-up to the "Issues in Contemporary Oil Paint" Conference held in Amersfoort in 2013 and its ensuing Proceedings (van den Berg et al, 2014).

The chapters in this volume cover a wide variety of topics including aspects of oil paint manufacture, artists' techniques and materials and degradation problems. These material aspects consider the conservation history of the works and the ethics, aesthetics and practical approaches to treatment. Where possible, the artists' viewpoints on the display and conservation of their works are considered. Research includes experiments on model systems as well as case studies of individual paintings and collections of paintings, using analytical techniques that range from the naked eye to advanced and state-of-the-art instrumental methods.

The volume begins with two introductory chapters that are at the heart of the CMOP project. The first chapter, "Challenges in Research: Connecting Scientific Analysis with Conservation Practice", discusses the need for conservators to evaluate scientific research to inform their conservation practice. "Evaluating Cleaning Systems for Use on Water-Sensitive Modern Oil Paints: A Comparative Study" is a keynote chapter summarising the approach and key findings of the CMOP project that relate to the practical surface cleaning of modern oil paint surfaces.

[2] https://www.tate.org.uk/about-us/projects/cleaning-modern-oil-paints-0

Paint Technology

The next six chapters discuss developments in technology of pigment and paint manufacture in the twentieth century. Chapter "Pigment Surface Treatments: Twentieth- and Twenty-First-Century Industrial Techniques and Strategies for Their Detection" presents information pigment surface modifications which have a profound influence on the production, rheology and probably ageing characteristics of paints. "Meet the Future: The Creation of New Pigments" presents the recent introduction of new pigments. The next four chapters provide insights into the paint manufacturer's considerations on the introduction of new pigments in oil paint ("Synthetic Organic Pigments in Royal Talens Oil Paint 1920–1950"), reactivity of pigments and oil and their stability ("Zinc Oxide in Oil-Based House Paint: Insights from a Paint Chemist's Notebook Dated 1949"), new oil-based emulsion paints ("Royal Talens' ETA (Emulsion) Paint") and the consequences of war-time shortage ("The House and Decorative Paint Market in Norway During World War I").

Case Studies: Artists Techniques and Materials, Degradation, Ethical Considerations and Treatment Options

This large section comprises a range of case studies of individual paintings and collections of paintings, describing with varying foci a combination of artists' biographies, techniques, materials and degradation problems, as well as conservation history of the works, ethics, aesthetics and practical conservation approaches.

The chapter "Ellsworth Kelly: The Studio and Beyond" discusses the artist's materials and working practices as well as his changing attitude to the conservation of his works and the consequences for their treatment and preventive conservation.

"Bocour Oil Paints and Barnett Newman", "Pigments of Soviet Artists in the 1950s–Late 1970s", "Alkyd-Oil Enamel Paints in Early Reliefs by Jan J. Schoonhoven", "Aspects of Lovis Corinth's Painting Technique in His Late Work" and "Art Materials and Techniques in Edvard Munch's *Old Man in Warnemünde* (1907)" describe the use of pigments and materials by the artists from different periods and geographic regions in the twentieth century.

"Modern Paints and Conservation at the Clyfford Still Museum" discusses the characteristic degradation and resulting conservation issues of paintings by the artist. Another study that takes into account the artist's voice as well as materials and techniques is "A Study of Knifer's Oil Paintings from the 1960s". "Preservation of Russian Abstract Art of the Second Part of the Twentieth Century" describes some issues of works of modern art that were kept under poor conditions and now pose serious conservation issues and also considers the artists' viewpoints.

A plethora of degradation problems related to materials and techniques used by a range of artists are discussed in "Oil Paint Straight from the Tube: Paint-Specific Deterioration in Works by Alexis Mérodack-Jeaneau (1910–1913)" and "Picasso

1917: Failure Mechanisms in the Paint Layers of Four Artworks". Mechanical failure in paint and ground layers and their causes are examined in "Delaminated Areas in a Black Oil Painting from 1960 by Pierre Soulages", "Paint Delamination in *C'est grace à nous* by Asger Jorn" and "Use and Degradation of Zinc White in *The Woodcutters* by Bart van der Leck" as well as in "Extreme Delamination on *Beach Scene* by J.E.H. Akkeringa (1861–1942). The deterioration of paintings characterised by materials protruding through the paint surface is discussed in "The *Speckles Syndrome* Affecting the Late 1920s Oil Paintings by René Magritte".

The degradation of paints and potential conservation solutions are discussed in "Efflorescence on Edwin Austin Abbey's *The Spirit of Light*" caused by metal soaps on the paintings' surface. "Softening and Liquefying Oil Paint on *Womanizer* by Alex Janvier" and "Deterioration and Treatment of Some Late Paintings by Paul-Émile Borduas" also discuss the effect of sometimes detrimental previous conservation treatments on the present condition.

The description of typical and conservation challenges and the way conservators may deal with the practical, aesthetic and ethical issues related to patina, surface dirt and/or degradation phenomena are discussed in "Some Considerations when Cleaning Robert Ryman's Oil Paint(ings)", "Challenges of Surface Cleaning Paintings by Asger Jorn (1914–1973)" and "Decision-Making Processes Regarding the Treatment of Modern Oil Paintings (1950s–present) exhibiting paint dripping and oil exudates".

Model Studies

Six chapters focus on analytical approaches to the study of the chemical and physical degradation of oil paint and the effects of conservation treatment. Chapter "Investigations into the Mechanical Properties of Commercial Artists' Oil Paint by Dynamic Mechanical (Thermal) Analysis (DMA), Nanoindentation (UNHT) and Dynamic Vapour Sorption (DVS)" examines physical/mechanical properties of small oil paint samples with a range of state-of-the-art techniques. "A Preliminary Investigation into the Behaviour of Modern Artists' Oil Paints in a Hot and Humid Climate" presents information on the influence of tropical and moderate climates on paints, while the influence of high temperatures on curing and ageing of oil paints is studied in "Inside the Forger's Oven: Identification of Drying Products in Oil Paints During and After Accelerated Drying with Increased Temperatures". Chapter "The Influence of Metal Stearates on the Water Sensitivity of Modern Oil Paints" describes some of the influence of metal stearate additives on the polymeric/ionomeric network of the oil paints and related solvent sensitivity of oil paints. Chapter "The Rate of Solvent Action on Modern Oil Paint" discusses fundamental aspects of solvent interaction with modern oil paint model systems.

Conservation Treatment

Amongst the most challenging practical conservation interventions is surface cleaning of solvent-sensitive paint surfaces. “Water Sensitivity in Modern Oil Paintings: Trends in Phenomena and Treatment Options” and “Modern Oil Paintings in Tate’s Collection: A Review of Analytical Findings and Reflections on Water Sensitivity” summarise the results of studies of the extent of water sensitivity of modern paintings to water and possible options for treatment. Potential alternative solutions are presented in chapters “Analytical Study to Monitor the Effectiveness of a Combined Liquid-Dispensing and Micro-aspiration System for the Cleaning of Modern Oil Paintings”, “Analysis of Cleaning Efficacy and Clearance of Silicone-Based Pickering-Type Emulsions Used in the Cleaning of Water-Miscible Oil Paints”, “Improving the Surface Cleaning of Water-Sensitive Oil Paint by Use of Alternative Application Methods” and “Designing Mock-Up Paintings for a Study of Novel Surface Cleaning Techniques for Munch’s Unvarnished Aula Paintings”. Chapter “Ketone Resins Varnishes on Canvas Paintings from the Collection of the Munch Museum” presents work on paintings by Munch, while “Evolon® CR Microfibre Cloth as a Tool for Varnish Removal” reviews a method for varnish removal without mechanical action with a novel microfibre cloth. Finally, chapter “Can Cellulose Beads Save the *Circling of the Planets*? Cellulose-Based Consolidating Filler to Stabilise Lifted Brittle Flakes on a Large Painting” presents a new filler composition to consolidate a fragile, extensively flaking painting by Giacometti.

This volume aims to combine technical evidence and the conservators’ perspective on the challenges related to conservation of modern oil paintings and to provide a basis for decisions related to treatment, storage and presentation of paintings in oil media.

We thank all our colleagues who contributed to this volume. We are grateful for the national agencies involved in the JPI Heritage Plus EU for funding the CMOP project and the Rijksmuseum for hosting the Conference on Modern Oil Paints. NWO, Rijksmuseum, AKZO Nobel and the Cultural Heritage Agency of the Netherlands are gratefully acknowledged for financially supporting the Conference on Modern Oil Paints and its Proceedings.

Amsterdam, The Netherlands — Klaas Jan van den Berg
Pisa, Italy — Ilaria Bonaduce
London, UK — Aviva Burnstock
London, UK — Bronwyn Ormsby
Caparica, Portugal — Leslie Carlyle
Köln, Nordrhein-Westfalen, Germany — Gunnar Heydenreich
Copenhagen, Denmark — Mikkel Scharff
Amsterdam, The Netherlands — Katrien Keune
17 April, 2019

Reference

Van den Berg KJ, Burnstock A, de Keijzer M, Krueger J, Learner T, de Tagle A, Heydenreich G Eds (2014) Issues in Contemporary Oil Paint, Springer International, Switzerland.

Contents

The original version of this book was revised: The affiliation of editor Prof. Gunnar Heydenreich has been updated. The correction to this chapter is available at https://doi.org/10.1007/978-3-030-19254-9_46

Chapter 1
Challenges in Research: Connecting Scientific Analysis with Conservation Practice

Aviva Burnstock, Klaas Jan van den Berg, Judith Lee, and Bronwyn Ormsby

1.1 Introduction

One of the key aims of the CMOP research program is to inform conservation practice for the treatment and display of modern oil paintings. The focus of research has been on the causes of water sensitivity based on molecular level studies of the curing and chemical changes on ageing of modern oil paints, and the removal of dirt from water sensitive unvarnished paintings. These studies have involved analysis of the composition and chemistry of manufactured modern oil paints and samples from paintings and testing some novel practical approaches to treatment. This paper aims to provide an introduction for conservators who are required to evaluate scientific research to inform their conservation practice. The range of methods and approaches to research in selected recent studies is highlighted together with suggestions about how the results of molecular level material studies may be used by conservators to guide more specific and appropriate treatment choices.

A. Burnstock (✉)
Courtauld Institute of Art, London, UK
e-mail: Aviva.Burnstock@courtauld.ac.uk

K. J. van den Berg
RCE, Utrecht, The Netherlands
e-mail: van.den.Berg@cultureelerfgoed.nl

J. Lee · B. Ormsby
Tate, London, UK
e-mail: Judith.lee@tate.org.uk; Bronwyn.ormsby@tate.org.uk

K. J. van den Berg et al. (eds.), *Conservation of Modern Oil Paintings*,
https://doi.org/10.1007/978-3-030-19254-9_1

1.2 Material Characterisation of Paintings

Fundamental research that aims to provide guidelines for conservation is challenged by the material complexity of paintings. The variety of materials, techniques and physical histories of works of art have a pronounced influence on their condition. Therefore, molecular characterisation of materials from paintings requires consideration of the presence of combinations of materials and their interactions, ageing properties and past and present chemical and physical environment. Despite these issues of interpretation of complex data, combined methods for technical research has led to significant advances in understanding a range of deterioration phenomena in paintings, notably the characterisation of metal soaps, salts and fatty acid efflorescence in oil paintings.

The most useful studies for conservators combine analytical characterisation of specific material deterioration based on molecular or spectral indicators linked with visual and microscopic evidence of the surface phenomena. The use of conventional light microscopy in these studies is critical for conservators who are familiar with this technique, while images of paintings or paint samples produced using Hirox, electron microscopy and other high-resolution techniques such as atomic force microscopy require more detailed interpretation of the practical implications of the results. Concomitantly, few conservators have access to resources for molecular analysis of organic materials, and thus rely on generalisations based on the results of analytical investigations that are often limited to case studies. The relationship between the results of material analysis and recommendations for treatment may influence conservation decision making and practical approaches to treatment. For example, deterioration of the paint surface that includes the accumulation of polar free fatty acids at the surface could indicate sensitivity to protic solvents [2, 8]. Alternatively, surface whitening caused by migration and aggregation of fatty acids to the paint surface that can take various forms is typically removable using solvents or if possible, dry polishing with a silk cloth [5]. Where surface crusts and whitening caused by the formation of metal soaps compromise reading of the composition, an approach may be to reduce the soaps using chelating agents with a preference for the metal to be removed and a low preference for chelation of the underlying paint [11]. Alternatively, soaps that have formed as part of the ageing of the work might be best left alone and the visual impact accepted or considered in the approach to retouching. In all these cases, molecular material characterisation has been critical for guiding treatment options, while the skill and judgment of the conservator remains key to the success of the treatment.

In relation to cleaning unvarnished paintings, fundamental analytical studies based on single case studies of paintings may provide data of limited general value, given the variation in artists' materials and technique, and the deterioration and condition of the works. Cleaning case studies typically report the use of an approach to treatment that removed unwanted surface material without removing paint, and few studies include a critical evaluation of the risks and choice of reagent and method of application that are important criteria for conservation practice.

1.3 Paint Samples and Reconstructions

Recent research on the chemistry of modern oil paints has benefitted from access to historical paint samples such as the set from the leading British brand Winsor & Newton that include paint-outs of every batch of oil paint made at the factory.[1] These samples are dated and naturally aged and provide a starting point for comparison with similar materials used in paintings. The composition and surfaces of the Winsor & Newton and other brands of manufacturers paint samples have been characterised using a combination of methods, including FTIR, GC-MS and electrospray MS and light microscopy combined with SEM-EDX for elemental analysis of the composition of oil paint and surface examination [6, 8].

The results of the analysis of manufactured paints and naturally aged paint outs has informed the production of new samples of known standardised composition, with the aim of identifying the aspects of paint formulations that affect curing, drying, changes on aging, and the cleaning of modern oil paints. As modern manufactured paints include additives to modify and homogenise the paints in a range, there are physical and chemical interactions that may contribute to their deterioration and resulting sensitivity. The interpretation of analytical studies of modern oil paints has been based on a model of oil paint that contains glycerides that oxidise and cross link to form a polymeric network, whilst in parallel, hydrolysis and oxidative scission reactions form free carboxylic acids, and saponification reactions form metal soaps [13]. These processes may be influenced by pigments that incorporate reactive or mobile metal ions, or those that do not contribute toward the formation of an ionomeric network. Based on a significant body of analytical research, this model was accepted some time ago and new analytical findings may help to refine the model.

To enable systematic study of the deterioration or the effects of different reagents on paint films, paint samples of homogeneous consistency have been produced and then artificially aged to try to mimic changes in the films that might occur under natural conditions. Some of the historical samples and experimental reconstructions of modern artists' oil paints produced for study have closely replicated the surface characteristics observed on paintings. For example, the formation of skins of medium and medium separation in ultramarine-containing paints has been consistently observed both in paintings and in experimental samples. These findings are useful to conservators together with the results of studies that medium skins may be associated with sensitivity to water and protic solvents caused by the formation of polar degradation products at the paint surface [2, 7, 8]. Some paints develop polar

[1] The paint swatches/samples were donated to Tate's science department by Ian Garrett, former Technical Director of Winsor and Newton, on the closure of the W&N factory at Wealdstone in North London in 2011. They include paint-outs of every batch of oil paint made at the factory applied to canvas or card. Each patch of paint is dated and the purpose of each paint out was to test for drying time. The painted areas are approximately 5 cm x10cm and therefore represent valuable historical records of paints made at the factory to be used intelligently and sparingly in analytical and cleaning investigations.

surfaces by hydrolysis and oxidation indicated by the formation of relatively high proportions of free diacids. Other polar species have been characterised by GCMS, which together explain the sensitivity of the surface to polar cleaning reagents [6].

Although it is important to consider just how similar artificially aged test films are to the materials used in works of art, this experimental methodology is useful for direct comparison and evaluation of different cleaning methods. Where cleaning studies aim to compare the efficacy of different reagents for dirt removal, some studies have included the application of a standard formulation of artificial surface dirt to paint samples, either before or after ageing [3]. Although this artificial dirt has a different relationship to the paint surface to dirt on paintings, the method provides the opportunity to compare the action of different reagents and means of application. Nonetheless, it is important to reflect on the scientific challenge of creating representative synthetic dirt and to consider how closely the artificial material resembles naturally acquired soiling of a range of materials that represent painted surfaces. There is no ideal or universal dirt and its deposition on paint of different media, pigmentation and texture may be the most significant variable in a cleaning study. It is therefore important to read the methods section as well as the conclusions of a research paper to usefully evaluate and contextualise the findings.

As historical samples of manufactured paint are a precious resource, the extent to which these samples can be used for testing is limited. In studies that employ these samples the small size of cleaning tests may not replicate the cleaning test methods used by conservators on paintings. An important factor that became clear in the CMOP studies is that while the choice of reagent and method of application was important in cleaning, the individual hand of the conservator was a critical variable in standardised cleaning studies.

1.4 Defining Aims for Treatment

Unvarnished paintings from the Tate collection examined as case studies were studied but not treated as part of the CMOP research, however the material analysis, together with consideration of the history and context of these paintings provide criteria that could inform their future treatment that may include surface cleaning [1]. One set of criteria may be considered for paintings at Tate, while works displayed or stored in a great range of other environments and contexts may have other criteria for treatment. Additionally, the condition of paintings made using similar materials will be influenced by their present context and physical history. An awareness of these variables that influence the condition of paint used for works of art is clearly essential in the interpretation of the results of material analysis based on artificially aged paint samples. To make the transition from experiment to practical reality it is important to characterise the surface of paintings to ascertain the condition of the paint surface, the presence of coatings, and the physical and chemical relationship between the paint and superficial or imbibed dirt.

Armed with an understanding of the materials and technique and condition of the painting the conservator can better evaluate the risks for treatment. First to decide whether cleaning is necessary. If so, then this may be followed by empirical testing to ascertain a method for effective and even removal of dirt from all passages of paint, that may necessitate working within the limitations of removing dirt from paint surfaces with varying degrees of sensitivity to water or other cleaning reagents. Central to the interpretation of published studies is the underlying assumption that before considering treatment it is important to ascertain whether the paint is bound in oil paint or other media. The condition of the paint surface – impasted, well bound, medium rich, porous or under-bound paints will require different methods and approaches to practical treatment. It may be an advantage to ascertain whether there are reactive metal ions or other additives that might affect the stability of the paint. Of critical importance is an assessment of the optical and physical condition of the surface and the risks posed by the treatment approach, including the choice of reagent and method of application.

1.5 Ethical and Practical Issues for Cleaning Unvarnished Water Sensitive Paintings

Removing surface dirt from water sensitive surfaces presents practical challenges that are influenced by the cause and extent of sensitivity, and the relationship between dirt and the paint surface. Restricting cleaning to a partial removal of the dirt imbibed in a sensitive polar medium skin may be considered where loss of original paint or changes in surface gloss are a risk. The separation of dirt from paint in which highly water-soluble magnesium sulphate heptahydrate (epsomite) crystals have formed presents different practical challenges. Epsomite has been characterised in both historical and aged paint samples and in oil paintings containing a range of pigments that were probably exposed to sulphurous environmental pollutant gasses [12]. The identification of epsomite in modern oil paintings is significant for conservators. The crystals are highly soluble an affected by damp swabs of water or saliva; disruption of a few microns of medium skin around them exposes pigment to mechanical pick up.

Treatment of *Still life with a Classical Bust* d.1936 by Mark Gertler[2] provides an example of how cleaning practice has been informed by fundamental research. The presence of a partial varnish coating, embedded dirt, extremes of paint texture, and passages of varying sensitivity to water and the formation of epsomite are all represented in this case study. The surface was covered with particulates and embedded dirt, the result of over 60 years of unglazed display. Visually disturbing white crusts of epsomite had formed at the perimeters of compositional elements, and in the skin

[2] Painted on paper adhered to Masonite, Southampton City Art Gallery

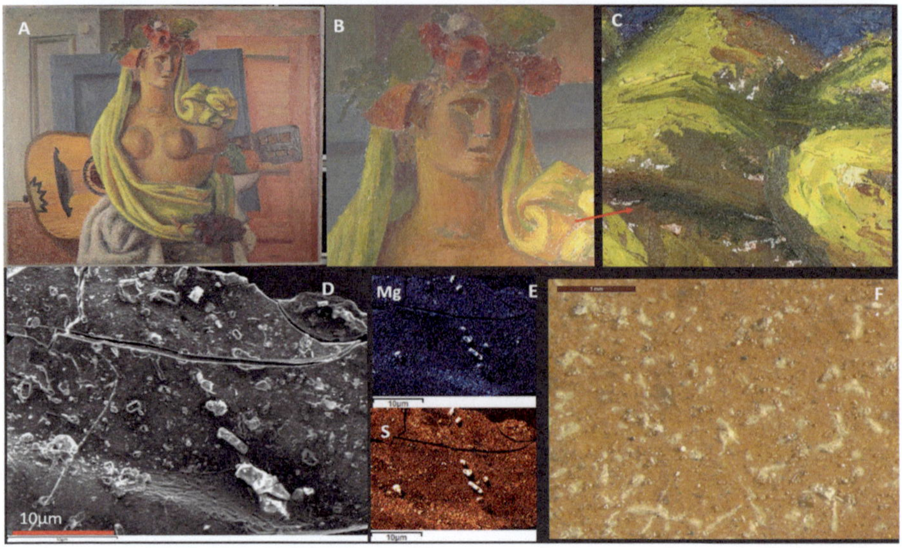

Fig. 1.1 (**a**) *Still life with a Bust 1936* by Mark Gertler; (**b**) Detail of head; (**c**) detail of drapery and white epsomite crusts, (**d**) SEM image of surface of yellow paint; (**e**) Element maps for Mg and S indicative of epsomite crystals at the paint surface; (**f**) light micrograph of yellow paint surface with epsomite crystals

of medium at the surface of vermilion, viridian and cadmium yellow paints (Fig. 1.1).

The perimeter epsomite crusts were relatively easy to remove with small tightly rolled locally applied saliva dampened swabs. The crystals embedded in the paint surface were large enough to be visible using the light microscope. The paints showed slightly different degrees of sensitivity to swab rolls of water; the vermilion containing red was the most sensitive, the yellow slightly less, and the green passages were slightly more tolerant to water before pigment was picked up.

Defining the goals for treatment was important in this case, given the challenges of removing surface dirt evenly from areas of different sensitivity. Consideration of the current condition of other contemporary works by the artist provided an indication of Gertler's intention, exemplified by his colourful works being well preserved by glazed framing in other collections. This raises the possibility that dirt can exacerbate the deterioration of his paint surfaces[3]. So, a decision was taken to find a way to address the practical challenge of evenly removing surface dirt from the whole painting. This was achieved using a range of reagents and means of application to clean different passages of paint.

[3] This has not been discussed widely in the literature, however the presence of carbon in surface dirt that may absorb gaseous agents of deterioration together with deposits of other reactive species in surface dirt present plausible factors for triggering surface deterioration in paintings.

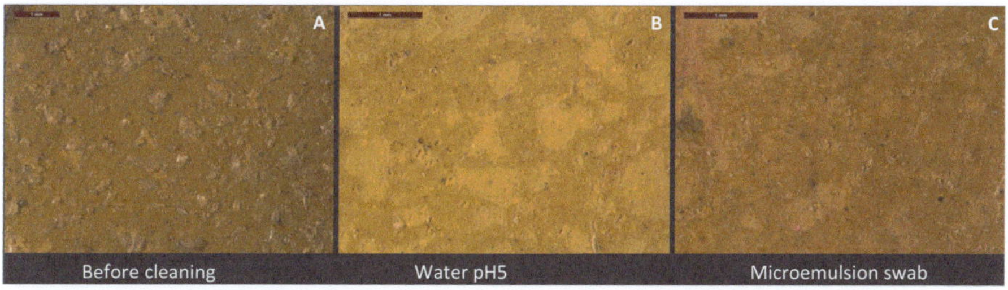

Fig. 1.2 Details of paint from the yellow drapery (**a**) before, (**b**) after cleaning with water at pH 5 using a swab which removed dirt and epsomite; (**c**) after cleaning with microemulsion ME3b-14 D1 H2O applied by swab, which removed some dirt and left epsomite on the surface

Details of the surface before treatment (Fig. 1.2) show the presence of epsomite embedded in a medium skin. The SEM image (Fig. 1.1d) shows that the surface dirt covers the crystals and the paint surface. While the dirt and epsomite were effectively removed using a swab of water adjusted to pH 5, holes in the skin of medium were left.

Another area cleaned with a swab and a microemulsion[4] removed some dirt and left most of the crystals in situ. The microemulsion was selected from a range of reagents used for cleaning acrylic paintings that are vulnerable to swelling in water [9]. These cleaning systems, which offer controlled action of water in a non-polar environment, have been tested for cleaning water sensitive oil paint samples [3] and paintings [4]. The microemulsion could be applied using a pre-rinsed make up sponge. The sponge was useful for cleaning ridged impasted paint where cotton swab snagging was a problem. The microemulsion was cleared from the paint surface with Shellsol D40, which that was slow to evaporate, but other non-polar low aromatic petroleum distillate solvents could be used, such as Shellsol D38 or iso octane, although their use has not yet been systematically studied[5]. The effectiveness of clearance of series 3b micro emulsions [9] will depend on the solubility of the NaDOSS [footnote 4] in the solvent[6]. The general recommendation is that the surfactant phase solvent and the clearance solvent should be the same. However, the effective clearance of the surfactant in the formulation is not known, nor have the consequences of leaving the surfactant on the paint surface been investigated for oil paints. Important challenges for the safe use of non-volatile reagents for cleaning are to evaluate the effects of residues on the painting in both the short and long term. This requires an experimental approach that can identify chemical changes that may occur even where small volumes of residue are present, and an assessment of the deterioration of the reagent in combination with a range of complex materials that

[4] ME3b-14 D1 H2O. This contained by weight: 40% Shellsol D40, 20% pH adjusted water 4.5 and 40% of a 65% solution of sodium dicotyl sulfosuccinate (Otherwise known as NaDOSS) in Shelsol D40.

[5] Shellsol D38 and D40 suggested by Alan Phenix.

[6] Ormsby suggested that the effectiveness of clearance may be related to the rate of dissolution of NaDOSS in the solvent, however this may be influenced by paint/pigment specific interactions.

are used for painting. The effectiveness of clearance of the reagent from the paint and the possibility of capillary penetration of cleaning reagents must also be considered. There are many practical variables associated with the application of gels and micro emulsions to a range of painted surfaces that require special consideration and focussed work with experimental designs and methods of analysis that are capable of specific characterisation of small amounts of material.

The *Still Life* by Gertler included areas of unvarnished paint that were not affected by epsomite formation but were nonetheless sensitive to water swabbing, suggestive of the formation of diacids characteristic of polar paint surfaces. The green passages of viridian-containing paint were patchily coated with a thin resin varnish, that afforded the paint surface some protection. It was possible to clean these passages with pH-adjusted water, using a swab or, in more textured areas of paint, a sponge. It is common to find that different passages of oil paint in a single work show different physical and chemical degradation phenomena. For example, epsomite formation, increased polarity of the paint surface or under-bound pigment, each of which demand a range of approaches to treatment. Other concerns for the choice of reagents and methods for surface cleaning include evaluation of changes in surface gloss and the effect of reagents including clearance solvent, as well as residues, on the appearance and condition of the original paint after treatment.

A significant number of cases of water-sensitivity have been studied, including large works comprising oil-rich paints, sometimes applied thickly, that lack metal ions which could react with fatty acids to form soaps, thus stabilising the paint film. In these and other works, glossy medium-rich surfaces present the greatest challenge for surface cleaning and the greatest risk of gloss change whatever reagent is used. In these extreme cases prudent goals for preventive conservation might include display in glazed frames, though this may challenge the wishes of both the living artist and the curator. Varnishing to protect the paint, which has been done in a few cases, can radically change the play of texture of the surface and create further problems for cleaning in the future.

CMOP studies have investigated the materials and methods for practical surface dirt removal from naturally aged unvarnished British oil paintings that exhibit a wide range of techniques and reasons for their sensitivity to water [4]. This included where appropriate, polysaccharide gum and rigid gels and microemulsions used for cleaning acrylic painted surfaces. While a case study approach may be limited in terms of the inferences it can yield for the treatment of other works, it has afforded the opportunity to compare realistic practical cleaning tests used by conservators. Results suggested that the choice of reagent and application method was important, but the most critical variable in the success of a cleaning treatment is the individual hand and judgment of the conservator.

1.6 Conclusions and Further Research

Research on the treatment of modern unvarnished oil paintings has characterised some of the causes of water sensitivity and provided guidelines for methods for surface cleaning [4]. However, the challenges posed by the removal of aged polar coatings from water sensitive oil paints remains to be investigated, for example for works that are varnished with polar resin coatings over passages of epsomite or polar paint surfaces. The need to address the non-uniformity of painting surfaces in cleaning and other treatments is underlined in these studies which have included oil paintings exhibiting combinations of degradation phenomena such as water sensitive oil paint coated with oil varnish, metal soaps, fatty acid efflorescence and salts together with the deposition of dirt.

While oil paints manufactured by Winsor & Newton and Royal Talens have been investigated in some detail, the chemistry of other brands of oil paint has received less attention in relation to trends in water sensitivity and the formation of brand-specific deterioration. The identification of characteristic markers for paints could be useful for the prediction of patterns of deterioration in paintings.

Where the use of cleaning systems that include surfactants and other non-volatile materials are the only option for treatment, it is important to investigate the consequences of reagent residues on artists' oil paints, in the short and longer term. While residues of microemulsions have been investigated for acrylic paints [10] further study is critical for the provision of guidelines for their safe use to clean modern oil paintings.

The underlying principle of the research to date is that an understanding of the materials is needed and is essential to inform conservation practice. While regular access to analytical resources is still limited for most conservators, the new studies in this volume that provide links between phenomena that can be identified visually or empirically and their analytical characterisation, demonstrate that the goal to connect scientific analysis with conservation practice is essential for those responsible for the conservation and display of modern oil paintings.

Acknowledgements CMOP partners, Lindsey Doyle, Tomasz Goral Natural History Museum, London; William Luckhurst Kings College London, Jennifer Gonzales Corujo.

References

1. Bay L, Burnstock A, Lee J, Ormsby B, Van den Berg KJ (2017) Water sensitivity of modern oil paintings in ICOM-CC 18th Triennial Conference Preprints, Copenhagen, 4–8 September 2017, ed. J. Bridgland, art. 1302. Paris: International Council of Museums pp 1–12.
2. Burnstock A, Lee J, Van den Berg KJ, Ormsby B (2016) Water Sensitivity of Modern Oil Paint Films, *Colour and Conservation*, Milan congress November 2015 *Association CESMAR7 with Il Prato Publishing,* VII Congress C & C2015 Proceedings, both in printed and eBook versions, *(www.salonedelrestauro.com)* pp 66–76.

3. Chung JY, Ormsby B, Lee J, Burnstock A, Van den Berg KJ (2017) An investigation of options for surface cleaning unvarnished water-sensitive oil paints based on recent developments for acrylic paints in ICOM-CC 18th Triennial Conference Preprints, Copenhagen, 4–8 September 2017, ed. J. Bridgland, art. 1302. Paris: International Council of Museums pp 1–12.
4. Gillman M, Lee J, Ormsby B, Burnstock A (2019) Water-sensitivity in modern oil paintings: Trends in phenomena and treatment options, These proceedings, Chapter 38
5. Hinde L, Van den Berg KJ, de Groot S and Burnstock A (2011) Characterisation of surface whitening in twentieth-century European paintings at Dudmaston Hall, United Kingdom in *ICOM-CC International Council of Museums Conservation Committee 16th Triennial Conference* Lisbon: ICOM-CC paper 1310.
6. Lee J PhD thesis (2018) Water sensitivity in modern oil paints, Courtauld Institute of Art, London.
7. Lee J, Ormsby B, Burnstock A, Schilling M, Van den Berg KJ (2017) The chemical characterisation of water-sensitive modern oil paint swatches by Winsor & Newton in ICOM-CC 18th Triennial Conference Preprints, Copenhagen, 4–8 September 2017, ed. J. Bridgland, art. 1604. Paris, International Council of Museums pp 1–12.
8. Lee J, Ormsby B, Burnstock A, Van den Berg KJ (2019) Modern oil paintings in Tate's collection: a review of analytical findings and reflections on water sensitivity, These proceedings, Chapter 39
9. Ormsby B, Keefe MH, Phenix A, Von Aderkas E, Learner T, Tucker C, Kozac C (2016) Mineral spirits-based microemulsions: A novel cleaning system for acrylic and other modern artists' paints.' *Journal for the American Institute for Conservation* 55(1) pp 12–31.
10. Ormsby BA, Willneff, EA, Clark MB, Lopez J, Keefe MH, Phenix A (2016a) *Cleaning acrylic paints: mineral spirits-based microemulsions and the question of residues.* In: Selva Bonino VE (ed.) CESMAR7 – Colore y Conservazione 2015. 7th International Conference Color and Conservation 2015, 13–14 Nov 2015, Milan, Italy Il Prato. ISBN 978-88-6336-328-9, pp 91–100.
11. Sawicka A, Burnstock A, Van den Berg KJ, Izzo F, Keune K, Boon JRJ, Kirsch K (2014) Metal soap efflorescence in contemporary oil painting in *Issues in Contemporary Oil Paint*, Van den Berg KJ, Burnstock A, de Keijzer M, Krueger J, Learner T, de Tagle A, Heydenreich G Eds, Springer International Publns Switzerland (2014), ISBN978-3-319-10100-2 pp 311–332.
12. Silvester G, Burnstock A, Megens L, Learner T, Chiari G, Van den Berg KJ (2014) A cause of water-sensitivity in modern oil paint films: The formation of magnesium sulphate', *Studies in Conservation* Vol 59 (1) pp 38–51.
13. Van den Berg JDJ, Van den Berg KJ, Boon JJ (1999) Chemical changes in curing and ageing oil paints. In: Bridgland J (ed) ICOM-CC 12th triennial meeting. James and James, Lyon, pp 248–253.

Chapter 2
Evaluating Cleaning Systems for Use on Water Sensitive Modern Oil Paints: A Comparative Study

B. Ormsby, J. Lee, I. Bonaduce, and A. Lluveras-Tenorio

2.1 Introduction

Collaborative research into the cleaning of modern painted surfaces is beginning to inform approaches to soiling removal that may offer lower risk options for solvent and mechanically sensitive unvarnished paint films [3, 8]. Building on earlier research on acrylic paints [4] and the use of polysaccharide gels [25], this study evaluates adjusted pH and conductivity waters [6], rigid gels, silicone emulsifiers [10], and mineral spirits-based microemulsions [17], for the surface cleaning of prepared sensitive oil paint samples; with additional explorations into the effects of application methods. Complementary studies using similar methodologies on naturally aged oil paintings are detailed by Gilman et al. [7].

The combined results of systematic studies on prepared and accelerated aged Winsor and Newton (W&N) oil paints are summarised via cleaning system type, involving evaluations of cleaning tests using star diagrams to capture a range of key empirical observations, digital microscopy (Hirox) and infrared spectroscopy (ATR-FTIR). Potential residues of cleaning materials were also investigated via visual inspection, microscopy and ATR-FTIR analysis. A selection of cleaned samples were then further investigated for residues using a pyrolysis on-line micro-reaction sampler and X-ray Photo Electron Spectroscopy (XPS). These studies highlight pigment-dependent behaviours with respect to cleaning and offer potentially useful approaches to soiling removal for these often challenging paint surfaces.

B. Ormsby (✉) · J. Lee
Conservation Department, Tate, London, UK
e-mail: Bronwyn.ormsby@tate.org.uk

I. Bonaduce · A. Lluveras-Tenorio
Department of Chemistry and Industrial Chemistry, University of Pisa, Pisa, Italy

K. J. van den Berg et al. (eds.), *Conservation of Modern Oil Paintings*,
https://doi.org/10.1007/978-3-030-19254-9_2

2.2 Methodology

2.2.1 Paint Samples

Samples were prepared using a Sheen instruments [Surbiton UK] film caster to a dry thickness of 100–150 μm onto titanium white and chalk oil-primed canvas pieces [Belle Arti 536 medium fine linen, 508gm, Jackson's Art Supplies London UK]. Based on previous studies [13] three W&N artist quality oil paint colours known to form water-sensitive paint films were selected (French ultramarine, yellow ochre, cadmium yellow), with one non-water-sensitive paint (titanium white) also selected. All paints contained magnesium carbonate which is known to contribute to water sensitivity via the formation of water-soluble magnesium sulphate heptahydrate (epsomite) salts [21]. To induce water-sensitivity, the samples were cured under ambient conditions for 6 weeks prior to being accelerated light aged (Philips TL-D/840 super 80, 58-watt, long tubes, 4000 K) at elevated relative humidity (70–80%) and 30 degrees centigrade, under an average ~3500 lux illumination with the ultraviolet (UV) component filtered out. Samples were monitored using a deionised water-moistened standard swab to assess the development of water sensitivity, which in some cases required up to 2.5 months exposure.

After ageing, the French ultramarine and yellow ochre paints had developed thin UV-fluorescing medium skins (<5 microns thick) which proved sensitive to deionised water (i.e. pigment pickup was noted) within 3–8 lightly moistened swab rolls, resulting in the surfaces appearing more matte. After ageing, ATR-FTIR analysis showed that both the French ultramarine and yellow ochre paints had more intense and broad hydroxy infrared absorption bands at ~3400 cm^{-1} and a new band at 1330 cm^{-1} attributed to the O-H in-plane bending vibration of alcohols. This is indicative of the formation of polar lipid oxidation products [12] at the paint surface. A high humidity environment is known to promote oxidation of oils [9] and it has been noted that French ultramarine and iron-oxide based oil paints may become more oxidised than other colours [14]. This suggests that the artificial ageing conditions employed promoted oxidation, resulting in the formation of higher levels of polar hydroxy-containing species, which may – at least in part – explain the sensitivity of these two paints.

The titanium white and cadmium yellow paints did not develop visible medium skins. However, the cadmium yellow paints had a general matte appearance and were susceptible to subtle shifts in gloss after swab rolling. Pigment pickup was also observed for all water-sensitive colours after between 16–25 swab rolls. Epsomite was detected (only) on the cadmium yellow samples (which may account to some extent for the matte appearance of this paint), based on the presence of characteristic ATR-FTIR spectral absorptions: a broad and intense S-O stretch at 1096 cm^{-1} and other characteristic bands including 986, 1466 and 3390 cm^{-1}. Whilst the possibility of the formation of water-soluble cadmium sulphate was also considered [20], the closest reference IR spectrum match was to magnesium sulphate heptahydrate. After ageing, the cadmium yellow paints had developed infrared absorption bands at 1710 cm^{-1} and 1320 cm^{-1} which can be attributed to the formation of carboxylic acid groups. This occurred in conjunction with a reduction in the intensity of C=O ester absorption bands at 1745 and 1165 cm^{-1} [24], which suggests that the ageing conditions promoted the hydrolysis of glyceride ester bonds and the formation of free carboxylic acid groups. It is possible, though not yet explored, that the accelerated ageing conditions, in conjunction with the photo-oxidation of cadmium yellow pigment, may have resulted in the formation of sulphuric acid which catalysed the hydrolysis of the oil medium [20].

After ageing, the samples were soiled using an artificial soil mixture [16], containing insoluble macroscopic solids (iron oxide, silica, kaolin), microscopic particles (carbon black), water reactive solids (cement type I), water swellable- or soluble- organics (gelatine powder, soluble starch), and apolar soluble organics (olive oil, mineral oil). For this study, the mixture was modified by reducing the carbon black and oil components by ~50% to avoid creating too dark or thick oil layers on sample surfaces. The soil was suspended in Shellsol D40 mineral spirits [Kremer Pigmente, Germany], and applied using an airbrush [Badger 250®, Amazon UK], with an area masked off with Melinex® to create an unsoiled control. After soiling, the French ultramarine, yellow ochre and cadmium yellow samples were further aged under the same conditions apart from a reduction in RH (to 50%) for 3 weeks; and the titanium white samples for 7 weeks, to encourage stronger adhesion between the applied soil and paint surfaces.

2.2.2 Cleaning Materials

The cleaning systems were selected from both established [25] and recently introduced options [10, 17, 22]; which are being increasingly widely used in conservation [2].

The cleaning materials were grouped into classes as follows:

Free aqueous solutions: included as the most efficient solvent for removing soiling; primarily acting on the kaolin, cement, gelatine and starch portions of the artificial soil mixture. Options included pH and conductivity *adjusted waters* derived from research into acrylic paints [6]. Adjusted waters were set to a pH range of 4.0, 4.5, 5.0, 5.5 and 6.0, at a conductivity of 1.0 mS/cm using ammonium hydroxide and acetic acid, which are believed to form volatile residues.[1]

Previous work [4] suggested that higher conductivity solutions did not offer detectable advantages for soiling removal, and that pH ranges between 4 and 6 resulted in minimised surface changes to sensitive oil paint samples; though it was also noted that lower pH values could result in reduced cleaning efficacy. The addition of 0.5% w/w. nonionic alcohol ethoxylate surfactants such as ECOSURF™

[1]Acetic acid boiling point is 118 °C (https://pubchem.ncbi.nlm.nih.gov/compound/acetic_acid#section=Color-Form) and ammonium hydroxide solution is 38 °C (https://pubchem.ncbi.nlm.nih.gov/compound/ammonium_hydroxide#section=Taste). Accessed 21.3.19.

EH-6 and ECOSURF™ EH-9[2] [4, 16] enhance soiling pickup and may also interact with carbon black [10]. Chelators such as citric acid (pK_a values 5.62, 4.34, 2.91), ethylenediaminetetraacetic acid (EDTA) (pK_a values 10.27, 6.16, 2.67, 2.0) and diethylenetriaminepentaacetic acid (DTPA) (pK_a values 10.42, 8.76, 4.42, 2.56, 1.79) were chosen from the Modular Cleaning Programme[3] for their ability to sequester the metal ions present in the artificial soil, including Fe^{3+}, Mg^{2+}, and Ca^{2+}. Citric acid is a strong chelating agent for Fe^{3+} ($\log K_1 = 11.4$) and at the concentration used, would also chelate the Mg^{2+} ($\log K_1 = 3.4$) and Ca^{2+} ($\log K_1 = 3.6$) ions present [18, 19]. The stability constants ($\log K_1$) for Fe^{3+}, Mg^{2+}, and Ca^{2+} ions for EDTA are 25.1, 8.7, and 10.6 respectively, and for DTPA, 28.6, 9.3 and 19.1 respectively, making them stronger sequestering agents than citric acid for these ions. All chelators were used at 1% w/v. concentration, with the pH adjusted using 10% w/v. sodium hydroxide in deionised water. Where required, clearance was carried out using adjusted waters set to the same pH as the cleaning system, using dilute ammonium hydroxide and acetic acid.

Free non-polar solvents: primarily act on the apolar soluble organic components (added oils) of the artificial soil and possibly the carbon black; and included aliphatic hydrocarbon solvents such as Shellsol D40 used alone, and with up to 2% w/w. added anionic surfactant NaDOSS (dioctyl sulfosuccinate, sodium salt)[4] [4]; as well as silicone solvents including hexamethyldisiloxane (HMDS) and cyclopentasiloxane (D5) [10]. The action of these solvents was compared to deionised water to assess their effects as clearance solvents (data not shown). Any option with added surfactant was cleared with the corresponding neat solvent.

Water-in-oil (W/O, or reverse) microemulsions: were designed to offer enhanced cleaning efficacy through the suspension of aqueous dispersed phases within a non-polar (low-swelling) solvent continuous phase [17]. *Series 3b* systems based on Shellsol D40 and NaDOSS surfactant were selected as potential options, based on previous investigations [4]. These systems act on many aspects of the applied soil simultaneously through solvent-soiling interactions, however in addition, the microemulsion droplet size (below ~100 nm) enhances cleaning activity through increased surface area [17]. These systems were cleared using Shellsol D40 with 2.5% w/w added propan-2-ol [17].

Silicone-based Pickering emulsifiers: were introduced for the controlled aqueous cleaning of a range of sensitive surfaces [10], with earlier products such as Velvesil Plus and KSG-210/240 [Shin-Etsu, Japan] showing potential for clean-

[2] http://msdssearch.dow.com/PublishedLiteratureDOWCOM/dh_0932/0901b80380932539.pdf?filepath=surfactants/pdfs/noreg/119-02244.pdf&fromPage=GetDoc. Accessed 14.1.19.

[3] http://cool.conservation-us.org/byauth/stavroudis/mcp/. Accessed 14.1.19.

[4] https://en.wikipedia.org/wiki/Docusate. Accessed 14.1.19.

ing sensitive oil paints [4]. Shin-Etsu's KSG 350z (Kremer Pigmente, Germany), is used in this study with silicone solvents and optimised aqueous phases at varied proportions [10, 11, 15]. This material can offer advantages for sensitive paints through viscosity, its modifiability with respect to emulsifying both polar and non-polar solvents in addition to its inherent non-polarity. The aqueous phase (usually included between 10 and 50% w/w.) is emulsified into the system as the key active ingredient, though the silicone components will also act to some extent on silicone oils, and possibly the two oils included in the artificial soil. In addition to the % aqueous phase used, the emulsion droplet size also influences cleaning activity as per microemulsions.

Rigid hydrogels: Agar/agarose and gellan gum [Special Ingredients, UK] were prepared as 2–3% w/w. mixtures, heated to ~95 °C and uploaded with the best performing tailored aqueous systems. For these rigid hydrogels, the main advantage is reduced mechanical action [25], where the removal of soiling is achieved through dissolution and absorption via capillary action [2]. For these gels, citric acid was used as a non-volatile buffer, even where the associated chelating properties were not required. Where used, surfactants were added when the gel had cooled to a temperature below the surfactant cloud point.[5] The prepared gels were applied at room temperature, for up to 10 min, with excess water blotted from the gel surface prior to application. Clearance was carried out using either deionised water hydrogels or with adjusted water applied with a rinsed makeup sponge, which tended to minimise associated swelling.

Application methods: Hand rolled cotton swabs, brushes, soft cosmetic sponges – e.g. Muji™ (blotting excess solvent before use), application through wet strength tissue [Spider's wet-strength Tarantula tissue][6], application through silicone solvent barriers (D5), and combinations of these were also explored with the primary aim of reducing the mechanical action applied to these sensitive paint surfaces.

2.2.3 Evaluation Methodology

2.2.3.1 Star Diagrams

Star diagrams offer a systematic, comparative way of capturing the empirical parameters used to evaluate cleaning tests, which can then form part of a conservation treatment record. The parameters evaluated in this study are described in

[5] The effect of any added ionic materials on the surfactant cloud point was not considered.

[6] http://www.cxdglobal.com/productdetails.aspx?id=309&itemno=PASWTT0100. Accessed 6.2.19.

Table 2.1, which included: relative dirt removal efficacy (e.g. how soil-free the surface is after the cleaning and clearance steps had been performed), relative pigment/colour pickup (e.g. judged by inspecting the application tool(s) after use), gloss integrity (e.g. judging the dry paint surface by eye after cleaning relative to the unsoiled control), clearance (e.g. effectiveness of the cleareance step as judged by eye), relative application time (e.g. as determined for each method) and relative amounts of swelling/blanching noted (e.g. as judged by eye and/or using microscopy). Star diagrams (Excel radar charts) were first used for dry cleaning studies [5] and have been further refined for wet cleaning evaluations [4]. For this study, the

Table 2.1 Star diagram rating scales based on those published by Chung et al., [4] amended for 'application time' to accommodate the use of several application methods.

Parameter		Rating 1 (POOR)	2	3	4	5 (GOOD)
Dirt removal efficacy (as judged by eye on dried paint film)		no effect	little effect	moderate	slightly effective	complete removal
Lack of pigment/stain pickup (as judged by eye on cleaning materials)		pigment pick-up visible	minor pigment visible with stain	faint stain	very faint stain	no pigment/ stain
Gloss integrity (compared to unsoiled control area)		significant change	change visible in normal light	change only visible in raking light	slight shift	no visible change
Ease of clearance (as judged by eye on dried paint film)		ring of residue visible on paint in normal light	faint tide mark visible in normal light	tide mark visible in raking light	slight concern	no concern
Lack of swelling / blanching (as judged during cleaning and by eye during/after cleaning).		major	fairly significant	slight	very slight	no swelling noted
Appropriate application time (defined separately according application method)	Hand-rolled swabs (swab roll count)	41–50	31–40	21–30	11–20	0–10
	Rigid polysaccharide gels; Time (min:sec)	8:01–10:00	6:01–8:00	4:01–6.00	2:01–4:00	0:00–2:00
	Sponge (number of sponge strokes applied)	41–50	31–40	21–30	11–20	0–10
	Shin-Etsu KSG-350z emulsifiers; Time (min:sec)	1:30–2:00	1:00–1:30	0.30–1:00	0:10–0:30	0–0:10

application time parameter was expanded to accommodate the range of application methods used. It is important to note that for this complex study the results described are based on the observations of one trained individual only[7], hence the outcomes do not represent definitive guidance.

2.2.3.2 Instrumentation

Digital microscopy images were taken using a Hirox digital microscope (Hirox, Japan) mounted onto a photo stand with a ring-light diffuser. The microscope zoom lens (MXG-2500REZ) was set to 100 and 1000x magnification. Images were processed using Hirox software.

Gloss measurement was carried out using a Rhopoint Novo-Gloss Trigloss glossmeter, using a 60° angle, carried out in triplicate where possible.

ATR-FTIR spectroscopy was carried out using a Thermo Scientific Nicolet iZ10 system with bench diamond and germanium ATR crystal accessories. Samples were allowed to dry for at least a week post-cleaning prior to analysis. Data was processed using Omnic 9 software.

XPS analysis was carried out using monochromatised Al (Kα) X-rays (1486.6 eV) in a Thermo K-Alpha spectrometer. All samples were analysed using a 300–400 μm beam to ensure that only the paint surface was analysed. Analyses were performed in triplicate, around two weeks after treatment. The high intensity carbon 1 s and oxygen 1 s peaks were curve fitted using reference values to determine atomic %. Data were interpreted using CASA XPS software and reported as atomic % to remove measurement dependent signal variations from XPS peak intensities.[8]

Pyrolysis with on-line gas chromatography-mass spectrometry (PyGCMS) was performed using a micro-furnace Multi-Shot Pyrolyzer EGA/Py-3030D [Frontier Lab] coupled with a gas chromatograph 6890 Agilent Technologies [Palo Alto, USA] and a 5973 Agilent Mass Selective Detector [Palo Alto, USA], single quadrupole. Paint samples were heated to 150 °C for 10 minutes. An HP-5 ms capillary silica column (30 m, 0.25 mm, film thickness 0.25 μm, Agilent Technologies, USA), connected to a deactivated silica precolumn (2 m, 0.32 mm, Agilent Technologies, USA) was used with a split ratio of 1:10. The GC oven was set at 70 °C then ramped to 300 °C at 20 °C/min, with a 15-minute isothermal hold. Surface scrapings of a 1 cm^2 paint surface area were used to compare the peak areas of both cleared and non-cleared surfaces. Analyses were performed in triplicate, 2 weeks after treatment.

---

[7] The cleaning tests, star diagrams, FTIR-ATR and microscopy evaluations were carried out by Dr. Judith Lee as part of the Cleaning of Modern Oil Paints (CMOP) project at Tate, London.

[8] XPS analysis was carried out as part of the CMOP project at Winnats Scientific (UK) through the University of Pisa, Italy. PyGCMS analysis was carried out at the University of Pisa, Italy.

2.3 Results and Discussion

2.3.1 Free Aqueous Solvents

2.3.1.1 Adjusting pH and Conductivity (Adjusted Waters)

A series of cleaning tests were carried out using hand-rolled cotton swabs to explore the effect of pH adjusted waters. Across the range of samples, lowering the pH to 4.5–6.0, was found to reduce unwanted changes such as pigment pickup and gloss change on paints with medium skins (i.e. French ultramarine (Fig. 2.1) and the yellow ochre samples), which may relate to a reduction in the ionisation of acidic functionalities on the paint surface. It was noted however that using waters at the lower pH range tended to reduce cleaning efficacy, which may relate to changes in the interfacial tension of the cleaning solution, influenced by concentration and speciation of the ions present in solution, and the solution pH [1]. For example, the pK_a of acetic acid is 4.75, hence at pH values lower than the 4.75, acetic acid will generally not be ionised. Despite a reduction in cleaning efficacy noted for the yellow ochre (not shown) and French ultramarine paints, lowering the pH proved beneficial for reducing pigment pickup and gloss change. However, for the titanium white and

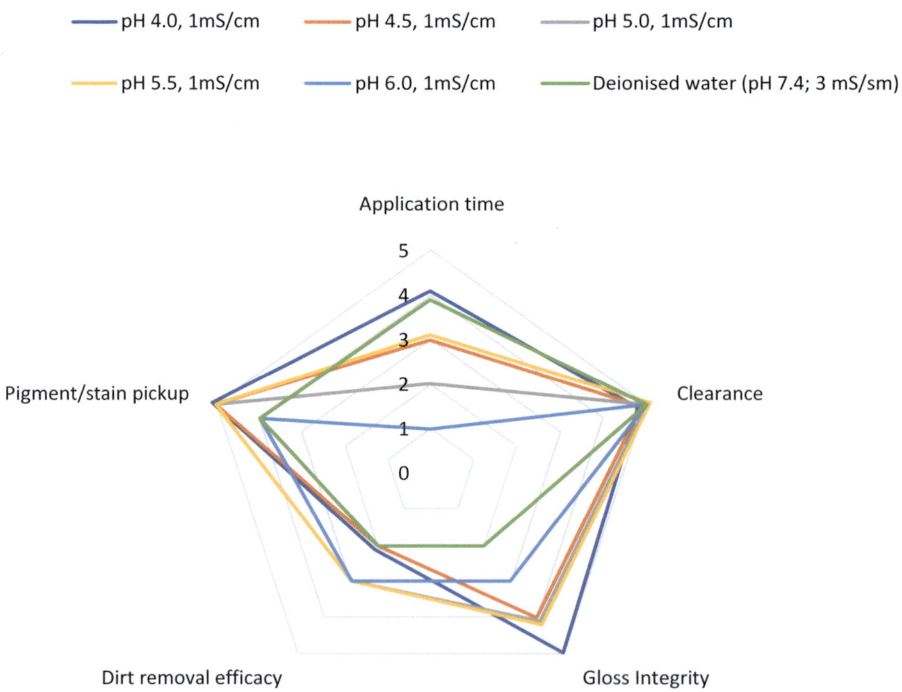

Fig. 2.1 Star diagram – adjusted waters swab applied to sensitive W&N French ultramarine blue paints

cadmium yellow paints there appeared to be no clear advantage to using adjusted waters set to below pH 5 or 5.5.

2.3.1.2 Adding Surfactants

As expected, the addition of 0.5% w/w. ECOSURF™ EH-6/EH-9 nonionic ethoxylate surfactant increased the soiling removal efficiency and reduced contact time. However, as the paint surface required a clearance procedure using the equivalent pH adjusted water, this occasionally negated any advantage. It was noted however that the addition of surfactant generally reduced the magnitude of gloss change when compared to adjusted waters alone. While the results naturally varied to some degree, the ECOSURF™ EH-6 (HLB 10.8), at 0.5% w/w. at pH 6.0 proved optimal for the cadmium yellow samples; 0.5% EH-9 (HLB 12.5), at pH 4.5/5 for the French ultramarine samples, and 0.5% EH-9 at pH 5/6 for the yellow ochre samples, which had relatively heavily imbibed soiling as shown in Fig. 2.2. This suggests that exploring the addition of low levels of nonionic surfactant at varying pH values may be useful; however it is noted that specific relationships between the paint surface, soiling layers and surfactant types have yet to be systematically explored.

2.3.1.3 Adding Chelators

Three chelators were evaluated including citric acid, EDTA and DTPA, used at 1% w/v. and tailored to the desired pH using a 10% w/v sodium hydroxide (NaOH) solution. Except perhaps for the titanium white samples, this group prove too active when applied using a swab. The addition of chelators increased cleaning efficacy as expected, and reduced contact time for the titanium white and French ultramarine paints; however, for the cadmium yellow and ochre paints, these systems were poor with respect to maintaining gloss integrity when compared to pH adjusted water and surfactant options. This may relate to the vulnerability of Cd^{2+} and Fe^{3+} ion-containing pigments to these chelating agents. In addition, due to the pK_a values of the chelators used, the cleaning efficacy of the solutions were noticeably reduced below pH 6. There appeared to be little difference in the overall performance of the chelators used in this study, however Gillman et al. [7] noted that EDTA options tended to cause less pigment pickup and facilitated a more even cleaning action on a group of matte, naturally aged paintings with varied soiling compositions. For the group of W&N samples the weaker binding citrate chelators set at pH 5–6 proved acceptable for the titanium white and French ultramarine samples, however this system resulted in unacceptable gloss changes for both the cadmium yellow and yellow ochre paints.

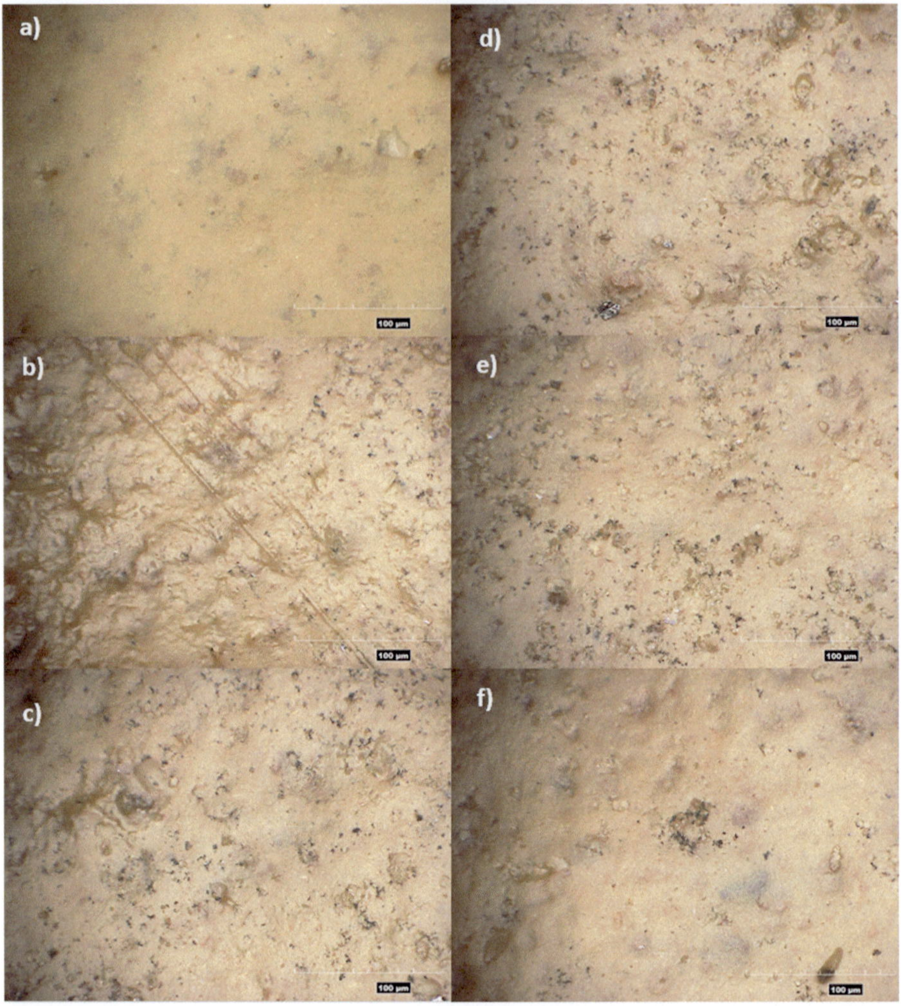

Fig. 2.2 Free aqueous systems swab-applied to a sensitive W&N yellow ochre oil paint sample evaluated using digital microscopy at 1000×. The 0.5% w/w. ECOSURF™ EH-9 option set to pH 5.5 proved optimal for minimising surface change on this sample (**a**) Unsoiled control (**b**) Deionised water (**c**) Adjusted water pH 5.0, 1 mS/cm (**d**) 0.5% w/w. ECOSURF™ EH-6, pH 5.0, 1 mS/cm (**e**) 0.5% w/w. ECOSURF™ EH-6, pH 6.0, 1 mS/cm (**f**) 0.5% w/w. ECOSURF™ EH-9, pH 5.5, 1 mS/cm (optimal)

2.3.1.4 Summary of Free-Aqueous Systems

These initial tests, carried out using liquid aqueous solutions applied using hand rolled cotton swabs, were designed to assess their effect on the four paint samples with the aim of determining the lowest-risk aqueous conditions for each paint. For this study, the 'most promising' options were determined via balancing soiling removal efficacy with swelling, pigment pickup and gloss change for each sample type. On occasion, reducing the solution pH resulted in less swelling of the paint film and less disturbance of medium skins, however this also reduced the soiling removal efficacy; hence in those cases surfactant was added and/or the pH was adjusted upwards to increase the cleaning efficacy. Gloss readings taken after cleaning suggested that the yellow ochre and French ultramarine paints (both with visible medium skins), were the most susceptible to gloss change, with the most significant reduction at a pronounced 30 units (Fig. 2.5). The titanium white and cadmium yellow paints appeared least affected, with less than 5 gloss unit shifts noted, which reflects (at least in part) an absence of medium skins on these two colours. However, pigment pickup was most pronounced for the cadmium yellow samples (epsomite-rich) with deionised water, which was minimised to some extent when the pH was reduced to 5.5. For the cadmium yellow paints, all of the free aqueous systems resulted in the removal of epsomite as determined through digital microscopy and/or ATR-FTIR analysis where removal was indicated via reduction in the broad absorption band at ~1095 cm^{-1} (data shown in Fig. 2.4). In essence, none of these sensitive paints were satisfactorily cleaned using the free-aqueous systems, which were typically too aggressive; however the best performing options taken forward for use in microemulsions and/or silicone emulsions are listed in Table 2.2.

2.3.2 *Rigid Hydrogels*

For this group of samples, the agar and gellan gels restricted visual access to the paint surface, did not remove soiling efficiently and tended to cause swelling, blanching, pigment pickup and gloss change. Similarly to the free aqueous systems, reducing the pH of these gels reduced paint swelling to some extent and facilitated slightly longer contact times with the paint surface; however soiling removal at low pH remained poor, and the addition of surfactants and chelating agents tended to make the action too aggressive. For example, the use of agar gel set to pH 4.5–5.5 proved somewhat useful for minimising swelling and gloss changes to the French ultramarine paints, however soil removal remained poor. It is noted however, that Gillman et al. [7] found both agar and gellan gels at 2–3% w/w. with 1% w/w. added EDTA useful for removing soil from naturally aged paintings; which may be due to the different nature of the soiling layers in addition to the older, less medium-rich paint layers.

2.3.3 *Non-polar Organic Solvents*

Compared to deionised water, the free-liquid aliphatic hydrocarbon Shellsol D40 and silicone solvents HMDS and D5 did not cause significant gloss changes or significant pigment pickup. However as expected, they did not function well as cleaning agents and required longer contact times. In Gillman et al. [7], the use of

Table 2.2 'Best performing' free aqueous systems, cotton swab applied, on four artificially aged and soiled W&N oil paints, with appropriate clearance step. (§) All chelators were pH adjusted using 10% w/w aqueous NaOH. All proportions are by weight

Paint colour with specific condition/ property	Groups of free aqueous systems that were most promising, and achieved similar cleaning outcomes. NB: none of the free aqueous systems proved optimal for these paints	Best performing Agar gel (3%), cleared with DI water hydrogel (Titanium white and Yellow ochre) and a sponge with adjusted water at same pH and 1mS/cm (French Uiltramarine Blue, Cadmium Yellow)
Titanium white (not water sensitive, moderate gloss, resistant to mechanical action)	Deionised water	
	pH6, 1mS/cm	Deionised (DI) water*
	0.5% ECOSURF™ EH-9, pH 6.0, 1 mS/cm	1% citric acid; 0.5% ECOSURF™ EH-9; pH 6.0[§]
	0.5% ECOSURF™ EH-6, pH 6.0, 1 mS/cm	1% citric acid; 0.5% ECOSURF™ EH-6; pH 6.0[§]
	1% citric acid, pH 6.0[§]	1% citric acid; pH 6.0*[§]
Yellow ochre (water-sensitive, thin medium skin, tended to swell, imbibed soil)	pH 5, 1mS/cm	Deionised water*
	ECOSURF™ EH-6, pH 5.0, 1 mS/cm	1% citric acid; pH 5* [§]
	ECOSURF™ EH-6, pH 6.0, 1 mS/cm	1% citric acid; 0.5% ECOSURF™ EH-6, pH 5.0[§]
	ECOSURF™ EH-9, pH 5.5, 1 mS/cm	1% citric acid; 0.5% ECOSURF™ EH-6, pH 6.0[§]
	ECOSURF™ EH-9, pH 6.0, 1 mS/cm	1% citric acid; 0.5% ECOSURF™ EH-9, pH 5.5[§]
		1% citric acid; ECOSURF™ EH-9, pH 6.0[§]
French ultramarine (water-sensitive, thin medium skin, vulnerable to mechanical action)	pH 5.5, 1 mS/cm	Deionised water*
	ECOSURF™ EH-9, pH 4.5, 1mS/cm	1% citric acid; pH 5.5*[§]
	ECOSURF™ EH-9, pH 5.5, 1 mS/cm	1% citric acid; pH 5.5[§]
	1% citric acid, pH 5.0[§]	1% citric acid; 0.5% ECOSURF™ EH-9, pH 4.5[§]
	1% citric acid, pH 6.0[§]	1% citric acid; 0.5% ECOSURF™ EH-9, pH 5.5[§]
		1% citric acid; pH 5.0[§]
		1% citric acid; pH 6.0[§]
Cadmium yellow (water-sensitive, matte surface, vulnerable to mechanical action, epsomite present)	pH 5.5, 1 mS/cm	Deionised water*
	0.5% ECOSURF™ EH-6, pH 5.0, 1 mS/cm	1% citric acid; pH 5.5*[§]
	0.5% ECOSURF™ EH-6, pH 6.0, 1 mS/cm	1% citric acid; 0.5% ECOSURF™ EH-6, pH 6.0[§]
	0.5% ECOSURF™ EH-9, pH 5.5, 1 mS/cm	1% citric acid; 0.5% ECOSURF™ EH-6, pH 5.0[§]
	0.5% ECOSURF™ EH-9, pH 6.0, 1 mS/cm	1% citric acid; 0.5% ECOSURF™ EH-9, pH 6.0[§]
		1% citric acid; 0.5% ECOSURF™ EH-9, pH 5.5[§]

iso-octane (2,2,4-trimethylpentane) afforded a slight reduction in the tendency to pick up pigment when compared to Shellsol D40 for some paint passages investigated.[9]

2.3.4 Water-in-Oil Microemulsions

In previous work [4], the Shellsol D40-based W/O (reverse) microemulsions [17] *Series 3b* group were noted as promising for sensitive oil paint surfaces. In this study, the effect of varying surfactant proportion, aqueous phase proportion and aqueous phase modifications (adjusting pH, adding surfactant) have been further explored. Cleaned samples were evaluated for soiling removal efficacy, surface change and surfactant residues. To compare the effects of the proportion and character of the aqueous phase on cleaning efficacy, several microemulsions were prepared[10] containing 40, 30, 20 and 10% w/w. aqueous content. Each was applied via systematic 10-swab roll applications, and cleared with the appropriate hydrocarbon solvent with 2.5% w/w. added propan-2-ol. The results, shown through two examples in Fig. 2.3, confirmed that decreasing the water content from 40% to 10% resulted in less pronounced changes to paint surfaces, and particularly when the aqueous content was ≤20%. Adjusted aqueous phases set to pH 4.5, 5.0, and 6.0 were also explored alongside deionised water and a solution containing 0.5% w/w. ECOSURF™ EH-6, at pH 6. As shown in Fig. 2.3, modifying the aqueous phase also had an effect; in some cases, lowering the pH reduced unwanted gloss change and pigment pickup. The effects appeared to be less noticeable when the aqueous content was 10% or less; and in all cases, abrasion was noted with cotton swab applications. A slight improvement was however noted in the gloss integrity where adjusted waters were used for the yellow ochre sample, where the deionised water appeared to slightly disturb the medium-skin (see Fig. 2.3).

Within a W/O microemulsion, modifying the aqueous phase with added surfactant, such as 0.5% w/w. ECOSURF™ EH-9 (or EH-6) did not appear to enhance cleaning efficacy or improve the overall cleaning outcome. Regarding the addition of chelators, initial indications were that citrates (Na-citrate or TAC) offer the most stable microemulsion options within the lower water content (~10%) options. It is possible to use EDTA or DTPA within these systems, however the region where stable microemulsions form on the phase diagram [17] may be somewhat reduced, and has yet to be mapped. It is also likely that a microemulsion incorporating stronger chelators would prove to be too active for these paint surfaces.

Comparisons were also made (not shown) between microemulsions with different levels of the NaDOSS surfactant phase [ME3b-18 (70% surfactant phase) and

[9] This has also been noted by the authors when using isooctane-based microemulsions.

[10] Series 3b-5 (30% D40, 40% aqueous, 30% NaDOSS), 3b-10 (50% D40, 30% aqueous, 20% NaDOSS), 3b-16 (60/20/20), and 3b-22 (60/10/30) or 3b-24 (80/10/10). The NaDOSS is made up as a 65% w/w. solution in Shellsol D40.

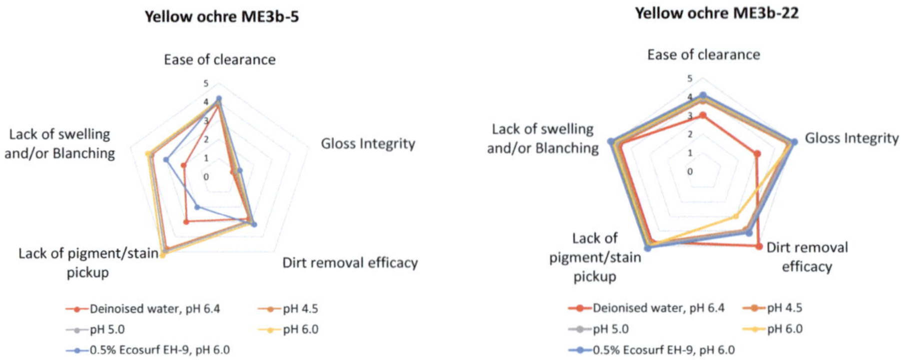

Fig. 2.3 Star diagrams for 10 roll swab-applied Shellsol D-40 based *Series 3b* microemulsions with different proportions and modified aqueous phases (e.g. ME3b-5 = 40% aqueous, ME3b-22 = 10% aqueous)

ME3b-24 (10% surfactant phase)] using the same quantity of aqueous phase (10% by weight). Promisingly, the higher % surfactant phase content did not appear to contribute significantly toward soil removal efficacy, suggesting that keeping surfactant levels low does not hinder the cleaning efficacy of these systems. The higher % surfactant ME3b-18 option was more viscous and had a smoother action but tended to leave visible surfactant residues in the form of glossy rings which required further clearance. Using ATR-FTIR analysis, residues were detected on samples where the surfactant proportion was 40% and above (see *Residues*). It is noted that the effects of the use of these and other microemulsions on the physical properties of oil paints has yet to be explored.

As some abrasion was visible when using swabs directly on these paints, the swabbing of W/O microemulsions through a wet-strength tissue barrier was also evaluated. However, in addition to preventing visual access to the paint surface, the tissue barrier did not prevent swelling and/or pigment loss and resulted in ineffective soil removal. Gillman et al. [7] however, reported some success using this method on older oil paint films. Using makeup sponges with microemulsions (while wearing gloves) resulted in a less abrasive action, however the sponge was eventually affected by the Shellsol D40 solvent, raising concerns around sponge residues. Loading microemulsions into soft sponges could be advantageous for these types of paints, should appropriate sponges become available.[11]

[11] http://www.nanorestart.eu/. Accessed 6.2.19. Note that trials of organo-gels and microemulsions are underway at Tate.

2.3.5 Silicone Emulsifiers

The initial set of Shin-Etsu KSG-350z silicone emulsifiers evaluated were based on published formulations using 10% KSG-350z (at 30–60% optimised aqueous phase) by weight [11]. For this study, a second group based on 25% KSG-350z were created to form options with reduced aqueous contents of 30–40%. All silicone emulsifier systems were made by weight, first by blending the emulsifier with D5 silicone solvent, followed by dropwise additions of the aqueous phase. The mixture was then stirred vigorously to create the emulsion, followed by being forced through blunt-end luer lock syringe needles of two different gauges (wider bore then narrow bore) prior to use [23]. In general, these systems were slow acting and required prolonged contact with paint surfaces. As expected, higher proportions of the aqueous phase enhanced the cleaning efficacy and reduced the working time required. It was noted that an aqueous content of 50 or 60% by weight significantly improved cleaning efficacy over the 30 or 40% equivalents; and that below 30%, the cleaning efficacy was poor. In general, formulations with moderate (30–50%) proportions of aqueous phase were chosen to help minimise paint surface change. Optimal application involved pre-wetting the surface with D5 solvent (which also enhanced cleaning efficacy), applying the prepared emulsifier by brush followed by light agitation on the paint surface. Clearance was carried out using D5-dipped cosmetic sponges as often as required (average 3 applications), then repeated using the faster evaporating HMDS solvent. For the paint samples evaluated, increasing the proportion of the aqueous phase proved more important to the final result than small adjustments to the character of the aqueous phase.

Figure 2.4 (upper) shows the star diagram results of the silicone emulsion applications on the epsomite-rich cadmium yellow paint film, where the optimal system was determined as a 10% KSG-350z emulsion with a 50% aqueous phase containing 1% citric acid and 0.5% ECOSURF™ EH-9, adjusted to pH 6 using an NaOH solution. Evaluation of the cleaned paint films using ATR-FTIR spectroscopy (Fig. 2.4, lower) revealed that epsomite was removed from the paint surface to a similar extent as a swabbed deionised water application. Adjusting the aqueous phase pH did not appear to make a significant difference to either the swelling behaviour, or paint gloss integrity for the cadmium yellow paints, though the addition of both the ECOSURF™ EH-9 surfactant and citric acid enhanced soiling removal and reduced application time.

For the particularly water-sensitive French ultramarine paint, the 25% KSG option with an aqueous phase at 40% modified to include 1% (w/w) citric acid and 0.5% (w/w.) ECOSURF™ EH-9, at pH 6.0 tended to reduce pigment loss and swelling, and minimised surface disruption (data not shown). Table 2.3 lists the optimised silicone emulsifier systems for each paint type. Differences in the optimal emulsifier formulation tended to reflect the relative degree of surface soil-adhesion. In general, reducing the aqueous phase content tended to reduce swelling, reduce gloss shift and decrease pigment pickup, while also increasing application time and

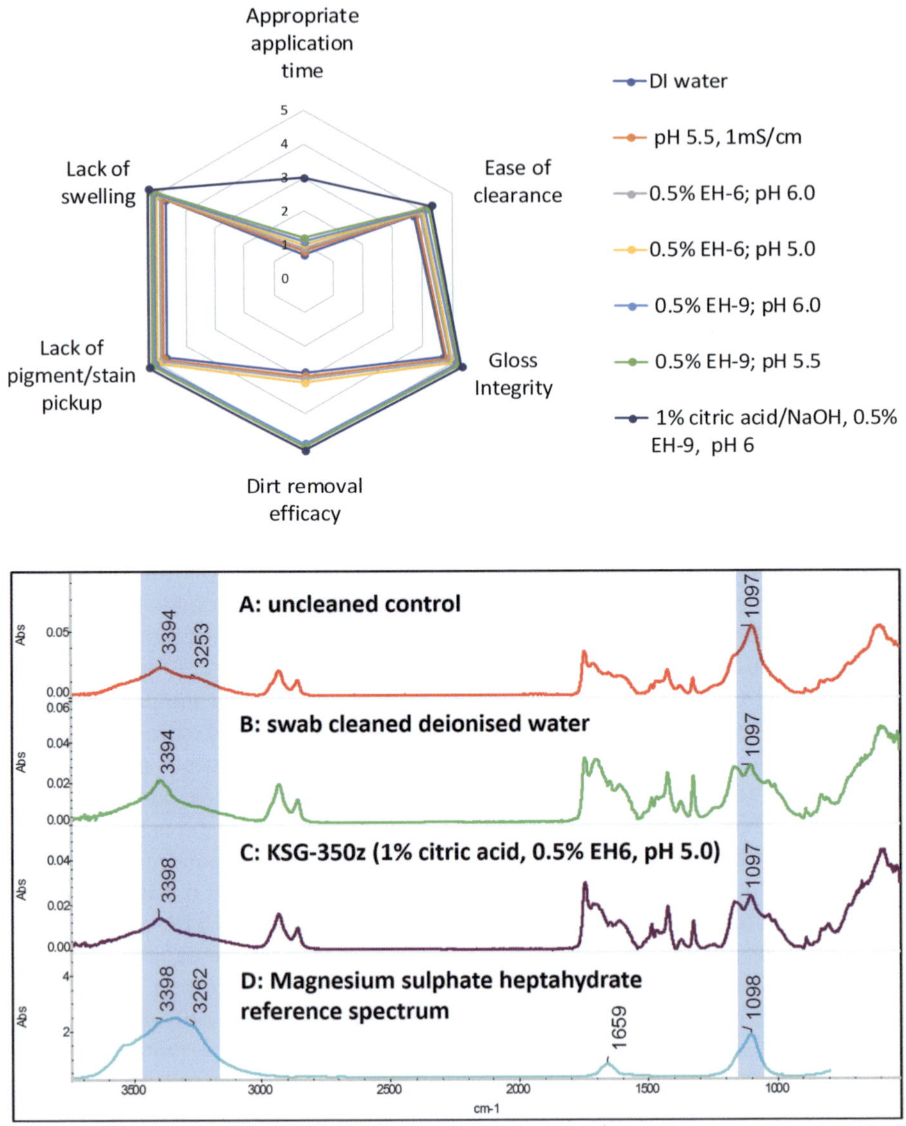

Fig. 2.4 Upper - phase diagram for KSG-350z emulsion applications on the sensitive W&N cadmium yellow (CY) paint using: 10% 350-z; 40% D5; 50% aqueous phase applied by brush, with 7 different aqueous phases. Lower – ATR-FTIR spectrum of the: soiled control CY paint top), the soiled sample after cleaning with deionised water (second from top), the soiled sample after cleaning and clearing using the optimised KSG-350z system (third from top) and a reference spectrum for magnesium sulphate heptahydrate/epsomite (bottom). Note that application time has now been incorporated into the star diagram (see Table 2.1)

Table 2.3 Optimal KSG-350z emulsion formulations derived for each paint type

W&N Paint type	KSG-350z (%w/w)	D5 (% w/w)	Aqueous phase % content	Aqueous phase Composition	Explanation for choice of cleaning system
Titanium White	10	40	50	0.5% w/w. ECOSURF™ EH-9, pH 6	The titanium white films had less heavily adsorbed soil than the other colours. The reduced water content necessitated the addition of surfactant to boost cleaning efficacy. For this paint, surfactant added to the aqueous phase was more effective than chelating agents.
Cadmium Yellow	10	40	50	1% w/w. citric acid, 0.5% w/w. ECOSURF™ EH-9, at pH 6	Long application times were required to achieve a moderate amount of soil removal, as the soil was adhered to the paint surface. The addition of both chelating agent and surfactant were required to boost the cleaning efficacy of the aqueous phase in order to reduce application time. Epsomite was reduced to a similar extent as swab-applied deionised water.
Yellow Ochre	10	40	50	Deionised water (no clear benefit to using adjusted water, due to heavily embedded soiling)	Yellow ochre paint films had highly imbibed soil; the use of chelating agents and surfactants did not improve the removal of the imbibed soil, hence deionised water was used as the aqueous phase to minimise prolonged clearance. NB: This sample was difficult to clean to an acceptable level with any of the systems evaluated.
French Ultramarine	25	35	40	1% w/w. citric acid, 0.5% w/w. ECOSURF™ EH-9, at pH 6	A lower % aqueous content was required as the surfaces of French ultramarine paints were particularly vulnerable to pigment removal. Chelating agents and surfactants were needed within the aqueous phase to achieve an acceptable level of soiling removal.

decreasing cleaning efficacy. Gloss changes resulting from the KSG systems were comparable to many of the optimal systems (see Fig. 2.5), however these systems did not completely prevent disruption to thin medium skins on the French ultramarine samples, due to the mechanical action required.

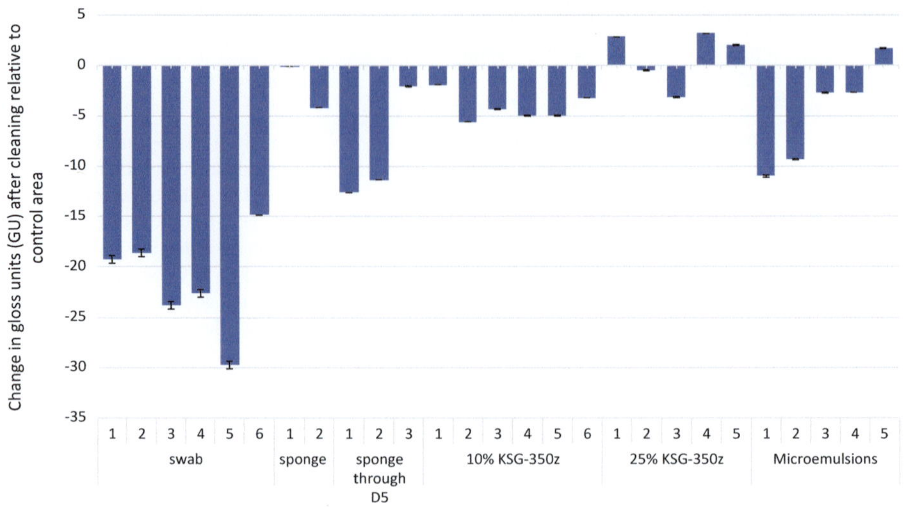

Fig. 2.5 Gloss change measured at 60 ° for sensitive French ultramarine samples – comparing application methods and cleaning materials

2.3.6 *Application Methods*

Table 2.4 summarises key observations made on the use of different application methods on these paint films. Reduced gloss change and pigment pickup was most evident when swabs were not used with all paint types apart from the Titanium white (data not shown) which appeared less affected by the application method. Figure 2.5 includes the gloss change data recorded for the French ultramarine samples where swabbing clearly resulted in the highest gloss change. Several of the sponge-applied systems; the systems applied through D5, the KSG-350z emulsifiers (12–22) and low water content microemulsions (25–27) all resulted in reduced gloss change when compared to the swabbed samples. Where present, epsomite was removed to some extent by all systems, and medium skins were also disrupted, albeit to a lesser extent when methods other then free-solvent swabbing were used. The occasional increase in gloss may indicate the successful retention of the thin medium skin, and/or the presence of residual cleaning materials such silicone emulsifiers or NaDOSS surfactant (see *Residues*).

2.3.7 *Residues*

Potential cleaning system residues were explored through visual observation, microscopy, ATR-FTIR, XPS and via a pilot study into the retention of D5 solvent using PyGCMS. Overall, there were no residues detected with microscopy or ATR-FTIR, except in cases where the measurement focussed on microemulsion

Table 2.4 Summary of notes on different application methods

Application method	General observations (NB: All epsomite-rich surfaces were affected by all systems)
Cotton swabs (hand-rolled)	Affects medium skin, can cause abrasion, tend to unravel with non-polar solvents and silicone emulsifiers. Not very effective for clearance. Removal of epsomite, gloss changes and pigment pickup risk most pronounced. Caused approximately double the magnitude of gloss change (reduction) than other systems.
Rigid polysaccharide hydrogels (room temp application)	Useful for delivering tailored aqueous systems, minimises abrasion, though can require pressure and repeated application. Difficult to assess surface during cleaning, some can be too active, causing the removal of medium skins, some paint swelling and some pigment pickup noted. Likely to be more useful for older paints/lighter soiling layers.
D5 silicone solvent barrier	Enhances cleaning efficacy of silicone emulsifiers, can limit penetration of liquid and aqueous cleaning systems into lean/porous paints and can slow down cleaning action of aqueous systems, which can be beneficial. Longer exposure times and mechanical action required (brush, swab, sponge); which may not reduce pigment pickup.
Tissue barrier (wet strength, non-woven)	Difficult to assess surface during cleaning and can slow down cleaning action. Can reduce pigment pickup and possibly useful for delivering W/O microemulsions (see *Gillman et al*, this volume). Minimal improvement in gloss change reduction.
Makeup sponges (e.g. Muji)	Requires assessment of softness, polymer type, additives and solvent compatibility prior to use. Useful for application of free aqueous systems and for clearance (free solvents and silicone emulsions). Repeated application often required, some gloss change noted. Some disintegration noted with W/O microemulsions.
Sponge through D5 barrier	Relatively successful on very sensitive paints. Repeated replenishments of D5 barrier required. Mechanical action required suggests this is unlikely to be suited to underbound/very lean surfaces. Gloss change similar with and without D5 barrier (due to mechanical action). Useful for reduction of pigment pickup and can enhance cleaning activity when used with silicone emulsions.

surfactant tidelines (visible to the naked eye) when the surfactant-phase was included above 40% w/w.[12] Additionally, no silicone residues were detected using ATR-FTIR for the KSG-350z systems with emulsifier contents of both 10% and 25% by weight. For the highly surface-sensitive XPS analysis [26], the quantitative elemental composition in atomic per cent and associated chemical states were determined at the surface of unsoiled cadmium yellow samples, including untreated control areas as well as areas cleaned with KSG emulsions and Shellsol-D40 microemulsions, with appropriate clearance steps.

[12] Tidelines are primarily a feature of the small cleaning tests carried out and are generally removed when cleaning over a larger scale.

In Fig. 2.6 (upper), the silicon (Si) concentration (relative to the control), appears enriched for many of the test sites. The Si binding energies detected were typical of siloxanes, indicating the presence of residues of silicone emulsifiers and/or silicone solvents on paint surfaces. This data also suggests that the ability to clear the surface of siloxanes (see columns 3-5 from the left) did not differ significantly with the use of D5, HMDS and D5/HMDS blends, as the Si concentrations remain similar. This was also supported by preliminary data obtained with PyGCMS (not shown) where D5 solvent was detected after 2 weeks within yellow ochre paint films cleaned with neat KSG 350z (with and without a D5 clearance step). Figure 2.6 (lower), indicates

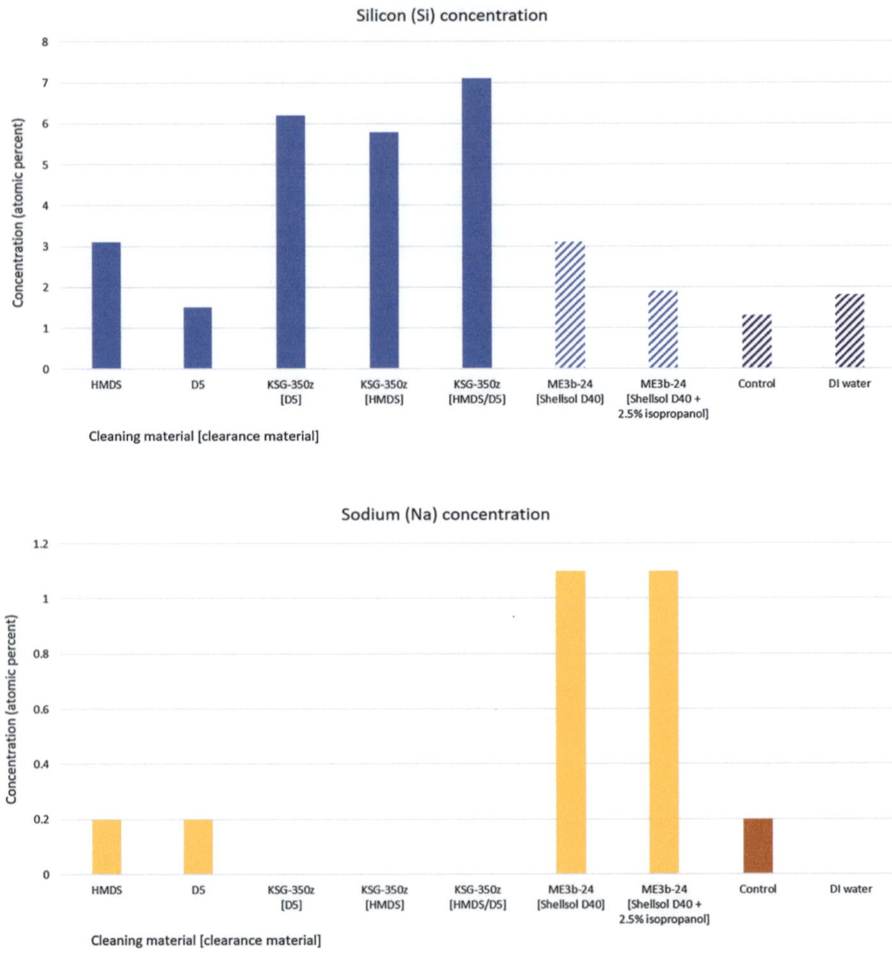

Fig. 2.6 XPS data in atomic % for various materials on cleaned cadmium yellow paints (upper) Si (from silicone [siloxane] solvents and emulsifiers) at the surface; (lower) Na (from NaDOSS surfactant) at the surface. The hatched bars on the Silicon graph indicate Si originating from the inorganic silica; the filled bars correspond to Si originating from siloxanes only

that deionised water and the (50% aqueous-phase containing) silicone emulsifiers removed any inherent sodium (Na) from paint surfaces during cleaning, whereas the HMDS and D5 silicone solvents did not (the levels appear similar to the control). The Series 3b-24 microemulsion (10% aqueous phase) appears to leave behind NaDOSS surfactant monolayers which, similarly to the silicone-treated surfaces, requires further exploration around the potential consequences of the use of these materials (and others) on the future soiling and cleaning behaviour of treated paint films. It is also noted that the atomic % sodium (Na) amount is significantly less than the siloxane (Si) remaining on the same paint film. The Na results also suggest there is little benefit to clearing the NaDOSS-based microemulsions using Shellsol D40 with added 2.5% w/w propan-2-ol when compared to using D40 alone. Similar to the FTIR data obtained (Fig. 2.4), reduced magnesium was also noted via XPS after cleaning (data not shown), confirming that epsomite was removed by all of the aqueous-containing systems evaluated, except where the HMDS and D5 silicone solvents had been applied neat. Although this data is preliminary, the XPS data results suggest that the 50% aqueous-phase silicone emulsifiers removed marginally more epsomite than the 10% aqueous-phase ME3b-24 microemulsion, presumably due to the higher aqueous content and extended application times required when using the KSG materials; whereas the epsomite reduction for the ME3b-24 and deionised water applications appeared comparable.

2.3.8 Optimal Systems for Each of the W&N Paints

The tenacity of the artificial soil varied significantly with paint colour, where the soil was noticeably less well-adsorbed for the non-water-sensitive titanium white paints than each of the more sensitive colours. This presumably reflects differences in the surface character of the paints, including softness, physical coherence, hydrophilicity, degree of oxidation, the presence of soluble salts, ionic character and other factors which require further investigation.

For the relatively robust titanium white sample(s) which did not have visible medium skins, the series 3b microemulsions (particularly 3b-5, 40% aqueous phase) resulted in optimum soil-removal efficacy, however other acceptable cleaning outcomes were also achieved using swab-applied aqueous options containing either citric acid chelating agent (1% w/w. at pH 6.0) or surfactant (0.5% w/w. ECOSURF™ EH-6 or 0.5% w/w. ECOSURF™ EH-9). For this paint, sponge application slightly reduced the cleaning efficacy, and the D5 silicone solvent barriers and/or KSG emulsifiers resulted in the least efficient soil removal. This could be enhanced by increasing the proportion of the aqueous phase and/or adjusting the application method/time.

The yellow ochre paints, which were characterised by medium-skins, enhanced oxidation and heavily imbibed soiling, could not be cleaned effectively with any of the systems evaluated, with only the lightly bound soiling being removed prior to pigment pickup and other undesirable changes becoming evident. As the medium-rich surface of this paint film was highly vulnerable to swelling and mechanical action, substituting swabs with cosmetic sponges and/or using KSG-350z silicone emulsifiers significantly minimised changes to these paint surfaces. Further work exploring the use of slightly raised pH aqueous phases within KSG emulsions could prove beneficial for encouraging the removal of imbibed soiling.

For the epsomite-rich cadmium yellow paints, optimal cleaning outcomes were achieved using the silicone emulsifier systems (50% aqueous phase with added 0.5% ECOSURF™ EH-6 or 0.5% ECOSURF™ EH-9 surfactant), cleared using sponges loaded with D5 then HMDS, as well as tailored aqueous systems applied using cosmetic sponges, e.g., 0.5% ECOSURF™ EH-9 pH adjusted to 5.5, cleared with adjusted water set to pH 5.5.[13] Both systems facilitated effective soil removal without causing visible changes to the surface; however, as stated earlier, for all the options evaluated, any epsomite present was at least partially removed during cleaning, and trace levels of silicone emulsifer residues were noted using XPS (Fig 2.6).

Similarly, for the highly water-sensitive French ultramarine blue paints, characterised by medium skins, enhanced oxidation and the risk of significant gloss change, the use of cosmetic sponges and silicone emulsifiers enabled a helpful degree of soil removal. Lowering the aqueous phase pH to 4.5 reduced undesirable effects further, and chelating agents offered an increase in soiling removal efficacy. Other slight enhancements were made where surfactants and chelating agents were used in combination. Optimal systems included sponge-applied 1% w/w. citric acid, pH 6; cleared with sponge-applied adjusted water at pH 6.0; and KSG-350z (40% aqueous phase, 1% w/w. citric acid, pH 6.0 with additional 0.5% w/w. ECOSURF™ EH-9). In addition to tailoring the cleaning options for each paint, further reductions in gloss change were achieved through the use of sponges, D5 barriers, KSG-350z emulsions and low % aqueous content microemulsions.

2.4 Conclusions

These combined evaluations may help guide conservators towards soiling removal strategies for water-, mechanical action- and solvent- sensitive painted surfaces. The approach used, involving exploring: the character of the paint surfaces, the effects of adjusting and optimising of the active aqueous phases prior to incorporation into gels and emulsions, as well as exploring the effects of application methods, proved beneficial to the cleaning outcomes achieved for each of the four paints evaluated. For the three sensitive paints in particular, this process improved the cleaning outcomes, including enhanced soiling removal, reduced pigment pickup and the reduction of undesirable surface changes.

[13] The cadmium yellow paints tended to be more vulnerable toward the use of chelating agents in comparison to the surfactants.

The moderately- and highly-sensitive paints benefitted from reduced mechanical action (using sponges and brushes), reduced aqueous exposure though controlling access and penetration (using D5 barriers, tissue, gels, microemulsions and emulsions), as well as from the tailoring of aqueous phases (via setting the pH to ~4.5–6, by adding 0.5% w/w. nonionic surfactant, and by adding up to 1% w/w. chelator). All of these modifications helped reduce pigment pickup, resulted in reduced gloss shifts and minimised surface (such as medium skin) disruption. Reducing the aqueous pH to 4.5–5 also resulted in lower levels of pigment pickup for the French ultramarine blue and cadmium yellow paints, however the cleaning efficacy also tailored off due to the pKa(s) of the acids used, which in some cases necessitated the addition of surfactants. The hydrogels evaluated offered poor cleaning efficacy and resulted in the swelling of the W&N paint films; however, agar and gellan gum options proved more useful for naturally aged oil paintings with lighter soiling layers and minimal medium skins [7].

The Shellsol D40 microemulsions proved most beneficial with a 10% w/w. aqueous phase, with additional gains made through adjusting the character of the aqueous phase. NaDOSS residues were confirmed in cleaning tidelines using ATR-FTIR when the surfactant phase content was above 40%. Trace levels of Na (from the NaDOSS surfactant) were detected on uppermost cleaned and cleared paint surfaces via XPS, most likely present as a monolayer. The silicone emulsifier KSG-350z proved beneficial at between 10–25% emulsifier content, with 35–50% aqueous phase adjusted with added surfactant and/or chelator, brush-applied and sponge cleared using neat silicone solvents. XPS analysis also confirmed the presence of siloxane residues on cleaned and cleared samples, and that the benefits of a double (D5, HMDS) clearance step were minimal. In all cases, the impact of trace cleaning system residues and solvent retention on the re-soiling and re-treatability of these paints is of key interest and remains to be explored.

Acknowledgements This research was undertaken as part of the Cleaning Modern Oil Paints project (2015–2018) supported by Heritage Plus, a Joint Programming Initiative of the European Commission. The authors would like to thank the following people and institutions: all CMOP partners and associate partners; Winnats Scientific, UK; Chris Stavroudis, US; Ian Garret, UK; and the Tate Conservation and Research Departments. The digital microscopy images were taken using a HIROX KH-8700 digital microscope, the purchase of which was supported in part by the Horizon2020 funded NANORESTART project (agreement 646063).

References

1. Álvarez F, Vázquez G, Sánchez-Vilas M, Sanjurjo B, Navaza JM (1997) Surface Tension of Organic Acids + Water Binary Mixtures from 20 °C to 50 °C. Journal of Chemical & Engineering Data 1997 42 (5) 957–960
2. Angelova L, Ormsby B, Townsend JH, Wolbers R Eds (2018) Gels in the Conservation of Art. Archetype Publications London: p. 381
3. Casoli A, Cremonesi P, Héritier P-A, Volpin S (2019) Analytical study to monitor the effectiveness of a combined liquid-dispensing and micro-aspiration system for the cleaning of modern oil paintings. These proceedings, Chapter 40

4. Chung JY, Ormsby B, Lee J, Burnstock A, van den Berg KJ (2017) An Investigation options for surface cleaning unvarnished water-sensitive oil paints based on recent developments for acrylic paints. In ICOM-CC 18th Triennial Conference Preprints, Copenhagen, 4–8 September 2017. Paris: International Council of Museums
5. Daudin-Schotte M, Bisschoff M, Joosten I, van Keulen H, van den Berg KJ (2013) Dry Cleaning Approaches for Unvarnished Paint Surfaces. New Insights into the Cleaning of Paintings: Proceedings from the Cleaning 2010 International Conference, Universidad Politecnica de Valencia and Museum Conservation Institute, edited by Mecklenburg, Marion F., Charola, A. Elena, Koestler, Robert J. Smithsonian Contributions to Museum Conservation. Washington, DC: Smithsonian Institution: 209–219
6. Dillon CE, Lagalante AF, Wolbers RC (2014) Acrylic emulsion paint films: The effect of solution pH, conductivity, and ionic strength on film swelling and surfactant removal. Studies in Conservation 59(1): 52–62
7. Gillman M, Lee J, Ormsby B, Burnstock A (2019) Water-sensitivity in modern oil paintings: Trends in phenomena and treatment options. These proceedings, Chapter 38
8. Hintz M, van den Berg KJ, Stols Witlox M, Steyn L (2019) Improving the surface cleaning of water sensitive oil paint by use of alternative application methods. These proceedings, Chapter 44
9. Labuza TP, Dugan Jr RL (1971) Kinetics of lipid oxidation in foods. C R C Critical Reviews in Food Technology 2:3: 355–405
10. Lagalante AF & Wolbers RC (2016) The cleaning of acrylic paintings: new particle-based water-in-oil emulsifiers. *Colore e Conservazione Proceedings*: 107–114
11. Lagalante A, Wolbers R (2017) Particle-based silicone cleaning emulsions: studies in model paint systems for the clenaing of water sensitive artworks. In Gels in the Conservation of Art, Lora Angelova, Bronwyn Ormsby, Joyce H Townsend and Richard Wolbers (Eds). Archetype Publications, London: 193–199
12. Lazzari M, Chiantore O (1999) Drying and oxidative degradation of linseed oil. Polymer Degradation and Stability 65:2: 303–313
13. Lee J, Bonaduce I, Modugno F, La Nasa J, Ormsby B, van den Berg KJ (2018) Scientific investigation into the water sensitivity of twentieth century oil paints. *Microchemical Journal*, Vol. 138: 282–295
14. Lee J, Ormsby B, Burnstock A, van den Berg KJ (2019) Modern oil paintings in Tate's collection: a review of analytical findings and reflections on water-sensitivity. These proceedings, Chapter 39
15. Myers JM, Cushman M (2019) Analysis of cleaning efficacy and clearance of silicone-based Pickering-type emulsions used in the cleaning of water-miscible oil paints. These proceedings, Chapter 43
16. Ormsby B, Soldano A, Keefe MH, Phenix A, Learner T (2013) An Empirical Evaluation of a Range of Cleaning Agents for Removing Dirt from Artists' Acrylic Emulsion Paints. AIC Specialty Group Postprints 23: 77–87
17. Ormsby B, Keefe MH, Phenix A, von Aderkas E, Learner T, Tucker C, Kozak C (2016) Mineral Spirits-based Microemulsions: A Novel Cleaning System for Acrylic and Other Modern Painted Surfaces. Journal for the American Institute for Conservation 55:1:12–31
18. Perrin D, Dempsey B (1974) Buffers for pH and Metal Ion Control. Springer Netherlands: 105
19. Phenix A, Burnstock A (1992) The removal of Surface dirt on Paintings with Chelating agents. The Conservator 16:1:28–38
20. Pouyet E, Cotte M, Fayard B, et al (2015) Appl. Phys. A 121: 967. https://doi.org/10.1007/s00339-015-9239-4
21. Silvester G, Burnstock A, Megens L, Learner T, Chiari G, van den Berg KJ (2014) A cause of water-sensitivity in modern oil paint films: the formation of magnesium sulphate. Studies in Conservation 59:1: 38–51
22. Stavroudis C (2012) More from CAPS3: Surfactants, silicone-based solvents, and microemulsions. WAAC Newsletter, Volume 34, Number 3

23. Stavroudis C (2016) Silicone-Based Solvents in Conservation. As free solvents and components of gel systems, and microemulsions. *Colore e Conservazione Proceedings*: 178–184
24. van der Weerd J, Van Loon A, Boon J (2005) FTIR Studies of the Effects of Pigments on the Aging of Oil. Studies in Conservation 50(1): 3–22
25. Volk A, van den Berg KJ (2014) Agar – A New Tool for the Surface Cleaning of Water Sensitive Oil Paint? In van den Berg K.J et al. (eds) Issues in Contemporary Oil Paint. Springer: 389–406.
26. Winnats Scientific Services, Unpublished report, *1D8UP02 XPS Analysis of Paints and Pigments* 2018

Part I
Paint Technology

Chapter 3
Pigment Surface Treatments: 20th and 21st Century Industrial Techniques and Strategies for their Detection

Carolina Salis Gomes, Cátia Ferreira, Brenda Rossenaar, Ineke Joosten, Inez van der Werf, Leslie Carlyle, and Klaas Jan van den Berg

3.1 Introduction

As pigments are not produced specifically for the artists' market, modern artist's oil paint contains pigments designed for industrial applications such as architectural coatings, automotive paint, plastics and cosmetics. These industrially produced pigments are supplied with surface treatments that can present different characteristics intended to improve performance in a range of binders[1], enabling manufacturers to provide a broad choice of pigment grades suitable for different applications [1, 32]. Pigment surface treatments can influence and/or control specific properties such as colour and batch reproducibility and ease of dispersion in the binder. In addition, they can be used to adjust paint consistency, improve compatibility with the binder, increase the chemical stability of the pigment in the binder, reduce pigment solubility, and enhance light stability and weather fastness [6].

Manufacturers such as BASF offer large pigment portfolios. Their data-sheets indicate that their pigments are appropriate for use in plastics, paints or textiles.

[1] Personal communication of CSG and CF with BASF representatives, October 2017.

The research for this paper was carried out by CSG and CF as part of an internship at the Cultural Heritage Agency of the Netherlands in the framework of the Cleaning of Modern Oil Paints.

C. Salis Gomes · C. Ferreira · L. Carlyle
Faculdade de Ciências e Tecnologia da Universidade NOVA de Lisboa, Lisbon, Portugal
e-mail: cb.ferreira@campus.fct.unl.pt

B. Rossenaar
Nouryon, Deventer, The Netherlands
e-mail: brenda.rossenaar@nouryon.com

I. Joosten · I. van der Werf · K. J. van den Berg (✉)
Cultural Heritage Agency of the Netherlands, Amsterdam, The Netherlands
e-mail: k.van.den.berg@cultureelerfgoed.nl

K. J. van den Berg et al. (eds.), *Conservation of Modern Oil Paintings*,
https://doi.org/10.1007/978-3-030-19254-9_3

Since they are intended to be multifunctional, not all treatments are necessarily ideal for a given binder or end use. As Levison points out, although there can be similarities between artists' and industrial paints, the requirements for stability, lightfastness and the pigment's contribution to paint durability are different [25]. Because they can confer a competitive edge, surface treatments are largely kept confidential ([9], p. 29)[2]. For conservators, the presence and composition of these treatments is of concern since detailed knowledge of the materials used in modern oil paints is important for understanding degradation phenomena [3] and for research calling for paint modelling using highly characterized historical reconstructions [8].

Part I of this paper provides a brief background on surface treatment and presents examples of materials and application methods. Key sources are Merkle and Schäfer [29], Hays [19], Schröder [32] and Bugnon [5] which review the surface treatment literature and provide background. Important sources in the technical literature (including patents) covering industrial pigment manufacturing are Lewis [26], Buxbaum and Pfaff [6] and Herbst et al. [21]. Part II reports on the results of the analytical approach used to detect the presence of these surface treatments with the electron microscope and hyphenated mass spectrometry techniques. The following pigments were selected for study: Cu-Phthalocyanine Blue and the Hansa yellows PY3 and PY74, Umber, Vine Black, Synthetic Ultramarine Blue, Lead White, Zinc White, and Titanium White. This broad pigment selection reflects their historical importance and use among modern-day artists. In addition, these pigments represent key examples of the rapidly developing pigment production industry in which the application of surface treatments is becoming increasingly important.

3.2 PART I: Background on Pigment Surface Treatments

As Schröder [32] explains, a surface treatment on an individual pigment particle consists of a "modification of the surface character of a pigment in a desired manner where the original pigment surface can either be modified or replaced by a completely new surface". In the literature consulted, terms such as *surface treatment*, *coating*, and *encapsulation* are often used interchangeably. However, as will be seen, a surface treatment does not always result in a coating, and the term encapsulation is distinguished from coating in some cases, where encapsulation describes a specific polymeric treatment involving total coverage of each pigment particle [5, 33][3].

There are very early references to pigment treatments prior to their use in paint. For example, the fourteenth century Montpellier manuscript *Liber diversarum arcium* directs that azure (presumably azurite pigment) should first be ground with

[2] For this study, CSG and CF interviewed representatives from BASF and Royal Talens. Details are found in their respective Masters Theses (Department of Conservation & Restoration, Universidade NOVA de Lisboa September 2018.

[3] Personal communication of CSG and CF with BASF representatives, October 2017.

fish glue, and cinnabar with egg and water prior to their use for painting on walls and ceilings ([10] p.143)[4].

Recent sources report that beginning in late nineteenth century and throughout the twentieth century (see [1, 14]) the patent literature and articles in technical journals reveal numerous efforts to improve the performance of pigments.

One of the oldest of the modern industrial treatments is resination, or rosination, which has been in use since at least the 1920's [29]. This treatment was adopted by the printing industry due to the need for greater transparency and smaller particle sizes. Treatments with rosin also reduces the generally undesirable characteristic of bronzing[5] ([24] p.335) Vernadakis reports that rosin is still "widely used today, as itis one of the least expensive surface treating agents" ([38], p.11). Natural rosin, used in organic pigments such as the azo and polycyclic pigments, is sensitive to oxidation and has an unfavourable effect on storage stability. Furthermore, due to its low melting point the mechanical properties of the paint film can also be affected [29].

Other early materials reported are waxes [11] and metal stearates [15, 28]. These latter materials were also used in oil paint as additives to correct and improve oil paint formulations [22].

3.2.1 *Materials and Applications*

The literature uses *surface treatment*, or *surface-modification*, as umbrella terms for various applications and processes which alter the surface character of pigments to improve their performance [33]. Depending on the nature of the pigment, organic or inorganic, and its intended use, these modifications are carried out during the synthesis of the particle and/or afterwards, in a post-synthesis phase [9, 32, 33, 38]. While the same materials can be used in both cases, they perform different functions.

3.2.1.1 During-Synthesis

Surface treatments carried out during synthesis are intended to control particle formation in terms of shape and size, and to modify the crystal phase. In the latter case substances such as rosin or Turkey red oil, which absorb selectively on different

[4] For a discussion on the role of pigment coating see [10] [§3.2.5] p.217). Von Baum et al. suggest that such protein coatings would confer advantages, for example for Azurite, coating could reduce susceptibility to yellowing, wrinkling or cracking during drying [39].

[5] Bronzing "is an effect of light reflection giving the print different shades depending on the observation angle (..)can be corrected by surface treatments or by other modifications of the pigment" ([24], p. 326).

crystal faces, are used to inhibit crystal growth (e.g. in case of Cu-Phthalocyanine pigments) see [14], p.241–242, and [32].

The materials applied during synthesis are varied and affect performance such as decreasing powder cohesion and enhancing pigment dispersibility [32]. They may be *surfactants;* Swarup and Schoff [34] report the use of long chain carboxylic acids (fatty acids). Both natural and synthetic *polymeric materials* are also commonly used and include compounds such as poly (12)-hydroxystearic acid [38]. Other treatments aim to decrease solubility and enhance light stability (see discussion on Titanium White pigments in [37]).

The materials listed above also affect the post-synthesis phase. That is, by modifying the surface character, active sites are created on the particle surface which aide further treatments; for example, by facilitating the adsorption of polymeric treatments thus proving an effective steric stabilization[6].

3.2.1.2 Post-Synthesis

After pigment particles are fully formed their surface character can be further modified. Treatments applied at this stage are commonly designated as *coatings*, or *after-treatments* ([6], p.10), also referred to as *post-treatments* [33]. A *pigment coating* is ideally a colourless or nearly colourless material which creates a physical and chemical barrier between the pigment and the binder. By adjusting the character and energy of the pigment surface, the coating influences pigment/binder interactions and improves weathering properties of the subsequent paint system (i.e. stability against ultraviolet light and humidity).

The coating creates affinity with the binder which improves pigment properties during paint manufacture such as wetting out of the particles, increased dispersion and anti-flocculation and during shelf-life, for example acting as suspension agents to prevent settling and separation ([6], p. 5). Coatings consist of a thin surface layer of "molecules that can be tailor-made through chemical modifications so that they have a higher affinity for selected interfaces, such as solid/liquid and coating/liquid" ([2], p. 70). They can be present in up to 10% of the total dry weight of the pigment particle, [5] and vary in thickness, porosity and density ([29], p. 157; Buxbaum and Pfaff, p. 67). The coating can be partial or total depending on the type of pigment and the nature of the application method [32].

A coating may consist of a monomolecular layer or a multilayer of inorganic compounds (e.g. silicium dioxide, aluminium oxide, zirconium dioxide), or organic materials (e.g. polyalcohols, siloxanes, organo-functional silanes or titanates) or a combination of both [6]. To receive the coating, an intermediate layer may first be added to the pigment surface [33]. Bugnon [5] writes that organic pigments which will be given an inorganic coating require such an intermediate layer with materials

[6] Ideally, the resulting system consists of well solvated and unfurled polymeric chains that should be highly compatible with the binder, allowing a durable and permanent adsorption of the polymeric dispersant onto the pigment surfaces [30].

such as pigment derivatives. Common substances which are closely related to the pigment both in terms of their constitution and hue are strongly adsorbed on the pigment surface to facilitate the adhesion of other polymeric treatments (also called *hyperdispersants*) [5, 38]. Cu-Phthalocyanine Blue pigments are usually treated by milling the crude with a pigment derivative such as sulfonated phthalimidomethyl phthalocyanine [38].

3.3 Part II: The Identification of Pigment Surface Treatments

3.3.1 *Pigment Selection*

A total of 31 samples of dry pigments were collected for investigation (Table 3.1).

The Hansa® Yellows, PY3 and PY74, were selected in order to establish whether possible surface treatments could contribute to the instability of the subsequent oil paint. Well-characterized Cu-Phthalocyanine Blue (CuPc Blue) pigments [13], were chosen since it was reported in the literature that this class of organic pigments were amongst the most surface treated pigments (PCI mag., 2005). For the same reason, synthetic Ultramarine Blue pigments were selected. (A. Gambardella, unpublished results). An Umber, a Vine Black, and two modern-period Lead Whites[7] used for historically accurate oil paint reconstructions [7] were selected to determine whether any of them had been treated during manufacture since surface treatments could potentially compromise their interaction in the oil binder.

Three other samples of Lead White pigments as well as the Zinc White pigments were collected from the RCE Reference Collection[8]. both Lead White and Zinc White have been implicated in paint degradation primarily due to their reactivity in oil media [23, 31]. Contemporary Titanium Whites included 3 anatase and 2 rutile pigments[9].

3.3.2 *Analytical Methodology*

Table 3.1 the analytical techniques employed for each pigment. Initial studies of the pigments were carried out in a conventional Scanning Electron Microscope (SEM, JEOL JSM5910LV) with imaging in low vacuum mode (LV, 30 Pa). Pigments were evaluated visually in terms of particle size and shape with backscatter imaging.

[7] These Lead Whites were acquired in 1999 from (respectively) Kremer Pigmente and the former Dutch Schoonhoven Company.

[8] One zinc white provided by Carlyle.

[9] Provided by Birgit van Driel [37].

Table 3.1 Summary of pigments investigated, the techniques used and observations

Pigment		Sample Codes, year of acquisition/ brand and/or donated by[a]	SEM-EDS	FEG-SEM[b]	(S) TEM-EDX[b]	EGA-MS[a,b]	Py-GC/MS[b]	Py-GC/MS[b] (TMAH)	Observation
Hansa® yellow	PY3	0526, n.d., RCE	C, N, O, Al, Si, Cl	n/a	n/a	Characteristic pigment fragments only.	n/a	n/a	Inconclusive results. Aluminium and silicon were identified, however needs further analyses using the other techniques.
	PY3	R.T., 2017, Royal Talens, CW	C, N, O, Cl	Coating indication	n/a	Characteristic pigment fragments only.	Characteristic pigment fragments only.	Characteristic pigment fragments only.	Inconclusive results. Needs further analyses using (S)TEM-EDX.
	PY74	0527, n.d., RCE	C, N, O	n/a	n/a	Characteristic pigment fragments only.	n/a	n/a	Inconclusive results. Needs further analyses using the other techniques.
	PY74	R.T., 2017, Royal Talens, CW	C, N, O	n/a	Si Coating	Characteristic pigment fragments only.	Characteristic pigment fragments only.	Characteristic pigment fragments only.	An inorganic coating is present.
Copper Pthalocyanine Blue	PB15:?	833, n.d., RCE	C, N, O, Al, Cu	Coating	Si Coating	Dehydroabietic acid (DHA)[d]	Diterpenoids (*pinaceae* resin): Dehydroabietic acid (DHA) (239, 240, 299) Methyl abietate (256, 342, 227) + characteristic pigment fragments.	Diterpenoids (pinaceae resin): Isopimaric acid methyl ester (241, 91, 105) Abietic acid methyl ester (213, 131, 273) 15-methoxydehydroabietic acid, methyl ester (329, 330, 73) 7-Oxo-dehydroabietic acid, methyl ester (253, 328, 187) + characteristic pigment fragments.	Both inorganic and organic coatings are present.

	PB15:?[c]	Monastral Blue 2755, n.d., Old Holland, RCE	C, N, O, Al, S, Cl, Cu	n/a	n/a	Characteristic pigment fragments only.	Characteristic pigment fragments only.	Characteristic pigment fragments only.	Inconclusive results. Needs further analyses using the other techniques.
	PB15:0	IRPA, 1956, The Royal Institute for Cultural Heritage of Belgium, CD	C, N, O, Al, S, Cu	n/a	n/a	Characteristic pigment fragments only.	Characteristic pigment fragments only.	Characteristic pigment fragments only.	Inconclusive results, however according to literature findings, PB15:0 is expected not to be surface treated. Needs further analyses using the other techniques.
	PB 15:0	CuPc, 1978, Brico, CD	C, N, O, Al, S, Cu	n/a	n/a	Characteristic pigment fragments only.	Characteristic pigment fragments only.	Characteristic pigment fragments only.	Inconclusive results. Needs further analyses using the other techniques.
	PB 15:2	W&N, 1980, Winsor&Newton, CD	C, N, O, Al, S, Cu, Ba	Coating	Si Coating	Characteristic pigment fragments. SO_2[d] (64, 48) 400–600 °C Characteristic fragment ions for alkane	Characteristic pigment fragments only.	Inconclusive.	Inorganic silica coating present in the pigment. Needs further analysis on organic treatments.
Synthetic ultramarine blue		Uncoated Ultramarine, 2015, Kokkie, AG'	C, O, Na, Al, Si, S, K, Ca	No coating present	No coating	Inconclusive	Inconclusive	Inconclusive	No evidence of surface treatment.

(continued)

Table 3.1 (continued)

Pigment	Sample Codes, year of acquisition/ brand and/or donated by[a]	SEM-EDS	FEG-SEM[b]	(S) TEM-EDX[b]	EGA-MS[a,b]	Py-GC/MS[b]	Py-GC/MS[b] (TMAH)	Observation
	Coated Ultramarine, 2015, Kremer Pigmente, BD	C, O, Na, Al, Si, S, K,	Coating	Si Coating	Inconclusive	Inconclusive	Methyl esters of fatty acids	Inorganic silica coating and possibly organic surface treatment present in the pigment. Needs further organic analyses.
	0867, n.d., RCE	C, O, Na, Al, Si, S	n/a	n/a	Inconclusive	Inconclusive	Inconclusive	Needs further analysis to confirm surface treatment.
	Lukas 2537, n.d., RCE	C, O, Na, Al, Si, S, K	n/a	n/a	Inconclusive	Inconclusive	Inconclusive	Needs further analysis to Confirm surface treatment.
Vine black	FCT-PNT MOLART 1999, Kremer Pigmente, LC	C, O, Al, Si, S	No coating present	No coating present	Inconclusve.	n/a	n/a	No evidence of surface treatment. Aluminium and silicon identified possibly related to traces from the combustion process during manufacture.

Umber	FCT-PNT – MOLART, 1999, Kremer Pigmente, LC	C, O, Mg, Al, Si, P, Ca, Mn, Fe	n/a	No coating present	Inconclusive.	n/a	n/a	No evidence of surface treatment. Aluminium and silicon identified possibly related to ore/ impurities.
Lead white	8233, 2004, Dr. A. Haagen, Roermond, RCE	C, O, Al, Pb, S,	n/a	Inconclusive.	SO_2[e] (64, 48)	n/a	n/a	Inconclusive results. Needs further analyses using the other techniques.
	9612, 2014 Scheveningen Classic, Old Holland, RCE	C, O, Al, Pb, S	n/a	n/a	Inconclusive	n/a	n/a	Inconclusive results. Needs further analyses using the other techniques.
	9641, 2014 Sikkens Chest Nr.1, RCE	C, O, Al, Pb, S	n/a	n/a	acetic acid[e] (60, 45, 43)	n/a	n/a	Inconclusive results. Needs further analyses using the other techniques.
	FCT-PNT – MOLART, 1999, Kremer Pigmente, LC	C, O, Al, Pb, S	Inconclusive.	Inconclusive	Inconclusive	n/a	n/a	Inconclusive results. Needs further analyses using the other techniques.
	FCT-PNT-MOLART, 1999, Schoonhoven, LC	C, OAl, Si, Pb, S, Cl	Inconclusive.	Inconclusive	Acetic acid[e] (60, 45, 43)	n/a	n/a	Inconclusive results. Needs further analyses using the other techniques.
Zinc white	0191, n.d., Kremer Pigmente, RCE	C, O, Zn, Al	n/a	Inconclusive.	Inconclusive.	Inconclusive.	Inconclusive.	Inconclusive results. Needs further analyses using the other techniques.

(continued)

Table 3.1 (continued)

Pigment	Sample Codes, year of acquisition/ brand and/or donated by[a]	SEM-EDS	FEG-SEM[b]	(S) TEM-EDX[b]	EGA-MS[a,b]	Py-GC/MS[b]	Py-GC/MS[b] (TMAH)	Observation
	8232, 2004, Old Holland-Classic colors, RCE	C, O, Zn, Al	n/a	Inconclusive.	Inconclusive.	n/a	n/a	Inconclusive results. Needs further analyses using the other techniques.
	9647, 2014, Sikkens, RCE	C, O, Zn, Al	N	Inconclusive.	Inconclusive.	Inconclusive.	Inconclusive.	Inconclusive results. Needs further analyses using the other techniques.
	9648, 2014, Sikkens, RCE	C, O, Zn, Al, S	n/a	n/a	SO_2[d] (64, 48) 500–700 °C	Inconclusive.	Inconclusive	Inconclusive results. Needs further analyses using the other techniques.
	Carlyle, Pre-1980's, LC	C, O, Zn	Inconclusive.	Possibly silica coating. Needs further analyses to confirm.	Inconclusive.	Inconclusive.	Inconclusive	Inconclusive results. Needs further analyses using other techniques.
Titanium white	A1, 2011, Hombitan LW	C, O, Al, Ti	No coating.	Previous results [37] showed no coting present.	Inconclusive.	Inconclusive.	Inconclusive.	Needs further analyses using other techniques.
	A2, 2014, Hunstman, A-HRF	C, O, Al, Si, Ti,	Inconclusive.	Previous results [37] showed no inorganic coating.	Inconclusive.	Inconclusive.	Trimethylolpropane (polyol)	Organic coating present in the pigment.

	A3, 2014, Hunstman, A-PP2	C, O, Al, Si,Ti	n/a.	Previous results [37] showed SiO and Al inorganic coating.	Inconclusive.	2-ethylacrolein Cyclopentanone	Trimethylolpropane (polyol) 2-ethylacrolein Products from Glycerols: Propane, 1,2,3, trimethoxy Diethylene glycol dimethyl ester	Inorganic (silica and alumina) and organic surface treatment present in the pigment.
	R2, 2011, Hunstamn HDCD	C, O, Al, Ti	n/a	Previous results [37] showed no inorganic coating.	Inconclusive	2-ethylacrolein	Phenol, 2, 4-bis(1,1-dimethylethyl) (antioxidant) 1,2-dimethyl phthalate (phthalate) Trimethylolpropane (polyol)	Organic coating is present.
	R3, 2011, Verfmo-len de Kat, BD	C, O, Al, P, Ti	n/a	Previous results [37] showed ZrO and Al inorganic coating.	Inconclusive.	2-ethylacrolein	(?) – Identified Polyol Fatty acids	Inorganic and organic coating present in the pigment.

Notes: 1:
[a]Pigment obtained from: RCE Reference Pigment Collection (RCE), Royal Talens, Mr. Coos van Waas, (RT), Catherine Defeyt (Liege Univ.) (CT), Alessa Gambardella (Rijksmuseum) (AG), Brigit van Driel (RCE/TU Delft) (BD), Leslie Carlyle/MOLART Fellowship (LC)
[b]n.d.: not dated, n/a: not applicable
[c]PB15 form is not known
[d]SO_2 – the presence of this compound possibly results from atmospheric contamination [31]
[e]Acetic acid possibly related to the conversion of metallic lead to lead carbonate [17]

Energy dispersive x-ray spectroscopy (EDX, Ultradry silicon drift detector with NSS software, Thermo Scientific) provided elemental composition and, where present in high concentration, could show traces of an inorganic treatment . A selection of pigments was studied with a FEG-SEM (NovaNanoSEM450 from FEI) at higher magnifications in high vacuum at low accelerating voltage (2 keV) showing greater detail in particle, texture and morphology[10]. (Scanning) Transmission Electron Microscope Energy Dispersive X-ray ((S)TEM-EDX) permitted the determination of the presence of inorganic coatings as well as providing information on possible pigment impurities ([37], p. 117).

The first step for detecting treatments with organic components was to employ Evolved Gas Analysis Mass Spectrometry (EGA-MS); thermal desorption desorption and pyrolysis in an inert atmosphere with direct MS. The resulting thermogram is interpreted using extracted ion chromatograms and average mass spectra ([35], p.5). For selected pigments, pyrolysis coupled to gas chromatography – mass spectrometry (Py-GC/MS), with thermally-assisted hydrolysis and methylation (THM-GC/MS) was performed.

3.4 Results and Discussion

Analytical results are summarised in Table 3.1. Seven of the 31 pigments showed clear evidence of treatment: 2 CuPc Blues, 1 of the Hansa Yellows PY74, and 4 Titanium Whites. (S)TEM-EDX and FEG-SEM were the most effective techniques for the detection of inorganic surface treatments.

For organic pigments as well as organic coatings, linking compounds to surface treatments proved to be more challenging. This was likely due to the very small amounts present and possible confusion with compounds related to pigment synthesis.

3.4.1 Hansa® Yellows – PY3 and PY74

SEM-EDX was useful for determining the size and morphology of the pigments but was limited in the identification of inorganic surface treatments since analysis is restricted to the elements present (Table 3.1). In addition, the small particle sizes of Hansa® Yellows make it particularly difficult to detect surface treatments with this method. However, there were interesting results: when the EDX results from PY3 and PY74 were compared, chloride (Cl) was found in both PY3 pigments, but not in any of the PY74 pigments (Table 3.1).

[10] https://www.jeolusa.com/DesktopModules/Bring2mind/DMX/Download.aspx?EntryId=610&Command=Core_Download&language=en-US&PortalId=2&TabId=320. Accessed 10 Sep 2018.

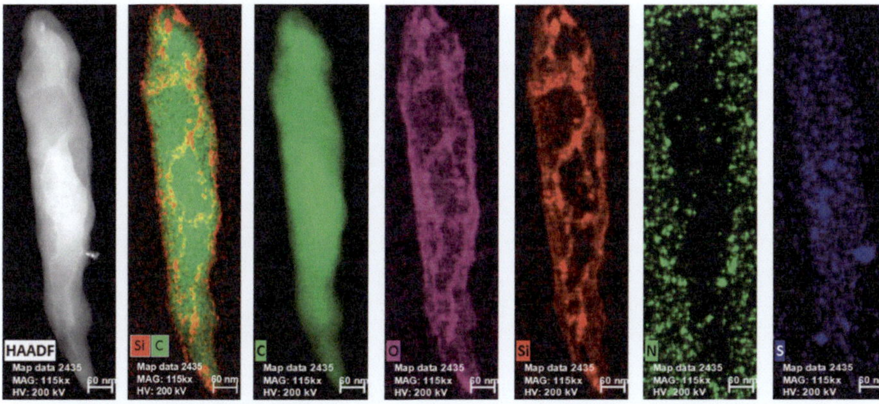

Fig. 3.1 (S)TEM-EDX phase-mapping of PY74 (Royal Talens) indicating the presence of silicon (Si) coating on the particle

With FEG-SEM it was evident from the pigment morphology that PY3 from Royal Talens consists of two different materials. In contrast with PY3, the PY74 from Royal Talens has irregular shaped particles of larger dimensions. (S)TEM-EDX analysis were performed on the Royal Talens PY74 pigment. The results suggest that parts of the pigment particles are covered with a silicon oxide coating (Fig. 3.1). The fact that only part of the pigment is coated raises the question of whether it is an intended partial coating or whether this represents an ineffective surface treatment application.

As for the detection of the organic coating, EGA-MS and Py-GC/MS, with and without TMAH were performed on PY3 and PY74. No organic surface treatment could be detected unequivocally.

3.4.2 Cu-Phthalocyanine Blue

SEM-EDX analysis of the Cu-Phthalocyanine Blue pigments did not demonstrate the presence of surface treatments. Two Cu-Phthalocyanine Blues, pigment 833 and the 1980 W&N pigment (Table 3.1) were analysed using FEG-SEM. CuPc 833 presented elongated particles with a distinctive surface appearance resembling "crocodile skin" (see Fig. 3.2a). While a similar surface was observed for the 1980 W&N pigment, the features were not as large (Fig. 3.2b).

Furthermore, the surface of the 1980 W&N pigment is irregular presenting a mixture of different particles and presumably different materials (see Fig. 3.3).

Regarding the analyses to identify organic materials, EGA-MS showed different masses related to dehydroabietic acid (DHA) (m/z 239, 285) which is observed in the mass spectra extracted around 260 ° C (Fig. 3.4a).

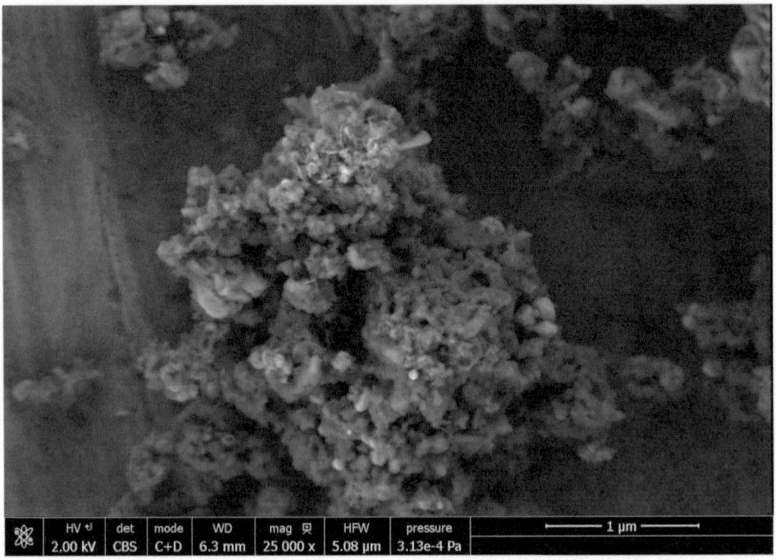

Fig. 3.2 Backscatter (BE) images using a circular backscatter detector (CBS), showing the visual appearance of (**a**) surface treated pigment form 1980 (W&N) and (**b**) surface treated CuPc 833 (RCE)

These characteristic ions evaporated early in the temperature program before the decomposition of the pigment and suggest that the resin was not chemically bound to the pigment's surface. To confirm these findings, pyrolysis gas chromatography mass spectrometry (Py-GC/MS) with a derivatization agent, TMAH (tetramethyl-ammonium hydroxide) was carried out. In Fig. 3.4b, very intense pigment-related peaks as well as abietic acid and its oxidized products such as DHA methyl ester, 7-Oxo-DHA, and oxidized abietane can be seen. These products are found in Pinacea diterpernoid resins. In addition, isopimaric acid (isomer of pimaric acid) was identified which confirms that rosin (pine colophony), and not only dehydroabietic acid, is present [36].

As for the organic treatment of the 1980 W&N pigment, although EGA-MS showed evidence of linear and branched alkanes, which might suggest the presence of a wax material, analysis with Py-GC/MS and Py-THAM-GC/MS did not allow further conclusions regarding the presence of these compounds.

3.4.3 Synthetic Ultramarine Blue

(S)TEM-EDX imaging of the synthetic Ultramarine Blue from Kremer Pigmente acquired in 2015, shows an inorganic silica coating. Using FEG-SEM, it was possible to observe full coverage of the pigment surface with the silica film. In Fig. 3.5a,

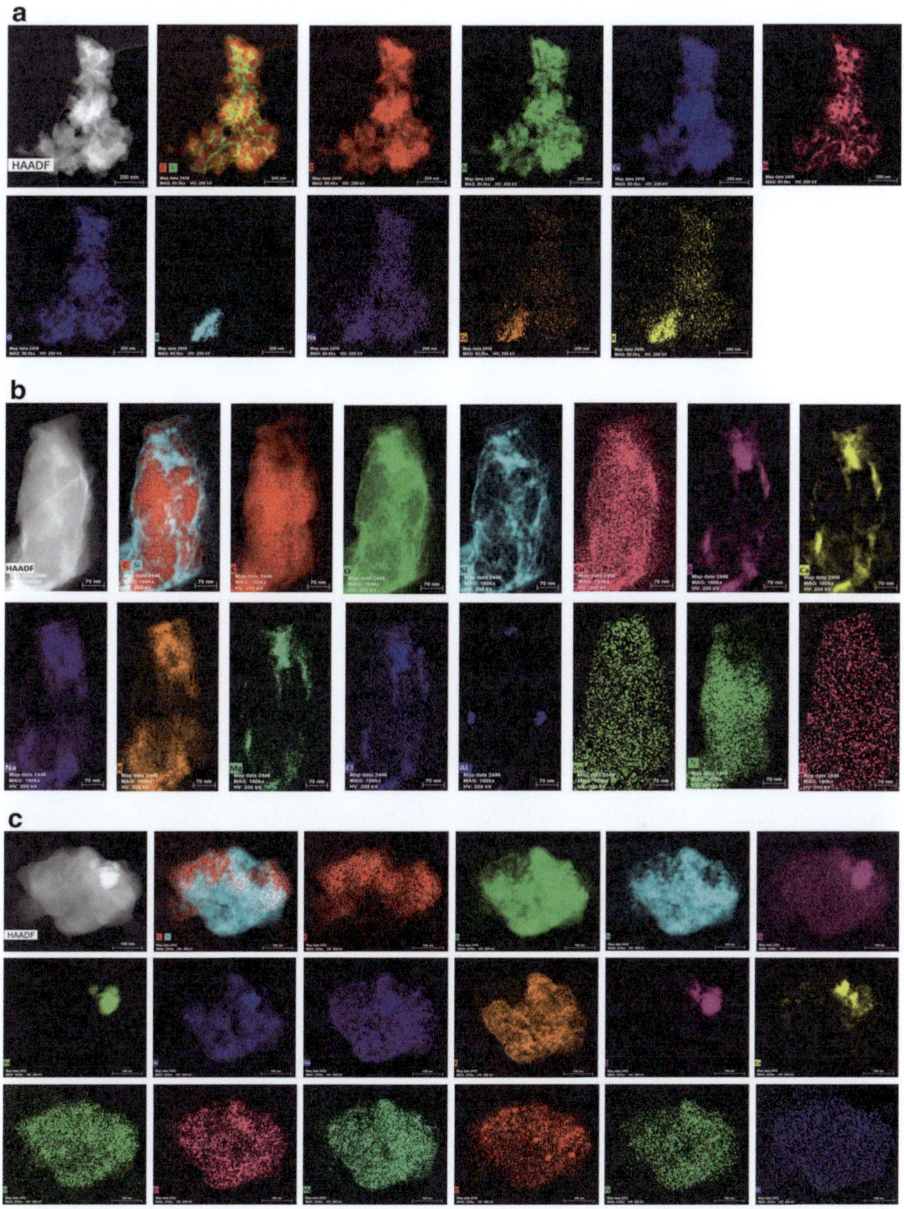

Fig. 3.3 (S)TEM-EDX phase mapping of pigment particles: (**a**) CuPc 833 (RCE) illustrating a non-homogenous silicon coating (Si), (**b**) 1980 (W&N) pigment showing silicon (Si) coating and (**c**) 1980 (W&N) pigment where the presence of silicon is identified, however, a surface coating is not clearly evident

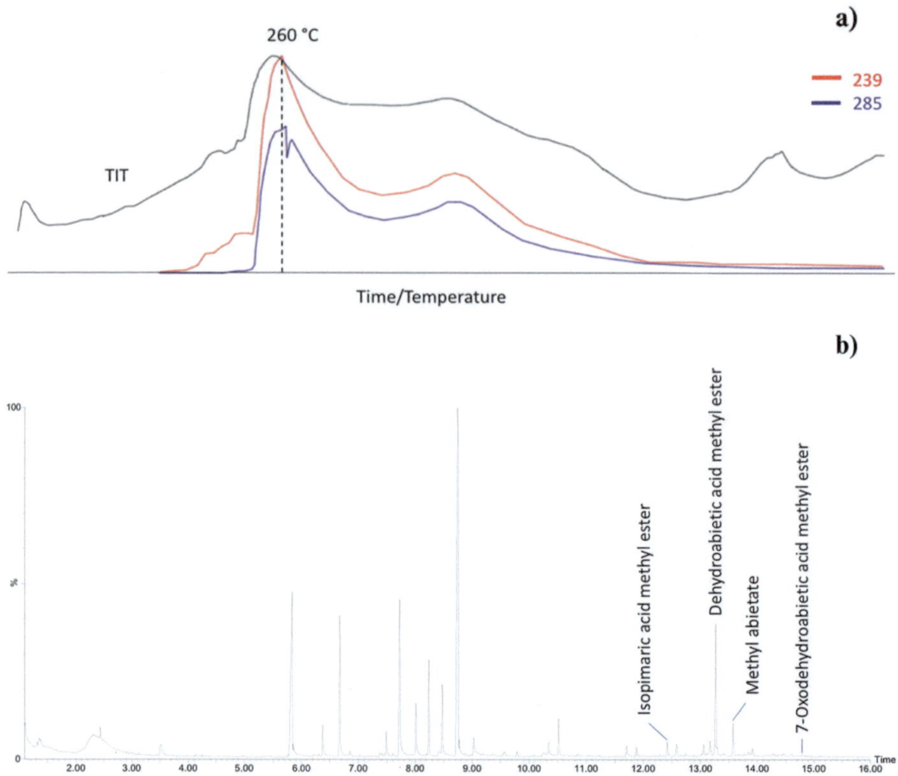

Fig. 3.4 (**a**) Total Ion Thermogram for CuPc 833 (RCE) showing the single ion chromatograms, m/z 239 and 285, extracted at 260 °C and (**b**) Pyrogram of CuPc 833 (RCE) with TMAH showing oxidation products of abietic acid: a high dominant peak of Dehydroabietic acid methyl ester and 7-Oxodehydroabietic acid methyl ester, and also Isopimaric acid methyl ester and methyl abietate

the increased particle size of this pigment when compared to an uncoated pigment (named 'Kokkie', Fig. 3.5b), is likely due to the addition of the coating material [27].

For the identification of organic treatments, the results were inconclusive and further studies are required.

3.4.4 *Vine Black*

Analyses did not detect the presence of inorganic treatments on the surface of the Vine Black pigment particles. The presence of aluminium and silicon from EDX results could represent trace materials from the combustion process during manufacture [16].

Results of EGA-MS analysis for organic surface treatments were inconclusive.

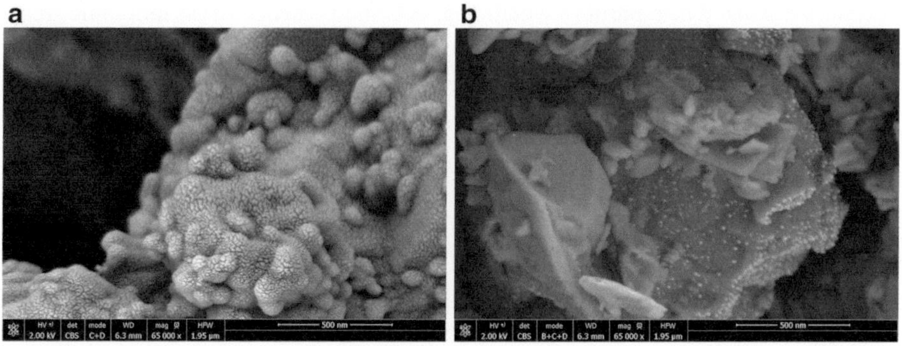

Fig. 3.5 BE images using a CBS detector at 2 kv showing, (**a**) coated Kremer synthetic Ultramarine Blue particle surface and (**b**) uncoated Kokkie synthetic Ultramarine Blue particle

3.4.5 Umber

Analyses of the Umber pigment did not determine the presence of inorganic materials that could be associated with a surface coating. Aluminium and silicon identified using EDX are possibly related to traces of impurities [20].

Results of EGA-MS analysis for organic surface treatments were inconclusive.

3.4.6 Lead White

The Lead White pigments studied showed no clear visual or analytical evidence of inorganic surface treatments. While EDX of the Schoonhoven pigment did determine the presence of ´silicon it was present in only trace amounts, which may be due to contamination [18]. No evidence of surface treatment was found using (S)TEM-EDX.

EGA-MS detected the presence of acetic acid on pigment 8233, and on the Schoonhoven and Kremer Pigmente; this may be associated with the manufacturing process [17].

3.4.7 Zinc White

Electron microscopic analyses revealed the morphology of the pigment particles but did not produce evidence of surface treatment on the Zinc White pigments studied, with the exception of (S)TEM-EDX analyses of the Zinc White supplied by Carlyle. Phase-map analyses revealed the presence of a silicon rich phase concentrated around the Zinc White particle which suggests the presence of an inorganic treatment, however, further analyses to confirm these findings are needed.

As for the detection of an organic coating using EGA-MS and Py-GC/MS no conclusive results were found.

3.4.8 *Titanium White*

The Titanium Dioxide pigments selected for this study (Table 3.1) were previously investigated by van Driel [37] for evidence of surface treatments. Van Driel reported (S)TEM-EDX results showing that pigments A3 and R3 had both received inorganic coatings with aluminium; silicon dioxide (SiO_2) was detected in A3; and zirconium dioxide (ZrO_2) was identified in R3.

Using ICP-OES phosphorous was found on R3 as well. Evidence of inorganic coatings on pigments A1, A2 and R2 was not found ([37], p.117–120).

In this study, the presence of carbon and aluminium was identified in all pigments, with silicon found in pigment A2, as well as A3. The presence of phosphorus in R3 was confirmed and may be related to the pigment manufacture [4].

FEG-SEM was carried out on sample A1, van Driel's "uncoated" reference, and pigment A3 which had been previously coated with gold (for SEM-EDX High Vacuum analysis). Due to the gold coating on A3 it was not possible to observe the pigment's surface to verify whether an inorganic treatment could be detected. However, in terms of morphology, the gold coated surface of pigment A3 appears to be very similar to the silicon coating detected on the synthetic Ultramarine Blue pigment from Kremer Pigmente.

As for the investigation of possible treatments with organic materials, in her research van Driel concluded that A1 was unlikely to be coated, A2, A3 and R2 showed the presence of what was "likely a polyol", and R3 was also likely to be coated ([37], p.120). The analysis of the thermal profiles of these pigments obtained by EGA-MS analysis confirmed van Driel's findings by showing that the pigment profiles contained organic material in all pigments except A1 (which presented only traces of common contaminants attributed to handling).

Further investigation with Py-GC/MS, with and without TMAH, enabled the detection and identification of polyols on pigments A2, A3, R2 and R3 which corroborated van Driel's conclusion that polyols were present. No attribution for the polyol in the R3 pigment could be made, however trimethylolpropane (TMP) was identified in A2, A3 and R2. According to the literature TMP is a common polyol used as a grinding aid [4]. The presence of 2-ethylacrolein was identified in A3, R2 and R3. Its presence was significant because it appears in the manufacturer's safety data-sheet, as a product of TiO_2 thermal decomposition (Huntsman datasheet 2008).

Glycerol pyrolysis products were found in A3 (Fig. 3.6) however more information is needed to draw conclusions. Analyses of R2 revealed the presence of an antioxidant which was unexpected and requires further study.

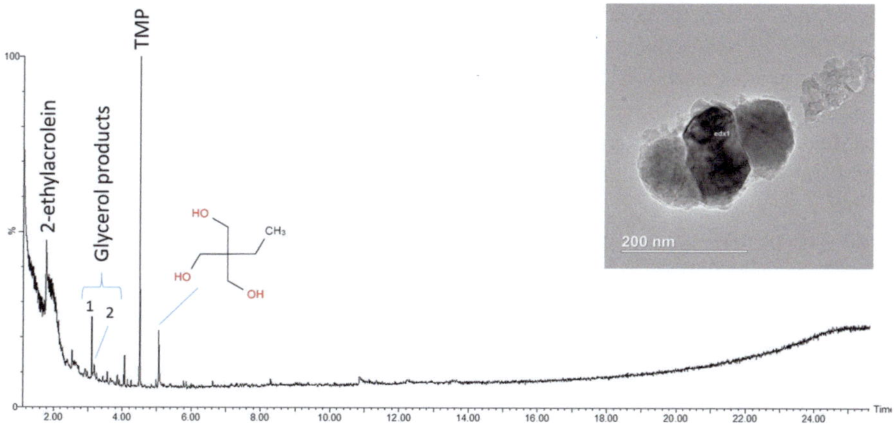

Fig. 3.6 GCMS trace of A3 with TMAH showing peaks related to 2-ethylacrolein, and Glycerol products (1: Propane, 1, 2, 3, trimethoxy and 2: Diethylene glycol dimethyl ester), and a high dominant peak which was identified as Trimethylolpropane (TMP). Insert: (S)TEM-EDX spectral image of A3- the inorganic coating (Zr, Al) is non-homogeneous, with some loose (coating) material (upper right)

3.5 Conclusions

For this study an investigation of the patent and technical literature provided an overview of pigment surface treatments. Interviews with industry specialists were very helpful[11] but also revealed the competitive nature of this field which leads to limited disclosure of industrial methods and materials.

An important finding regarding the technical literature was that material datasheets may provide information regarding the chemical composition of inorganic surface treatments, however, for organic treatments, information may not be as forthcoming.

For the detection of inorganic surface treatments FEG-SEM and (S)TEM-EDX were the most effective techniques. FEG-SEM enabled morphological and topographical imaging, while (S)TEM-EDX enabled the detection of inorganic treatments particularly through elemental phase mapping analyses. EDX provided possible indications of inorganic treatments through the detection of aluminium and silicon for example on a synthetic Ultramarine Blue pigment, however with this method it was not possible to identify inorganic treatments without further analyses using the other techniques.

Relating organic compounds to surface treatments can be more challenging since surface treatments are present in such small amounts. For the analysis of unknown organic compounds, EGA-MS was crucial for a first approach whereas pyrolysis methods such Py-GC/MS with and without TMAH were effective for the identifica-

[11] see footnote 1.

tion of anticipated materials (e.g. polyols on Titanium White and rosin on the 833 CuPc) as well as the discovery of unexpected compounds such as antioxidants.

For future research it will be important to explore methods to detect the presence of pigment surface treatments in paint systems and to research their possible impact in actual art works -particularly to evaluate whether they may play a role in the deterioration of artists' paint.

Acknowledgments The authors are grateful to Birgit van Driel for the introduction to her work on coated titanium dioxide; to Mark Clarke for pointing out that pigment coating was practiced as early as the fourteenth century and for providing the reference to the Montpellier Manuscript; to Matthijs de Keijzer for guidance during patent research; and to Luc Megens and Henk van Keulen for instructions and guidance with the analytical instruments. Sample materials were kindly provided by Catherine Defeyt, Birgit van Driel, Alessa Gambardella, Gilian Osmond and Royal Talens. Coos van Waas and Daphne van Mansom (Royal Talens); Arne Janssen and Tony Davies (AkzoNobel Chemicals); Conny te Walvaart and Tanja Takman (BASF) are gratefully aknowledged for their invitations to the companies and constructive suggestions.

Appendix

SEM was carried out on a JEOL JSM-5910LV Scanning Electron Microscope – Energy Dispersive X-ray spectroscope) with secondary (SE) and backscatter (BSE) detectors coupled to an energy dispersive X-ray spectrometer (EDX, Thermo Scientific Ultra Dry SDDS with NSS software). Analyses were performed under low vacuum conditions (30 Pa) at 15 kV.

TEM Bright Field (BF) and STEM-EDX were conducted at AkzoNobel Chemicals with a JEOL 2010F TEM operated at 200 kV and equipped with a Thermo Scientific Ultra Dry SDD for EDX analysis; EDX Spectrum Images (SI) were acquired with Thermo Scientific NSS Software or Velox, respectively; Principal Component Analysis (PCA) was applied for phase identification in the SI.

The Field Emission Gun Scanning Electron Microscope (FEG-SEM) used was a FEI Nova Nano SEM450, with concentric (insertable) higher energy electron (CBS) detectors. Analyses were carried out under high vacuum conditions at 2 kV.

Selected samples were analysed using thermally assisted hydrolysis and methylation (THM) gas chromatography-mass spectrometry (GC/MS) in combination with ultrafast thermal desorption (UTD; Decq et al. [12]), after adding tetramethylammonium hydroxide (TMAH) in methanol (5%) with tridecanoic acid (FA-C13) internal standard. UTD was performed in a Frontier Lab 3030D pyrolyser mounted on a Thermo Scientific Trace 1310 GC /ISQ mass spectrometer combination, by heating at 500 °C/min from 360 °C up to 700 °C. A SLB5 ms (Supelco) column was used (20 m, i. d. 0.18 mm, film 0.18 μm) with a helium carrier at 0.9 ml/min and 1:30 split ratio. Temperature program: 35 °C (1.5 min); 60 °C/min to 100 °C; 14 °C/min to 250 °C; 6 °C/min to 315 °C (1.5 min); interface and ion source temperatures 270 and 220 °C, respectively. Mass spectra were recorded from 29 until 600 amu with a speed of 7 scans per second. Xcalibur 2.1, AMDIS 2.7 and MassLynx 4.0 software were used for collecting and processing of the data.

EGA-MS analysis was performed on the same pyrolysis and mass spectrometer combination using a deactivated, uncoated capillary (1 m, i.d. 0.1 mm). Samples were directly introduced via a steel pyrolysis cup and heated 70 °C/min from 70 °C to 700 °C.. Carrier gas was Helium with a constant flow of 0.9 ml/min; split ratio 1:20. Mass spectra were recorded as above.

References

1. Barnett CE (1949) Physics and Chemistry of pigments. Ind Eng Chem 41 (2):272–279
2. Bieleman J (ed) (2000) Additives for Coatings. Wiley-VCH, Weinheim
3. Blumenroth D, Zumbühl S, Scherrer NC, Müller W (2014) Sensitivity of Modern Oil Paints to Solvents. Effects on Synthetic Organic Pigments. In: van den Berg KJ, Burnstock A, Keijzer M, Tagle A, Heydenreich G (eds) Issues in Contemporary Oil Paint, Springer, Cham, pp. 351–362
4. Braun JH, Baidins A, Marganski RE (1992) TiO_2 pigment technology: a review. Prog Org Coat 20(2):105–138
5. Bugnon P (1995) Surface Treatment of Pigments: Treatment with inorganic materials. Prog Org Pig 29(1–4):39–43.
6. Buxbaum G, Pfaff G (eds) (2005) Industrial Inorganic Pigments. 3rd ed. Wiley VCH, Weinheim
7. Carlyle L (2001) Molart Fellowship, "Historical Reconstructions of Artist's Oil Paint: an investigation of oil processing methods and the use of medium-modifiers", Canadian Conservation Institute, April 2000 (revised edition 2001), Report No.72894.
8. Carlyle L (2018) Towards historical accuracy in the production of historic recipe reconstructions. Tempera 2018 Conference, Doerner Institut, Munich, Germany, postprints (in press)
9. Carr W (1973) Improving the Physical Properties of Pigments: Pigment Elaborations. In: Patton TC (ed) Characterization and Physical Relationships. Pigment Handbook, vol 3. Wiley, New York, pp. 29–35
10. Clarke M (2011) Mediaeval Painters' Materials and Techniques: The Montpellier 'Liber diversarum arcium'. Archetype Publications, London
11. Coolidge C, Holt HS (1932) Noncaking pigment. US Patent 2,009,436, 24 May 1932
12. Decq L, Frederic L, Schilling M, Fremout W, Cattersel V, Steyaert D, Indekeu C, Van Binnebeke E, Saverwyns S (2016) The analysis of European lacquer: optimization of thermochemolysis temperature of natural resins. Applied Physics A. 122. https://doi.org/10.1007/s00339-016-0550-5.
13. Defeyt C (2013) Copper-phthalocyanine Blue Pigments in Conservation. PhD dissertation thesis, Université de Liège
14. Fisher EK (1950) Surface-active agents. In: Colloidal Dispersions, John Wiley & Sons, New York, pp. 224–257
15. Gardner HA (1933) Process of modifying the characteristic of pigments and pro-duct. US Patent 1,894,168, 22 Jan 1931
16. Gatta T, Campanella L, Coluzza C, Mambro V, Postorino P, Tomasetti, Vusco G (2012) Characterization of black pigment used in 30 BC fresco wall paint using instrumental methods and chemometry. Chem Cent J 6(Suppl 2): 1–9
17. Gettens, RJ, Kühn H, Chase WT (1993) Lead White. In: Roy A (ed) Artist's Pigments: A Handbook of Their History and Characteristics, vol. 2, Oxford University Press, Oxford, pp. 125–139
18. Gonzalez V, Calligaro T, Wallez G, Eveno M, Toussaint K, Menua M (2016) Composition and microstructure of the lead white pigment in Masters paintings using HR Synchrotron XRD. Microchem J 25: 43–49
19. Hays BG (1984) Surface Treatment of Organic Pigments for Printing. Am Inkmaker 62(6):28–50

20. Helwig K (2007) Iron oxide pigments natural and synthetic. In Berrie B (ed) Artist's Pigments: A Handbook of Their History and Characteristics, vol. 4, National Gallery of Art, Oxford, pp. 39–109
21. Herbst W, Hunger K, Wilkner G (eds) (2004) Industrial Organic Pigments: Production, Properties, Applications, 3rd edition. Wiley-VCH, Weinhein
22. Izzo FC, van den Berg KJ, van Keulen H, Ferriani B, Zendri E (2014) Modern Oil Pints – Formulations, Organic Additives and Degradation: some case studies. In: van den Berg KJ, Burnstock A, Keijzer M, Teagle A, Heydenreich G (eds) Issues in contemporary oil paint, Springer, Cham, 75–104.
23. Keune K, Boon J (2007) Analytical Imaging Studies of Cross-Sections of Paintings Affected by Lead Soap Aggregate Formation. Stud Cons 52(3):161–176
24. Lenoir J (1971) Organic Pigments. In: Venkataraman K (ed) The Chemistry of Synthetic Dyes, vol. 5, Academic, New York, pp. 313–474
25. Levison HW (1973) Pigmentations of Artist's Colors. In: Patton TC (ed) Applications and Markets. Pigment Handbook, vol 2. Wiley, New York, pp. 423–434
26. Lewis P (ed) (1988) Pigment Handbook: Properties and Economics, vol 1, Wiley, New York
27. Lia SF, Liu M (2011) Preparation of Acid-resistant Ultramarine Pigment by Dense Silica Process. Advanced Materials Research, 233-235:246–249
28. Meister, F Walter, Elizabeth NJ (1938) Pigments of improved wetting, dispersing, and nonsettling characteristics and the process of manufacturing same. US Patent 2,133,539, 7 June 1933.
29. Merkle K, Shäffer F (1973) Surface treatment of organic pigments. In: Patton TC (ed) Characterization and Physical Relationships. Pigment Handbook, vol 3. Wiley, New York, pp. 157–167
30. Napper D (1983) Stabilization by Attached Polymer: Steric Stabilization. In: Polymeric Stabilization of Colloidal Dispersions, Academic, London, pp. 18–30
31. Osmond G (2014) Zinc oxide-centred deterioration of modern artist's oil paint and implications for the conservation of twentieth century paintings". PhD dissertation, University of Queensland, Australia
32. Schröder J (1988) Surface Treatment of pigments. Prog Org Pig 16: 3–17.
33. Senna M (2006) Surface Modification. In: Yoshida H, Masuda H, Higashitani K (eds) Powder Technology: Fundamentals of Particles, powder beds, and particle 3rd ed. CRCPress, Boca Raton, pp. 485–491
34. Swarup S, Schoff CK (1993) A survey of surfactants in coatings technology. Prog Org Coat 23(1):1–22
35. Tsuge S, Ohtani H, Watanabe C (eds) (2018) Pyrolysis - GC/MS Data Book of Synthetic Polymers: Pyrograms, Thermograms and MS of Pyrolyzates, Elsevier New York
36. Van den Berg KJ, Boon JJ, Pastorova II, Spetter LF (2000) Mass spectrometric methodology for the analysis of highly oxidized diterpenoid acids in Old Master paintings. J Mass Spectrom 35:512–533.
37. Van Driel B (2018) Titanium White: Friend or Foe? Understanding and predicting photocatalytic degradation of modern oil paintings. PhD thesis, Delft University of Technology
38. Vernardakis T (2007) Chapter 31 – Pigment Dispersion. In: Tracton AA (ed) Coating Materials and Surface Coatings, CRC Press, California, pp. 1–19
39. Von Baum K, Dietemann P, Fischer U, Klein P, Krischel R, Neuhoff T, von Saint-George C, Schaefer I, Schawe M, Stege H, Walcher J (2014) Painting Materials, Paint Application and Painting Techniques. In: Let the Material Talk: Technology of Late-Medieval Cologne Panel Painting. Birdcage Walk, London, pp. 145–148

Chapter 4
Meet the Future: The Creation of New Pigments

Matthijs de Keijzer

4.1 Introduction

In the world of colour science, the creation of new pigments is extremely rare, but in the beginning of the twenty-first century chemists were successful in discovering four pigments. The synthetic inorganic yellow pigments rutile tin zinc oxide and niobium tin pyrochlore can be of special interest to replace the chrome yellows (PY 34) and cadmium yellows (PY 35) in the near future. The pigment YInMn blue was discovered accidentally and could be an alternative to the existing blue inorganic pigments, because the properties of this modern blue pigment are much more durable. The most recent discovery is Vantablack, whose use as an artists' paint will be very limited, due to its limited availability and extraordinarily high price.

As these pigments are hardly described in the published literature, this article has been compiled on the basis of numerous websites; the most important ones are cited in the text. This contribution is meant to be a follow-up on two articles published by the author on twentieth century synthetic inorganic [5] and synthetic organic artists' pigments [6].

M. de Keijzer (✉)
Cultural Heritage Laboratory, Cultural Heritage Agency of the Netherlands, Amsterdam, The Netherlands
e-mail: m.de.keijzer@a1.net

K. J. van den Berg et al. (eds.), *Conservation of Modern Oil Paintings*,
https://doi.org/10.1007/978-3-030-19254-9_4

4.2 Chrome- and Cadmium Yellows: Legislation, Prohibition and Alternatives

Stable yellow and orange-yellow pigments are highly valued in the art world. Environmental and health concerns, market demands and regulating requirements have placed increasing pressure on some pigment groups to find alternatives, especially for the chrome – and cadmium yellows.

The toxicity of the element lead and the carcinogenicity of hexavalent chromium are of particular concerns for the chrome yellows. For several decades the use of chrome yellows (PY 34) has not been permitted in artists' pigments, in coatings for toys and consumer goods, including food contact applications and packaging and have been labelled as 'only for industrial purposes'. Since 2015 its industrial production and use has ceased worldwide.

Yellow pigments with a high chromaticity are made from cadmium sulphide (PY 35). Since the 1980s the use of cadmium in containing consumer goods has been forced back. In 1981 Sweden prohibited the use of the cadmium pigments for ecological reasons and around 1990 the European Committee and Switzerland banned them for their main use, the colouring of plastics. As a result, many industries have cut down or entirely eliminated the use of these colorants. Even in the automobile industry their use declined. In the industry the yellow, orange and red cadmium pigments have already been largely replaced by synthetic organic pigments discovered in the second half of the twentieth century, like the perylenes, quinacridones, diketopyrrolo-pyrroles, the improved Hansa-Yellows and benzimidazolones. These synthetic organic pigment groups offer suitable alternatives, because their properties are sufficiently similar to cadmium pigments in some respects, and even superior in others [6, 8].

For artists' colours and ceramic products the cadmium pigments are still permitted. Only 5–7% of the total cadmium pigment production is used in the manufacture of artists' materials. If these trends continue, application of cadmium pigments in artists' colours will become difficult, and may be impossible due to the lack of availability [1]. However, substitutes for the pigment for use in artists' paints have recently become available.

4.3 Substitutes for Cadmium Pigments in Artists' Paints

The orange and red diketopyrrolo-pyrroles (DPPs) (PO 71, PO 73, PR 254, PR 255, PR 264, PR 272), discovered in the early and produced in the middle of the 1980s, are suitable alternative options for the cadmium reds (PR 108) and cadmium oranges (PO 20) as artists' pigments. They possess excellent properties, but disadvantages are their transparency and their high price. However, after the expiration of the first DPP patent in 2004, several products of European and Asian competitors caused a drop in prices since.

Substitutes for the cadmium yellows are harder to find. The synthetic inorganic yellow pigment bismuth vanadate (PY 184), introduced to the artists´ palette in the middle of the 1990s, is a candidate, but due to the toxicity of the metals bismuth and vanadium it offers little benefit over the cadmium yellows.

A potentially good substitute is from the synthetic organic pigment group of improved Hansa-Yellows, namely Permanent Yellow GX (PY 74), discovered in 1957. Its hue is between cadmium yellow light and cadmium yellow medium and has a relatively high opacity and an excellent light-fastness.

The best alternative nowadays seems to have been the synthetic organic pigment Benzimidazolone Yellow H3G (PY 154). Since it entered the pigment market in 1975, its use is reflected in many international artists' paint catalogues. The only disadvantage is its high price.

4.4 New Developments in the Twenty-First Century

Still, there is need for yellow substitutes that meet the colour characteristics of both (lead) chromate and cadmium yellows. Two developments in the inorganic pigment field can be of interest in replacing these pigments, the first one is the recent commercialization and improvement of the rutile tin zinc oxides (PY 216 and PO 82). The other is the new patented complex inorganic colour pigment class of yellow pigments named niobium tin pyrochlore (NTP). NTP Yellow (PY 227) has recently gained the regulatory approval for commercial sale.

4.5 Rutile Tin Zinc Oxides (RTZ)

Rutile tin zinc oxide is a mixed crystal compound of oxides of titanium, tin and zinc with the chemical formula $(Sn,Zn)Ti_xO_y$. Synthetic oxide pigments containing two or more different metals are named *complex inorganic colour pigments* (CICPs). Coloured pigments are created by substituting titanium atoms in the rutile crystals with different metal atoms. The discovery of the CICPs dates back to 1934. Nickel titanium yellow (PY 53) and chrome titanium yellow (PBr 24), are the most important pigments of this class [23].

Rutile tin zinc oxide (PY 216) was developed in the 1980s. This yellow complex pigment was patented by the chemists John Ward Jenkins and John Wolstenholme of the British Johnson Matthey Public Limited Company (see Table 4.1). In 2003 this firm, now called Rockwood Pigments, introduced three yellow pigments under their trade name Solaplex®: Rockwood Solaplex® Yellow 34H1001R, Rockwood Solaplex® Mid Yellow 34H1002 and Rockwood Solaplex® Orange 34H1003.

In the period 2006–2010 improvements have been brought to the product class by increasing the colour from a red-shade yellow to a more orange hue. Recent modifications have enabled the achievement of a true orange shade (PO 82), such as

Table 4.1 Patents and applications of rutile tin zinc oxide

Yellow pigment	Discoverers and patents		Industrial production and *occurrences as an artists' paint*	
Rutile tin zinc oxide PY 216 Rockwood Solaplex® Yellow 34H1001R	1984	John W. Jenkins, John Wolstenholme; Johnson Matthey Public Limited Company, Great Britain, U.S. Pat. 4.448.608, Colored inorganic complex for use as a pigment and compositions containing it	2003	Introduction to the pigment market by Rockwood Pigments, trade name: Solaplex®
Rockwood Solaplex® Mid Yellow 34H1002 Rockwood Solaplex® Orange 34H1003	1987	John Ward Jenkins, John Wolstenholme; Johnson Matthey Public Limited Company, Great Britain, European Pat. 0.113.229, Farbiger anorganischer Komplex welcher als Pigment benutzt wird und Zusammenstellungen die ihn enthalten	2009	Yellow shades (PY 216) oil, watercolour paints Orange shades (PO 82)
Rutile tin zinc oxide PO 82 Rockwood Solaplex® 34H1004	1987	John Ward Jenkins, John Wolstenholme; Johnson Matthey Public Limited Company, Great Britain, German Pat. 3.373.488, Coloured inorganic complex for use as a pigment and compositions containing it	2009	*H. Schmincke & Company, Germany, Norma catalogue* *Chrome yellow hue light,* *Chrome yellow hue middle,* *Chrome yellow hue deep,* *Schweinfurt green hue*
BASF Sicopal® Orange L2430 Shepherd's RTZ Orange 10C341	1987	John Ward Jenkins, John Wolstenholme; Johnson Matthey Public Limited Company, Great Britain, Austrian Pat. 29.511, Farbiger anorganischer Komplex welcher als Pigment benutzt wird und Zusammenstellungen die ihn enthalten	2015	*Winsor & Newton, UK, watercolour catalogue: Turner's Yellow hue*
	2006	Ghila Lanzendörfer, Heidi Riedel, Lynn Viehmeyer, Volker Kallmayer, Sophie Viala, Nicole Mocigemba; Beiersdorf AG, Germany, German Pat. 102005007467, Kosmetische Zubereitungen zur dekorativen Anwendung enthaltend anorganische Gelbpigmente auf der Basis von Titan-, Zinn- und Zinkmischoxiden	2016	*Kremer Pigmente, Germany, Hokkaido-Orange, PY 216*
	2006	Ghila Lanzendörfer, Heidi Riedel, Lynn Viehmeyer, Volker Kallmayer, Sophie Viala, Nicole Mocigemba; Beiersdorf AG, Germany, European Pat. 1.690.523, Kosmetische Zubereitungen zur dekorativen Anwendung enthaltend anorganische Gelbpigmente auf der Basis von Titan-, Zinn- und Zinkmischoxiden		
	2008	Norbert Mronga, Kirill Bramnik; BASF SE, Germany, WO 2008/083897, Pigments containing titanium, zinc, tin, rare earth and optionally alkali earth oxides		
	2009	Kirill Bramnik, Norbert Mronga; BASF SE, Germany, Chinese Pat. 101583567		
	2009	Norbert Mronga, Kirill Bramnik; BASF SE, Germany, European Pat. 2.125.622, Titan-, Zink-, Zinn-, selterend-, sowie wahlweise Erdalkali-Oxide enthaltende Farbkörper		
	2010	Norbert Mronga, Kirill Bramnik; BASF SE, Germany, U.S. Pat. 7.837.781, Colorants comprising tin and rare earth elements		

BASF's Sicopal® Orange L2430, Rockwood's[1] Solaplex® Bright Orange 34H1004 and Shepherd's Orange 10C341.

4.6 Properties and Applications of the Rutile Tin Zinc Oxides

The colours of the rutile tin zinc oxides are ranging from bright yellow to orange. The pigments possess an excellent light- and weather-fastness, a high colour strength and a high hiding power. They present an excellent chemical resistance, are extremely heat-resistant, being stable to c. 320 °C. The yellows are already established in road marking, interior and exterior architectural paints and suitable for food packaging and toys. The orange ones like BASF's Sicopal® Orange K2430 and Shepherd's Orange 10P340 are applied in plastics, outdoor applications, such as garden tools, patio furniture and playground equipment. BASF's Sicopal® Orange

[1] In 2014 Rockwood Holdings was bought by the Huntsman Corporation.

L 2430 and Shepherd's Orange 10C341 are especially recommended for coating and paint applications. The difference in these pigments lies most probably in the different surface treatments [15].

4.7 Rutile Tin Zinc Oxide in Art

The Solaplex pigments have already been introduced in the oil – and watercolour techniques as alternatives for the chrome yellows. Rutile tin zinc oxide (PY 216) has been listed in 2009 in the finest artists' oil colours catalogue Norma® of Schmincke as a substitute for different hues of chrome yellow. *Chrome yellow hue light* is made by mixing rutile tin zinc oxide with nickel titanium yellow (PY 53); *chrome yellow hue middle* and *chrome yellow hue deep* only contain the yellow RTZ pigment. Schmincke's *Schweinfurt green hue* is a mixture of rutile tin zinc oxide (PY 216) with cobalt blue (PB 28), cobalt green (PG 19), viridian (PG 18) and nickel titanium yellow (PY 53). Winsor & Newton uses rutile tin zinc oxide (PY 216) in their watercolour product *Turner's Yellow hue*. In 2016 the firm Kremer Pigmente puts PY 216 on their pigment list as *Hokkaido-Orange*, named after the orange Hokkaida pumpkin (*Cucurbita maxima*).

4.8 Niobium Tin Pyrochlore (NTP)

The most recent innovation in the yellow field is niobium tin pyrochlore (PY 227) with the chemical formula $Sn_2Nb_2O_7$. It is a mixed crystal compound of oxides of niobium and tin produced by high temperature calcination. Research in the early 1950s by the scientists W.R. Cook Jr. and Hans Jaffe raised interest in the pyrochlores [3]. Some niobates exhibit ferroelectric properties, but unfortunately the best one contains the element lead in the pyrochlore structure. Since the industrial use of lead poses environmental problems, it is of interest to replace this element by a less toxic one.

In 2000 the Portuguese chemists Cruz, Rocha and Pedrosa de Jesus of the University of Aveiro together with the French chemists Savariant of the CEMES-CNRS in Toulouse and Jumas of the ESA 5072 CNRS at the University of Montpellier attempted to replace lead by tin. Several lead niobates with the pyrochlore structure are known, while only two tin niobates $SnNb_2O_6$ and $Sn_2Nb_2O_7$ have been reported. Their research gave an insight into the synthesis, the structural characterisation of the tin niobates and that the colour of tin niobate, nominally $Sn_2Nb_2O_7$, can range from yellow to red depending on the Sn: Nb – ratio [4].

Between 2005 and 2010, Shepley Booth of the Johnson Matthey Public Limited Company together with Dann and O'Brien of the Loughborough University patented production methods of yellow to red hues based on a modified pyrochlore structure (see Table 4.2). However they had limited success by improving the products of the Portuguese and French scientists and they found it difficult to produce reliable coloured compounds.

In the U.S. patent 7.594.961 the chemists proposed and disclosed the use of this NTP class as a pigment and reported that the invention could be of importance for replacing the cadmium sulphides (cadmium yellows) or cadmiumsulfoselenides (cadmium reds and cadmium oranges). The pigments are particularly suitable for the coloration of glasses and plastics, but they can also be used in paints and inks.

Table 4.2 Patents and applications of niobium tin pyrochlore

Yellow pigment	**Discoverers and patents**		**Industrial production and *occurrences as an artists' paint***	
Niobium tin pyrochlore PY 227 NTP Yellow 10C151	2005	Jonathan Charles Shepley Booth; Johnson Matthey Public Limited Company; Sandra Elisabeth Dann and Duncan Lee	2012	NTP Yellow has gained regulatory approval for commercial sale
	2006	John O´Brien, Loughborough University; Great Britain, WO 2005/052068, Inorganic Pigments	-	*Future: Replacing the chrome yellows and cadmium yellows*
	2007	Jonathan Charles Shepley Booth, Sandra Elisabeth Dann and Duncan Lee John O´Brien; Loughborough University Enterprises Limited, Great Britain, European Pat. 1.694.779, Inorganic Pigments		
	2007	Jonathan Charles Shepley Booth, Sandra Elisabeth Dann and Duncan Lee John O´Brien; Loughborough University Enterprises Limited, Great Britain, Chinese Pat. 1.954.036, Inorganic Pigments		
	2008	Jonathan Charles Shepley Booth; Henley-Upon-Thames; Sandra Elisabeth Dann and Duncan Lee John O´Brien; Loughborough, US 2007/0144402, Inorganic Pigments		
		Jonathan Charles Shepley Booth, Sonning Common; Sandra Elisabeth Dann and Duncan Lee John O´Brien; Loughsborough University, Loughborourgh University Enterprises Limited, Great Britain, German Pat. 602004011107, Anorganische Pigmente		
	2009	Jonathan Charles Shepley Booth, Sonning Common; Sandra Elisabeth Dann and Duncan Lee John O´Brien; Loughsborough University, Loughborourgh University Enterprises Limited, Great Britain, U.S. Pat. 7.594.961, Inorganic Pigments		
	2011	Simon K. Boocock; Shepherd Color Company, USA, Canadian Pat. 2.801.260, Tin Niobium Pyrochlore and Tin Niobium Oxide Pigments		
	2011	Simon K. Boocock; Shepherd Color Company, USA, WO 2011156362, Pigments substitutes à base d'oxyde double d'étain et de niobium		
	2011	Simon Boocock; USA, US 2011/0297045, Substituted Tin Niobium Pyrochlore and Tin Niobium Oxide Pigments		
	2012	Simon Boocock; Shepherd Color Company, USA, U.S. Pat. 8.192.541, Substituted Tin Niobium Pyrochlore and Tin Niobium Oxide Pigments		
	2012	Simon K. Boocock; Shepherd Color Company, USA, Australian Pat. 2011264994, Substituted Tin Niobium Oxide Pigments		
	2013	Simon K. Boocock; Shepherd Color Company, USA, Chinese Pat. 103097299, Substituted Tin Niobium Oxide Pigments		
	2013	Simon K. Boocock; Shepherd Color Company, USA, Korean Pat. 20130083431, Substituted Tin Niobium Oxide Pigments		
	2013	Simon K. Boocock; Shepherd Color Company, USA, European Pat. 2.580.163, Substituted Tin Niobium Oxide Pigments		
	2013	Simon Boocock; Shepherd Color Company, USA, Japanese Pat. 2013528156, Substituted Tin Niobium Pyrochlore and Tin Niobium Oxide Pigments		
	2014	Simon K. Boocock; Shepherd Color Company, USA, Spanish Pat. 2.522.215, Substituted Tin Niobium Pyrochlore and Tin Niobium Oxide Pigments		
	2015	Simon Boocock; USA, USRE 45382, Substituted Tin Niobium Pyrochlore and Tin Niobium Oxide Pigments		
	2015	Simon K. Boocock; Shepherd Color Company, USA, Japanese Pat. 5.778.264, Substituted Tin Niobium Pyrochlore and Tin Niobium Oxide Pigments		

In 2010, the chemist Simon K. Boocock of the American Shepherd Color Company came up with an alternative, slightly different from the original, which he called niobium tin pyrochlore (NTP Yellow). In 2012 the company covered the pigment NTP Yellow by a number of patents around the world (see Table 4.2). In the U.S. patents 7.594.961 and 8.192.541 production methods for niobium tin pyrochlore are described. The products range from yellow through orange to red.

4.9 Properties and Applications of NTP Yellow

NTP Yellow is a bright clean yellow similar in shade to the middle chrome yellows. The colour lies between bismuth vanadate (PY 184) and the orange version of rutile tin zinc oxide (PY 216) and is close to the redder shades of Benzimidazolone Yellow H3G (PY 154). It possesses a higher colouring power than nickel titanium yellow (PY 53) and chrome titanium yellow (PBr 24). NTP Yellow provides high chroma, opacity and durability, does not fade, is of excellent resistance to acids and alkalis, fast to weathering, chemically inert, heat resistant, stable to ultraviolet light, non-migratory and does not contain toxic and carcinogenic materials [16]. The Shepherd Color Company specifically developed NTP Yellow 10P150 for plastics applications, while NTP Yellow 10C151 is tailored for coatings and paints. Hopefully they provide viable replacements for the chrome yellows and cadmium yellows on the artists' palettes in the near future.

4.10 YInMn Blue

YInMn blue (PB 86) is a mixed crystal compound of oxides of yttrium, indium and manganese and its name – pronounced as yin-min blue – is derived from these three chemical elements. The chemical formula of the pigment is $YIn_{1-x}Mn_xO_3$ and is the result of the crystal structure, known as trigonal bipyramidal coordination. It is obtained when Mn^{3+} is introduced into the trigonal bipyramidal sites of the metal oxides [12, 17]. By varying the In: Mn – ratio the intensity of the colour can range from light to dark blue and black (see Table 4.3).

In 2009 a team of chemists, led by Professor Munirpallam A. Subramanian in the materials science laboratory at the Oregon State University (OSU) tried to find new materials with electromagnetic properties that could be used in electronics applications, such as computer hard drives. For one series of tests they mixed black manganese(lll) oxide with other chemicals and heated the grey mixture in a furnace to nearly 1200 °C. Instead of a new, high-efficiency electronic material one of the samples turned out to be a brilliant blue.

The scientists Subramanian, Arthur W. Sleight and Andrew E. Smith of the OSU patented the invention in October 2012 (see Table 4.4). The U.S. patent 8.282.728

Table 4.3 Different colours of YInMn blue

Composition	Colour
$YInO_3$	White
$YIn_{0,95}Mn_{0,05}O_3$	Sky blue
$YIn_{0,90}Mn_{0,10}O_3$	Bright blue
$YIn_{0,85}Mn_{0,15}O_3$	Bright blue
$YIn_{0,80}Mn_{0,20}O_3$	Bright blue
$YIn_{0,75}Mn_{0,25}O_3$	Royal blue
$YIn_{0,70}Mn_{0,30}O_3$	Royal Blue
$YIn_{0,65}Mn_{0,35}O_3$	Navy blue
$YIn_{0,60}Mn_{0,40}O_3$	Navy blue
$YIn_{0,55}Mn_{0,45}O_3$	Navy blue
$YIn_{0,50}Mn_{0,50}O_3$	Dark navy blue
$YIn_{0,25}Mn_{0,75}O_3$	Black
$YIn_{0,10}Mn_{0,90}O_3$	Black

Table 4.4 Patents and applications of YInMn blue

Blue pigment	Discoverers and patents		Industrial production and *occurrences as an artists' paint*	
YInMn blue PB 86 Blue 10G513	2012	Munirpallam. A. Subramanian, Arthur W. Sleight, Andrew F. Smith; Oregon State University, USA, U.S. Pat. 8.282.728, Materials with trigonal bipyramidal coordination and methods of making the same	2011	*Rebecca Shapiro: Purify*
			2011	*Rebecca Shapiro: Summers Eve*
			2015	*Madelaine Corbin: OSU Memorial Union Facade*
	2015	Exclusive license between the Oregon State University and the Shepherd Color Company, USA	2015	*Carol Chapel: CaMas001*
			2017	*Derivan, Australia, Matisse acrylic paint: Oregon Blue*
	2017	The Shepherd Color Company has received regulatory approval for commercial sale of YInMn blue, known as Blue 10G513	2017	*David Brinsden painted his guitar with Oregon Blue*
			2018	*Crayola LLC, USA, blue crayon named Bluetiful inspired by YInMn blue*
			2019	*Kremer Pigmente, Germany, YInMn Blau*

describes the synthesis, and the individual processes for the manufacture of YInMn blue, such as drying, mixing, pelleting and calcining.

Three years later, in 2015, the OSU reached an exclusive licensing agreement with the Shepherd Color Company for its industrial manufacture and since 2017 the blue pigment is commercially known as Blue 10G513 used in industrial coatings and plastics [13].

4.11 Properties and Applications of YInMn Blue

YInMn blue possesses unique characteristics, which makes it a special material for commercial companies and artists. The synthetic inorganic pigment has a brilliant blue colour, a high colour strength, a very good covering power and is non-toxic. The pigment is more durable than the existing blue inorganic pigments, as it is stable, lightfast, resistant to UV-light, organic solvents and has a good acid, alkali, and heat resistance.

One of the interesting characteristics of YInMn blue is that it reflects a large amount of infrared radiation. The pigment has an infrared reflectivity of about 40 percent, which is much higher compared to other blue pigments, such as synthetic ultramarine blue (PB 29) and cobalt blue (PB 28). The heat-reflecting properties indicate that the pigment is a suitable candidate for energy efficiency and contributes to energy savings. It can find use in a variety of exterior applications, like outdoor paints to paint roofs to keep buildings and cars cool [18].

4.12 YInMn Blue in Art

Before 2017 YInMn blue was not on the market as an artist's pigment, because the pigment had not received regulatory approval for commercial sale. Nevertheless, it has already found its way into the art world. Subramanian has sent out samples to artists, allowing them to experiment creatively with the vivid blue in various techniques, e.g. oil painting, watercolour and dry-point (a form of engraving). In 2011 Rebecca Shapiro used YInMn blue in her paintings *Purify* and *Summers Eve*. Four years later, in 2015, Madelaine Corbin applied it in many shades in her artwork *OSU Memorial Union Façade* and in the same year Carol Chapel made her art-object *CaMas 001* with YInMn blue using dry-point printing. In the aquarel *Mount Hood* by Rajeevi Subramanian except the white, all blue and black colours are derived from the YInMn blue family [2, 10]. Even for art restorers YInMn blue can be of interest, because while the colour is very close to synthetic ultramarine blue (PB 29) and cobalt blue (PB 28), the pigment is more durable, especially with regards to acid resistance.

One problem is the price of YInMn blue which is approximately six to seven times higher than the already expensive pigments cobalt blue (PB 28) and ceruleum blue (PB 35). This is largely due to the rare chemical element indium. The Shepherd Color Company offers test samples of YInMn blue at 10 US dollars per 10 grams. For small quantities, the price level is 1000 US dollars per kilogram.

In 2017 the Australian paint supplier Derivan was the first company in the world to make YInMn Blue commercially available to artists. Derivan named its new blue acrylic paint *Oregon Blue*. The Australian David Brinsden used the blue paint to adorn his handmade guitar. In March 2017 the firm Crayola LLC announced that the yellow Dandelion crayon will be retiring and will be replaced by a crayon inspired by YInMn blue. After a contest the new crayon became the name *Bluetiful* and is since 2018 available and expands the Crayola assortment in the USA. From early 2019, YInMn blue is also available by Kremer Pigmente GmbH & Co. KG. The blue pigment can be used in acrylic-, oil-, tempera-, watercolour and gouache techniques.

4.13 Vantablack

So called super-blacks, including Vantablack, consist of carbon nanotubes (CNTs). The discovery of the carbon nanotubes goes back to 1952, when the Russian scientists L.V. Radushkevich and V.M. Lukyanovich of the Physics and Ecology Department published transmission electron microscopy (TEM) images of 50 nanometer diameter tubes made of carbon in the Soviet Journal of Physical Chemistry. This discovery was largely unnoticed, as the article was published in the Russian language, and Western scientists' access to Soviet press was limited during the Cold War [14]. The most famous historical discoveries on the carbon nanotubes between 1952 and 1993 are listed in Table 4.5. In the period 1990–2010, extensive research on the CNTs led to tens of thousands articles.

Table 4.5 Patents and applications of Vantablack

Black pigment		Discoverers and patents		Industrial production and *occurrences as an artists' paint*
		Discovery of carbon nanotubes		
	1952	L.V. Radushkevich, V.M. Lukyanovich; Physiscs and Ecology Department, Russia, Article: O strukture ugleroda, obrazujucegosja pri termiceskom razlozenii okisi ugleroda na zeleznom kontakte. *Zurn Fisic Chim* 26, 88 - 95, 1952		
	1976	M. Endo, A. Oberlin, T. Koyama; CNRS Orléans, France, Article: Filamentous Growth of Carbon Through Benzene Compostion, Journal of Crystal Growth 32, 335 - 349, 1976		
	1979	J. Abrahmson, P.G. Wiles, B.L. Rhoades; Department of Chemical and Process Engineering, University of Canterbury, Christchurch, New Zealand, Abstract in proceedings of the 14th Biennial Conference on Carbon, June 1979, Reprinted in Carbon 37, 1873 - 1874, 1999		
	1991	S. Iijima; Fundamental Research Laboratories, Nippon Electric Company Corporation (NEC), Miyukigaoka, Tsukuba, Japan, Article: Helical Microtubes of Grahitic Carbon, Nature 354, 56 - 58, 1991		
	1993	S. Iijima, T. Ichihashi; Fundamental Research Laboratories, Nippon Electric Company Corporation (NEC), Miyukigaoka, Tsukuba, Japan, Article: Single-shell carbon nanotubes of 1-nm diameter, Nature 363, 603 - 605, 1993		
	1993	D.S. Bethune, C.H. Kiang, M.S. de Vries, G. Gorman, R. Savoy, J. Vazquez; IBM Almaden Research Center, San Jose, USA, Article: Cobalt-catalysed growth of carbon nanotubes with single-atomic-layer walls, Nature 363, 605 - 607, 1993		
Vantablack Vantablack Vantablack 2.0. Vantablack S-VIS		**Discovery of super blacks, including Vantablack**	2010	*Frederik De Wilde: Hostage, use of NASA's black coating*
	2007	John Hagopian; National Aeronautics and Space Administration Goddard Space Flight Center, Greenbelt, USA, start of the experiments with carbon nanotubes	2010	*Frederik De Wilde: Nanoblack-Sqr#1, use of NASA's black coating*
	2008	Shawn-Yu Lin; Rensselaer Polytechnic Institute, Troy, USA, claiming to have made the blackest black	2014	Surrey NanoSystems Ltd., Newhaven, UK, start of producing Vantablack
	2009	Sembukutiarachillage Ravi Silva, Ben Paul Jensen, Guan Yow Chen; Surrey Nanosystems Limited, Great Britain, U.S. Pat. 2009/0061217 A1, Nanostructure production methods and apparatus	2014	Presentation of Vantablack during the Farnborough International Air Show
			2014	*Anish Kapoor starts to use Vantablack*
			2015	Increase in production of Vantablack due to orders from the military and aerospace
	2010	Ben Poul Jensen, Guan Yow Chen; Surrey NanoSystems Ltd., United Kingdom, UK Pat. 2.467.320, Providing gas for use in forming a carbon nanomaterial	2016	Surrey NanoSystems Ltd., Newhaven, UK, start of producing Vantablack 2.0.
		Ben Jensen; Surrey Nanosystems Limited, United Kingdom, UK Pat. 2.478.269, Nanomaterials growth system and method	2016	Surrey NanoSystems Ltd., Newhaven, UK, start of producing the spray-on version Vantablack S-VIS
	2011	Ben Poul Jensen, Guan Yow Chen; Surrey NanoSystems Ltd., Great Britain, U.S. Pat. 2011/0311724 A1, Providing gas for use in forming a carbon nanomaterial	2016	*Anish Kapoor secured the exclusive rights for the artistic use of Vantablack*
	2013	National Physical Laboratory, Enersys' ABSL Space Products Division and Surrey NanoSystems, Great Britain, change in th production process from silicon to aluminium		
	2016	Sembukutiarachillage Ravi Silva, Ben Poul Jensen, Guan Yow Chen; Surrey Nanosystems Ltd., Great Britain, U.S. Pat. 9.334.167 B2, Nanostructure production methods and apparatus		

The search for 'world's blackest black' started in 2007, when a team of ten technologists, led by the physicist John Hagopian, at the NASA Goddard Space Flight Center in Greenbelt, Maryland, USA, attempted to develop a super-black coating for aerospace applications by nanotechnology. In 2010 the team succeeded in growing CNTs for stationed optical instruments in space. A breakthrough was the discovery of a method to grow CNTs on a catalytic layer of iron to heat-resistant materials such as silicon, diamond and sapphire. This material was heated in an oven to 750 °C introducing a carbon-containing gas to form the CNTs. In 2014, a first sample of NASA's super-black nano coating was sent to the International Space Station to test its resistance to the harsh space environment.

Unbeknownst to the NASA group, the New York-based Rensselaer Polytechnic Institute (RPI) had initiated a similar effort. RPI announced in 2008 the development of the darkest carbon nanotube-based material ever made by its researchers professor Shawn-Yu Lin et al.. The material had a reflection of 0.045 percent and the Guinness World Records acknowledged the material as the ´blackest black´.

After a few years the technology company Surrey NanoSystems Ltd. (SNS Ltd.) in Newhaven, Great Britain, introduced a new and better manufacturing process, developed by Ben Jensen and his technical department technical department. The new process entailed a high-tech chemical vapour deposition (PECVD) at a relatively low temperature, patented in 2010 and 2011 (see Table 4.5) following this development, a two-year research program (2012–2014) to create a super-black for space applications led to the development of Vantablack in December 2013. In the summer of 2014, SNS Ltd. announced that it had transferred its low-temperature manufacturing process from silicon to aluminium structures [21].

The Vantablack name is trademarked by SNS Ltd. and this complicated production process is registered by several British and American patents (see Table 4.5). The term VANTA comes from the words **V**ertically **A**ligned **N**ano**T**ube **A**rrays and was mentioned for the first time in an article published in 2012 [20]. Vantablack was launched at the 2014 Farnborough International Air Show and first orders were delivered in July 2014. In the following year the production was scaled up for the aerospace and defence sectors. In March 2016, improvements in the process allowed for the introduction of the even darker Vantablack 2.0. In the same year, the coating became also available in a spray-on version, namely Vantablack S-VIS.

The coating is not available to individuals. SNS Ltd. restricted the use of Vantablack exclusively to selected institutions. Only verified companies, research facilities and educational institutions can order a sample of Vantablack. Press releases from SNS Ltd. emphasized that Vantablack has a minimum reflection of 0.035 percent. Although the difference with RPI nano-black is small, soon, Guinness World Records officially announced: Vantablack is the new blackest black.

4.14 Properties of Vantablack

In addition to its exceptionally high light absorption 99.965 percent of light at 750 nm, Vantablack possesses a number of other excellent properties according to the manufacturer. Vantablack is extremely light and hydrophobic. When submerged in water, its optical properties are not influenced. The material has the highest thermal conductivity and conducts heat seven times more effectively than copper and has ten times the tensile strength of steel. Further, it has virtually undetectable levels of outgassing and particle fallout.

It has an exceptionally high degree of elasticity and is robust against heavy mechanical shocks, loads, long-term vibrations and exposure to high vacuum. It has also a very high thermal shock resistance. Extreme changes in temperature (temperature cycling) (in liquid nitrogen from −196 °C to +300 °C) do not affect the Vantablack properties. Vantablack appears to be sensitive to touch and scratches, but the coating adheres well to the substrate [19].

4.15 Other Applications of Vantablack and Vantablack S-VIS

Vantablack has interesting potential applications in a number of fields ranging from aerospace and military sectors to design. The material was originally developed to increase the sensitivity of precision optical systems, such as optical satellites, imaging and calibration systems for earth observation. The super-black coating is particularly effective in the coating on the inside of telescopes absorbing stray light and obtaining a sharper image.

Further on, the application of Vantablack may be of interest for scientific instruments, such as infrared cameras and sensors. Due to its unique physical and optical properties, the product can be used for cinema projectors and lenses. Aesthetic effects can be obtained for luxury products and design goods, such as watches, jewelry, dashboard displays and smartphones. The ability to absorb light and to convert it into heat is also important for collector elements in solar energy systems, which leads to the creation of more efficient solar panels and solar cells.

Its light absorbing properties make Vantablack applicable for stealth applications for the military in e.g. aircraft, helicopters, and armoured vehicles as well as thermal camouflage, making uniforms appear virtually invisible during nighttime.

The spray-on version Vantablack S-VIS is particularly suitable to cover larger surfaces and complex three-dimensional objects. The main condition is that the surface of the material to be sprayed is resistant to temperatures up to 100 °C, which means that this super-black coating is also suitable for polymers and composite materials.

4.16 Artists' Use of Super-Black Pigments

4.16.1 Frederik De Wilde

The super-blacks not only aroused interest in science, aerospace and defence, but also in the art world. NASA's coating was so black that it attracted the interest of the Belgian artist Frederik De Wilde, who persuaded the NASA group to allow him to use their super-black coating in some of his projects, such as 3D printed sculptures, in the works *Hostage* (2010) and *Nano Black-Sqr # 1* (2014). After the discovery of Vantablack, De Wilde also approached the firm Surrey NanoSystems Ltd. with a proposal for artistic collaboration. However, the company refused, because it had already signed an exclusive agreement with the British-Indian artist Anish Kapoor [7, 22].

4.16.2 Anish Kapoor

Since the summer of 2014 the artist Anish Kapoor has been working with Vantablack. In the beginning of 2016 he secured, for a very high amount of money, the exclusive rights for the artistic use of Vantablack. This led to a great outrage among a number of artists, such as the Belgian Frederik De Wilde, the British Christian Furr and Stuart Semple and the British-Indian Shanti Panchal. They think that Kapoor has given a monopoly on a material which he did not discover and/or patent, so actually it should belong to everyone. Furr, largely known for being the youngest artist to portrait the Queen of England, stated: 'I have never heard of an artist monopolising a material. We (artists) should be able to use it – it is not right that it belongs to one man.' Panchal also criticized Kapoor: 'I have not known of anything so absurd – in the creative world, artists, nobody should have a monopoly.' To date, Kapoor is the only artist in the world given permission to paint with the blackest black [9].

4.16.3 Yves Klein

It is not the first time that an artist has claimed a unique bond with a pigment. In 1960 the French artist Yves Klein (1928–1962) took out the French patent 1.258.418 for his International Klein Blue (IKB) [24]. He developed in collaboration with the Parisian art paint supplier Edouard Adam a deep matt shade of blue, which he used in a series of monochrome blue paintings. Klein experimented with synthetic ultramarine blue (PB 29) and the synthetic resin binder polyvinyl acetate. The French chemical company Rhone-Poulenc developed this synthetic binder and marketed it under the name Rhodopas M60A. The paint mainly consists of synthetic ultramarine blue (PB 29), but it is the matt binder which makes the IKB colour. After drying the paint appeared to hover over the surface creating a rich velvety texture and an unusual appearance of depth. The difference between Klein and Kapoor is that the latter did not invent or patent a material, but just bought the right to use a colour. So, he does not actually own the colour black, he has just an exclusive legal right to that particular material for artistic purposes.

4.16.4 Stuart Semple

In 2016 Stuart Semple has addressed this situation as an art project in the internet by selling his own pigments, like PINK, Diamond Dust and Black 2.0. To buy the pigments the customer is required to make a declaration stating that he or she is not Anish Kapoor, is in no way affiliated to Anish Kapoor, is not purchasing this item on behalf of Anish Kapoor or is an associate of Anish Kapoor. The ongoing colour battle over the artistic rights to Vantablack continues. Semple will ony stop his project, when Kapoor wants to share his super-black with everyone [11].

References

1. Berger G, Endriss H (1993) Colored Pigments, *Industrial Inorganic Pigments*, edited by G. Buxbaum, VCH Verlagsgesellschaft mbH, Weinheim, pp. 110–114.
2. Cascone S (2016) The Chemist Who Discovered the World's Newest Blue Explains Its Miraculous Properties, https://news.artnet.com/art-world/yinmn-blue-to-be-sold-commercially-520433.
3. Cook WR Jr., Jaffe H (1952) Ferroelectricity in Oxides of Fluorite Structure, *Physical. Review* 88 (6), p. 1426.
4. Cruz LP, Savariault J-M, Rocha J, Jumas J-C, Pedrosa de Jesus JD (2001) Synthesis and Characterisation of Tin Niobates, *Journal of Solid State Chemistry* 156, pp. 349–354.
5. De Keijzer M (2011) A choice of colour: modern synthetic inorganic artists' pigments, *Restauratorenblätter*, Band 30 zum Thema: Kunst des 20. und 21. Jahrhunderts, Probleme und Perspektiven zur Erhaltung, Verlag Stift Klosterneuburg, Klosterneuburg, pp. 33–42, Table pp. 184–192.
6. De Keijzer M (2014) The Delight of Modern Inorganic Pigment Creations, *Issues in Contemporary Oil Paint*, *Restauratorenblätter*, K.J. van den Berg et al. (eds), Springer International Publishing, Switzerland, pp. 45–73.
7. Fabian (2015) Meet Scientist and Artist Frederik De Wilde, Maker of the Blackest Black in the World, https://i.materialise.com/blog/meet-scientist-and-artist-frederik-de-wilde-maker-of-the-blackest-black-in-the-world/.
8. Golden Artists Colors (1996) Will Cadmium always be on the Palette?, *Just Paint*, Issue 4.
9. Hammoudi J (2017) Anish Kapoor, Vantablack And Manufactured Moral Outrage, http://artreport.com/anish-kapoor-vantablack-manufactured-moral-outrage/.
10. Herreria C (2016) New Shade of Blue Set to Change our Homes for the Better, http://www.huffingtonpost.com/entry/newest-shade-of-blue_us_577321d9e4b0d1f85d47c1da.
11. Holmes K (2016) Anish Kapoor is banned from buying the world's pinkest paint, https://creators.vice.com/en_us/article/4xqzb3/the-worlds-pinkest-paint.

12. Ocaña M, Espinós JP, Carda JB (2011) Synthesis, through pyrolysis of aerosols, of $YIn_{1-x}Mn_xO_3$ blue pigments and their efficiency for colouring glazes, *Dyes and Pigments* 91, pp. 501–507.
13. Oregon State University, Department of Chemistry, The Story of YInMn Blue, htpp://chemistry.oregonstate.edu/content/story-yinmn-blue.
14. Radushkevich LV, Lukyanovich VM (1952) O strukture ugleroda, obrazujucegosja pri termiceskom razlozenii okisi ugleroda na zeleznom kontakte, *Zurn Fisic Chim* 26, pp. 88–95.
15. Salis Gomes C, Ferreira C, Rossenaar B, Joosten I, van der Werf ID, Carlyle LA, van den Berg KJ (2019) Pigment Surface Treatments: 20th and 21st century industrial techniques and strategies for their detection. These proceedings, Chapter 3.
16. Shepherd Color Company, New NTP Yellow and RTZ Orange, www.shepherdcolor.com/new-ntp-yellow-and-rtz-orange.
17. Smith AE, Mizoguchi H, Delaney K, Spaldin NA, Sleight AW, Subramanian MA (2009) Mn in Trigonal Bipyramidal Coordination: A New Blue Chromophore, *Journal of the American Chemical Society* 131, pp. 17084–17086.
18. Smith AE, Comstock MC, Subramanian MA (2016) Spectral properties of the UV absorbing and near-IR reflecting blue pigment, $YIn_{1-x}Mn_xO_3$, *Dyes and Pigments* 133, pp. 214–221.
19. Surrey Nanosystems, Vantablack, www.surreynanosystems.com/de/vantablack
20. Theocharous SP, Theocharous E, Lehman JH (2012) The evaluation of the performance of two pyroelectric detectors with vertically aligned multi-walled carbon nano coatings, *Infrared Physics & Technology* 55, pp. 299–305.
21. Theocharous E, Chunnilall CJ, Mole R, Gibbs D, Fox N, Shang N, Howlett G, Jensen B, Taylor R, Reveles JR, Harris OB, Ahmed N (2014) The partial space qualification of a vertically aligned carbon nanotube coating on aluminium substrates for EO applications, *Optics Express* 22, 6, pp. 7290–7307.
22. Thomas G (2014) Vantablack might not be the world's blackest material, http://www.dazeddigital.com/artsandculture/article/22302/1/vantablack-might-not-be-the-worlds-blackest-material.
23. White J (2002) Complex Inorganic Colour Pigments: An Overview, *High Performance Pigments,* editor H.M. Smith, Wiley-VCH Verlag GmbH, Weinheim, Germany, pp. 41–51.
24. Yves Klein: French patent 1.258.418, 1960. Procédé de décoration ou d'intégration architecturale et produits obtenus par application dudit procédé.

Chapter 5
Zinc Oxide in Oil-Based House Paint: Insights from a Paint Chemist's Notebook Dated 1949

Anne Carter, Gillian Osmond, Paula Dredge, and Bruce Leary

5.1 Introduction

Commercial paint company British Australian Lead Manufacturers (BALM) Pty. Ltd. was established in Sydney in 1918. The company became Dulux Australia Ltd. in 1971, renamed after its well-known alkyd paint. The Dulux Archives in Melbourne hold a set of four personal laboratory notebooks written by Brian Roberts while a trainee paint chemist in the BALM *Recruit Training Scheme* of 1949 (Fig. 5.1). Nothing more is known about Roberts, except for these handwritten records which detail industry insights, research observations and recipes that are not published elsewhere.

The notebooks detail nitrocellulose, oil and alkyd paints being produced by BALM. This paper focuses on information contained in the notebooks applicable to BALM's management of known zinc oxide reactivity in their oil-based house paint formulations. Roberts describes pigment grades, driers, oil modifications and custom methods for production of durable zinc oxide-based oil paint which complements historical paint industry research by other authors [2, 5, 14, 16]. Zinc oxide reactivity in oil is still not completely understood and this historical information is useful for continuing research by conservators and artist paint manufacturers concerned with deterioration mechanisms of zinc oxide in oil-based paints.

In the first half of the twentieth century, zinc oxide was a frequent white pigment used in oil-based house paint formulations including paints based on lead white. The BALM notebooks date from a time of post WWII supply restrictions and

A. Carter (✉) · G. Osmond
Queensland Art Gallery|Gallery of Modern Art, Brisbane, Australia
e-mail: anne.carter@qagoma.qld.gov.au; gillian.osmond@qagoma.qld.gov.au

P. Dredge
Art Gallery of New South Wales, Sydney, Australia

B. Leary
Dulux History Project, Melbourne, Australia

K. J. van den Berg et al. (eds.), *Conservation of Modern Oil Paintings*,
https://doi.org/10.1007/978-3-030-19254-9_5

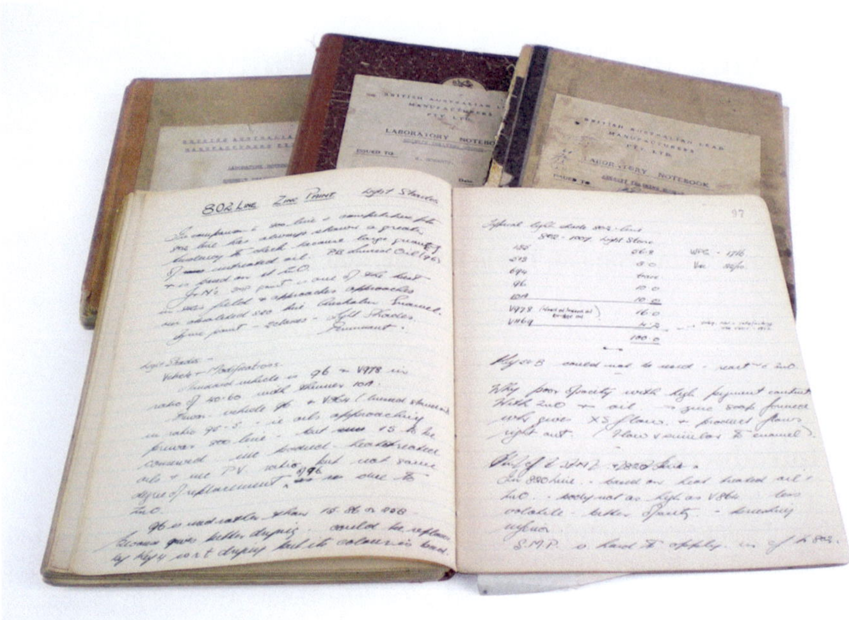

Fig. 5.1 Brian Roberts [13] BALM Laboratory Notebooks. Photograph Trudy Scott

tightening regulations around the use of toxic lead-based pigments. Innovations of rutile titanium dioxide and pentaerythritol alkyd production were anticipated, but were not yet locally viable. In contrast to artist paint formulations, zinc oxide and oil-based formulations would soon be displaced from commercial Australian house paint products, however in 1949 zinc oxide continued to play a significant role.

Zinc oxide reactivity in oil to form 'soaps' was well known to commercial paint chemists in 1949, although it was not entirely understood ([13]:L83). Zinc oxide is acid reactive and complexes with long chain fatty acids available in drying oils to form soaps. In paint chemistry, zinc soaps imparted beneficial properties for both paint application and weathering, but could also cause problems as detailed below. Controlling zinc soap formation in oil paint was an industry priority at this time and an important symposium on zinc oxide was held in Melbourne on 5 June 1949, organised by the Victorian branch of the Oil and Colour Chemists Association (OCCA) [14].

Current understanding of zinc oxide reactivity in oil distinguishes between network co-ordinated and crystalline structures of zinc carboxylates. This is reflected in FTIR spectra of zinc oxide containing oil paint where a broad peak at 1590 cm^{-1} indicates network co-ordinated zinc carboxylates while a sharp peak at 1540 cm^{-1} indicates crystalline phases [4]. Crystalline forms have been found to be most problematic for paint film stability [11, 12].

5.2 Mid Twentieth Century Australian Paint Industry

BALM's 1949 paint testing included assessment of dry film characteristics as well as stability/settling of contents and viscosity in the paint can ([13]:L79a). From this testing Roberts documented three primary factors which contributed to a durable, high grade zinc oxide oil-based paint – good quality pigment at correct pigment volume ratio, appropriate driers and a high proportion of stand oil.

These issues are discussed below in relation to two of BALM's oil-based paint lines: *Pure Prepared Paint* (800 line) typically containing 65% lead white/35% zinc oxide – which was the main house paint line and the only paint product still using lead white; and *Zinc Paint* (802 line) a lead free interior paint. Only the pale shades of the 800 and 802 paint lines are applicable to this discussion as the darker 'permanent colour' products using iron oxides, lead chromates and Prussian blue had transitioned to Glyptal alkyd vehicles by 1949 ([13]:L71a).

5.3 Zinc Oxide Pigment

Australia commenced production of Indirect (French) process zinc oxide in 1947 which was of higher purity and finer particle size than Direct (American) grades previously produced [10, 14]. Particle size is known to influence zinc oxide reactivity in oil, with finer grades having increased surface area per volume and thus higher oil absorption and reactivity [6, 10]. The 1949 Melbourne OCCA zinc oxide symposium was prompted by a sudden and marked increase in problems associated with this local switch to finer particle size zinc oxide [10].

The notebooks refer to fine or coarse pigment particle size, but not to production methods. BALM's in-house research found that finer particle size zinc oxide produced paint with excess zinc soaps ([13]:L82). Durability required a certain amount of soap. Too much soap resulted in a paint film with cracking, excess checking (paint failure by fine cracking) and chalking (progressive powdering of the dry film). Too little soap resulted in paint with no body and quicker checking ([13]:L82). Rogala [14] cites paint industry research from the 1940s which also found that continued soap production beyond that required for paint film formation had negative consequences for film strength and moisture sensitivity.

Roberts documents that BALM had been experiencing issues with the supply of 'poor' quality zinc oxide – with 'poor' referring to paint working properties, not purity ([13]:L80). This prompted thorough testing of all incoming pigment and 'poor' grades were excluded from use in finishes by 1949 ([13]:L80a).

BALM's preferred pigment choice was Lysaght brand zinc oxide ([13]:L80, R28) with two highly rated pigment grades (BALM Key 103 Green Seal ZnO and

135 Red Seal ZnO) having respective specific gravities of 5.28 and 5.95 and oil absorptions of 12.9 and 18.5 (wt%) ([13]:L83a, L107a). Roberts also records respective data of 56 and 59 lbs./gallon for these pigments, with a note that "other brands are approximately 50 lbs./gallon" (1949:107a, 80). Unfortunately, Roberts does not explain what these lbs./gallon measurements refer to. The wide variation in zinc oxide properties recorded in the notebooks predates determination of optimum pigment specifications and developments in production technology which enabled industry standards for zinc oxide pigments [10].

BALM used pigment combinations to improve the durability of paints. The best combination for oil-based paints was considered to be 65% lead white/35% zinc oxide as used in *Pure Prepared Paint*. However, anywhere between 20 and 80% zinc oxide in combination with lead white was considered to produce a good paint – minimising the tendency of lead white to chalk and zinc oxide to check ([13]:L83a).

An additional pigment option was lithopone (30% zinc sulphide/70% barium sulphate) – used for specialised paints requiring a non-acid reactive zinc pigment. Lithopone was reported to have similar opacity and tinting strength to zinc oxide ([13]:R6) however was used for applications where exterior durability was not required ([13]:R6).

5.4 Pigment Volume Ratio

Roberts discusses pigment volume ratio as an important parameter in paint formulation ([13]:L80). Zinc oxide has a relatively high oil requirement in paint formulation, however requires sufficient pigment concentration and a coarse particle size to maintain hiding power (opacity) [6].

Samples of zinc white paint dating prior to 1950 from the highly regarded oil-based Ripolin brand were found to contain 63.5 wt% pigment and 36.5 wt% binder ([9]:6). Roberts records that BALM's recipe for pale shade *Zinc Paint-Light Stone* (Fig. 5.2) contained similar content of 63.1 wt% zinc pigment, 3.3 wt% other pigment and 33.5 wt% non-volatile binder. No fillers are listed in this recipe for *Zinc Paint-Light Stone* which is consistent with analysis of Ripolin paints by Muir et al. [9]. The recipe contains no resin additives unlike some Ripolin paints [3].

5.5 Driers

Driers work to accelerate oxidation, polymerisation or gelation of the oil medium ([15]:50). In BALM's testing, cobalt napthenate used in lead based paints was found to lose effectiveness within 2 weeks when used with zinc oxide ([13]:L99a). This was suspected to result from formation of zinc napthenate which retarded drying. Cobalt resinate or pale terebine (metal octoates) were used in zinc

BALM *Zinc Paint* 802-1007 Light Stone		
Key		%w/w
135	Lysaght ZnO pigment	56.8
513	yellow ochre	3.0
694	red ochre	trace
96	pale boiled linseed oil	10.0
10A	mineral turpentine	10.0
V978	Bridge oil	16.0
V1169	Pale terebine drier	4.2
		100.0

Fig. 5.2 Recipe for *Zinc Paint, Light Stone (802–1007)* ([13] L:97)

paint pale shades (Fig. 5.2). Manganese napthenate or terebine were used for colours. Lead napthenate was also effective as a drier in zinc paints (but not for lead) ([13]:L99a).

5.6 Stand Oil

Stand oil is a general term given to heat-bodied (thermally polymerised) oil. The stand process is used to modify viscosity and improve weathering properties of the drying oil. Either a 'non-breaking' or an 'alkali refined' oil must be used to produce a stand oil which is heated at high temperature (above 270 °C) to enable thermal polymerisation ([15]:51).

The BALM notebooks attribute the superiority of their oil paints to the use of a vehicle with a high percentage of linseed stand oil, together with 'good' zinc oxide ([13]:L80a). Roberts specifies preference for up to 85% linseed stand oil and never less than 55% for *Pure Prepared Paint*. Inclusion of heat-treated oil is noted to curtail checking of zinc paints ([13]:L83) compared to untreated oil (raw, acid and alkali refined or boiled) ([13]:L97a).

This is consistent with other documentary sources which identify a high content of linseed stand oil as a marker of high quality oil-based/enamel paint [16]. Analysis of zinc oxide in oil-based Ripolin paints has similarly identified a high degree of thermal polymerisation of the oil medium suggesting the use of stand oil [3, 5].

Stand oils can be made from various oils, but linseed oil is common. BALM specified Key 15 linseed oil as their preferred raw material to produce stand oil. Roberts does not include properties of this particular linseed oil in his notes, however BALM specifies Key 15 as a raw (refined) linseed oil with a relatively low maximum acid value of 3 (Dulux Archives, 1956 specification).

Due to postwar shortages of Key 15 linseed oil for making stand oil, the notebooks advise substitution with Bridge Oil (BALM Key V978) as an alternative heat-bodied oil for *Zinc Paint* (Fig. 5.2). Bridge Oil is a "heat bodied mixture of two oils – 10–15% wood oil, with the remainder linseed" ([13]:L71). Wood oil is also known as Tung oil ([8]:36). To form a binder, Bridge oil was combined with pale-boiled linseed oil in preference to raw refined linseed oils for better drying ([13]:L97a, Fig. 5.2). Boiled oil incorporates driers and has been oxidised under controlled conditions (including heat) to a defined relative density and drying time ([15]:51–64).

Reactions occurring during stand oil production are influenced by the presence or absence of oxygen ([8]:40). In 1949, stand oils required considerable skill to produce and were usually made in open varnish pots (Fig. 5.3). Processing was carried out to a specified viscosity, with acid value and colour controlled as required. Enclosed vessels fitted with stirring and vacuum facilities could be used where paler colours and lower acid values were required. The additional cost of vacuum processing was rarely justified in terms of finished product performance ([15]:66). It is unclear from the 1949 notebooks whether any of the stand oils used by BALM were produced in the absence of oxygen.

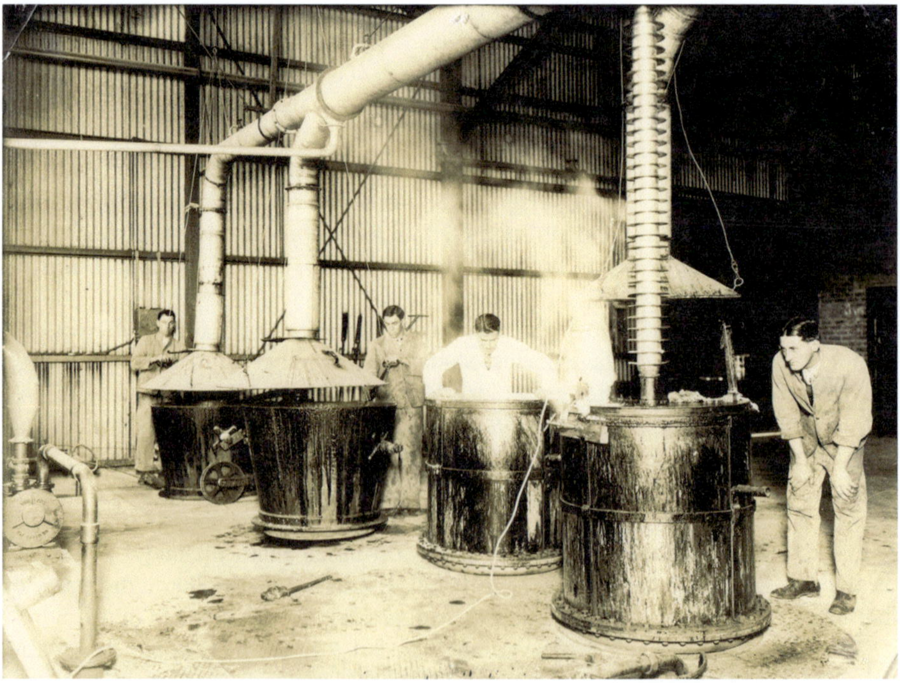

Fig. 5.3 Dulux Archives image 170 HIS1037. Heat treatment of oils and varnishes in pots over gas flames at Cabarita Resin and Varnish plant c.1950. Image courtesy Trudy Scott

5.7 Oil Acid Value

Acid value (AV) is a measure of the free acids present in oil (mg KOH/g). As acid value increases there are more available acids and consequently more zinc soap formation. BALM used different types of oil and pre-processing methods to control acid value and influence the level of soap formation. Linseed oils with acid value above 5 were found to cause excess soap and rapid bodying. Too low an acid value, for example, alkali refined linseed oil with AV 0.3–0.6, produced too little soap, no bodying and quicker checking ([13]:L82,84).

The well-known significance of acid value is discussed by Roberts but with the additional suggestion that soap formation is influenced by the type of acid an oil contains. Zinc paints made with relatively high acid value linseed stand oil (up to AV 10) were found to be viable in contrast to high acid value untreated oils, possibly due to stand oil containing 'different' acids. Roberts ([13]:L84) writes:

> …with heat treated oils – different soap – formed by different type of acid encountered. Thus it is not the acid value, but type of acid which is important in paints containing ZnO (explanation unknown here).

Roberts suggests that the stand reaction imparts qualities which reduce problematic soap formation. This premise is supported by recent research comparing zinc oxide reactivity in a range of drying oils where crystalline zinc carboxylate formation was found to be lowest in stand oil (and Tung oil) [1].

Heating linseed oil promotes rearrangement of some unconjugated double bonds in the constituent unsaturated fatty acid chains [17]. An FTIR spectrum of fresh linseed stand oil when compared to cold pressed linseed oil has an additional peak characteristic of *trans*- unsaturation (peak at 970 cm^{-1}) and reduced *cis*- absorption (peak at 720 cm^{-1}) [7]. The presence of both *cis*- and *trans*- fatty acid isomers during linseed stand oil production enables Diels-Alder addition reactions to occur, forming cyclic dimer acids (Fig. 5.4). Cyclised monomeric acids may also result [8]. These cyclic carboxylic acids have potential to react with zinc ions, however it is possible that the resulting cyclic structures are less problematic in paint films than zinc carboxylates of the straight chain fatty acids in untreated linseed oil due to their molecular organisation. This hypothesis requires further research.

5.8 Conclusion

There is valuable historic paint industry information in the 1949 BALM notebooks. Amongst details of many paint ranges and concerns, the notebooks record research and practical testing on a commercial scale of zinc oxide oil paints at a time when lead white was being phased out due to its toxicity. Zinc oxide in oil was later displaced by rutile titanium dioxide and pentaerythritol alkyd in commercial house paint but remains in artist paint formulations.

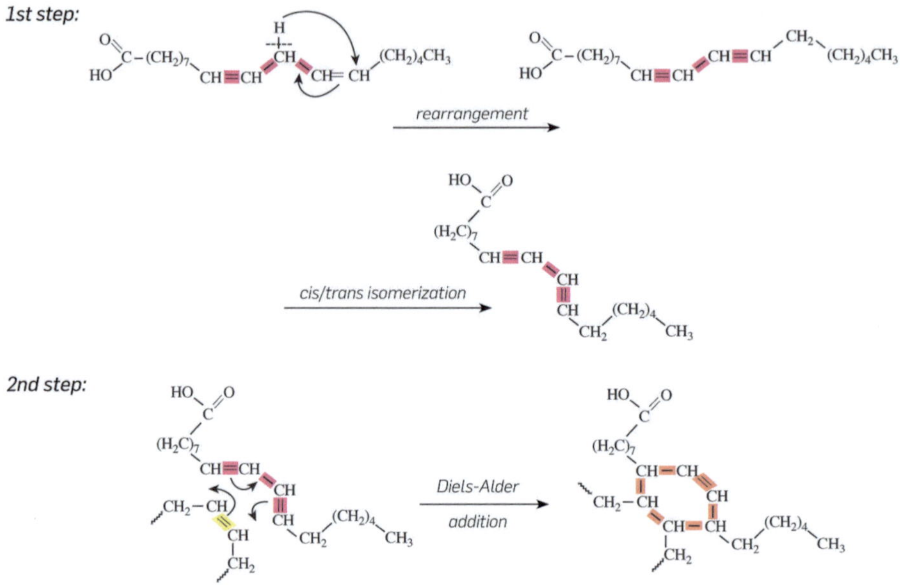

Fig. 5.4 Possible mechanism during stand reaction: Representative Diels-Alder addition involving 9-*trans*, 12-*trans* isomer of linoleic acid. (Adapted from [17])

In 1949, zinc oxide was known to be reactive in oil and management of zinc soap formation was the subject of experiments and variations to commercial paint formulation. The most durable zinc-based oil paint required use of good quality, large particle size zinc oxide at correct pigment volume ratio, incorporation of appropriate driers and use of a heat-bodied oil. The notebooks provide a prompt to consider the effect of the stand reaction on linseed oil acids and subsequent soap formation. Although stand oils can contain a high proportion of carboxylic acids available for reaction with zinc ions, following the stand reaction, these are likely cyclic. It is proposed that subsequent cyclic zinc carboxylates may inhibit aggregation and crystallisation and could be the 'different soap' suggested by Roberts, which renders zinc oxide in linseed stand oil more stable.

Acknowledgements Trudy Scott, Archivist, Dulux Archives, Melbourne.

References

1. Casadio F, Bellot-Gurlet L, Paris C (2019) Factors affecting the reactivity of zinc oxide with different drying oils: a vibrational spectroscopy study. In: Casadio F et al. (eds) Metal Soaps in Art: Conservation and Research, Springer International Publishing, Cham.
2. Dredge P (2012) A history of Australian housepaint technology from the 1920s to the 1950s, with reference to its use by Australian artists, particularly Sidney Nolan. AICCM Bulletin (33):53–61.

3. Dredge P, Schilling MR, Gautier G, Mazurek J, Learner T, Wuhrer R (2013) Lifting the lids off Ripolin: a collection of paint from Sidney Nolan's studio. JAIC (52):213–226.
4. Hermans JJ, Keune K, van Loon A, Iedema PD (2015) An infrared spectroscopic study of the nature of zinc carboxylates in oil paintings. Journal of Analytical Spectrometry, vol. 30, 1600–1608.
5. Kokkori M, Casadio F, Boon J (2014) A comprehensive study of early twentiethcentury oil-based enamel paints: Integrating industrial technical literature and analytical data. ICOM-CC 17th Triennial Conference Pre-prints, Melbourne.
6. Kühn H (1986) Zinc white. In: Feller R (ed) Artists' pigments vol 1. Cambridge University Press, p 169–186.
7. Meilunas RJ, Bentsen JG, Steinberg A (1990) Analysis of aged paint binders by FTIR spectroscopy. Studies in Conservation (35):33–51.
8. Mills J, White R (1994) The organic chemistry of museum objects. second edn. Routledge.
9. Muir K, Gautier G, Casadio F, Vila A (2011) Interdisciplinary investigation of early house paints: Picasso, Picabia and their 'Ripolin' paintings. ICOM-CC 16th Triennial Conference Preprints, Lisbon.
10. Osmond G (2012) Zinc white: a review of zinc oxide pigment properties and implications for stability in oil-based paintings. AICCM Bulletin (33):20–29.
11. Osmond G (2019) Zinc Soaps: An overview of zinc oxide reactivity and consequences of soap formation in oil-based paintings. In: Casadio F et al. (eds) Metal Soaps in Art: Conservation and Research, Springer International Publishing, Cham.
12. Raven LE, Bisschoff M, Leeuwestein M, Geldof M, Hermans JJ, Stols-Witlox M, Keune K (2019) Delamination due to Zinc Soap Formation in an Oil Painting by Piet Mondrian (1872–1944): Conservation Issues and Possible Implications for Treatment. In: Casadio F et al. (eds) Metal Soaps in Art: Conservation and Research, Springer International Publishing, Cham.
13. Roberts B (1949) BALM Laboratory Notebooks, Recruit training scheme. No.G 7/2/49 HIS2041, No.L 2/5/49 HIS2044, No.R 23/6/49, HIS2039, No.K 5/12/49 HIS2043. Dulux Archives, Melbourne.
14. Rogala D (2019) Everything old is new again: revisiting a historical symposium on zinc oxide paint films. In: Casadio F et al. (eds) Metal Soaps in Art: Conservation and Research, Springer International Publishing, Cham.
15. Surface Coatings Association Australia (SCAA) (1983) Surface Coatings. Raw materials and their usage. The New South Wales University Press.
16. Standeven H (2011) House Paints, 1900–1960: History and Use. Getty Conservation Institute.
17. Zovi O, Lecamp L, Loutelier-Bourhis C, Lange CM, Brunel C (2011) Stand reaction of Linseed oil. Eur J Lipid Sci Technol 113:616–626.

Chapter 6
A Glimpse into the House and Decorative Paint Market in Norway During World War I (1914–1918)

Barbro Wedvik

6.1 Introduction

Norway did not participate in the First World War (WWI); however, the paint market was strongly affected in the war's closing years. During the war, the cost of linseed oil—the primary binding medium used in house paint, decorative paint and artists' paint in Norway—rose dramatically (Fig. 6.1), and then it became unavailable [13]. The aim of this study is to investigate the consequences of the restricted access to linseed oil and what other binding media were available that could have been used as substitutes. The research gives background information for further material investigation and treatment of paint layers from the period. The paper aims to present changes in the production and use of house paints in Norway during WWI and gives an overview of the binding media and paint products presented and discussed in the journal *Maleren*.

6.2 Method and Sources

The research is based on the analysis of written sources. The main sources are reports, articles and advertising published in *Maleren (The Painter 1908–present)*, a biweekly Norwegian journal for house and decorative paint, with painters as the main target group. Primary and secondary sources on governmental directives and committee work have been studied to understand the wartime situation and the import and distribution of raw materials (*Maleren 1908–present*, [10, 11, 13]). Primary and secondary sources of industry development and industry manufacturing

B. Wedvik (✉)
Conservation Department, Norwegian Institute for Cultural Heritage Research (NIKU), Oslo, Norway
e-mail: barbro.wedvik@niku.no

K. J. van den Berg et al. (eds.), *Conservation of Modern Oil Paintings*,
https://doi.org/10.1007/978-3-030-19254-9_6

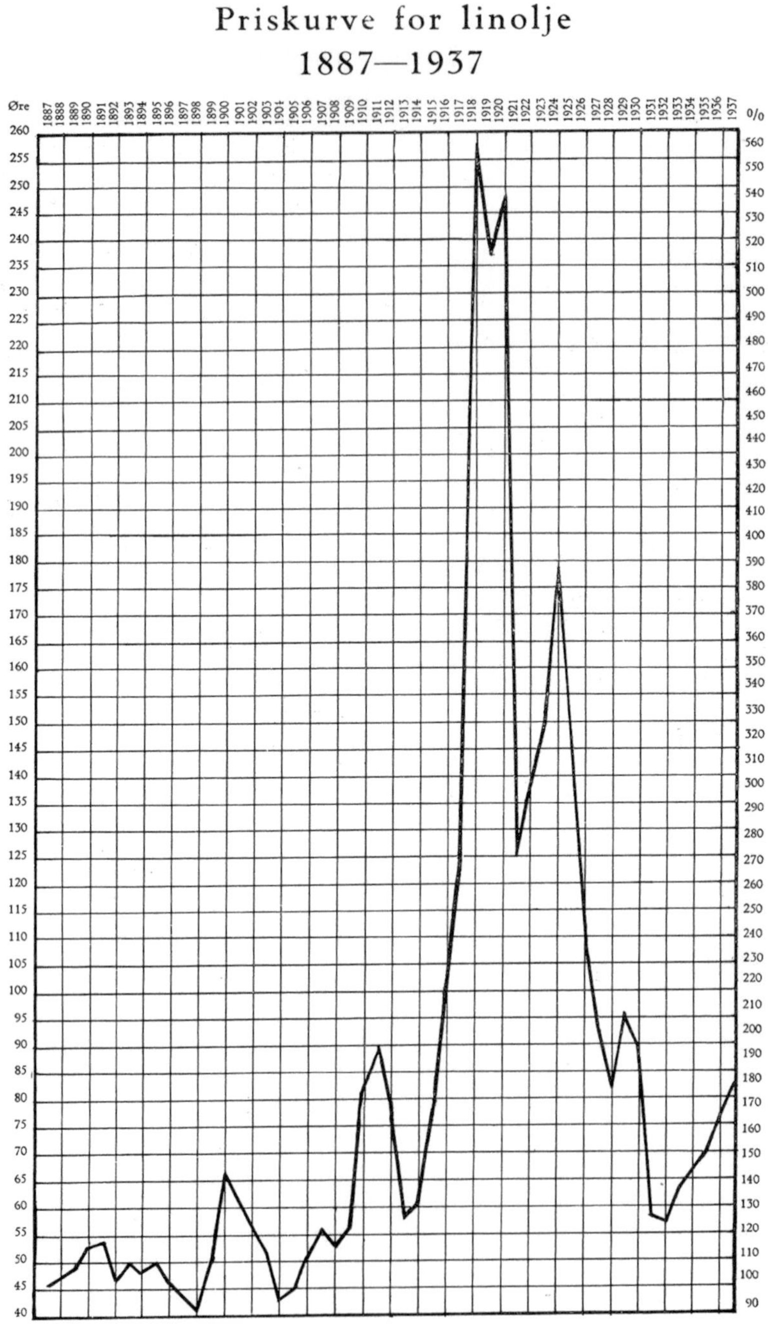

Fig. 6.1 Linseed oil price curve, 1887–1937 (Reproduced from [13])

processes have also been studied to obtain insight into the available materials [1, 2, 8]. For an understanding of the international paint development context, the book *House Paint 1900–1960* by Harriet Standeven [15] is an important source. Norwegian painters' books from the time, paint material courants and brochures were also studied.

6.3 Oils Used in Paint in the Early Twentieth Century

Linseed oil was the most important of the drying oils used as binders in house paint, decorative paint and artists' paint in the early twentieth century. Linseed was imported to Norway both as oil and as linseeds for grinding in domestic mills. From 1914 to 1916, linseed oil was mainly imported from the United Kingdom [10, 13]. There were also other drying oils used in paint. For example, tung, sunflower and hemp oil were used in the commercial fabrication of oil varnishes in the early years of WW I [1]. In addition to linseed oil, poppy and nut oils were used for home-made decor paint and artists' paint ([6], *Maleren* 1911 nr. 13). For outdoor paint, there was a long tradition along the coast of using fish oil [12].

6.4 Ready-Made House Paints

In the late nineteenth century, in the United States and the United Kingdom, ready-mixed house paints were introduced, both oil- and water-based, and by the first decades of the twentieth century, ready-made paints had been formulated for all purposes, including interior and exterior woodwork, interior walls and masonry, and these paints were available in flat, gloss and eggshell finishes [15]. According to Dredge [7], all oil-based ready-made paints were called enamels. Kokkori et al. [9] write that the term 'oil enamel' was employed to describe ready-to-use paints which contained linseed oil or a blend of oils with resin and pigments, and this term connoted a hard, glossy surface similar in appearance to that of porcelain enamel. Another important development was the industrial production of so-called 'washable water paints.' These were early type emulsions [4], meaning mixtures of aqueous and oleaginous binding media [5], often with a content of resin and the white pigment lithophone. Both enamels, such as Moline (Netherlands), Decorin and Japonol (local), and 'washable water paints', such as Duresco (UK) and Temperatin (local), were available ready-made paints in Norway before WWI (*Maleren* 1908–1913).

6.5 Linseed Oil Regulations in Norway During World War I

By 1917, access to linseed oil had become very limited, and the Norwegian Ministry of Industrial Supplies implemented regulations for its use for paint (*Maleren* 1917 nr. 23). As far as possible, only glue, limewash, casein and tar were to be used as well as emulsion paints and surrogate oils. Linseed oil was only permitted in the following circumstances:

- New woodwork, such as panels, doors, windows and floors, should be oiled only once to preserve the wood
- The flooring of old buildings should be oil painted only when absolutely necessary to preserve health and for home care purposes. The walls, ceilings, panels, doors and windows of old buildings should not be painted with oil
- Plaster walls painted with water-based paint should not be repainted using oil paint
- Hospitals, and thereafter residential houses, should be given preference before other buildings for treatment with oil paint.

In *Maleren,* the house painters were advised to keep painting (*Maleren* 1918 nr. 8). Many paint recipes were given in *Maleren,* featuring traditional binders, such as animal glue, fish liver oil, starch, flour, gum and casein. Many pieces of advice on how to economize on the use of linseed oil were also given, including adjustment of the recipes and the mixing ratios. The painters were advised to use distemper paint for interior paint on walls and ceilings and to use ready-made enamels on surfaces which had to be washable (*Maleren* 1918 nr. 8). The extent to which the advice was followed has yet to be verified.

6.6 Surrogate Products

The frequent use of the terms 'surrogate oil' and 'surrogate paint' corresponds, not surprisingly, to times with limited access to linseed oil. From 1917 to 1919, when linseed oil prices rose very high (Fig. 6.1), there were several ads and articles in *Maleren* where the terms were used. The term 'surrogate oil' seems to have included any product which could be used as a binder except pure linseed oil, while 'surrogate paint' could be any paint except pure or high-content linseed oil paint.

The products labelled 'surrogate oil' and 'surrogate paint' advertised in *Maleren* were both Norwegian and imported. The content was not described, and the manufacturers often remained anonymous, perhaps to avoid repercussions owing to the poor quality of the products sold. Lower quality surrogate paints were described as often being brittle and having a resin and mineral oil content (*Maleren* 1911 nr. 8). In a review of *The exhibition of surrogates* (*Maleren* 1918 nr. 8), current surrogate paint products were presented, and their suitability was discussed. Many of the paints were presented as distemper paints. They were described as matte paints which could be lacquered upon; these were washable and suitable for indoor use and could also be used outdoors with the addition of oil. One paint description sought to reassure the consumer that the paint did not contain sulphite lye or water and that the raw material needed for production was available domestically. A later article from the same year stated that many of the surrogates for linseed oil were emulsions (*Maleren* 1918 nr. 15).

In the spring of 1918, the need for an evaluation of the paint products became apparent. There were surrogate paints in the market, but some of them were described as useless (*Maleren* 1918 nr. 9). Consequently, a governmental committee named the Surrogate Committee was established to conduct product tests and provide recommendations for the surrogate products (*Maleren* 1918 nr. 10). The committee was composed of a colour trader, a chemist, a paint manufacturer and a master painter. Through testing, the committee made recommendations, and the approved product's manufacturer could then apply to the government for raw materials for production.

6.7 Availability of Binders

What type of binding media could be present in a paint which was produced in Norway during WWI? The need for binding media supplies inspired the creative application of local and national resources. According to an article in *Maleren*, "Most of the mentioned surrogate oils and paints are to a greater or lesser degree made of sulphite lye" (see p. 66 in *Maleren* 1918 nr. 8). Large quantities of sulphite lye from the pulp and paper industry were readily available, representing a cost-effective material. A by-product from the sulphite process was large quantities of terpenes, which could be distilled into wood turpentine with a residue of resin [8]. There was possibly also the production of ethanol from sugar and the production of a fine lignin precipitate for dyeing, from neutralization with lime stone [2]. It has yet to be elucidated which ingredients of sulphite lye were wanted and how the lye was used—as solvent, binder or dye.

According to the advertisements in *Maleren*, products from the modern paint industry, imported and domestic ready-made enamels, were continuously available throughout the war (*Maleren* 1914–1918). Examples of available imported ready-made enamels include Robbialac (UK) and Teoline (the Netherlands). Enamels could be made of pure oil binders or binders with a resin content [15]. Alternative oils were also imported to Norway. Tung oil-based products manufactured by the Alf Bjerckes paint factory became very popular in the years before the war. Those, among others, included an enamel paint called Decorin and a varnish called Oxan oil (Fig. 6.2). However, in 1918, the ads for Decorin disappeared, and in the same year, when Oxan oil was called for to substitute for linseed oil in interior paint, it was not available (*Maleren* 1918, nr. 13).

All the binders used in the traditional recipes, such as animal glue, starch, flour, gum and casein, could be present in a WWI paint. Recipes distributed in *Maleren*

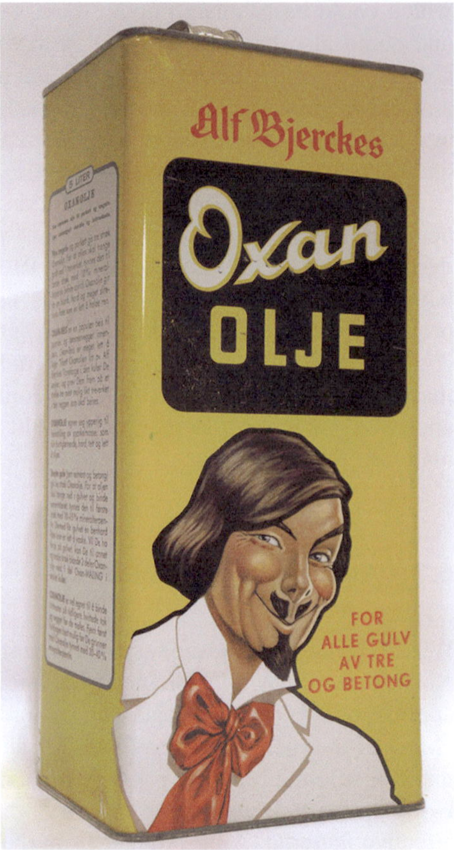

Fig. 6.2 Oxan oil based on tung oil made by Alf Bjerckes Factories. (Photo: Jotun Museum 2017)

during the war were both domestic and from the neighbouring countries of Sweden and Denmark. Locally produced ready-made emulsion paints, such as Temperatin and Hydratin from the Alf Bjerckes factory, were available (*Maleren* 1914–1918). In Norway, raw material from the sea was also important. *Tran* or fish liver oil could be sourced from industrial suppliers and local small-scale producers (Fig. 6.3, see p. 13 in [14]). In northern Norway, the use of paint based on oil from fish livers is documented from about 1860 until after WWII, and the paint was probably used frequently all along the coastline [12]. Its primary use as a binding medium was for outdoor paint. It could be employed on its own or in combination with another oil, such as tung oil, or in emulsion paints (*Maleren* 1918 nr. 11).

Whaling was one of the largest Norwegian industries before WWI, and whale oil as a binder for paint had been accessible. Norwegian whaling activity abroad decreased during the war and whaling along the Norwegian northern coast had been prohibited. In 1917, to meet people's need for meat and fat, whaling along the northern coast became legal again, and in 1918, by law, whale oil became available to produce paint ([10], *Maleren* 1911 nr. 8). However, even though whale oil was

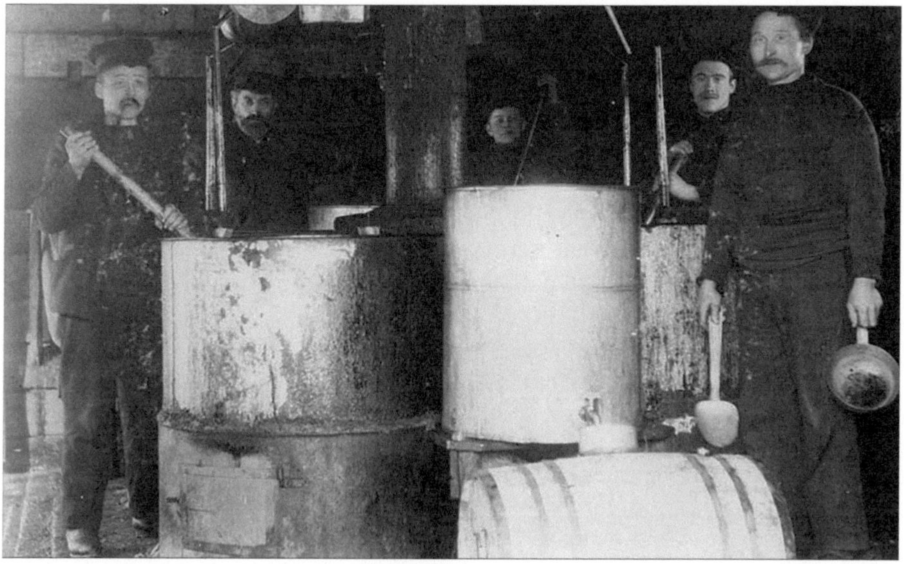

Fig. 6.3 Damp boiling of fish liver oil at Hamarøy, ca. 1910. (Reproduced from [14])

now legally available, it is not clear whether whale oil-based paint was manufactured during and after WWI.

Based on information in *Maleren,* characteristics of the house paint market in the closing years of WWI could be summarized this way:

- Many newly developed paint products with a content of sulphite lye, a by-product from industrial wood processing, became available. The use of sulphite lye as an ingredient in paint is not described further in *Maleren* and has yet to be elucidated
- Extension of the use of available modern paint materials, especially ready-made enamels
- Revival or extended use of distemper and emulsion paints based on traditional recipes
- Available traditional binders were mixed with modern products; for example, fish liver oil was mixed with tung oil.

6.8 Consequences for Today

In the articles in *Maleren*, some of the surrogate paints were reported to be non-drying and running, while others were found to flake, lack adhesion and become like powder. Yet others were reported to be well-functioning paints. No comments or discussions have been found in *Maleren* on the shortage of linseed oil as a binder in artists' paint. However, an archived letter from the well-established, productive

Norwegian fine art painter and academic professor Christian Krogh testifies to the lack of oil in the final year of the war:

> Concerning paint, one can buy dry pigments, grind them and spread them with a palette knife. Admittedly one needs a little oil for this, which is nowhere to be had. But at Andersen & Engebretsen or Sudmann in Lille Grendsen one can buy a 'surrogate' for oil. I have tried it and it is very good. (See p. 225 in [3])

Apparently, the presence of substituting binding media can be expected not only in house and decorative paint, but also in fine art paints from this period. This may have influenced painting techniques, optical effects and the drying of the paint. For the conservation and treatment of paint layers from WWI, the diversity of binders has important implications. Not only will all the different types of binders used in paint display unique sets of properties and responses to treatment, but a conservator should also expect unpredictable variations on every painted object, due to situational paint mixtures and economizing mixed ratios.

For architectural paint researchers the research findings show features to look for, such as the absence of linseed oil-painted surfaces—on panels, doors, windows, walls and ceilings—and just one allowed oil layer on new woodwork—such as panels, doors, windows and floors. Such qualities are important lead treads in the interpretation and dating of historic interiors. The extended use of both home-made and ready-made distemper and emulsion paints on interiors as well as of ready-made enamel paint on interiors which had to be cleanable was an exception to usage in normal times, and thus, it was also characteristic of the period.

6.9 Conclusion

This study gives a glimpse into the paint market in Norway from 1914 to 1918. The restricted access to linseed oil during WWI caused important changes in the house painters' market. A space opened for an extended use of modern paint products, such as ready-made enamel paints—both locally produced and imported—and of tung oil-based products; for the development of new paint products; and for the revival of traditionally used binders in paint production. Emulsion paints, both as traditional home-made paints and as ready-made paints, seem to have been used frequently. By-products from modern industry, such as sulphite lye, also came into use.

Norwegian house paints from the period 1914–1918 may be based on a wide range of binding media, including fish liver oil. This may have influenced the optical effects and drying properties of the paint. To what extent the same binding mediums were used for easel paintings has not yet been elucidated. The very wide range of binders which can be present in paint from WWI must be taken into consideration when planning conservation treatment.

Acknowledgments Grateful thanks are extended to the Norwegian Research Council for the provision of funding.

References

1. Alf Bjercke Fernisfabrik (1915) Alf Bjerckes fernisser. Pamphlet in *Maleren* nr. 7. Alf Bjercke Fernisfabrik, Oslo
2. Andersen CC (1949) Utnyttelse av sulfittavluten. Teknisk ukeblad 24:378–383
3. Aslaksby TE (2012) Christian Krogh and photography: some observations and reflections. In: Hansen et al (eds) Christian Krohg: pictures that captivate. Nasjonalmuseet for kunst, arkitektur og design, Oslo
4. Baty P (2017) The anatomy of colour: the story of heritage paints and pigments. Thames & Hudson, London
5. Beltinger K, Nadolny JM (2016) Painting in tempera, c. 1900. Schweizerisches Institut für Kunstwissenschaft in association with Archetype, London
6. Brænne J (1998) Dekorasjonsmaling: marmorering, ådring, lasering, patinering, sjablondekor, strukturmaling. Teknologisk forlag, Oslo
7. Dredge P (2012) A history of Australian housepaint technology from the 1920s to the 1950s, with reference to its use by Australian artists, particularly Sidney Nolan. AICCM Bulletin, 33(1):53–61
8. Halse OM, Samuelsen S (1918) Undersøkelse av en norsk træterpentinolje. Tidsskrift for Kemi 11:177–184
9. Kokkori M et al (2014) Charting the development of oil-based enamel paints through the correlation of historical paint technology manuals with scientific analysis. In: van den Berg KJ et al (eds) Issues in contemporary oil paint. Springer International Publishing, Cham, p 117–125
10. Larsen LPH (2012) Fett -og Oljeforsyningen under 1. verdenskrig: En analyse av statens reguleringer og Statens Fettdirektorat. Master thesis, Norges teknisk-naturvitenskapelige universitet, Trondheim.
11. *Maleren* (1908–) Malermestrenes landsforbund in association with Skarland Press AS, Oslo
12. Meissner K, Brekke A (2017) Tranprosjektet: produksjon og maling med tran fra seilever: sluttrapport. Byggkult AB, Öland
13. Schiørn F (1937) Olie- og farvehandlerforeningen i Oslo gjennem 50 år 1887–1937. Olie- og farvehandlerforeningen i Oslo in assosiation with Emil Moestue AS, Oslo
14. Schiøtz O (2010) A thousand years of Lofoten cod-liver oil: Lofoten, a global manufacturer. Museum Nord—Norsk fiskeværsmuseum, Å in Lofoten
15. Standeven HAL (2011) House paints, 1900–1960: history and use. Getty Conservation Institute, Los Angeles

Chapter 7
Approaches to the Identification of Royal Talens ETA (Emulsion) Paint in Objects of Art

Lisanne van den Heuvel, Inez van der Werf, Coos van Waas, and Klaas Jan van den Berg

7.1 Introduction

The Dutch manufacturer of paints and art supplies, Royal Talens, was founded in Apeldoorn in 1899. At the beginning of the twentieth century, Talens started producing artists oil paints, such as the high quality Rembrandt oil paint [4]. Another paint Talens produced was ETA paint, which is a relatively inexpensive casein in oil emulsion paint. This rainwater resistant decoration paint was produced for in- and outdoor embellishments and commercial design. After this paint was discontinued around 1980, the knowledge of its composition and history disappeared.[1]

Although many other Talens paints were specifically targeted towards artists, ETA seems also to have been used. For example, a famous still from the documentary 'De werkelijkheid van Karel Appel'[2] shows the artist in an atelier with many tins of ETA in the background.

A collaborative agreement between the Cultural Heritage Agency of the Netherlands (RCE) and Royal Talens provided access to the formulations and production of ETA from the Talens archive. These included recipes, notes on manufacture, price lists and colour charts with paint outs. The sources enabled to establish a time line of the developments in formulation of each colour in the ETA range.

[1]At the moment of writing, Talens' website still states that ETA was introduced in the 1960s: https://www.royaltalens.com/about-us/history/.

[2]'Karel Appel's reality'. Documentary by Jan Vrijman, 1961.

L. van den Heuvel
Independent Researcher, Utrecht, The Netherlands

I. van der Werf · K. J. van den Berg (✉)
Cultural Heritage Agency of the Netherlands, Amsterdam, The Netherlands
e-mail: k.van.den.berg@cultureelerfgoed.nl

C. van Waas
Royal Talens, Apeldoorn, The Netherlands

K. J. van den Berg et al. (eds.), *Conservation of Modern Oil Paintings*,
https://doi.org/10.1007/978-3-030-19254-9_7

In this paper, the context in which Talens' ETA was developed will be presented together with a discussion on how these painting materials fit into the product range supplied by Talens. Then, the sources that were used in the construction of a chronological database will be described, with information about the development in the paint formulation. Based on these outcomes, links between the recipes and examples of analysis of paint from works of art are presented. The results can inform identification and subsequent dating of ETA media in objects (of art) with analytical techniques based on the ingredients used in the production of the paint at a specific time.

7.2 The Life of ETA: From a New Product to Discontinuation

'A matte, opaque casein emulsion paint, especially intended for the production of 'etalages' (shop windows), for in- and outside commercials, wall and theatre decorations, etc. ETA is lightfast and does not rain off.'[3] This is how the paint was introduced in Talens' 1933 brochure or 'Prijscourant' (price list). The reason why ETA was added to the Talens existing ranges of artist' paints and what motivated the production of the new medium was investigated from written sources, including the surviving catalogues from 1899 until the discontinuation of ETA in the 1980s.

Talens was initially established for the production of office supplies, but soon Talens started offering a wide range of products for use by artists. In their 1902 catalogue, for example, Rembrandt oil paints and water paints were introduced.[4] The first two Talens paints that provided a matt, opaque and cheaper alternative to the artist quality paints were *Talens Rembrandt Tempera Colours* (*Rembrandt Tempera Verven voor Kunstschilders)* and *Showcard Tempera Colours* (*Plakkaat Tempera Verven* or *Etalagekaartenverf*).[5] In the price catalogue from 1933 ETA is advertised right above the *Glass Paint*,[6] which was developed for commercial advertising purposes on glass (illuminated signs, decorations on glass windows, etc). ETA, probably an abbreviation for 'etalage' (shop window) paint, could be used to decorate outdoor and indoor objects on all kinds of supports and grounds. The paint resembling ETA most closely in 1933 was the *Showcard Tempera Colours* paint, but this paint was not suitable for outdoor embellishments. ETA was available in containers of several sizes including large quantities,[7] and it was relatively cheap. With the

[3] 'Een matte, dekkende caseïne-emulsieverf, speciaal bestemd voor het vervaardigen van ETAlages, binnen- en buitenreclame, muur- en theaterdecoraties, enz. ETA is lichtecht en regent niet af.' Translation by author (LvdH). Prijscourant 1933.

[4] Prijscourant 1902. Recent research shows that oil paints were already produced as early as 1900 (Van den Berg and Pause, unpublished results).

[5] Prijscourant 1922, pp. 114–116.

[6] Price catalogue 1933, p. 59.

[7] In 1933, ETA was supplied in 0,5, 1, 2.5 and 5 kg tins, at the price of 32,50–47,50 French Francs per 1 kg.

introduction of ETA, Talens therefore added to their line of products a cheap, rain-water resistant paint that could also be used outdoors. After the first advertisement for ETA in 1933, the product description in the catalogues and folders did not change much until the discontinuation of the range, with the introduction of new colours as the most remarkable variation.

In a personal recipe book by P.J. Berends, a Talens employee in the 1970 and '80s, there is a description of the search for another recipe of ETA paint without casein in the binding medium. Apparently, shelf life of ETA was problematic because after a while the casein would start to rot, causing the paint to thicken and the binding medium to separate from the pigment ('uitzakken'). In the beginning of the 1980s, a new product, TalETA, based on a synthetic water soluble polymer instead of casein, was introduced to replace ETA. However, this product was discontinued after a few years. Probably because the advantages - brilliant colours, covering power, matte surface, water-resistance and possible application to a variety of surfaces - could also be found in more commonly used, more stable paints such as acrylic paints.

7.3 Available Sources in the Royal Talens Archive

Talens' archives contain catalogues, price lists, production books, lab journals of chemists, ledgers, colour charts with original paint outs and recipe cards in a contemporary analogue back-up system (Fig. 7.1).

The catalogues provided insight in the continuation and extensions of the colour ranges of ETA paint that were available on the market in the Netherlands and abroad between 1933 and 1980. In the catalogues from 1933 to 1976, 26 up to 40 colours were presented. These catalogues seem to have been distributed with separate

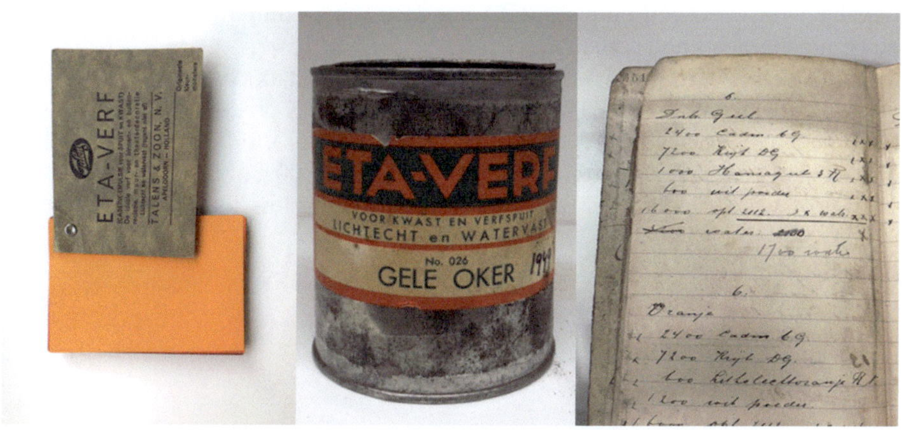

Fig. 7.1 Colour chart dated 1946–1951, ETA tin dated 1949 and part of a production book dated 1932

corresponding price lists for the paints, which were more regularly updated than the catalogues themselves and provide information about the cost of the paints.

Furthermore, the logbooks from 1932 to 1946 archive the daily production: for each colour the materials, their quantities and sometimes the the machinery used. These books also contain recipes for the binding medium and, sometimes, the companies from which pigments were purchased are mentioned.[8] While no logbooks after 1958 were found, records of recipes were typed on separate cards [4]. Between 1946 and 1958 there are recipe cards from an old analogue back-up system. Furthermore, Talens' chemists kept personal lab journals in which they described experiments with new materials and other observations and challenges they still needed to find solutions for. In addition, Talens also kept ledgers which provide information on the companies where Talens bought its pigments and their costs. A material source found in the archive are colour charts with paint outs. They provide a visual representation of the colours.

The evolution of the formulation, composition and material properties of ETA can be established through productions records which give an overview of its production. An Excel© database was created that incorporated the written sources. This was used to highlight the production and availability of Talens' ETA paints.[9]

7.4 The Composition of ETA

The database of production recipes provided an overview and a method for obtaining information of how the paint's ingredients changed over time. The changes in ingredients will facilitate the identification and dating of ETA in historical painted objects and may help explain the paint's characteristics such as: the colour, viscosity, gloss, drying time etc.

ETA's special property was that it could be used as an emulsion paint that was mixable with water and oil paint. ETA consists of a binding medium, organic and inorganic pigments and certain additives.

As an example, in Fig. 7.1, the production log (1936) of a batch of 28,9 kg dark yellow ('d(o)nk(er) geel') ETA paint is reported. It was produced with 2400 g of Cadm. 6G (cadmium yellow?), 7200 g of Krijt DG (chalk), 1000 g of Hansa Yellow 3R, 600 g of witpoeder (a different grade of chalk), 16,000 g of the medium. To this, 1700 g of water was added, which was for this batch less than the usually used 3200 g.

[8] See also Pause et al., these Proceedings, Ch. 8.

[9] By LvdH. Database accessible via KJvdB.

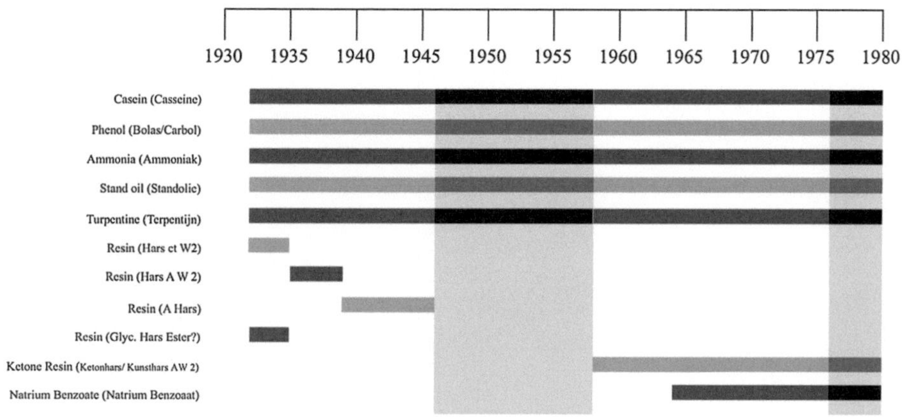

Fig. 7.2 Time line of ingredients in ETA binding medium 1932–1946 and 1958–1976 (with Dutch original text). Grey columns represent time slots from which no information is available

7.4.1 *The Binding Medium*

Figure 7.2 provides a summary of the key components of the binding medium for the ETA paints based on reports in the production books from 1932 to 1946 and recipe cards from 1958–1976. Between 1958 and 1976 seven recipe cards relating to the formulation of the binding medium were found. These list a mixture of casein, water, ammonia, phenol and solution 'E30/100'. In a note of another recipe card from 1976 it is stated that solution E30/100 contains (stand) oil, ketone resin, and white spirit. In another binding medium recipe from that time instead of 'white spirit', 'turpentina' was noted. It seems more likely that oil of turpentine was meant and not 'white spirit'. The most significant changes in the binding medium were the addition of different kinds of resin and the addition of natrium benzoate in 1964.

7.4.2 *Pigments and Additional Materials*

Generally, ETA paints contain organic pigment and/or inexpensive inorganic pigment, which concurs with the low price of the product. The inorganic pigments included ultramarine (for ultramarine and cobalt blue), yellow (probably synthetic[10]) iron oxide and lithopone and titanium oxide whites. Remarkably, the still relatively new titanium white pigment was, with few exceptions, used only as an additive to other colours [3].

Figure 7.3 plots the composition of the colour 'Donkergeel' (Yellow Deep), from the first production in 1932 until the last production in 1980, during which some

[10] Rika Pause, personal communication 18 April 2019.

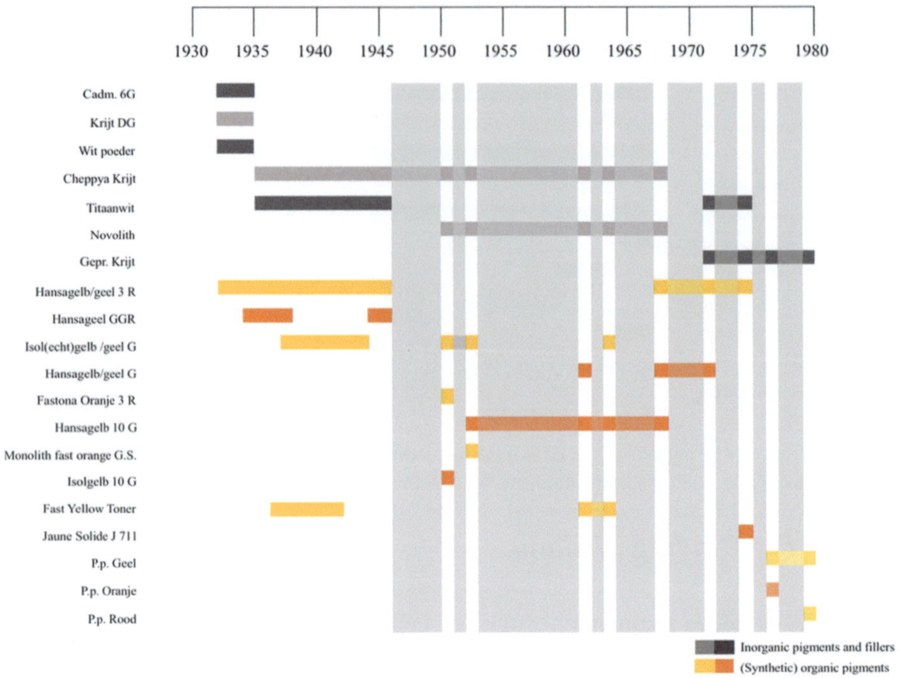

Fig. 7.3 Overview of developments in organic and inorganic pigments and fillers of Yellow Deep. P.p. refers to pigment paste, it remains unclear what ingredients these pigment pastes contained. Grey columns represent time slots from which no information is available

significant changes occurred. For example, in 1935 Hansageel GGR was added to the Hansageel/gelb 3R. The additions or omissions of certain organic pigments could affect the appearance and the properties of the paint. Also inorganic materials used in Yellow Deep were used for different batches over time, such as the alternative use of lithopone (Novolith) or titanium white. "White powder" most likely refers to chalk, which is a common extender in Talens paints [4].

7.5 Analysis of ETA in Historical Paint Samples

Contemporary colour charts, remnants of ETA paints from old tins and 'flacons' (flasks) were investigated as well as some works of art.[11]

[11] All results are available in L. van den Heuvel, internal RCE/CMOP report, 2018.

7.5.1 X-Ray Fluorescence Spectroscopy

Analysis of three Talens' ETA colour charts using X-Ray fluorescence spectroscopy (XRF) provided elemental analysis that could be used to infer similarities between the paints and corresponding recipes.

For example, analysis of the colour chart in Fig. 7.1 (*left*) demonstrated that the colour 'white' in this booklet contains titanium and calcium. Results from analysis could correspond to two ingredients reported in a recipe from 1946: titanium white and bentonite. This chart, for example, could be dated before 1952, because it does not contain the colour 'Parelgrijs' (Pearl Grey) yet, which was introduced in a catalogue from 1952. Considering this and other results of analysis, we suggest that this booklet dates between 1946 and 1951.

Where the coloured components of the ETA formulations comprised synthetic organic pigments, other analytical techniques were used for analysis.

7.5.2 Raman Spectroscopy

Paint samples were taken from the early, undated ETA colour chart (Fig. 7.1 *left*) to identify organic pigments with Raman spectroscopy (785 nm laser). Three colours were chosen for analysis: Donker geel (Yellow Deep), Oranje (Orange) and Vermiljoen (Vermillion).

The analysis of Yellow Deep is discussed as an example. In the recipe from 1946, Yellow Deep was produced with Hansageel GGR and Hansagelb 3 R, corresponding respectively to Pigment Yellow 2 (PY2) and Pigment Orange 1 (PO1), as defined by the Colour Index. The Raman spectrum of Yellow Deep showed clear peaks that matched PO1, pointing to the presence of Hansagelb 3 R from the 1946 recipe (Fig. 7.4a). The presence of this pigment was confirmed with Py-GC/MS (see Fig. 7.4b).

In a case study of the painting *Playmate of the year* (1964), painted with ETA by Dutch artist Wim Gijzen (1941), a sample taken from a purple passage was analysed using Raman spectroscopy. The results correspond to a *Blauwviolet* ETA recipe from 1958 listing water, binding medium E 30/99, Cheppyakrijt (chalk), Geprec. Krijt (precipitated chalk) and the two synthetic organic pigments Fanal Rose B Supra and Fanal Violet R Supra. The Raman results of the purple paint sample showed a match with Fanalviolet R (PV3) and some possible matches with Fanalrosa.

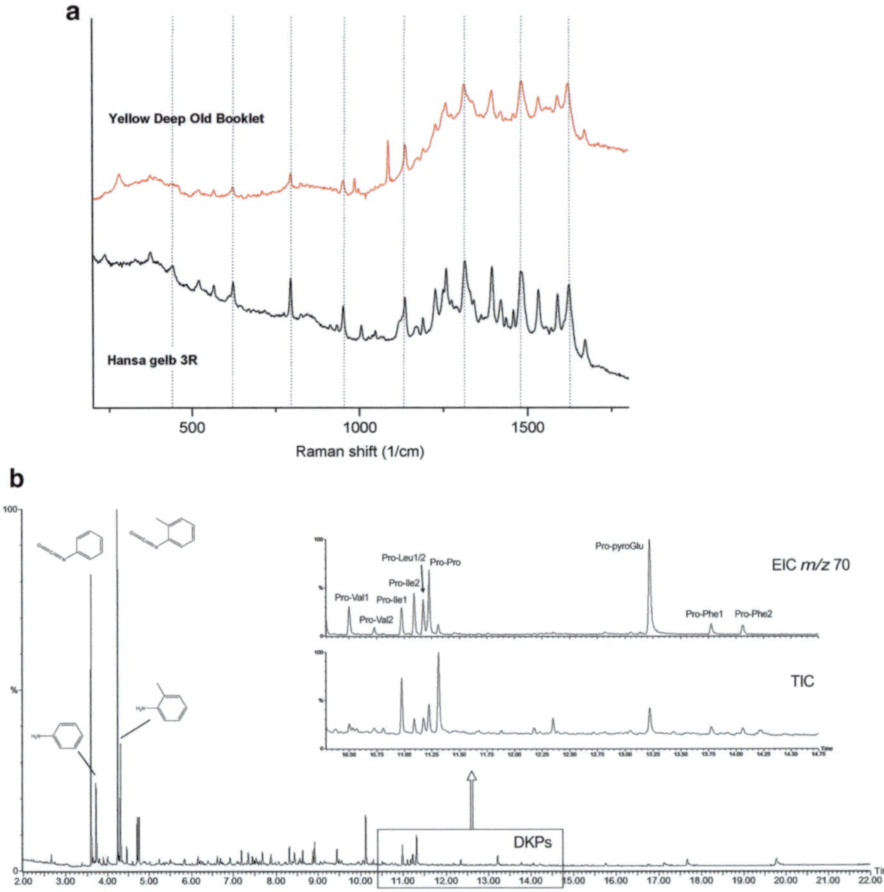

Fig. 7.4 Analysis of Yellow Deep of the old ETA booklet (**a**) Raman spectrum compared to a Hansagelb 3R (PO1) reference (courtesy of Rika Pause). (**b**) Py-GC/MS trace (Total Ion Chromatogram – TIC) (T_{Py} 350 °C). Very intense peaks are assigned to pyrolysis products of the main pigment (PO1). Inset: TIC and Extracted Ion Chromatogram (m/z 70) of the 2,5-diketopiperazines (DKPs) related to the pyrolysis of casein proteins. The latter three compounds are markers for casein

7.5.3 Pyrolysis-Gas Chromatography/Mass Spectrometry

The ETA binding medium consistently contained casein, water, ammonia, phenol, stand oil and turpentine. A range of different resins was used; sometimes even adjusted to a specific colour. In 1964 natrium benzoate was added. With the aim to identify binders and organic additives in ETA paint, samples from yellow, ultramarine and vermillion were taken from four historic colour charts and 3 ETA flasks. All samples were analysed using thermally assisted hydrolysis and methylation (THM) GC/MS with ultrafast thermal desorption (UTD) or with pyrolysis (Py) – GC/MS.

The results showed that all ETA paints contain proteins, which were confirmed to be casein on the basis of various specific pyrolysis markers (2,5-diketopiperazines – DKPs; [2]). These could be efficiently identified by performing Py-GC/MS analysis at a rather low temperature (T_{Py} 350 °C) without TMAH (Fig. 7.4b).[12] Although glycerol and dicarboxylic acids are generally low in intensity, the use of stand oil was ascertained by the identification of specific markers. Concerning the additives, phenol was detected in high amounts only in the paints from the pots, which is probably due to evaporation with time. The addition of natrium benzoate was clearly attested for the ETA paints from the flacons. Small amounts of *Pinus* diterpenoids point to the use of turpentine – except for the ETA paint from the pots - and ketone resins were also found.

Eugène Brands (1913–2002) revealed in an artist interview with IJsbrand Hummelen (RCE) in 1998 that he painted with ETA from Talens. There are two pictures of Brands in his studio surrounded by tins of ETA to confirm this statement. A comparison between the colour charts and some of his works on paper, and Py-GC/MS analysis support the artist's use of ETA for the work. Py-GC/MS[13] analysis of the black paint from Brand's *Surreëel Landschap* (Surreal Landscape, 1979) shows the presence of casein and other materials. It was concluded that this work was made with ETA and another kind of water borne paint (plakkaat verf).

Similarly, photos of Karel Appel in his studio with tins of ETA paint suggest that he too used this medium for painting. However, to date we have not found analytical evidence for its use in Appels paintings.

In 1959 the artist Lex Metz (1913–1986) painted a composition on two walls in the canteen of the former Dutch Red Cross Central Laboratory in Amsterdam [1, 5]. The possible use of ETA is suggested by a picture where a flacon of ETA is visible next to him. Examination on location and comparison of the mural with the ETA colour charts revealed that some colours corresponded, while others did not. Py-GC/MS of some samples allowed to identify casein, oil and some benzoic acid (likely related to natrium benzoate additive), which seems to confirm the presence ETA paint. Apparently, the mixing of colours before application and perhaps some discolorations have led to some colours of the painting not corresponding to the colours of the colour chart.

---

[12] Three ETA paint samples were also analysed with liquid chromatography – tandem mass spectrometry (LC-MS/MS) and the occurrence of casein could be confirmed by the presence of β-casein, α-S1-casein and α-S2-casein proteins (analyses performed by Cosima Damiana Calvano – SMART Centre, University of Bari, Italy).

[13] Performed by Henk van Keulen.

7.6 Conclusion

ETA is a relatively cheap casein oil emulsion paint with bright colours, matte appearance and quick drying time that was produced and used in the Netherlands between 1932 and 1985.[14] In this research project a database was constructed consisting of written sources about the formulation of ETA paint with the purpose of providing a chronology of the materials used to make the paint in this period. This was used as a reference for identification of the paint in works of art.

Identification of ETA in works of art is important, since some pigments seem to be very prone to discolouration and its specific composition may require conservation treatments different from those used for oil paintings. The database provided a means to obtain an overview of how the paint formulation changed over time.[15] This information will be crucial for the identification of ETA from an object.

The analysis on the Talens colour charts and old ETA paints provided data that will be useful for the identification of ETA in works of art, exemplified by case studies of works by three Dutch artists. Visual comparison with Talens colour charts can provide a first indication of the presence of pure ETA colour, or a mix with other ETA or paints with different media. XRF might then be used as a first non-destructive analytical technique providing elemental information indicating a possible presence of typical inorganic pigments and additives in ETA (chalk, lithopone, titanium dioxide). Subsequently, non- or micro-destructive analysis of synthetic organic pigments with Raman spectroscopy can give additional information on the organic pigments' use.

If sample-taking is possible, Py-GC/MS provides an overview of all organic ingredients and might allow for identification of casein and other organic components from the binding medium. All information can be checked with the database.

Future research is intended to complete the paint formulation database, particularly for 1946–1958. Furthermore, fading studies of the sometimes fugitive pigments could inform conservation and restoration practice.

References

1. Beljon, JJ, et al (1970) Dialoog kunst en gebouw, voorbeelden uit de praktijk, Bouwcentrum, pp 167–171
2. Orsini S, Parlanti F, Bonaduce I (2017) Analytical pyrolysis of proteins in samples from artistic and archaeological objects, J Anal Appl Pyrol 124, 643–657
3. Phenix A, Soldano B. Van Driel B, van den Berg KJ (2017) 'The Might of White': formulations of titanium dioxide-based oil paints as evidenced in archives of two artists' colourmen, mid-twentieth century. In Bridgland J (ed) ICOM-CC 18th Triennial Conference Preprints, Copenhagen, 4–8 September 2017. Paris: International Council of Museums. Article 0104

[14] To date, no information is available on its use abroad.

[15] Accessible via RCE: KJvdB.

4. Van den Berg KJ, Bayliss S, van Gurp F, Burnstock A, Klein Ovink B (2016) Making paint in the 20th century: the Talens Archive. In Eyb Green S et al. (eds) Sources on Art Technology; Back to Basics. Proceedings of the 6th symposium of ATSR,16 and 17 June 2014. Archetype, pp 43–50
5. Van der Linden JJ (1962) Gebouwencomplex voor Bloedtransfusiedienst te Amsterdam, in: Bouwkundig Weekblad 80 (1962) nr 22, pp 453–459

Chapter 8
Synthetic Organic Pigments in Talens Oil Paint 1920–1950 – The Case of Vermillion *imit*

Rika Pause, Johan G. Neevel, and Klaas Jan van den Berg

8.1 Introduction

After the discovery of mauve in 1856 [11], the emerging chemical industry invented and developed many new synthetic organic pigments (SOPs). However, the extent of application in early twentieth century artists' oil paint remains largely unclear and is thus an important area for art technological and conservation research. Production archives from this area are very rare. The *Royal Talens*[1] archive, which has preserved most of these references is a unique source for the investigation of modern paint production.[2]

The conducted research project "Synthetic organic pigments in Talens oil paint" focused on *Talens'* interest and championing of the use of SOPs from 1922[3] onwards and *Talens'* acquisition of raw materials from pigment suppliers and producers.

This paper is based on this project and will first illustrate *Talens'* stance towards SOPs. It then outlines the methodology of linking production recipes of the acquired logbooks with the investigation of oil paint tubes and illustrates this approach by presenting the representative analysis of a red Vermilion imitation paint and pigment.

---

[1] From hereon called *Talens*. The company names changed over time: 1899 *Talens & Co*; 1908 *Talens & Zoon*; 1949 *Koninklijke Talens*.

[2] The research agreement between *Royal Talens* and the *Cultural Heritage Agency of the Netherlands* (RCE) allows researchers to gain insight in the early oil paint production of this globally represented company in the beginning of the twentieth century [14].

[3] In 1912, the Talens & Zoon catalogue distinguishes in the fine oil paint range between natural and synthetic alizarin, for the first time.

R. Pause (✉) · J. G. Neevel · K. J. van den Berg
Cultural Heritage Agency of the Netherlands, Amsterdam, The Netherlands
e-mail: r.pause@cultureelerfgoed.nl; k.van.den.berg@cultureelerfgoed.nl

K. J. van den Berg et al. (eds.), *Conservation of Modern Oil Paintings*,
https://doi.org/10.1007/978-3-030-19254-9_8

8.2 Talens' Historical Dealings with SOPs

In the catalogue of 1931, *Talens* officially announced the use of SOPs for the production of fine quality oil paint. One year later, *Talens* published the book *Kunstschildersmaterialen en Schildertechniek* (The Artists' Materials and Painting Techniques) authored by Frits Kerdijk (1879–1972), which was followed by several editions. The book refers to *Talens* oil paints, belonging to the *Rembrandt* fine quality range. Kerdijk addresses SOPs in an individual chapter. In the chapter's first sentences he refers to the often negative attitude towards these pigments in artists' material at the time: "Their rare beauty of tone unfortunately deceived many manufacturer and self-producers of artists' paints, to use first coal-tar dyes and to damage their artworks by painting with them which damaged the reputation of coal-tar dyes. And so it always goes: once good reputation is lost, it is hardly possible to reacquire it, even if the former culprit now shows an excellent character."[4] Accordingly, he implies that SOPs were potentially good pigments. From 1932 onwards, Kerdijk devotes a full colour chart to coal-tar paints, emphasising their high quality by naming the chart "Eerste groep: betrouwbaar, Teerkleurstoffen" (First group: reliable, coal-tar colourants). In the second edition, published in 1937, he first lists the lightfastness standards for all oil colour samples in his book. Here, only three coal-tar colourants samples on the colour chart are listed as having the lowest grade of lightfastness and nineteen the middle grade. Three pigments received the highest rating. These three belonged to *Talens'* in house brand, introduced in 1931, which contained oil paints only consisting of SOPs. Hence, their lightfastness was indicated as higher than that of the synthetic alizarin containing oil paints (logbook 1, 1922; Fig. 8.1). This is interesting, since synthetic alizarin was generally considered as the best reference for lightfastness tests due to its high light stability [10]. Accordingly, *Talens* demonstrated trust in the high quality of their oil paints containing SOPs, which were by no means intended only for the production of inexpensive, student quality paints.

The importance of the use of SOPs for artists' oils paints by *Talens*, is illustrated in the following statement: "Recent years have enriched us with various new coal-tar dyes of splendid colour shades, which, in all respects, can gloriously pass the fastness tests and become a part of the artists' palette, without any objection."[5] Hence, Kerdijk openly and enthusiastically confirms *Talens'* dealing with SOPs, which formed an important part in their fine and student quality oil paint production since at least 1922.

Talens' open endorsement of these pigments as adequate and reliable artist's material since 1931 is remarkable since the general opinion concerning coal-tar based colourants was still rather critical in the 1930s [10]. Interestingly, *Talens* had used SOPs since at least 1922.

---

[4] Original quote in Kerdijk, (1937, p.65) (Translation R. Pause).

[5] Original quote in in Kerdijk [8]. (Translation R.Pause).

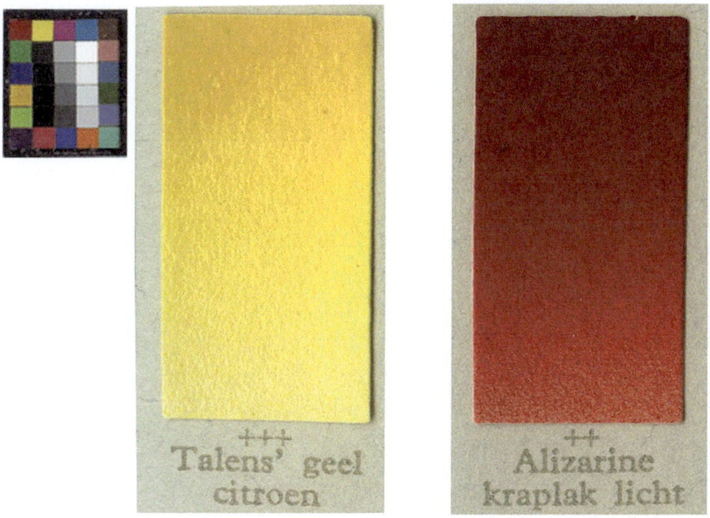

Fig. 8.1 Light stability of *Talens' geel citroen* (Talens' yellow lemon; *Talens'* in house brand) vs. *Alizarine kraplak licht* (Alizarine madder light; lightfastness reference) [8]. Paint outs in varying thickness

8.3 Methodology

8.3.1 Logbooks and Paint References

Fourteen production logbooks that document *Talens'* oil paint production from 1922–1950 were examined with a focus on SOPs. The recipes document the manufacturing process for every paint produced by *Talens*, listing the date of production, quality level and all ingredients and amounts [14]. In total, recipes related to more than 3000 batches of oil paints and 60 different colours. The registered ingredients also provide new insights into the origin of the pigments. They are listed with their commercial name, often followed by an abbreviation, linking the pigments to producers and suppliers.

Subsequently, corresponding paint samples that were expected to contain the SOPs listed in the respective recipes were selected for chemical analysis. At least one reference paint was chosen for the identification of each SOP listed in the recipes. It was important that these references contained the SOP in a pure form and not in a mixture with other pigments, to provide clear analytical results and establish a reference database. For this selection the paint samples were tentatively dated according to their paint tube shape and label design ([9]; see Fig. 8.2).

Furthermore, the listed SOPs were set into their art-technological context through examining contemporary specialist literature. These books were primarily selected from the historical library of the *Royal Talens* archive, marked with the stamp

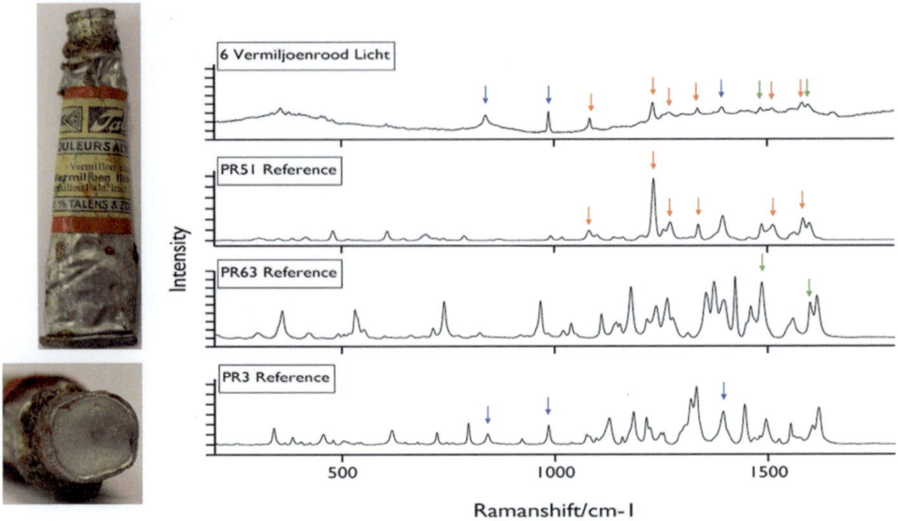

Fig. 8.2 Left: Vermilion licht imitation sample (No.6). Right: Raman spectrum of sample number 6; KIK-IRPA reference spectra of PR51, PR63 and PR3; range 200 cm^{-1} to 1800 cm^{-1}

"*Talens & Zoon*". It can therefore be assumed that these works were used by *Talens* at the time.

8.3.2 *Analyses*

Through the pre-dating and the elemental analyses of the oil paints, it was possible to match the results with original production recipes and limit the number of possible organic pigments, and thus refine dating. Elemental analyses were conducted with SEM/EDX (Scanning Electron Microscopy/Energy Dispersive X-ray spectroscopy) or handheld XRF (X-Ray Fluorescence). Tube paints were analysed with SEM/EDX and colour charts were analysed with handheld XRF to avoid destructive sampling. For the investigation of some samples both methods were necessary.

Analyses of the organic pigments were then conducted with micro-Raman spectroscopy. When Raman was not conclusive, SOP analysis was performed using Py-GC-MS (Pyrolysis-Gas Chromatography-Mass Spectrometry).

Raman spectra were compared to KIK-IRPA's spectral library ([5]; https://modern.kikirpa.be/).

8.4 Case Study: Vermiljoenrood Licht Imitatie – Vermilion Red Light Imitation

8.4.1 Recipes in Talens' Logbooks

A *Vermiljoen licht imitatie* paint was firstly offered in *Talens'* 1918 sales catalogue for the student quality range 101, with an annotation that the imitation form has no inferior characteristics in comparison with the original vermilion pigment [1].

In 1922, the production of vermilion light red imitation for student quality paint was registered for the first time. Table 8.1 presents the developments. Apparently, its composition changed several times within the investigated time frame, using six different SOPs. Comparing the different proportions of weight of the SOPs, it is striking that the pigments *Vermiljoen licht imit L 1484* and *Vermiljoen licht imit A arzb* required almost 10 times weight percentage as *Permanentrood 2G extra* and *Siegle oranje 2S*, probably either due to a lower tinting strength or the presence of extenders, added by the pigment supplier.

From 1928 to 1930 the main chromophore SOP *Vermiljoen licht imit A arzb* was no longer used in a pure form, but with the addition of Hansa Yellow pigments.[6] Both Hansa yellows were only used in small amounts in comparison to the vermilion imitation pigment, which demonstrates their high tinting strength [7].

Three different extenders were used in the *Talens* paint production process. Gypsum was replaced with chalk from 1925 onwards. Likewise, the additive *F vers* was replaced by *Nieuw Bindmiddel 2187*, a zinc stearate, in 1925 and by N.B., a sodium stearate in 1930 (see Table 8.1).[7] Zinc-white was added from 1930. Until 1947, the same recipe of 1936 was used for the paint production.

The pigment *Vermiljoen licht imit L 1484* was purchased from *Haarlemsche Stoomverffabriek*, the pigment *Vermiljoen licht imit A arzb* from *Arzberger, Schöpff & Co.*, *Siegle Oranje 2S* from *G. Siegle & Co*, Stuttgart and *Permanentrood 2G extra* from *N.V. Defa Arnhem*. According to the Colour Index [2], *Permanentrood 2G extra* and *Siegle Oranje 2S* are both PO5.

8.4.2 Vermilion Substitution in the Art-Technological Context

To substitute inorganic vermilion, Gettens et al. [6] mention possible organic replacements such as "monolite red, lithol reds and other synthetic dyes of appropriate hue [...], called 'vermillionettes'". According to Heaton (1928, cited in

[6] Hansa Yellow GR (PY2) and Hansa Yellow 3R (PO1), [3], p.49.

[7] The additives NB and 'Nieuw bindm(iddel)' are used frequently in Talens oil paint production from the 1920s on [14]. Confusingly, the name describes two different stearates, sodium (NB only) and zinc stearate (2187). Work in progress.

Table 8.1 Transcription of paint recipes for *Vermiljoen licht imitatie studie* (1922–1947). From 1936 to 1947, the recipe did not change. (Information concerning the extenders, driers, dispersion agents and stabilizers: [14])

Year	Pigment 1	Pigment 2	Extender 1	Extender 2	Additive 1	Additive 2	Binding Media	Comment
1922	6000 *Vermiljoen licht imit L 1484*		4000 *Gips* (gypsum)		600 *F vers* (unknown)		1100 + 1700 *Lijnolie* (linseed oil)	Maaltijd 14uur (grinding time 14 h)
1925	3000 *Vermiljoen licht imit A arzb.*		2500 *Witpoeder* (chalk)		360 *Nieuw Bindm* (zinc stearate)		600 + 1000 Lijnolie (linseed oil)	Maaltijd 6 uur (grinding time 6 h)
1928	4000 *Vermiljoen licht imit A arzb.*	60 *Hansageel 3R* (Hansa yellow 3R)	3400 *Witpoeder* (chalk)		480 *Nieuw Bindm* (zinc stearate)		800 + 1600 *L olie* (linseed oil)	Maaltijd 7 uur (grinding time 7 h)
1928/1929	4000 *Vermiljoen licht imit A arzb.*	25 *Hansageel 3R* (Hansa Yellow 3R)	3400 *Witpoeder* (chalk)		480 *Nieuw Bindm* (zinc stearate)		1500 + 300 *L olie* (linseed oil)	Maaltijd 6 uur (grinding time 6 h)
1930	600 *Permanentrood 2G extra*		3000 *Zinkwit* (zinc white)	3000 *Witpoeder* (chalk)	300 *NB vers* (sodium stearate)	130 *Pale Terebine* (unknown)	1600 + 1000 *Lijnolie* (linseed oil)	6 × malen (grinding 6 h)
1935	600 *Siegle oranje 2S*		3000 *Zinkwit* (zinc white)	3000 *Witpoeder* (chalk)	300 *NB* (sodium stearate)	100 *Pale Terebine* (unknown)		Proef (Test) 4 × malen (grinding 4 h)
1936	600 *Permanentrood 2G extra*		3000 *Zinkwit* (zinc white)	3000 *Witpoeder* (chalk)	300 *NB* (sodium stearate)	100 *Pale Terebine* (unknown)	2250 + 850 *Lijnolie* (linseed oil)	

Table 8.2 Results of analysis of sample 6 (Vermilion Red Light imitation)

Sample	SEM/EDX	Micro Raman	GC-MS
No 6	Zn, S, Ba, Ca	PR 51 + PR3 + PR63	PR51 + PR3 + PR63 or PR49
	Gips (*$CaSO_4(2H_2O)$*) *or Witpoeder* (*$CaCO_3$*) Ba, S: *Part of PR51*	*Vermiljoen licht imitation L 1484 or Vermiljoen licht imitatie A Arzb.*	

[4]), this term refers to "dyes such as eosin [that] were precipitated onto 'orange lead'[...] rather than a white base" [4].

Trillich [13] describes similar practice in the third volume of *Das deutsche Farbenbuch,* elaborating that these reds can also consist of artificial, organic coal-tar based colourants, like triphenylmethane-dyes, azo dyes, alizarin-dyes, anthracene-dyes, or indanthrene-dyes laked on substrates such as lead compounds or barium sulphates. However, Trillich does not refer to these pigments in artist paint and calls the organic imitations a falsification of original cinnabar.

Wagner [15] also defines Vermilion imitation pigments as coal-tar dyes, laked on lead-substrates, like eosin laked on red lead.

Schultz refers to both an inorganic type of Vermilion, which is basic lead chromate $Pb_2(OH_2)CrO_4$, and an organic one that consists of various mixed dyes. These may be eosin and other aniline colourants, laked on lead substrates, which is in agreement with other authors. The organic imitation form can also be mixed with red lead [12].

The Colour Index [2] also describes inorganic and organic imitation forms, mentioning PR 103 (CI 77601; basic lead chromate) and PR50:1 (CI 15500:1), which is a ß-naphthol, laked on a barium salt.

8.4.3 Analyses

The paint tube selected for the identification of "Vermiljoen licht imitatie" was tentatively dated around the 1920s, according to its ribbed cap shape and paper label design [9], which are shown in Fig. 8.2.

Results of the analyses are presented in Table 8.2. Calcium and sulphur were detected by SEM-EDX, which could be assigned to gypsum, used before 1925, or chalk[8] that was used after 1925.[9] Complementary analysis conducted with Micro Raman Spectroscopy and GC-MS[10] identified a mixture of three SOPs (see Fig. 8.2).

[8] The SOP itself contains sulphur atoms in its structure.

[9] Since the Micro Raman Spectrum includes the spectra of three different synthetic organic pigments (see Fig. 8.2), no specific peaks for gypsum or chalk were found.

[10] Conducted by Henk van Keulen.

The vermilion imitation pigment, consists mostly of PR51, with PR3 and PR63 being smaller additions.[11]

PR51 is a ß-naphthol pigment lake on barium salt [7] and therefore barium and sulphur identified by SEM-EDX form part of the organic pigment. This pigment was developed in 1920 and was used in *Talens* oil paint almost immediately. PR63 belongs to the group of monoazos and is a BONA pigment. It can be laked on salts based on sodium, calcium, barium or manganese and can contain additives such as calcium or manganese, depending on its shade [7]. It was not determined onto which salt the BONA pigment was laked due to its lower Raman activity in comparison with PR51. The monoazo PR3 is also a ß-naphthol pigment.

8.4.4 Discussion

Through analysis, a mixture of the three SOPs PR51 + PR3 + PR63, was identified. It remains unclear which of the two possible Vermilion imitation pigments is contained in paint sample no. 6 and thus if it was delivered either by *Haarlemsche Stoomverffabriek* or *Arzberger, Schöpff & Co*. As a mixture of the three pigments is not listed in any of Talens recipes, it is likely that the pigments were mixed by the pigment supplier itself.

When buying a paint called Vermilion imitation, artists presumably expected either an inorganic pigment or a synthetic dye laked on a lead-based substrate. The mixture of the analysed three SOPs is not mentioned explicitly.

Consequently, *Talens'* usage of SOPs for the production of Vermilion imitation does not correspond to the suggestions found in the selected contemporary literature and artists would not have been aware of the presence of the included SOPs.

8.5 Conclusion and Future Work

This project generated relevant art-technological information about the commercial origin and chemical composition of 34 SOPs, used in the production for *Talens* oil paint in the first half of the twentieth century. The detected pigments mostly corresponded to those in the production recipes and logbooks, which allowed for more precise dating of the *Talens* paint samples.

Further material- and technical investigations will determine possible causes of paint degradation which may be used for targeted conservation approaches. Since the composition of the *Talens* product lines before 1950 changed frequently, detecting *Talens'* paints in artwork may give relatively precise dates of their manufacture. In time, this may also become possible for other brands. This is useful for the exact dating of paintings and can also be used to identify counterfeits. In

[11] Reference for PR51, PR3 and PR63 by KIK/IRPA. https://modern.kikirpa.be/PHP/spectrum.php

addition, investigations into the longevity of the examined tube colours would be relevant to conservation.

Acknowledgements Prof. Dr. Christoph Krekel, Daphne van Mansom, Coos van Waas, Ruud Knol, Jan Slief, Matthijs de Keijzer, Suzan de Groot, Dr. Luc Megens, Dr. Ineke Joosten, Dr. Inez van der Werf, Art Ness Proãno Gaibor, Henk van Keulen, Dr. Wim Fremout, Dr. Steven Saverwyns.

Appendix

Raman spectra were recorded with a combined *RamanMicro 300* and *RamanStation 400F (Perkin Elmer);* 785 nm laser, maximum laser power 100 mW. Spectra were recorded with a wave speed range from 200 – 2000 cm^{-1} with integration times between 6 to 15 seconds.

SEM/EDX was performed with a *JEOL JSM-5910 LV* variable pressure Scanning Electron Microscope; setting "point and shot" in low vacuum SEM, 22 kV.

References

1. Catalogue (1918): Catalogus 1918 Fabriek van fijne Verven en Inkten $^{N}/_{V}$ Talens & Zn. Apeldoorn, Royal Talens Archive, Apeldoorn
2. Colour Index (1971): The Society of Dyers and Colourists; Lund Humphries, Bradford and London GB, 3rd edition
3. De Keijzer M (2014): The Delight of Modern Pigment Creations, in K.J van den Berg, A. Burnstock, M. de Keijzer, J. Krueger, T. Learner, G. Heydenreich and A. de Tagle (eds), *Issues in Contemporary Oil Paint*, pp 45–73 Cham: Springer
4. Eastaugh N, Walls V, Chaplin T, Sidall R (2008): Pigment compendium: A dictionary of historical pigments, Elsevier Butterworth-Heinemann, 1st edition, pp 387
5. Fremout W, Saverwyns S (2012): Identification of synthetic organic pigments: the role of a comprehensive digital Raman spectral library, *Journal of Raman Spectroscopy*, Wiley Online Library, pp 1536–1544
6. Gettens R J, Feller R L, Chase W T (1993): Vermilion and Cinnabar, *Artists' Pigments A Handbook of Their History and Characteristics*, Volume 2, Ashok Roy, National Gallery of Art, Oxford University Press, pp 159–180
7. Herbst W, Hunger K (1993): Industrial Organic Pigments – Production, Properties, Application, VCH Verlagsgesellschaft mbH Weinheim (Federal Republic of Germany), 1st edition, pp 226, 228, 319, 334
8. Kerdijk F (1932, 1937): Kunstschildersmaterialen en Schildertechnik – Kleurstoffen, bindmiddelen en vernissen in de olieverftechniek, Talens & Zoon, Apeldoorn, 1st and 2nd edition, respectively pp 64–68
9. Pause R (2018): Synthetic Organic Pigments in Talens Oil paints, Masterthesis Staatliche Akademie der Bildenden Künste, Stuttgart
10. Schäning A (2010): Synthetische organische Farbmittel aus einer technologischen Materialsammlung des 19./20. Jahrhunderts: Identifizierung, Klassifizierung und ihre Verwendung sowie Akzeptanz in (Künstler)Farben Anfang des 20. Jahrhunderts, Dissertation Akademie der Bildenden Künste Wien, pp 177–240

11. Scherrer N C, Zumbühl S, Delavy F, Fritsch A, Kühnen R (2009): Synthetic organic pigments of the 20th and 21st century relevant to artist's paints: Raman spectra reference collection, *Spectrochimica Acta Part A 73 Molecular and Biomolecular Spectroscopy*, published in Elsevier B.V., pp 505–524
12. Schultz G (1931): Farbstofftabellen (siebente Auflage, neu bearbeitet und erweitert von Dr. Ludwig Lehmann), Leipzig akademische Verlagsgesellschaft MBH, pp 670–672
13. Trillich H (1926): Das deutsche Farbenbuch – III. Teil Die Anstrichfarben und Lacke, Verlag B. Heller/München, pp 62
14. van den Berg K J, van Gurp F, Bayliss S, Burnstock A, Klein Ovink B (2016): Making paint in the 20th century: the Talens Archive, *Sources on Art Technology: Back to Basic*, Archetype Publications Ltd., pp 43–50
15. Wagner H (1928): Die Körperfarben in *Chemie in Einzeldarstellungen, Ban XIII*, wissenschaftliche Verlagsgesellschaft m. b. H., Stuttgart, pp 328

Part II
Case Studies: Artists Techniques and Materials; Degradation; Ethical Considerations; Treatment Options

Chapter 9
Ellsworth Kelly: The Studio and Beyond

Mary H. Gridley

9.1 Introduction

The painting career of Ellsworth Kelly (American 1923–2015) started its steady upward rise in the 1950s and 1960s. Kelly's singular style of abstraction was a complete departure from gestural and personal mark-making, embracing instead a highly refined reduction of form and surface, expressed in planes of single colors.

Over the seven decades of his working career, Kelly expressed a keen interest in maintaining the pristine surfaces of his paintings. As a result, he engaged with conservators from early on. His artistic maturation coincided with changes in conservation training: a gradual shift from the apprentice system to one of university post-graduate degree programs, with a concurrent standardization of protocols and ethics. Additionally, austere artworks like Kelly's set up new challenges for conservators, forcing a break from traditional Old Master painting conservation techniques. To review the conservation treatment of Kelly's paintings over the last 70 years is to see the development of conservation treatment and prevention strategies unfold in real time.

We at Cranmer Art Group (CAG) in New York have had the opportunity to examine or treat over 110 of Kelly's paintings, ranging in date from the 1950s to the 2010s. While many had never been treated, others provided an opportunity to examine and assess prior treatments.

Information about Kelly's work is scarce in the conservation literature, although art historical and critical writings abound. This paper will first describe Kelly's materials and working methods–drawn from observation, discussion with the artist and his studio, and scientific analyses—and then review the kinds of condition

M. H. Gridley (✉)
Cranmer Art Group, LLC, New York, NY, USA
e-mail: mary@cagsoho.com

K. J. van den Berg et al. (eds.), *Conservation of Modern Oil Paintings*,
https://doi.org/10.1007/978-3-030-19254-9_9

issues that affect his paintings (both common and anomalous). Finally, it will address some of the history of the structural treatment of his works over several decades and his relationship to conservation and conservators.

9.2 The Artist

Kelly served in the US Army's Camouflage Unit in Europe from 1943 until the end of the war, when he returned to Boston. After 18 months of art school, Kelly moved to France in 1948, initiating a fertile period during which he began to move away from the figurative painting style in which he had been trained toward the kind of abstraction that would become his legacy.

Returning to the States in 1954, he settled in New York City, occupying several studios and interacting with fellow artists (Fig. 9.1). He began to exhibit more regularly, in both solo and group shows. The 1960s was a decade that brought

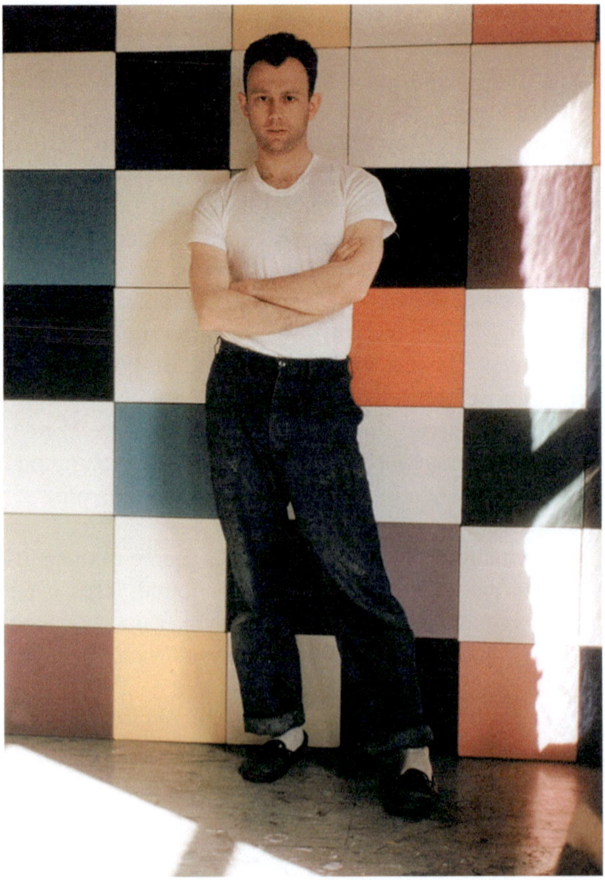

Fig. 9.1 Ellsworth Kelly with *Colors for a Large Wall*, 1951 (EK46) Broad Street studio, New York 1954
© Ellsworth Kelly Foundation
Image: courtesy Ellsworth Kelly Studio

commissions, awards, and solid recognition, with regular sales of work in both the US and abroad. In 1970 he moved to Spencertown, New York, where he bought a farmhouse and eventually built a large studio. He lived and worked there until his death in 2015.

Kelly was a formalist. His artistic inspiration came from things seen in the real world; juxtapositions of shapes, colors, and found objects that he drew or collaged, and which were then abstracted, flattened out, magnified, and rendered in paint. While he created many sculptures, drawings, and prints, paintings form a substantial portion of his oeuvre.

Many of his early works were small in scale, often limited to two or three colors, and sometimes relief-like, made of wood, Masonite, or cardboard. He also began making joined paintings, in which two or more stretched canvases were displayed as one work. He experimented throughout his career with blurring the line between relief and painting and with the multi-panel format. Although the scale of the work increased as he moved to larger studio spaces, the use of expanses of a single color and the reduction of forms remained a constant. In the mid-1960s, Kelly began experimenting with shaped canvases, using curves, triangles, and irregular polygons. Together these simple formats—alone or in combination—occupied Kelly for the rest of his career.

In addition to the finished works, there are small oil studies, working drawings, collages, sketchbooks, watercolors, and even found objects which exist as archival material, illuminating the thought processes behind the finished paintings.[1]

9.3 Studio Practice

A unique and very helpful feature of Kelly's working practice is the EK number system. The artist started numbering his works very early on. This numbering is invaluable when talking about paintings that are titled only by color and year. A full description of the system is available in the first volume of the catalogue raisonné.[2] For the purpose of this paper, I will just note that the numbering is sequential and that paintings and sculptures are one set of numbers. The earliest numbered painting CAG has examined is *Deesis I,* 1950 (EK7) and the latest is *Four Panels*, 2012 (EK1038). The numbering stops with Kelly's last, unfinished work, *Four Panels (Blue Black Green Red)*, made in 2015 (EK1090).

Kelly was a traditionalist in the studio, and took care to use the best materials he could afford. Through a lifetime of contact with conservators and museum professionals, he also adopted some of their 'best practices' for storage, handling, and transport.

[1] Kelly, Tablet.

[2] Bois [2], 14.

While he employed studio assistants[3] to help in a general way, Kelly enjoyed the whole process of stretching, priming, and painting. Before he began painting, a long, temporary wooden batten was screwed to the back of the stretcher so the canvas could be handled without touching the sides or face. Entire tubes of paint were squeezed into white enameled bowls so Kelly could judge the color. Measuring was by the tube, and the final color was recorded as the ratio of tubes used. He diluted the oils with turpentine to get the desired consistency. Kelly retained jars of each color of each layer he used on a painting and recorded in notebooks descriptions of the proportions and manufacturers. A few of the paintings even have printed and typewritten labels listing their materials affixed to the backboards (Figs. 9.2 and 9.3).

Some of the early paintings, made with two colors, may have very distinct surfaces for each color (for example, smooth for the 'ground' and more vigorously brushed for the 'figure'), but the later paintings generally have a uniform surface achieved by using a three-inch brush working in bands a foot or so wide. The strokes are often visible in raking light.

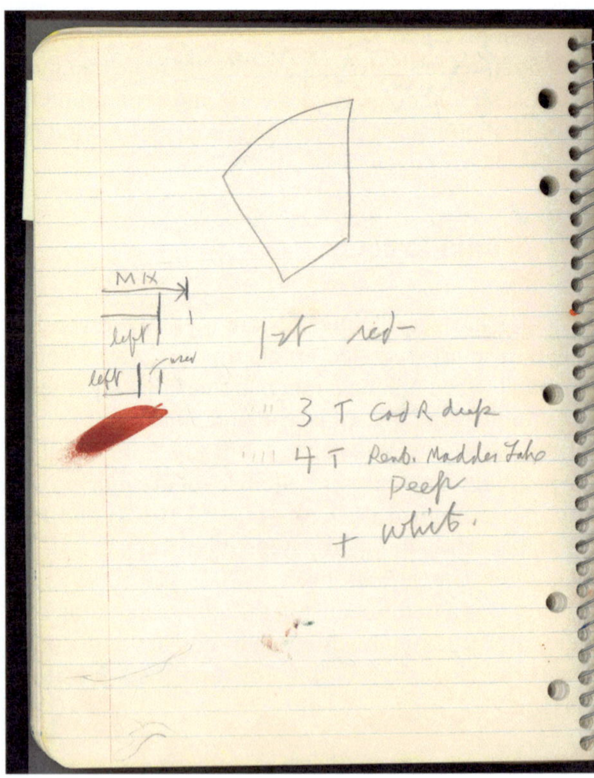

Fig. 9.2 Ellsworth Kelly's notes for *Documenta Panel III: Red Panel with Curve* 1991, pencil and oil on spiral bound lined paper
A page from one of Kelly's studio journals. The entry for each painting includes a thumbnail sketch of the shape of the painting, measurements made by dipping a stick into the bowl of mixed paint before and after each layer, indicating how much paint was left in the bowl, and the paint color names and proportions (T = tube). Generally the EK number was added later
© Ellsworth Kelly Foundation
Image: courtesy Ellsworth Kelly Studio

[3] Henry Persche (1966–1993), Nick Walters (1993–present), Joe Yetto (2009–present).

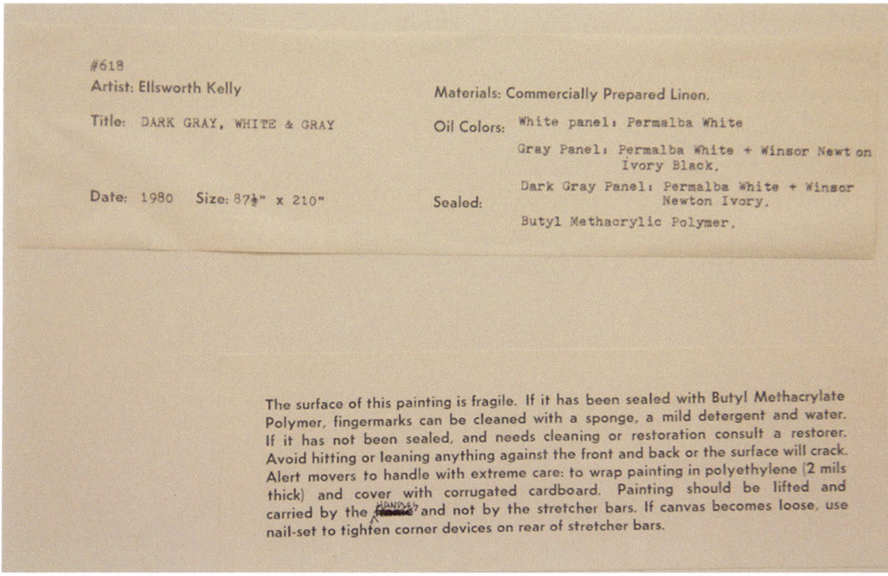

#618
Artist: Ellsworth Kelly
Title: DARK GRAY, WHITE & GRAY
Date: 1980 Size: 87½" x 210"

Materials: Commercially Prepared Linen.
Oil Colors: White panel: Permalba White
Gray Panel: Permalba White + Winsor Newton Ivory Black.
Dark Gray Panel: Permalba White + Winsor Newton Ivory.
Sealed: Butyl Methacrylic Polymer.

The surface of this painting is fragile. If it has been sealed with Butyl Methacrylate Polymer, fingermarks can be cleaned with a sponge, a mild detergent and water. If it has not been sealed, and needs cleaning or restoration consult a restorer. Avoid hitting or leaning anything against the front and back or the surface will crack. Alert movers to handle with extreme care: to wrap painting in polyethylene (2 mils thick) and cover with corrugated cardboard. Painting should be lifted and carried by the HANDLES and not by the stretcher bars. If canvas becomes loose, use nail-set to tighten corner devices on rear of stretcher bars.

Fig. 9.3 Labels from the back of *Dark Gray, White & Gray*, 1980 (EK618) indicating materials and instructions for handling
© Ellsworth Kelly Foundation

9.4 The Materials

9.4.1 Stretchers

In Paris, Kelly's stretchers were made by a local carpenter or he purchased student-grade stretchers. These stretchers (and occasionally strainers) had a very thin profile (3/4 in. or 2 cm). In the early 1960s, Kelly started using expansion-bolt stretchers, which had been developed in the late 1950s by James Lebron.[4] Kelly used this type of stretcher almost exclusively until the end of his life, even for his shaped paintings. They were first purchased directly from Lebron, then Tom Zoufaly, and later from Simon Liu. Most of the stretchers incorporate a hanging bar.

Having heard that stretchers should be coated in order to inhibit acid migration into the fabric, Kelly began to coat his stretchers with shellac sometime in the mid 1970s. At first it was a clear coat, but in 1984 he switched to white pigmented shellac, applied in the studio (Fig. 9.4).[5]

True to the nature of artists' studios, though, things are never quite so cut and dried. We have run across stretchers that are hybrids of clear and white shellac as a result of later upgrades.

[4] Buckley [4], 157.

[5] Brand name: Zinsser B-I-N Shellac-based primer, colorant: titanium dioxide.

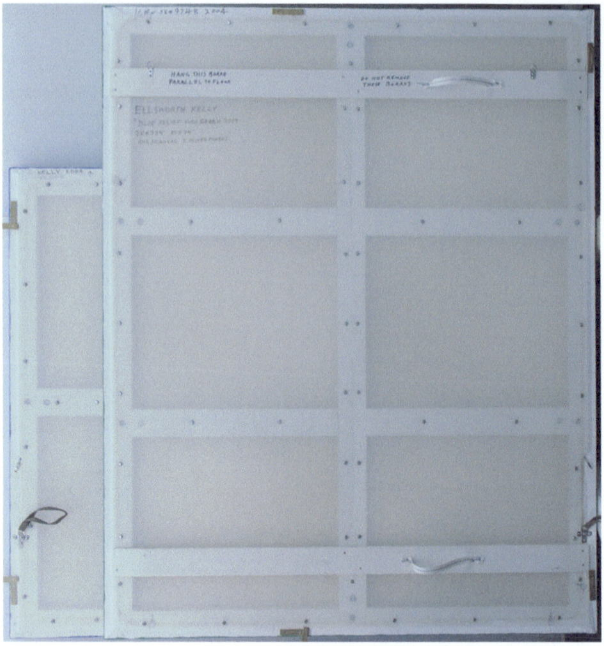

Fig. 9.4 Verso of *Blue Relief over Green,* 2004 (EK974) with white shellacked stretchers, corrugated polypropylene backing board, and artist's notations © Ellsworth Kelly Foundation

9.4.2 Fabric Supports

Kelly painted frequently on linen, from 1949 while in France through the mid-1960s, with sporadic use thereafter. His use of cotton duck began in the mid-1950s. He found that the linen reacted too readily to humidity fluctuations in his studios and the galleries (which were without climate control in that era), but he also preferred the texture of the cotton duck.[6] The fact that the fabric was available in large widths may also have been a factor in this choice. The cotton duck was ordered in bolts. Working from his detailed notes, the fabric was carefully cut to fit the shapes of the previously ordered stretchers in the studio. The procedure for stretching was to stretch once, wet the whole canvas, and then stretch again to ensure a taut surface. This ritual was an opportunity for Kelly to get a sense of the proportions and to begin to scale the paintings up from his preparatory drawings and studies.

[6] Wijnberg et al. [12], 6.

9.4.3 Grounds

Kelly's early linen supports during the mid-1950s were often commercially pre-primed. The limited analyses we were able to perform indicate these are white alkyd or oil-based grounds (Table 9.1). With the switch to cotton duck, Kelly began priming the canvases himself with acrylic gesso. Winsor & Newton acrylic gesso was

Table 9.1 Analysis of ground layers

Title	Date	EK#	Bottom ground	Top ground	Test date	Instrumentation
Leaf (Castor bean)	1962	278	Alkyd, $BaSO_4$, TiO_2, $CaSO_4$	Oil, $BaSO_4$, TiO_2, metal soap	2010	FTIR[a]
Green blue red	1963	310	*Presence undetermined*	Lead	2012	Lead test kit
Blue green red III	1965	347	*Presence undetermined*	Oil, $PbCO_3$ (metal soap), ZnO, TiO_2	2010	FTIR[a], Raman Microscopy[b]
White brown	1968	394	Alkyd, $BaSO_4$, TiO_2, $CaCO_3$, metal soap, $CaSO_4$	Oil, $BaSO_4$, TiO_2, metal soap	2010	FTIR[a]
Blue green yellow orange red (yellow panel)	1968	390C	Alkyd, $BaSO_4$, TiO_2, $CaSO_4$	Oil, TiO_2, $CaCO_3$, kaolin	2010	FTIR[a]
Black white	1968	396	*Presence undetermined*	Alkyd, TiO_2, $CaSO_4$	2011	FTIR[a]
Gray blue	1980	601	*Presence undetermined*	Lead	1980	Printed label on reverse
Blue panel	1999	887	*Presence undetermined*	Ethylacrylate/ methylmethacrylate copolymer, TiO_2, $CaCO_3$	2017	FTIR[c], SEM-EDX[d]

[a]FTIR: Thermo Continuum microscope fitted with a nitrogen-cooled wide-band MCTB detector (4000-450 cm^{-1}), coupled to a Thermo FTIR Avatar 360 spectrometer. Samples were compressed on single Spectra-Tech diamond cells and analyzed in transmission at 4 cm^{-1} resolution for 128 sample and background scans, using Happ-Genzel apodization

[b]Raman: Thermo DXR dispersive Raman microscope fitted with a 785 laser (power 1–14 mA) and high-resolution grating (1800-50 cm^{-1})

[c]Bruker Vertex 70 FTIR equipped with a mid-infra-red source, a potassium bromide (KBr) beamsplitter, a HeNe laser and a deuterated triglycine sulfate detector. Spectrometer equipped with a diamond Pike GladiATR accessory. IR spectra were collected between 4000 and 600 cm^{-1} using 64, 128 and 256 sample scans and a spectral resolution of 4 cm^{-1}

[d]Hitachi TM-1000 scanning electron microscope, equipped with Oxford Instruments energy dispersive X-ray spectrometry system. Accelerator voltage was 15 kV. Samples mounted on conductive adhesive pads applied to stubs

used until the late 1990s, at which point he switched to Golden Acrylic Gesso,[7] and for the late black canvases, Kelly used Golden Black Acrylic Gesso,[8] both in a methylmethacrylate/butylacrylate copolymer dispersion. Kelly experimented with a variety of diluents for the first coat, including water and Liquitex medium, eventually settling on Golden Artist Color's GAC 100, a methylmethacrylate/butylacrylate aqueous copolymer which is chemically identical to Golden Acrylic Gesso. Two or three coats of full-strength gesso followed, sanded in between with 220 grit sandpaper. The final coat was lightly sanded.

9.4.4 Paint

Kelly painted in oil, and his manufacturer of choice was Winsor & Newton. Although he occasionally used Old Holland or Williamsburg paints in an effort to arrive at the color or effect he wanted, or even another brand if it was recommended to him, he remained loyal to Winsor & Newton. The major exception to this was his consistent use of Permalba® white, manufactured by Martin F. Weber Company.[9]

Kelly generally mixed paints: he told us once that he added Permalba® white to all his colors, as well as using it for all of his pure white paintings or panels. Blacks were often problematic. They painted out well when used straight from the tube, but the addition of white sometimes caused separation, or changed the gloss to something irregular or too glossy for Kelly's taste (he preferred matte surfaces). Changes in paint formulations by manufacturers were a constant irritation for Kelly, and he established communication with Winsor & Newton both to share his frustration and to search for solutions.

Each work usually consists of two or three layers of paint, and they can be of a single, repeated color or of slightly different hues. In order to determine how much paint would be needed for a single coat, Kelly would dip a small measuring stick into the bowl of paint he had mixed up, then dip it again after painting one coat, so he knew how much paint had been used, and therefore how much was needed for the next coat. (Fig. 9.2).

Kelly did experiment with acrylic paints on his canvas works. According to studio records, in the 1960s this was limited to 24 paintings (including 2 floor pieces[10]) and two very large paintings commissioned for public spaces.[11] He did not use acrylic paint again until the very end of his life, returning to it for four paintings in 2015.

---

[7] Titanium PW6.

[8] Carbon black PBk7 and Mars black PBk11.

[9] Primarily titanium white.

[10] Canvas stretched over wood panels.

[11] *White over Blue*, 1967 (EK382) for World's Fair in Montreal (Norton Simon Collection); *Blue Green*, 1969 (EK424) for Unesco.

9.4.5 Varnish

Kelly liked to have his paintings varnished—not necessarily to saturate the colors or give them an even gloss, but because he felt that a layer of varnish would protect the surface of his paintings and make them easier to clean. It is unclear whether Kelly came to the decision to varnish on his own, or whether it was suggested to him by other artists or by a well-meaning conservator. In truth, his oil paints have a gloss and saturation of their own, as decided by the artist. Our experience of these coatings is that they are seldom a cohesive film, normally presenting as a uniform scatter of resin droplets.

From the mid-1960s up until the late 1990s, the New York City-based restorer Daniel Goldreyer, with whom Kelly worked on restoring damaged paintings as well, would come up to the Spencertown studio and varnish the paintings in batches. According to labels on the reverse of several of Kelly's paintings, Goldreyer's preferred varnish was butyl methacrylate. From the mid-1990s, the paintings were spray-varnished by Nick Walters, Kelly's studio assistant, with Golden MSA (Mineral Spirit Acrylic), a solution of isobutyl methacrylate and *n*-butyl methacrylate with HALS and substituted benzotriazole, with amorphous silica as the matting agent (Satin & Matte only). All three varieties, gloss, satin and matte, were used in varying combinations.

After 2006, varnish was abandoned altogether, as an uptick in the frequency of exhibitions did not leave enough time in the studio for the paintings to adequately cure before being varnished. It was also, from a practical standpoint, very difficult to achieve an even spray coat with the desired gloss on paintings of such a large scale.

9.4.6 Edges

The artist's approach to the edges of his paintings varied over the years. The linen canvases were commercially primed, and therefore included white edges, which sometimes Kelly painted over to match the face color, to emphasize the 'object' quality of the work. Initially, the edges of the cotton duck paintings remained unprimed and unpainted. However, after it was suggested to Kelly that priming the edges made them sturdier, this became a regular part of his studio practice. We have seen edges without primer, primed with white gesso, or painted to match the face color, and occasionally colored edges that have been painted over with white. Sometimes there is evidence that Kelly changed his mind at a later date. For example, staples (or the tacks used in the early years) were later removed, the holes spackled, and locally repainted, often only on the sides or any edge visible from a normal viewing perspective. This painting or repainting of the edges was an aesthetic refinement Kelly developed over time. He told us that he came to regard colored versus white edges as an indication that a work was more—or less—sculptural,

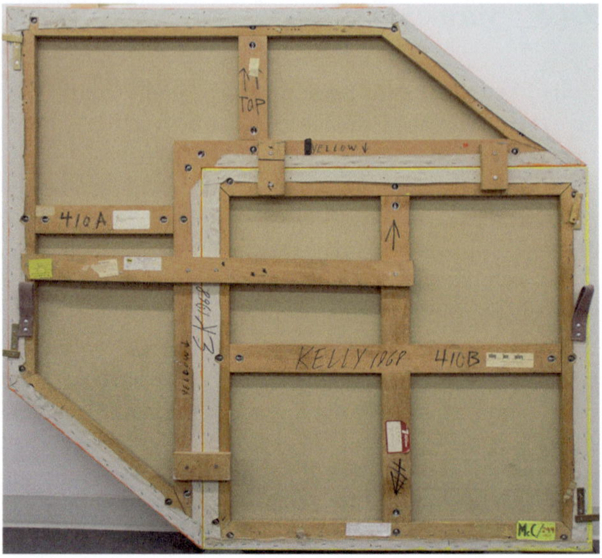

Fig. 9.5 Verso of *Yellow Orange*, 1968 (EK410). The expansion bolt stretchers, joining bars, and notations in the artist's hand are typical of this period

functioning as either an object or a window. It was also true that edges could be repainted at a later date just because Kelly saw them with fresh eyes, noticing paint splatters or other distracting marks.

Framing was another option the artist employed, although irregularly. Early on Kelly might use simple wood stripping to describe a line around the edge. In the 1960s, he started having aluminum shadow box frames made for the larger panels, but he considered these to be for travel and handling purposes only, not for display. However, this request was not always adhered to, and consequently many of these paintings are displayed in these frames.[12] The aluminum frames were abandoned by the late 1980s and generally replaced by wooden travel frames and crates to minimize handling of the paintings.

9.4.7 *Backs*

Because so many of Kelly's paintings are multi-panel works, the backs have some rather distinctive features. In order to hold the canvases together, Kelly screwed permanent, long battens across the elements to keep them attached and in plane (Fig. 9.5). These wooden bars became a regular feature in Kelly's works, at first

[12] In a letter to Peter Cochrane of Arthur Tooth & Sons Ltd., dated 15 April 1962, Kelly said "The larger pictures will be sent by the end of this week, I hope. I am having aluminum strip frames made for them for protection while moving them. However I prefer exhibiting them without the frames."

serving the dual function of joining multiple panels and providing a hanging bar for installation. Over time, these bars morphed into an aesthetic choice, appearing even on paintings that comprised a single canvas. Kelly discovered he liked to have the paintings float slightly off the surface of the wall.

Kelly always provided a lot of written information on the backs of his paintings: signatures (in many formats), the EK number, titles, dimensions, orientation, joining instructions and so forth, usually written in black paint or marker.

Borrowing best practice ideas from conservators, Kelly also began putting backing boards and handles on his works. This was done for new paintings, but also for any older ones that left the studio for exhibition or sale. Although they appear regularly starting in the late 1970s, the earliest painting we have seen with a foamcore backing board was painted in 1963. The early backboards of foamcore had cutouts or long narrow slits for air circulation and thick knotted white plastic coated cord threaded through eyehooks for handles. Later, air holes were drilled in a grid pattern. In about 2004, the foamcore was replaced by corrugated polypropylene, with a similar upgrade in handles. Of course there are hybrids, as older paintings came back to the studio and were retrofitted.

9.5 Condition Issues

9.5.1 Handling Damage

By far the most common condition issue encountered in Kelly's work is the cracking that results from handling, impacts, or contact with stretcher rails and bars. It is clear from the types of cracks we have seen that–with the exception perhaps of a few very early works–these are not a result of faulty technique or materials, but rather of handling or environment (Fig. 9.6). What may go unnoticed in a painting with vigorous brushwork or complex surface iconography becomes instantly obvious on the pristine and monochromatic surfaces of these paintings. Second to the cracks are localized irregularities in surface color or gloss caused by handling, abrasions, or splatter. Black is a color that had unpredictable results; in the studio Kelly found when mixing black with other paints the result of the finished surface could vary; and sometimes the black paint changed as it aged, exhibiting an uneven saturation.

9.5.2 Material and Environmental Degradation

In addition to these common damages, a very small subset of paintings we have seen exhibited more global surface issues, some of unknown origin and others undoubtedly linked to environmental degradation or previous conservation.

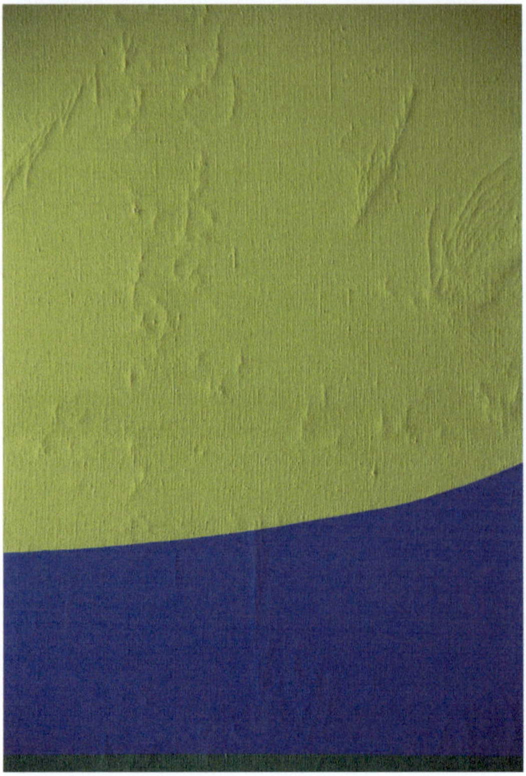

Fig. 9.6 Raking light detail of handling damages on an untreated 1960 painting (photographed 2016)
© Ellsworth Kelly Foundation

The recent focus in conservation research on the question of surface efflorescence has encouraged us to rethink assumptions made about paintings examined many years ago. While the occurrence of a whitish haze on Kelly's paintings has usually turned out to be dust or dirt, we have noted a few instances in which the haze was being generated by the painting's materials in conjunction with environmental stressors or previous conservation.

Curiously, on three of the paintings with a whitish haze that was not dust, there was a negative correlation with the stretcher rail– that is, the stretcher rail protected the paint surface from developing efflorescence (Fig. 9.7).

Among the 17 red paintings that have passed through the studio, three have had whitish hazes on the surface, appearing anywhere from 20 to 50 years after the painting was made.

Black paintings, too, have exhibited various types of surface change, including efflorescence. Out of 29 paintings examined where black was a predominant color, one painting from 1968 (examined in 2012) had an area of zinc stearates on the surface, with a negative correlation to the stretcher rails.

One yellow painting (dated 1963, examined 2014) exhibited all-over chalking and browning, and another (dated 1963, examined 2010), hung for decades over an

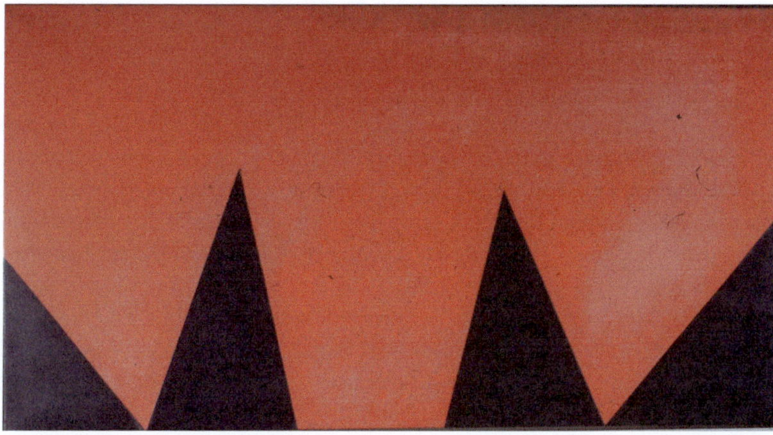

Fig. 9.7 Efflorescence, with some negative correlation to the stretchers bars, on *Red Black (Tanager)*, 1957 (EK128). The painting had been previously lined, and, in the blacks only, varnished at an unspecified date (photographed 2001)

air vent (where it essentially functioned as an air filter for the room), exhibited powdering and darkening in both the yellow and the white passages.

And finally, there was the appearance of epsomite (hydrated magnesium sulphate) on the surface of one painting (dated 1965, examined 2010) that had been comprehensively treated with a wax-resin lining, repainting of all three colors, and a final varnish.[13] The presence of surface accretions of epsomite on oil paintings, particularly in reference to Winsor & Newton paints, was first published in 2014,[14] and is referenced throughout this volume. Because of the extensive past conservation treatment of this particular Kelly, however, it is impossible to know what triggered epsomite formation in this instance.

Going forward, the combination of aging varnish and changes in paint solubility brought about by various degradation phenomena will no doubt further complicate, perhaps even prohibit, the cleaning of select surfaces.

[13] FTIR identified epsomite on the surface of the green and blue, $MgCO_3$ in the body of all three paint colors (red, green, and blue), and $BaSO_4$ in the body of the green and blue. See Table 9.1, footnote [a] for instrumentation.

[14] Silvester et al. [11].

9.6 Historical Structural Treatments

Given that handling and impact cracks are the most common form of damage to Kelly's paintings, most treatments–done with or without the knowledge or approval of the artist–have been undertaken to address this issue. It cannot be stated strongly enough that the austere format of Kelly's paintings render even small damages unacceptable. They impact the aesthetics as well as the financial value of the work, so there has always been a strong desire by the artist and his collectors to treat these kinds of damages.

Traditionally, the remedy for this kind of damage was lining, and Kelly was no stranger to this process. We do not know how many of his paintings have been lined, but the previously lined paintings we have seen (15 in total) demonstrate almost every popular combination of adhesives and supports our profession has used over the last 70 years, with the mixed success one would expect.[15]

One painting had a glue-paste lining, which effectively eliminated structural distortions, but which created weave interference and subtle irregularities caused by the uneven adhesive layer (dated 1980, examined 2009).

The wax-resin lining mentioned above satisfactorily took care of the single large impact crack but contributed to dramatic changes in the paint sensitivity (dated 1965, examined 2011).

Beva linings generally failed to hold the cracks in plane over the long term. One example of such a lining exhibited bubble-like distortions in the surface caused by localized delamination (dated 1959, examined 2013).

Another Beva lining, employing a stiff Mylar release layer, improved the profile of the cracks somewhat, but effectively fixed them in place. Removal of the Mylar, now the most rigid element of the laminate, would entail flexing of the paint film, possibly inducing further damage (dated 1959, examined 2013).

One five panel painting had three of the elements lined with a flocked PVA. The thick adhesive layer conformed to the distorted cracks, and rather than bringing them back into plane, effectively locked them in. The adhesive's resistance to solubility or mechanical removal prevents any further treatment to the reverse (dated 1968, examined 2008).

As conservators began to move away from linings—realizing that they were creating as many problems as they solved—local treatments started to appear. One painting we examined had numerous thin paper patches coated with PVA resin applied to the reverse of cracked areas. They were not stiff enough to solve the problem but—happily—were easily reversible (dated 1963, examined 2010).

However, as sophisticated collectors have become more focused on previous conservation treatments—with mostly a negative view—these lined paintings are problematic both for the artist's legacy and the market. What was once considered a

[15] Hartin et al. [9].

traditional, acceptable treatment has become a diminishment to the artwork's value, even when the treatment is successful or when it was sanctioned by the artist.

Kelly was well aware of this changing value judgment, and began bringing damaged paintings to CAG, where, over several years, we had developed an approach of treating the cracks locally from the reverse, without lining or patches. We use water, alcohol, low heat, and pressure to relax the ground and paint layers, adding gelatin to stiffen the fabric locally on the back. The process reduces or visually eliminates the cracks and does not prohibit future treatment.[16]

A second way in which the structural vulnerability of these paintings has been addressed is by attaching them to new sub-supports. As a preventive measure, strong sub-supports with adequate bevels or keyable solid panel supports can improve a painting's stability. Goldreyer often replaced original stretchers with sturdier models and backing boards. At CAG we have occasionally replaced original stretchers with panel stretchers, made inserts of foamcore or Mylar, or added cami-linings, as well as providing backing boards and handles, all of which reduce the chances of torquing, stretcher bar cracks, and other damages going forward.

9.7 Conservators

Like many artists, Kelly was willing to oblige a collector, or rescue a favorite painting, by repairing it himself or supervising its restoration. Many of the lined paintings are signed on the back, so we can only assume that Kelly had a hand in the treatment or at least approved it. Over the course of his life, he engaged in different types of collaborations with conservators whom he liked and trusted. A lot of the early work was done by Goldreyer, with whom Kelly had a close relationship. Goldreyer got great results, not just for Kelly's paintings but also for works by Mark Rothko, Jackson Pollock, and Barnett Newman, among others, primarily by lining and varnishing. When this level of intervention began to be viewed as undesirable, Kelly turned to younger conservators in New York, CAG included. Aside from undertaking stand-alone treatments, we have occasionally prepared parts of paintings so that he could repaint them. He followed the ongoing treatments with interest, often checking in on the progress. He also formed close relationships with conservators at institutions across the globe, notably SFMOMA, Stedelijk Museum Amsterdam, Fondation Beyeler, Whitney Museum of American Art, MoMA and Kröller-Müller Museum, collaborating with them on both treatments and display questions.

The Ellsworth Kelly Foundation, established in 1991, has made conservation a priority in its mission. Kelly's leadership in contributing resources to museum conservation departments in the U.S. and abroad was a natural extension of his own

[16] Gridley [7], Gridley and Cranmer [8].

interest and experience. This recognition that stewardship of cultural heritage is an ongoing process is a perspective that few collectors, foundations or fellow artists have acknowledged in such a proactive way.

9.8 Conclusion

While a review of Kelly's materials and techniques is interesting in and of itself, what was striking about this research was the historical overview of structural conservation—how and why our materials and techniques have changed–seen through the lens of a single artist with an unwavering aesthetic. Over the course of his career, Kelly had almost every kind of interaction with conservation that one could imagine. Because his notion of what his surfaces should look like remained constant over seven decades, a study of his treated works provides a very good snapshot of the ways his practice and ours have intertwined within the broader context of a changing market and changing attitudes towards conservation.

Kelly wanted his surfaces to be pristine, and there is no ambiguity in that. Beyond stating this, he even took part in efforts to bring damaged works back to that state, setting a clear baseline for an acceptable appearance, to which collectors and institutions try to adhere. So when a work is damaged, there is a concerted effort to have it treated. Everyone can agree on what is a damage, but what constitutes a successful treatment has been a bit of a moving target, and reflects changes not only in conservation approaches, but also changes in the market's perception of what is an acceptable conservation intervention.

At CAG, where we were fortunate enough to have had an ongoing dialogue with Kelly, our approach to treating his work has been to get as close as possible to the ideal through localized treatments and preventive measures. We have often been the first to work on a given painting, but we know we will not be the last. Our experience with previously treated works reminds us that it is critical that what we do remains re-treatable so that we are not closing doors against future advances.

Acknowledgments Dana Cranmer, Kristin Robinson, Eva Walters, Nick Walters, Jack Shear, and Yves-Alain Bois.

References

1. Abraham L, Wijnberg L, and Sauerberg ML (2005) The conservation of monochrome paintings. In: Hummelen I et al (eds) Modern art: who cares, Archetype Publications Ltd., London, pp 362–388
2. Bois Y (2015) Ellsworth Kelly, Catalogue raisonné of paintings, reliefs, and sculpture, volume one, 1940–1953. Cahiers d'Art, Paris
3. Bracht E (2005) The restoration of Kelly's works at the Stedelijk Museum. In: Hummelen I et al (eds) Modern art: who cares, Archetype Publications Ltd., London, pp 250–253

4. Buckley B (2012) Stretchers, tensioning, and attachments. In: Stoner JH, Rushfield R (eds) Conservation of easel paintings. Routledge, London, pp 128–160
5. Clark R, Barger, M (2016) The artist initiative at San Francisco Museum of Modern Art. In: Studies in conservation vol 61 no. 2-Supp: 24–28 https://doi.org/10.1080/00393630.2016.1193692
6. Coplans J (1971) Ellsworth Kelly. Harry N. Abrams, Inc., New York
7. Gridley MH (2017) Notes on the treatment of cracks in canvas paintings. In: American Institute for Conservation of Historic and Artistic Works Paintings Specialty Group Postprints 2014 vol 27: 1–20.
8. Gridley MH, Cranmer D (2006) Unforgiving surfaces: treatment of cracks in contemporary paintings. In: Learner TJS et al (eds) Modern paints uncovered. Getty Conservation Institute, Los Angeles, pp 143–148
9. Hartin DD et al (2018) Overview of the CCI lining project: do linings prevent cracking and cupping in paintings? In: American Institute for Conservation of Historic and Artistic Works Paintings Specialty Group Postprints 2015 vol 28: 137–150.
10. Kelly E (2002) Ellsworth Kelly tablet 1948–1973. The Drawing Center, New York and Musée Cantonal des Beaux-Arts, Lausanne
11. Silvester G et al (2014) A cause of water-sensitivity in modern oil paint films: the formation of magnesium sulphate. In: Studies in Conservation vol 59, no. 1: 38–51.
12. Wijnberg L et al (2011) A study of the grounds used by three post-war American artists (1954–1974): Barnett Newman, Ellsworth Kelly and Brice Marden. In: Bridgland J (ed) ICOM-CC 16th triennial conference preprints Lisbon, 19–23 September 2011. Critério–Produção Grafica, Lda., Lisbon, pp 1–10

Chapter 10
A New Vocabulary of Color: Bocour Oil Paints and Barnett Newman

Corina E. Rogge and Bradford A. Epley

10.1 Introduction

The twentieth century was a time of great competition in the artists' paints market as established companies such as Winsor & Newton faced new upstarts including Shiva, Grumbacher and Bocour. The latter company, founded by Leonard Bocour and Leonard Lefcourt in 1932 and joined by Sam Golden in 1936, was one of the most well-known and innovative in the U.S.A. [14]. Although perhaps best known for their introduction of *Magna Plastic Paints* in 1947 and *Aqua-tec Acrylics* in 1963–1964, the Bocour Company also made three different lines of oil paint (Fig. 10.1) [14]. *Bocour Hand Ground Oil Colors* were first available in the mid-1930s; the paints were originally sold under the name 'Lefcourt' when introduced in 1932 [14]. Initially available in about 18 colors, the line later expanded to 37 colors [9, 18]. In 1947, advertisements include mention of the *Bellini Artist's Colors* line, which was a less expensive, machine-ground 'student' line [7–9]. The introduction date of the third line, *Bocour Artist Oil Colors*, is more difficult to establish as the name of the paint line and the name of the company are identical. In 1954 the company began to run magazine advertisements touting "Bocour the finest hand-ground colors ever made!" [2], but it is unclear if this refers to the *Bocour Hand Ground* colors or *Bocour Artist Colors* paint line. In American Artist Magazine Buyer's Guides from 1956 and 1957, *Bocour Hand Ground Professional Oil Colors* are listed, but in 1958 the listing changes to *Bocour Artist Oil Colors* [3–5], suggesting that the *Bocour Hand Ground* line may have been phased out (or

C. E. Rogge (✉)
The Museum of Fine Arts, Houston, Houston, TX, USA

The Menil Collection, Houston, TX, USA
e-mail: crogge@mfah.org

B. A. Epley
The Menil Collection, Houston, TX, USA

K. J. van den Berg et al. (eds.), *Conservation of Modern Oil Paintings*,
https://doi.org/10.1007/978-3-030-19254-9_10

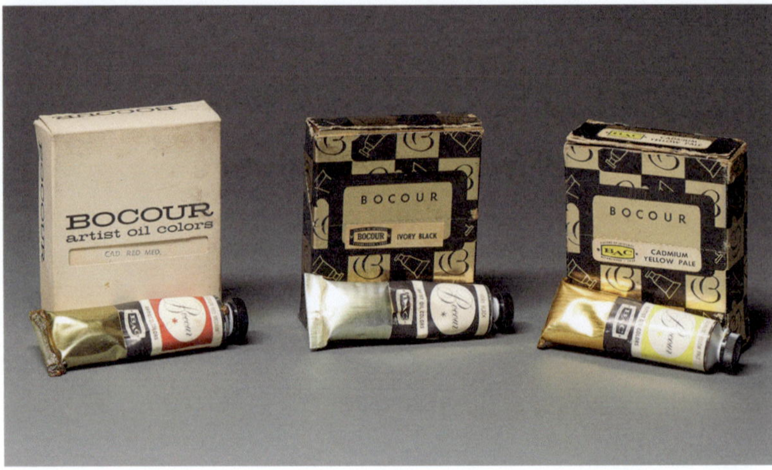

Fig. 10.1 Bocour oil paints from Barnett Newman's studio, gifted to the Menil Collection by Robert Murray

renamed) in favor of the *Artist Oil Colors* line between 1957 and 1958. The first advertisement found to date showing tubes of the *Artist Oil Colors* line is from 1959, providing a terminus ante quem for the introduction date [6].

Because their company was situated in the heart of Manhattan close to many artists' studios, Bocour and Golden developed close relationships with artists including Morris Louis, Ad Reinhardt, Jackson Pollock and Barnett Newman, using them as 'guinea pigs' for new products lines and sometimes creating bespoke paint formulations on their behalf [10, 14, 23]. Bocour paints would likely have been utilized by many of the artists working in New York from the 1940s onward, but apart from *Magna,* which contained a relatively easily identifiable poly(n-butyl methacrylate) binder [11, 12], Bocour paints have not been positively identified in artworks. A recent study of historic Bocour oil and acrylic paints using X-ray fluorescence (XRF), Raman and Fourier transform infrared spectroscopies to characterize pigments, fillers/extenders, and binders provides a known set of reference materials with which to compare data from artworks [18] to address this lacuna.

Barnett Newman (1905–1970) was one of the many artists known to have purchased paints from the Bocour Company. Receipts held in the Barnett Newman Foundation Archives record frequent purchases of *Aqua-tec Acrylic* paints, and photographs of Newman's studio from the mid-1960s show the presence of *Magna, Bellini, Bocour Hand Ground* and *Bocour Artist Oil Colors* paints [19, 20, 22]. After Newman's death in 1970, Annalee Newman, Barnett's widow, gave away many of his paints to the public schools of New York, but also to Newman's friends and colleagues [16]. Robert Murray, a sculptor and friend of Newman's, received some of this material and in 2015 donated it to the Menil Collection. Paints present in the donation materials include all three of the oil color lines produced by Bocour.

As it is apparent that Newman purchased Bocour oil colors and had them in his studio, his works present an ideal opportunity to utilize our database of historical paints to attempt to identify Bocour oil paints.

In 2007, Annalee Newman presented the Center for the Technical Study of Modern Art (CTSMA) of the Harvard University Art Museums with a major gift of Newman's studio materials and ephemera such as canvas fragments, paint brushes, rollers, straw hats, drawings, and sketches. Amongst these materials is an object described as a paint trial on stretched canvas (ARCH.2007.9.33) measuring 130.81 × 17.15 cm consisting of a piece of canvas, painted recto and verso, roughly tacked to a hand-built wooden strainer (Fig. 10.2a, b). The object is undated but the proportions and size of the strainer suggest that it may have been created in 1950, as it clearly relates to a series of five untitled paintings dated 1950. In particular, the dimensions are similar to: *Untitled* 2 (1950) (121.9 × 13 cm) held in the Menil Collection and *Untitled 3* (142.2 × 7.6 cm) at the Art Institute of Chicago. The red tones of the verso composition also correspond well with his 1950 Untitled series, so it is reasonable to hypothesize that Newman had originally intended this work to be part of the series, but was dissatisfied with it and at some point unstretched the painted canvas and flipped it over for reuse. The canvas appears to be a cotton duck sized with animal glue; Newman switched his sizing material from animal glue to Rivett's glue, which is poly(vinyl acetate) based, in 1958 [13].

There are 12 clearly identifiable 'tests' on the recto of the canvas that are evocative of extant works of art (Fig. 10.2c). Tests 1–7 contain red, yellow and blue

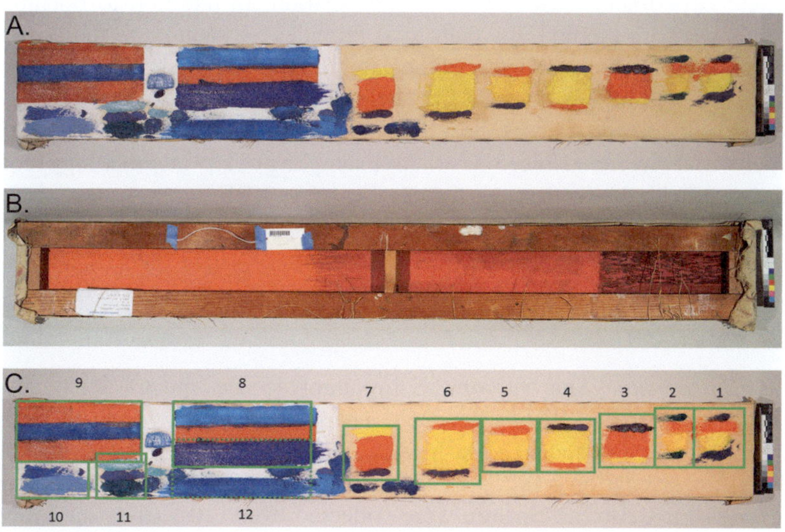

Fig. 10.2 (**a**) Recto and (**b**) verso of *Paint trial on stretched canvas* (Twentieth c), oil paint on canvas, 130.81 × 17.15 × 4.13 cm, Harvard Art Museums/Center for the Technical Study of Modern Art, Gift of the Barnett and Annalee Newman Foundation (2007.9.33). Images taken by Ellen Davis, © President and Fellows of Harvard College. (**c**) Assigned numbers to individual paint tests referred to in the text

in five different compositions painted directly onto the sized canvas. The color palette of these suggests an obvious association with the four paintings of the *Who's Afraid of Red, Yellow and Blue* series, created between 1966 and 1970. Test 3 in particular bears close resemblance to the layout of *Who's Afraid of Red, Yellow and Blue I* (1966). Tests 8–12 were painted over a white ground layer. The colors and composition of test 8 are evocative of *Voice of Fire* (1967), and test 12 resembles *Profile of Light* (1967). The blue colors present in test 10 are reminiscent of two lithographs, *Canto VII* and *Canto VIII* (1963–1964), and a screenprint on plexiglass, *The Moment* (1966). The greenish-blue tones of test 11 evoke those in a series of paintings from 1954 including *The Word, White Fire I*, *Primordial Moment*, *The Gate*, and *Right Here.* Thus, the colors on the trial seem to relate to paintings dating from 1954 onwards.

The paint trial represents one of the few opportunities to observe Newman's creative processes. Newman insisted on solitude while painting, perhaps because he desired the creation of a work to be a dialogue between himself and the canvas, devoid of outside distractions or influences. He also rarely chose to describe his own methods, so this object is one of the few things that bears witness to his practice [13]. Newman stated that "… I am an intuitive painter, a direct painter. I have never worked from sketches, never planned a painting, never 'thought out' a painting. I start each painting as if I had never painted before." [17] However, among the materials bequeathed to the Harvard University Art Museums are sketches related to *Chartres* (1969) with notation describing proportions. The sketches and this paint trial suggest that, contrary to his statement, Newman did sometimes plan out his works and experiment with color and composition. In addition to documenting Newman's working method, the paint trial provides an ideal opportunity to try to identify the paints present, as many of the paints appear to have been applied straight from the tube, although in a few of the tests the blue paints are layered (tests 3 and 4), or obviously mixed (tests 10 and 11). As the fragile state of the paint on the verso of the paint trial precluded micro-destructive sampling or any methodology requiring direct contact with the recto, analysis was limited to XRF spectroscopy.

10.2 Methods of Analysis

X-Ray Fluorescence Spectroscopy X-ray fluorescence spectra were collected using a Bruker Tracer III-SD handheld energy dispersive X-ray spectrometer equipped with a Peltier cooled XFlash silicon drift detector (SDD) with a resolution of 145 eV. The excitation source was a Rhodium (Rh) target X-ray tube, operated at 40 kV and 10 μA current. Spectra were collected over 120 s (live time). Spectral interpretation and peak integration was performed using the Bruker Artax Spectra 7.4.0.0 Software. This is the same instrument and settings previously used to analyze historic Bocour materials [18].

10.3 Results and Discussion

As heavy elements present in the paints on the verso may be detected from the recto, we collected XRF spectra at multiple locations on the verso to identify elements that might contribute to recto measurements. Visually, the surface paints on the verso composition do not appear to be oil paints, and so although their elemental signatures must be interrogated we do not seek to compare them to our database of Bocour oils. Lead is detected in all areas of the verso, and the Lα to Lγ peak intensity ratio suggests the signal derives from underlying paint layers [15]. Visual inspection confirms that the red paints on the verso appear to have been applied over a lead white ground. Three visually distinct red paints are present on the verso. The elemental signatures indicate that the dark red paint may contain cadmium red (Cd, S, Se), red ochre (Fe), and bone/ivory black (Ca, P). The medium red probably also contains cadmium red and red ochre, and peaks indicative of barium sulfate are detected (Ba, S, Sr). The bright red paint likely contains cadmium red, red ochre, and barium sulfate; elements consistent with the presence of lead chromate (Pb, Cr), barium sulfate (Ba), and a molybdate or molybdenum based heteropolyacid pigment (Mo) are also present.

Heavier elements in paint on the verso of the canvas are clearly observable from the recto: areas of bare canvas show lead, cadmium, and iron peaks deriving from the verso (Fig. 10.3), indicating that the verso paint will contribute to the spectra of paints on the recto. The lead signal in particular is quite strong: the lead L_α peak is attenuated fourfold less on the recto side compared to the verso. The peaks for iron

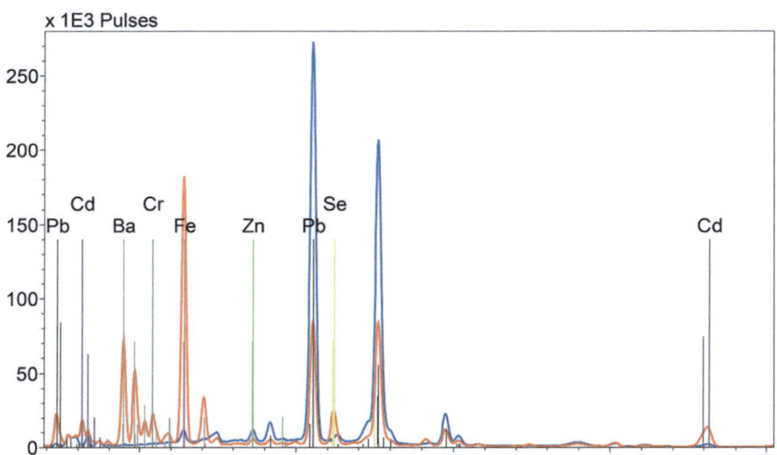

Fig. 10.3 XFR spectra of the medium red paint on the verso (orange), and the bare canvas on the recto (blue). The lower energy peaks due to Cd, Ba, and Fe are largely attenuated in the recto measurement, but the lead peaks from the ground on the verso appear more intense and less attenuated when measured from the recto

and cadmium are more significantly attenuated (18 fold and 7 fold, respectively) due to the intervening lead white ground of the verso. Therefore, elemental ratios of cadmium red and yellow paints and iron containing paints on the recto may be slightly distorted by the inadvertent co-analysis of the verso paint.

XRF spectra of red paints on the paint trial recto indicate that all are cadmium red (Cd, S, Se) paints with barium sulfate (Ba, S, Sr). We assessed the similarity of the elemental compositions at these sample sites by the ratios of the intensities of Se Kα and Cd Kα peaks and the Ba Lγ and Cd Lα peaks. These peaks were chosen to avoid interference from other elements and to have similar energies, which will minimize differences in attenuation. The average Se Kα/Cd Kα and Ba Lγ/Cd Lα ratios are 2.55 ± 0.4 and 0.36 ± 0.03, respectively. The greater variability in the Se/Cd ratio value may arise from a combination of effects including self-attenuation (paint thickness) and contribution from the verso paint. The similarity in ratios and lack of any other detectable elemental differences lead us to conclude that these red paints are all the same. Comparison of the elemental ratios of the paint trial reds with historic samples of Bocour oil paints indicate that the paint is most similar to *Bocour Artist Oil Colors Cadmium Red Light* paint (Fig. 10.4, see Table 10.1 for elemental comparison between paint trial paints and Bocour paints). *Bocour Artist Oil Colors* and *Bellini Cadmium Red Light* paints have similar Se Kα /Cd Kα ratios, but the *Bellini* paint contains significantly more barium sulfate, consistent with a less expensive student grade paint.

XRF suggests the presence of four blue paints on the paint trial (Fig. 10.5a), which occur in 'from the tube' form, mixed with white, mixed with each other, or layered. The two most frequently used blues appear to be a cobalt blue (Co, Al) paint with barium sulfate (Ba, S, Sr) (blue type 1); and phthalocyanine blue (Cu, Cl) (blue type 2). Identification of phthalocyanine blue by XRF can be difficult when the high tinting strength pigment is present in low amounts. However, in the paint trial the paint is only minimally mixed with other paints or applied in nearly pure strength over another blue paint. As a result, a significant copper peak is detected (Fig. 10.5b) that indicates the presence of a copper-based pigment; in a modern, intensely blue colored paint this strongly implies a copper phthalocyanine species, most likely phthalocyanine blue (PB15). The absence of elements commonly associated with other blue pigments supports this identification. A cobalt stannate cobalt chromate cerulean blue (Co, Cr, Sn), occurs in two areas of test 11 (blue type 3), and a cobalt blue without any detectable barium sulfate occurs in the blue shared between tests 8 and 12 (blue type 4). The blue type 4 has no analog amongst the Bocour materials analyzed to date, as the *Bocour Artist Oil Colors Cobalt Blue* paints analyzed contain barium sulfate and the *Bellini Cobalt Blue* paints contain ultramarine as the colorant [18]; therefore blue type 4 may derive from a different paint manufacturer.

The cobalt blue paint with barium sulfate (type 1) is positively identified as *Bocour Artist Oil Colors Cobalt Blue*: the spectra are nearly identical (Fig. 10.5c), the Al K/Co Kα ratios are the same (0.0017 ± 0.0002 and 0.0017 ± 0.005 for the Bocour paint and the blue type 1, respectively) as are the Ba Lα/Co Kα ratios (0.44 ± 0.02 and 0.44 ± 0.06 for the Bocour paint and the blue type 1, respectively).

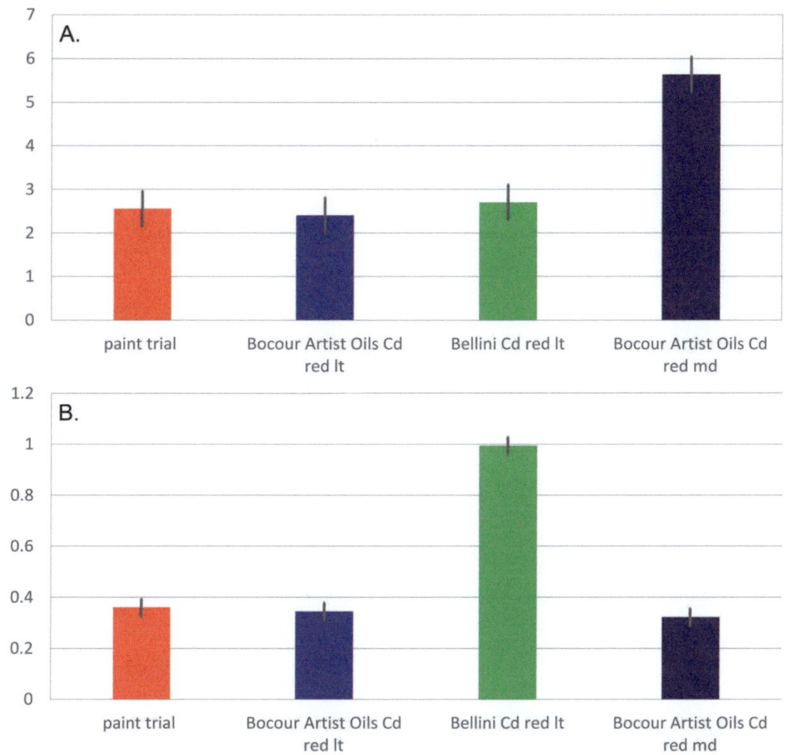

Fig. 10.4 (**a**) Se Kα/Cd Kα ratios for the red areas from the painting trial and historic samples of *Bocour Artist Oil Colors Cadmium Red Light* paint, *Bellini Cadmium Red Light paint, and Bocour Artist Oils Cadmium Red Medium* paints. (**b**) Ba Lγ/Cd Lα ratios for the red areas from the painting trial and historic samples *of Bocour Artist Oil Colors Cadmium Red Light* paint, *Bellini Cadmium Red Light* paint, and *Bocour Artist Oils Cadmium Red Medium* paints. The elemental ratios of the paint trial red closely resemble those of the Bocour Artist Oils Cadmium Red Light paint

The phthalocyanine blue paints on the paint trial (blue type 2) are either mixed with white paint or mixed with or layered over another blue paint, complicating the analysis. This material cannot be a *Bellini Blue* paint, as the *Bellini* line contains high levels of barium sulfate. XRF analysis of the *Bocour Artist Oil Colors Bocour Blue* paints detects only copper, and small amounts of calcium and chlorine. Although chlorine is often associated with phthalocyanine green (chlorinated copper phthalocyanine), it can also be present if the phthalocyanine blue pigment was synthesized from $CuCl_2$ or CuCl rather than metallic copper [1]. The Cl/Cu ratio of the *Bocour Blue* paint is variable (0.2 ± 0.1) but similar to the ratio for the phthalocyanine paint on the trial (0.25 ± 0.08) (Fig. 10.5b). We conclude that the paint trial blue type 2 paint is consistent with the *Bocour Blue* paint.

The cobalt stannate/cobalt chromate cerulean blue (type 3) may be a *Bocour Artist Oil Colors Cerulean Blue*, as the XRF spectra (Fig. 10.5d) and elemental

Table 10.1 Elemental comparison of paint trial recto paints and Bocour paint analogues

Paint trial recto paints	Elements detected on paint trial by XRF[a, b]		Commercial paint analogue	Elements in commercial paint detected by XRF[a, c]	
	Major	Minor		Major	Minor
Red	S, Cd, Ba, Pb, Se	Si, Ca, Fe, Zn, Sr	*Bocour Artist Oil Color Cadmium Red Light*	S, Cd, Ba, Se	Fe, Zn, Sr
Orange	S, Cd, Cr, Zn, Pb, Se	Ba, Ca, Fe	*Bocour Artist Oil Color Cadmium Orange*	S, Cd, Cr, Pb, Se	Al, Ca, Ba, Fe, Cu, Zn
Yellow type 1	S, Cd, Zn, Pb	Ba, Fe, Se	*Bocour Artist Oil Colors Cadmium Yellow Pale*	S, Zn, Cd	Ba, Fe, Pb
Yellow type 2	S, Cd, Ba, Zn, Pb	Fe, Se, Sr	None identified		
Blue type 1	S, Ba, Co, Pb, Sr	Al, Si, K, Ca, Fe, Zn, Mo	*Bocour Artist Oil Colors Cobalt Blue*	S, Ba, Co	Al, Si, S, Ti, Cr, Cu, Zn, Sr
Blue type 2	Ti, Cu, Zn, Pb	S, Cl, K, Ca, Cr, Cd	*Bocour Artist Oil Colors Bocour Blue*	Cu	S, Cl, Ca, Ba, Cr, Fe
Blue type 3	Sn, Co, Pb	Al, Si, S, Ba or Ti, Cr, Fe, Cu, Zn, Cd	*Bocour Artist Oil Colors Cerulean Blue*	Sn, Co	Al, Si, S, Ti, Cr, Cu, Zn
Blue type 4	Ti, Co, Pb	Al, Si, K, Ca, Fe, Cu, Zn	None identified		
White	Ti, Pb	Si, K, Ca, Fe, Zn, Cd	None identified		

[a]Elements deemed major had peak heights higher than the rhodium lines due to Rayleigh scattering of the primary x-rays. Those deemed minor had peak heights lower than the Rh signal
[b]X-ray penetration depth means the analysis area includes paints on both recto and verso
[c]Data taken from Ref. [18]

ratios are very similar. The *Cerulean Blue* paint has a Sn Lα/Co Kα ratio of 0.332 ± 0.008 and a Cr Kα/Co Kα ratio of 0.059 ± 0.01 and the type 3 paint has ratios of 0.33 ± 0.06 and 0.060 ± 0.007.

All the yellow paints tested contain Cd, Zn, and S, at levels consistent with the presence of cadmium zinc sulfide or cadmium sulfide and zinc oxide. Two distinct yellow paint types are present in tests 1–7. Type 1 (found in tests 1, 3, 4, 6 and 7) contains very low levels of barium, while type 2 (found only in tests 2 and 5) contains high levels of barium and strontium, consistent with the presence of barium sulfate fillers/extenders (Fig. 10.6a). The only samples of historic Bocour cadmium yellow paints available for comparison are *Bocour Artist Oil Colors Cadmium Yellow Pale* and *Cadmium Yellow Deep*. Yellow type 1, with relatively low levels of barium, displays an elemental signature similar to that of the *Cadmium Yellow Pale* paint (Fig. 10.6b). However, the elemental ratios do not match precisely: the *Cadmium Yellow Pale* paint has a Zn Kα/Cd Lα ratio of 7.2 ± 0.1 and S Kα/Cd Lα ratio of 0.54 ± 0.01, and yellow type 1 has values of 8.6 ± 0.9 and 0.37 ± 0.07,

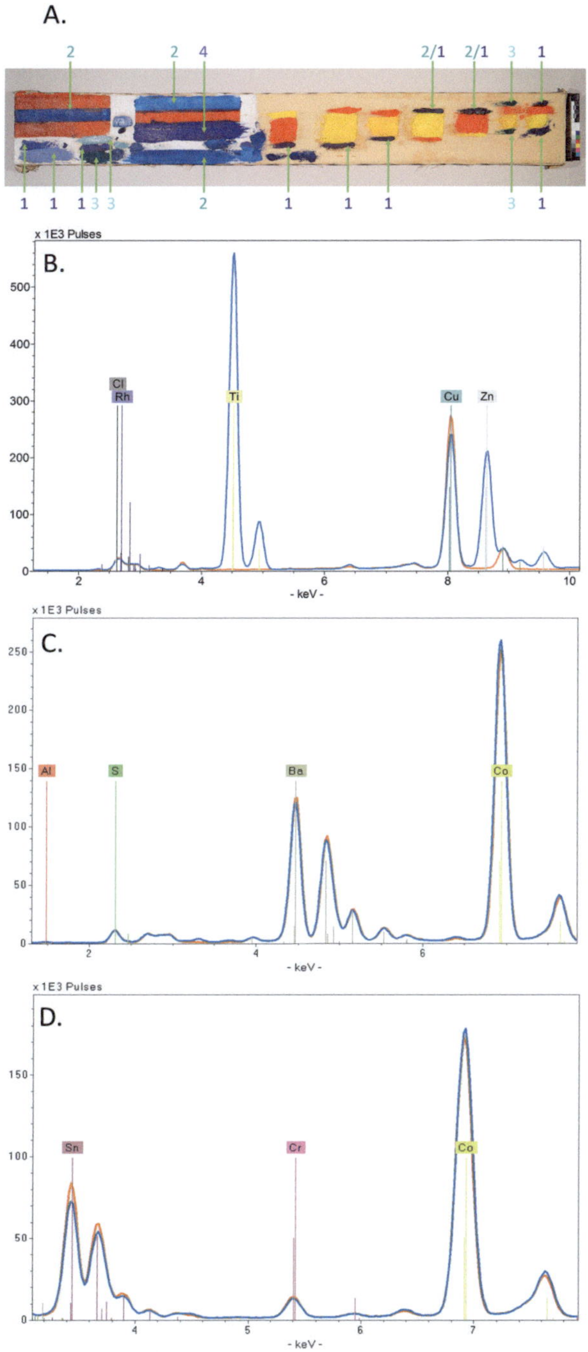

Fig. 10.5 (**a**) The location of the four different blue paints identified. Type 1 is the cobalt blue with barium sulfate; type 2 is the phthalocyanine blue; type 3 is the cobalt stannate/ cobalt chromate cerulean blue; and type 4 is a cobalt blue without detectable levels of barium sulfate. (**b**) XRF spectra of *Bocour Artist Oil Color Bocour* Blue paint pigmented with phthalocyanine blue (orange) and the type 2 paint from the paint trial (orange). The paint trail paint displays a chlorine to copper ratios similar to the Bocour blue paint. (**c**) XRF spectra of *Bocour Artist Oil Color Cobalt Blue* paint (orange), and the type 1 paint from the paint trial (blue). The spectra of the two materials are nearly identical. (**d**) XRF spectra of *Bocour Artist Oil Color Cerulean Blue* paint (orange), and the type 3 paint from the paint trial (blue). The spectra are nearly identical with only slight differences in the tin to cobalt ratios

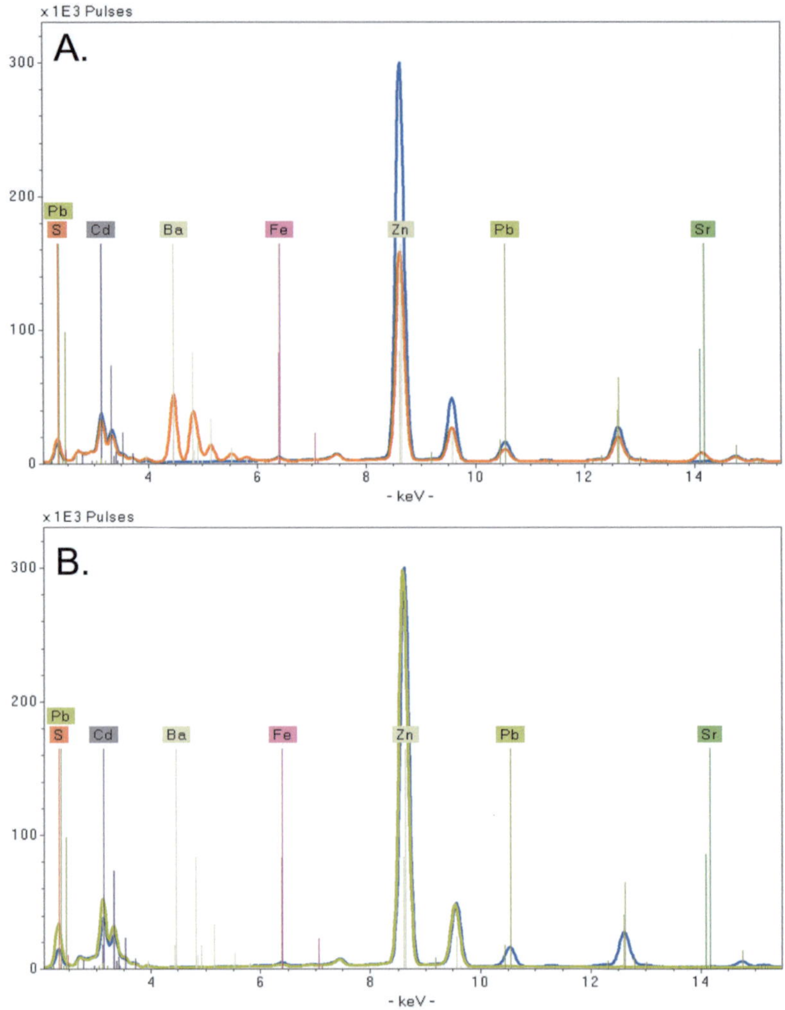

Fig. 10.6 (**a**) XRF spectra of yellow type 1 (blue) and type 2 (orange), on the paint trial. The two types are clearly distinguishable based upon relative barium and zinc contents. (**b**) XRF spectra of yellow type 1 from the paint trial (blue) and a historic sample of *Bocour Artist Oil Colors Cadmium Yellow Pale* paint (yellow), which show a similar cadmium to zinc ratio

respectively. The differences could be due to batch-to-batch variation in the pigment supplied by manufacturers, but the identification of the paint trial yellow type 1 as *Bocour Artist Oil Colors Cadmium Yellow Pale* paint must be classified as tentative. The yellow type 2 on the paint trial, with high levels of barium, matches neither historic paint, although the high levels of barium suggest that this could be a *Bellini* cadmium yellow paint as that product line tends to have higher amounts of fillers and extenders than the other lines [8, 18].

A white ground present on a portion of the test panel appears to be titanium white based, although the presence of some lead white cannot be excluded as the paint on the verso provides a high background signal for this element. The historic *Bellini* titanium whites analyzed to date all contain zinc in addition to titanium, and therefore do not correspond to this paint. Unfortunately, no historic examples of *Bocour Artist Oil Colors* titanium white paints are available for comparison, so the source of this paint remains unidentified.

One swatch of an orange paint is present on the test panel between tests 1 and 2; it contains cadmium sulfoselenide (Cd, S, Se) with small amounts of barium sulfate (Ba, S) and perhaps lead chromate given the presence of Cr. The low levels of barium exclude the possibility that this is a *Bellini Cadmium Orange*. However, the similarity of the Se Kα/Cd Kα ratios (1.7 ± 0.3 and 1.5 for the Bocour paint and the paint trial, respectively) and the Cr Kα/Cd Lα ratios (0.80 ± 0.1 and 0.82 for the Bocour paint and paint trial, respectively) suggest that this paint could be *Bocour Artist Oil Colors Cadmium Orange*.

We conclude that six paints on the paint trial may be from *Bocour Artist Oil Colors* paint line: *Cadmium Red Light, Cadmium Orange, Cobalt Blue, Cerulean Blue, Phthalocyanine Blue*, and *Cadmium Yellow Pale* although the latter two are a more tentative identification. These six paints were among the Bocour materials in Newman's studio at the time of his death, samples of which were given to Robert Murray by Annalee Newman. Only three paints, the white, the cobalt blue without barium sulfate (blue type 4), and the cadmium yellow with high levels of barium sulfate (yellow type 2) do not correspond to a *Bocour Artist Oil Colors* paint. Due to the lack of historical samples of *Bocour Artist Oil Colors* titanium white and *Bellini Cadmium Yellow Pale* paints, we cannot positively assert that the white and yellow type 2 paints are not Bocour paints. However, the blue type 4, cobalt blue without barium sulfate, has no analog within the samples of *Bellini* and *Bocour Artist Oil Colors Cobalt Blue* paints and barring a significant change in formulation by the Bocour Company is likely from a different manufacturer. The preponderance of *Bocour Artist Oil Colors* paints suggests that this paint trial, in addition to revealing an exploration of color and composition, might document Newman's explorations of a new paint line. He was notoriously sensitive to color differences; once, when the Bocour Company had difficulties purchasing cadmium red and substituted other colorants, Newman detected the exchange and phoned them to complain about the substitution [13]. It reasonable to suppose that when presented with a new product line, Newman would carefully evaluate it.

The colors on the paint trials visually relate to extant works dating from 1954 onwards, earlier than the availability of the *Bocour Artist Oil Colors* line is firmly documented (1958–1959). The blue-green colors in test 11 likely contain *Bocour Artist Oil Colors Cerulean Blue* paint; but the light bluish green of *Primordial Light* (1954) contains viridian, not a cobalt stannate/ cobalt chromate cerulean blue. Accordingly, the paint trial may not be a preparatory study for the 1954 series of paintings containing the aqua color (*The Word, White Fire I, Primordial Moment, The Gate*, and *Right Here)* but instead a post facto exploration of a similar color. The red, yellow and blue swatches of tests 1–7 evoke the *Who's Afraid of Red,*

Yellow and Blue series, the first of which from 1966 is in oil. Although it would be of great interest to compare the paints on this painting with those on the paint trial to determine if they match, that work is in a private collection and inaccessible.

Newman began experimenting with acrylics in 1964 [13] but continued using oils in paintings up to his death in 1970, with *Shimmer Bright* (1968) being his last painting executed entirely in oil. Although tests 8, 9 and 12 are painted in oils, they are evocative of the color and general format of two 1967 acrylic paintings, *Voice of Fire* and *Profile of Light*. The paint trial may predate these works, so perhaps Newman initially planned to execute the two 1967 paintings in oil but decided to switch to faster drying acrylics. It is also possible that he used previous ideas as sources of inspiration for his later works and looked backwards to an idea sketched out in oils for an acrylic painting made at a later date. Newman may not have felt constrained to use the same materials in tests as he did in the final works. The blues in test 10 seemingly relate to lithographs and screenprints, which would not utilize Bocour oil paints.

Unfortunately, due to Newman's intense desire for privacy, we do not have a good understanding of how long the creation of a given work took, from inception of idea to his final painting of his signature, or what processes it entailed. It may be that similar to the 'incubation period' when he spent 1 year contemplating *Onement I* (1948), many of his paintings had a long creation process, a hypothesis that is perhaps supported by the relatively low number of paintings he produced in his lifetime [21]. These considerations mean that the date of the paint trial cannot be firmly tied down through association with extant paintings, and our conjecture that the trial documents Newman's exploration of a new paint line introduced in 1958–1959 remains unproven. It is of course also possible that the individual trials are not all from the same date; a canvas used for one test may have been a natural choice for subsequent trials.

10.4 Conclusions

Non-contact, non-destructive XRF analysis reveals that the elemental signatures of six paints on an undated paint trial created by Barnett Newman show close correspondence to *Bocour Artist Oil Colors Cadmium Red Light, Cadmium Orange, Cadmium Yellow Pale, Cobalt Blue, Cerulean Blue,* and *Phthalocyanine Blue* paints, all of which were present in Newman's studio at the time of his death in 1970. This is the first time that Bocour oil paints have been identified in a work, proving the value of the analytical database of historic Bocour paints created previously and proffering hope that these results can be translated into future identification of Bocour paints on paintings. The preponderance of *Bocour Artist Oil Colors* paints suggests a deliberate exploration of that paint line, but the presence of one paint that does not appear to be a Bocour paint, a cobalt blue without barium sulfate, indicates that Newman was actively seeking a precise blue color and comparing the Bocour paints with an as-yet-unidentified other paint line.

Acknowledgements The authors thank many individuals and institutions including: Dr. Narayan Khandekar, Kate Smith, Mary Lister, and Elizabeth Sirrine of Harvard Art Museums, for allowing access to the Annalee and Barnett Newman Foundation materials held in the Center for the Technical Study of Modern Art; Ellen Davis, for digital photography of the paint trial; Dr. Barbara Berrie and Michael Skalka of the National Gallery of Art (Washington D.C.), for providing access to Bocour Bellini paints held in the Art Materials Research and Study Center; Robert Murray, artist and colleague of Barnett Newman, for sharing his knowledge and providing samples of Newman's materials; and the Cecily Horton Professional Development Fund and the Andrew W. Mellon Foundation for supporting conservation science at the Museum of Fine Arts Houston and the Menil Collection.

References

1. Aravindakshan A (2005) Copper phthalocyanines. Paints and coatings industry. https://www.pcimag.com/articles/83452-copper-phthalocyanines. Accessed 16 May 2018
2. Bocour Artist Colors (1954) untitled advertisement. Art News 52(9): 76
3. Bocour Artist Colors (1956) Oil Colors. American Artist Buyers Guide 20(1): 45
4. Bocour Artist Colors (1957) Bocour Hand-Ground Artist Colors. American Artist Buyers Guide 21(1): 47
5. Bocour Artist Colors (1958) Bocour Artist Colors. American Artist Buyers Guide 22(1): 12
6. Bocour Artist Colors (1959) For the creative artist who uses only the best. American Artist 23(5): 50–51
7. Bocour Colors (1947) New! Casein colors by Bocour Colors. American Artist 11(10): 55
8. Bocour L (1960) The permanent palette. In: Chaet B (ed) Artists at work. Webb Books Inc., Cambridge, p 89–92
9. Bocour L (1978) Oral history interview with Leonard Bocour, 1978 June 8, Archives of American Art, Smithsonian Institution. http://www.aaa.si.edu/collections/interviews/oral-history-interview-leonard-bocour-12884 Accessed 10 May 2017
10. Golden TH (2012) Recollections by a paintmaker's son. Golden Paints. http://www.goldenpaints.com/blog/2012/general/recollections-by-a-paintmakers-son-new-show-at-sagg/ Accessed 10 May 2017
11. Learner T (2000) A review of synthetic binding media in twentieth-century paints. *The Conservator* 24 (1): 96–103.
12. Lodge RG (1988) A history of synthetic painting media with special reference to commercial materials. *AIC preprints. American Institute for Conservation 16th Annual Meeting, New Orleans, Louisiana*: American Institute for Conservation, Washington D.C., p 118–127.
13. Mancusi-Ungaro C (2004) The Paintings of Barnett Newman 'involved intuition on the highest level.' In: Schiff R, Mancusi-Ungaro C, Colsman-Freyberger H (eds) Barnett Newman: A Catalogue Raisonné. Yale University Press, New Haven, p 116–146
14. Marontate JLA (1996) Synthetic media and modern painting: Case study in the sociology of innovation. Dissertation, Université de Montréal
15. McGlinchey C (2012) Handheld XRF for the examination of paintings: proper use and limitations. In: Shugar AN, Mass J (eds) Studies in Archaeological Sciences. Vol. 3. Handheld XRF for art and archaeology. Leuven University Press, Leuven, p 131–158
16. Murray R (2015) Personal communication. Sculptor.
17. Newman B (1992) "Frontiers of Space" interview with Dorothy Gees Seckler. In: O'Neill JP (ed) Barnett Newman: Selected Writings and Interviews, University of California Press, Berkeley, p 248

18. Rogge CE, Epley BA (2017) Behind the Bocour label: identification of pigments and binders in historic Bocour oil and acrylic paints. Journal of the American Institute for Conservation 56(1): 15–42
19. Rosenberg H (1978) Barnett Newman. Harry N. Abrams, Inc. Publishers, New York, p 87
20. Temkin A (ed) (2002) Barnett Newman. Philadelphia Museum of Art, Philadelphia, p 113
21. Schiff R, Mancusi-Ungaro C, and Colsman-Freyberger H (eds) (2004) Barnett Newman: A Catalogue Raisonné. Yale University Press, New Haven
22. The Barnett Newman Foundation Archives (1969–1970) Numerous receipts from the artist's purchases of paint from Bocour Artists Colors Incorporated
23. Upright D (1985) Morris Louis: the complete paintings. Harry N. Abrams, New York

Chapter 11
Analysis of Modern Paints and Conservation at the Clyfford Still Museum

James Squires and Pamela Skiles

11.1 Introduction

Clyfford Still was an integral member of the uniquely American Abstract Expressionist art movement of the mid-twentieth century. He developed his form of abstraction earlier than his contemporaries, such as Jackson Pollock, Mark Rothko, Willem de Kooning, and Barnett Newman, and is considered by scholars as the father of the movement. When viewing Still's paintings, Jackson Pollock suggested as much, stating, "…Still makes the rest of us look academic" [3]. Still held strong convictions about the nobility and purpose of art and art making; it was an expression of ideas, purpose, and meaning, not a form of commerce or means to wealth and fame. In her unpublished manuscript, *Clyfford Still: A Critical Study*, Betty Freeman wrote that to Still, painting was "an instrument of thought." This viewpoint drove his falling out with gallerists, curators, critics, and even other artists of the movement. Eventually, he left the art world behind just when Abstract Expressionism was gaining critical acclaim.

After Still withdrew in the early 1960s, he continued to paint and create meaningful art until his death in 1980. He strongly believed that the only way to truly understand an artist was through viewing their collective body of work en masse and without distraction of other artists. As such, from early in his career he sought to maintain possession of as much of his artwork as possible (even copying paintings he sold), so that one day his life's work could be housed and exhibited as a single collection. In 2005, Still's wife, Patricia, agreed to bequeath the entirety of his work to the City and County of Denver upon construction of his namesake

J. Squires (✉)
Clyfford Still Museum, Denver, CO, USA
e-mail: jsquires@clyffordstillmuseum.org

P. Skiles
Denver Art Museum and Clyfford Still Museum, Denver, CO, USA
e-mail: pskiles@denverartmuseum.org

K. J. van den Berg et al. (eds.), *Conservation of Modern Oil Paintings*,
https://doi.org/10.1007/978-3-030-19254-9_11

museum. The Clyfford Still Museum opened in 2011 and houses approximately 95% of the artist's body of work, including 836 canvas paintings, 2573 works on paper, 3 sculptures, and his extensive personal archive. Most of the collection, spanning his career from the 1920s to 1980, has rarely or never been exhibited publicly.

11.2 Artist Materials, Technique, and Storage

Clyfford Still considered himself a traditional artist, primarily using oil paints and traditional techniques. Still's library contained Max Doerner's *The Materials of the Artist and Their Use in Painting with Notes of the Techniques of the Old Masters*, with notations indicating he referenced the text. His routine artistic process included sizing cotton duck fabric with rabbit skin glue prior to painting by utilizing a loose mopping technique. Dry pigments were hand-mixed with boiled linseed oil on his palette and applied directly to the sized canvas with palette knives. On the rare occasion, brushes were used and late in his career, he added small touches directly from tube paints. With the hand-mixed paints, the consistency of each batch varied, in part so Still could vary the optical effects of the paint, as well as its handling properties.

Due to fiscal economy and space constraints in his studios, once the paint was dry to the touch, Still removed the canvases from their stretchers and rolled them (paint-side out) onto tubes of varying diameters and materials. On average, he rolled four to five paintings per tube, though some tubes had as many as 13 paintings or as few as one. The rolls were then stored in his house and studio; many were only unrolled again for inventory photography.

11.3 Condition Issues, Interpretation, and the Role of the Conservator

When the Clyfford Still Museum received the collection, approximately 85% of the paintings were still rolled. In general, there are very few instances of structural damage (flaking, paint loss, cracking, etc.) to the paint films within the collection. Instead, the majority of the visual changes to the paint films are physical changes associated with paint that was not fully cured and rolled prematurely. These range from the superficial (detritus embedded in the paint, cotton canvas fibers) to permanent alterations to the paint surface (flattened impastos, canvas weave impressions, interleaving material impressions). Additional visual transformations, like uneven surface sheen, reveal the fact that the paint is often still in the process of drying. Metal soap formation, in some cases coinciding with other reactions, and the breakdown of binder indicate chemical changes that are not yet fully characterized nor understood. Oil bleed and darkening of the medium are also regularly observed.

Conservators can recognize many of the shifts that have inevitably occurred to the artworks over time. However, the lack of a clear documentation, both written and photographic, calls into question our understanding of these changes. This is particularly true with abstract, experimental work where artists were challenging many of the academic painting conventions. The traditional "signifiers of age" [2] (i.e. craquelure, discolored coatings, loss of a recognizably painted form) used by museum professionals may or may not be applicable to modern paintings.

Still's documented re-engagement with works during his lifetime adds another level of complexity and ambiguity to interpreting his work and original intent. When he observed visual surface changes to a painting he did not appreciate or no longer desired, Still would intervene. After his death in 1980, his wife would "restore" his paintings per his written specifications. They both "oiled out" passages of paintings with linseed oil or spar varnish when they thought the surfaces appeared too "dry" or uneven. In some cases, additional passages were oiled out that had not originally been coated, transforming and reinterpreting the visual relationships within the picture. Over time these additions have also aged and changed, further altering their appearance and increasing the complexity of treatment avenues.

Understanding Clyfford Still's intent is paramount to the Museum's conservation efforts and can only be fully understood through technical analysis. The Museum has reached out to both art historians and conservation scientists to develop this body of research. As part of this outreach and sharing of information, the Museum has partnered with the Getty Conservation Institute to begin this much needed analysis. This research has already begun to empower and both curatorial understanding and interpretation of Still's works, along with rationales for conservation decisions. Significantly, these collaborations will aid in establishing new parameters around why and how these artworks age, along with a potential range of acceptable aesthetic shifts within the works.

11.4 Case Studies

The ongoing analytical research has been used as an aid in conservation decisions, but factors such as artist intent and current cultural aesthetics must also be considered. The following case studies, and the condition issues they represent, highlight the institution's commitment to better understanding the artworks and the interplay between clinical assessment and contemporary cultural standards with interpretation.

11.4.1 PH-1142 *(1957)*

The first case study investigates one of several paintings that all exhibit the same discoloration phenomenon to varying degrees, predominantly occurring in a particular orange paint that is present throughout much of the collection. *PH-1142* is one of the most extreme occurrences identified thus far, due to the broad swaths of black discoloration that have formed within the bright orange passage (Fig. 11.1a, b). Most often the black color develops around the edges of the painted forms. There have also been a handful of instances where it appears in another color, such as white, or less often, yellow.

This peculiar phenomenon has been an area of interest in the Museum's collaboration with the GCI. Samples were taken from several paintings and underwent analysis that included optical microscopy, GC-MS, Raman microscopy, and SEM-EDS to begin characterizing these significant visual changes. Analysis of the orange pigment with Raman microspectroscopy identified it as red lead. Cross sections reveal that the discoloration is an extremely thin (<5 μm) and superficial layer of black crystalline material (Fig. 11.2). SEM-EDS analysis confirmed it is a lead-rich material with a spectrum closely matching the bulk of the paint film. Fluorescence of the samples prevented positive identification of the black material with Raman microspectroscopy, however, preliminary XRD results indicate the presence of galena and lead formate.

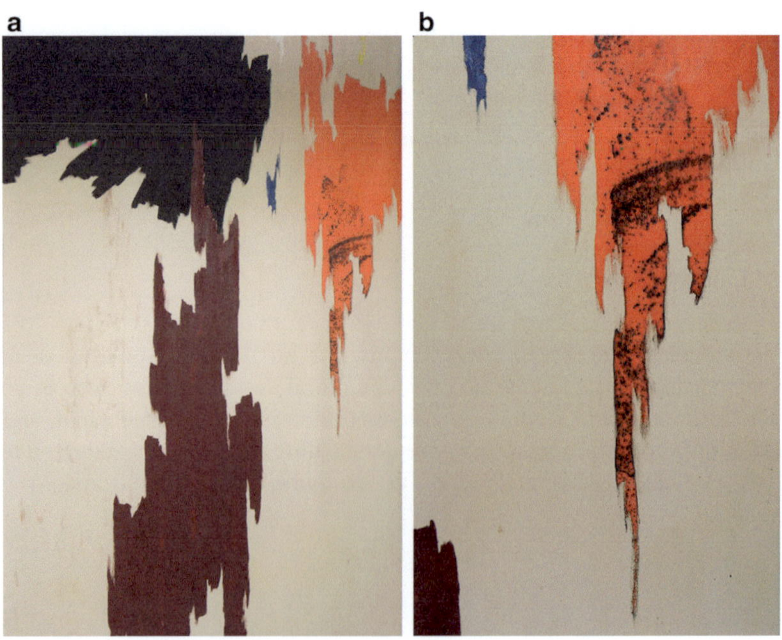

Fig. 11.1 *PH-1142* (1957) (**a**) Overall; (**b**) Detail of black discoloration

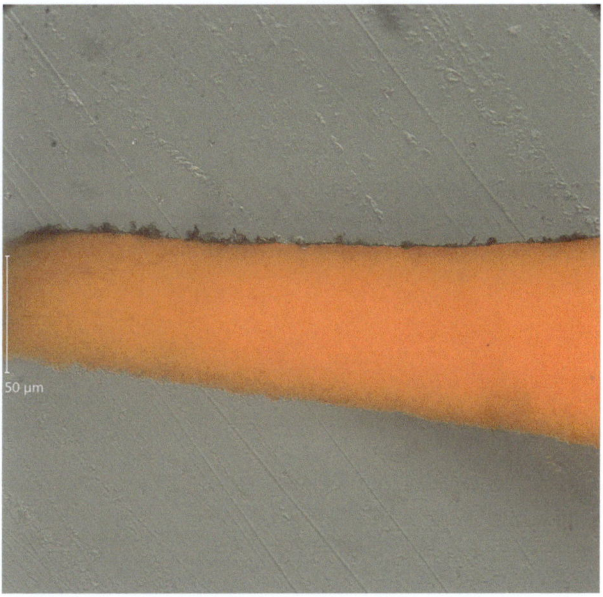

Fig. 11.2 *PH-1142* (1957) 50× photographic magnification, © 2018 The J. Paul Getty Trust, Image: Alan Phenix

Visual analysis of the cross sections with ultraviolet optical microscopy revealed the presence of horizontal medium-rich bands devoid of pigment particles. These bands develop within the body of the paint film in a layered structure throughout the paint film. Preliminary thinking suggests these bands are the formation and migration of lead soaps within the paint film. A possible hypothesis of the mechanism is the accumulation of lead soaps on the paint surface react with atmospheric sulfur to form the superficial black crystalline material. Completion of the technical analysis, including identification of the black material and confirmation of lead soap presence, is essential in understanding the observed phenomenon and developing a treatment strategy.

11.4.2 PH-247 *(1951)*

Known colloquially as "Big Blue," *PH-247* is actively undergoing an uneven color shift throughout the fields of synthetic ultramarine blue (Fig. 11.3). A photograph in the Clyfford Still Museum Archive from 1959 shows the painting hanging in the Albright-Knox Art Gallery in Buffalo, New York. The color slide itself has shifted in color with age, but it is still evident that the painting once had a more even, saturated appearance.

This phenomenon has been studied for the forthcoming book, *Clyfford Still: The Artist's Materials* (Lake et al), from the Getty Conservation Institute (GCI). Analysis of the paint film and hypotheses as to the causes of these visible changes have resulted. ATR-FTIR mapping of the cross sections carried out at the GCI has

Fig. 11.3 *PH-247* (1951) Overall

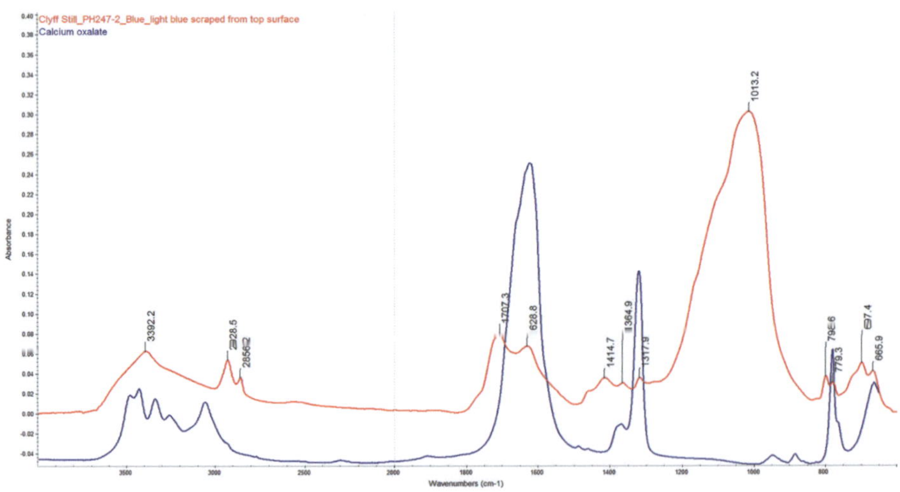

Fig. 11.4 *PH-247* (1951) FTIR spectrum of ultramarine paint sample, © 2017 The J. Paul Getty Trust. Analysis: Herant Khanjian

determined that Still built up the paint film with increasingly medium rich layers to create a deep, transparent effect. Transmitted FTIR determined the presence of oxalates, which may indicate that the whitening is due to the degradation of the linseed oil binder, not a breakdown of the pigment particles themselves as with ultramarine disease (Fig. 11.4). Additional studies have also pointed to a photocatalytic degradation process in synthetic ultramarine paint films [1].

The theorized breakdown of the binding medium in *PH-247* causes microfissures, which scatters light and creates the visual appearance of uneven whitening (Fig. 11.5). Application of a varnish would fill these voids, both re-saturating the

Fig. 11.5 *PH-247* (1951) Detail, bottom right edge

color and dampening color contrasts within the ultramarine blue field. However, the light scattering effect gives the painting a luminosity and range of color that is universally admired by museum visitors and staff alike. Arguably, the artist's intent was for a more saturated, even field of color. This presumed intent conflicts with the current aesthetic preference of the painting and the treatment choices of conservators play an integral role in changing the appearance. For the present time, no treatment will be carried out, instead monitoring the progression of visual changes through photographic documentation.

11.4.3 PH-891 *(1972)*

The example of *PH-891* addresses treatment of disfiguring efflorescence on paint surfaces, one of the most common issues in the collection needing mitigation. *PH-891* had a heavy build-up of white efflorescence, giving the raw umber paint a mottled light gray appearance, broken up by the red paint that had little to no efflorescence (Fig. 11.6a). Additionally, the orientation in which the painting had been rolled led to an unusual concentration of the efflorescence in the top quarter section of the umber and red field, creating a block-like composition that was unrelated to the painting's composition. The other significant condition issue seen in the image are small fragments of masking tape adhered to the paint surface.

Both curatorial and conservation staff determined that extent and uneven concentration of the efflorescence obscured the color and changed the compositional relationships to the degree that it no longer represented Still's original intent, necessitating intervention. Transmitted FTIR spectroscopy carried out prior to treatment confirmed the efflorescence was comprised primarily of stearic acid (Fig. 11.7). The paint layer proved to be too sensitive for any solvent tested, so treatment began

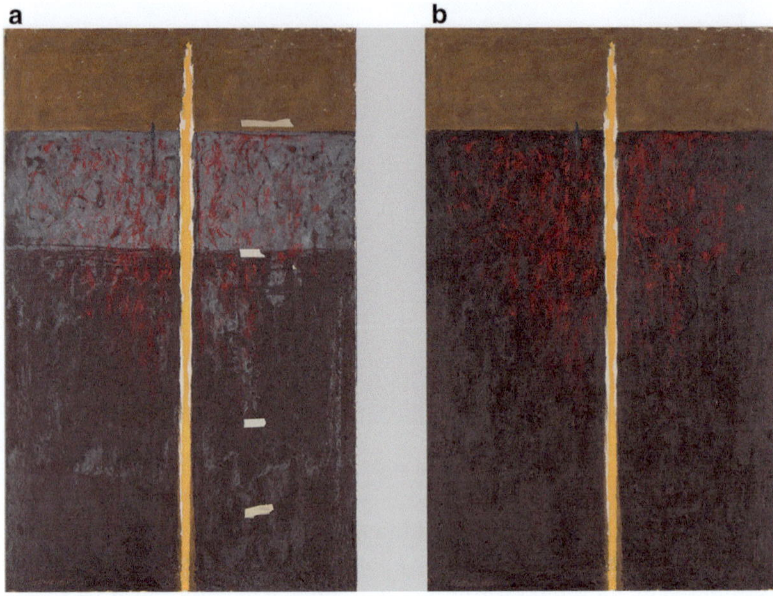

Fig. 11.6 *PH-891* (1972) (**a**) Before treatment; (**b**) After treatment

with mechanical removal using a combination of latex sponges and non-vinyl erasers. The surface texture of the painting did not allow for complete or uniform mechanical removal and the efflorescence remained distracting. Tests ensued with a hot air pen to reduce the remaining efflorescence crystals and thus visually improve the appearance. Since this method does not actually remove the migrating materials, instead driving them back into the paint film, there was concern the technique would alter the surface in unintended ways, either deepening saturation and unifying gloss or creating disparities therein. It was also understood the efflorescence may return in the future and this possibility was accepted. Upon further consultation with the curatorial team it was determined to move forward with treatment. The resulting surface revealed subtle differences in sheen within the umber field and is believed, based on archival images, to be closer to the artist's original intent (Fig. 11.6b).

11.4.4 PH-254 *(1945)*

The final example presents *PH-254*, which considers the artist's intended appearance of the artwork. The painting came into the collection in 2011 with the composition virtually unreadable due to an uneven and significantly darkened varnish (Fig. 11.8a). It presently appears as if some of the compositional elements have been preferentially varnished to a very high gloss. Early in Still's career, generally the 1930s and early 1940s, he seems to have explored varying the gloss of different

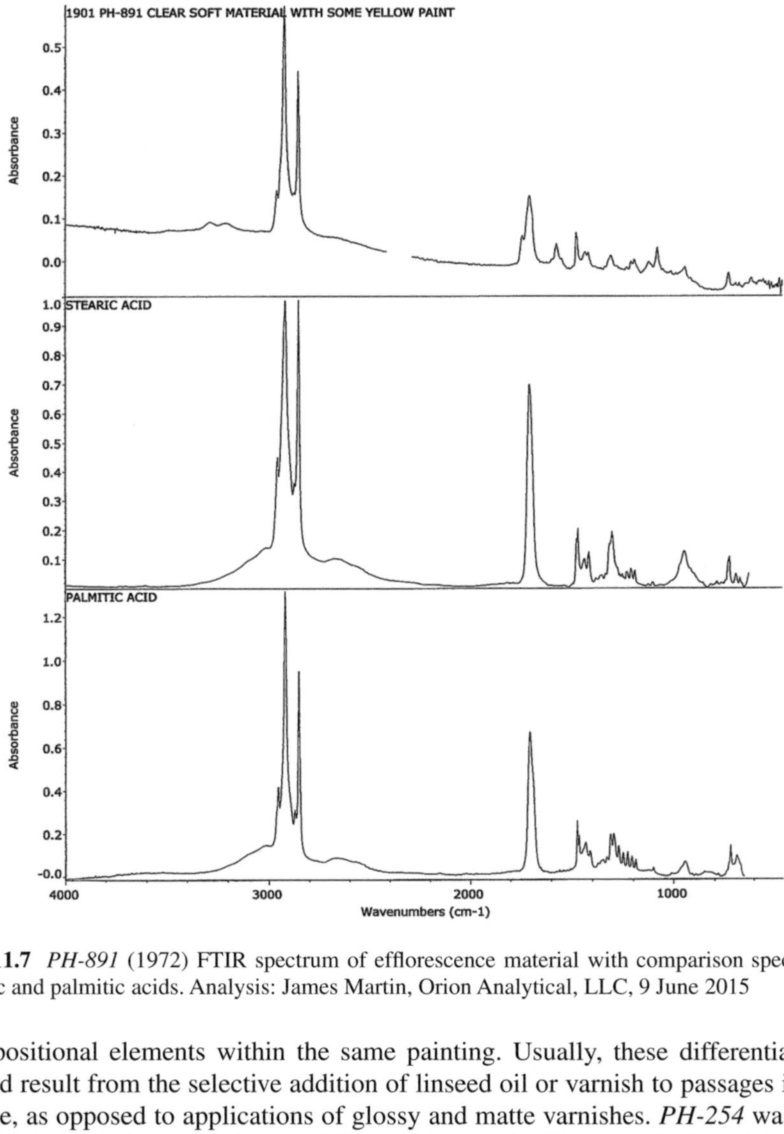

Fig. 11.7 *PH-891* (1972) FTIR spectrum of efflorescence material with comparison spectra of stearic and palmitic acids. Analysis: James Martin, Orion Analytical, LLC, 9 June 2015

compositional elements within the same painting. Usually, these differentiations would result from the selective addition of linseed oil or varnish to passages in the image, as opposed to applications of glossy and matte varnishes. *PH-254* was created during the time period when Still experimented with this technique, however, it has a continuous varnish layer which ranges from a high gloss to matte by compositional element.

An undated archival image, likely from the early 1960s, shows the painting with a very legible composition (Fig. 11.8b). The artwork appears evenly saturated with a relatively unified gloss. In addition, there are several other historical images where a more even varnish can potentially be seen. One such photo dates from 1946 when it hung in Still's first solo exhibition at Peggy Guggenheim's *Art of This Century* gallery. Notably, this photograph provides an indication of the painting's appearance

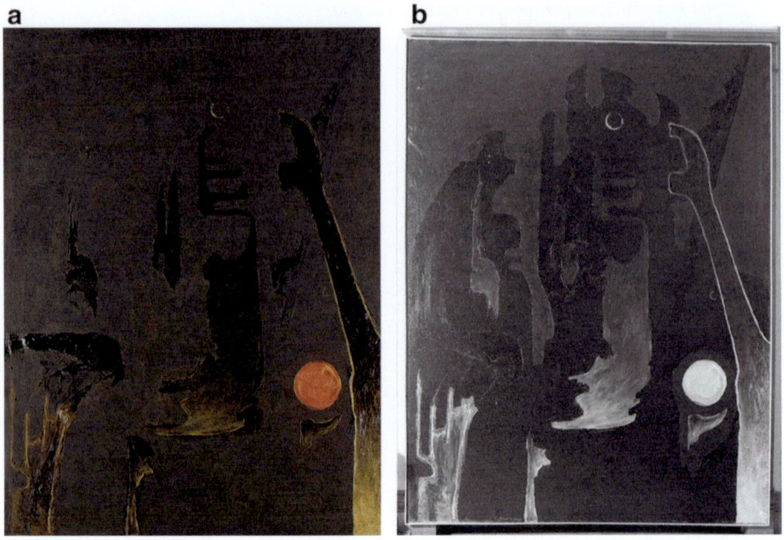

Fig. 11.8 *PH-254* (1945) (**a**) Overall; (**b**) Photograph of *PH-254* (1945), circa 1964. (Photo by Sandra Still. Courtesy of Clyfford Still Archives © Sandra Still)

soon after it was created. *PH-254* was also photographed in 1962 at his Maryland studio during his inventory. By this point in time the varnish appears to be sinking in unevenly across the painting.

The archives also contain a series of images of *PH-254* with several different lighting angles. This group also appears to have originated some time during the 1960s when the bulk of this documentation was happening, but they are not necessarily part of the formal inventory. It is uncertain what the intent of the various lighting angles was, but uneven gloss within the painting appears to be evident here, as well. Comparing the myriad of archival images with the current appearance of the painting and not knowing their exact dates, Still's true intent remains unclear.

Investigating the archival photographs and records led to documentation of the painting's acquisition in 1969 by the Marlborough-Gerson Gallery in New York City. It remained in their possession until December 1974 when Still traded the Gallery a new painting for *PH-254* saying in his diary, "I always regretted letting this picture go. When I found Marlborough still had it, I requested an exchange for a more recent work. Marlborough was happy to make the exchange – obviously understanding little of the history or significance of the work."

Clearly this painting was of great importance to Still, so when he re-acquired it in 1974, it is possible that he creatively engaged with the painting again. If he did, he could have been attempting to re-saturate the painting to make it more readable and the varnish has aged poorly, resulting in unintended variations of gloss. Perhaps he was re-visiting his earlier technique of selectively highlighting compositional forms.

Analysis thus far includes optical microscopy of cross sections and transmitted FTIR spectroscopy. Two discrete layers of varnish were evident in the cross sections, as well as significant discoloration in the bottom layer. FTIR confirmed the presence of an alkyd resin varnish.

Any number of scenarios are possible for these layers and when they may have been applied. But the crux of the question remains unanswered – what was his artistic intent? In order to treat the painting and return it to an exhibitable state, clarification is necessary.

11.5 Conclusion

Unrolling Clyfford Still's paintings, many of which have yet to be viewed publicly, revealed several interesting technical and philosophical problems. How his artworks should be maintained and interpreted as they age are at the forefront of discussion between the Museum's conservation and curatorial staff. Within documents in the archive, Still stated:

> Gradual oxidation of the linseed binder to some degree is unavoidable and the additional potent chemicals in the atmosphere of cities act to hasten the process. Therefore, when the effects become visible, an application of varnish or restoration of the linseed binder by cloth or brush is advisable. Especially where matte effect was not the original intention.

Though acknowledging changes will occur and authorizing intervention, Still unrealistically assumes a universal understanding of his artistic intent. However, the natural aging of his works coupled with a modern societal perspective makes interpretation ambiguous. It is therefore necessary to engage in significant technical and art historical research to guide the interpretation of his aging life's work. The differing treatment decisions highlighted in the case studies reflect not only our present state of research and understanding, but the interplay of technical art history and cultural standards in the interpretation of aging artworks.

References

1. De la Rie R, Michelin A, Ngako M, Del Federico E, Del Grasso C (2017) Photocatalytic degradation of binding media of ultramarine blue containing paint layers: a new perspective on the phenomenon of "ultramarine disease" in paintings. Polymer degradation and stability 144: 43–52
2. Hedley G (1990) Long lost relations and new found relativities: issues in the cleaning of paintings. In: Ramsay-Jolicoeur BA, Wainwright INM (eds) Shared responsibility: proceedings of a seminar for curators and conservators, National Gallery of Canada, Ottawa 26–28 October 1989. Canadian Conservation Institute, Ottawa, p 159–169
3. Hunter, S (1985) Masters of the fifties: american abstract painting from Pollock to Stella. Marisa del Re Gallery, New York

Chapter 12
Pigments of Soviet Artists in the 1950s – Late 1970s

Irina Kadikova, Svetlana Pisareva, Irina Grigorieva, and Maria Lukashova

12.1 Introduction

Nowadays, Socialist Realism Art is becoming increasingly popular in Russia because it depicts a certain period of Soviet history not taking politics in consideration. The growth in popularity of this style in art has led to the establishment of new museums, such as the Institute of Russian Realist Art (IRRA), a private museum and an exhibition centre opened in December 2011. It represents the collection of Russian and Soviet realistic Art of the twentieth century, including works of renowned and little-known Soviet artists. It also led to opening of numerous exhibitions, such as 'Thaw' (Tretyakov Gallery, Moscow, 2017), 'Nissky. Horizon' (IRRA, Moscow, 2018), 'Andrei Mylnikov (1919–2012). The 100th Birth Anniversary' (The Russian State Museum, St. Petersburg, 2019), and others that have become important events in national cultural life. These paintings now require careful examination, detailed description and conservation treatment.

The official Soviet Art of the 1950s – late 1970s has never been studied thoroughly. The particular feature of the Soviet palette is a restricted number of pigments compared to those available to contemporary European and North American artists. This is related to the Iron Curtain era, as the artists had access mainly to paints manufactured in the USSR. New and modified pigments produced in the USSR after the 1950s when the Soviet paint industry was rapidly developing are of

I. Kadikova (✉) · S. Pisareva
The State Research Institute for Restoration, Moscow, Russia
e-mail: kadikovaif@gosniir.ru

I. Grigorieva
The State Hermitage Museum, St. Petersburg, Russia

M. Lukashova
TESCAN Ltd., St. Petersburg, Russia

K. J. van den Berg et al. (eds.), *Conservation of Modern Oil Paintings*,
https://doi.org/10.1007/978-3-030-19254-9_12

particular interest to researchers. Cobalt blue spectral ($Co_2SiO_4{\cdot}Zn_2SiO_4$), a new pigment synthesized in 1956, was introduced as an oil paint pigment for the first time in the Soviet market in 1961. Some of the pigments, such as anatase titanium white (TiO_2), blue and green phthalocyanines and Pigment Green B ($C_{20}H_{12}N_2O_4Fe$) were produced by Soviet industry from the late 1960s, much later than in Europe and North America. At this time some well-known pigments with modified chemical composition, such as cobalt green (e.g. cobalt dark green, $CoO{\cdot}15ZnO{\cdot}0.03Al_2O_3{\cdot}0.26MgO$; cobalt light green, $CoO{\cdot}50ZnO{\cdot}0.03Al_2O_3$, etc.) and chromium aluminium cobalt oxide (e.g. chromium aluminium cobalt oxide greenish-blue, $CoO{\cdot}2Al_2O_3{\cdot}Cr_2O_3{\cdot}ZnO$; chromium aluminium cobalt oxide bluish-green, $CoO{\cdot}1.4Cr_2O_3$, etc.) were produced [1, 2].

The aim of this research was to examine important paintings from the 1950s – late 1970s to ascertain the extent to which artists used the new pigments produced by the Soviet industry, such as titanium white, phthalocyanines blue and green, Pigment Green B, manganese blue ($Ba_3(MnO_4)_2{\cdot}nBaSO_4{\cdot}mBaO$, where $n = 10$–16, $m = 1$–3), cobalt blue spectral, and cobalt ammonium phosphate ($CoNH_4PO_4{\cdot}H_2O$) (Table 12.1). Juxtaposing the obtained results with published archival data[1] allow us to distinguish between artists' usage of foreign paints before their manufacturing was established in the USSR. This data may be important in determining forgeries from original works and later copies.

12.2 Materials and Methods

The pigments and grounds from 60 paintings produced from the 1950s to the late 1970s from Russian museums and private collections were examined using a range of analytical methods. The authenticity of the paintings was verified by documented provenance provided by museums, and results of artwork expertise. Most paintings have a signature and a date. The results of 60 works listed in Table 12.1 are given here.

The majority of examined works were painted by little-known artists, such as Aleksei Ratnikov, Vasilii Basov, Boris Vitollo, and others. The paintings of renowned Soviet artists (People's Artists of the USSR), such as Aleksandr Gerasimov, Andrei Mylnikov, Porfirii Krylov, Yurii Pimenov, and Sergei Gerasimov were also studied. This limited number of important artists of the Academy of Arts had status that allowed them the opportunity of travelling overseas and bringing back paints from

[1] The archival data on chemical contents and the beginning of the Soviet art materials manufacturing in the period of the 1950s–1990s were firstly reviewed and systematized by Yu. I. Grenberg [1]. While working with the Archives of the Leningrad Artistic Paints Plant (LAPP, the largest manufacture of art paints in the USSR) and the Expert Quality Council of Art Materials of the Culture Ministry of the USSR, he gathered and analysed technical specifications (TS) and state standards, or GOST standards, which regulated paints composition for a certain period as well as feedbacks from the professionals about paints quality.

Table 12.1 Ground content and pigment composition of the examined paintings. The works of representatives of unofficial Soviet Art are indicated with *

	Author/Painting	Ground	White pigments	Blue pigments	Green pigments	Red/Orange pigments	Yellow/ Brown pigments	Violet/ Metallic pigments
1	Nikolai Kozochkin (1899–1961). *The Shed.* 1950. Oil on cardboard, 19 × 14.2 cm. Moscow Region, Museum-Reserve named after F. Tyutchev *Muranovo*	No primer	Zinc white	Ultramarine, Cobalt blue		Red organic pigment, Mars red	Cadmium yellow, Mars yellow	
2	Aleksandr Buzovkin (1886–1966). *Portrait of Valia Korneeva*. 1950. Oil on canvas, 30 × 35 cm. Serpukhov History and Art Museum	Zinc white with a small amount of chalk	Zinc white	Ultramarine	Cobalt green, Volkonskoite	Red organic pigment, Red ochre, Mars orange	Strontium yellow, Cadmium yellow, Zinc yellow, Mars yellow	
3	Aleksei Zhibinov (1905–1955). *Portrait of a Woman*. 1950. Oil on canvas, 47.6 × 40.7 cm. Irkutsk Regional Art Museum	Zinc white with barium sulphate	Zinc white	Cobalt blue	Viridian, Cobalt green	Red organic pigment, Red ochre	Strontium yellow, Chrome yellow, Yellow organic pigment, Yellow ochre	

(continued)

Table 12.1 (continued)

	Author/Painting	Ground	White pigments	Blue pigments	Green pigments	Red/Orange pigments	Yellow/Brown pigments	Violet/Metallic pigments
4	Konstantin Dorokhov (1906–1960). *The Lake*. 1950. Oil on canvas, 30 × 74 cm. Solnechnogorsk, The museum and an exhibition centre *Putevoi dvorets* (En Route Imperial Palace)	Zinc white with a small amount of chalk	Zinc white	Ultramarine, Cobalt blue, Manganese blue	Viridian, Volkonskoite	Cadmium sulfoselenide,[a] Red organic pigment, Mars orange	Strontium yellow, Mars yellow	
5	Ivan Blokhin (1887–1954). *On the Veranda*. 1952. Oil on canvas, 52.3 × 66.8 cm. Private collection, Moscow	Primed by artist. Calcite with a small amount of kaolinite	Zinc white	Ultramarine, Cobalt blue	Viridian, Chrome green, Volkonskoite	Cadmium sulfoselenide, Red organic pigment, Mars orange	Strontium yellow, Cadmium yellow, Mars yellow	
6	Sergei Gerasimov (1885–1964). *The Swamp in the Forest*. 1952. Oil on cardboard, 69 × 49 cm. Serpukhov History and Art Museum	Zinc white with a small amount of lead white, and chalk	Lead white	Ultramarine, Manganese blue	Viridian	Cadmium sulfoselenide, Mars orange, Chrome orange	Mars yellow	

7	Ol'ga Kolmakova (1905–1999). *Portrait of Horse-breeder A.I. Gorbenko*. 1952. Oil on canvas, 100 × 80 cm. Solnechnogorsk, The museum and an exhibition centre *Putevoi dvorets* (En Route Imperial Palace)	Zinc white with chalk	Zinc white	Ultramarine, Cobalt blue	Viridian, Cobalt green	Cadmium sulfoselenide, Red ochre, Mars orange	Strontium yellow	
8	Aleksandr Buzovkin (1886–1966). *Self-portrait*. 1953. Oil on canvas, 43 × 55 cm. Serpukhov History and Art Museum	Chalk	Zinc white	Ultramarine	Cobalt green, Volkonskoite	Red organic pigment, Red ochre, Mars orange	Strontium yellow, Cadmium yellow, Zinc yellow, Mars yellow, Brown ochre	
9	Aleksandr Gerasimov (1881–1963) with the students. *The Statement by Iosif Stalin*. 1953. Oil on canvas, 645 × 1100 cm. Moscow, The VDNKh (Expo center), pavilion #1	Primed by artist. Zinc white with chalk	Zinc white, Lead white with zinc white	Ultramarine, Cobalt blue	Viridian, Cobalt green, Volkonskoite, Emerald green	Cadmium sulfoselenide, Red organic pigment, Mars red and orange	Strontium yellow, Cadmium yellow, Mars yellow and brown	
10	Aleksei Zhibinov (1905–1955). *The Teacher*. 1954. Oil on cardboard, 49.5 × 62 cm. Irkutsk Regional Art Museum	Zinc white	Zinc white	Ultramarine	Cobalt green	Vermilion, Red organic pigment, Mars orange	Strontium yellow, Cadmium yellow	

(continued)

Table 12.1 (continued)

	Author/Painting	Ground	White pigments	Blue pigments	Green pigments	Red/Orange pigments	Yellow/Brown pigments	Violet/Metallic pigments
11	V. Nikolaev (1920–?). *The Harvesters*. 1955. Oil on canvas, 90 × 50 cm. Serpukhov History and Art Museum	Zinc white	Zinc white	Ultramarine, Cobalt blue	Viridian, Cobalt green, Volkonskoite	Cadmium sulfoselenide, Red organic pigment, Mars orange	Strontium yellow, Cadmium yellow	Cobalt ammonium phosphate
12	Viktor Ferber (1908–1965). *The Sea*. 1955. Oil on canvas, 135 × 98.5 cm. Private collection, Moscow	Zinc white	Zinc white, Lead white (with large crystals of cerussite),[b] **	Ultramarine, Cobalt blue	Viridian, Volkonskoite	Cadmium sulfoselenide, Red organic pigment, Mars orange	Strontium yellow, Cadmium yellow	
13	Vasilii Basov (1918–1962). *The Volga near Kerzhenets*. 1955. Oil on canvas, 41 × 81 cm. Serpukhov History and Art Museum	Zinc white	Zinc white	Ultramarine, Cobalt blue	Viridian, Cobalt green, Volkonskoite	Cadmium sulfoselenide, Red ochre, Mars orange		Cobalt ammonium phosphate
14	Nikolai Nikonov (1889–1975). *Flowers*. 1956. Oil on canvas, 59.5 × 79.5 cm. Solnechnogorsk, The museum and an exhibition centre *Putevoi dvorets* (En Route Imperial Palace)	Zinc white	Zinc white	Ultramarine, Cobalt blue	Cobalt green, Volkonskoite	Cadmium sulfoselenide, Red organic pigment, Mars red and orange	Strontium yellow, Brown ochre	

15	Yevgenia Maleina (1903–1984). *On the River.* 1956. Oil on canvas, 80 × 41 cm. Solnechnogorsk, The museum and an exhibition centre *Putevoi dvorets* (En Route Imperial Palace)	Zinc white	Zinc white	Ultramarine	Viridian, Volkonskoite	Red organic pigment, Mars orange	Strontium yellow	
16	Konstantin Maksimov (1913–1994). *The Lotuses.* 1956. Oil on canvas, 100 × 150 cm. Private collection, Moscow	Chalk with zinc white	Zinc white, Lead white**	Ultramarine	Viridian, Volkonskoite, Emerald green	Cadmium sulfoselenide, Red organic pigment, Mars orange, Chrome orange	Cadmium yellow, Mars yellow	
17	Vladimir Gavrilov (1923–1970). *The Bridge.* 1956. Oil on canvas, 83 × 64 cm. Solnechnogorsk, The museum and an exhibition centre *Putevoi dvorets* (En Route Imperial Palace)	Zinc white	Zinc white	Ultramarine, Cobalt blue	Viridian, Cobalt green, Volkonskoite	Red organic pigment, Mars orange	Strontium yellow, Yellow organic pigment, Brown ochre	
18	Andrei Vasnetsov (1924–2009). *Still-life with the Thistle.* 1958. Oil on canvas, 88.5 × 110.5 cm. Moscow Region, The State Museum-Reserve *Abramtsevo*	Primed by artist. Gypsum with chalk	Zinc white	Ultramarine	Viridian, Cobalt green, Volkonskoite	Mars red and orange	Brown ochre	

(continued)

Table 12.1 (continued)

	Author/Painting	Ground	White pigments	Blue pigments	Green pigments	Red/Orange pigments	Yellow/Brown pigments	Violet/Metallic pigments
19	Dmitrii Tegin (1914–1988). *The Shepherd.* 1958. Oil on canvas, 46 × 36 cm. Solnechnogorsk, The museum and an exhibition centre *Putevoi dvorets* (En Route Imperial Palace)	Zinc white with barium sulphate	Zinc white with a small amount of lead white	Ultramarine, Cobalt blue	Viridian, Cobalt green, Volkonskoite	Cadmium sulfoselenide, Red organic pigment, Mars orange	Cadmium yellow, Mars yellow	
20	Baqi Urmanche (1897–1990). *Vintaging*. 1958. Oil on canvas, 143 × 195 cm. Kazan, The State Museum of Fine Arts of the Republic of Tatarstan	Lead white with zinc white	Zinc white, Lead white with zinc white	Ultramarine	Viridian, Cobalt green, Chrome green, Volkonskoite, Phthalocyanine green	Cadmium sulfoselenide	Strontium yellow	Cobalt ammonium phosphate, Violet organic pigment
21	Rostislav Galitskii (1920–1979). *The Head.* 1960. Oil on cardboard, 47 × 35 cm. Rostov, The State Museum-Reserve *Rostov Kremlin*	Zinc white	Zinc white	Ultramarine, Cobalt blue	Viridian, Volkonskoite	Cadmium sulfoselenide, Red organic pigment, Mars orange	Strontium yellow	
22	Anatolii Timofeev (1928–?). *The Walking Excavator*. 1961. Oil on cardboard, 37.8 × 51.5 cm. Irkutsk Regional Art Museum	Zinc white	Zinc white	Ultramarine, Cobalt blue, Chromium aluminium cobalt oxide	Cobalt green, Volkonskoite	Cadmium sulfoselenide, Mars orange	Strontium yellow, Cadmium yellow, Mars yellow	

23	Lev Viatkin (1931–?). *Autumn*. 1961. Oil on cardboard, 11.5 × 16.7 cm. Moscow, Russian National Museum of Music	No primer	Zinc white	Ultramarine, Prussian blue	Cobalt green	Red organic pigment, Cadmium orange	Strontium yellow, Brown ochre	
24	Viktor Perel'man (1892–1967). *A Hot Day*. 1961. Oil on canvas, 93 × 60 cm. Private collection of the artist's heir, Moscow	Zinc white with calcite	Zinc white	Ultramarine, Cobalt blue	Volkonskoite	Vermilion, Red organic pigment, Mars orange	Cadmium yellow, Yellow and brown ochres	
25	Aleksandr Zaitsev *Latin Cathedral in L'vov*. 1961. Oil on canvas, 49 × 36.5 cm. Moscow, Russian National Museum of Music	Lead white (with large crystals of cerussite) with chalk	Zinc white, Lead white (with large crystals of cerussite)**	Ultramarine, Cobalt blue	Cobalt green, Chrome green	Cadmium sulfoselenide, Red organic pigment, Mars orange	Strontium yellow, Mars yellow	
26	Viktor Perel'man (1892–1967). *Mimosa*. 1962. Oil on canvas, 64.2 × 79.8 cm. Private collection of the artist's heir, Moscow	Zinc white	Zinc white	Ultramarine	Chrome green, Volkonskoite	Red organic pigment, Mars red, Orange ochre	Strontium yellow, Cadmium yellow, Brown ochre	
27	Nadezhda Poluektova (1900–?). *Seliger*. 1962. Tempera on canvas, 49.8 × 50 cm. Solnechnogorsk, The museum and an exhibition centre *Putevoi dvorets* (En Route Imperial Palace)	Primed by artist. Zinc white with chalk	Zinc white	Ultramarine, Phthalocyanine blue	Phthalocyanine green	Red organic pigment, Mars orange	Yellow organic pigment	

(continued)

Table 12.1 (continued)

	Author/Painting	Ground	White pigments	Blue pigments	Green pigments	Red/Orange pigments	Yellow/ Brown pigments	Violet/ Metallic pigments
28	Vsevolod Filimonov (1930–1980). *Portrait of Volodia, the Carpenter*. 1962. Oil on canvas, 74.4 × 56 cm. Solnechnogorsk, The museum and an exhibition centre *Putevoi dvorets* (En Route Imperial Palace)	Was not examined	Zinc white	Ultramarine, Cobalt blue	Viridian, Volkonskoite	Cadmium sulfoselenide, Red organic pigment, Red ochre	Mars yellow, Strontium yellow, Yellow and brown ochres	
29	Miron Lukianov (1936–2007). *A Rainy Sunday* movie poster. 1963. Gouache on paper, 150 × 90 cm. Private collection of the artist's heir, Moscow	No primer	Zinc white	Ultramarine, Phthalocyanine blue	Viridian, Phthalocyanine green			
30	Andrei Polyushenko (1911–1985). *Spring in the Collective Farm*. 1963. Oil on canvas, 60 × 73 cm. Serpukhov History and Art Museum	Primed by artist. Chalk with zinc white	Zinc white, Lead white (with large crystals of cerussite)**	Ultramarine, Cobalt blue, Chromium aluminium cobalt oxide, Cerulean	Viridian, Cobalt green	Cadmium sulfoselenide, Mars red and orange	Strontium yellow	Cobalt phosphate, Cobalt ammonium phosphate
31	Lado Gudiashvili (1896–1980). *Bride and Bridegroom*. 1964. Oil on canvas, 55.6 × 42 cm. Private collection, Moscow	Primed by artist. Chalk with orange ochre, and kaolinite	Zinc white, Lead white with admixture of titanium white (rutile)	Prussian blue, Manganese blue	Viridian, Cobalt green	Red organic pigment	Cadmium yellow	

32	Yefrem Zver'kov (1921–2012). *The Road in the North Village*. 1964. Oil on cardboard, 51 × 59 cm. Serpukhov History and Art Museum	Chalk	Zinc white	Ultramarine, Cobalt blue spectral	Viridian, Chrome green	Red organic pigment, Mars orange	Yellow and brown ochres	
33	Boris Vitollo (1908–1992). *The Guitarist*. 1965. Oil on canvas, 87 × 67 cm. Moscow, Russian National Museum of Music	Zinc white	Zinc white, Lead white (with large crystals of cerussite)	Ultramarine, Chromium aluminium cobalt oxide, Cerulean, Cobalt blue spectral	Viridian, Cobalt green, Chrome green	Cadmium sulfoselenide, Red organic pigment, Mars red, Cadmium orange	Cadmium yellow, Mars yellow, Yellow ochre	
34	Aleksandr Zaitsev *The View of the Dante Primary School*. 1967. Oil on canvas, 51 × 37.5 cm. Moscow, Russian National Museum of Music	Calcite with zinc white	Zinc white, Lead white (with large crystals of cerussite)**	Cobalt blue spectral	Viridian, Cobalt green	Cadmium sulfoselenide, Red organic pigment	Strontium yellow, Cadmium yellow, Mars yellow	
35	Yurii Pimenov (1903–1977). *The Urban Landscape*. 1967. Oil on canvas, 45 × 44.7 cm. Private collection, Moscow	Zinc white with a small amount of chalk	Zinc white	Cobalt blue, Chromium aluminium cobalt oxide, Cerulean		Mars red and orange		
36	Sergei Kolybanov (1898–1975). *The Share of Lake Baikal*. 1967. Oil on canvas, 55 × 120 cm. Private collection, Moscow	Primed by artist. Zinc white with a small amount of calcite	Zinc white	Ultramarine, Cobalt blue, Phthalocyanine blue, Cobalt blue spectral	Cobalt green, Phthalocyanine green	Red organic pigment, Red ochre, Cadmium orange	Cadmium yellow, Yellow ochre	Cobalt ammonium phosphate

(continued)

Table 12.1 (continued)

	Author/Painting	Ground	White pigments	Blue pigments	Green pigments	Red/Orange pigments	Yellow/Brown pigments	Violet/Metallic pigments
37	Alexandr Tyshler (1898–1980)*. *A Caryatid with a Ladder*. 1968. Oil on canvas, 84 × 68.5 cm. Moscow, The museum-panorama *Borodinskaia Bitva* (The Battle under Borodino)	Zinc white	Zinc white	Ultramarine, Cobalt blue, Chromium aluminium cobalt oxide, Manganese blue, Cobalt blue spectral	Viridian, Phthalocyanine green	Cadmium sulfoselenide, Red organic pigment	Strontium yellow, Mars yellow	Manganese phosphate
38	Tatiana Radimova (1916–2000). *Kakushkino*. 1969. Oil on canvas, 47.3 × 78.9 cm. Solnechnogorsk, The museum and an exhibition centre *Putevoi dvorets* (En Route Imperial Palace)	Zinc white	Zinc white	Ultramarine, Cobalt blue	Viridian	Cadmium sulfoselenide, Red organic pigment	Strontium yellow, Cadmium yellow	
39	Aleksei Ponsov (1920–2009). *The Forest Edge*. 1969. Oil on cardboard, 49.4 × 35 cm. Private collection of the artist's heir, Moscow	Zinc white	Zinc white	Ultramarine	Viridian, Cobalt green, Volkonskoite	Red organic pigment, Mars orange	Strontium yellow, Yellow ochre	
40	Porfirii Krylov (1902–1990). *Malakhovka*. 1970. Oil on canvas, 49 × 34 cm. Private collection, Moscow	Titanium white (anatase) with barium sulphate	Zinc white, Titanium white (anatase) with zinc white, and barium sulphate	Ultramarine, Manganese blue, Cobalt blue spectral	Viridian, Chrome green	Cadmium sulfoselenide, Red organic pigment, Mars orange	Cadmium yellow, Yellow organic pigment	Cobalt phosphate

41	Zinaida Kovalevskaia (1902–1979). *Still-life with Peonies*. 1971. Oil on canvas, 79 × 60 cm. Private collection, Moscow	Zinc white	Zinc white	Ultramarine, Cobalt blue, Prussian blue, Chromium aluminium cobalt oxide, Phthalocyanine blue	Cobalt green, Chrome green, Phthalocyanine green	Red organic pigment, Orange ochre	Cadmium yellow, Zinc yellow, Brown ochre	
42	Mikhail Volodin (1912–1987). *The Last Hydrofoil.* 1972. Oil on cardboard, 33 × 97 cm. Solnechnogorsk, The museum and an exhibition centre *Putevoi dvorets* (En Route Imperial Palace)	Zinc white	Zinc white	Ultramarine, Cobalt blue	Viridian, Cobalt green, Volkonskoite	Cadmium sulfoselenide, Red organic pigment, Mars orange	Strontium yellow, Cadmium yellow, Mars yellow	
43	Vladislav Zubarev (1937–2013)∗. *Reveller. Oleg Sataev*. 1972. Mixed technic on canvas, 66 × 79 cm. Moscow, Russian Abstract Art Foundation	Zinc white	Zinc white	Ultramarine, Cobalt blue	Cobalt green	Red organic pigments	Strontium yellow, Yellow organic pigment	Aluminium powder
44	Alexandr Tyshler (1898–1980)∗. *A Caryatid with Musicians*. 1972. Oil on canvas, 69.5 × 60 cm. Moscow, The museum-panorama *Borodinskaia Bitva* (The Battle under Borodino)	Zinc white	Zinc white	Ultramarine, Cobalt blue, Manganese blue	Viridian, Phthalocyanine green	Cadmium sulfoselenide, Red organic pigment	Strontium yellow, Cadmium yellow	

(continued)

Table 12.1 (continued)

	Author/Painting	Ground	White pigments	Blue pigments	Green pigments	Red/Orange pigments	Yellow/Brown pigments	Violet/Metallic pigments
45	Alexandr Tyshler (1898–1980)∗. *A Caryatid with Houses*. 1972. Oil on canvas, 84.5 × 75.6 cm. Moscow, The museum-panorama *Borodinskaia Bitva* (The Battle under Borodino)	Zinc white with a small amount of chalk	Zinc white	Ultramarine, Manganese blue, Phthalocyanine blue	Phthalocyanine green	Cadmium sulfoselenide, Red organic pigment	Strontium yellow, Cadmium yellow	
46	Yurii Pimenov (1903–1977). *The Wardrobe of the Bolshoi Theatre*. 1972. Oil on canvas, 87 × 81.7 cm. Private collection, Moscow	Zinc white	Zinc white, Lead white		Viridian, Volkonskoite, Glauconite	Mars red	Mars yellow, Brown ochre	
47	Alexandr Tyshler (1898–1980)∗. *Galatea*. 1974. Oil on canvas, 74 × 59 cm. Moscow, The museum-panorama *Borodinskaia Bitva* (The Battle under Borodino)	Zinc white	Titanium white (rutile, anatase) with zinc white	Ultramarine, Cobalt blue, Chromium aluminium cobalt oxide, Cobalt blue spectral	Viridian, Cobalt green, Phthalocyanine green	Cadmium sulfoselenide, Red organic pigment	Cadmium yellow	
48	Leontii Denisov (1904–?). *Karelia* 1974. Oil on canvas, 38 × 63 cm. Solnechnogorsk, The museum and an exhibition centre *Putevoi dvorets* (En Route Imperial Palace)	Zinc white	Zinc white	Ultramarine, Cobalt blue	Viridian, Cobalt green, Chrome green, Emerald green, Phthalocyanine green	Cadmium sulfoselenide, Red organic pigment, Mars orange, Chrome orange	Cadmium yellow, Yellow and brown ochres	Cobalt phosphate

49	Aleksei Ratnikov (1926–1984). *Us't-Ilim. Mira Avenue*. 1975. Oil on canvas, 15.5 × 21.5 cm. Serpukhov History and Art Museum	Zinc white with a small amount of quartz	Zinc white	Ultramarine	Cobalt green, Pigment Green B	Cadmium sulfoselenide, Red organic pigment, Orange ochre	Cadmium yellow, Chrome yellow, Yellow organic pigment, Mars yellow	Cobalt ammonium phosphate
50	Aleksei Ratnikov (1926–1984). *Bratsk Timber Industry*. 1975. Oil on canvas, 15.5 × 21.5 cm. Serpukhov History and Art Museum	Zinc white	Zinc white	Ultramarine, Cobalt blue		Cadmium sulfoselenide, Red organic pigment, Orange ochre	Chrome yellow, Yellow ochre	Cobalt ammonium phosphate
51	Aleksei Ratnikov (1926–1984). *Wood of Us't-Ilim*. 1975. Oil on canvas, 15.5 × 21.5 cm. Serpukhov History and Art Museum	Zinc white	Zinc white	Ultramarine, Cobalt blue	Chrome green	Red organic pigment, Mars red, Orange ochre	Yellow ochre	
52	Aleksei Ratnikov (1926–1984). *Bratsk Hydroelectric Power Station*. 1975. Oil on canvas, 15.5 × 21.5 cm. Serpukhov History and Art Museum	Zinc white	Zinc white	Ultramarine	Cobalt green, Pigment Green B	Red organic pigment, Orange ochre, Mars orange	Cadmium yellow, Chrome yellow, Yellow organic pigment, Mars yellow	

(continued)

Table 12.1 (continued)

	Author/Painting	Ground	White pigments	Blue pigments	Green pigments	Red/Orange pigments	Yellow/ Brown pigments	Violet/ Metallic pigments
53	Andrei Mylnikov (1919–2012). *Early Spring*. 1976. Oil on canvas, 70 × 88 cm. Private collection, Moscow	Titanium white (rutile) with chalk	Zinc white, Titanium white (rutile) with zinc white, and barium sulphate	Ultramarine, Phthalocyanine blue	Phthalocyanine green	Red organic pigment, Mars red	Yellow organic pigment	
54	Pavel Nikonov (1930). *The First Snow.* 1977. Oil on canvas, 38.9 × 43.6 cm. Moscow Region, The State Museum-Reserve *Abramtsevo*	Primed by artist. Zinc white with a small amount of lead white	Zinc white	Ultramarine	Cobalt green	Red organic pigment, Mars red and orange		
55	Yevgenii Tsei (1926–1982)∗. *The Bad Dreams*. 1979. Oil on canvas, 55.5 × 63.5 cm. Krasnodar Region Art Museum	Primed by artist. Titanium white (rutile) with zinc white, and calcite	Zinc white	Ultramarine, Prussian blue, Chromium aluminium cobalt oxide	Viridian, Phthalocyanine green	Red organic pigment	Yellow organic pigment	
56	Anatolii Talalaev (1929–1989). *The Wife Portrait.* 1979. Oil on canvas, 85 × 60 cm. Private collection of the artist's heir, Reutov	Titanium white (anatase) with zinc white	Zinc white	Ultramarine	Cobalt green	Red organic pigment, Mars orange	Strontium yellow, Mars yellow	

57	Porfirii Krylov (1902–1990). *On the Carpet.* 1980. Oil (?) on cardboard, 37.5 × 51.7 cm. Private collection, Moscow	Titanium white (rutile) with barium sulphate	Titanium white (rutile, anatase) with zinc white, and barium sulphate	Ultramarine, Cobalt blue, Cerulean, Cobalt blue spectral, Phthalocyanine blue	Viridian, Cobalt green, Phthalocyanine green	Vermilion, Red organic pigment, Mars orange	Cadmium yellow, Yellow organic pigment, Brown ochre	Cobalt ammonium phosphate
58	Yurii Darii (1932–2001). *The Olympic Sketch.* 1980. Oil on canvas, 67 × 77 cm. Private collection, Reutov	Zinc white	Zinc white	Ultramarine, Cerulean, Manganese blue	Viridian, Cobalt green	Mars orange	Cadmium yellow	Cobalt phosphate
59	Vadim Baldin (1932–1992). *Danilovo.* 1980. Oil on cardboard, 78 × 50 cm. Private collection, Reutov	Zinc white with chalk, and kaolinite	Zinc white	Ultramarine, Cobalt blue	Viridian, Cobalt green	Red organic pigment	Yellow ochre	Cobalt ammonium phosphate
60	Viktor Panfilov (1924–2009). *Spring in the Mountains.* 1980. Oil on canvas. Private collection, Reutov	Zinc white	Zinc white	Ultramarine, Cobalt blue, Cerulean, Chromium aluminium cobalt oxide	Cobalt green, Volkonskoite, Phthalocyanine green	Cadmium sulfoselenide, Red organic pigment, Mars orange		

[a]According to the TS of the LAPP from 1950s the oil paints based on cadmium sulfoselenide contained 50% of barium sulphate. As a rule, such mixture was identified in red paint layers, except two paintings by Sergei Gerasimov (*The Swamp in the Forest*, 1952) and Konstantin Maksimov (*The Lotuses*, 1956) where barium sulphate was not found in red paint layer

[b]In five paintings lead white was identified only in coloured paint layer in a very small amount (indicated with **), so we may conclude that it was an accidental admixture to some coloured paint

abroad. In addition, this study included six works by representatives of unofficial Soviet Art, such as Aleksandr Tyshler, Vladislav Zubarev, and Yevgenii Tsei.

Paint samples were studied by an optical microscope (MSP-1, LOMO) and examined with polarizing microscopy (POLAM L-213M microscope, LOMO). Scanning electron microscope (JSM 5610LV, JEOL) with energy-dispersive X-ray spectrometer (SEM-EDX) was used to determine the elemental composition and to characterize the morphology of some pigments. Analysis of pigments and fillers was performed using micro-Fourier transform infrared (μ-FTIR) spectroscopy (LUMOS, Bruker). The molecular and phase composition of the samples was studied by means of μ-Raman-spectroscopy (SENTERRA, Bruker) and electron backscatter diffraction method (EBSD) on a scanning electron microscope (MIRA 3 LMU, Tescan with AZtecEnergy Advanced energy dispersive microanalysis system with the X-Max 80 spectrometer and EBSD with NordlysNano detector, Oxford Instruments). Cross-sections were prepared to study the paint layer structure of some paintings, as well as to study the elemental and molecular composition layer-by-layer using SEM-EDX, μ-FTIR and μ-Raman-spectroscopy [6].

12.3 Results and Discussion

12.3.1 Ground Content

Soviet artists often used commercially-prepared canvases and cardboards primed with white emulsions (typically a mixture of oil and protein) or oil grounds, which had good adhesion to the support (Fig. 12.1a, b). In contrast, the adhesion between the paint layers and ground was often found to be very weak.

Analysis showed that supports were mainly primed with zinc white (ZnO) or zinc white with various additives, such as lead white ($2PbCO_3{\cdot}Pb(OH)_2$), barium sulphate ($BaSO_4$), calcite or chalk ($CaCO_3$), kaolinite ($Al_4[Si_4O_{10}](OH)_8$), and quartz (SiO_2). Lead white and a mixture of lead white and zinc white were detected in only two paintings. Titanium white and a mixture of titanium white and zinc white were identified in the ground of five examined works (Table 12.1).

Artists' primed canvases were much less common – there were only eight works with white ground and one work with a coloured yellow ground that were thought to be primed by the artists (Fig. 12.1c, d) (Table 12.1).

12.3.2 Pigments

Zinc white was the most widely used pigment identified in 57 paintings, while lead white was identified in only 12 paintings. Lead white was the main white pigment in six works and an additional component to zinc white in another painting. In five

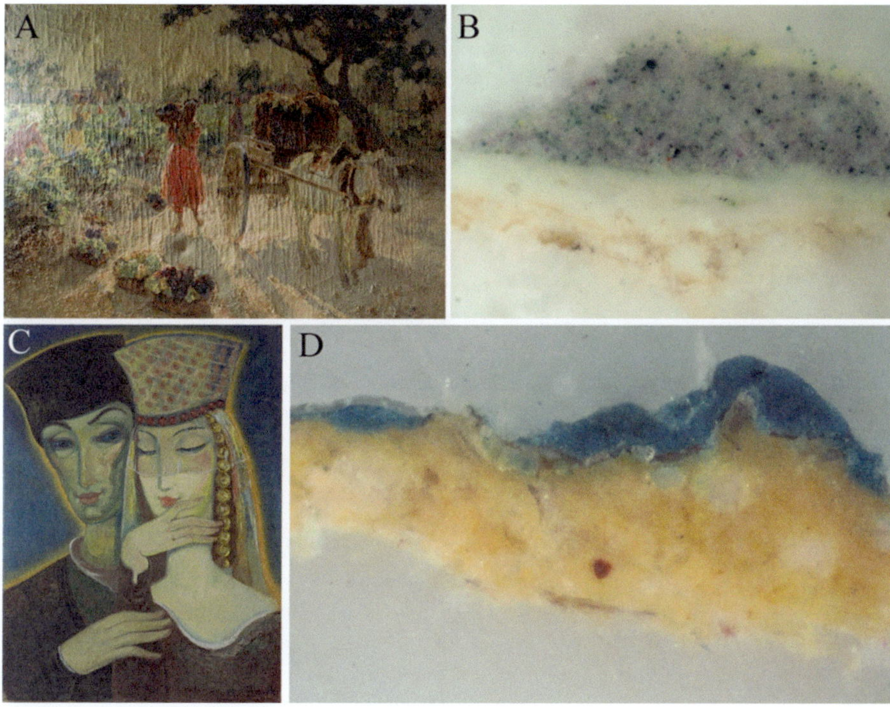

Fig. 12.1 (**a**) Baqi Urmanche *Vintaging*, 1958, oil on canvas, 143 × 195 cm, Kazan, The State Museum of Fine Arts of the Republic of Tatarstan; (**b**) cross-section of a sample from the grape in the basket showing a white commercially-prepared emulsion ground, consisting of lead white with zinc white and brownish glue layer; the violet paint layer consists of zinc white, violet organic pigment, ultramarine, cobalt green, and phthalocyanine green; (**c**) Lado Gudiashvili *Bride and Bridegroom*, 1964, oil on canvas, 55.6 × 42 cm, private collection, Moscow; (**d**) cross-section from the blue background showing coloured glue primer, consisting of chalk, orange ochre (an iron oxide-rich earth), and kaolinite; the blue paint layer consists of lead white with admixture of titanium white (rutile), Prussian blue, and a little amount of manganese blue, cobalt green and viridian

paintings lead white was identified in the paste of different colours in a small amount (Table 12.1). Relatively rare usage of lead white is consistent with the restriction of lead-based pigment production [1, p. 31].

A study of the morphology of lead white particles indicated the presence of unusually large (up to 8 μm) crystals of cerussite ($PbCO_3$) (Fig. 12.2) [3, p. 91, 7, pp. 123–146]. Same crystals were found in five paintings (Table 12.1). The large size of the crystals is characteristic of the method of manufacture of the pigment.

Titanium white was a new pigment for the Soviet paint industry. According to the technical specification (TS) of the Leningrad Artistic Paints Plant (LAPP), in 1972 a mixture composed of titanium dioxide in the form of *anatase* and 25% of zinc white was used as white paint [1, pp. 49–60] (Table 12.2).

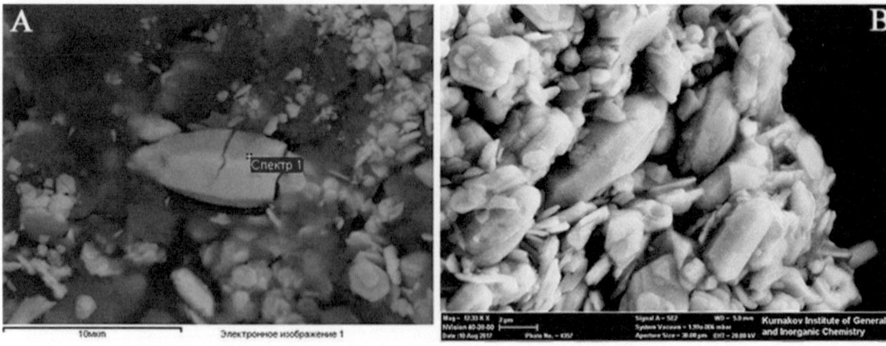

Fig. 12.2 SEM-images of dispersed particles of hydrocercussite and unusually large crystals of cerussite in the paintings by: (**a**) Aleksandr Zaitsev *Latin Cathedral in L'vov*, 1961, oil on canvas, 49 × 36.5 cm, Moscow, Russian National Museum of Music; (**b**) Boris Vitollo *Guitarist*, 1965, oil on canvas, 51 × 37.5 cm, Moscow, Russian National Museum of Music

Table 12.2 Manufacturing of a number of paints in Europe and in the USSR

Pigment	Beginning of manufacturing as an artists' paint		The first appearance in the examined works (see Table 12.1)
	in Europe [3–5]	in the USSR [1]	
Pigment	1921	1955 (water-soluble paints)	
Green B		1971 (oil paints)	1975, Aleksei Ratnikov
Titanium white	1925 (anatase)	1972 (anatase)	1964 (rutile), Lado Gudiashvili
	1938 (rutile)		
Phthalocyanines blue and green	1937–1939	1955 (water-soluble paints)	1962, Nadezhda Poluektova
		1967 (oil paints)	1958, Baqi Urmanche
Manganese blue	1942	1948	1950, Konstantin Dorokhov
Cobalt ammonium phosphate	1943	1966	1955, Vasilii Basov
Cobalt blue spectral	Was not produced	1961	1964, Yefrem Zver'kov

Titanium white was identified in five paintings (Table 12.1). Its rare use is consistent with a conviction of Soviet specialists that the paints were of low quality [8, pp. 186–188]. In 1962, Vasilii Tyutyunnik claimed that titanium oil paints *become yellowish,* noting their *negative influence on the oil film strength and poor miscibility with some other paints* [9, p. 108].

The earliest example of titanium white was detected in a painting dated 1964 (Fig. 12.1c) (Table 12.1). Titanium dioxide in the rutile phase was present in a small quantity (Fig. 12.3a). Since the production of titanium white in the USSR started in 1972, it was concluded that the paint used was from abroad.

Surprisingly, in the painting by Porfirii Krylov from the1980s, titanium dioxide was identified as a mixture of anatase and rutile (Fig. 12.3b, c). Also, zinc white and barium sulphate were identified in the white paint layer (Table 12.1). This mixture of pigments was used in house paint production in the USSR from 1961 [1, p. 53].

The most commonly used blue pigments were ultramarine ($Na_7Al_6Si_6O_{24}S_3$) and cobalt blue ($CoO{\cdot}Al_2O_3$) – there were identified in almost all examined paintings. Chromium aluminium cobalt oxide was found in nine paintings, cerulean blue

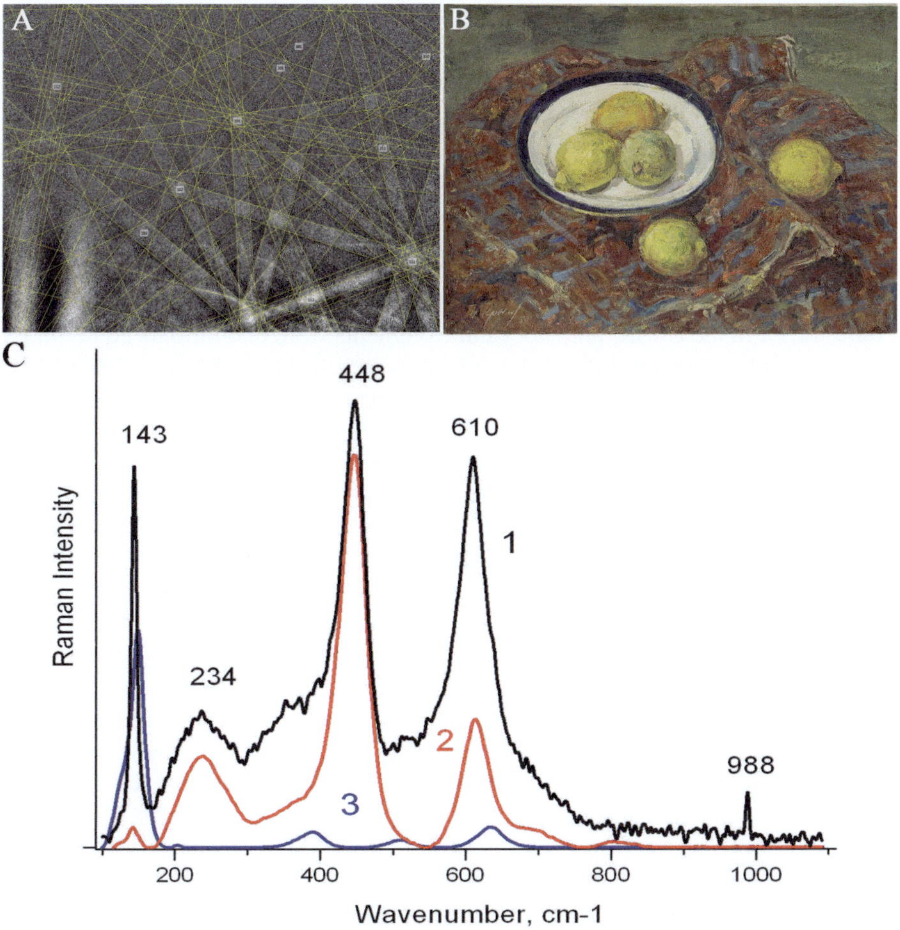

Fig. 12.3 (**a**) the EBSD-pattern of studied crystal TiO_2 (rutile in tetragonal crystal system) from the white sample of the painting by Lado Gudiashvili, 1964; (**b**) Porfirii Krylov *On the Carpet*, 1980, oil on cardboard, 37.5 × 51.7 cm, private collection, Moscow, (**c**) (1) μ-Raman spectrum from white paint layer from the dish (λ_0 = 785 nm, power 1 mW), (2) rutile, RRUFF database, R060493 (λ_0 = 780 nm); frequencies are 143 (mode A_{1g}), 447 (E_g) and 612 (B_{1g}) cm^{-1} and (3) anatase, RRUFF database, R120013 (λ_0 = 780 nm); frequencies are 143, 395, 612 cm^{-1} [10, 11]

($CoO \cdot nSnO_2$) – in six paintings, and Prussian blue ($Fe_4[Fe(CN)_6]_3$) – in four paintings. The latter two were used less frequently than others (Table 12.1).

Manganese blue was first mentioned in the patent literature in 1935, however the start of its production as a paint in Europe dates from 1942. In the USSR the TS for the paint was introduced from 1948 to 1950s, nevertheless this paint was first mentioned in professional literature on art materials only in 1964 [1, p. 90] (Table 12.2). Manganese blue was used quite rarely and identified in only eight paintings, presumably due to insufficient information about the pigment and the earliest dates of pigment detection are 1950 and 1952 (Table 12.1).

Another important blue pigment was initially developed as a colouring agent for ceramics at the Dulevo porcelain factory in 1956 [12]. However, it was not mentioned as an artists' oil paint until the early 1960s. The pigment was introduced to the TS as oil paint in 1961 under the name of *Cobalt Blue Spectral* [1, pp. 93–96] (Table 12.2). The pigment was detected in eight paintings, created after 1964 (Table 12.1).

Oil paints based on phthalocyanine blues and greens were widely used only after the World War II, although their production in Europe commenced in 1937 and 1939 respectively [4]. Interestingly, phthalocyanine green pigment was found in the work *Vintaging* by Baqi Urmanche painted in 1958, when the Soviet paint industry had not yet produced this pigment, and was therefore acquired from a foreign source (Table 12.1). In the USSR, phthalocyanine pigments were first introduced to the TS as water-soluble paints in 1955 and as oil paints in 1967 [1, pp. 85–89] (Table 12.2). Phthalocyanine blue was identified in seven works, and phthalocyanine green in 14 works (Table 12.1).

Amongst the green pigments, viridian ($Cr_2O_3 \cdot 2H_2O$), volkonskoite ($Ca_{0.3}(Cr,Mg,Fe)_2\ (Si,Al)_4O_{10})(OH)_2 \cdot 4H_2O$) [1, pp. 73–75, 3, p. 392], cobalt green, and chrome green (Cr_2O_3) were often identified in the paintings. Emerald green ($Cu(CH_3COO)_2 \cdot 3Cu(AsO_2)_2$) and glauconite ($(K,Na)(Mg,Fe^{2+},Fe^{3+})(Fe^{3+},Al)(Si,Al)_4O_{10}(OH)_2$) were infrequently detected (Table 12.1). In Europe, emerald green was used in painting until the 1960s despite its toxicity [5, pp. 219–271]. The pigment was identified in three works, including the painting by Leontii Denisov *Karelia,* 1974 (Fig. 12.4). For this work, more than one green pigment was identified in samples from different areas of the paint layers, including viridian, cobalt green, chrome green, and phthalocyanine green (Table 12.1). Since there is no information about emerald green production in the USSR, it can be concluded that it was obtained from an external source.

Pigment Green B was also a new pigment for the USSR paint industry. In Europe, the pigment was used in the manufacture of water-soluble paints in the 1930s, and in the early 1950s it was replaced by a phthalocyanine pigment. In the USSR it first appeared in 1955 in watercolour paints (mixture of Pigment Green B and blanc fixe in a ratio of 1:1) and in 1971 as an oil paint *Viridonovaia Green* [1, pp. 83–85] (Table 12.2). Pigment Green B is an extremely rare pigment in the Soviet artists' palette – it was spotted only in two works by Aleksei Ratnikov: *Us't-Ilim. Mira Avenue* and *Bratsk Hydroelectric Power Station* painted in 1975 (Table 12.1).

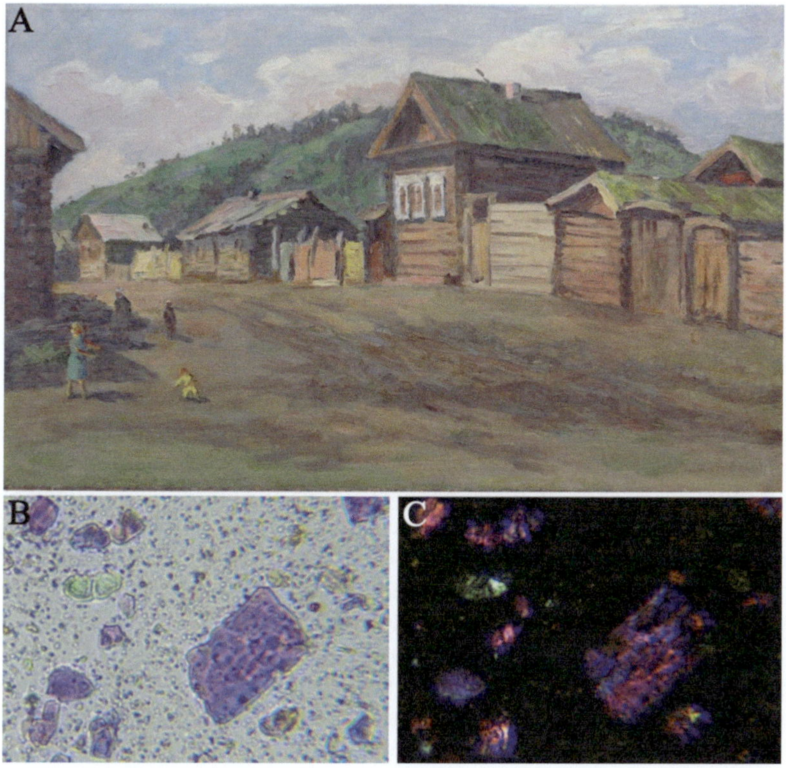

Fig. 12.4 (**a**) Leontii Denisov *Karelia,* 1974, oil on canvas, 38 × 63 cm, Solnechnogorsk, The museum and an exhibition centre *Putevoi dvorets* (En Route Imperial Palace); particles of emerald green and cobalt phosphate from the violet area of the painting: (**b**) transmitted light and (**c**) the same specimen between crossed polars

Violet pigments are represented very poorly in the paintings examined. Among them cobalt phosphate ($Co_3(PO_4)_2$) was detected in four works, manganese phosphate ($MnNH_4P_2O_7$) and purple organic pigment[2] were found only once. Surprisingly, cobalt ammonium phosphate, a new for the Soviet industry pigment, was more popular than other violet pigments. It was identified in nine works after 1955, however the official production of the pigment in the USSR started in 1966 [1, pp. 99–101] (Tables 12.1 and 12.2).

The common red pigments of Soviet artists were red organic pigments,[3] cadmium sulfoselenide red (CdS·CdSe), and Mars red and orange (iron oxide). Vermilion (HgS) and chrome orange ($PbCrO_4$·PbO) were used less frequently than others (Table 12.1).

[2] The exact chemical composition was not identified.

[3] The exact chemical compositions were not identified.

Soviet artists used a variety of yellow pigments, such as strontium yellow ($SrCrO_4$), cadmium yellow (CdS), yellow organic pigments[4], Mars yellow, and yellow ochre. Chrome yellow, which was very popular in the first half of the twentieth century and zinc yellow ($ZnCrO_4$) were detected infrequently (Table 12.1).

12.4 Summary

Socialist realist artists had a preference for traditional pigments. These included zinc white, ultramarine, cobalt blue, viridian, cobalt green, volkonskoite, chrome green, cadmium sulfoselenide red, strontium yellow, cadmium yellow, various ochres and Mars pigments. A number of pigments, which are more characteristic for use in the first half of the twentieth century, such as emerald green, chrome orange, vermillion, chrome yellow, and zinc yellow were found infrequently.

New pigments, such as titanium white (anatase), manganese blue, cobalt ammonium phosphate, cobalt blue spectral, blue and green phthalocyanines, and Pigment Green B had gradually emerged in the Soviet artists' pallet in the period 1950s – early 1970s. However, research shows that some of these pigments were not very common. Probably, their rare application is consistent with a conviction of Soviet specialists that the paints were of low quality.

Comparison of the obtained results with the published archival data allowed the authors to determine rare cases when Soviet artists used certain paints prior the implementation of their full-scale production in the USSR, and therefore – manufactured elsewhere.

References

1. Grenberg Yu.I., Pisareva S.A. (2010) Maslianye kraski XX veka i expertiza proizvedenii zhivopisi. Sostav, otkrytie, kommercheskoe proizvodstvo i issledovanie krasok (Oil paints of the XX century and examination of paintings. Composition, invention, commercial production and research of paints). Zerkalo mira, Moskva. S. 194.
2. Belenkii E.F., Riskin I.V. (1974) Khimiia i tekhnologiia pigmentov (Chemistry and technology of pigments). Khimiia, Leningrad. S. 761.
3. Eastaugh N., Walsh V., Chaplin T., and Siddall R. (2004) The Pigment Compendium. A Dictionary of Historical Pigments. Elsevier, Oxford. P. 499.
4. Keijezer M.De (2002) The history of modern synthetic inorganic and organic artists'· pigments discovered in the twentieth century. In: Mosk J.A., Tennent N.H. (eds.) Contribution to Conservation Research from the ICN. James & James, London, pp 42–54.
5. FitzHugh E.W. (ed.) (1997) Artists' Pigments. A handbook of their history and characteristics, vol 3. National Gallery of Art, Washington. P. 364.
6. Kadikova I.F., and Grigorieva I.A. (2017) Investigation of multilayer painting by means of molecular spectroscopy. EPJ Web of Conferences (132): Art. No 03015.
7. Grenberg Yu.I., Pisareva S.A., and Kadikova I.F. (2017) Anatomy of Russian Avant-garde. View from the Lab. Tri Quadrata, Moscow. P. 292.

8. Vinner A.V. (1950) Materialy maslianoy zhivopisi (The art materials of oil painting). Iskustvo, Moskva. S. 523.
9. Tyutyunnik V.V. (2009) Materialy i tekhika zhivopisi (Materials and technology of paintings). V. Shevchuk, Moskva. S. 214.
10. Pasteris J.D. (1998) The laser Raman microprobe as a tool for the economic geology. In: McKibben M.A., Shanks W.C., and Ridley W.I. (eds.) Application of microanalytical techniques to understanding mineralizing processes, Reviews in economic geology, Society of Economic Geologists, Socorro, pp. 233–250.
11. Lafuente B., Downs R.T., Yang H., and Stone N. (2015) The power of databases: the RRUFF project. In: Armbruster T., and Danisi R.M. (eds) Highlights in Mineralogical Crystallography. W. De Gruyter, Berlin, Germany, pp. 1–30.
12. Pisareva S.A. (2005) A note on cobalt spectral. Studies in Conservation (50):190–192.

Chapter 13
A View Through the Meander: A Conservation Study of Knifer's Oil Paintings from the 1960s

Mirta Pavic and Martina Munivrana

13.1 Introduction

Julije Knifer (1924–2004) was one of the most prominent Croatian contemporary artists, whose unconventional approach to painting marked the second half of the twentieth century. When he said, "I have probably already made my final paintings, but possibly I haven't made my first paintings yet" [5], Knifer expressed the very essence of his art; one to which conventions did not apply, both in terms of the perception of his work as such and in the approach to the preservation of the artistic material itself. Knifer's entire oevre was characterized by variations on a single idea. This idea is very precise and strictly defined, and does not allow for superficiality or improvisation, in the processes of either creation or subsequent conservation, the latter being unavoidable given the reality of transience.

Julije Knifer's Meanders, named after the motif with which the artist continued to play using various techniques, are a first-rate part of the collection of the Museum of Contemporary Art in Zagreb (MSU), and appear in the collections of the greatest museums in the world. Knifer had been painting them in oil until the late 1960s; he later turned to acrylics, with which it was easier to achieve the desired intensity of black and white.

> The colours were more or less converging into gray, until there was only black in the end. Within a few months I had reached the end, so to speak, i.e. a black and white painting... the meander. [3]

Still, his first oil paintings of meanders are interesting not only for their historic and artistic roles, but also because of their physical characteristics, which greatly affect their conservation, preservation and presentation. By analyzing Knifer's oil

M. Pavic (✉) · M. Munivrana
Museum of Contemporary Art, Zagreb, Croatia
e-mail: mirta.pavic@msu.hr

K. J. van den Berg et al. (eds.), *Conservation of Modern Oil Paintings*,
https://doi.org/10.1007/978-3-030-19254-9_13

paintings through their material aspect, one can learn many details and facts that greatly contribute to the understanding and interpretation of his art.

Knifer created his Meanders with many thin layers of paint, most often in only two chromatic extremities – black and white – and without varnish, which makes them extremely sensitive to external influences [4]. This sensitivity is particularly characteristic of white surfaces which were not painted with classic oil paint like the black ones, but were made with a traditional chalk-glue ground to which a white pigment, usually zinc white, was added. This fact explains the gradual natural darkening of the white zones and the high sensitivity to moisture which complicates all phases of the conservation process.

The five most important characteristics of all the selected oil paintings by Knifer from the period 1959–1970 are:

- numerous thin paint layers
- generally two chromatic extremities: black and white
- black and white surfaces painted with oil and traditional chalk-glue ground
- no varnish
- Knifer's meditative repetition of the process has a wider conceptual context

In accordance with these characteristics, the following arise in the conservation process:

- sensitive and fragile surfaces
- high sensitivity to moisture
- white surfaces react faster to external influences, and are hygrophilic
- every painting, despite identical techniques and technologies, responds differently during the conservation treatment, demanding an individual approach
- challenging procedures for the removal of overpaint and varnish

Due to their sensitivity and age, a number of changes are already visible in most of the paintings. We shall describe four examples of conservation, referring to several different problems: removing surface dirt, unwanted overpaint and varnish.

13.2 Conservation Methods Applied to Selected Paintings

Even though oil painting is one of the oldest techniques and artistic expressions that museum experts encounter, contemporary oil paintings are constantly raising new issues and therefore introducing new conservation methodologies and materials into practice. The oil paintings by Julije Knifer are an inexhaustible source of new information, but they also require extreme caution because no two paintings from the same period will respond identically to the methods applied. This is why every problem has to be approached as if we are facing it for the first time. However, certain steps have proved to be adequate for each of the four selected paintings.

The white surfaces on most of Knifer's paintings display minute losses of paint on the ends of the warp and weft of the canvas that the human eye perceives as impurities. The thin layer of paint is also porous. Cleaning tests have shown that for each painting a different system is effective.

Aquazol 500 yielded largely good results in local consolidation. Even though it is applied as a water-based solution, its precise and controlled introduction with a thin brush to the target areas does not jeopardize the painted layer. The consolidation of the entire surface is a more complex problem due to its high sensitivity to moisture, but also to most solvents.

Paintings which need to be removed from the stretcher to be straightened are most commonly applied to a working frame using the strip lining method, allowing for safer handling and a constant tension in the canvas. This method was applied to all four of the selected paintings.

Most losses in the painted layer caused by mechanical damage were filled with a poly(vinyl alcohol) putty; 7% Mowiol 4-88 [2] solution in distilled water with Bologna chalk as a filler. Its advantages are the ability to adjust viscosity, ease of application and molding, flexibility and reversibility. The retouching was done with pigments, and the binder was selected according to the intensity of the gloss of the original paint surface.

A comparison of the cleaning methods was carried out on two paintings created within a five-year period: *M19* (1965) and *Untitled* (1960–1961). Nevertheless, they remain different and the same method was not effective for both paintings.

13.3 Methods of Removing Dirt, Overpaint and Varnish

13.3.1 Removing Dirt

The painting **M19** from 1965 (oil/canvas 48 × 75.3 cm) (Fig. 13.1) was found to be in very poor condition. Although various types of damage were present, we shall focus on the cleaning phase because the surface of the painting was virtually

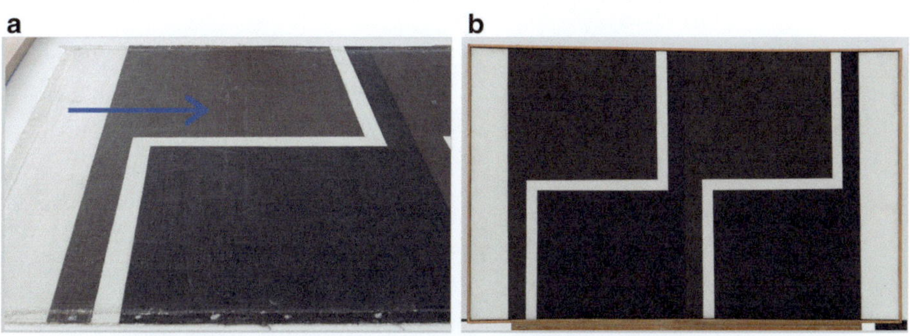

Fig. 13.1 Julije Knifer, ***M19,*** 1965, oil on canvas, 48 × 75.3 cm; (**a**) detail: surface dirt and traces of a fluid as indicated by the arrow (the painting stretched onto a strainer), (**b**) after the conservation treatment

unrecognizable due to various stains and traces of a fluid. The first step was to establish the pH of the original surface and to provide tests while controlling factors such as pH, conductivity [7] and method of application. The original pH of all the painted areas is neutral (pH 7). The entire surface was first dry cleaned with No-Smear drafting powder and a Magic Rub eraser. The tests have shown that wet cleaning has to be applied with different products on dark and light surfaces. Dirt was removed from the black and brown painted layers using a gel based on a non-polar solvent (100 ml Shellsol T, 2 g Carbopol, 20 ml Ethomen C-12). The white layer, a chalk-glue ground with zinc white, was cleaned using a Pemulen gel (pH 7), but it was applied through a membrane of Shellsol T to prevent the gel penetrating the layers of the painting. Although the advantage of gels is in their consistency, the chalk-glue ground's hygroscopicity and easily soluble binders needed further protection from moisture.

After filling up the damaged parts with 7% Mowiol putty, the reconstitution of the painted layer was performed with pigments bound with a 5% Klucel EF [2] solution in distilled water. Klucel creates a matte coat that integrates into the original painted layer.

The painting **Untitled** from 1960–1961 (oil/canvas, 48 × 65 cm) (Fig. 13.2) required an entirely different cleaning method. The condition of the painting demanded complete conservation: straightening, consolidation, filling up of the damaged parts in the ground layer and colour, cleaning and retouching. Due to the risk of dissolving the white painted layer, the painting was consolidated with an 8% solution of Plexisol P 550 in the non-polar solvent Shellsol T, with a small addition of aromatic Shellsol A, which facilitated the homogeneous decomposition of the resin in the solvent.

Although many of Knifer's oil paintings from the same period have been successfully cleaned with microemulsions and gels [2], these cleaning systems did not

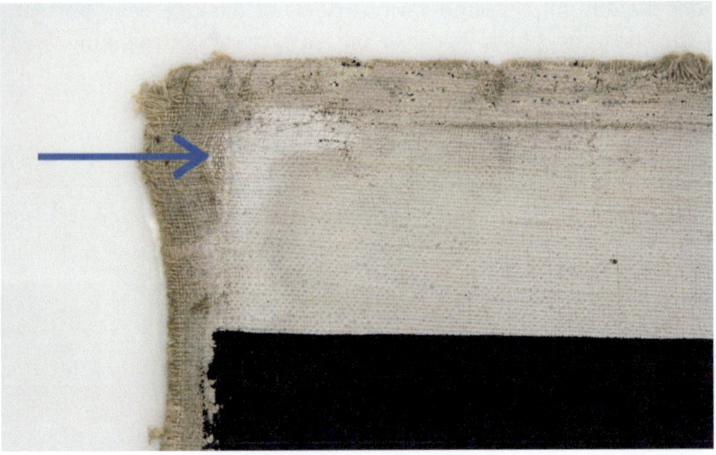

Fig. 13.2 Julije Knifer, ***Untitled*****,** 1960–1961, oil on canvas, 48 × 65 cm; detail: area of cleaning test

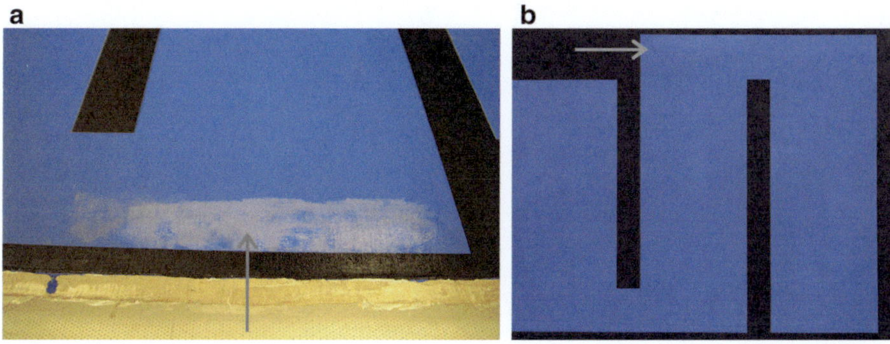

Fig. 13.3 **Julije Knifer,** ***MNB,*** 1970, emulsion on canvas, 89.7 × 109.6 cm; (**a**) found condition: overpainted area (indicated by the arrow), (**b**) after the conservation treatment (treated overapainted area indicated by the arrow)

yield satisfactory results with this painting. Interestingly, neither agar gel [1] nor xanthan gel yielded good results. Agar, even with the addition of surfactants, could not remove the dirt from the surface of the painting. Xanthan gel was too moist for the sensitive and hygroscopic white surface. Furthermore, agar was unsuccessful, leaving behind a faint trace on the matte black surface [8].

The pH of the original paint layer was 8, and tests have shown that the impurity was successfully eliminated with a pH neutral solution of citric acid in distilled water [10], which was applied indirectly to the surface previously coated with Velvesil silicone gel to make a water-tight barrier. In this way, the solution picked up dirt without moisture penetrating the layers of the painting.

13.3.2 Removing Overpaint

The painting **MNB** from 1970 (emulsion/canvas, 89.7 × 109.6 cm) (Fig. 13.3) had not been exhibited for many years due to the overpaint applied in a prior intervention, which had impaired the perception of the monochrome surface. The overpaint had been applied to the upper left edge of the painting to cover damage in the original paint layer.

Although the museum documentation lists the technique as acrylic on canvas, FTIR analysis of the original colour sample showed that it is actually an emulsion. The blue paint exhibited very strong positive reactions for both a protein (+AlexaFluor 488) and an oil component (+RHOB). The FTIR spectrum of the blue paint also suggests an emulsion composition (a blend of oil and protein) rather than an acrylic binding system. This may explain the water sensitivity observed during

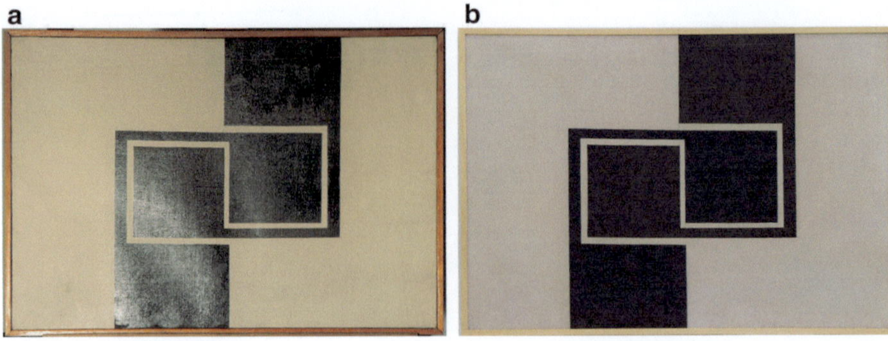

Fig. 13.4 **Julije Knifer,** ***Meander 20,*** 1965, oil on canvas, 48.4 × 69 cm; (**a**) photo of found condition under raking light, (**b**) after the conservation treatment with the new frame

the cleaning tests. The ground also exhibited a similar staining pattern (positive for both protein and oil).[1]

A gel was prepared according to a recipe by Professor Richard Wolbers, and the amounts of individual components were adjusted through testing until the appropriate ratio was reached. A gel composed of 35 ml of Xylene, 7.5 ml of benzyl alcohol, 9.5 ml of Shellsol T, 1 g of Carbopol and 10 ml of Ethomeen C12 was gradually applied, and the overpaint was slowly removed. In determining the composition of the overpaint, FTIR spectroscopy[2] showed the presence of an alkyd resin in the sample.

The retouching was performed as before, with a blue pigment[3] which Knifer experimented with in an attempt to get the Yves Klein Blue.

13.3.3 Removing Varnish

The layer of varnish found on the 1965 painting **Meander 20** (oil/canvas, 48.4 × 69 cm) (Fig. 13.4) is probably the result of a previous unprofessional intervention. Observing the stratigraphic layers of Knifer's paintings reveals that a finishing layer of varnish was not originally present on them. FTIR spectroscopy[4] showed that it is a ketone resin-based varnish, whose composition is much stronger than the composition of the paint, and requires aromatic solvents to dissolve. It was removed with an emulsion of xanthan gel with the addition of a 5% xylene and benzyl alcohol mixture, which enhanced its ability to disintegrate and remove the

[1] From the report by Prof. Richard Wolbers, who carried out the analysis.

[2] The analysis was conducted at the Natural Science Laboratory the of the Croatian Conservation Institute in Zagreb by Dr. Domagoj Mudronja.

[3] Kremer 45,010 Ultramarinblau dunkel and 45,020 ultramarinblau rotstichig.

[4] The analysis was conducted at the Naturwissenschaften und Technologie in der Kunst, Akademie der Bildenden Künste, Wien by Dr. Dubravka Jembrih-Simbürger.

varnish from the surface. The penetration of the varnish into the layers of the painting was further prevented by a barrier of a biodegradable silicone D5 solvent applied to the surface prior to the application of the gel. The painting was retouched with pigments[5] bound with a 5% Klucel EF solution, under a stereo microscope, with thin retouch brushes to precisely fill only those areas where the colour was missing. Such damage is almost impossible to make absolutely invisible, which means that when exhibiting **Meander 20**, as well as **MNB**, diffuse, properly selected lighting is important. This can greatly assist the presentation of the paintings to the public in a way that will not alter the visual appearance of the original surface.

13.4 Concluding Thoughts

This research demonstrates that every painting responds to physical interventions into its structure in its own specific way. Works by the same artist will rarely respond to treatments in the same way; in the case of Knifer, the only common feature is the paintings' unpredictability. One needs to approach each of them as if one has never worked on a painting by the same artist before. It is also important to mention green chemistry, which has been increasingly discussed of late. The tendency is to replace solvents and similar ingredients with biodegradable and less harmful ones. Leading conservation scientists are trying to turn to natural ingredients and water-based methods. In encountering paintings which raise questions, as the oil paintings by Knifer do, it is not easy to discard materials which still cannot be adequately replaced in specific situations such as these. This is why Richard Wolbers explains that basic Carbopol/Ethomeen solvent gels can increase dwell time, slow down solvent evaporation, and carry out other functions [9].

As challenges arise, so the competencies of conservators increase through trial and error; but they nevertheless yield useful conclusions. With Knifer's oil paintings, careful research, testing and cautious action have been conducted with the aim of maintaining their materiality and presentability. These procedures are intended to protect the physical structure as well as the significance of Knifer's Meanders. Although Knifer himself stated that, if you look at them a little closer, his paintings are not really perfect,[6] the impression they leave on the observer is one of purity and infinity. It is this quality and this message that we wish to preserve, but harmonizing ethical and aesthetic standards is not always easy. By looking at his oil paintings at several levels – visual appearance, inner structure and artist's message – we tried to find the best solution for each one individually. In a controlled museum environment, this challenge may be easier to overcome, especially when the decisions are made by a team.

---

[5] Kremer 47,000 Rebschwarz, 48,420 Eisenoxidschwarz 306, 47,150 Elfenbeinschwarz and 41,000 Kasslerbraun

[6] Documentary by Dunja Blažević "Pet umjetnika i Gorgona" (Five artists and Gorgona) 1986, Radio Television of Belgrade (RTB).

References

1. Cremonesi P (2016) Surface cleaning? Yes, freshly grated Agar gel, please. In: Studies in conservation, vol 61, no 6, IIC, Lindon, pp 362–367
2. Hill Stoner J, Rushfield R eds. (2012) Conservation of Easel Paintings, Routledge, London and New York, pp 248–250, 260–263, 516–523, 532–548, 604–606, 628–634, p 624, 626
3. Knifer J (1983) Zapisi, In: Život umjetnosti, (Writings, In: Life of art publication) No. 35, Zagreb, pp 28–34
4. Burnstok A, van den Berg KJ, de Grool S, Wijnberg L (2007) An Investigation of Water-Sensitive Oil Paints in Twentieth-Century Paintings. In: Modern Paints Uncovered, proceedings from the Modern Paints Uncovered Symposium. Getty Conservation Institute, Los Angeles, http://hdl.handle.net/10020/gci_pubs/paints_uncovered, pp 177–188
5. Maković Z (2002) Knifer. Meandar & Studio Rašić, Zagreb
6. Matičević D (2011) Prilozi periodizaciji opusa Julija Knifera, In: Suvremena umjetnička praksa Ogledi 1971–1993, (Contributions to the Periodization of Julije Knifer oeuvre, In: Contemporary Art Practice Assays 1971–1993). Diurieux/Muzej suvremene umjetnosti/ Hrvatska sekcija AICA, Zagreb, pp 125–143
7. Soldano A, van den Berg JK (2014) Investigation into the Surface Conductivity of Wather-Sensitive Modern Oil Paints. In: Issues in Contemporary Oil Paint, proceedings from the international conference ICOP 2013, held in Amersfoort. Springer, Switzerland, pp 407–417
8. Volk A, van den Berg JK (2014) Agar – A New Tool for the Surface leaning of Water Sensitive Oil Paint. In: Issues in Contemporary Oil Paint, proceedings from the international conference ICOP 2013, held in Amersfoort. Springer, Switzerland, pp 389–406
9. Wolbers R (2017) Gels, green chemistry, gurus and guides. In: Gels in the Conservation, proceedings from the international conference 2017, held in London, Archetype Publications, London, pp 3–8
10. Wolbers R (2000) Cleaning painted surfaces, Aqueous Methods. Archetype Publications, London

Chapter 14
Preservation of Russian Abstract Art of the Second Part of the Twentieth Century

Anastasia V. Yurovetskaya, Maria S. Churakova, and Irina Kadikova

14.1 Introduction

Recently the Russian museum and heritage scientific communities have turned their attention to the abstract paintings of the second part of the twentieth century. A prominent part is formed by theoretical and artistic works of group organized by Soviet artist Ely Bielutin in 1950s.

After graduating from the Fine Arts faculty of the Moscow Pedagogical Institute, where he studied under Aristarkh Lentulov, Nikolay Kuznetsov and Lev Bruni, Bielutin started developing and imparting his *Theory of Universal Contact* [8] to students. He formulated its main laws as an example of generalization of universal notions. The first law is the law of equilibrium of the emotional supraconsciousness[1] as an intuitive component, the highest degree of personal psychic activity. The second law of trusting the intellectual supraconsciousness explains that a person is turning to art motivated by the aim to establish an emotional and intellectual connection with another person. The third law of creative potential states that each man has a character and individuality, just like any particle has its own properties. Ely Bielutin understood how to affect the emotional sphere aesthetically by means of a metalanguage, making the purified feeling the subject of a portraiture [6].

Having a developed theoretical platform puts the *New Reality* group closer to avant-garde art of the first part of the twentieth century than to contemporary artistic movements in the Soviet Union, that often confined their mission statements just to manifestos (e.g. *the New Realism*). Followers of Bielutin were active since 1950s till 2000s, forming a group, whose main aim was to express their creativity with the help of abstract artistic vision. But as they still had some thingness in their works

[1] https://en.oxforddictionaries.com/definition/supraconsciousness, accessed February 17, 2019.

A. V. Yurovetskaya (✉) · M. S. Churakova · I. Kadikova
The State Research Institute for Restoration (GOSNIIR), Moscow, Russian Federation

K. J. van den Berg et al. (eds.), *Conservation of Modern Oil Paintings*,
https://doi.org/10.1007/978-3-030-19254-9_14

the New Reality artists stayed aside from abstract expressionism or lyrical abstraction, even though their expressive means had much in common.

The historical background of the *New Reality* movement is discussed in this paper, together with some insight into the, often inferior, materials' use of the artists. The problems related to the use of materials and/or techniques are illustrated within the context of a research and conservation project of *Composition* (1962) by Lucian Gribkov.

14.2 Historical Context

Bielutin's theories that propose the existence of artistic consciousness in every person were very popular among young painters and graphic artists. As a result, at the very beginning of 1960s his studio tallied roughly 300 students [6]. New ideas and expressive means of Ely Bielutin and his students initially totally belonged to the spirit of the *Thaw* period of Soviet culture, but very soon they had to meet with the misoneism and imperception of the authorities. On the first of December 1962, Nikita Khrushchev, Chairman of the Council of Ministers of the Soviet Union, was taken to the Manege Gallery to view an exhibition "30 Years of the Moscow Union of Artists" which included a number of avant-garde and abstract works next to official art of representationalism. On seeing the latter, Khrushchev exploded with anger, proclaiming that "*a donkey could smear a better art than this with its tail*" [12]. Members of the Soviet government lost their temper and would shout at the artists that they had to be arrested or even executed [11]. One of the members of the studio *the New Reality,* Leonid Rabichev describes this moment in his memoirs as a very impressive one. In the end no punitive measures were taken other than public and official censure. After this event, and for the next 30 years, government officials didn't express anything but disparagement of abstract art.

The artists who stayed devoted to their teacher Ely Bielutin and his theories formed the core of the studio *the New Reality* (Fig. 14.1). Being an underground part of Soviet culture they never had a chance to have official exhibitions or be published in books or magazines. Creating their art pieces for self-expression, they were indifferent to the preservation of their works. Some of them were kept in their studios, but mostly the art works were stored in lumber rooms or upper-storeys of their summer houses. Some were literally kept out-of-doors, suffering from rain and snow.

In the past few years specialists from the State Research Institute for Restoration (GOSNIIR) started to collaborate with Russian Abstract Art Foundation. The members of this foundation strive to arrange appropriate storages for the art objects and work on their introduction into scientific discourse. The collection of the foundation comprises about 1000 paintings and works on paper, sculptures and installations by artist members of *The New Reality* studio at different stages of their artistic careers. Working as a team, collaborating scientists -and conservators of GOSNIIR support exhibition activities of the foundation.

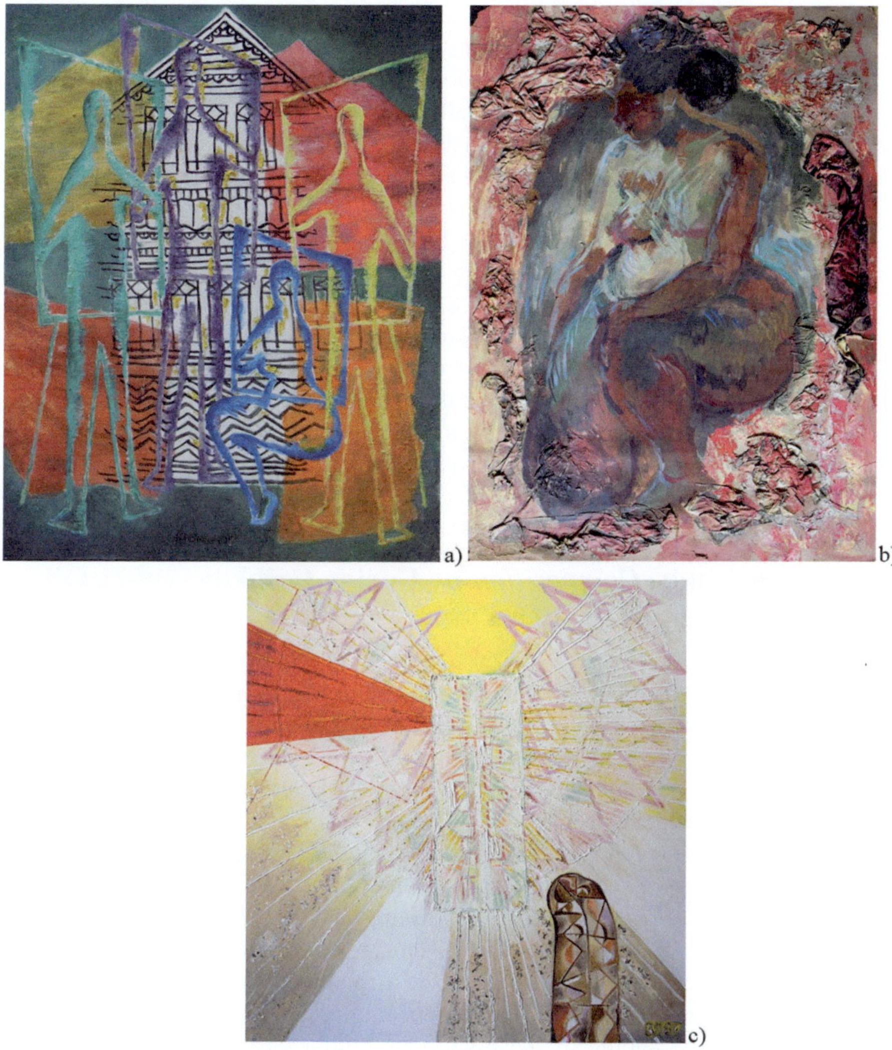

Fig. 14.1 Paintings by members of *the New Reality Studio*: (**a**) Natalia Levyant. Summer in Abramtsevo (Leto v Abramtsevo). 1977. Oil on canvas. 90 × 70 cm (**b**) Anatolij Safokhin. Nude (Obnazhennaya). 1961. Mixed technique on cardboard. 97 × 70 cm. (**c**) Vera Preobrazhenskaya. Annunciation (Blagoveshenie). 1997. Oil on canvas. 125 × 115 cm

14.3 Art Technique

At the present time, as paintings and installations, created by members of *The New Reality* group, are becoming a part of state museums and private collections, conservators face serious challenges treating them. Aside from very bad storage conditions in the past, preservation of abstract pieces of art depends a lot on the painting materials and techniques that were used. As each work was created as a purely artistic and emotional experiment, the variety of materials can really be impressive. Fibreboard, canvas, cardboard, paper, wood or even foam plastic could be used as a support. And the primer layer, if used at all, could be covered by oil, pentaoil[2] [1, 7] or acrylic paints, tempera, water colour, gouache, encaustic and drawing ink. Furthermore, many different materials may have been mixed and combined in one object depending on the artist's idea. Most pictures were not varnished.

As mentioned by *New Reality* artist Inessa Ryabinina, the quality of artistic materials was far from perfect.[3] Working together as a big group in Ely Bielutin's house or workshop, artists of *the New Reality* studio always needed large quantities of painting materials, especially when working on large scale projects. As professional artists' colors of high quality were luxury goods in the Soviet period, it was absolutely impossible to get them in the desired quantity, so often industrial oil or even alkyd paints would be used. Solvents and mediums were also chosen on the basis of their availability and affordability, but not their properties.

14.4 Case Study. *Composition* by Lucian Gribkov (1962). Oil on Canvas, 173 × 70 cm

When the Russian Abstract Art Foundation asked GOSNIIR for a conservation treatment of the painting *Composition* (Kompoziciya) by Lucian Gribkov (Fig. 14.2), conservators first proposed to perform technical examination and research in order to learn as much as possible about the technique of the art work.

The painting was created in a very expressive manner. The texture of painting layers varies from very thin, almost transparent, to very thick, formed with spatula. An interesting effect of shrinkage of the paint created a very specific relief. This effect can also be seen on some other paintings of that time by the artist (e.g. *Composition.* 1969. Oil on canvas. 100 × 83 cm, private collection, Moscow). Inessa Ryabinina confirmed the supposition that the artist was using paint shrinkage as a decorative technique.[4] He achieved this beautiful impression in certain pastose areas of white oil paint by adding of siccative to the colour paste. The picture was not varnished, its surface was covered with a thick layer of dust, oil exudations can be seen in the areas with a shrunk paint (Fig. 14.3).

[2] Pentaoil colours were firstly suggested by Soviet paint manufactures in 1959 to improve the paints' quality. The paints on the basis of this medium were first produced in 1963. In contrast to alkyd medium, pentaoil is medium based only on pentaerythritol esters of fatty acids of linseed oil and doesn't contain any synthetical modification including styrene, vinyl toluene, etc. The paints on the basis of alkyd resins were firstly introduced to Soviet paint market only in 1976 [1].

[3] Personal communication of AVY with *New Reality* artist Inessa Ryanbinina, August 2016.

[4] Personal communication with AVY, August, 2016 and January 2019.

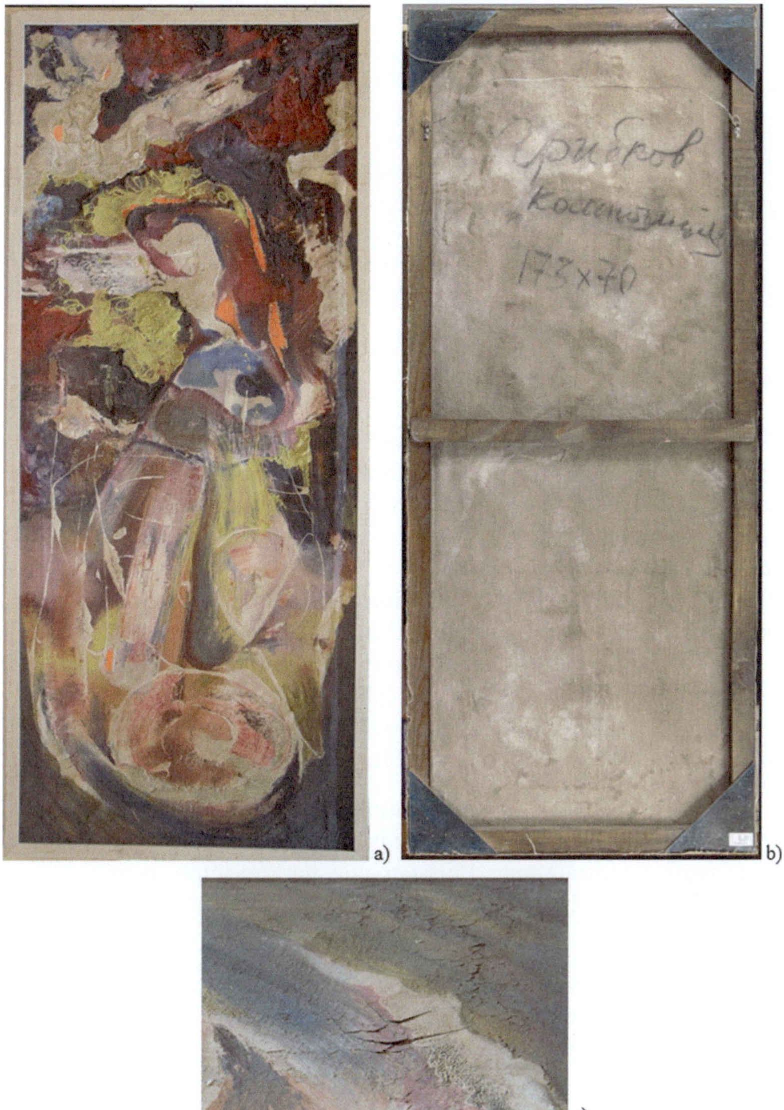

Fig. 14.2 *Composition* by Lucian Gribkov (1962). Oil on canvas. 173 × 70 cm. Front (**a**) and back (**b**) sides before treatment; (**c**) fragment with flaking oil paint

Paint samples taken from the painting were studied with[5] an optical microscope (MSP-1, LOMO) and examined using polarizing microscopy (POLAM L-213 M microscope, LOMO). Scanning electron microscope (JSM 5610LV, JEOL) with energy-dispersive X-ray spectroscopy (SEM-EDX) were used to determine the

[5] See for a description of the techniques Kadikova and Grigorieva [5].

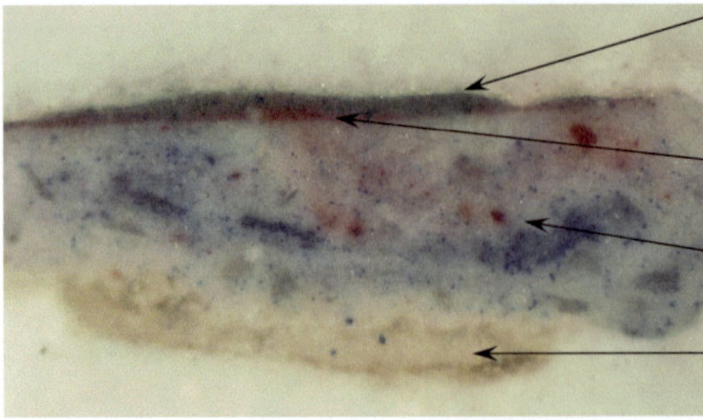

Fig. 14.3 *Composition* by Lucian Gribkov (1962). Oil on canvas, 173 × 70 cm. Cross-section in reflected polarized light, sea-green, the upper right corner

elemental composition of pigments. Analysis of the binder medium, as well as a number of pigments and fillers, was performed by using micro- Fourier transform infrared (μ-FTIR) spectroscopy (LUMOS, Bruker). The molecular and phase composition of pigments was studied by means of BRAVO Handheld Raman Spectrometer (Bruker).

Stratigraphic study showed that the painting is multilayered. Gribkov used a self-made priming of oil and natural chalk that demonstrates a good adhesion to the canvas support in contrast to the adhesion between the paint layers and priming, which is not too strong and lead to the local flacking of the paint. Clots of oil were found between the canvas and the priming layer.

Composition was created with quite a limited number of colours:

White: zinc white with talc and barium sulphate;
Dark red: red organic pigment, mars brown, talc;
Dark blue: synthetic ultramarine, Prussian blue, barium sulphate, talc;
Light brown: yellow ochre, mars orange, lead white, synthetic ultramarine, zinc white, strontium yellow.

Although the painting surface had specific relief and varying gloss, analyses showed that oil was the binding medium. Unfortunately, μ-FTIR and EDX did not provide identification of the siccative that was used by the artist due to the very low quantities.

14.5 Conservation Issues

The painting arrived in the conservation studio of GOSNIIR in a poor condition and in need of consolidation, strip lining, stretching on a new pine frame and surface cleaning. It contained areas with flaking paint, strains along the frame, both the front and the back side of the picture were covered with thick layer of dust. The stretcher had edges fastened with nails and very thin wood strips that distorted as a result of its weakness and big and unbalanced load of canvas with prime and color paste. The painting looked as if it had never experienced conservation before, probably even including dusting.

Russian traditional conservation methods include protein glue (e.g. sturgeon glue, rabbit skin glue) for consolidation and surfactants or ox bile solution for removing soil from the surface of the paint layer. But these methods were developed decades ago for treating pictures created in classical painting techniques. However, Russian abstract art of the second part of the twentieth century is drastically different in all ways and needs alternative remedial conservation approaches.

As already mentioned, an art piece created in Bielutin's studio can contain various materials at the same time. An example is a painting by Vladislav Zubarev, *Landscape. Thickets* (*Pejzazh. Zarosli*), that was created with gouache colors on the paper saturated with wax. But even if the painting technique is relatively straightforward, resembling more classical oil on canvas, traditional conservation methods might still be inappropriate. Throughout the twentieth century, artistic materials have survived a lot of changes, connected with the development of technologies or efforts to cost-cut the manufacturing, many natural mediums were replaced by synthetic ones. Mirta Pavić [9] turns conservators' attention to the problem of treating *objects in traditional or similar technique,* emphasizing that "*days when oil on canvas led to known and tested conservation treatments are long gone, despite of the fact that the technique has been known for centuries*".

Structural conservation of contemporary pieces of art usually implies usage of either water- or alcohol-based synthetic adhesives for consolidation [4]. But the choice of cleaning methods can be much more complicated. As a matter of fact the decision making in this case is absolutely impossible without technological examinations. Since most paintings of *the New Reality* group are not varnished, knowing the binding media of the paint layers are key to inform the choice of a cleaning formulation and application method. Furthermore, since sensitivity to water and the brittleness of the paint layer may play an important role, careful examination and testing is an essential part of the preparation of a cleaning intervention.

Treatment of *Composition* started with consolidation in the areas with lifted paint surface. As an adhesive for the flaking paint conservators used 10% AK211 (copolymer of butyl acrylate with methyl methacrylate) in isopropanol. Alcohol made the brittle painting layer a little softer and more flexible which made the consolidation process more effective. Strip lining was performed using Beva®371 film and the picture was stretched on a new pine wood stretcher, that replaced the old one, condemned as defective because it had deformations and rigidly fixed wooden strips.

On stretching the painting, the subject of surface cleaning came to the fore. Here, the key problem was that the relief varied on different parts of the art work. Series of cleaning tests were established in order to find an appropriate cleaning approach. It was found that while even ingrained dust could be gently removed with distilled water, this could only be done with cotton swabs on a very limited part of the surface (Fig. 14.4).

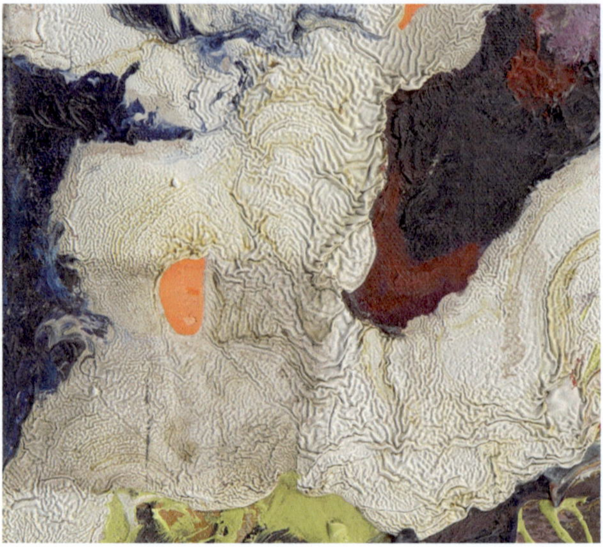

Fig. 14.4 Removal of surface dirt on a pastose area. *Composition* by Lucian Gribkov (detail). Oil on canvas. 173 × 70 cm

Based on tests with different types of sponges [2, 10], it was decided to clean the most prominent parts of the paint surface with a latex-free high density polyurethane sponge produced by Deffner&Johann (Germany). The deepest and the hardest to access parts of the impasto were carefully treated with a bristle brush after slightly loosening the dirt with water. The process was then finished by wiping with a sponge for the second time in order to absorb the dissolved dirt. During the surface cleaning process it turned out that on some areas the soil was more resistant. In these areas conservators would use a small amount of Ph-neutral soap dissolved in distilled water, which was further rinsed. It was decided to leave the medium exudations on the surface [3].

As a result of the treatment conservators of GOSNIIR returned the brightness of colors to the picture so that it could be displayed during the next exhibition arranged by Russian Abstract Art Foundation.

14.6 Conclusion

Painters, who belonged to the underground part of Soviet culture and were rejected by the official artistic community, are rapidly becoming popular in Russia at present time. Their art works are displayed in state museums and respectable galleries, art connoisseurs are competing to get them for their collections. Painting conservators are more and more often coming across a problem of treating pictures created in complex mixed techniques, including works of Russian abstract art of the second part of the twentieth century. In such situations choosing a conservation methodology becomes a serious challenge. The case study of cleaning *Composition* by Lucian Gribkov provides conservators with useful experience of working with different reliefs of the paint layer. It shows that it's not only the cleaning agent itself, but also the method of its application and further rinsing that can become a key to successful cleaning of surfaces of paintings by the *New Reality* group.

Acknowledgements The authors thank Marina Voronina and Elena Iurovetskaia from the State Research Institute for Restoration and Anna Karganova from Russian Abstract Art Foundation for help and support in practical and research stages of working on the project.

References

1. Grenberg Y I , Pisareva S A (2017) Maslyanie kraski XX veka i ekspertiza proizvedeniy zhivopisi. Sostav, otkrytie, kommercheskoe proizvodstvo I issledovanie krasok. (Oil paints of the 20th century and examination of works of art. Composition, disclosure, commercial production and research of paints). 2nd edition. Lan'. THE PLANET OF MUSIC. Saint-Petersburg
2. Daudin-Schotte M, Bisschoff M, Joosten I, van Keulen H, van den Berg K J (2013) Dry cleaning approaches for unvarnished paint surfaces. In: Mecklenburg M F , Charola A E , Koestler R J (eds). New Insights into the Cleaning of Paintings Proceedings from the Cleaning 2010 International Conference Universidad Politécnica de Valencia and Museum Conservation Institute. Smithsonian Institution Scholarly Press, Washington D C, pp 209–219
3. De Groot J, Steyn L, Stols-Witlox M, van den Berg KJ (2019) Decision-making processes regarding the treatment of modern oil paintings (1950's – present) exhibiting paint dripping and oil exudates. These proceedings, Chapter 30.
4. Fedoseeva T S , Belyaevskaya O N , Gordyushina V I , Malachevskaya E L , Pisareva S A (2016) Restavracionnie materialy. Kurs lekciy. (Conservation materials. Course of lectures). Indrik. Moscow
5. Kadikova IF, Grigorieva IA (2017) Investigation of multilayer painting by means of molecular spectroscopy. EPJ Web of Conferences (132): Art. No 03015
6. Karganova A , Volkova V (2016) The New Reality Studio (1958–1991). Transformation of Consciousness. Album. MAIER. Moscow
7. Katalog-spravochnik. Khudozhestvennie kraski, masla, laki, razbavitely. Leningradskiy zavod khudozhestvennih krasok. (1964) (Reference book. Artistic colors, oils, varnishes, mediums. Leningrad factory of artistic colors). Leningrad
8. Moleva N (2013) Ely Bielutin. Teoria Vseobshej Kontaktnosti. (Ely Bielutin. Theory of Universal Contact). Novaya Reaslost'. Moscow
9. Pavić M (2014) Non-traditionally painted oil painting: how to treat it properly? Josip Vaništa's Cakes (1955). In: Van den Berg, K J Burnstock, A , de Keijzer, M , Krueger, J , Learner, T , de Tagle, A , Heydenreich, G , (eds) Issues in Contemporary Oil Paint. Springer, pp 159–166
10. Schorbach S (2009) Reinigungsschwämme in der Restaurierung - Vergleichende Untersuchung zu Material, Wirkung und Rückständen. (Cleaning Sponges in Restoration—A Comparative Study with Regards to Material, Effectiveness and Conditions.) Zeitschrift für Kunsttechnologie und Konservierung, № 23 (1), pp 41–54
11. Talochkin L P , Alpatova I G (1991) Moskva 1956–1976. K hronike hudozhestvennoy zhizni. (Moscow 1956–1976. To the chronicle of artistic life). In two volumes. SP Interbook. Moscow
12. Taubman W (2003) Khrushchev: The Man and His Era. W.W Norton & Company. New York

Chapter 15
Investigating Colour Changes in Red and Blue Paints – A Preliminary Study of Art Materials and Techniques in Edvard Munch's *Old Man in Warnemünde* (1907)

Jin Strand Ferrer, Irina Crina Anca Sandu, Terje Syversen, Ana Margarida Cardoso, António Candeias, and Camelia N. Borca

15.1 Background

This paper presents a study on deterioration phenomena observed in a painting in the Munch Museum collection. A preliminary visual survey of a selection of paintings revealed that specific blue and red paints appear to suffer from various forms of deterioration and chromatic alteration. These colour changes are observed on several paintings in the collection, and are evident in works from throughout Edvard Munch's oeuvre. From a preliminary visual investigation, these phenomena appear to occur frequently in the works that Munch painted in the period 1907–1908, when the artist lived and worked in Warnemünde, Germany. The *Catalogue Raisonné* of Munch's work attributes 69 paintings to this period, of which 56 are in the Munch Museum collection [1]. The work selected for this study, *Old Man in Warnemünde*, is an oil painting on a primed canvas (Fig. 15.1).

Characteristic of the artistic practice of Edvard Munch (1863–1944) was his experimentation, which early on developed into a deviation from traditional painting approaches, both in terms of visual representation and of paint handling [2]. The paintings dating from the period 1907–1908 form an exception in Munch's oeuvre in terms of their style and the artist's application of paint, showing a tendency towards a lesser degree of manipulation of the painting materials. As such, they are distinct from Munch's later paintings, in which the artist usually applied the oil paint in a more diluted and thin, nearly watery, paint consistency, as well as mixing

---

J. S. Ferrer (✉) · I. C. A. Sandu · T. Syversen
Conservation Department, Munch Museum, Oslo, Norway
e-mail: jin.ferrer@munchmuseet.no

A. M. Cardoso · A. Candeias
Hercules Laboratory, Évora University, Évora, Portugal

C. N. Borca
Paul Scherrer Institute – Phoenix beamline, Villigen, Switzerland

K. J. van den Berg et al. (eds.), *Conservation of Modern Oil Paintings*,
https://doi.org/10.1007/978-3-030-19254-9_15

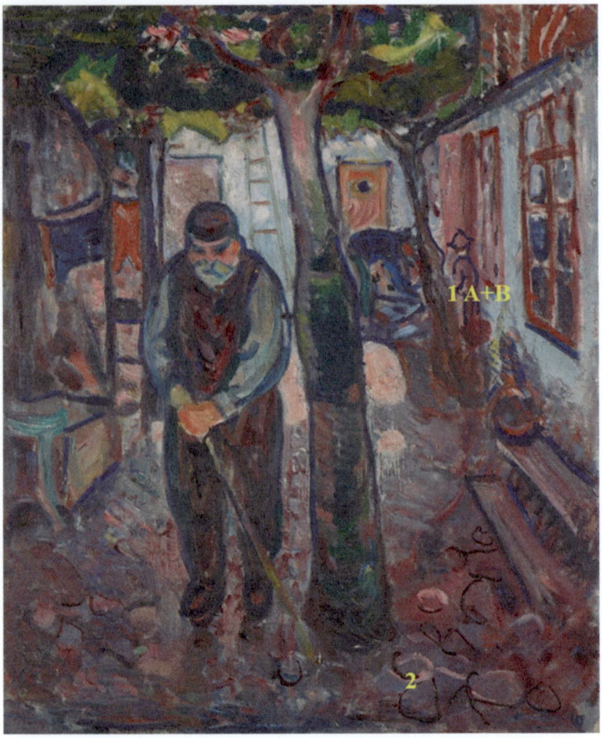

Fig. 15.1 E. Munch, Old Man in Warnemünde (1907). Inv.nr. Woll M 755/M 491. Oil on canvas 110 × 85 cm. Location of sampling of blue and red paint are indicated and numbered in yellow. (Photo credits: Ove Kvavik, 2015 © Munch Museum)

paints with different binding media within a single work [3]. Typical for paintings produced in this period is textural surface variation, with thickly applied paint, often in the outlines of the figurative elements, leaving the surface covered with impasto paint and pronounced brush strokes. *Old Man in Warnemünde* (1907) is an illustrative example, exhibiting areas where paint appears to have been squeezed directly out of the tube [4]. The blue and red paints used in this painting show colour changes in addition to other signs of deterioration. The paint used in rendering shapes and outlines shows irregular colour variations. The red paint Munch used for outlining the cobblestones in the foreground exhibits phenomena such as darkening, deep cracks and local areas with yellow crust on the surface. This red paint partially retains its brilliant red colour, albeit with areas of deep brown, lending several areas of colour a patchy appearance. The darkening in the blue paint seems to occur on the paint surface, as well as on the ridges of impasto. Preliminary analysis of these painted areas using portable X-ray fluorescence spectroscopy (pXRF) indicated that ultramarine, and possibly red lake, were used.

15.2 Sampling and Analytical Techniques

In order to describe and characterise the blue and red paints, samples were taken from the painting and analysed using different techniques. These included digital and optical microscopy (HIROX RH-2000 and OM) and physical chemical characterisation by means of portable X-ray Fluorescence (pXRF), μ-Fourier Transform Infrared spectroscopy (μ-FTIR), Scanning Electron Microscopy coupled to Energy Dispersive X-ray spectrometry (SEM-EDX), X-ray Absorption Near Edge Spectroscopy (XANES) and Pyrolysis Gas Chromatography Mass Spectrometry (Py-GC/MS). These analyses were carried out in three laboratories (Swiss Light Source in Switzerland, Hercules Center at University of Évora in Portugal and in the Conservation Department at Munch Museum in Norway).

Three samples (Table 15.1) were collected from the painting. Two from a blue (S1 A + B), and one from a red paint area (S2) (Figs. 15.1 and 15.2). The blue paint sample was taken from an area that exhibited both darkening and bright blue colour, and was divided in two (hence A + B). It was taken from an area where the paint layer had previous losses. The red paint sample was extracted from impasto, cracked red paint showing both darkening and bright red colour.

Preliminary pXRF measurements were performed on the painting with a portable Niton XL3t spectrometer equipped with a Geometrically Optimized Large Area Drift Detector (GOLDD™), using the Mining calibration setting to collect XRF spectra over a spot size of 3 mm. Four spectra were collected (Main, Light, High, and Low elemental range) for 30 s each, amounting to 120 s total collection time, employing a 50 kV current up to 40 μA. A Bruker ArtTAX with rhodium tube and polycapillary focusing optics, using 50 kV and 600 μA data collection parameters, and data collection time of 100 s was used for measuring the paint samples.

OM on samples and cross-sections was performed at different magnifications (from 100× up to 500×) using a LEICA DM2500M stereo-zoom microscope, coupled with a digital camera. Examination using visible light (illumination position for dark field observation, position 1) was performed in reflection geometry and a UV block filter was used for examination using UV fluorescence (position 3).

μ-FTIR measurements were carried out using a Bruker Tensor 27 Mid-IR (MIR) spectrometer coupled with a HYPERION 3000 microscope controlled by the OPUS 7.2 software. The microscope is equipped with a Mercury Cadmium Telluride (MCT) detector cooled with liquid nitrogen. Spectra were acquired from different points of the sample. The samples were analysed in transmission mode using a 15× objective and an EX'Press 1.6 mm diamond compression microcell, STJ-0169. The IR spectra were plotted in the region of 4000–600 cm^{-1}, with 64 scans and 4 cm^{-1}

Table 15.1 Samples and analyses

Sample ID	Surface colour	Analyses	Laboratory ID
M491 S1 A	Blue	OM, pXRF, XANES	MUM + SLS
M491 S1 B	Blue	OM, pXRF, FTIR, Py-GC/MS	MUM + Hercules
M491 S2	Red	OM, pXRF, FTIR, SEM-EDX	MUM + Hercules

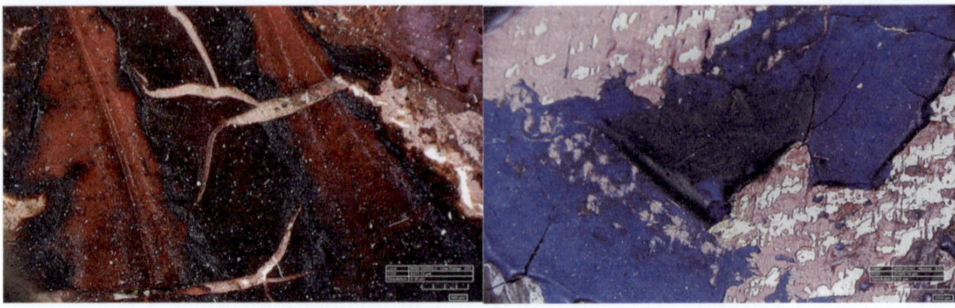

Fig. 15.2 HIROX RH-2000 images of area marked 1 (right), and 2 (left image) on Fig. 15.1. (Photo credits: Emilien Leonhardt, 2017)

spectral resolution. An Omnic software database archive (both built in house and from collections of reference materials from industry and chemical applications) was used to compare the spectra from the samples with reference materials.

A Hitachi S-3700N scanning electron microscope coupled with a Bruker XFlash 5010 SDD energy dispersive X-ray spectrometer was used with an accelerating voltage of 20 kV for SEM-EDX analysis of the samples. Cross-sections were analysed in variable pressure to preserve the samples for subsequent analysis and images were collected in Back Scattering Electron mode (BSE).

Py-GC/MS was performed with a Frontier Lab PY-3030D double-shot pyrolyser attached to a Shimadzu GC2010 gas chromatographer and to a Shimadzu GCMS-QP2010 Plus mass spectrometer. Paint layers were separated from ground layers using a stereomicroscope. Samples of about 200 μg were firstly derivatised in a 50 μl Eco-cup capsule and then placed in the double-shot pyrolizer for a 2 min helium purge. Each sample was pyrolised with a single-shot method at 500 °C for 12 s. The interface was kept at a temperature of 280 °C. A capillary column Phenomenex Zebron-ZB-5HT (30 m length, 0.25 mm internal diameter, 0.50 μm film thickness) was used for the separation, with helium as carrier gas, which was adjusted to a flow of 1.5 ml/min. The splitless injector was operated at a temperature of 250 °C. The GC temperature programme was set at 40 °C during 5 min, ramp until 300 °C at 5 °C/min, with an isothermal period of 3 min. The source temperature was placed at 240 °C and the interface temperature was maintained at 280 °C. The mass spectrometer was lined up to acquire data among 40 and 850 m/z. Software AMDIS and NIST and Wiley databases of reference materials were used for compound identification.

The micro-XANES investigations have been performed at the PHOENIX beamline from the Swiss Light Source, using a 5 × 5 microns beam size and an energy dispersive fluorescence detector (4-element Vortex, USA) with an energy resolution of 150 eV. The samples originating from the bright blue and darkened blue areas were placed in a vacuum chamber (10^{-6} mbarr) at 45° incident angle with respect to the incoming beam, in order to minimise the scattering contributions. After scanning each sample for elemental composition, several micro-XANES spectra have been recorded at both Al (1560 eV) and S (2472) K-edges on the surface of bright blue and darkened blue regions. Using this setup geometry the interaction depth of the X-rays with the sample surface was limited to a few microns. No time-dependent changes of the XANES spectra were detected for the different locations on the sample, indicating that the sample was not damaged by the extended exposure to the high flux X-ray beam. All XANES spectra were normalised to a post edge jump of 1 for comparison using the software ATHENA [5].

15.3 Results

In Table 15.2 a summary of the analytical results obtained from the three samples is given.

The pXRF measurements on the dark blue, blackened edges in the paint used for (the measurements were taken in the area of sampling 1 A+B) indicated the presence of the following elements: Pb, Si, Fe, and Al, suggesting the use of an ultramarine blue pigment. The detection of lead is interpreted as related to the white ground layer, which is likely to contain lead white. Measurements on the red paint detected Ca, Pb, Fe, Zn, Hg, Cl, Si, and P. The detected elements are likely to be related to the presence of a calcium-based compound, lead white and clay minerals (a red ochre or burnt umber). Zinc white and vermilion in the underlying paint layers, as confirmed by SEM-EDX and OM of the cross-section. SEM-EDX analysis of cross-section S2 from the red paint used in the outlines of the cobblestones confirmed the detection of P and Ca, along with other elements (Al, Sn, Na, C, O), suggesting the presence of the substrate for an organic lake (Fig. 15.3).

The ultramarine blue pigment in sample S1B was identified in the μ-FTIR spectrum (Fig. 15.4a) by the strong absorptions around 1150–950 cm^{-1} with overlapping stretching bands for Si-O-Si and Si-O-Al. The lesser intense bands in the region of 800 cm^{-1} may be related to the Si-OH stretch, whilst the weak band in the region of 1600 cm^{-1} is most likely due to O-H bending. The aluminosilicates $(Al_6Si_6)\,O_{24}$ exhibit Al-O vibrations which are visible in the 750–650 cm^{-1} region. The spectrum also shows a broad absorption around 3417 cm^{-1} [6].

The presence of oil is indicated by two sharp peaks at 2918 and 2850 cm^{-1}, which relate to the C-H stretching, and by the C=O stretching band at 1721 and

Table 15.2 Summary of results

Sample ID	Binding media	Other organic compounds	Pigments	Fillers, extenders
M491 S1 A			Synthetic ultramarine	
M491 S1 B	Heat-bodied linseed oil	Beeswax	Synthetic ultramarine	
M491 S2			Alizarin (upper layer), Vermilion (underlying layer)	Aluminum/calcium and/or tin/calcium based substrate (Al, P, Ca, Sn, Na)

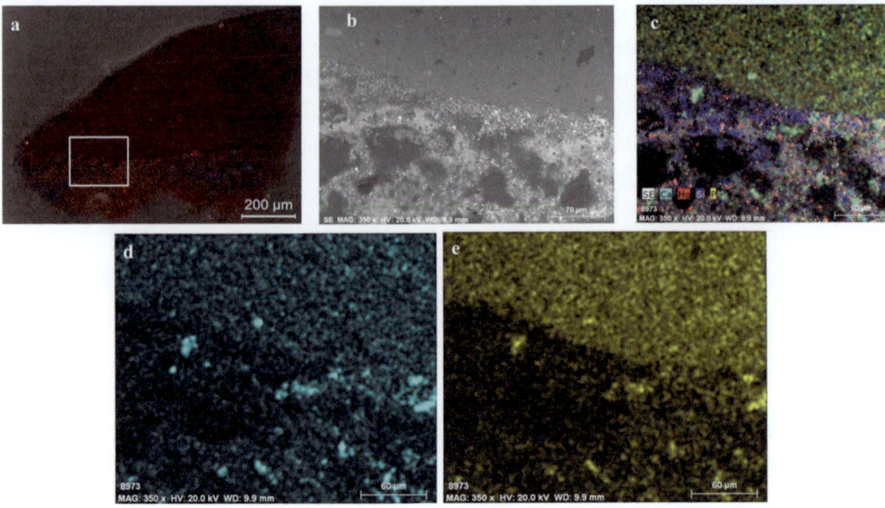

Fig. 15.3 (**a**) Cross-section of S 2 (red paint) in optical microscopy. (**b**) Backscattered electron image of outlined area in Fig. (**a**). (**c**) Combined elemental distribution of Ca, Hg, Si and P in the red paint sample. (**d**) Shows the elemental distribution of Ca. (**e**) Shows the abundantly and evenly distribution of P

1709 cm^{-1}. The fingerprint region for linseed oil shows bands at 1463 and 1380 cm^{-1} due to the C-H bending and at 718 cm^{-1} due to the C-H rocking. The results from Py-GC/MS analysis of this sample indicated the fatty acid ratios A/P 0.81 and P/S 1.24 with a suberate/azelate ratio of 0.37, suggesting the presence of a heat-bodied linseed oil [7]. Analysis with GC/MS detected the presence of natural wax (beeswax) in addition to the drying oil.

The µ-FTIR spectrum in Fig. 15.4b exhibit the characteristic absorption band of alizarin with the hydroxyl functional group at 3389 cm^{-1}, attributed to water molecules and -OH groups. The C-H stretching bands appears at 2929 and 2855 cm^{-1} and the C=O stretching bands at 1712 cm^{-1}. The band at 1638 cm^{-1} are characteristic of the C=O functional group and the bands at 1590 and 1468 cm^{-1} corresponds to the C=C functional group, also belonging to the anthraquinone molecules.

A total of five fragments of the paint sample (S 1A) have been investigated using spatially resolved XANES at the Al and S K-edges. The elemental distribution of the fragments have been mapped at 2500 eV incident radiation, just above the S K-edge. The Na, Si, Al, S fluorescence lines were distributed uniformly across the blue and dark fragments (maps not shown). Figure 15.5 shows the micro-XANES spectra on only two representative samples corresponding to a blue and a dark region of the ultramarine pigment. The S K-edge spectra show the presence of several sulphur oxidation states including sulphide (2470.8 eV), elemental sulphur (2472.9 eV), sulphite (2477.3 eV) and sulphate (2481.6 eV) in both the pristine, bright blue area and in the darkened regions. The pre-edge at 2468.0 eV was reported to relate to the presence of the S^{-}_{3} chromophore [8]. The sulphate (SO_4^{2-}) peak at

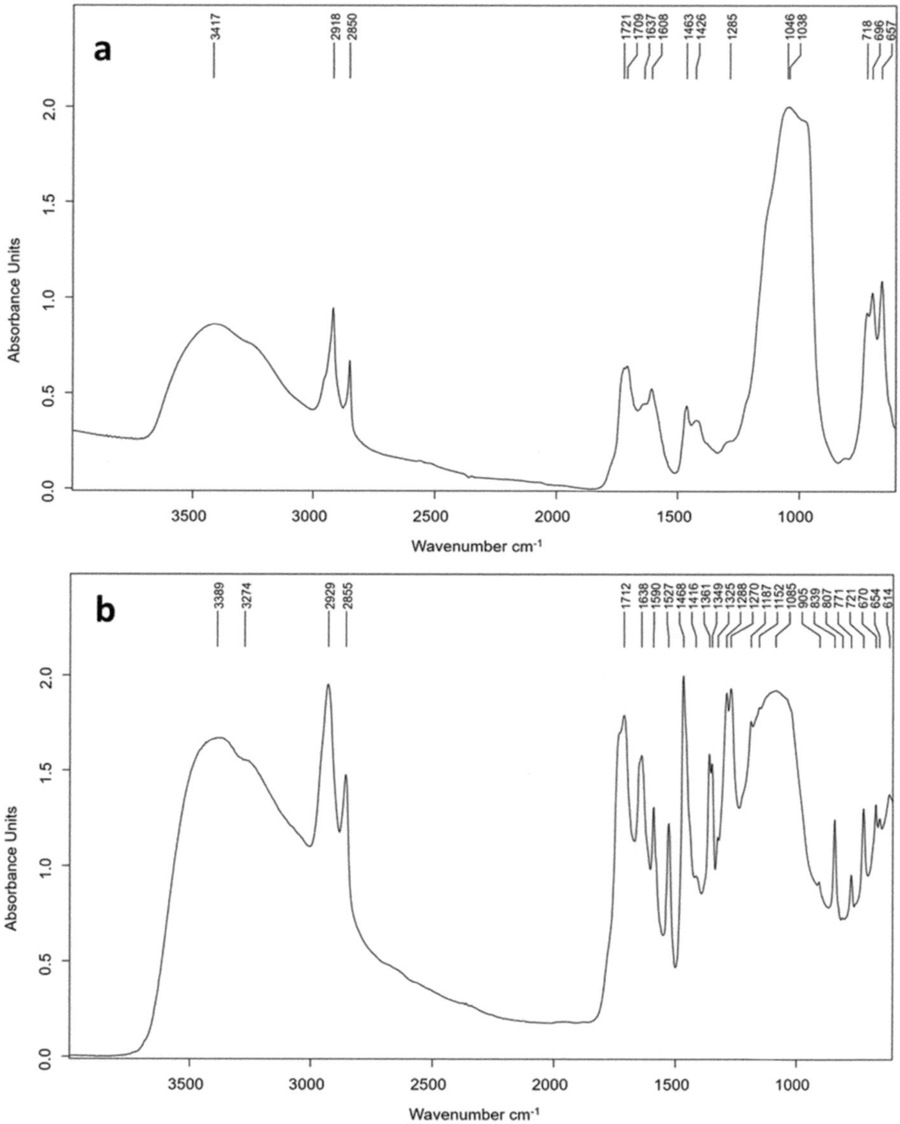

Fig. 15.4 Micro-FTIR spectra of (**a**) S1 B, ultramarine blue and oil. (**b**) S2, alizarin

2481.7 eV is stronger in position 1 (blue area), but the rest of the peaks do not differ significantly. The Al K-edge spectra are almost identical for positions 1 (bright blue) and 2 (darkened blue) indicating the presence of mainly tetrahedral aluminium (AlO_4 units) [9].

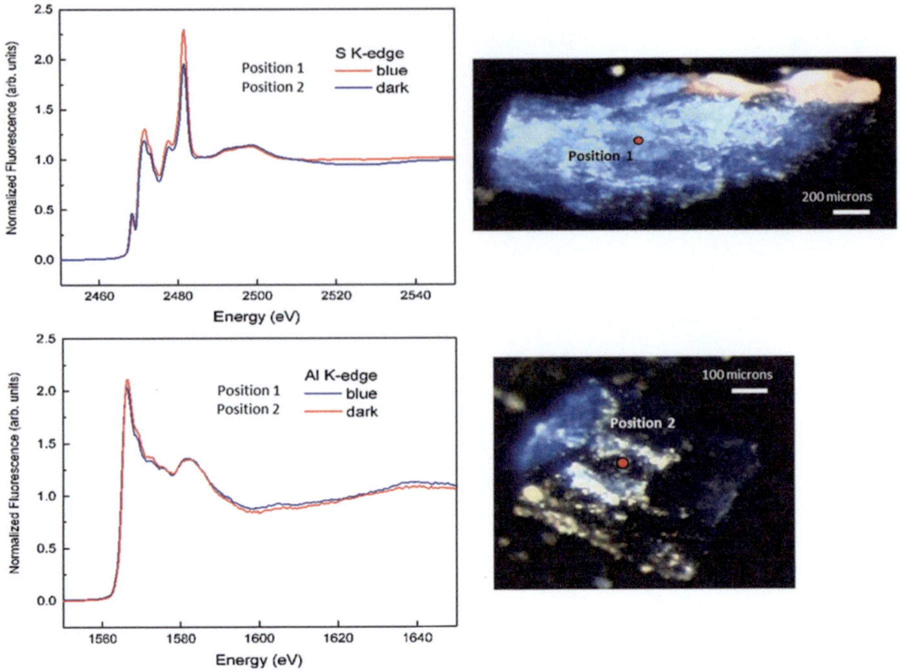

Fig. 15.5 The micro-XANES spectra recorded on v S1 A with micro-beam on the surface of blue (position 1) and dark (position 2) areas of ultramarine pigment at the S (top) and Al (bottom) K-edges

15.4 Discussion of Results

Ultramarine blue paints are known for having poor drying properties in oil. Although the pigment is noted to be comparatively permanent, there have been reports of changes in colour due to instability of the pigment [10]. Recent research has linked the discolouration to the degradation of the binding medium, understood to be a result of a of pigment–binder separation [11]. The phenomenon of blackening of deep blue paint observed in *Old Man in Warnemünde*, as well as in other paintings in the Munch Museum collection, may be seen in relation to early reports on similar phenomena, dating from the period of early manufacturing of synthetic ultramarine. As synthetic ultramarine became available on the market, it was noted that the paint had issues of blackening, at the time attributed to impurities or additions of iron, lime, magnesia and potash, or to the paint being mixed with white lead [12]. The elemental distributions measured with the synchrotron micro-beam in fluorescence mode show that the composition of the ultramarine pigment used in *Old Man in Warnemünde* has a uniform distribution of Al, Si, S, Na across the surface, in both dark and brighter blue areas. This suggests that darkening is likely due to the organic components rather than to the pigment itself. The micro-XANES analysis indicates no significant change of the mineral structure in the darkened blue compared to the bright blue regions. The Al K-edge spectra correspond to Al tetrahedral structure, while the S K-edge spectra show a slight decrease of the S^{-}_{3} chromophore (pre-edge peak) and an increase of the sulphate peak SO_4^{2-} in the dark blue regions. In order to gain further understanding of the causes of the darkening, further analysis should be performed to obtain comparative data on the binding media. The presence of natural wax in the paint sample is interpreted as a probable additive from the manufacturers. Such additions have been identified in ready-made paint from the period [13], and have also been indicated by FTIR/ATR analysis of certain paint tubes in the Munch Museum's reference collection [14].

The instability of the red lake paint has been subject to studies on paintings made by Munch's contemporary, Vincent van Gogh [15]. Certain types of organic red paint degradation on paintings by Van Gogh have been linked to paint formulations containing starch and specific types of mordant–colourant. The deterioration characteristics on Munch's paintings may be compared to these findings. The degradation phenomena described in red lake paints in van Gogh's *Glass with Yellow Roses*, and *Portrait of an Old Woman* [16], exhibits visual surface characteristics similar to the red lake paint in Munch's *Old Man in Warnemünde*. The red lake in Munch's painting exhibits both pronounced cracks and spotting, in addition to browning. In the red sample from *Old Man in Warnemünde*, alizarin was identified as the colourant, with a possible tin/calcium-based and/or aluminium/calcium based-substrate, which might suggest a mixture of two types of red lake. The deterioration, resulting in a brown discolouration of the red lake used by Van Gogh, was, as noted above, connected with the inclusion of starch. The absence of starch from the sample from *Old Man in Warnemünde* notwithstanding, the similarities of paint deterioration merits a further investigation and comparative analysis on the uses of red lake paint by these two artists. The presence of a phosphorous compound is a topic for further investigation. Moreover, specifying the other minor constituents in the paint, as well as the binders used, would contribute to our understanding of the deterioration phenomena.

15.5 Conclusions

The results produced by these investigations may furnish more comparative information useful in research on binder–pigment–additive issues affecting these problematic paints. Furthermore, analyses on paint tubes used by Edvard Munch covering the two typologies of colours (blue–ultramarine and red–alizarin) will be undertaken to establish whether a correlation exists between the composition of such paints and the ageing/darkening phenomena observed on the paintings. Analysis and identification of binding media, pigments and additives from Munch's paint tubes are on-going, and the results will be published in a future article [17].

References

1. Woll G (2008) Edvard Munch Samlede malerier, Catalogue raisonné. Cappelen Damm, Korotan
2. Bruteig M (2008) Umalte tegninger. In: Ydstie I and Guleng M B (eds) Munch blir «Munch» Kunstneriske strategier 1880–1892, Munchmuseet, Oslo, p 63
3. Singer B, Aslaksby T, Topalova-Casadiego B, Tveit ES (2010) Investigation of materials used by Edvard Munch. Studies in Conservation 55 (4): 274–292. Sandbakken EG, Tveit ES (2012) Edvard Munch's monumental sketches (1909–1916) for the Aula of Oslo University, Norway: Conservation issues and treatments. Studies in Conservation 57: 258–267
4. Eggum A (1999) Munch og Warnemünde. In: Biørnstad S (ed) Munch og Warnemünde. Labyrinth Press, Drammen, p 26
5. Ravel B, Newville M (2005) ATHENA, ARTEMIS, HEPHAESTUS: data analysis for X-ray absorption spectroscopy using IFEFFIT. J Synchrotron Radiat. 12: 537–541. https://doi.org/10.1107/S0909049505012719
6. Derrick MR, Stulik D, Landry J (2004) Infrared Spectroscopy in Conservation Science. The Getty Conservation Institute, Los Angeles. Learner T (2004) Analysis of Modern Paints. The Getty Conservation Institute, Los Angeles
7. Mills J, White R (1994) The Organic Chemistry of Museum Objects, 2nd edn. Butterworth – Heinemann, Oxford, pp 170–172
8. Fleet M, Liu X (2010) X-ray absorption spectroscopy of ultramarine pigments: A new analytical method for the polysulfide radical anion S_3- chromophore. Spectrochimica Acta Part B 65(1):75–79
9. Ildefonse PH, Cabaret D, Sainctavit P, Calas G, Flank AM, Lagarde P (1998) Aluminium X-ray absorption Near Edge Structure in model compounds and Earth's surface minerals. Phys Chem Minerals 25(2):112–121
10. Cato E, Borca C, Huthwelker T, Ferreira ESB (2016) Aluminium X-ray absorbtion near-edge spectroscopy analysis of discoloured ultramarine blue in 20th century oil paintings. Microchemical J, 126:18–24
11. Wijnberg L (2014) Do we see what we know or do we know what we see? Conservation of Oil Paintings in the Stedeljik Museum Amsterdam. In: van den Berg KJ, Burnstock A, Kreuger J, Learner T, de Tagle A, Heydenreich G (eds) Issues in Contemporary Oil Paint, Springer, Switzerland pp 21–32
12. Carlyle L (2001) Artist's Assistant:oil painting instruction manuals and handbooks in Britain, 1800-1900, with reference to selected 18th century sources. Archetype, London, p 473
13. Laar M, Burnstock A (1997) With paint from Claus & Fritz: A study of an Amsterdam painting materials firm (1843 – 1931). JAIC 36(1):1–16
14. Ferrer JS, Hull A, Syversen T (2016) Investigating surface phenomena in paints and paintings by Edvard Munch (1863 – 1944) in the context of the artist's painting materials. In: Clarricoates R, Dowding H, Gent A (eds) Colour Change in Paintings. Archetype, London, pp 126–130
15. van Bommel M, Geldof M, Hendriks E (2005) An investigation of organic red pigments used in paintings by Vincent van Gogh (November 1885 to February 1888). ArtMatters: Netherlands technical studies in art 3:111–137
16. Burnstock A, Lanfear I, van den Berg KJ, Carlyle L, Clark M, Hendriks E, Kirby J (2005) Comparison of the fading and surface deteriorating of red lake pigments in six paintings by Vincent van Gogh with artificially aged paint reconstructions. In: Verger I (ed) ICOM 14th Triennial Meeting The Hague, 12–16 September 2005, Committee for Conservation:Preprints. James and James, London, pp 459–466
17. Syversen T, Hull A, Sandu IC, Ferrer JS, Mazurek J, Schilling M, Mass J, Miliani C, Monico L, Casini A (2019) The Paint tubes collection at Munch Museum – an integrated disciplinary approach. Studies in Conservation (under revision).

Chapter 16
Aspects of Lovis Corinth's Painting Technique in His Late Work: Working Process and Colour Structure Captured in Paint and on Film

Thomas Krämer

16.1 Introduction

Lovis Corinth (1858–1925) was a painter, graphic artist and illustrator. He is regarded as one of the leading German Impressionists. His late work was influenced by the consequences of a stroke, which he suffered in 1911.[1] Based on visual analysis a comparative art-technological study of the brushwork and the working process of Corinth's late work is attempted. In the first part of the contribution, the painting *Walchensee, Landschaft mit Kuh* (*Lake Walchensee, landscape with cow*), dated 1921 and part of the collection of the Neue Galerie (Museumslandschaft Hessen Kassel), is presented (Fig. 16.1). Focussing on two spots the paint surface is described and a reconstruction of the painting process is attempted. The second part deals with a documentary film made by Hans Cürlis (1889–1982) in 1922/23, which captured Corinth working on a small-scale oil study. The art-technologically relevant shots are presented and described. The analysis of the painting process and characteristics of the paint application are described through examples. The findings are correlated with selected written sources by Lovis Corinth, his wife Charlotte Behrend-Corinth (1880–1967) and their son Thomas (1904–1988) concerning aspects of painting technique, material and execution.

[1] Bärnreuther [1] pp. 18–19.

T. Krämer (✉)
Conservation Department, Museumslandschaft Hessen Kassel, Kassel, Germany
e-mail: t.kraemer@museum-kassel.de

K. J. van den Berg et al. (eds.), *Conservation of Modern Oil Paintings*,
https://doi.org/10.1007/978-3-030-19254-9_16

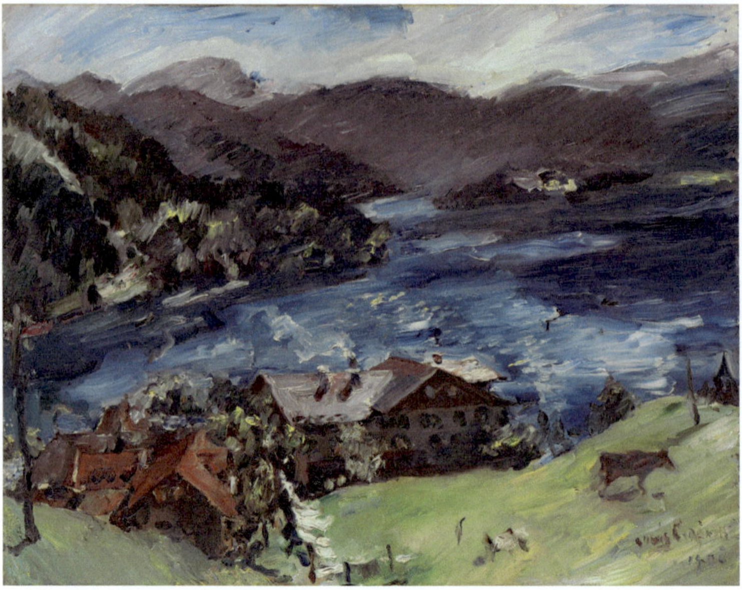

Fig. 16.1 Lovis Corinth, *Walchensee, Landschaft mit Kuh*, canvas, 95 × 120 cm, signed and dated 1921, collection Neue Galerie, Museumslandschaft Hessen Kassel. (Photo U. Brunzel)

16.2 The Painting *Walchensee, Landschaft mit Kuh*, Part of the Series of *Lake Walchensee* Landscapes

From 1918 onwards, Corinth often spent time in the village of *Urfeld* on the shore of Lake Walchensee, a mountain lake in the alps. From 1918 to 1924, he created 78 paintings and graphic works of art at this location. They mostly show the lake landscape in different perspectives and in changing times of day and seasons. These works have, from a present day perspective, tendencies towards Expressionism and abstract painting.[2]

The painting *Walchensee, Landschaft mit Kuh,* is signed and dated. At 95 × 120 cm it represents one of the large-scale *Lake Walchensee* landscape paintings. From an elevated spot the viewer looks down onto the lake landscape in a glaring midday summer light. In the foreground there is a meadow, steeply sloping towards the houses of the village of *Urfeld.* Laundry hung out to dry is fluttering in the wind, a cow stands to the right. The light is reflected on the lake's waves, the slightly clouded sky is mirrored in the water. A broad view unfolds onto the shore and wooded hills. Behind them the mountain ranges can be made out.

[2]Timm [9]; Heinz [6] pp. 12–37. Based on the catalogue raisonné by Berend-Corinth, Heinz quoted '*approx 60'* paintings and a time frame from 1918 to 1925.

16.2.1 Techniques of Painting

The canvas used was ready-primed with a white ground. In a few spots the ground layer and in one spot the underdrawing are exposed. Corinth preferred painting *'with pure colours without any additives'*, i.e. without linseed oil, siccative or turpentine, but he didn't rule out their use generally.[3] Brushmarks on this painting indicate the stiffness of his oil paint, but also the addition of some oil or siccative.[4] In addition, they prove that a flat hogs' hair brush with square-cut tip was used.[5] The width of the strokes is up to 2.5 cm. Corinth, however, argued that the type of brush used is not crucial for the execution of a painting.[6] For example, in the mountain ranges long brushstrokes run in a straight and diagonal direction; in the wooded hills the strokes are noticeable zig-zagged. Corinth applied numerous dabs of varying size and shape and occasionally calligraphic strokes. For example Fig. 16.2 shows a detail of the nearby shore on the left of the painting. Arrow 3 points to the beginning of the stroke, which was first drawn with the narrow side of the brush, barely touching the surface; at arrow 4, the brush was turned to its wide side and drawn with increasing pressure. The paint layer was largely, but not entirely worked wet in wet. We can see a brushstroke applied wet on dry at one spot of the water surface near the shore on the right (Fig. 16.3, arrow 3). By his own admission Corinth painted mostly *alla prima*. He wrote: *'[I] have […] observed as much as*

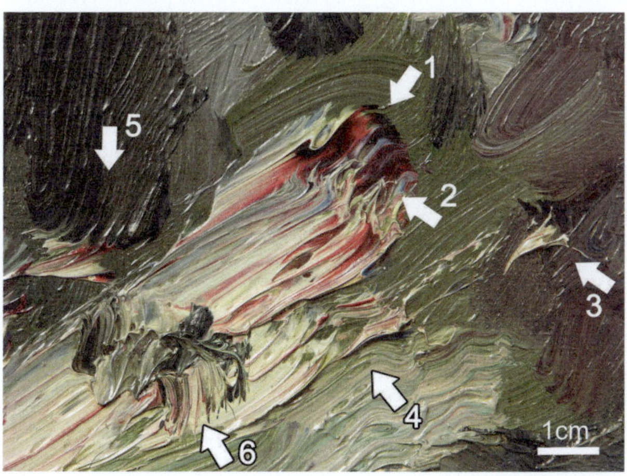

Fig. 16.2 *Walchensee, Landschaft mit Kuh*, detail of the lakeshore, brushstroke with heterogenous colour mixing (arrows 1–6)

[3] Corinth [3] p. 192. Transcription. *'mit reinen Farben ohne jede Zutat'*.

[4] Except the dark blue colour, which formed tears in some places.

[5] Callen [2] p. 175.

[6] Corinth [3] p. 193.

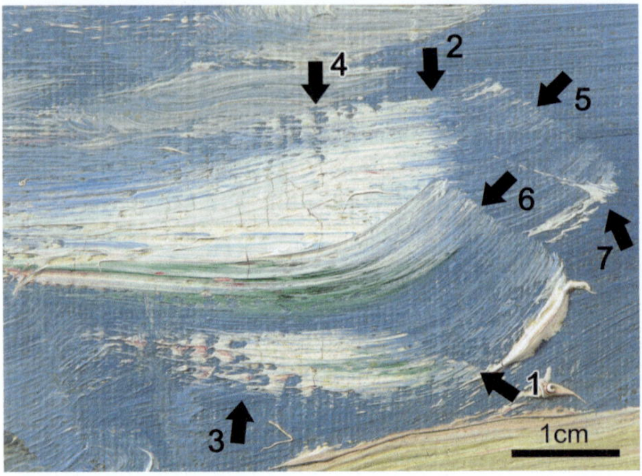

Fig. 16.3 *Walchensee, Landschaft mit Kuh*, detail of light reflections on the water surface, dragged brushstrokes (arrows 1–7)

possible the main law for every painting: wet in wet.'[7] Subsequently he added: '*[A]s long as the paint is wet, you can go over again and correct, as often as it's deemed necessary.*'[8] The painting under discussion may have been created outdoors. Traces on the paint layer indicate that the folding mechanism of the stretcher, which was developed for *plein air* painting, was used.[9] As the paint layer shown in Fig. 16.3 (arrow 3) indicates, Corinth worked in more than one session. With regard to other paintings of the *Lake Walchensee* series Behrend-Corinth reported single sessions of up to half an hour, but also repeated sessions during several days.[10] The painting is covered with a thin varnish. It is not clear whether the varnish is original or not.[11]

16.2.2 Brushstrokes with Heterogenous Colour Mixing

The above-mentioned Fig. 16.2 taken from one spot of the nearby shore shows various mixed and overlapping brushstrokes. First, consider the sequence of three diagonal strokes. Arrow 1 indicates the beginning of the first stroke, arrow 2 the end of the second one in opposite direction. The third, already described stroke (arrows 3 and 4) runs to the left again. The strokes merge different colours, which partially lie next

[7] Corinth [3] p. 188. Transcription. *'[I]ch habe […] soviel wie möglich das Hauptgesetz für jede Malerei beobachtet [sic!]: naß in naß zu malen.'*

[8] Corinth [3] p. 189. Transcription. *'[S]olange die Farben naß sind, kann man so oft wieder rübergehen und korrigieren, wie man es für notwendig hält.'*

[9] Heinz [6] p. 17; Stöbe [8] p. 122.

[10] Timm [9] pp. 40–41.

[11] There is no information avaible about the conservation history of the painting.

to each other in unmixed strands. In the present contribution this striking appearance is named as 'heterogenous colour mixing'.[12] Firstly, it was developed by painting with a brush, which was loaded with different (white, dark red and light blue) colours. Secondly, fine strands of light green paint in the tips (arrow 2) emerged through a wet in wet mixture with the green paint layer beneath. A dark green stroke touched the already applied stroke (arrow 5). Probably with the same paint Corinth applied a dab and created further heterogenous colour mixing (arrow 6).

16.2.3 Dragged Brushstrokes

The second area mentioned above features the water surface near the shore, where Corinth created the illusion of light, reflected on the moving water (Fig. 16.3). The underlying light blue paint layer shows slight brushmarks. Several dragged brushstrokes in white or light colour and with some slight heterogenous colour mixing can be discerned.[13] Again, the strokes were applied on top of each other, but at this spot both wet in wet and wet on wet. Arrows 1 and 2 point at the beginning of the strokes beneath. Small patches of paint adhere to vertical threads of the canvas (arrows 3 and 4). However, the dragging effect of the overlying brushstrokes (arrows 5 and 6) is less noticeable, probably due to the less loaded brush or reduced pressure with the brush. Finally, on one spot the flanking impasto of one stroke (arrow 7) was dragged by a slight touch of the brush.

16.3 Corinth Filmed at Work

16.3.1 Hans Cürlis and His Cinematographic Work 'Creating Hands'

The German art historian, film director and producer Hans Cürlis was a pioneer of cultural and documentary film. Best known are his portraits of artists dating from 1922 to 1967. The first artist portrayed by Cürlis was Lovis Corinth, in 1922 or 1923. Cürlis' compilation entitled *Maler bei der Arbeit* (*Painters at Work*) was first presented in 1923, featuring not only Corinth, but also Max Liebermann (1847–1935) and Max Slevogt (1868–1932) among others. This work later became known under the title *Schaffende Hände* (*Creating Hands*). It was continuously expanded and presented in different compilations.[14]

---

[12] Stöbe [8] p. 126. Stöbe called this *'Rudimente' [rudiments']* of the initially used colours.

[13] Callen [2] p. 81, Fig. 118, pp. 158, 217.

[14] Döge [5] pp. 27–31, 57–58, 65, 68, 72, 80–83, 88, 93, 96, 121; Ziegler [10] pp. 200, 223–225; Kirchmayer [7].

For the following analysis, a publicly available source was selected, a documentary by Josef Kirchmayer in 1989 about Cürlis and his cinematographic work.[15] In this documentary the historical film by Cürlis starts at 3:10. It is a silent black-and-white film with a running time of about 5 min. In this version the film contains 14 shots. In shot 1 we see the artist *en fac*e (close up), in the shots 2, 4, 7, 8 and 11 in profile at work (close up) and in shot 13 (medium long shot) sitting at the easel working at a painting. In the shots 3, 5, 6, 9, 10, 12 and 14 the camera positioned to the artist's right focusses on the painting. The upper and lower parts of the painting are outside the camera's focus. In addition, Cürlis showed Corinth's right hand holding the brush. A title sequence or credits are not available or not preserved.[16]

16.3.2 The Film Footage as an Art-Technological Source

Although the documentary generally claims to depict reality, the documentary value of the depiction has to be questioned. For recording and conceptual reasons, Cürlis preferred to film small studies or excerpts from paintings.[17] Besides that Corinth represented himself as a successful artist in various photographs. He can be seen in his studio with brush and palette in hand in front of already framed and obviously finished paintings.[18] This exemplifies an intervention in film production and the influence of narrative elements.

Corinth painted his oil study especially for the movie shot sitting in the front garden of his home and studio in Berlin.[19] This situation was staged too, but Charlotte Berend-Corinth reported years later: '*He painted […] just like he always did.*'[20] According to her, the painting was finished in a few minutes.[21] However, the painting process is not fully documented in the film. Surprisingly, shot 9 is repeated in the longer shot 11. As the painting has been lost, no comparison is possible.

---

[15] Kirchmayer [7]; Youtube, https://www.youtube.com/watch?v=pujVYyCRBDU. Accessed 2 Jan 2019. The film can be watched in a TV broadcasting by the channel 3sat in 1994.

[16] Döge [5] pp. 28, 57; Ziegler [10], pp. 221, 223–225. According to Ziegler an early version had contained intertitles.

[17] Ziegler [10] p. 223.

[18] Bärnreuther [1] pp. 17, 21, 22, Fig. 16, 24, 28; Corinth [3] pp. 191–192. Corinth reported that he didn't frame paintings until he had finished them.

[19] Ziegler [10] p. 225; Corinth [4] pp. 339–400.

[20] Berend-Corinth: Letter addressed to Cürlis, dated 1926. In: Ziegler [10] p. 225. Transcription. '*Er malte* […] *genauso, wie er stets malte.*'

[21] Ziegler [10] p. 225.

16.3.3 Aspects of the Filmed Working Process

Based on the film shots, the size of the painting is estimated as 60 × 40 cm. Corinth's son assumed that it was on cardboard.[22] The artist also used a flat, square-ended hogs' hair brush (shot 5, 3:40–4:34). In his left hand he hold further brushes and his palette. No vessels or brush-dippers can be discerned (shot 13, 8:00–8:10). The colours can only be characterised as light or dark.

Shot 3 gives an insight into the first stage of the painting process (3:22–3:39). At first we see a white blocking-in of the upper part of the painting. The rough brushstrokes indicate the stiffness of the paint. Corinth dispensed with an underdrawing. In an expressive manner circular and hatched strokes were applied. In later stages, for example captured in shot 9 (4:56–5:10) and in shot 12 (7:15–7:36), Corinth moved between different areas, alternating between extended paint application, accentuating dots and reworking areas.

It can be seen how the artist touched the paint surface in different ways, either sweeping so slightly that only the brush hair tips touched the painting surface, or with heavy pressure and bending the brush hairs (shot 5, 4:20–4:28 and shot 12, 6:30–6:48). He drew strokes also with the narrow side of the brush, turned the brush in the hand and varied the width of his strokes by a curved motion (shot 9, 4:56–4:59 and 5:20–5:25, shot 12, 7:40–8:00). Surprisingly, before Corinth applied the paint, he first moved the brush over the painting without touching it.

16.3.4 Brushstrokes with Heterogenous Colour Mixing

In shot 12 (6:05–6:22) an example of heterogenous colour mixing can be observed. Corinth drew the strokes with heavy pressure. Although the brush was loaded with different colours the first strokes applied appear light. Then he worked wet in wet in a circular motion and by touching the dark paint layer on the left. The result can be seen in Fig. 16.4 (arrow 1). The further painting process led to a more homogenized colour mixing by working over areas already painted and probably reducing the pressure with the brush (arrow 2).[23] Furthermore, Corinth applied this colour elsewhere (shot 12, 6:22–6:25).

16.3.5 A Dragged Brushstroke

In the further course of shot 12 (6:50–7:02), a wet on wet applied and partially dragged brushstroke leading from left to right can be observed. From its beginning on (Fig. 16.4, arrow 3) the layer thickness decreased. Light streaks running along

---

[22] Corinth [4] p. 400.

[23] Ziegler [10] p. 225.

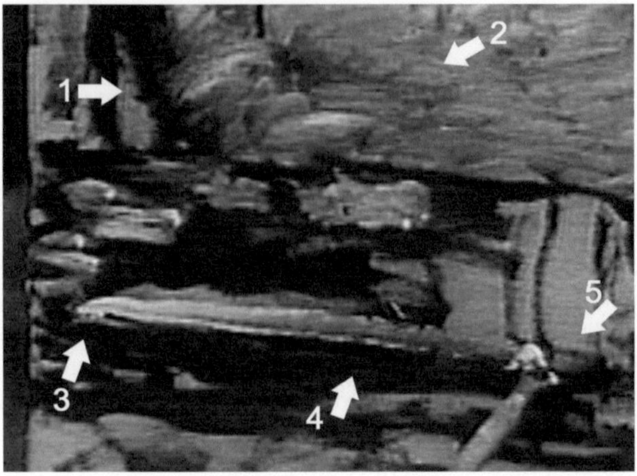

Fig. 16.4 Painting process captured in a documentary film by Hans Cürlis in 1922/1923, film still, detail, reproduced from Kirchmayer [7], © Peter Cürlis Film-Produktion, Berlin (arrows 1–5)

the edges of the brushstroke clearly show the '*broken colour effects*' of the dragging, probably contributing to the halting motion (arrow 4).[24] We see the brush at the end of the stroke, arrow 5 points to its tip loaded with light colour. Subsequently, Corinth broke this eye-catching brushstroke with crossing marks (shot 12, 7:20–7:21, 7:33, 7:54–7:58).

16.4 Conclusion

The two works examined from Corinth's Late Period, a painting *in natura* on the one hand and on the other a small-scale oil study that today only survives on film, were created under different conditions and with different intentions. Nevertheless, the results of the visual analysis focussing on the application of paint and the working process are confirmed and complemented in essential aspects. These aspects include turning the flat brush used during the stroke, drawing strokes with the narrow or the wide side of the brush, varying pressure, and alternating movement between fast and slow. Heterogeneous colour mixing was the result of both the brush loading with different colours and the wet in wet paint application. Dragged brushstrokes could be applied over wet or dry paint. The film source gives an insight into Corinth's *alla prima* painting technique and suggests that he partly developed his composition in dialogue with the painting. Although the written sources mentioned are informative, important technical aspects are not mentioned. Both the painting by Corinth and the cinematographic work by Cürlis provide a basis for further art-technological research.

[24] Callen [2] p. 217.

References

1. Bärnreuther A (1996) Biographie. In: Schuster P-K, Vitali C, Butts B (eds): Lovis Corinth. Prestel, München, New York, pp. 11–24
2. Callen A (2000) The Art of Impressionism. Yale University Press, New Haven, London
3. Corinth L (1979) Das Erlernen der Malerei. Ein Handbuch. 3rd edition [1920], Gerstenberg, Hildesheim
4. Corinth T (1979) Lovis Corinth. Eine Dokumentation. Wasmut, Tübingen
5. Döge U (2005) Kulturfilm als Aufgabe. Hans Cürlis (1889–1982). Filmblatt-Schriften, vol 4, CineGraph Babelsberg, Berlin
6. Heinz M (2008) Die wahre Kunst ist, Unwirklichkeit zu üben. Lovis Corinth, „Walchensee, Landschaft mit Kuh". Deutscher Kunstverlag, München, Berlin
7. Kirchmayer J (1989) Aus den Ateliers der 20er Jahre. Der Filmpionier Hans Cürlis und seine Malerporträts. Produktion Hessischer Rundfunk (© Hessischer Rundfunk, www.hr.de)
8. Stöbe E (2009) Bemerkungen zur Maltechnik von Lovis Corinth. In: Husslein-Arco A, Koja S (eds): Lovis Corinth. Ein Fest der Malerei. Prestel, München, Berlin, London, New York, pp. 121–131
9. Timm W (1986) Lovis Corinth. Die Bilder vom Walchensee. Vision und Realität, Exhibition catalogue. Ostdeutsche Galerie Regensburg, Kunsthalle Bremen. Garamond, Regensburg
10. Ziegler R (2005) Schaffende Hände. Die Kulturfilme von Hans Cürlis über bildende Kunst und Künstler. In: Kreimeyer K, Ehmann A, Goergen J (eds): Geschichte des dokumentarischen Films in Deutschland, vol 2, Weimarer Republik 1918–1931. Reclam, Stuttgart, pp. 219–227

Chapter 17
Oil Paint Straight from the Tube: Paint-Specific Deterioration in Works by Alexis Mérodack-Jeaneau, 1910–1913

Agata Graczyk, Pauline Hélou-de La Grandière, Alan Phenix, and Sigrid Mirabaud

17.1 Introduction

Alexis Mérodack-Jeaneau has nowadays sunk into oblivion despite his renown and his influence on painters at the early beginning of the twentieth century, particularly painters from the first École de Paris. The preparation for the retrospective exhibition scheduled for 2019 for the centenary of his death at the musée des Beaux-Arts in Angers gave the opportunity to study eight of his paintings. The present study[1] focuses on paintings between 1910 and 1913 when Mérodack-Jeaneau adopted a novel technique whereby paint was squeezed directly from the tube onto the pre-primed canvas. The paintings were examined and compared to better understand the painter's palette, tools and work habits. The technique and the associated degradation phenomena were thoroughly studied to design a conservation methodology and strategy.

[1] The project is the continuation of a master thesis in paintings conservation at the Institut national du patrimoine focusing on *La Créole au perroquet* [13].

A. Graczyk (✉)
Independent paintings conservator, Paris, France
e-mail: agata.graczyk.fr@gmail.com

P. Hélou-de La Grandière
Independent paintings conservator, Nîmes, France
e-mail: lagrandiere@gmail.com

A. Phenix
Getty Conservation Institute, Los Angeles, CA, USA
e-mail: alan.phenix_2@outlook.com

S. Mirabaud
Head of the laboratory, Deputy Head of Studies, Department of restorers, Institut national du patrimoine, Paris, France
e-mail: sigrid.mirabaud@inp.fr

K. J. van den Berg et al. (eds.), *Conservation of Modern Oil Paintings*,
https://doi.org/10.1007/978-3-030-19254-9_17

17.2 Alexis Mérodack-Jeaneau (1873–1919)

Alexis Mérodack-Jeaneau was born in Angers (western France). The starting point of the development of his personal style he later calls *Synthétisme* (synthetism)[2] occurs around 1901. His manner of painting results in simplified and clear compositions where pure colours settle in fields and lines, and where the materiality of paint is celebrated with gestural and thick paint layers: "Seize life, note the movement, render the atmosphere. I try to do my paintings as my sketches, with the simple necessary touches of colour".[3] At the time, Guillaume Apollinaire (1880–1918), one of the foremost critics of the era, had described Mérodack-Jeaneau as "bold" and acknowledged his influence on important painters of the time: "[…] it is true that he has had a great influence on van Dongen[4] and others […]" ([1]: 350).

17.2.1 Corpus

The project focused on the "Synthetic paintings" purchased by the musée des Beaux-Arts these last two decades (Table 17.1). They are inspired by exotics subjects, as the *Lady with a dish of oranges* (*Femme à la coupe d'oranges*), the *Creole with a parrot* (*La Créole au perroquet*) and subjects concerning the circus the artist particularly enjoyed, as the *Blue horsewoman* (*L'Écuyère bleue*), and the so-called Triptych *"the Circus"* re-composed today[5] of the *Monkey Perfomer* (*La Montreuse de singe*), the *Yellow dancer* (*La Danseuse jaune*) and the *Green horsewoman* (*L'Écuyère verte*). Those paintings met with great success in Paris when created.

17.3 Materials

17.3.1 Supports

During the first years of his career, Mérodack-Jeaneau used various substrates such as metal plates (zinc and brass boards), cardboard, cotton canvases, etc., frequently applying a hand-made preparation. In the years 1910–1913 he focused more on the

[2] See [18], pp. 9–33.

[3] (Personal translation): "Saisir la vie, noter le mouvement, rendre l'atmosphère. J'essaie de faire des grands tableaux comme mes croquis rapides, avec les simples touches nécessaires de couleur" [20].

[4] Indeed, close stylistic links between Mérodack-Jeaneau and van Dongen can be drawn especially when looking at *Femme au fond blanc* at the Art Institute of Chicago (1910–1914) by van Dongen.

[5] This ensemble was separated at an unknown date. The three paintings in the museum's collection have been called a "triptych" according to their similar formats, but as a painting with a similar subject and format exists in a private collection, this denomination could be incorrect.

Table 17.1 Supports used by Mérodack-Jeaneau, 1910–1913 © La Grandière / Graczyk

Title	*La Tour à plomb, Angers*	*Femme à la coupe d'oranges*	*La Créole au perroquet*	*La [petite] Montreuse de singe*	*L'Écuyère bleue*	*La Montreuse de singe*	*La Danseuse jaune*	*L'Écuyère verte*
Year	Before 1910 (?)	Before 1910	c. 1910	1911	1911	1911	1912	1913
Size	81 × 65 cm	81 × 65.3 cm	130.5 × 97.5 cm	81 × 65.3 cm	80 × 61 cm	253.5 × 154.5 cm	253.5 × 93.5 cm	253.5 × 93.5 cm
Front		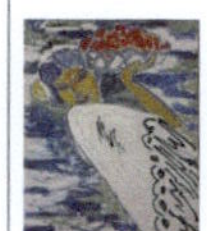	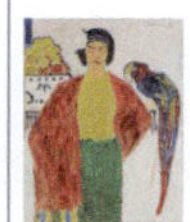	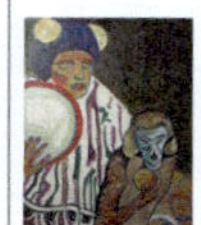	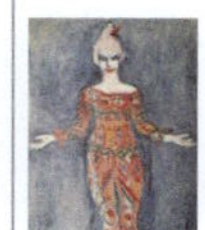	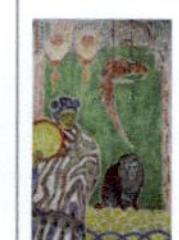	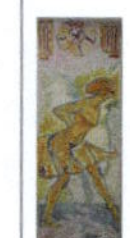	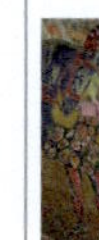
Reverse	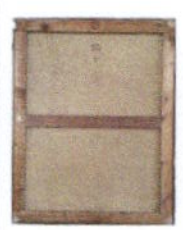	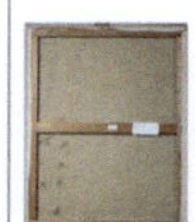		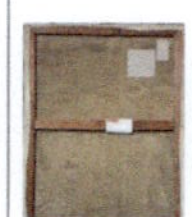	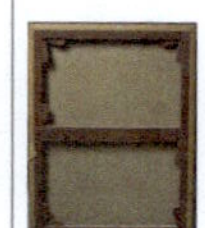	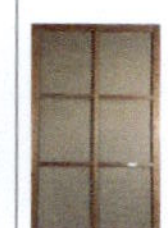	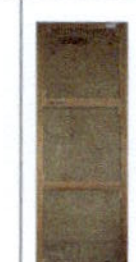	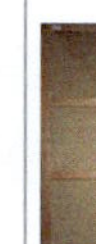
Canvas: Thread count per cm^2	13 by 16	13 by 16	17 by 13	13 by 16	12 by 15	16 by 12	17 by 13	16 by 12
Presence of colourmen's stamp or sticker	Y [stamp]	Y [sticker]	N	N	N/A	N	N	N

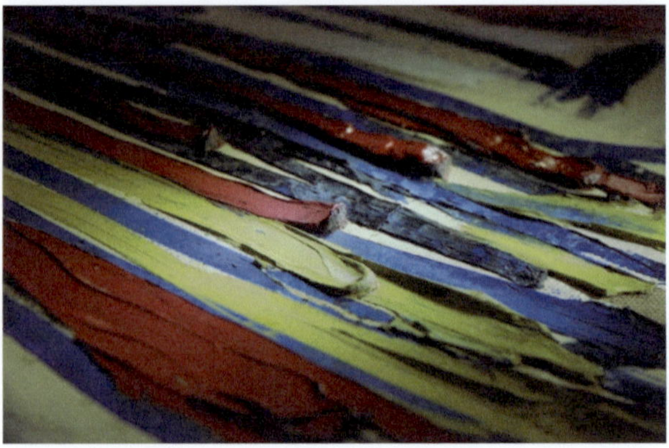

Fig. 17.1 The tube: an unexpected tool for painting c. 1910 on *La Créole au perroquet*; detail from the parrot's tail © Graczyk

oil-paint matter itself, thereby using ready-made canvases of standard sizes (Table 17.1). Parisian colourmen's stamps are frequently visible on the reverses (Bourgeois Aîné, Lefranc & Cie), revealing his suppliers. This recurrence allows us to differentiate which paintings were made in his Parisian workshop, and which were made in Angers. He often himself stretched the pre-primed canvas on the strainer or stretcher, as suggested by the marks left by the stretching pliers on the edges on the canvas.

Mérodack-Jeaneau chose generally fine weave linen pre-primed canvases with a white, beige, or grey oil ground, evenly spread to the edges of the canvas, available per square-meter or per roll in art suppliers' shops at the time. The canvas used for *La Créole* probably comes from such a roll as there are parallel bands visible in the colour of the canvas suggesting a prior rolling located at regular intervals on the reverse of the canvas.[6]

Directly on the ground, the painter began with a charcoal preparatory sketch and/or a blue coloured pencil (sometimes both), delineated later with blue oil paint (mainly cobalt blue) which consists of steady, confident lines.

17.3.2 Paint Layer

Keeping the sketch's marks visible, Mérodack-Jeaneau applies pure oil colours in fields and lines, leaving the ground visible. Paint is laid by brush but also by squeezing paint directly from the tube (Fig. 17.1), creating impastos up to 0.8 cm thick.[7]

[6] This type of marks is often observed on the reverse of oil-bound pre-primed canvases of more recent paintings by Pierre Soulages and Georges Mathieu [15] and can be related to a *couche grasse* added in the oily ground, or to the glue from the sizing (personal discussion with Leslie Carlyle).

[7] Generally speaking all the elements of the painting except the background are realized with this technique.

The latter is characterized by a ribbon-like shape, of uniform width with a slight indentation following the center of the stroke that results from the pressure of the rim of the tube. Mérodack-Jeaneau seems to be one of the first artists who painted directly with the tube in such an extensive manner. This technique gives his paintings a bold and striking appearance where the materiality of the painting is crudely visible. As far as our investigation goes, only Van Gogh, Kandinsky, Vlaminck, and van Dongen might have painted in a similar fashion at the beginning of the twentieth century,[8] but less extensively than Mérodack-Jeaneau.

The application of the paint directly from the tube onto canvas provides evidence of tube manufacturing around 1910. The widths of the tube strokes on the studied artworks measure 1 cm, 0.8 cm, and 0.3 cm, which corresponds to the diameters of the different sizes of paint tubes illustrated in the catalogues of the main art manufacturers of the time, such as Lefranc & Cie and Bourgeois Aîné ([16]: 5; [4]: 152).

Reconstructing the tube's painting technique gave very interesting details on the mode of application. One surprising detail was to find out that the black particles visible on paint impasto were charcoal and not dust or soil. When paint is applied using a tube, the charcoal particles from the underdrawing stick to the paint at the edge of the tube, so that it is lifted from the canvas. Then a subsequent tube stroke brings the charcoal to the top of the impasto.

On *La Créole,* tide-lines can be observed in visible and ultraviolet light around the impastos applied directly with the tube. This suggests that, because of the paint direct application, excess oil is not removed, and the resulting tide-lines are due to the oil diffusion into the ground. Interestingly, this excess oil has resulted in the preservation of the charcoal preliminary sketch on the painting, which was erased probably accidentally during the history of the artwork where it was not located in a tide-line area.

Mérodack-Jeaneau's paintings are thereby constituted of pure colours, with a recurrent and simple palette composed mainly of vivid mineral synthetic colours, applied by brush or tube. As an exception, mainly only two hues were applied solely with the brush: carbon black, probably to avoid siccative issues linked with the slow or poor drying or because of a particular rheology; and lead white, perhaps because of its specific consistency. This direct technique gives a rare opportunity to study the natural ageing of tubed-on impastos after a 100 year period.

---

[8] Van Gogh appears to be the first artist to use the tube to apply paint ([23]: 30) followed by Kandinsky at the beginning of his career, notably on *Landschaft bei Regensburg*, (1903) as discussed with Iris Schaefer (Wallraf-Richartz-Museum). Vlaminck discusses the use of the tube to paint in his writings [6, 12] and it was also observed on some of his paintings.

17.4 Degradation Phenomena

The development of tube-paint[9] was generally considered a major advancement, rendering "(…) bad paintings as sustainable as good ones".[10] The development of tube paints led to significant changes in paint formulation ([5]: 178) as various additives,[11] such as metal soaps, additional oils, beeswax begun to be used in order to stabilize paints for long-term storage and handling purposes. Already at that time some painters and chemists expressed doubts about the quality of manufactured paints ([29]: 97–292, [3]: 217–219, [27]: 6, [11]: 35–36, 107–108) owing to the observation of some unusual early deterioration phenomena such as discolourations, darkening of colours, and drying cracks.

Coming back to the study, the most important damages can be observed in some of the tubed-on colours, mainly cobalt blue, ultramarine blue, chrome yellow and organic lakes. In those areas, there is a severe lack of adhesion between the ground and canvas, and also between the paint and ground. The delaminating paint flakes lift up to 0.5 cm from the canvas and have resulted in paint losses. The ultramarine blue, whether applied thick or thin, is strongly underbound: the superficial blue crust fell off in several places leaving visible the interior of the paint, a phenomenon that has been reported previously [21].

Regarding the surface, cobalt blue, when applied thick, appears black although the inside of the paint is blue. Under the microscope, white protrusions, foam-like formations and medium-exudates can be seen on almost all of the colours. Discolouration of an organic lake, efflorescence and deposits (Fig. 17.2) were observed on the surface, as well as black spots and discolourations on cerulean blue, and specific white spots on a pink organic lake. In addition, cobalt blue, vermilion red, chrome yellow, and ultramarine blue are water-sensitive, and it was observed by application of a localized humid chamber that water vapour induces whitening of the surface of the ultramarine blue.

17.5 Analysis and Diagnosis

Scientific analyses were conducted on three paintings: at the Institut national du patrimoine (*La Créole au perroquet*), the Getty Conservation Institute (ibid.), and the Centre de Recherche et de Restauration des Musées de France (*La Montreuse de singe*, *L'Écuyère verte*). Results provided evidence to support hypotheses about the processes involved in the degradation phenomena.

---

[9] Patented in 1841 by the American John G. Rand, the tin-iron-tin laminate collapsible tube was invented precisely to contain paint.

[10] (personal translation); "Je crains qu'avec les couleurs de la Maison Lefranc, les mauvais tableaux soient aussi durables que les bons" ([17], np., cited by [25]: 320).

[11] Authors mention additives since the middle of the nineteenth century ([7]: 194).

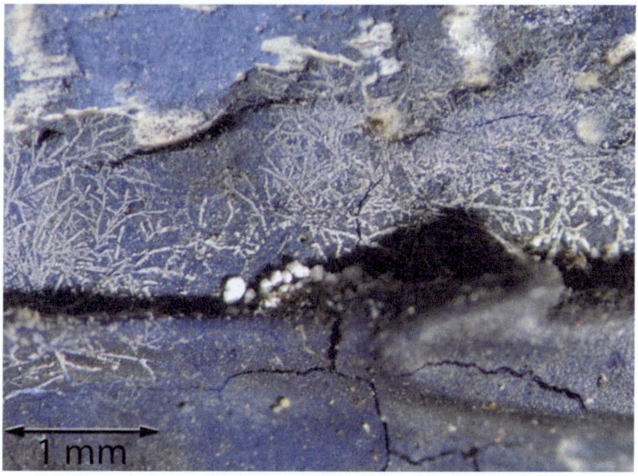

Fig. 17.2 Efflorescence on ultramarine blue on *La Créole au perroquet* © Graczyk

17.5.1 Grounds

Grounds are manufactured, oil bound, and composed of lithopone with lead white, or solely lead white (Table 17.2). The ground is more degraded under tube application areas, as observed under the microscope on *La Créole* (Figs. 17.3 and 17.4). In those areas the ground has yellowed and appears swollen. Observation of the ground under ultraviolet light reveals a strong fluorescence which is different from the unpainted ground which has no specific fluorescence, suggesting a particular deterioration of the ground in the tubed-on areas that might be linked with medium excess. Underneath the microscope, the ground layers under those areas appear to consist of several layers. The upper ground layer is quite transparent and yellow-brown with a distinct amorphous, lead-rich domain in the lower part. The ESEM-BSE image shows high atomic mass particulate substances, probably lead compounds, especially in the upper portion of the upper layer and in the interface between the two layers (Fig. 17.3). On *La Montreuse*, the ground consists again of two layers with a strong concentration of lead at the interface between the ground and the paint. On the two paintings analyzed, dissolution of pigment and metal soap formation are suspected in both ground layers, with upper layer being especially severely affected.

17.5.2 Colours

The deteriorated colours were analysed (ESEM-EDS, GC-MS), in particular the blues, their discolourations, and some surface efflorescence. The probable presence of additives was investigated. GC-MS analyses were performed on surface and bulk samples of ultramarine and cobalt blue (*La Créole)*. Beeswax was found to have accumulated toward the bottom of the paint on the ultramarine blue sample.

Table 17.2 Ground analyses

Painting	Colour	Elements	Interpretation	Analysis	Laboratory
La Créole au perroquet (c. 1910)	Beige: 2 layers.	Upper layer:	Lead compounds, both particulate and amorphous (organo-lead substance); barium sulphate[a], zinc compounds (lithopone?)	XRF	Inp
		Al, **Pb**/S, *(Ba), (Fe), (Zn)*		ESEM-EDS	GCI
		Lower layer: (3 sub-layers):			
		1c: **C**, O, *Al*, **Pb**/S, *(Ba)*, Zn			
		1b: **C**, O, *Al, Si*, **Pb**/*S*, *Ba, Zn*			
		1a: **C**, O, *Al, Si*, **Pb**/*S*, *Ba, Zn*			
La Montreuse de singe (1911)	Grey	Pb	Two layers of lead white different in grain size or concentration (?)	ESEM-BSE	C2RMF

Notes to table:
Key to notation: **bold** = major, normal = minor, *italic* = trace, *(italic)* = very trace
[a]Inferred by coincident abundance of Ba and S

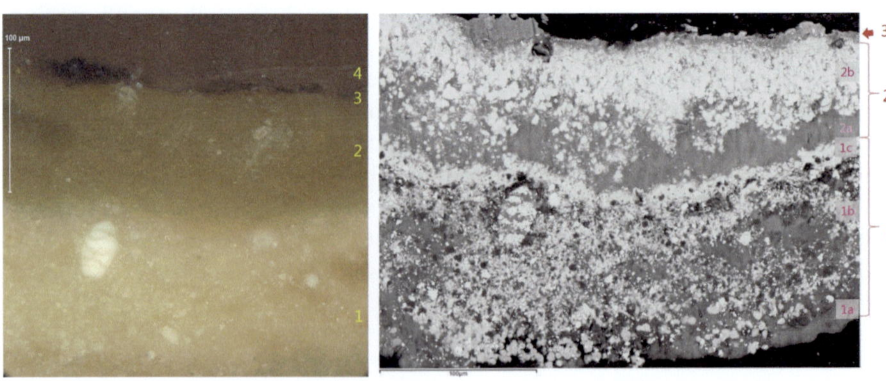

Fig. 17.3 Ground under cobalt blue paint from the parrot's wing (tubed-on paint) on *La Créole au perroquet*. *1*- Lower ground: lead compounds, both particulate and amorphous (organo-lead substance); barium sulphate, zinc compounds (lithopone?), *2*- Upper ground: lead compounds, both particulate and amorphous (organo-lead substance), *3*- Cobalt blue, *4*- Organic coating. (Photos by Alan Phenix © The J. Paul Getty Trust 2018)

Beeswax was used in paint industry as a thickener for ultramarine blue in oil, as the pigment alone in oil otherwise may lack body. It could be one of the components of the powdery deposits visible on the paint surface, in this regard qualified as efflorescence ([26]: 311–332; [22]), but no beeswax could be found on the surface.

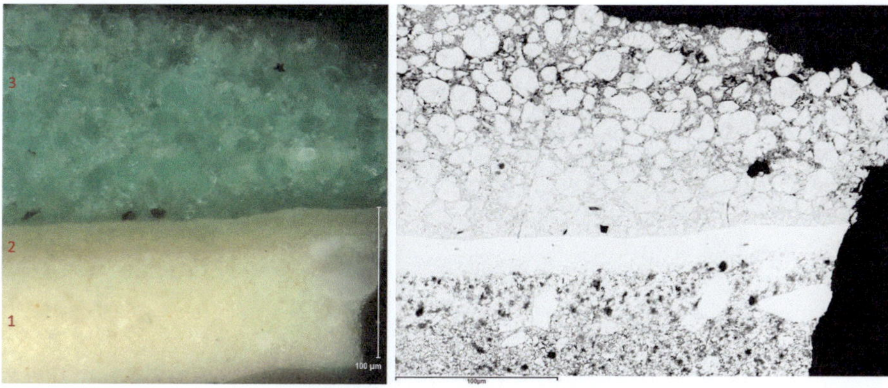

Fig. 17.4 Ground under green paint from the skirt (brushed-on paint) on *La Créole au perroquet*. *1*- Lower ground: lead compounds, barium sulfate, zinc compounds (lithopone?), *2*- Upper ground: predominantly lead compounds, barium sulfate, zinc compounds (lithopone?), *3*- Green paint: Copper-arsenic green. (Photos by Alan Phenix © The J. Paul Getty Trust 2018)

On the surface and in the cracks of the ultramarine blue *(La Créole),* a very thin layer of pigment-free material was observed at high magnification. It is quite brownish in colour in visible light and fluoresces weakly under UV. Selected elemental maps obtained by electron microscopy of a cross-section revealed ribbon-like crystals in the cracks in which sodium, oxygen and sulfur were present in coincident abundance, suggesting the presence of sodium sulfate in the ultramarine blue (Fig. 17.5). The most common mineral phases of sodium sulfate are thenardite, heptahydrate, and mirabilite or Glauber's salt. The latter is deliquescent over 80% RH ([30]: 324–326, [19]: 3) which means it would dissolve when breathed on.[12] Their crystals ([10]: 51) look quite similar to those observed on the surface of the painting (Fig. 17.2) which are deliquescent. Mirabilite is one of the most destructive salts for stone and other porous materials because it exhibits very pronounced and rapid phase changes in response to indoor climate variations (a whisker of mirabilite can grow in less than a day). Occurrence of highly water-soluble sodium sulfate in the ultramarine blue paint could contribute to the observed water sensitivity[13] and underbinding.[14] A known cause of the water-sensitivity of some artists' oil paints is the presence of a different soluble sulfate salt, magnesium sulfate heptahydrate (Epsom salt), formed by reaction of a magnesium carbonate extender with airborne sulfuric acid pollutant [28]. What is uncertain is whether the sodium sulfate is a

[12] "The commercial product forms large, transparent crystals of the monoclinic system, with 55–76% of water of crystallization, readily efflorescing in the air" ([31]: 317).

[13] High azelaic acid content was also noted (Mirabaud and Genachte-Le Bail, Inp, report Inp-DR.16–11; [2]).

[14] Crystal formation within the paint creates stresses that can alter the binding network of the paint.

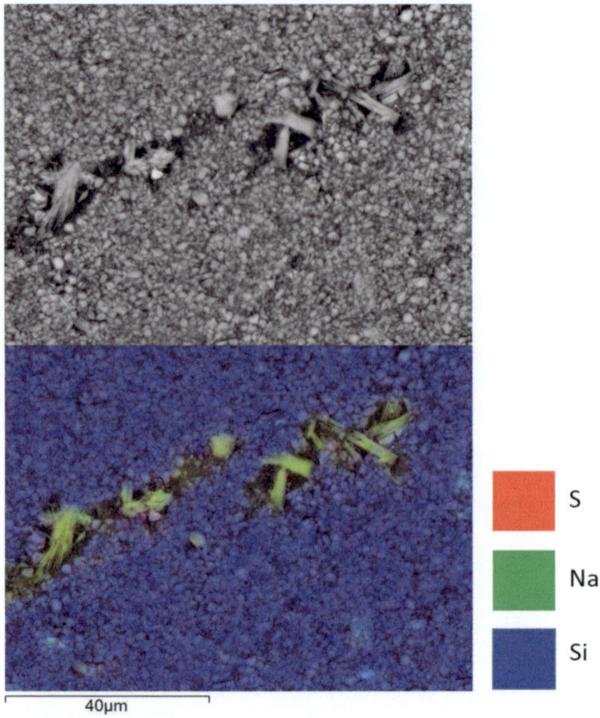

Fig. 17.5 *Top*: Backscattered electron image of ribbon-like crystals formed in cracks in the ultramarine blue paint sample from *La Créole au perroquet*. *Bottom*: Coloured 'phasemap mix' overlaid on BSE obtained using Oxford INCA SEM-EDS software phase mapping tool, with red for sulfur, green for sodium, and blue for silicon. The ribbon-like crystals render as yellow (red + green) and are inferred as sodium sulfate. (Photos by Alan Phenix © The J. Paul Getty Trust 2018)

pigment degradation product or residual raw material from manufacture[15] that was not properly removed by washing prior to incorporation in the paint.[16]

Concerning the black spots on the cerulean blue areas on *L'Écuyère verte*, ongoing analysis seems to point out the presence of lead sulphide (Nathalie Balcar, C2RMF, report n°39,019) of unidentified origin.

In closing, analyses of deteriorated areas painted with the tube show interaction with the ground, metal soaps and dissolution of pigment, which may support the link between paint application [direct with the tube] and deterioration. Sodium sulfate formation may contribute to water-sensitivity, underbinding of paint and possibly constitutes the white deliquescent deposits on the surface. The residual internal stresses of the layer, due to its thickness, might have led to the early formation of

[15] Artificial ultramarine is prepared by heating china clay with sulphide, soda, carbon, and Glauber's salt (sodium sulfate), of which it should be free because this may cause efflorescence ([9]: 78, [24]: 55–65).

[16] It might also depend on the grade of the ultramarine ([31]: 178).

cracks and to delamination of the paint [14]. It is likely that environmental conditions have had a significant impact on the deterioration of the paintings due to temperature and relative humidity variations.

17.6 Conservation Treatment[17]

The heavy impastoed lifting flakes required a demanding consolidation process (*La Montreuse de singe* and *La Créole au perroquet)*. In order to flatten the lifting flakes, two different treatments were processed depending on the presence of varnish or not, as the surface reacted differently. On the varnished painting (*La Montreuse*), several layers of facing paper strips with glue paste were temporarily applied and allowed to dry between each layer of paper on the varnished surface prior to consolidation with Paraloïd B72® 10% in ethyl acetate (w/v). On the unvarnished one (*La Créole*), concentrated Funori glue was introduced under the flakes with a syringe on a small suction table located at the back of the canvas. Local heating with an infrared lamp or a hot air gun and several applications of glue were necessary to return the raised flakes into place.

The underbound unvarnished ultramarine blue was consolidated using a 1:1 mixture of sturgeon glue and Funori at 0.5% final concentration in water (w/v) applied with a brush through hemp paper, the whole on a suction table located at the rear of the canvas. This process was chosen because it also saturated the efflorescence present on its surface by action of the aqueous solution without affecting the aspect of the surface.

The non-original PVA-varnish,[18] applied probably around the 1980–1990's on three of the paintings, gave them a very glossy appearance whereas Mérodack-Jeaneau left them unvarnished in the first place.[19] In discussion with curator Christine Besson from the musée des Beaux-Arts of Angers, it was decided to remove them. A mixture of organic solvents (xylene / methylethyl ketone 1:3), as no water-based or gelled method was satisfactory despite extensive tests, was found to dissolve the resin fast enough without inducing solvent sensitivity of the paint. On *La Danseuse Jaune*, it was found that another solvent mixture (white spirit / butyl acetate 4:1) heated to 40° Celsius was satisfactory (the varnish could be removed without affecting the paint layer only at this temperature). The treatment had to be adapted on the yellow colour with the aspiration-suction machine adapted for conservation ([8]: 315–325), vermilion red, ultramarine and cerulean blues because of pronounced solvent sensitivity.

[17] The structural treatments carried out on some of the paintings to stabilize tears and deformations of the canvas are not presented here.

[18] FTIR analysis on varnish samples identified polyvinyl-acetate resin internally plasticized with VeoVA (Nathalie Balcar, C2RMF, reports n°39,018 and n°39,019).

[19] Three out of the eight paintings are originally unvarnished.

17.7 Conclusion

This project was a great opportunity to get to know Mérodack-Jeaneau's novel technique and work habits of applying paint directly from and with the tube. His paintings constitute today 100 year-old examples of the degradation of tube-applied paint impastos of commercially manufactured paint, suggesting a link between paint application and deterioration. An alteration phenomenology showing that severe mechanical and chemical alterations occurred in areas painted directly with the tube was established on all of the paintings of the corpus. Occurrence of highly soluble sodium sulfate in ultramarine paint may be one factor contributing to its water-sensitivity. Coherence in the conservation treatment was ensured thanks to scientific analysis and knowledge about the materials and their sensitivity.

Acknowledgments This project was realized with the scientific input of Nathalie Balcar (C2RMF), conservation support of Gilles Barabant (C2RMF), and the confidence of Christine Besson (musée des Beaux-Arts, Angers).

Appendix: Experimental Details for Analysis

GC-MS performed by Sigrid Mirabaud, Inp, was used for the separation and identification of organic components in the paint layers of samples from *La Créole au perroquet*. A trace GC Ultra gas chromatograph (Thermo Scientific) and an ISQ Single quadrupole Mass Spectrometer (Thermo Scientific) were used with the following specifications: non-polar fused silica capillary column (Rxi 5Sil MS, 20 m, 0.18 mm di, 0.18 mm df Restek); carrier gas H2 at 1.5 ml/min; temperature program: 50 °C for 1 min, ramp of 20 °C/min until 320 °C, hold 320 °C for 5 min; split injector set at 290 °C with a total flow of 50 mL/min and a split ratio of 15 (except sample P03 which was injected in the splitless mode); automated injection; interface temperature set at 320 °C; electronic impact ionisation mode at 70 eV, ion source temperature 220 °C. The samples required a derivatisation step prior to injection, which included mixing fragments of the samples with 50–100 μL of Meth-Prep II (Grace Alltech) and heating at 60 °C for 30 min prior to injection. This transmethylation is needed to hydrolyse the glyceride and terpenoid components of the binder, and convert the polar groups (acids and alcohols) into the corresponding methyl esters and ethers.

ESEM-EDS performed by [A] Alan Phenix, GCI, and [B] Nathalie Balcar, C2RMF

[A] samples from *La Créole au perroquet*: cross-sections by embedded in acrylic resin Technovit 2000 LC, first examined microscopically using a Leica DM4000 microscope, and analyzed by Environmental Scanning Electron Microscopy with Energy-Dispersive X-ray Spectroscopy (ESEM-EDS) using a

Philips XL30 ESED-FEG instrument with Oxford INCA EDS analysis software. ESEM-EDS conditions: H_20 mode, 10 mm working distance, 20 kV accelerating voltage, and 1.0 torr water vapour pressure in the chamber.

[B] samples from *La Montreuse de singe* and *L'Écuyère verte*: the identification of pigments and degradation products were carried out on samples embedded in polyester resin, after a wet polishing and a carbon coating. The instrument used for elemental analyses was a scanning electron microscopy (SEM, Philips XL30CP) with energy dispersive spectroscopy (EDS analysis, SDD Oxford S-Max 50); acceleration voltage of 15 kV, software AZTEC.

XRF performed by [A] Anne Genachte-Le Bail, Inp, and [B] Eric Laval, C2RMF

[A] In situ analysis on *La Créole au perroquet* with portable x-ray fluorescence spectrometer Bruker ARTAX800 (excitation source: molybdenum, voltage: 50 kV, intensity 600 μA, duration of analysis: 100 s, atmosphere: helium (1 L/min)), number of analysed spots: 21.

[B] In situ analysis on *L'Écuyère verte* [1] and *La Montreuse de singe* [2] with portable x-ray spectrometer Elio from XGLab (excitation source: transmission X-Ray generator, 5–200 μA, 10–40 kV, Rh anode (Ag, Au, Mo, W)), number of analysed spots: 14 [1], 16 [2].

FTIR spectroscopy performed by Nathalie Balcar, C2RMF

Samples of varnish from *La Montreuse de singe* and *L'Écuyère verte* were analyzed with a handheld AGILENT FTIR Exoscan 4100 spectrometer with ATR Diamond (4000-and 650 cm^{-1} range at a resolution with 8 cm^{-1} and by summing 32 scans), MicroLab software.

References

1. Apollinaire G (1960) Chroniques d'art (1902–1918). Gallimard, Paris, p 350
2. Banti D, La Nasa J, Lluveras Tenorio A, Modugno F, van den Berg KJ, Lee J, Orsmby B, Burnstock A, Bonaduce I (2018) A molecular study of modern oil paintings: investigating the role of dicarboxylic acids in the water sensitivity of modern oil paints. RSC Adv., 8, pp 6001–6012
3. Bellanger C (1907) Le peintre : traité usuel de peinture à l'usage de tout le monde. Garnier, Paris, p 217–219
4. Bourgeois Aîné (1913) Catalogue illustré Couleurs fines et matériel d'artistes. Paris, p 152
5. Burnstock A, van den Berg KJ, Groot S, Hinde L (2007) An Investigation of Water-Sensitive Oil Paints in Twentieth-Century Paintings. In: Learner T, Smithen P, Krueger J, Schilling M (eds) Modern Paints Uncovered. 2006, Getty Publications, Los Angeles, p 178
6. Campa L (2013) Guillaume Apollinaire. Gallimard, Paris. https://books.google.fr/books?id=jQ9AAQAAQBAJ&pg=PT198&dq=derain+tube+cobalt&hl=fr&sa=X&redir_esc=y#v=onepage&q=touches%20mobiles&f=false. Accessed 25 Sept 2018
7. Carlyle L (2001) The artist's assistant: oil painting instruction manuals and handbooks in Britain, 1800–1900, with reference to selected eighteenth-century sources. Archetype Publications, London p 194

8. Cremonesi P (2018) Combination of a liquid-dispensing and micro-aspiration device for the cleaning of sensitive painted surfaces. Studies in Conservation 63, 6, p 315–325.
9. Doerner M (1934) The materials of the artist and their use in painting. Harcourt, Brace and Company, New York, p 78
10. Denecker M (2014) Le rôle des sulfates de sodium dans l'altération des pierres du patrimoine bâti: méthodes indirectes d'identification pour l'approche expérimentale. PhD thesis, Université de Cergy Pontoise, p 51
11. Dinet E (1926) Les fléaux de la peinture, Moyens de les combattre. Laurens, Paris, pp 48–49
12. Franck D (2015) Le temps des bohèmes. Grasset, Paris. https://books.google.fr/books?redir_esc=y&hl=fr&id=SDeHCgAAQBAJ&q=%C3%A9crasant+ses+tubes#v=snippet&q=%C3%A9crasant%20ses%20tubes&f=false. Accessed 25 Sept 2018
13. Graczyk A (2016) A pleins tubes ! Le tube de peinture utilisé comme outil de mise en œuvre dans les années 1910 pour *La Créole au perroquet* par Alexis Mérodack-Jeaneau (musée des Beaux-Arts, Angers). Postgraduate diploma in the Conservation of easel paintings, Institut national du patrimoine, Paris
14. Hélou-de La Grandière P, Le-Hô AS, Mirambet F (2008) Delaminating paint films at the end of 1950s: a case study on Pierre Soulages. In: Joyce H. Townsend (ed) Preparation for Painting: the artist's choice and its consequence, ICOM interim meeting on paintings. Archetype Publications, London, pp 156–162
15. Hélou-de La Grandière P (2018) A Montparnasse disease? Severe Expressions of Metal Soaps in Paintings by Pierre Soulages from around 1959–1960. In: Casadio F, Keune K, Noble P, van Loon A, Hendriks E, Centeno S, Osmond G (eds) Metal Soaps in Art: Conservation and Research. Springer (in press)
16. Lefranc et Compagnie [1910] Catalogue Couleurs fines pour la peinture à l'huile, la gouache, l'aquarelle, le pastel [...]. Paris, p 5
17. Lefranc (1925) Les peintures et la couleur. Paris, 1925, np.
18. Le Nouëne P (1997) Alexis Mérodack-Jeaneau dans les collections du musée des Beaux-Arts d'Angers. Musées d'Angers, Angers, pp 9–33
19. Lopez-Arce P, Doehne E (2006) Kinetics of Sodium Sulfate as observed by humidity cycling with ESEM. In: Heritage, Weathering & Conservation Conference, Madrid 21–24 June, Consejo superior de investigaciones cientificas, Madrid, p 3
20. Mérodack-Jeaneau A (1912) L'Ouest, 2 août 1912, pp 1–2
21. Mills L, Burnstock A, Duarte F, de Groot S, Megens L, Bisschoff M, van Keulen H, and van den Berg KJ (2008) Water sensitivity of modern artists' oil paints. In: Bridgland J (ed) ICOM Committee for Conservation, 15th triennial conference, New Delhi, 22–26 September 2008: preprints. ICOM Committee for Conservation, Paris, pp 651–659
22. Ordonez E, Twilley J (1998) Clarifying the haze, efflorescence on works of art. WAAC Newsletter, vol. 20 / 1. https://cool.conservation-us.org/waac/wn/wn20/wn20-1/wn20-108.html. Accessed 27 Sept 2018
23. Peres C, Hoyle M, van Tilborgh L (1991) A Closer look: technical and art-historical studies on works by Van Gogh and Gauguin. Waanders Publisher, Zwolle, p 30
24. Plesters J (1968) Ultramarine blue, natural and artificial. In: Roy A, Artists' Pigments: a Handbook of their history and characteristics. New York, Oxford, pp 55–65
25. Roth-Meyer C (2004) Les marchands de couleurs à Paris au XIXème siècle. PhD thesis, Université Paris IV, p 320
26. Sawicka A, Burnstock A, Izzo FC, Keune K, Boon J, Kirsch K, van den Berg KJ (2014) An Investigation into the Viability of Removal of Lead Soap Efflorescence from Contemporary Oil Paintings. In: van den Berg KJ, Burnstock A, Keijzer M, Krueger J, Learner T (eds) Issues in Contemporary Oil Paint. Springer International Publishing, 2014, pp 311–332
27. Sennelier G (1912) Chimie des couleurs : Fabrique de couleurs fines pour les arts. Paris, p6

28. Silvester G, Burnstock A, Megens L, Learner T, Chiari G, and van den Berg KJ (2014) A cause of water-sensitivity in modern oil paint films: the formation of magnesium sulphate. Studies in Conservation 59, 1, pp 38–51
29. Vibert JG (1981) La science de la peinture. Gutenberg reprint, p 97–292
30. Zehnder K, Schoch O (2009) Efflorescence of mirabilite, epsomite and gypsum traced by automated monitoring on-site. Journal of Cultural Heritage, 10, pp 324–326
31. Zerr G, Rubencamp R (1908) A treatise on colour manufacture: a guide to the preparation, examination, and application of all the pigment colours in practical use. C. Griffin, London, p 178–317

Chapter 18
Picasso 1917: An Insight into the Effects of Ground and Canvas in the Failure Mechanisms in Four Artworks

A. Vila, A. Murray, C. K. Andersen, F. C. Izzo, L. Fuster-López, E. Aguado-Guardiola, R. Jiménez-Garnica, and A. Scharff

18.1 1917 Paintings

Four paintings by Picasso from the Museu Picasso collection (Barcelona) painted in 1917 were examined: *Seated Man*[1] (Fig. 18.1a), *Man with Fruit Bowl*[2] (Fig. 18.1b), *Woman on an Armchair*[3] (Fig. 18.1c) and *Blanquita Suárez*[4] (Fig. 18.1d) [7]. Despite presenting similarities in materials, technique and concept and having been displayed together in Picasso's family residence in Barcelona from 1917 to 1970, the paintings show very different degrees of cracking. The hypothesis is that differences in the composition of canvas and ground layers have contributed to the degree of cracking, from one painting with deep cracks throughout to another with no cracks. Fortunately, according to the principle of proofed fluctuations [11], it is unlikely that the fracture damage will increase in the present environment since the paintings have already been through stress relaxation and cracking caused by the less

[1] *Home Assegut,* MPB 110.005, oil on canvas, 104 × 54 cm.

[2] *Home amb Fruiter*, MPB 110.006, oil on canvas, 100 × 70 cm.

[3] *Dona en Butaca,* MPB 110.007, oil on canvas, 92 × 64 cm.

[4] *Blanquita Suárez,* MPB 110.013, oil on canvas, 73 × 47 cm.

A. Vila (✉)
Art Exhibition Department. Culture and Scientific Dissemination Area, La Caixa Foundation, Barcelona, Spain
e-mail: anna.vila@fundaciolacaixa.org

A. Murray
Art Conservation Program, Queen's University, Kingston, ON, Canada
e-mail: alison.murray@queensu.ca

C. K. Andersen · A. Scharff
Institute of Conservation, The Royal Danish Academy of Fine Arts, Schools of Architecture, Design and Conservation, Copenhagen, Denmark
e-mail: cka@kadk.dk; abs@kadk.dk

K. J. van den Berg et al. (eds.), *Conservation of Modern Oil Paintings*,
https://doi.org/10.1007/978-3-030-19254-9_18

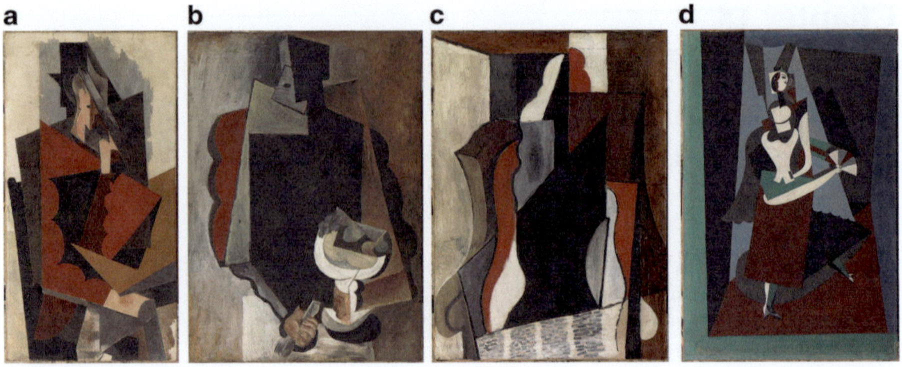

Fig. 18.1 (**a**, **b**, **c**, **d**). (**a**) *Seated Man;* (**b**) *Man With A Fruit Bowl* (**c**) *Woman In An Armchair* (**d**) *Blanquita Suárez*. Pablo Picasso, Barcelona, 1917, Oil on canvas

controlled climate of the Picasso residence. They have in other words become less vulnerable due to this initial response to former climate fluctuations.

All four paintings have thin paint layers on top of a thick ground and a cotton canvas. *Seated Man* is, however, more affected by cracks than the other paintings in the series. A deep crack network covering the whole painted surface was observed in the paint, regardless of colour. A less dense crack pattern was evident in the areas coincident with the stretcher bars, suggesting that protection of the back of the canvas was afforded by the stretcher [8]. *Blanquita Suárez* showed similar phenomena while within the boundaries of the stretcher bars, locally extensive, deep cracks can be observed. *Man with Fruit Bowl* has no significant cracks whereas *Woman on an Armchair* exhibited varied surface phenomena and the paint shows a few cracks in the areas that are not coincident with the stretcher. The film appears more compact

F. C. Izzo
Dipartimento di Scienze Ambientali, Informatica e Statistica, Università Ca' Foscari Venezia, Venice, Italy
e-mail: fra.izzo@unive.it

L. Fuster-López
Instituto Universitario de Restauración del Patrimonio, Universitat Politècnica de València, Valencia, Spain
e-mail: laufuslo@crbc.upv.es

E. Aguado-Guardiola
Escuela Superior de Conservación y Restauración de Bienes Culturales de Aragón/Andrew W. Mellon Conservation Researcher for the Barrié Foundation's Santiago Cathedral Program, Huesca, Spain
e-mail: eaguado@escyra.es

R. Jiménez-Garnica
Museu Picasso, Barcelona, Spain
e-mail: rjimenezga@bcn.cat

than in the other three paintings studied. *Seated Man* and *Woman on an Armchair* each showed evidence of local water stains on the back of the canvas, but this cannot explain the differences in overall conditions.

18.2 Materials and Methods

Canvas analysis was carried out by taking close-up photos of the back of the original canvases, including a scale bar, and analysing them using the open-source software ImageJ. The weave and spinning geometry were measured on ten different threads, the width of the threads and degree of spinning were determined, and the cover factor of the fabric was calculated in both weave directions. Cover factor is the combined coverage of an area with threads [9]. Identification of fibres was performed with polarized light microscopy.

Cross-section samples from the edge of each painting were taken and embedded in Technovit 2000 LC in an EasySection®. The first part of the polishing was carried out wet. The last part of the polishing was done dry using Micromesh with grits up to 12,000.

Cross-section micrographs were taken with a Leica DM 4000 M microscope with Leica Application Suite 4.1 software. Incident light was used with the filter settings Dark Field (DF), Bright Field (BF), Differential Interference Contrast (DIC), and ultraviolet (UV) filters A and I3.

A JEOL JSM 6300 scanning electron microscope operating with a Link-Oxford-Isis energy-dispersive X-ray microanalysis system was used to observe surface and cross sections of the films. Data was collected with 20 kV accelerating voltage, 2.10–9 A beam current, and 15 mm working distance.

A selection of the collected samples was analyzed by micro-Fourier transform infrared spectroscopy (μ-FTIR) in attenuated total reflectance (ATR) and transmittance mode with a Thermo Fisher Scientific Nicolet iN10 MX Ultra Fast Motorized stage with MCT Detector. The spectra were acquired between 4000 and 600 cm^{-1}, 128 scans, with a resolution of 4 cm^{-1}.

Due to the complex and different nature of organic materials present in modern paint samples, two different analytical procedures were used:

Procedure 1 -Identification of lipidic, waxy and resinous materials

The samples were trans estherified using (trifluoromethylphenyl) trimethylammonium hydroxide, overnight reaction. The procedure and the details are described in several publications [2–4].

Procedure 2 -Identification of proteinaceous and polysaccharidic materials

The samples were analysed as N,O-acetyl methyl esters after hydrolysis with trifluoroacetic acid and derivatisation. GC-MS analysis was performed using an Agilent 6890NGC instrument with a capillary Carbowax Column, 30 m, 0.20 mm, 0.5 μm interfaced with an 5973NetworkMS. The temperature program was: 80–250 °C, 20 °C/min until 210, then 2 °C/min until 250 °C. The MS was run in Full Scan mode (m/z 40–500). Mannitol and Norleucine were used as internal standards for sugars and amino acids, respectively.

18.3 Results and Discussion

Analyses have revealed minor but important differences related to the specific use of some materials in both canvases and grounds.

18.3.1 Analysis of Canvas

The canvases used for all four paintings were identified as mercerized cotton. The weave geometry showed a significant difference between the canvases (Table 18.1). Threads showed similar thickness and degree of spinning, while the canvas in *Seated Man* was more tightly woven with cover factors of 71% and 85% in each weave direction, whereas the other paintings only had cover factors of around 50% in each weave direction. This signifies a tight weave in the *Seated Man*, which is indicative of canvases that are prone to shrinkage [1]. Moreover, it has been suggested that a hygroscopic layer containing animal glue combined with a tightly woven canvas is more prone to develop stress and possibly cracking than if the canvas is a more open weave [5].

18.3.2 Analysis of Ground Layers

Table 18.2 shows optical microscopy (OM) and SEM backscattered electron (BSE) cross-section images of samples taken from the edges where mainly the ground layers are present. All four paintings have two ground layers: the OM cross-sections show a dark brown layer beneath the double ground that is highly fluorescent in UV light and includes small particles. In all four paintings the upper ground appeared more compact with slight variations in tonality and elemental composition characterised using SEM-EDX as summarised in Table 18.2.

The lower ground layer of *Seated Man* contains lead (Pb) and/or sulfur (S) together with relatively lower amounts of calcium (Ca), barium (Ba), potassium (K)

Table 18.1 Study of support (weave geometry) of the four case studies

"Seated man" (MPB 110.005)		*"Woman on an armchair"* (MPB 110.007)		*"Man with fruit bowl"* (MPB110.006)		*"Blanquita Suárez"* (MPB 110.013)	
		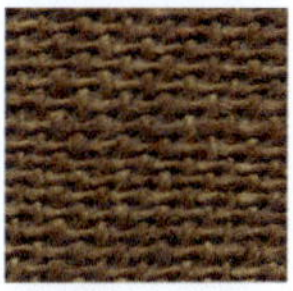		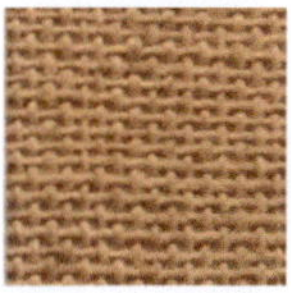		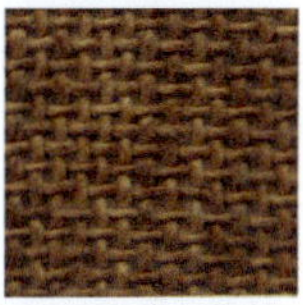	
Thread count per cm (vertical x horizontal)	22 × 21	Thread count per cm (vertical x horizontal)	13 × 15	Thread count per cm (vertical × horizontal)	16 × 17	Thread count per cm (vertical x horizontal)	13 × 14
Vertical threads relative cover factor	0,71	Vertical threads relative cover factor	0,52	Vertical threads relative cover factor	0,53	Vertical threads relative cover factor	0,57
Horizontal threads relative cover factor	0,85	Horizontal threads relative cover factor	0,50	Horizontal threads relative cover factor	0,45	Horizontal threads relative cover factor	0,51
Width of threads in mm, vertical	0,323	Width of threads in mm, vertical	0,403	Width of threads in mm, vertical	0,31	Width of threads in mm, vertical	0,435
Width of threads in mm, horizontal	0,403	Width of threads in mm, horizontal	0,330	Width of threads in mm, horizontal	0,28	Width of threads in mm, horizontal	0,341

and chlorine (Cl). The upper ground layer presents relatively high amounts of Pb and/or S and relatively lower amounts of Ba, Ca, Si and strontium (Sr). The results suggest the presence of lead white as a major component in the upper layer of this painting. This was confirmed by the micro-FTIR results that identified basic lead carbonate in the white preparation layer together with calcium carbonate and barium sulphate.

Elemental analysis of the preparation ground layers from *Woman on an Armchair* found that the lower layer contains Pb/S and Ca with traces of magnesium aluminium, sodium, phosphorous, potassium and chlorine while the upper ground layer contained relatively high amounts of lead and calcium and magnesium with lower proportions of silicon and sodium. Micro-FTIR analysis of the lower brown layer identified both a lipidic and proteinaceous organic compounds and inorganic materials, such as $CaCO_3$ and silicates. In the upper whitish layer, micro-FTIR analysis characterised basic lead carbonate and calcium carbonate mixed with a lipidic binder. Thanks to the typical IR absorptions at around 1510 and 1460 cm^{-1}, the presence of lead carboxylates was also observed, that may have affected the mechanical properties of the ground layers [13].

Table 18.2 Optical microscopy (OM), SEM backscattered electron (BSE) cross-section images and SEM-EDX results of samples taken from the edges. In MPB 110.007, part of the top surface of the upper layer of this stratigraphy shows the presence of a thin white layer of a denser nature and a slightly different fluorescence, as seen in the OM and SEM BSE images. The study of the surface might indicate that it is part of the whitish paint layer of the painting

	"Seated Man" (MPB 110.005)	*"Woman on an armchair"* (MPB 110.007)	*"Man with a Fruit Bowl"* (MPB110.006)	*"Blanquita Suárez"* (MPB 110.013)
Optical microscopy (OM)	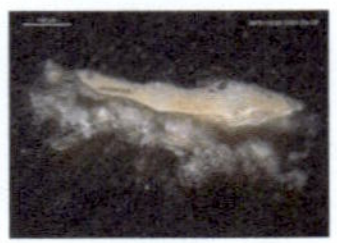	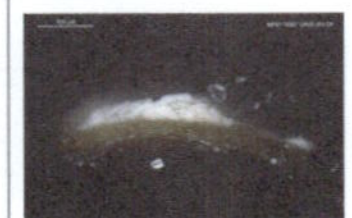	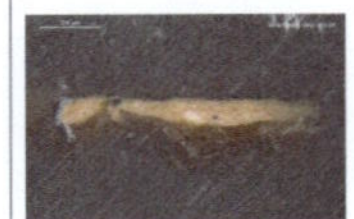	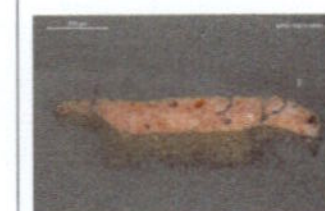
	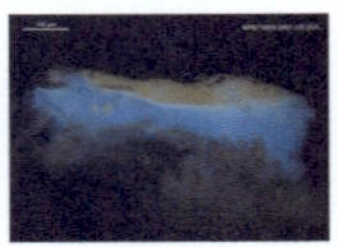	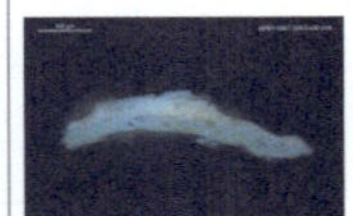	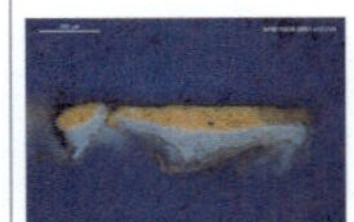	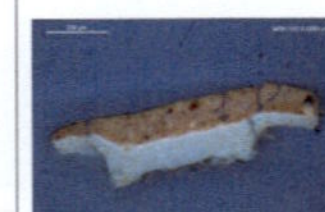
SEM-BSE	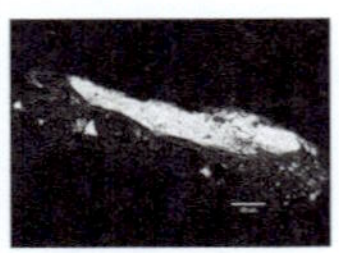	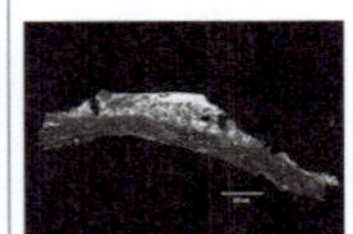	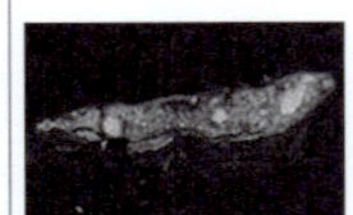	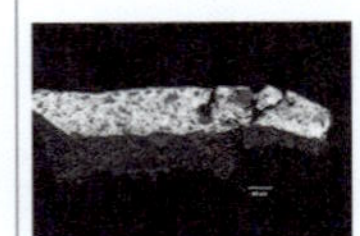
SEM-EDX	Upper layer			
	Pb/S (Ba, Ca, Si, Sr)	Pb (probably from the white paint layer close by)	Pb/S, Ca, Ba, Zn (Si, Sr, K, Mg)	Pb, Ca, Fe (Mg, Al, Si, Ti?)
		Pb, Ca (Mg, Si, Na) + carboxylates FTIR		
	Lower layer			
	Pb/S (Ca, Ba, K, Cl)	Pb/S, Ca (Mg, Al, Na, P, K, Cl)	Ca, S, Mg (Si, Al, Na, Cl)	Pb, Ca (Si, Al, Mg)

Elemental analysis of the ground layers of *Man with Fruit Bowl* showed that the lower ground included Ca, S, and Mg and relatively small amounts of Si, Al, Na and Cl and an ochre upper ground layer composed of relatively high amounts of zinc (Zn), Ba and Ca, together with Pb and/or S, and relatively lower amounts of Si, Sr, K and Mg. Micro-FTIR has also confirmed the presence of calcium carbonate and barite. It might be concluded that the upper ground contains chalk and lithopone and/or zinc white (ZnO), zinc sulfide (ZnS) and barium sulfate. ZnO containing paint layers in oil media are known to be stronger but also brittle, having a tendency to delaminate rather than to crack [12]. Although no cracks were found in Man with Fruit Bowl, no delaminations were evident either.

Different elements were identified in the lower and upper ground layers of *Blanquita Suárez*: in the lower brownish coloured ground layer, lead, calcium and relatively lower amounts of Si, Al and Mg were found. Furthermore, in the orange upper ground layer, relatively high amounts of Pb, together with iron (Fe) and Ca, and relatively lower amounts of Mg, Al, Si and possibly Ti were identified. Micro-FTIR confirmed the presence of lead and calcium carbonates together with alumino-silicates. These silicates (together with traces of titanium) are characteristic of natural iron oxides responsible for the orange colour of the upper ground layer.

Based on GC-MS results, in all the case studies, the grounds contained a proteinaceous binder, which was identified as animal glue indicated by the amino acidic profile and the detection of hydroxyproline. This is the specific biomarker for animal glue, typically used as a binder in the execution of ground layers. This finding points to the presence of animal glue, particularly in *Seated Man* and *Blanquita Suárez*, which exhibited characteristic fluorescence under UV illumination and the typical IR absorption bands of a proteinaceous compound. Animal glue has been previously recognized as responsible for stress in ground and paint layers due to its high response to relative humidity [6, 10]. Furthermore, a lipidic fraction (containing both mono- and di-fatty acids) was detected in all the ground layers, possibly due to the presence of a drying oil. In *Man with Fruit Bowl*, the oil fraction was high and its presence was also corroborated by micro-FTIR where the binder of the ground was identified as a mixture of lipidic and proteinaceous compounds, characterised by the typical IR absorption at 1740–1460–1373-1090-720 cm^{-1} and 3390–1620 cm^{-1} respectively. Finally, and considering that the paintings have not been lined, the presence of polysaccharidic compounds (glucose and galactose derivatives) identified in the grounds of all four paintings was possibly from starch that could have been mixed into the grounds or from a pre-treatment of the canvas as part of the manufacture proceedings.

18.4 Conclusion

This paper presents the results of the first stage of a project where the canvas and ground layers of four paintings by Picasso were analysed. The visual observations associated with diverse kinds of mechanical deterioration in the paintings suggests that the material differences in the works created by Picasso in Barcelona within the same year have caused various degrees of vulnerability to environmental and chemical deterioration.

Material characterization has revealed some key differences between the grounds of the paintings. Three of the four paintings included an upper ground layer composed mainly of lead white and other materials that result in slight differences in the tonality of the layer. *Man with a Fruit Bowl* is the only painting in this group that contains a Zn compound in the upper ground layer. Lead carboxylates were identified in *Woman on an Armchair*. All four paintings had grounds that contain animal glue, a possible drying oil, and polysaccharides as detected by GC-MS. In particular cases, some compounds are in relatively higher amounts, such as the polysaccharide compounds in *Blanquita Suárez* that could be related to the sizing recipes.

It is proposed that material differences in the canvas and grounds can be related to the degree of cracking observed in the paintings. For example, *Seated Man* exhibited the most extensive cracking, which could be explained by a high proportion of animal glue in or below the ground that responded rapidly to fluctuations in moisture while the paint was stiffer and slower to respond. Animal glue is known to shrink both at low and very high relative humidity and furthermore the tightly woven canvas support used for this painting rendered it prone to shrinkage at high humidity. *Woman on an Armchair* has large, shallow cracks that could be explained by the combination of animal glue and possibly the carboxylates that can make a film stiffer and somewhat brittle. *Blanquita Suárez* also has locally extensive cracks, again probably due to the relatively high amount of animal glue and impact from handling. *Man with a Fruit Bowl* has no cracks. The zinc compound detected in the upper ground layer may indicate a material that has a tendency to delaminate rather than to crack, but no evidence was found of delaminations. Further research to identify the Zn-containing compound will help to better understand the condition of this painting.

In conclusion, the mechanisms behind the cracking observed in each painting is complex and warrants further analysis. Future research will include chemical characterization of the paint layers as the combination of specific pigments and binders is known to result in paint layers with different mechanical properties, moisture sensitivities, and tendencies for cracking. The influence of fillers will also be explored.

It is unlikely that these paintings will suffer in the controlled environment of the Picasso Museum. However, if the paintings are moved to other locations or if the works are treated by conservators, the resulting response to the environment should be monitored. This would be especially important for the paintings that contain higher amounts of animal glue and the one that is painted on closely woven canvas.

Acknowledgements This study has been developed within the framework of the R + D Project HAR2016-75131-P '*Study of the mechanical and dimensional properties of commercially manufactured paint films and their influence in the physical and chemical degradation of modern and contemporary paintings*' (2017–2019) granted by the Ministerio de Economía, Industria y Competitividad de España (MICINN/FEDER) -Programa Estatal de Fomento de la Investigación

Científica y Técnica de Excelencia, Subprograma Estatal de Generación del Conocimiento. Authors acknowledge preliminary studies carried out by Andrés Sánchez Ledesma (ARTE-LAB) and the high resolution images of Humberto Durán. Authors also thank Marion F. Mecklenburg (Smithsonian Institution) for interesting suggestions and technical advice.

References

1. Bilson T (1996) Canvas shrinkage: a preliminary investigation into the response of a woven structure. IN: J. Bridgeland (ed.) ICOM Committee for Conservation, 11th Triennial Meeting; Edinburgh, September 1996, James & James Ltd, pp 245–252.
2. Izzo FC (2011) 20th century artists' oil paints: a chemical-physical survey, PhD Dissertation, Ca' Foscari University.
3. Izzo FC, Van den Berg KJ, Van Keulen H, Ferriani B, Zendri E (2014) 20th century artists' oil paints: the case of the Olii by Lucio Fontana. Journal of Cultural Heritage 15: 557–563.
4. Izzo FC, Zanin C, van Keulen H, da Roit C (2017) From pigments to paints: studying original materials from the atelier of the artist Mariano Fortuny y Madrazo. International Journal of Conservation Science 8: 547–564.
5. Fuster-López, L et al. (2017). Glue-Paste Linings: an evaluation of some biological, chemical and mechanical aspects of a traditional technique. ICOM-CC 18th Triennial Conference Preprints, Copenhagen, International Council of Museums.
6. Hedley G (1988) Relative Humidity and the Stress/Strain Response of Canvas Paintings: Uniaxial Measurements of Naturally Aged samples. Studies in Conservation 33: 133–148.
7. Gual M, Jiménez R, Bru R (2017) 1917: Picasso en Barcelona, Barcelona: Fundació Museu Picasso.
8. Ligterink FJ, di Pietro G (2007) Canvas paintings on Cold walls: relative humidity differences near the stretcher. Museum Microclimates, T. Padfield & K. Borchersen (eds.) National Museum of Denmark, pp. 27–38.
9. Mathur, K, Seyam AF (2011). Color and weave relationship in woven fabrics. Advances in Modern Woven Fabrics Technology, pp. 129–150. IntechOpen.
10. Mecklenburg M (1982) Some aspects of the mechanical behaviour of fabric-supported paintings. Smithsonian year: annual report of the Smithsonian Institution. Washington D.C., unpublished, Smithsonian Institution, Washington DC.
11. Michalski, S. (2014). The power of history in the analysis of collection risks from climate fluctuations and light. In: J. Bridgland (ed), 17th Triennial Conference, Preprints, Melbourne, Australia, International Council Of Museums.
12. Rogala D, Lake S, Maines C, Mecklenburg M (2010) Condition Problems Related to Zinc Oxide Underlayers: Examination of Selected Abstract Expressionist Paintings from the Collection of the Hirshhorn Museum and Sculpture Garden, Smithsonian Institution, Journal of the American Institute for Conservation, 49:2, 96–113.
13. van den Berg JD, van den Berg KJ, Boon JJ (1999) Chemical Changes in Curing and Ageing Oil Paints. In: J.Bridgland & J. Brown, eds. ICOM Committee for Conservation, 12th Triennial Meeting, Preprints. James & James (Science Publishers, Ltd.), London, 1: 248–53.

Chapter 19
Investigation on the *Speckles Syndrome* Affecting Late 1920s Oil Paintings by René Magritte

Catherine Defeyt, Francisca Vandepitte, Joy Mazurek, Elodie Herens, and David Strivay

19.1 Introduction

As part of the research project *Magritte on Practice* [1], visually disturbing networks of dark speckle-like stains, referred to here as *speckles syndrome*, were noted on the surface of six early oil paintings conserved at the Royal Museums of Fine-Arts of Belgium (RMFAB, Brussels): *La voleuse*, *Le démon de la perversité*, *L'homme du large*, *Le joueur secret, Découverte* and *Personnage méditant sur la folie*. All the affected pictures, shown in Fig. 19.1, were executed in 1927, except the last one, dated 1928. *Speckles syndrome* appears to be a typical degradation feature of Magritte's oil paintings from the late 1920s, as the proliferation of such dark stains were also reported for *L'assassin menacé* from the Museum of Modern Art (MoMA, New York) and *Le sens de la nuit* from The Menil Collection (Houston) which are both dated to 1927.

The examination of the Brussels paintings through digital microscopy revealed the presence of the same symptoms as described for *L'assassin menacé* by Duffy et al. [2], i.e., dark stains formed by radiating micro-cracks with whitish protrusions and/or translucent droplets. A subsequent analytical study conducted on *L'assassin menacé* outlined the probable link between the spotted surface and the formation of metal soaps. However, the exact metal(s) and the layer(s) involved in the saponification reaction, as well as the contribution of past environmental conditions in the degradation process remain unresolved. This comparative study of the

C. Defeyt (✉) · E. Herens · D. Strivay
Centre Européen d'Archéométrie, U.R. Art, Archéologie Patrimoine, Université de Liège, Liège, Belgium
e-mail: Catherine.Defeyt@uliege.be

F. Vandepitte
Royal Museums of Fine-Arts of Belgium, Brussels, Belgium

J. Mazurek
Getty Conservation Institute, Los Angeles, CA, USA

K. J. van den Berg et al. (eds.), *Conservation of Modern Oil Paintings*,
https://doi.org/10.1007/978-3-030-19254-9_19

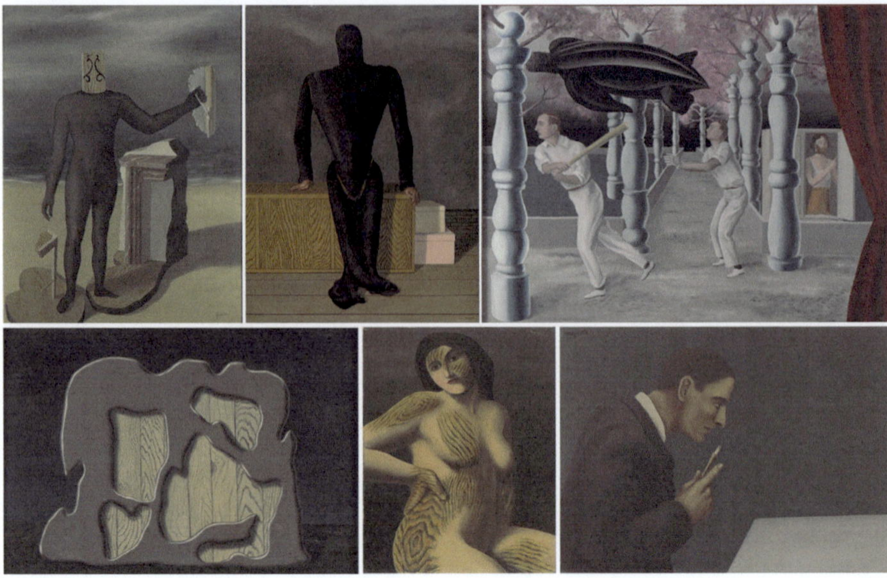

Fig. 19.1 From left to right: *L'homme du large*, 1927, 139 × 105 cm; *La voleuse*, 1927, 101 × 72 cm; *Le joueur secret*, 1927, 150 × 195 cm; *Le démon de la perversité*, 1927, 81 × 116 cm; *Découverte*, 1927, 65 × 50 cm; *Personnage méditant sur la folie*, 1928, 54 × 73 cm. © Ch. Herscovici

RMFAB paintings offers a unique opportunity to elucidate the mechanism(s) and the compound(s) responsible for the *speckle* appearance. Indeed, by comparing the phenomena, the painting materials and techniques and by exploring past environmental conditions, common features shared by these speckled artworks may be determined.

To this end, digital and optical microscopy, Fourier-Transform Infrared spectroscopy (FTIR), Raman Spectroscopy (RS), Scanning Electron Microscopy with Energy Dispersive X-ray analysis (SEM-EDX) and X-Ray Fluorescence in macro scanning mode (MA-XRF) techniques were employed. This paper presents an overview of the results obtained and discusses the most relevant findings regarding contributions from past environmental conditions and the constitutive materials identified in the speckle phenomena.

19.2 Experimental

MA-XRF was used with the Centre Européen d'Archéométrie (CEA) XRF system [3–5]. Scanning step was set to 1 mm, speed to 3 mm/s. The X-ray tube was set to a voltage of 35 kV and a current of 120 µA. Spectra were treated in batch mode using PyMCA [6] to obtain elemental images. Raman analyses were conducted using the

Enwave Optronics Raman analyzer I-Dual-G, with a laser at 785 nm [7]. Three Raman spectra were acquired for each site using low laser power (*ca.* 30 mW) and treated with GRAMS AI™. All MA-XRF and Raman data were acquired on the museum site with CEA portable instrumentation.

In addition, speckle specimens scraped from painting surfaces were investigated using optical microscopy, micro-FTIR and SEM-EDX techniques. Photomicrographs were acquired with Dino-Lite and Olympus BX51M microscopes. The FTIR spectra were collected using a Thermo Nicolet Avatar 360 FTIR spectrometer. The samples were placed on a Diamond window and individually analyzed in transmittance mode in the 4000 to 650 cm^{-1} range (4 cm^{-1} resolution). SEM-EDX data were acquired with a Phillips FEI-XL-30 Environmental SEM at 20 kV.

19.3 Insight into Past Environmental Conditions

Duffy et al. [2] pointed out the possible influence of long-term exposure to high humidity on the dark stained appearance of *L'assassin menacé*. Indeed, for approximately 10 years *L'assassin menacé* and *Le joueur secret*, which exhibit identical symptoms, were displayed on either side of the main staircase of a seaside casino in Knokke (Belgium) [8]. However, among the six Brussels paintings with spotted surfaces, only *Le joueur secret* and *L'homme du large*, were at this location [9]. By contrast, five out of the six works belonged, for more than 20 years, to the Belgian art dealer E.L.T. Mesens, who acquired the works in 1932–1933 via the stock liquidation of several Brussels art dealers.

On this occasion, Mesens purchased more than 150 works by Magritte [10]. This substantial collection was stored for decades in a container rented to the Brussels company La Continentale, at a storage site that was totally inappropriate for preserving artworks[1]. Interestingly, Mesens also owned MoMA's painting *L'assassin menacé* from the early 1930s. Although it remains uncertain, *Le démon de la perversité*, which does not have a fully traced provenance, may have also been owned by Mesens, possibly as the unidentified work of its size Magritte sent to him in 1927 [8].

19.4 Speckles Syndrome

La voleuse is the one work where *speckles syndrome* is evident in normal light conditions. Here dark stain networks can be readily observed in the body suit, particularly at the level of the female thief's waist (Fig. 19.2b). Although much less prominent, they can also be seen in the grayish background and the other colored

[1] Personnal communication with Isy Brachot formerly Magritte's dealer who personnaly knew E.L.T. Mesens

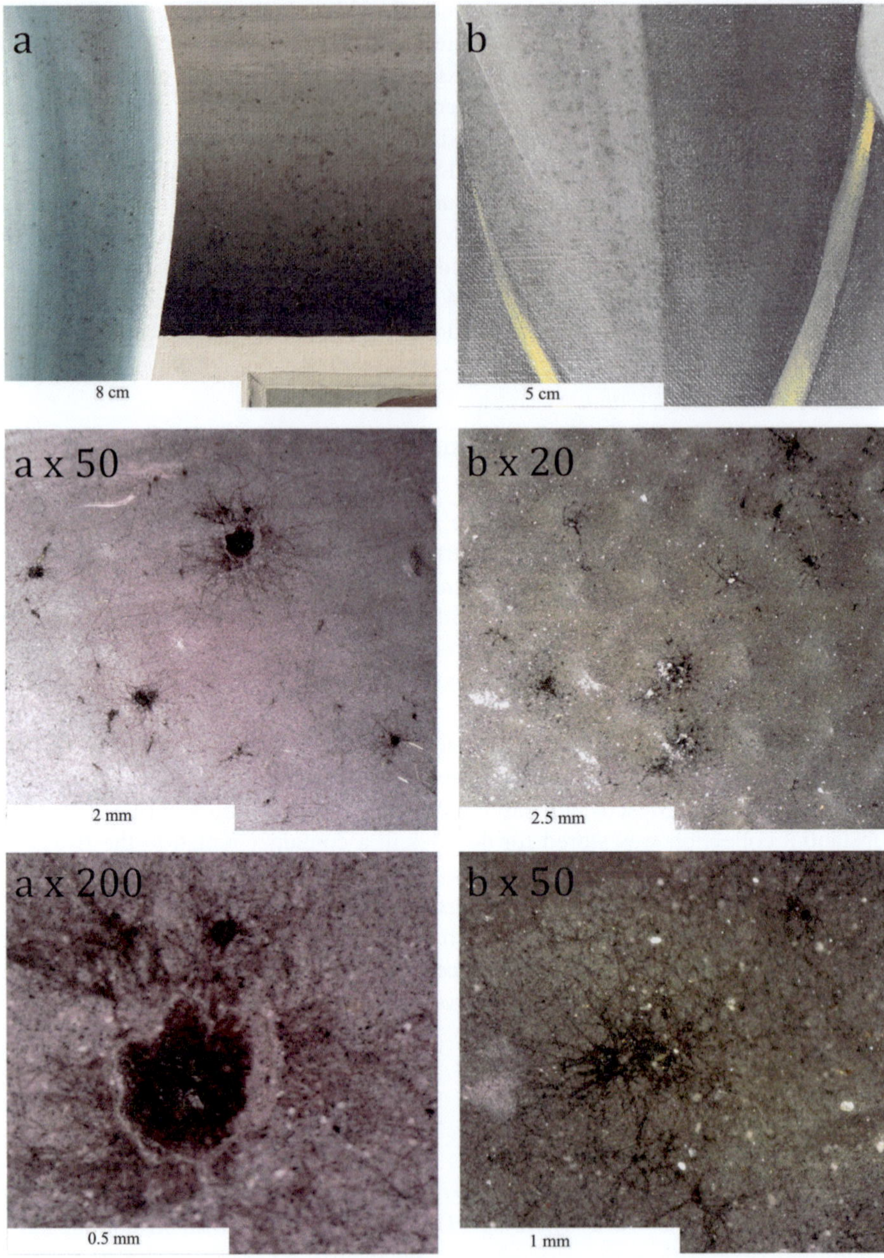

Fig. 19.2 (**a**) Detail of *Le joueur secret* before the 1995 restoration campaign showing the speckles are more prominent in the darker shades, and microphotographs of dark stains currently visible on the painting surface ×50 and ×200, (**b**) Detail of the *La voleuse* showing the speckles currently visible in the body suit gray and microphotographs of related dark stains ×20 and ×50

areas do not appear to be affected. There is no mention of such features in past condition and conservation treatments reports. Under the digital microscope, the speckles appear as dark accretions or protrusions surrounded by micro-cracks forming cobweb-like networks. Disparate whitish aggregates and translucent droplets located around and/or above the protrusions and accretions were also observed on the top of the painted surface.

The speckled surface of *Le joueur secret*, is no longer evident under visible light. Indeed, most of the visually disturbing dark stains were mechanically removed before the RMFAB acquired the work in 1995. Close up views showing spotted areas before cleaning provide clear proof of *speckles syndrome* (Fig. 19.2a). When examined under UV light however, widespread networks of dark halos became immediately apparent, irrespective of the paint colors. Digital microscopy revealed the presence of radiating micro-cracks with residual dark accretions and the remains of whitish protrusions at related sites. Alongside the partly removed stains, speckles that had not been scraped became apparent (Fig. 19.2a), which were similar in appearance to those noted in *La voleuse*. This syndrome is also present in the four other paintings, although the accretions and protrusions occurring in *Le demon de la perversité* do not seem to be correlated with microcracks.

19.5 Painting Materials

The tacking edges of *La Voleuse* and *Le joueur secret* show discontinued priming layers on the canvas fringes, which suggests that both canvas supports were likely to be artist-primed. Moreover, the uneven UV luminescence observed on the tacking edges of *Le joueur secret* and *L'homme du large* suggests the presence of separately applied ground layers. Indeed, the cross sections investigated facilitated the identification of two separate ground layers for four of the five paintings (note that no cross sections were taken from *La Découverte*). As shown in Fig. 19.3, these two layers appear to be consistently implemented: the first is relatively thick, consisting of zinc white and chalk, and encloses isolated tiny blue particles, the second is much thinner and is based on lead white. No other pigments or fillers were identified in the second ground layer. UV microphotographs show the characteristic luminescence of zinc white in the first layer. Note that kaolin was identified by FTIR from the tacking edges of *La Voleuse* and *Le joueur secret* however the layer it originates from is unclear. In *Le demon de la perversité*, a single lead white-based ground layer (without additional pigment/filler) was identified. For *La Découverte*, the RS and FTIR analyses performed on the primed tacking edges led to the identification of titanium anatase white and barites. Anatase titanium white was introduced onto painters' palettes in the early 1920s, and its presence in paintings dating from 1927 and 1928 is an early use of this novel white pigment by Magritte. However, the MA-XRF results including Zn, Sr and Pb (L-shell) suggest the possible presence of another ground layer underneath.

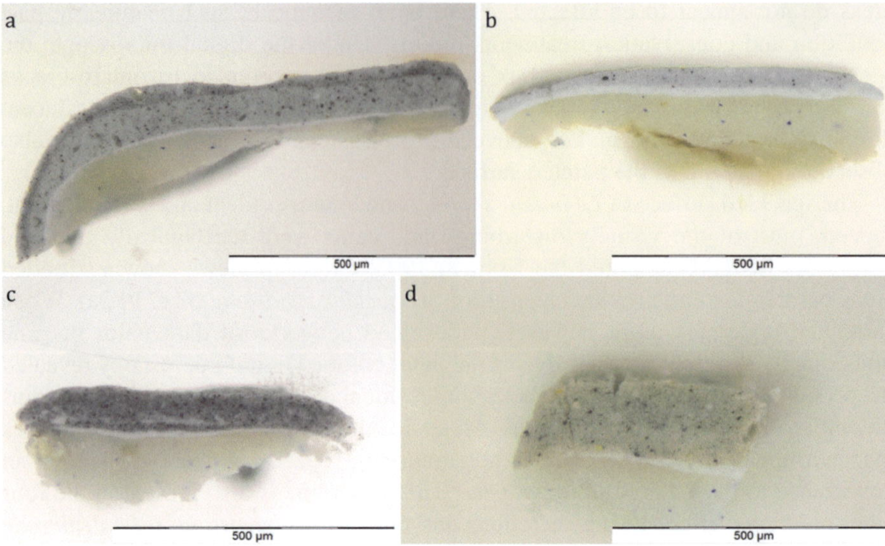

Fig. 19.3 Cross sections from (**a**) *Personnage méditant sur la folie* (**b**) *L'homme du large* (**c**) *Le joueur secret* (**d**) *La voleuse* showing two separate ground layers identically applied (A. Dupuis)

The *speckle syndrome* paintings were characterized through non-invasive analytical techniques (XRF and RS) and contain the following pigments in common: lead white, zinc white, ivory black and iron oxide/hydroxide pigments.

The palette used for *La voleuse* incorporates lead white, zinc white, ivory black, titanium white, Prussian blue, chromium pigment(s) and yellow ochre with kaolinite. In this painting, the speckle networks are only visible in the gray body suit and in the grayish background. The corresponding colors primarily include lead white and zinc white admixed with ivory black. The MA-XRF scans recorded in the waist region of the bodysuit showed that the medium gray, where the speckles are the most prominent, is richer in lead white than the surrounding darker gray shades, which contain ivory black and zinc white. The pigment content of the painted surfaces that do not display spotted features differ from the speckled surfaces through the identification by RS of anatase titanium dioxide.

The palette identified for *L'homme du large* is mainly composed of lead white, ivory black, zinc white, chromium pigment(s), Prussian blue, ocher yellow and further iron oxide pigment(s). Although speckles (retouched and not retouched) were observed in all the colors, they are more evident in the gray bodysuit. The MA-XRF distribution maps of P, Ca, Zn and Pb (L-shell), reproduced in Fig. 19.4, indicate that the related gray colors primarily contain lead white, zinc white and ivory black mixtures. As in *La voleuse*, the medium gray of the bodysuit corresponds to the color most affected by *speckle syndrome*. Indeed, UV light illumination showed that the retouched speckles sites are more abundant in the medium gray region of the bodysuit (see Fig. 19.4). From the MA-XRF results, this color differs

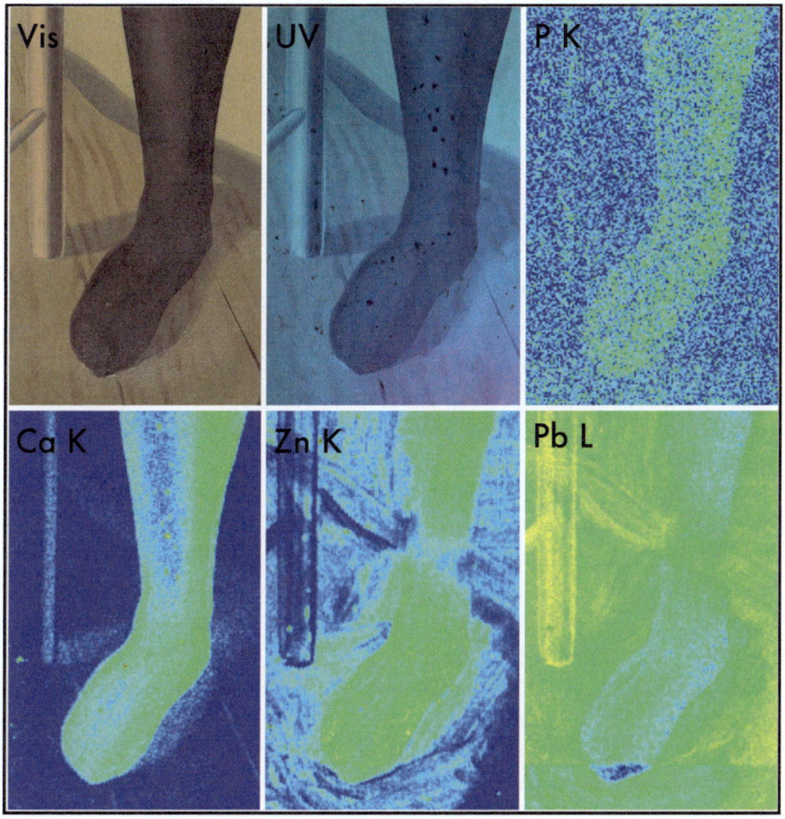

Fig. 19.4 Detail of *L'homme du large* in visible and UV light, and MA-XRF distribution maps of P (K-shell), Ca (K-shell), Zn (K-shell), Pb (L-shell). Retouched speckles are more concentrated in the gray bodysuit where lead white and ivory black are both dominant. The Ca signals, correlated with speckles, are associated with the iron oxide pigment(s) used for retouching the bodysuit

from neighboring darker gray shades through a higher lead white content and a lower zinc white content.

The gray jacket in *Personnage méditant sur la folie* and the gray of the biomorphic figure in *Le demon de la perversité*, both consisting of lead white and ivory black, also show the highest concentration of *speckles*. By examining the UV image of *Le joueur secret* it appears that the *speckles syndrome* is more severe in the black turtle, in the grayish background colors and in the blackish blue used for the shadows of the bilboquet-trees. These areas are rich in lead white and ivory black. As reported above, in *Découverte* the only spotted surface visible is the dark background. This color consists of a combination of Prussian blue, cadmium yellow, lead white, zinc white and ivory black. Except for the Prussian blue, the same group of pigments plus titanium white (anatase) were identified in the flesh tones and the wood 'rendered' surfaces of the female figure.

The characterization of the pigments used in these six paintings has demonstrated that lead white and ivory black are common to all of the altered paint surfaces. Interestingly, the speckle networks appear to be consistently more present in paint layers where both pigments dominate. In contrast, as noted in *Découverte* and *La voleuse,* the paint layers involving anatase titanium white do not exhibit a speckled appearance.

19.6 Degradation Products

The *speckle syndrome* microphotographs showed protrusions made of whitish mineralized aggregates, covered by yellowish/ brownish translucent crusts, most often enclosing disparate black particles. The micro-FTIR analyses performed on speckle specimens sampled from *La voleuse*, *Le joueur secret*, *L'homme du large* and *Démon de la perversité* revealed the presence of the characteristic IR absorption bands for metal carboxylates at *ca.* 1550 cm^{-1} [11]. Metal soap identification in seven out of the ten samples implies a connection to *speckles syndrome*. The deleterious environmental conditions induced by a long-term storage of the paintings in a container, especially with high relative humidity appear to have promoted the formation of metal soaps [12].

The diagnostic metal soap IR band(s) occur in the spectra taken from the white aggregates and translucent crusts; suggesting that both types of degradation products involve metal soaps. The identification of hydrocerussite, irrespective of the analyzed spot, and the position of the broad IR band arising between 1560 and 1520 cm^{-1} (appearing as a doublet in some spectra) indicate that the metal involved in the saponification process is lead [12]. Hence the presence of an amorphous lead soap or a lead ionomer may be responsible for the broad IR absorption in that region. The predominant occurrence of lead, carbon and oxygen in all the SEM-EDX spectra recorded from different spot sites of the speckles also confirmed this interpretation.

The micro-FTIR and SEM-EDX results also suggest that the translucent crusts lying on the top of the protruding whitish aggregates, generally comprise additional pigments particles originating from the surface paint layer, most often ivory black. An increased rate of lead soap formation within micro-cracks may explain the accumulation of black pigment grains at the painting surface, which is partly responsible for the darkness of the efflorescent degradation products [13]; and the growth of protruding aggregates on the subsurface could be the origin of the radiating micro-cracks associated with the *speckles*.

19.7 Conclusions

The serial approach used in this study has resulted in an enhanced understanding of the *speckles syndrome* affecting Magritte's early oil paintings. Common features to the six paintings include deleterious environmental conditions induced by long-term storage in a container, and the use of lead white based-surface layers involving ivory black. Painting degradation related to lead soaps, with or without carbon black, are relatively common, with many case studies reported in the literature. However the visual changes common to these six paintings should be contextualised within Magritte's technical practice, which incudes the recurrent use of ivory black. Indeed, the accumulation of black particles around and on the top of the protruding aggregates greatly increases their visibility, especially in the light and medium colors, and is in part responsible for the darkness of the speckles.

Acknowledgments The authors sincerely thank Michel Draguet, RMFAB General Director, for his commitment, which greatly assisted this research. The authors also express their gratitude to Ludovic Godfrin (RMFAB Modern painting collection keeper) and Alix Dupuis (ULiège Master student) for their assistance.

References

1. Defeyt C, Herens E, Leen F, et al (2018) Discovery and multi-analytical study of the last missing quarter from René Magritte's La pose enchantée. Heritage Science 6: 33–41
2. Duffy M et al (2014) Metal Soaps and Visual Changes in a Painting by René Magritte – The Menaced Assassin, 1927. In : van den Berg KJ et al (eds) Issues in contemporary oil paint, Springer : 197–203
3. Strivay D, Clar M, Rakkaa S, et al (2017) Development of a translation stage for in situ noninvasive analysis and high-resolution imaging. Applied Physics A 122 : 950–953.
4. Hocquet FP, Garnir HP, Marchal A, et al (2008) A remote controlled XRF system for field analysis of cultural heritage objects. X-Ray Spectrom. 37: 304–308
5. Hocquet FP, Calvo del Castillo H, Xicotencatl A et al (2011) Elemental 2D imaging of paintings with a mobile EDXRF system. Anal. Bioanal. Chem. 399(9): 3109–3116
6. Solé VA, Papillon E, Cotte M et al (2007) A multiplatform code for the analysis of energy-dispersive X-ray fluorescence spectra. Spectrochim Acta B, 62(1): 63–68
7. Lauwers D, Garcia Hutado A, Tanevska V, et al (2014) Characterisation of a portable Raman spectrometer for in situ analysis of art objects. Spectrochim Acta B 118, p 294–301
8. Sylvester D (1992) René Magritte – Catalogue Raisonné, tome 1: Oil paintings 1916–1930. Catalogues Raisonnés, Flammarion
9. Sylvester D (2009) Magritte. Actes Sud Beaux-Arts
10. L 'œuvre de René Magritte, XVe Festival belge d'été, Knokke. Editions de la Connaissance, Bruxelles (1962)
11. Robinet L, Corbeil MC (2003) The Characterization of Metal Soaps. Studies in Conservation, 48(1) :23–40
12. Keune K (2005) Binding medium, pigments and metal soaps characterised and localised in paint cross sections. Dissertation, University of Amsterdam
13. Hermans JJ, Keune K, van Loon A, et al (2016) The crystallization of metal soaps and fatty acids in oil paint model systems. Phys. Chem. Chem. Phys., 18:10896–10905

Chapter 20
Examination of Paint Delamination in *C'est grace à nous* by Asger Jorn

Ida Antonia Tank Bronken, Jaap J. Boon, and Calin Constantin Steindal

20.1 Delamination or Cleavage Linked to Soft Paint

Previous published research pointed out repeatedly that there are some conservation issues related particularly to paintings dated after second world war. Already in 1994 Moffatt and Miller describe lead soaps in connection to cracking soft paint [13]. In 2001, it was suggested that lead soaps in the ground might be a factor for cracks in paintings by Borduas [15]. The connection between soap formation and delaminations have since been described by several authors [9, 12, 14, 17].

One of the earliest of several published findings of high concentrations of fatty acids in soft paint is from 1993 [16]. Later research has shown that the increased levels of polar fatty acids are the cause of the softness, and are the main constituent of the drip material [2, 6].

C'est grace à nous by Asger Jorn (1957) shows delaminations visually linked to the upper, soft paint layer. The simple build-up consists of a white ground with a pastose, partly soft, black paint layer on top which in local areas exhibits a strong efflorescence. SEM-EDX shows that the white ground has a high concentration of zinc, indicating the use of zinc white. The black pigment of the surface paint is bone black.

I. A. T. Bronken (✉)
Department of Collection Management, National Museum of Art, Architecture and Design, Oslo, Norway
e-mail: ida.bronken@nasjonalmuseet.no; http://www.nasjonalmuseet.no

J. J. Boon
JAAP Enterprise for Art Scientific Studies, Amsterdam, The Netherlands
e-mail: info@jaap-enterprise.com; http://www.jaap-enterprise.com

C. C. Steindal
Museum of Cultural History, University of Oslo, Oslo, Norway
e-mail: c.c.steindal@khm.uio.no; http://www.khm.uio.no/

K. J. van den Berg et al. (eds.), *Conservation of Modern Oil Paintings*,
https://doi.org/10.1007/978-3-030-19254-9_20

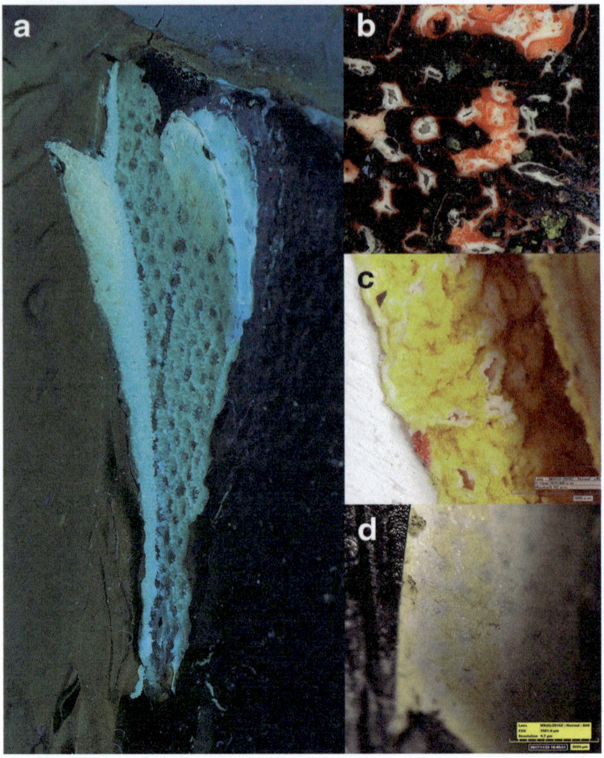

Fig. 20.1 (**a**) Delamination rupture in zinc containing ground under soft black paint in *C'est grace à nous* by Asger Jorn (1957) under UV light. (**b**) Delamination in an area of paint loss showing dome shaped deformations. *Composition* by Jean Paul Riopelle 1952. (**c**) Delamination in an area of soft yellow paint over a white paint layer. Libellules blesses by Karel Appel 1961. (**d**) Delamination in Pierre Soulages *Peinture* from 1954. (Photo **a**: The National Museum, Børre Høstland. Photos **b–d**: Jaap Boon)

The zinc white paint layer has split in two parts and is presently brittle and solid. The surfaces show a ductile looking split and break pattern that hints to a former mobility and fluidity inside the layer (Fig. 20.1a), although the white paint in the area was not soft or malleable when examined. Additionally, the peaks formed indicate that this change in condition must have happened over time as there are no clean breaks or ruptures visible in the SEM-BSE image in Fig. 20.2. Small, peak-like deformations of the ground layer exist in all the delaminating areas of the painting.

The research focusses on the differences in chemical composition between the ground still connected to the canvas (B), and the part of the ground connected to the soft black layer that curves away from the canvas (A) (Fig. 20.4). For sampling, an area that had not been treated with heat or solvents was chosen. Samples were analysed by SEM-EDX, ATR-FTIR and DTMS.

20.2 Results

After storage on a glass slide for several months, sample A, black paint without ground, showed some degree of phase separation of the black paint and a clear medium. DTMS analysis of the black paint sample after homogenization (Fig. 20.3a

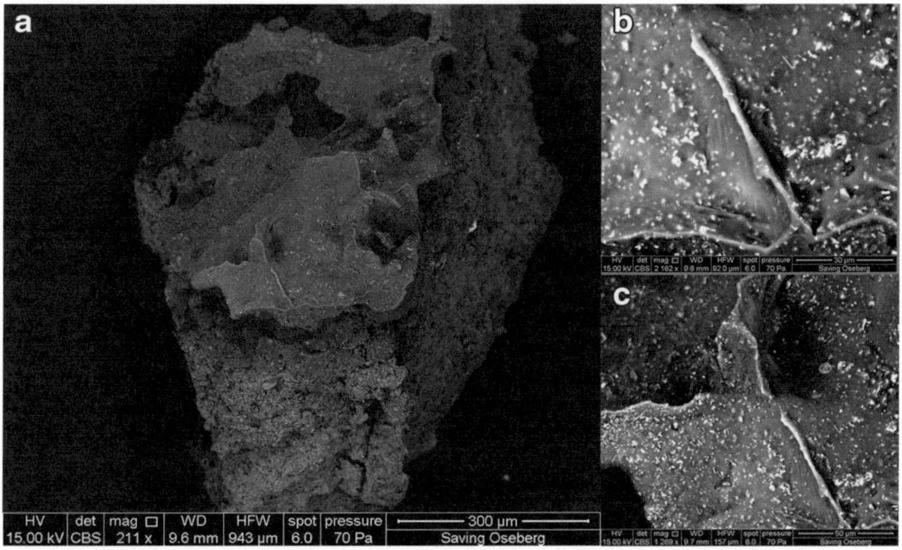

Fig. 20.2 (**a**) The sample upside down black down and white up. (**b**) The edge of the white layer. (**c**) Details of the sample in electron backscatter image from the point of delamination, showing formed ridges in the ground left on the underside of the black paint layer. (Photos: Calin Constantin Steindal)

and b) shows m/z of 62 indicative of phosphate (from the bone black), lead (m/z 207, 208, 209), zinc (m/z 66, 67, 68) at higher temperature. The spectra in the broad desorption profile of the total ion current profile, labelled X, are presented as a summed mass spectrum (Fig. 20.3b). Peaks indicative of fatty acids (palmitic (m/z 256) and stearic (m/z 284) acids), features indicative of azelaic acid (m/z 152) and various oxygenated (keto, hydroxyl and epoxy) stearic acids (m/z 155, 171, 280, 292) are observed. The intensity of features of azelaic acid and the oxygenated stearic acids are high relative to the features of palmitic and stearic fatty acid. The ratio of polar to apolar fatty paint constituents links to the degree of softening and the tendency to form exudates [6]. The desorption profile in DTMS points to a relatively large amount of fatty components. It can be expected that this excess of fatty acids does contribute to the cohesive failure of the paint layer structure.

The white ground was analyzed by SEM-EDX. The ground shows a high content of zinc relative to calcium, supporting that the paint is zinc white with some calcium carbonate filler. The black paint layer's main elemental peaks were phosphorous and calcium indicating the use of bone black pigment (unpublished results). When analyzing the delaminated part of the white ground adhering to the black paint layer, the main elements were zinc and calcium. The EDX identified L-line and both K-lines for zinc, and also K and L-lines for calcium. So, the results from the SEM-EDX strengthens the initial examination indicating that the delaminations occurs inside the white ground layer.

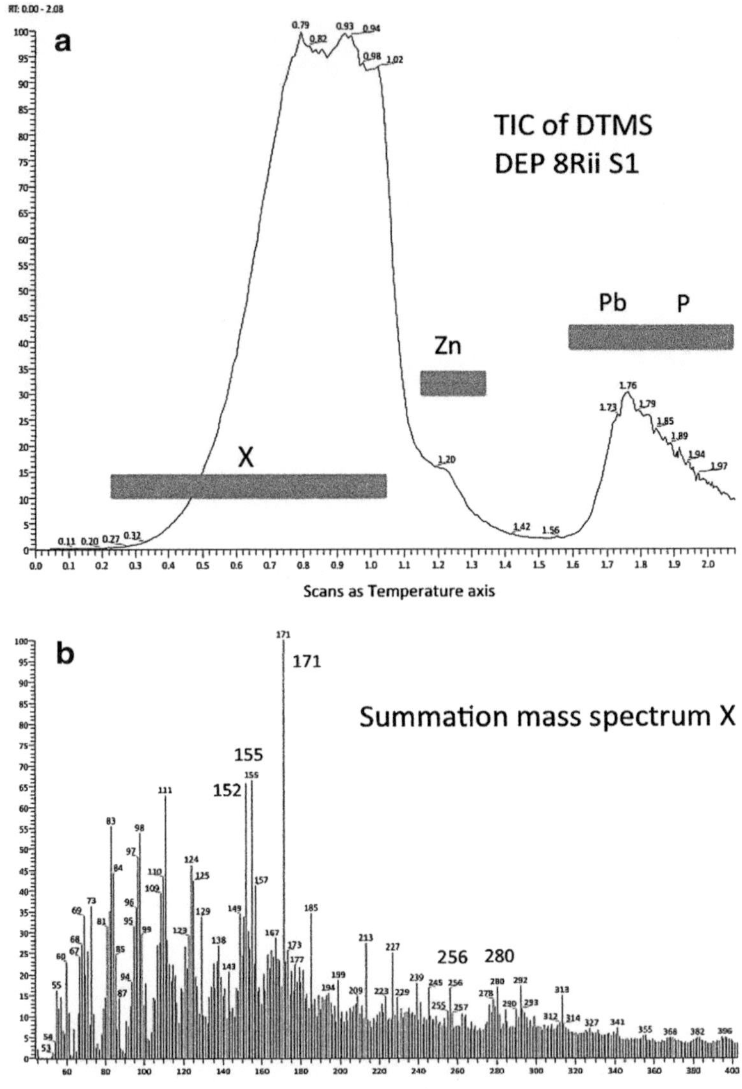

Fig. 20.3 DTMS data of soft black paint from *C'est grace à nous* by Asger Jorn (1957). (**a**) Total ion current profile. (**b**) Mass spectrum summed over band width X. Analysis: Jaap Boon and Karin Wyss SIK-ISEA, Zurich

The ATR-FTIR spectrum of the white ground attached to the black paint shows a strong, sharp band at 1528 cm^{-1}, with a shoulder at 1547 cm^{-1} and a weak band at 1592 cm^{-1}. The band positions are consistent with the presence of zinc fatty acid soaps (salts). Straight chain fatty acid soap shows a band near 1528 cm^{-1}, while diacids absorb near 1546 cm^{-1}. The band at 1590 cm^{-1} is exhibited by the more

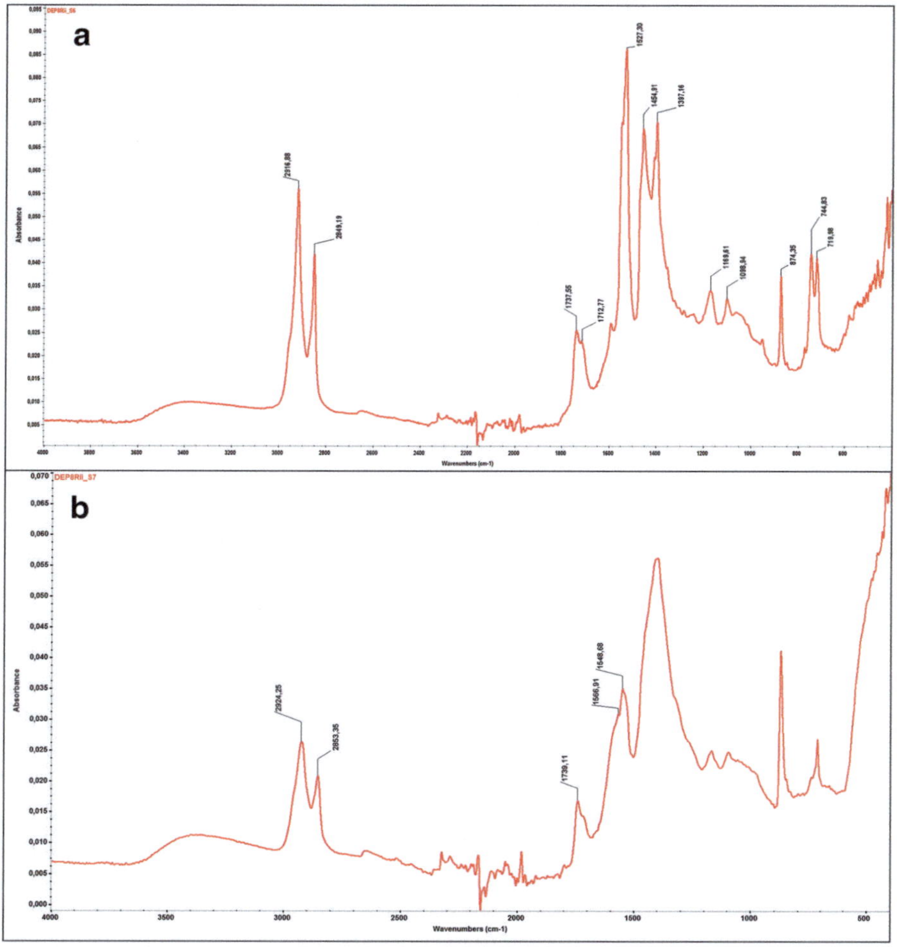

Fig. 20.4 (**a**) FTIR-ATR spectra from sample of white ground taken from the detached delaminated area of black paint. The peaks at 1527–1546 cm^{-1} indicate zinc fatty acid soaps. (**b**) FTIR-ATR spectra from sample of white ground still attached to the canvas. The peak profile at 1548–1590 cm^{-1} is indicative of the carboxylate absorption of zinc fatty acid soaps with a more amorphous structure. Spectra: Calin Constantin Steindal

complex zinc carboxylates observed in zinc white paint [10, 11, 14]. The spectrum also shows weak bands from ester and free fatty acid components at approximately 1730 and 1705 cm^{-1}.

The ATR-FTIR spectrum of the white layer still connected to the canvas Fig. 20.4b shows peaks at 873 and 713 cm^{-1} confirming the presence of calcium carbonate. The strong, broad band beginning at 500 cm^{-1} is consistent with zinc white vibrations. The bands of the zinc carboxylate region are shifted compared to those exhibited by the sample of white still attached to the black paint (Fig. 20.4a).

The spectrum shows a broad band between 1540 and 1590 cm^{-1}, suggesting more complex and amorphous zinc soap structures [8, 11].

The lower part of the ground exhibits a broad band belonging to zinc soaps, while near the interface the zinc soap bands are stronger and sharper. This suggests both a different structure of zinc soaps at the interface and an increased soap concentration. The fact that the zinc soaps at the interface are structurally different from those in the bulk of the ground indicates specific reactions in the contact zone. The additional zinc soaps in the delamination plane must have mechanically weakened the ground layer.

20.3 Discussion

The zinc soaps found in the white layer still connected to the black paint show that the interface between the soft layer and the ground is chemically active. It is inferred that, due to phase separation, the mobile components of the oil in the black paint not only move to the surface, but also diffuse into the underlying layers, thus reacting and creating a layer of zinc fatty acid soaps with problematic physical properties. It has been well documented that free carboxylic acid groups increase the reactivity for soap formation [1, 3, 14]. Zinc soaps have the tendency to increase in volume [14, 18]. High humidity promotes both soap formation and the increase in volume of the paint [14]. Zinc soaps are described as more mobile than lead soaps [18]. Mobile fractions in the structure most likely contribute to the formation of the peak-like ridges in the delaminated areas. Analysis of the spectra from the ground layer in contact with the bone black paint shows peaks from zinc carboxylates. The black paint in *C'est grace à nous* was found to have a high abundance of azelaic acid and oxygenated stearic acid moieties. This is consistent with previous studies of soft and dripping paints ([2], Unpublished, Bronken PhD thesis due 2021 [5]). The part of the zinc white ground layer attached to the delaminated black paint contains a high concentration of zinc fatty acid soaps. The combination of volume increase of zinc soaps with the malleable properties of the black paint is inferred to create delaminations.

Similar cases of interlayer cleavage between layers where one is zinc containing have been described in literature ([10, 12, 17] etc). Especially, the study presented in Maines et al. [12], where the description of cleavage in a Hans Hofmann painting bears a strong resemblance to this case study. The cadmium red paint in question was delaminating from areas in contact with zinc white. Moreover, GC-MS analysis of the red paint shows a distinct high peak for C9 diacid ([12], in Fig. 20.2 in this article) similar to the other case studies of soft paint. Hence, our interpretation is that the combination of zinc containing layers and their reaction with phase separating carboxylic acids from adjacent layers contribute to the severe delamination.

The presence of zinc soaps complicates any future treatments, but removal of the soap containing ridges in the areas of delamination seems not plausible for practical reasons (Fig. 20.1). If the painting material to be consolidated lacks stability, there will be extra demands for re-treatability.

20.4 Treatment, Tests and Choices

In the 1950s and 1960s, many of the artists used paint colours more directly, without extensive use of wet-in-wet, and often partly in thick layers. The structure is often found to be a mix of paint layers with varying physical properties: some solid and very stiff, others soft malleable and solvent/heat sensitive. This creates practical complications for consolidation of delaminations for paintings with soft and dripping paint.

Several of the paintings in the ongoing PhD project *Investigation of soft and dripping paint in selected paintings from 1946–1972* show delamination similar as the case study presented here and, in some cases, consolidation treatments were carried out (unpublished, Bronken PhD thesis due 2021 [5]). These treatments have given indications to issues of importance when choosing a treatment strategy for *C'est grace à nous* in the future. Considering the treatment reports from other institutions and through treatments of similar paintings, water soluble glues are so far showing to be ineffective [4]. An obstacle for consolidation is the heat and solvent sensitivity of the soft paint layers. Several earlier treatments have shown that Beva 371 works quite well when the area is easy to access with a brush. In practice, consolidants with low viscosity like Medium for consolidation help to ensure that the consolidating substance moves by capillary action as deep as possible into the delaminated area. Modifications of other consolidants may also achieve this. One drawback is that retreating consolidated areas often require solvents and heat. So no perfect treatment option seems to be apparent. Specific choices must be made for each individual case.

Heat will be needed for successful consolidation. In some treatments of other paintings, a hot air gun was used. Instead of a metal spatula, a malleable silicon tool was used for consolidation of the soft layers, minimizing the risk of flattening the structure when applying pressure in combination with heat.

Consolidation with a heat spatula on soft paint can be difficult as the surface is obstructed by the spatula and different paint layers, and different areas can have very different sensitivity to heat [7]. The silicon tool made it easy to work directly on the surface. Often consolidation treatments are done through a layer of melinex on top, but also this approach should be used with caution as it can stick to the surface and give gloss changes or surface deformations if left over time.

20.5 Conclusion

Analysis of an area of delamination from an Asger Jorn painting shows the result of reactivity of zinc ground, and the availability of fatty carboxylic acids in the superimposed black paint layer. Soft paints often have a high level of mobile free fatty acids, and can form fatty acid soaps with available metal ions. So, it is important to analyse the entire paint build-up as layers will chemically interact. The physical properties of zinc soaps could predict that future delaminations would be impossible to avoid. When planning to consolidate soft paint layers where soaps are present or suspected, conservators should inform the owners that the need for re-treatment of the paintings is probable. Furthermore, caution should be taken in transportation and during storage.

The possibility of removing original material is controversial and something conservators are trained to avoid. Removing the excess fatty acids when forming a pools or drips would be more advantageous for long-term preservation as the diacids and diacyl-glycerides have the potential to react with metal ions from surrounding paint layers. Many damages in paintings are related to soap formation, but this study supports the idea that mobile fatty acids are also a potential reason for future damages.

20.6 Analytical Techniques

20.6.1 Scanning Electron Microscopy Energy Dispersive X-Ray Spectroscopy (SEM-EDX)

A FEI Quanta 450 scanning microscope coupled to an X-Max^N Oxford analyser 50 mm^2 was used for SEM–EDX. Paint samples were placed directly on carbon tape.

20.6.2 Fourier-Transform Infrared Spectroscopy (FTIR)

FTIR spectroscopy measurements at the University of Oslo facilities were carried out in attenuated total reflectance (ATR) with an IS50 Thermo scientific Nicolet continuum FTIR microscope with a resolution of 4 cm^{-1}.

20.6.3 Direct Temperature Resolved Mass Spectrometry (DTMS)

DTMS was performed in a DSQ-II Thermo Quadrupole MS-instrument (EI-mode, 16EV, source T 250 °C, m/z 42-1050). The sample was homogenized and applied as methanol droplet on Rhenium filament (ramp 5 mA/s to 1000 mA). X-calibur software was used for data analysis.

References

1. Boon J, Hoogland F and Keune K (2007) Chemical Processes in Aged Oil Paints affecting Metal soap Migration and Aggregation. In: AIC paintings specialty group postprints: papers pres. at the 34th annual meeting of the AIC of Historic & Artistic Works providence, Rhode Island, June 16–19, 2006 /ed. H. Mar Parkin, AIC. Pp. 16–23
2. Boon J and Hoogland F (2014) Investigating Fluidizing Dripping Pink Commercial Paint on Van Hemert's Seven-Series Works from 1990–1995. In: Issues in Contemporary Oil Paints. Proceedings from the Symposium, 28 and 29 March 2013. eds. van den Berg KJ, Burnstock A, de Tagle A, de Keijzer M, Heydenreich G, Krueger J and Learner T. Amersfoort: Springer. Pp. 227–246
3. Boon J, Keune K and Zucker J (2005) Imaging analytical studies of lead soaps aggregating in preprimed canvas used by the Hudson River School painter F.E. Church. In: Microscopy and Microanalysis vol. 11 (Supplement 2). Pp. 444–445
4. Bronken I (2009) Undersøkelse og dokumentasjon av *Composition,* 1952, Jean-Paul Riopelle. Master thesis, University of Oslo
5. Bronken I (unpublished, due 2021) Categorization of soft and dripping paint. Examination of the condition issues for 25 paintings dated between 1946 and 1971 with soft, exuding and dripping oil paint. PhD thesis, University of Amsterdam
6. Bronken I and Boon J (2014) Hard dry paint, softening tacky paint, and exuding drips on Composition 1952. In: Issues in Contemporary Oil Paints. Proceedings from the Symposium, 28 and 29 March 2013. eds. van den Berg KJ, Burnstock A, de Tagle A, de Keijzer M, Heydenreich G, Krueger J and Learner T. Amersfoort: Springer. Pp. 247–62
7. Bronken IAT and Boon JJ (2015) Conservation challenges and physical changes in Composition (1952) by Jean-Paul Riopelle: a case study. In: Public Paintings by Edvard Munch and his Contemporaries: Change and Conservation Challenges, ed. Frøysaker T, Streeton NLW, Kutzke H, Hanssen-Bauer F, Topalova-Casadiego B. London: Archetype publications. Pp 334–348
8. Corbeil M-C, Helwig K, Poulin J (2011) Jean Paul Riopelle: the artist's materials. Getty Publications, Los Angeles
9. Helou-de la Grandière P, Le Hô A-S and Mirambet F (2008) Delaminating paint films at the end of 1950s: a case study on Pierre Soulages. In: Preparation for Painting. The artist's choice and its Consequences. London: Archetype publications. Pp. 156–162.
10. Helwig K, Poulin J, Corbeil M-C, Moffatt E and Duguay D (2014) Conservation Issues in Several Twentieth-century Canadian Oil Paintings: The Role of Zinc Carboxylate Reaction Products. In: Issues in Contemporary Oil Paints. Proceedings from the Symposium, 28 and 29 March 2013. eds. van den Berg KJ, Burnstock A, de Tagle A, de Keijzer M, Heydenreich G, Krueger J and Learner T. Amersfoort: Springer. Pp. 167–184
11. Hermans JJ (2017) Metal soaps in oil paint, structure mechanisms and dynamics. PhD Thesis University of Amsterdam

12. Maines CA, Rogala D, Lake S, Mecklenburg M (2011) Deterioration in Abstract Expressionist Paintings: Analysis of Zinc Oxide Paint Layers in Works from the Collection of the Hirshhorn Museum and Sculpture Garden, Smithsonian Institution. In: Materials Research Society Symposium Proceedings. Materials Issues in Art and Archaeology IX. Volume 1319
13. Moffatt EA and Miller D (1994) Spectroscopic and Chromatographic Analysis of Selected Paintings from the Parisian Periode of Paul-Émil Bourdas. In: Journal of the International Institute for Conservation-Canadian Group, vol. 19, 1994. Pp 3–13
14. Osmond G (2014) Zinc oxide-centered deterioration of modern artists' oil paint and implications for the conservation of twentieth century paintings. PhD thesis, University of Queensland
15. ÒMalley M and Moffatt EA (2001) Paul-Émile Bourdas: the blacks and whites of the later paintings (1955–1960). In: Deterioration of Artists' Paints: Effects and Analysis. London: UKIC and British Museum. Pp 84–87
16. Pfandlbauer S (1993) Leinöl in der modernen, zeitgenössischen Malerei. Oberflächenphänomene und Schadensbilder. Diplomarbeit. Fachhochschule Köln
17. Rogala D, Lake S, Maines C, Mecklenburg M (2010) Condition problems related to zinc oxide underlayers: examination of selected abstract expressionist paintings from the collection of the Hirshhorn Museum and Sculpture garden, Smithsonian institution. In: Journal of the American Institute for Conservation 49, volume 2 winter/fall. Pp: 96–113
18. Van der Weerd J, Geldof M, Struik van der Loeff L, Heeren RMA and Boon J (2004) Zinc soap aggregate formation in 'falling leaves' (Les Alyscamps)' by Vincent van Gogh. In: Z. Kunsttechnol. Konserv. 17. Pp. 407–416

Chapter 21
A Synchrotron Photoluminescence Microscopy Study into the Use and Degradation of Zinc White in 'The Woodcutters' by Bart van der Leck

Selwin Hageraats, Katrien Keune, Mathieu Thoury, and Ruth Hoppe

21.1 Introduction

Bart van der Leck, who—together with Piet Mondrian and Theo van Doesburg—co-founded the artistic movement *De Stijl,* stood at the forefront of early abstract art. Van der Leck's works are constructed from a profound, well-documented vision on color, space, geometry and light, resulting in a style in which composition is key, and arrangements are often subject to revision. Between 1916 and 1918, Van der Leck's works became increasingly more abstract. He made series of so-called *compositions*, in which the starting point is often a sketch or study with a figurative depiction of his subject. In subsequent steps he then reduces the composition into an increasingly

S. Hageraats (✉)
Department of Conservation and Science, Rijksmuseum, Amsterdam, The Netherlands

IPANEMA, CNRS, Ministère de la Culture, UVSQ, USR3461, Université Paris-Saclay, Gif-sur-Yvette, France

Van 't Hoff Institute for Molecular Sciences, University of Amsterdam, Amsterdam, The Netherlands
e-mail: s.hageraats@rijksmuseum.nl

K. Keune
Department of Conservation and Science, Rijksmuseum, Amsterdam, The Netherlands

Van 't Hoff Institute for Molecular Sciences, University of Amsterdam, Amsterdam, The Netherlands
e-mail: k.keune@rijksmuseum.nl

M. Thoury
IPANEMA, CNRS, Ministère de la Culture, UVSQ, USR3461, Université Paris-Saclay, Gif-sur-Yvette, France
e-mail: mathieu.thoury@synchrotron-soleil.fr

R. Hoppe
Gemeentemuseum Den Haag, Den Haag, The Netherlands
e-mail: rhoppe@gemeentemuseum.nl

K. J. van den Berg et al. (eds.), *Conservation of Modern Oil Paintings*,
https://doi.org/10.1007/978-3-030-19254-9_21

more abstract representation of the subject, in which the forms are transformed into rectangles or other geometric shapes floating on a white background, constituting the essence of the subject. In these abstract compositions he would only work with primary colors, black, and grey [1, 2]. *The Woodcutters*, (oil on canvas, 157 × 182.5 cm, Fig. 21.1a), painted in 1928 and depicting two woodcutters cutting down a tree, is representative for his work from 1919 onwards; when Van der Leck broke with the *De Stijl* movement. It was in this period that he moved away from highly abstracted forms to compositions in which the subject is yet again more readable.

In a recently published report of a technical study of *The Woodcutters,* it is described that under ultraviolet (UV) light the white background exhibits green luminescence in three distinctly different intensities [3]. As can be seen in Fig. 21.1b, the lowest intensity is predominantly visible in the lower left part of the painting, while an intermediate intensity can be observed in the top half, and the brightest luminescence dominates in the lower half of the painting. Although most of the differently luminescing paint areas were impossible to differentiate without the use of UV-light, in some areas of the background, such as those directly surrounding the color planes, differences in luminescence could be observed to correspond to slight variations in color and surface texture. It was found that the sole pigment used in the top layer of this white background is zinc white, which is most likely mixed with linseed oil as a binding medium. Although it must be noted that both zinc white and linseed oil are known to exhibit green luminescence [4, 5], in this report, the differences in appearance of the background under UV were ascribed to the use of different types of zinc white. Considering the known variation in luminescence behavior among different batches of zinc white [6], and the established use of UV photography for the identification of zinc white in oil paintings [7, 8], it is indeed presumed likely that the appearance of the background under UV is primarily related to the use of zinc white.

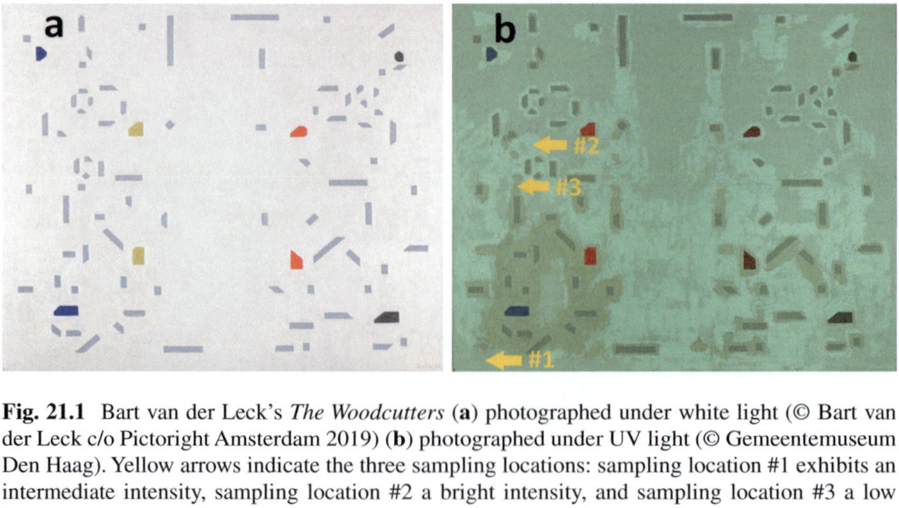

Fig. 21.1 Bart van der Leck's *The Woodcutters* (**a**) photographed under white light (© Bart van der Leck c/o Pictoright Amsterdam 2019) (**b**) photographed under UV light (© Gemeentemuseum Den Haag). Yellow arrows indicate the three sampling locations: sampling location #1 exhibits an intermediate intensity, sampling location #2 a bright intensity, and sampling location #3 a low intensity of green luminescence

The luminescence behavior of zinc white (ZnO) originates from its semiconductor properties. ZnO is a (direct) wide band gap II-IV semiconductor with a band gap of 3.3 eV (~375 nm) at room temperature [9]. Defects in the ZnO crystal structure, such as vacancies and interstitials, introduce energy levels within the band gap. These energy levels are often called trap states, and result in additional radiative relaxation pathways [9, 10]. Although the green trap state emission (~500–550 nm) is the most prominent and most often observed, emissions in the blue and red are also reported. Differences in the relative intensities of these trap state emissions and the band gap emission have been reported for different batches of zinc white, and allow discrimination between zinc whites based on their photoluminescence behavior [6].

In order to examine whether different types of zinc white are indeed present in the white background, it was proposed to study microscopic cross-sections using the same novel photoluminescence micro-imaging method as used by Bertrand et al. [6], and employ a newly developed computational method to detect statistical differences between zinc whites in different paint layers. The aforementioned photoluminescence micro-imaging technique makes use of synchrotron-generated deep-UV excitation and a multispectral acquisition approach [11, 12]. Besides its ability to resolve submicrometric particles and examine their luminescence behavior, its deep-UV excitation capabilities allow access to an emission range with energies exceeding the ZnO band gap. It has been shown in a previous study [13] that in this range, zinc white exhibits no emission and more weakly luminescing compounds, such as zinc white degradation products, can be detected. As the (surface) degradation of zinc white may also have an effect on the macroscale luminescence, using the novel photoluminescence micro-imaging technique, two aspects that influence this phenomenon can be examined simultaneously. Yet, since photoluminescence micro-imaging does not provide a chemically specific probe for all pigments and degradation products, complementary methods of analysis need to be employed in parallel. In this study, optical microscopy and SEM-EDX were used to complement the photoluminescence data in helping to resolve the painting build-up and identifying degradative surface layers.

21.2 Experimental

21.2.1 Sample Preparation

All analyses were performed on three samples taken from the locations indicated in Fig. 21.1b. The regions from which the paint samples were taken were chosen so as to represent each of the three different intensity levels of green luminescence. Paint samples were embedded in Technovit® 2000 LC resin, and dry polished with silicon carbide polishing cloths (Micro-Mesh®, final step 12.000 mesh) to be prepared as cross-sections.

21.2.2 Instrumental Analysis

Optical Microscopy Cross-sections were analyzed using a Zeiss Axio Imager. A2m optical microscope. The microscope was equipped with visible and UV LEDs, to provide white and UV light respectively. Images under UV light were recorded by filtering the excitation beam with a Zeiss G 365 band-pass filter and diverting the lower energy emission using a beam splitter with a cut-off wavelength of 395 nm. Emission was subsequently filtered with a 420 nm low-pass filter and recorded using a AxioCam MRc5 camera.

SEM-EDX Scanning electron microscopy images were recorded using a Thermo Fisher Nova Nano SEM 450 electron microscope. Backscattered electron images were taken at a 15 kV accelerating voltage and eucentric working distances between 5 and 6 mm. For the purpose of pigment identification, EDX analysis was performed using the coupled Thermo Fisher EDX system. Cross-sections were gold coated (3 nm thickness) in a Quorum Technologies SC7640 gold sputter coater prior to SEM-EDX analysis in order to improve surface conductivity.

S-DUV-PL Micro-Imaging Synchrotron deep-UV multispectral photoluminescence micro-imaging was performed at the DISCO beamline of the SOLEIL synchrotron. The optical set-up has been described in detail in Thoury et al. [12], so the focus here will be on the used parameters and the choice of exchangeable optics. All three cross-sections were analyzed with an excitation energy of 4.43 eV (280 nm), and a dichroic mirror with a cut-off wavelength of 300 nm. Four emission filters were used to record photoluminescence images: 327–353, 370–410, 412–438, and 499–529 nm. The three longer wavelength filters are meant to specifically transmit ZnO band gap emission, blue ZnO trap state emission, and green ZnO trap state emission respectively. It is worth noting that it is exactly this green ZnO trap state emission that gives the painting its green luminescence on the macroscale. Variable exposure times were used so as to ensure the highest possible signal, without reaching saturation. For the 327–353 nm emission filter, exposure times were set between 70 and 370 seconds, whereas for the remaining filters, due to the bright ZnO emission, exposure times could be set between 2.5 and 22 s. A Zeiss Ultrafluar 100 × UV-transmitting immersion objective was used, capable of resolving the individual submicrometric zinc white particles.

21.2.3 Image Processing & Statistical Analysis

Due to the limited field of view of the 100 × objective used in the S-DUV-PL micro-imaging set-up, composite images were recorded in a mosaic-like manner. Individual tiles were stitched together by optimizing the cross-correlation of adjacent tiles and subsequent linear blending. In order to highlight micrometric and submicrometric structure, reduce scattering effects, and filter out spatial features caused by inhomogeneous illumination, an algorithm based on scale space theory [14] was applied to the individual tiles. The algorithm works by separately applying two Gaussian filters with $\sigma = 35$ (yielding F_{35}) and $\sigma = 150$ (yielding F_{150}). From the original tile image I, the scale-space filtered tile image I_{SSF} is then obtained through:

$$I_{SSF} = I - F_{35} + F_{150} \tag{21.1}$$

This reduces each image to scales $150 < \sqrt{t} < 35$. The scale-space filtered composite images were then intensity stretched and—in case of the 370–410, 412–438, and 499–529 nm emission filters—combined into false color images.

In parallel, numerical analysis of the recorded images was performed to detect statistical differences in luminescence behavior between the different zinc white paint layers. These statistical differences could then be used as evidence for the use of different batches of zinc white throughout the painting, which can help explain the macroscale luminescence of *The Woodcutters* and provide information about its production process.

First, particle sizes—expressed as σ_p—were estimated based on the 2D autocorrelation matrix of $\sqrt{t} < 35$ scale-space filtered images. Then, a set number of particles in each image were integrated within a $2\sigma_p$ radius, of which the median integrated intensity was taken as its metric $M_{L,i}$. Here, the subscripts L and i indicate the paint layer number and emission filter number respectively. To avoid experimental reproducibility issues, ratios were determined from the calculated metrics of each cross-section:

$$R_{L,i,j} = \frac{M_{L,i}}{M_{L,j}} \tag{21.2}$$

Uncertainties in the particle size estimation and the median calculation were combined with the errors resulting from the quantized integration boundaries to yield the overall uncertainty in ratios $R_{L,i,j}$:

$$u_{R_{L,i,j}} = \sqrt{u^2_{p_{L,i,j}} + u^2_{med_{L,i,j}} + u^2_{quant_{L,i,j}}} \tag{21.3}$$

The dissimilarity between two paint layers L and L' is then measured by its combined zeta-score [15]:

$$\zeta_{L,L'} = \frac{\Sigma_j \left| R_{L,i,j} - R_{L',i,j} \right|}{\sqrt{\Sigma_j \left(u^2_{R_{L,i,j}} + u^2_{R_{L',i,j}} \right)}} \tag{21.4}$$

Here, the denominator gives the standard deviation of the distribution of the numerator, given that paint layers L and L' have the same chemical composition. The score $\zeta_{L,L'}$ then returns the number of standard deviations that the numerator deviates from its expected value of zero. Purely by chance alone, a score of 1.96 or higher is expected for only 5% of all measurements. When such scores are found, it is regarded as statistical evidence for the dissimilarity of paint layers L and L', which, in this context, is attributed to the use of different batches of zinc white.

21.3 Results

21.3.1 Instrumental Analysis

Optical Microscopy Optical microscopy images, recorded using visible light and UV illumination, are shown in Fig. 21.2. The three cross-sections show a similar stratigraphy, comprising at least three white paint layers. Layer 3 shows a bright green luminescence under UV illumination, which is predominantly emitted by submicrometric particles. This indicates that the green luminescence on the macroscale—as was hypothesized—indeed originates from the pigment, rather than the binding medium, degradation products, or possible additives. Interestingly, on top of layer 3 in cross-section #3 an additional paint layer appears to be present (layer 4), which does not show the same brightness as layer 3. The inset of Fig. 21.2f shows this layer in detail. In the lower part of each cross-section, a thin translucent layer of glue is visible, which can be seen in cross-section #1 & #3 to be directly applied onto the canvas support.

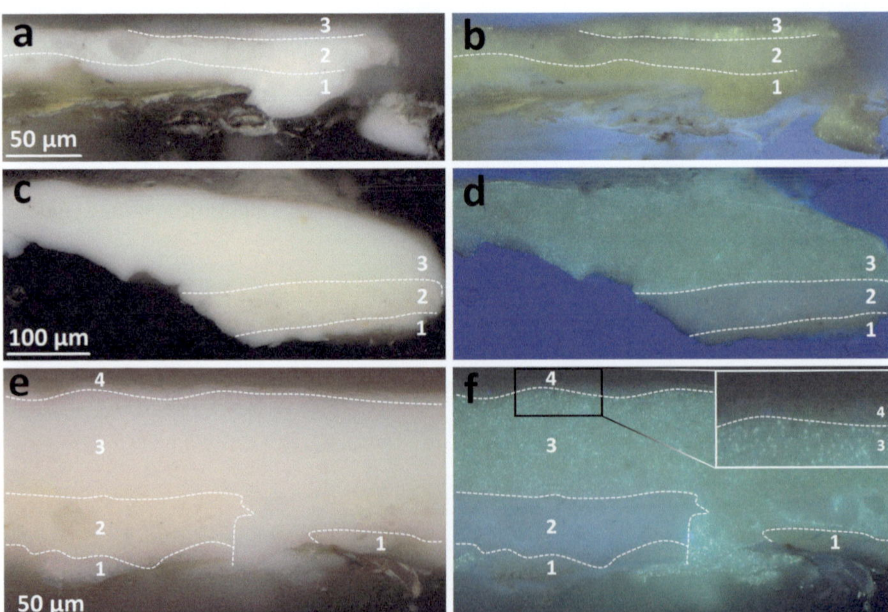

Fig. 21.2 Optical microscopy images of cross-sections #1 (**a, b**, intermediate intensity), #2 (**c, d,** bright intensity), and #3 (**e, f**, low intensity on macroscale). Figures **a, c** & **e** show the images recorded using white light illumination, while figures **b, d** & **f** show the images recorded using UV illumination. The inset in figure f shows a magnification of the region indicated by the black rectangle

Scanning Electron Microscopy SEM backscatter images for the three cross-sections, as well as details of the top layers, are shown in Fig. 21.3a–f. The paint layers that have been discussed in the previous paragraph can be clearly distinguished in each cross-section based on the differences in particle size and electron density. Using EDX analysis (data not shown), layer 1 was determined to contain lead, zinc, and sulfur, layer 2 mostly lead, and layer 3 and 4 only zinc. In an ongoing micro X-ray diffraction study, the lead pigment in layer 1 was identified as lead sulfate (Flemish white), and the lead pigment in layer 2 as lead white. Furthermore, the zinc pigment in layer 1 was identified as zinc white, even though this layer does not show the characteristic green luminescence under UV (Fig. 21.2b, d, and f) that is seen in layers 3 & 4.

On top of each cross-section, a thin unpigmented surface layer was noted (Fig. 21.3b, d, and f). This is especially apparent in cross-sections #1 & #2, which both exhibit a surface layer with an average thickness of 1–2 µm. In cross-section #3, the surface layer has a similar appearance, but was found to be much thinner. The apparent electron density of this thin surface layer is remarkably high, indicating the presence of a substantial inorganic fraction. This presumption was confirmed by EDX analysis, which revealed the presence of mostly zinc, sulfur, and chlorine.

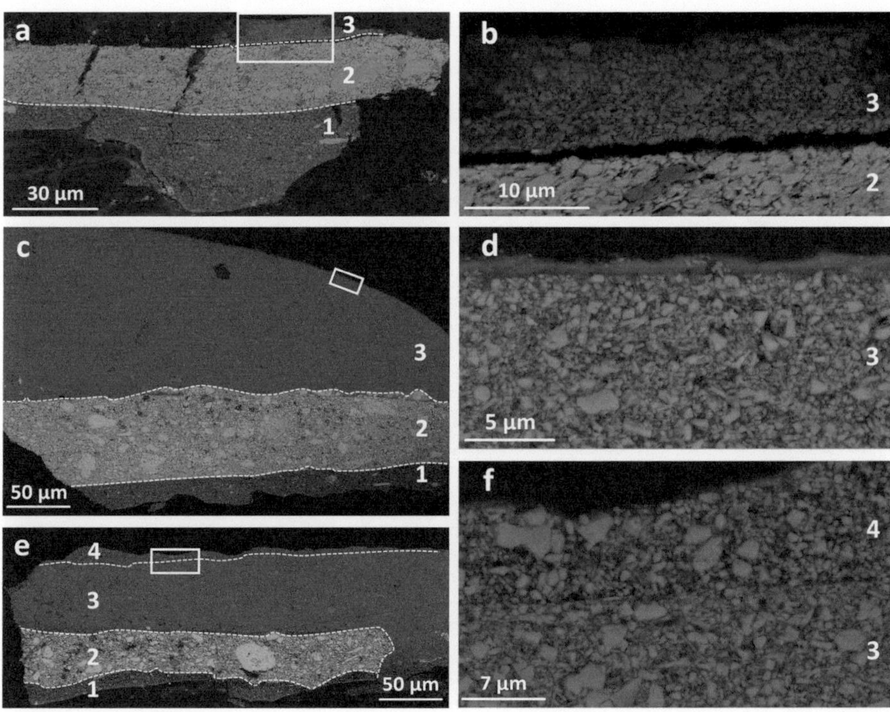

Fig. 21.3 SEM backscatter images of sample #1 (**a, b**, intermediate intensity on macroscale), #2 (**c, d**, bright intensity on macroscale) and #3 (**e, f**, low intensity on macroscale). White rectangles in figures **a, c** & **e** indicate regions of interest of which magnified images are shown in figures **b, d** & **f**

S-DUV-PL Micro-Imaging Figures 21.4a–c show the false color images of the three cross-sections, using the 499–529 nm filter for the red channel, the 412–438 nm filter for the green channel, and the 370–410 nm filter for the blue channel. White rectangles indicate regions of interest which are magnified in Fig. 21.4d and e. In the full-scale false color images, zinc white layers 3 & 4 can be clearly distinguished from the other paint layers, due to the vibrant colors which correspond with the differently luminescing zinc white particles. In each cross-section, layer 3 shows a similar color distribution; particles luminescing predominantly in the blue, green, and red emission channels each present in nearly equal concentrations. However, in layer 4 in cross-section #3 (Fig. 21.4e), particles luminescing in the red channel (499–529 nm) are nearly absent.

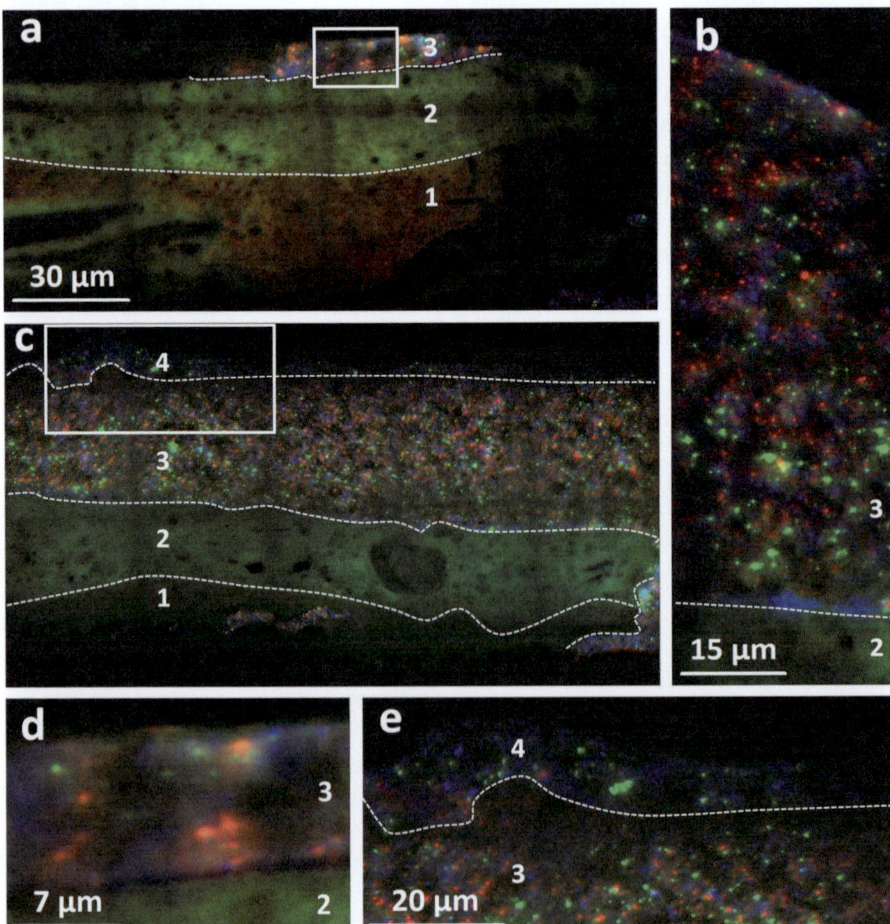

Fig. 21.4 S-DUV-PL false color images of samples #1 (**a, d**), #2 (**b**), and #3 (**c, e**). RGB values were assigned to the following emission filters: red: 499–529 nm, green: 412–438 nm, and blue: 370–410 nm

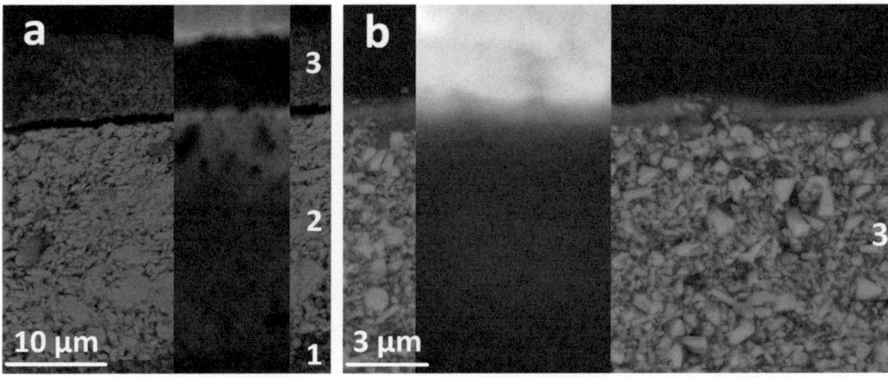

Fig. 21.5 Details of the SEM backscatter images of samples #1 & #2 (Fig. 21.3a and d), superimposed with the photoluminescence images recorded in the 327–353 nm emission range

The presence of zinc white in layer 1 is only evident from some weak luminescence in the 412–438, and 499–529 nm emission ranges. Oddly enough, no signal in the band gap emission range (370–410 nm) could be ascribed to zinc white particles. Although this makes the results obtained from this layer unfit for statistical analysis, this fact alone already shows that this zinc white must come from a different source than the zinc whites used in layers 3 and 4.

Figures 21.5a and b show details of the top of cross-sections #1 & #2, in which thin unpigmented surface layers could be observed in the SEM backscatter images (Fig. 21.3b and d). Here, the backscatter images are overlapped with those obtained using S-DUV-PL micro-imaging through the 327–353 nm emission filter. In both cross-sections, the thin unpigmented surface layer appears to exhibit some luminescence in the UV.

21.3.2 Image Processing & Statistical Analysis

According to the principles outlined in Sect. 2.3, values of $R_{L,i,j}$ and their uncertainties were calculated for all three ZnO emission ranges: 370–410, 412–438, and 499–529 nm. Even though for the interpretation of the macroscale luminescence of *The Woodcutters* only the 412–438 and 499–529 nm emission filters are directly relevant, the 370–410 nm emission is also included, as the resulting multivariate method of analysis has a higher discriminatory power. Due to the previously mentioned lack of signal from layer 1 in the 370–410 nm emission range, statistical analysis was only performed on zinc white layers 3 and 4.

The results are listed in Table 21.1, where the labels $\#1_{L3}$, $\#2_{L3}$, $\#3_{L3}$, and $\#3_{L4}$ are used to distinguish between layer 3 in cross-sections #1, #2, and #3, and layer 4 in cross-section #3 respectively. The zeta-scores and corresponding p-values, meant to quantify the differences between each set of R-values, are listed in Table 21.2. These values can be interpreted as the probability that two sets of

Table 21.1 Values of $R_{L,\alpha,\beta}$ and $R_{L,\alpha,\gamma}$ calculated for layer 3 in cross-sections #1, #2 and #3 ($\#1_{L3}$, $\#2_{L3}$, $\#3_{L3}$) and layer 3 in cross-section #3 ($\#3_{L4}$)

L	$R_{L,\alpha,\beta}$	u	$R_{L,\alpha,\gamma}$	u
$\#1_{L3}$	3.4	0.7	0.59	0.1
$\#2_{L3}$	2.5	0.2	0.35	0.04
$\#3_{L3}$	2.4	0.3	0.42	0.07
$\#3_{L4}$	3.2	0.5	1.4	0.3

Table 21.2 Zeta-scores and corresponding p-values calculated by comparing the R-values from Table 21.1 according to Eq. 21.4

L vs. L'	$\zeta_{L,L'}$	p
$\#1_{L3}$ vs. $\#2_{L3}$	1.6	0.11
$\#1_{L3}$ vs. $\#3_{L3}$	1.5	0.13
$\#1_{L3}$ vs. $\#3_{L4}$	1.1	0.27
$\#2_{L3}$ vs. $\#3_{L3}$	0.38	0.70
$\#2_{L3}$ vs. $\#3_{L4}$	3.0	0.0025
$\#3_{L3}$ vs. $\#3_{L4}$	2.8	0.0048

R-values—$\{R_{L,\alpha,\beta}, R_{L,\alpha,\gamma}\}$ and $\{R_{L',\alpha,\beta}, R_{L',\alpha,\gamma}\}$—are found given that paint layers L and L' have the same composition.

21.4 Discussion

21.4.1 *Explaining Differences in Luminescence on the Macroscale*

Instrumental analysis of the three cross-sections obtained from Bart van der Leck's *The Woodcutters* suggests that the layer build-up of the white background is similar throughout the painting, despite the distinctly different luminescence behavior that can be observed in the three sampling regions when the painting is viewed under UV light (Fig. 21.1b). In each of the three cross-sections, a lead sulfate/zinc white layer (1) and a lead white containing layer (2) can be distinguished, of which layer 1 was identified as the ground layer present on the commercially prepared canvas [3]. On top of layer 2, each cross-section shows at least one layer of zinc white. In cross-section #1 (intermediate luminescence on macroscale), this layer is relatively thin (~10 μm), whereas in cross-sections #2 & #3 (bright and low intensity respectively on macroscale), the zinc white-containing layer is much thicker: 150 μm and 100 μm respectively.

From the zeta-scores listed in Table 21.2, conclusions can be drawn on the use of zinc white throughout Bart van der Leck's *The Woodcutters*. First of all, it is evident from the very low value of $\zeta_{\#2_{L3},\#3_{L3}}$ (0.38) that the zinc white particles visible in layer 3 in cross-section #2 and cross-section #3 are very similar. Comparison of these two layers with the thin zinc white layer in cross-section #1 returns zeta-scores $\zeta_{\#1_{L3},\#2_{L3}}$ and $\zeta_{\#1_{L3},\#3_{L3}}$ of 1.6 and 1.5 respectively. Although the probability of finding these values or higher is relatively small (0.11 and 0.13 respectively), as per convention, this is regarded as insufficient evidence for dissimilarity. Moreover, it must be noted that the number of particles that could be analyzed in cross-section #1 was about an order of magnitude smaller than on cross-section #2 & #3, which makes a statistical comparison challenging to begin with. The remaining set of zeta-scores $\{\zeta_{\#1_{L3},\#3_{L4}}, \zeta_{\#2_{L3},\#3_{L4}}, \zeta_{\#3_{L3},\#3_{L4}}\}$ expresses the statistical significance of the apparent dissimilarities in photoluminescence behavior between layer 4 in cross-section #3, and the other zinc white paint layers analyzed so far. $\zeta_{\#2_{L3},\#3_{L4}}$ and $\zeta_{\#3_{L3},\#3_{L4}}$ were calculated to be 3.0 and 2.8 respectively, indicating significant differences, with probabilities of only 0.0025 and 0.0048 to occur when the same zinc white was used. At 1.1, $\zeta_{\#1_{L3},\#3_{L4}}$ returns a value that indicates no significant dissimilarities. However, as the number of particles that could be analyzed was small for *both* paint layers, this value is not taken into account in the following discussion.

Examining the *R*-values listed in Table 21.1, the statistical dissimilarity between layer 4 in cross-section #3 and layer 3 on cross-sections #2 & #3 seems mostly due to the much higher value of $R_{\#3_{L4},1,3}$, as compared to $R_{\#2_{L3},1,3}$ and $R_{\#3_{L3},1,3}$. This was in turn found to be caused by a significantly weaker signal measured through the 499–529 nm emission filter on layer 4 in cross-section #3 as compared to layer 3 on cross-sections #2 & #3. These results show that the zinc white used in layer 4 in sampling region #3 differs from the zinc white used in layer 3 in sampling regions #2 & #3. It also demonstrates that the significant differences can be ascribed to the weaker luminescence in the green spectral range of the zinc white used in layer 4, which explains the darker appearance of sampling region #3 under UV at the macroscale.

Based on these statistical results and those obtained from the instrumental analyses, it is now possible to formulate a hypothesis as to how the luminescence behavior of *The Woodcutters* on the macroscale is related to the use of pigment throughout the painting and what purpose the corresponding paint layers likely would have served. From the analysis on cross-sections #1 & #2 it could be concluded that the intermediate and bright intensity of green luminescence (Fig. 21.1b) originate from the same zinc white pigment. Differences in luminescence intensity on the macroscale are therefore ascribed to differences in the thickness of the applied paint layer. It can be noted that the intermediate intensity luminescence is most dominantly present in the regions between color planes, whereas the bright intensity luminescence is often present directly around color planes. This suggests that the thicker zinc white layer was applied as a means to modify the composition of the color planes, hereby possibly overpainting the thinner zinc white layer. The absence of a distinct interface within layer 3 in cross-sections #2 & #3 indicates that, if the thicker zinc white layer was indeed applied on top of the thinner zinc white layer, this application was done wet-in-wet.

Just like the bright intensity luminescence, the low intensity luminescence is also found predominantly in the direct vicinity of color planes. This indicates that layer 4 was also applied with the purpose of modifying the composition of the painting. The interface between layer 3 and 4 in cross-section #3—which is clearly visible in the SEM backscatter image shown in Fig. 21.3f—reveals that layer 4 was only applied when layer 3 was already dry. This means that at least several days had passed since the application of layer 3 before these final adjustments to the composition were made.

21.4.2 Assessing Surface Degradation Phenomena

As can be seen in Fig. 21.5a and b, cross-sections #2 & #3 show a thin, unpigmented zinc-rich surface layer that luminesces in the 327–353 nm emission range. Due to the presence of sulfur in this surface layer, its formation is thought to be a result of the reaction of zinc white with atmospheric sulfur compounds, such as sulfur dioxide (SO_2) and/or hydrogen sulfide (H_2S), producing zinc sulfide (ZnS). ZnS, like ZnO is a II-IV semiconductor with a similarly wide band gap of 3.54 eV (351 nm) at room temperature [16]. Therefore, if ZnS crystallites were indeed formed on the paint-air interface, this would explain the luminescence of the surface layer in the 327–353 nm emission range. Yet, when the signal intensities measured on these thin surface layers through the 327–353 nm emission filter are compared to the intensities of the band gap emission of ZnO measured on the same cross-sections, it must be noted that the ZnO band gap emission is over two orders of magnitude stronger than the emission measured on the surface layers. It is thought that this major discrepancy in band gap emission intensity can be attributed to the nanometric size the ZnS crystallites would have, given that they cannot be resolved using SEM at 5000x magnification (Fig. 21.3d). Such nanoparticles, formed in a complex chemical environment, are expected to have severely distorted electronic band structures due to high concentrations of surface states, impurities, and other crystal defects. These severe distortions can make non-radiative relaxation and radiative relaxation through trap-states dominant, thereby reducing the observed band gap emission. If the hypothesis of the formation of ZnS nanoparticles on the painting surface is correct, emission at longer wavelengths is therefore also expected. Such emission can most prominently be seen in the false color image in Fig. 21.4d, where the thin surface layer has a distinct green appearance, indicating emission in the 412–438 nm spectral region. It has been shown previously that ZnS nanoparticles with an average particle size of 1.24 nm exhibit an emission band in this exact spectral region, which is ascribed to the nanoparticles' abundant surface states [17].

The likely degradative origin of the surface layer and its absence in cross-section #3 suggests that there are differences in the chemical properties of these two different varieties of zinc white. Whereas the one used in layer 3 seems very susceptible to reaction with atmospheric sulfur compounds, the variety used in layer 4 appears far less prone to be converted to ZnS. As the only noticeable differences between the two varieties of zinc white show up in their photoluminescence behavior, it is hypothesized that the reactivity of zinc white and the concentrations of certain crystal defects are related.

21.5 Conclusion

Using a combination of instrumental and statistical analysis, the use and degradation of zinc white in Bart van der Leck's *The Woodcutters* was investigated. The results provide an explanation as to why the white background of the painting has such an inhomogeneous appearance under UV illumination (Fig. 21.1b). It could be concluded that the intermediate intensity luminescence—that appears in the painting for instance in sampling region #1—indicates the presence of a thin zinc white layer that was applied in the background, most likely in an early stage of the painting process. The bright intensity luminescence (e.g. sampling region #2) was found to correspond with a more thickly applied paint layer that contains the same variety of zinc white. These thick layers were applied to modify the composition, possibly by overpainting the previously applied thin zinc white layer wet-in-wet. The low intensity luminescence (e.g. sampling region #3) appears to relate to compositional adjustments that were made by Van der Leck in a later stage using a different variety of zinc white pigment. It is possible that these changes were made several days, weeks, or even months later, after a major part of the composition was already finished. These results demonstrate how—using a multispectral photoluminescence approach—three different varieties of zinc white can be distinguished in a set of historical paint cross-sections.

A thin, unpigmented surface layer was observed in cross-sections #1 and #2. Based on its luminescence in the 327–353 and 412–438 nm spectral regions and its relatively high electron density, it is hypothesized that this surface layer is the result of a reaction between zinc white and atmospheric sulfur compounds, forming nanometric ZnS crystallites. The nearly unaffected appearance of layer 4 in cross-section #3 suggests that significant differences in reactivity exist between differently luminescing varieties of zinc white. Given the well-documented relation between the photoluminescence behavior of zinc white and its crystal defects, it is proposed that a link may exist between defects in the ZnO crystal structure and its reactivity. Further research into this relationship and how it extends to its reactivity towards paint components may help explain why zinc white degradation phenomena vary so widely among paintings. Furthermore, revealing these possible relationships could pave the road for non-invasive risk assessment that is based on the luminescence properties of the paint surface.

Acknowledgments The authors would like to express their gratitude to Matthieu Réfrégiers & Frédéric Jamme (DISCO beamline, SOLEIL synchrotron) for their assistance at the beamline and to Victor Gonzalez for providing the μ-XRD maps used for pigment identification. The authors gratefully acknowledge financial support from The Bennink Foundation and the Access to Research Infrastructures activity in the Horizon 2020 Programme of the EU (IPERION CH Grant Agreement n. 654028).

References

1. Hilhorst C, Wijnia, L (2017) Chronologie. In: De uitvinding van een nieuwe kunst, Piet Mondriaan & Bart van der Leck, Laren 1916–1918, Gemeentemuseum Den Haag/WBooks, Den Haag/Zwolle, p 11–57
2. Janssen H (2017) Inleiding. In: De uitvinding van een nieuwe kunst, Piet Mondriaan & Bart van der Leck, Laren 1916–1918, Gemeentemuseum Den Haag/WBooks, Den Haag/Zwolle, p 7–10
3. Albrecht M, Koenen J (2017) Meer dan het oog ziet: een analyse van drie schilderijen van Bart van der Leck. In: De uitvinding van een nieuwe kunst, Piet Mondriaan & Bart van der Leck, Laren 1916–1918, Gemeentemuseum Den Haag/WBooks, Den Haag/Zwolle, p 147–156
4. De La Rie E R (1982) Fluorescence of Paint and Varnish Layers (Part I). Studies Conserv 27:1–7
5. De La Rie E R (1982) Fluorescence of Paint and Varnish Layers (Part II). Studies Conserv 27:65–69
6. Bertrand L, Réfrégiers M, Berrie B H, Échard J-P, Thoury M (2013) A multiscalar photoluminescence approach to discriminate among semiconducting historical zinc white pigments. Analyst 138:4463–4469
7. De La Rie E R (1982) Fluorescence of Paint and Varnish Layers (Part III). Studies Conserv 27:102–108
8. Broadway M, Sutherland K, Biolcati, V, Casadio, F (2018) Mastery of materials: Sargent's watercolors at the Art Institute of Chicago. In: John Singer Sargent & Chicago's gilded age, Art Institute of Chicago, Chicago, p 188–190
9. Özgür, Ü Alivov, Y I, Liu, C, Teke, A, Reshchikov, M A, Dogan, S, Avrutin, V, Cho S-J, Morkoc, H J (2005) A comprehensive review of ZnO materials and devices. Appl Phys 98:041301
10. Djurisic A B, Leung Y H, Tam K H, Hsu Y F, Ding L, Ge W K, Zhong Y C, Wong K S, Chan W K, Tam H L, Cheah K W, Kwok W M, Phillips D L (2007) Defect emissions in ZnO nanostructures. Nanotechnology 18:095702
11. Giuliani A, Jamme F, Rouam V, Wien F, Giorgetta J-L, Lagarde B, Chubar O, Bac S, Yao I, Rey S, Herbeaux C, Marlats J-L, Zerbib D, Polack F, Réfrégiers M (2009) DISCO: a low-energy multipurpose beamline at synchrotron SOLEIL. J Synchrotron Radiat 16:835–841
12. Thoury M, Echard J-P, Refregiers M, Berrie B H, Nevin A, Jamme F, Bertrand L (2011) Synchrotron UV-Visible multispectral luminescence microimaging of historical samples. Anal Chem 83:1737–1745
13. Hageraats S, Keune K, Réfrégiers M, Van Loon A, Berrie B, Thoury M (2019) Synchrotron Deep-UV Photoluminescence Imaging for the Submicrometer Analysis of Chemically Altered Zinc White Oil Paints. Anal Chem DOI: 10.1021/acs.analchem.9b02443
14. Lindeberg T J (1994) Scale-space theory: A basic tool for analysing structures at different scales Appl Stat 21:225–270
15. Analytical Methods Committee (2016) z-Scores and other scores in chemical proficiency testing-their meanings, and some common misconceptions. Anal Methods 8:5553–5555
16. Yongnian Y, Hickey C F, Gibson U J (1987) ZnS molecular beam epitaxy on silicon substrates. Thin Solid Films 151:207–214
17. Chen W, Wang Z, Lin Z, Lin L (1997) Absorption and luminescence of the surface states in ZnS nanoparticles J Appl Phys 82:3111–3115

Chapter 22
'Breaking Waves'. The Relation Between Zinc-Oxide Degradation and Extreme Delamination from the Panel Support of Beach Scene, by J.E.H. Akkeringa (1861–1942)

Alma Jongstra, Klaas Jan van den Berg, Ella Hendriks, Suzan de Groot, Henk van Keulen, and Maartje Stols-Witlox

22.1 Introduction

Johannes Evert Hendrik Akkeringa (1861–1942) belonged to the second generation of the Hague School of painters. He worked in The Hague for most of his life, where he worked closely together with contemporaries, such as Hague School painter Willem de Zwart [3]. Akkeringa's production of paintings was large and he used both canvas and panel supports. 'Children playing on the beach' was a frequent subject in his works.

The undated panel painting *Beach Scene* (Fig. 22.1a) was donated in 2012 to the Programme of Conservation and Restoration of Cultural Heritage at the University of Amsterdam from a private collection, where it had been kept in an uncontrolled climate. Severe delamination in the form of cleavage, tenting and flaking affected a large part of the painted surface. This phenomenon mainly occurred in the areas directly corresponding to the widest year rings in the tangential grain of the panel and a large filling with retouching suggested that some of this damage had been treated in the past (Fig. 22.1b). Due to this delamination the painting was unexhibitable and difficult to handle. *Beach Scene* had been framed behind glass, without a backing board and was tightly held in place with nails, which only allowed for

A. Jongstra · E. Hendriks · M. Stols-Witlox
University of Amsterdam, Amsterdam, The Netherlands

K. J. van den Berg (✉)
University of Amsterdam, Amsterdam, The Netherlands

Cultural Heritage Agency of the Netherlands, Amsterdam, The Netherlands
e-mail: k.van.den.berg@cultureelerfgoed.nl

S. de Groot · H. van Keulen
Cultural Heritage Agency of the Netherlands, Amsterdam, The Netherlands

K. J. van den Berg et al. (eds.), *Conservation of Modern Oil Paintings*,
https://doi.org/10.1007/978-3-030-19254-9_22

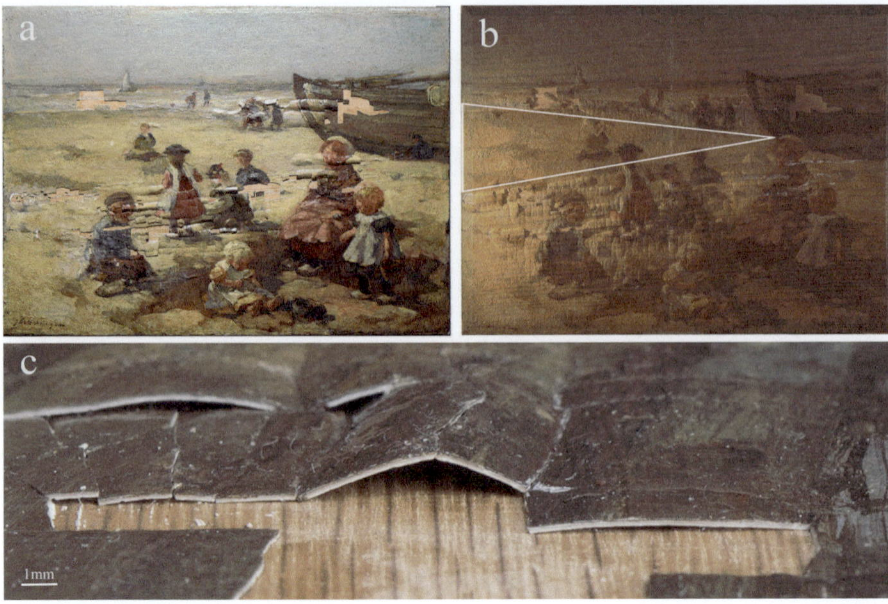

Fig. 22.1 J.E.H. Akkeringa, *Beach Scene*, unframed, date unknown, oil on panel, 42.0 × 32.5 × 1.1 cm. Raking light from the top showing extent of delamination (**a**) overlay of reverse (mirrored) and raking light image. The white triangle indicates the location of old filling and retouching. Note the correspondence of the year rings and delamination (**b**) Macroscopic detail of tenting paint showing that no residue was left on the panel (**c**)

in-plane shrinkage of the panel. The reverse of the panel was covered with a thick coating layer.

The adhesion failure occurred at a distinct interface between the preparation layers and the panel, where little or no residues of ground layers were visible after delamination (Fig. 22.1c). A semi-translucent, brownish material was visible on the reverse of delaminated paint flakes. As this could be connected in some way with the delamination of the ground from the panel, the materials in the preparatory layers were investigated. The aim of this diagnostic study was to acquire understanding of the observed phenomena as a basis towards finding a suitable treatment method.

To determine which materials J.E.H. Akkeringa used to prepare his wooden support, samples for instrumental analysis were taken and cross-sections prepared to visualise the build-up of the paint film and the delamination interface. The cross-sections and individual samples were used to analyse their contents with Scanning Electron Microscopy with Energy Dispersive X-ray Spectrometry (SEM-EDX), Fourier Transform Infrared Spectroscopy-Attenuated Total Reflection (FTIR-ATR) and Thermally assisted Hydrolysis and Methylation Gas Chromatography/Mass Spectrometry with Pyrolysis (THM-Py-GC/MS), respectively (see Appendix).

22.2 Results

Cross-sections were prepared from delaminated samples and microscopic analysis under ultraviolet (UV) and visible (VIS) light revealed three individual preparatory layers (Fig. 22.2a). The lowest layer, layer 1, filled the grain of the wood. At the bottom of this layer a semi-translucent brownish zone was visible. The brownish part of this layer did not fluoresce under UV light.

SEM-EDX analysis revealed that layer 3 (the upper priming) contained lead, zinc, barium and calcium, indicating the presence of lead and zinc white pigments, calcium carbonate or sulphate and barium sulphate extender. Layer 2 contained calcium and zinc and layer 1 contained mostly zinc, with a little calcium. A similar combination of elements to layer 1 was found in the brownish region. Elemental mapping revealed that zinc was evenly distributed throughout both the semi-translucent brownish zone and the white portion of the zinc containing layer 1 (Fig. 22.2b). However, there was a difference in morphology between the two parts:

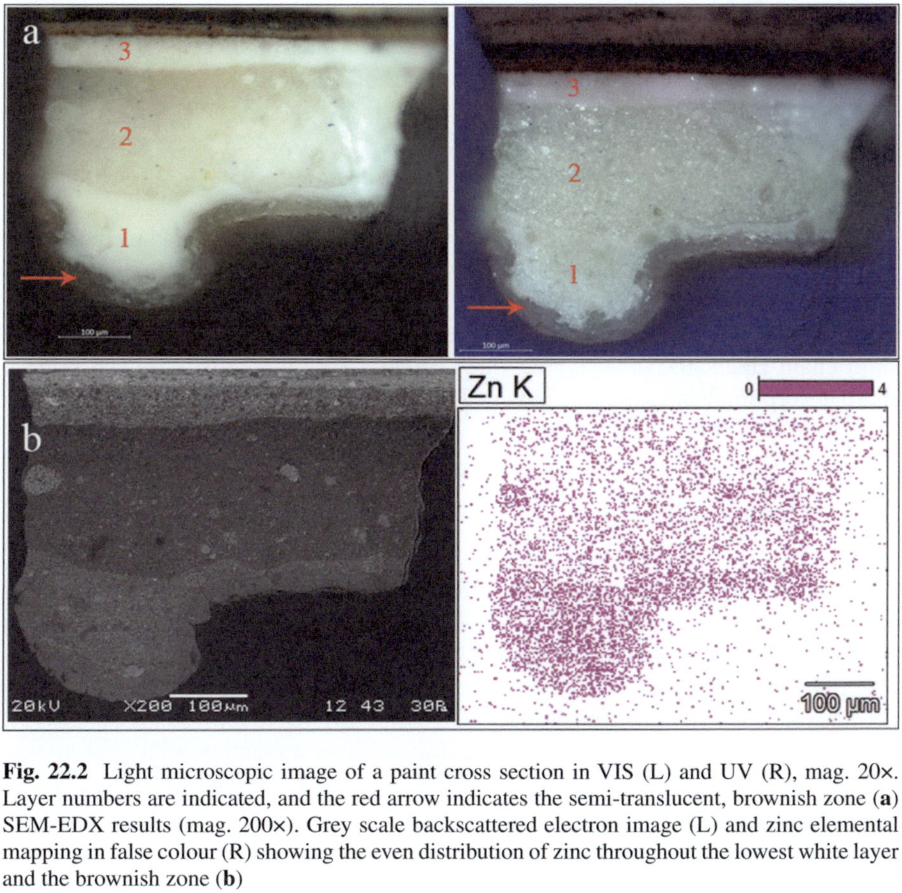

Fig. 22.2 Light microscopic image of a paint cross section in VIS (L) and UV (R), mag. 20×. Layer numbers are indicated, and the red arrow indicates the semi-translucent, brownish zone (**a**) SEM-EDX results (mag. 200×). Grey scale backscattered electron image (L) and zinc elemental mapping in false colour (R) showing the even distribution of zinc throughout the lowest white layer and the brownish zone (**b**)

the 'intact' ground (layer 1) contained densely packed zinc white particles, and the brownish zone below was homogeneous (Fig. 22.2b).

A sample of the surface of the wooden support was taken with a hollow needle, diameter 1 millimetre. The wood sample was analysed by THM-Py-GC/MS, and the pyrogram showed the presence of glycerol and a series of di- and long chain mono-carboxylic acids. The peak area ratios for palmitic to stearic acid, azelaic to palmitic, and suberic to azelaic acid were 1.3, 1.1 and 0.15 respectively. These values were comparable with the THM-Py-GC/MS analysis of a dried, not-pre-polymerised, linseed oil [10]. The pyrogram also showed the presence of extractable materials from the oak wood, such as acetic and gallic acid [2]. This agreed with visual observations that the panel was oak. The THM-Py-GC/MS analysis showed no sign of protein marker compounds [15], indicating that the presence of a glue-based size layer was not plausible, which would fit with traditional methods for panel preparation.

THM-Py-GC/MS analysis of scrapings from the brownish material showed the presence of glycerol and di- and mono-carboxylic acids, with peak area ratios for palmitic to stearic acid, azelaic to palmitic acid and suberic to azelaic acid, of 1.2, 0.9 and 0.3 respectively. All those values suggested the presence of a drying oil [10].[1] Next to a drying oil, trace amounts of beeswax and starch were identified [15]. The analysis of the brownish material furthermore showed the presence of acidic species possibly originating from the oak panel [4].

FTIR-ATR imaging of an embedded cross-section resulted in an average spectrum of the brownish zone (Fig. 22.3). The average spectrum of the white layer (Fig. 22.3b) showed absorption bands ascribable to chalk and a drying oil as well as to zinc soaps (1558 cm^{-1}). However, the relative intensity of the band at 1558 cm^{-1} was significantly smaller than that observed in the spectrum of the brownish material. Since oak-related acidic components were identified using TMH-Py-GC/MS it was also possible that the absorption band at 1558 cm^{-1} was (partially) caused by short chain zinc salts, for example zinc acetate. Crystalline zinc soaps showing a sharp absorption band around 1536 cm^{-1} [6] were not identified.

22.3 Discussion

The experimental results indicated that the brownish, semi-translucent zone was not an individual layer applied by the artist. Instead it appeared to consist of conversion products from zinc oxide in the ground layer that was in direct contact with acidic materials from the panel and the binding medium.

Zinc oxide has been proven to react with acidic species to form salts or soaps [9]. In oil paintings on panel this could manifest itself as a reaction with acid-groups

[1] Remarkable in the analysis was a series of short chain fatty acids, with a carbon number of C6 to C9, and the presence of a rather high amount of oleic acid and some linoleic acid. No explanation for these presences was formulated yet.

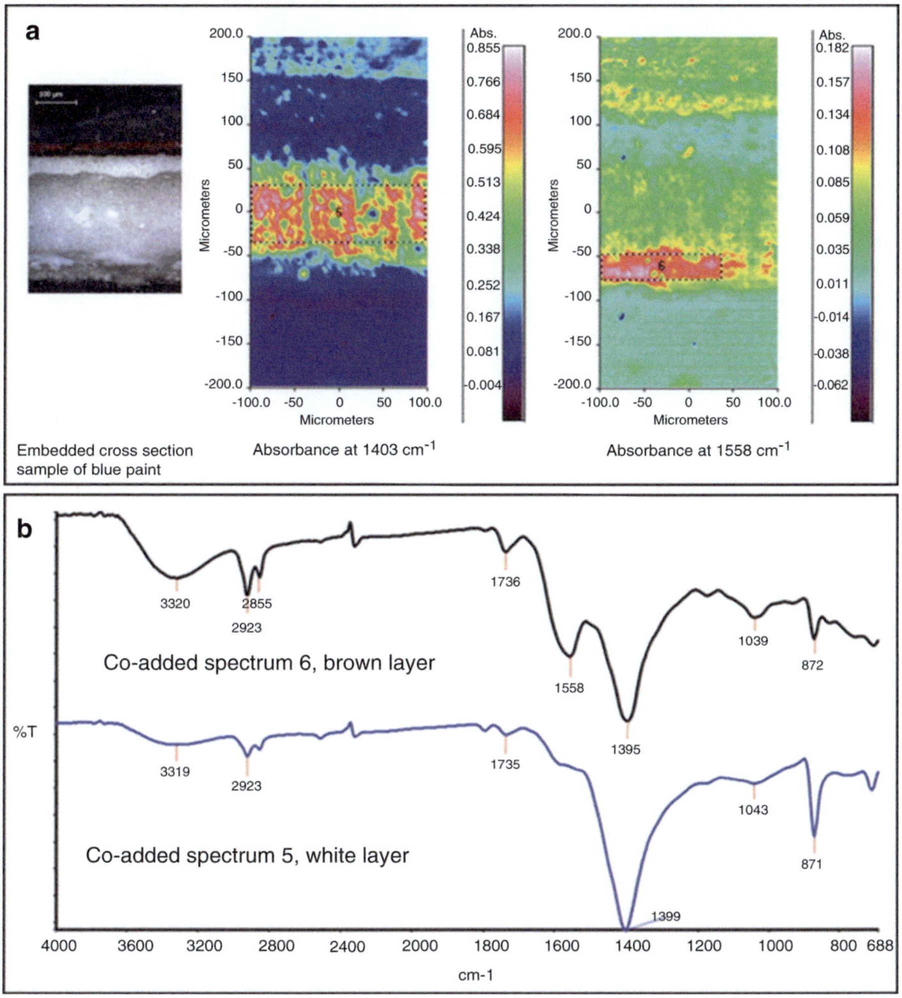

Fig. 22.3 False colour-plots FTIR-ATR images at 1403 cm^{-1} and 1558 cm^{-1} (**a**) Co-added FTIR spectra derived from of the degraded, brownish layer (top) and the 'intact' white layer (below) (**b**)

present in components of a drying oil, or with acids originating from the wooden support. In the case of a reaction with fatty acids, amorphous or crystalline zinc carboxylates are formed [6, 11].

When zinc oxide reacts with short chain acids, such as acetic acid or formic acid (present in oak wood), this can result in the formation of zinc acetate dihydrate or zinc formate dihydrate, which were observed on *Couple with Clouds in Their Heads* by Salvador Dalí [8]. The authors described the degradation of a zinc white-containing oil paint layer that manifested itself as a brownish, semi-transparent zone. This layer had formed under the joint influence of organic acids emitted from the wooden frame, the 'greenhouse' effect caused by the glazed framing system and exposing the painting to heat and high humidity in the past. No (crystalline) zinc soap aggregates were observed in the Dalí painting[2] [8] and similarly in the brownish zone in *Beach Scene* no crystalline zinc carboxylates were detected. These were, however, implicated as responsible for zinc soap-related delamination in other case studies.[3]

It was hypothesised that the thick coating on the panel's reverse may have reduced moisture evaporation from the wood, possibly leading to an increased transport of acidic components to the panel-ground interface. It might also be possible that in fluctuating relative humidity related to *Beach Scene*'s former uncontrolled storage conditions, its restricting frame had led to compression set in the panel and subsequent compression shrinkage as the RH dropped. This would have significantly compressed the paint film and caused it to delaminate.

22.4 Conclusion

Analytical research gave insight into Akkeringa's painting technique and the degradation occurring at the bottom of the first ground layer. With the help of SEM-EDX, FTIR-ATR and THM-Py-GC/MS it was shown that Akkeringa did not use a proteinaceous material to prepare his panel for painting, instead he applied a zinc white oil paint layer directly onto the oak panel. The zinc white pigments at the bottom part of this ground layer reacted with acidic species originating from the wood and fatty acids in the oil binder.

It was hypothesised that under the influence of the coating on the panel's reverse the transport of acidic material to the ground-panel interface was promoted. In addition, the restrictive framing might have caused compression set in the panel and significant compression shrinkage in response to thermo-hygrometric fluctuations. The bond between the deteriorated ground and the panel then proved to be the weakest under the suspected compression on the paint film exerted by the shrinkage of the wooden support.

This study helped to provide insight into the combination of factors that might explain the observed degradation phenomena. It also helped to focus paths of further research, for instance, into the influence of *Beach Scene*'s restricting frame on the compression shrinkage of the tangential panel. Possible continued susceptibility of the panel to climate fluctuations also made the design of an appropriate framing system and preventive conservation measures of prime importance for *Beach Scene* as well as for the preservation of Akkeringa's other panel paintings.

---

[2] This absence of zinc soaps was detected in Male Figure from *Couple with Clouds in Their Heads* by Salvador Dalí, 1936.

[3] Examples are: [1, 5, 7, 11–14].

Acknowledgements Maarten van Bommel, Rene Peschar, Rosa Boute, Marjolein Hupkes, Noa Kollaard, Madeleine Vaudremer and Sepha Wouda (University of Amsterdam).

Appendix

Paint Cross-Sections

Mounted in Technovit® 2000 LC (Heraeus Kulzer, Germany) cured in Technotray POWER– Light Polymerization Unit® (20 min). Wet ground on Labopol® polishing machine with MOPAS CPH7 XS cross-section polishing holder. Dry polished with Micro mesh® polishing cloths. Photography on Zeiss Axioplan 2 optical microscope with incident polarised light. Xenon-lamp for bright field and dark field illumination, and mercury short arc photo optic lamp HBO for UV-fluorescence. The filter-set for UV-fluorescence: excitation BP 365/12, beam splitter FT 395 and emission LP 397.

SEM-EDX

JSM-5910-JSM-5910LV Scanning Electron Microscope – Energy Dispersive X-ray spectroscopy (JEOL Technics Ltd., Japan) with secondary (SE) and backscattering (BSE) detectors coupled to an energy dispersive X-ray spectrometer (EDX).

FTIR-ATR

Perkin Elmer Spectrum 100 FTIR spectrometer in combination with a Spectrum Spotlight 400 FTIR microscope equipped with a 16 × 1 pixel linear Mercury Cadmium Telluride (MCT) array detector. ATR imaging was performed using a Perkin Elmer ATR imaging accessory with a germanium crystal.

THM-Py-GC/MS

Sample material was made into suspension with few drops of a 5% solution of tetra methyl ammonium hydroxide in methanol and the solvent/suspension was transferred to a pyrolysis cup and pyrolyzed using a high-speed temperature range starting at 360 °C, raised to 660 °C. The pyrolysis unit was a Frontier Lab 3030D pyrolyser on a Thermo Scientific Focus GC / ISQ mass spectrometer. Separation on a SLB5 ms (Supelco) column with a length of 20 meters, an internal diameter of 0.18 mm and a film thickness of 0.18 μm. Helium was used as the carrier with a constant flow of 0.9 ml/min. The temperature program was 35 °C (1) – 60 °C /min – 110 °C – 14 °C/min – 240 °C – 5 °C/min – 315 °C (2). The column was directly coupled to the ion source of the mass spectrometer. Interface temperature was 270 °C, ion source temperature was 220 °C. Mass spectra were recorded from 29 until 600 amu (5 scans per second).

References

1. Barnes S (2014) An Investigation of Extreme Delamination in Nu Féminin (1967) by Montréal Artist Jori Smith. Poster presented at the 2014 Annual Student Conference hosted by Buffalo State College (ANAGPIC), Kingston (ON), Queen's University
2. Challinor JM (1996) Characterisation of Wood Extractives by Pyrolysis-Gas Chromatography/Mass Spectrometry of Quaternary Ammonium Hydroxide Extract. Journal of Analytical and Applied Pyrolysis 37:1–13
3. Clercq S de (2010) Johannes Evert Hendrik Akkeringa (1861–1942) Schilder van het Onbezorgde Leven. Scriptum Art, Schiedam
4. Gibson LT, Watt CM (January 2010) Acetic and formic acids emitted from wood samples and their effect on selected materials in museum environments. Corrosion Science, volume 52:172–178
5. Helwig K, Poulin J, Corbeil MC et al (2014) Conservation Issues in Several Twentieth-Century Canadian Oil Paintings: The Role of Zinc Carboxylate Reaction Products. In: Van den Berg KJ, Burnstock A, De Keijzer M, Krueger J, Learner T, Tagle A de, Heydenreich G (eds) Issues in Contemporary Oil Paint. Springer International Publishing/Cultural Heritage Agency, New York/Dordrecht/London, p 167–184
6. Hermans JJ (2017) Metal Soaps in Oil Paint. Structure, Mechanisms and Dynamics. PhD Thesis. Amsterdam: University of Amsterdam
7. Keune K (2005) Binding Medium, Pigments and Metal Soaps Characterised and Localised in Paint Cross-sections. PhD Thesis, MOLART 11, FOM AMOLF, University of Amsterdam
8. Keune K, Boevé-Jones G (2014). It's Surreal: Zinc-Oxide Degradation and Misperceptions in Salvador Dalí's Couple with Clouds in Their Heads, 1936. In: Van den Berg KJ, Burnstock A, De Keijzer M, Krueger J, Learner T, Tagle A de, Heydenreich G (eds) Issues in Contemporary Oil Paint. Springer International Publishing/Cultural Heritage Agency, New York/Dordrecht/London, p 283–294
9. Kühn H (2012) Zinc white. In: Feller RL (ed), Artists' Pigments: A Handbook of their History and Characteristics, vol I. National Gallery of Art in association with Archetype Publications Ltd, London, pp 169–186
10. Lee J, Bonaduce I, Modugno F et al (2018) Scientific investigation into the water sensitivity of twentieth century oil paints. Microchemical Journal 138:282–295
11. Osmond G (2012) Zinc white: a review of zinc oxide pigment properties and implications for stability in oil-based paintings. AICCM Bulletin, no. 33:20–29
12. Osmond G, Keune K, Boon J (2005) A study of zinc soap aggregates in a late 19th century painting by R.G. Rivers at the Queensland Art Gallery. AICCM Bulletin, nr. 29:37–46
13. Raven L, Bisschoff M, Leeuwestein M et al (2016) Delamination due to Zinc Soap Formation in an Oil Painting by Piet Mondrian (1872–1944). Conservation Issues and Possible Implications for Treatment. Paper presented at the conference on Metal Soaps in Art: Conservation and Research, Rijksmuseum, Amsterdam, 14–15 March 2016
14. Rogala D, Lake S, Maines C et al (2010) Condition problems related to zinc oxide underlayers: examination of selected abstract expressionist paintings from the collection of The Hirshhorn Museum and Sculpture Garden, Smithsonian Institution. Journal of the American Institute for Conservation, vol. 49:96–113
15. Schilling MR, Heginbotham A, Van Keulen H et al (2016) Beyond the basics: a systematic approach for comprehensive analysis of organic materials in Asian lacquers. Studies in Conservation, 61, sup. 3:3–27

Chapter 23
Efflorescence on the Paintings of Edwin Austin Abbey: Examination, Analysis, and Cleaning of Surface Bloom on *The Spirit of Light*

Kelsey Wingel, Richard Hark, Katherine Schilling, Cynthia Schwarz, and Anikó Bezur

23.1 Introduction

The American artist Edwin Austin Abbey (1852–1911) began his career in New York City as an illustrator for *Harper's Weekly*, and first began oil painting in 1889 after relocating permanently to England. Passionate about the intersection of architecture, painting, and sculpture, Abbey completed several mural commissions throughout his 20-year painting career, most notably for the Delivery Room of the Boston Public Library and the rotunda of the Pennsylvania State Capitol Building in Harrisburg. Abbey painted these mural commissions in his purpose-built studio in Gloucestershire, England, which he shared for several years with fellow American painter John Singer Sargent [13].

A crucial component of Abbey's artistic process was the production of numerous sketches both before and during the painting of the final composition. More than 600 of Abbey's paintings, mainly oil sketches, were given to the Yale University Art Gallery (YUAG) in 1937 along with the contents of his studio. Many of these paintings are unvarnished or selectively varnished preparatory studies for mural commissions and, rarely exhibited or treated, over half of these paintings exhibit efflorescence.

This study examines the relationship between Abbey's materials and efflorescence formation on the oil-on-canvas study *The Spirit of Light* (Fig. 23.1a). This work was painted in England between 1902–1908 as a preparatory study for a lunette in the rotunda of the Pennsylvania State Capitol Building. The lunette depicts several

K. Wingel (✉) · C. Schwarz
Yale University Art Gallery, New Haven, CT, USA
e-mail: kelsey.wingel@yale.edu; cynthia.schwarz@yale.edu

R. Hark · K. Schilling · A. Bezur
Yale University Institute for the Preservation of Cultural Heritage, New Haven, CT, USA
e-mail: richard.hark@yale.edu; katherine.schilling@yale.edu; aniko.bezur@yale.edu

K. J. van den Berg et al. (eds.), *Conservation of Modern Oil Paintings*,
https://doi.org/10.1007/978-3-030-19254-9_23

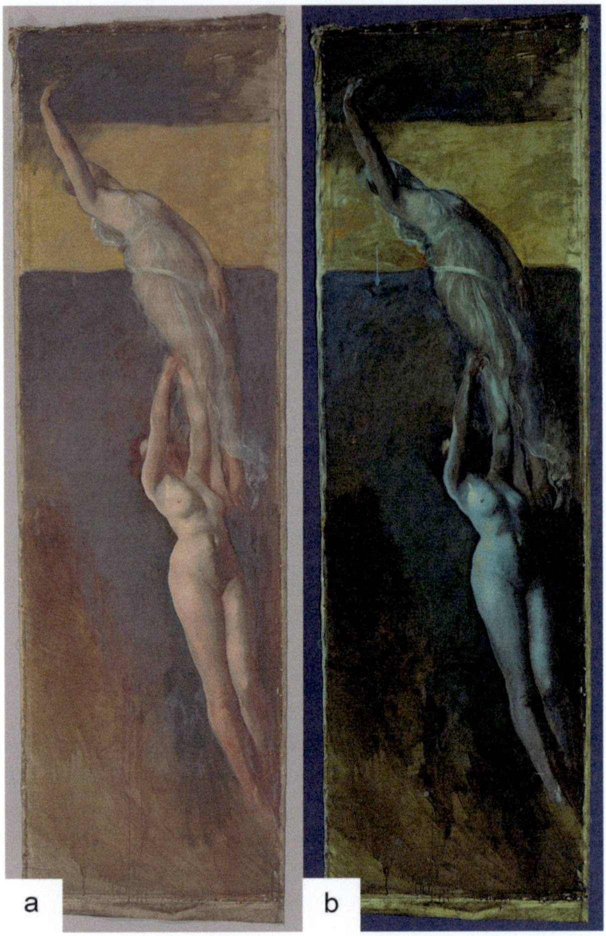

Fig. 23.1 *The Spirit of Light* before treatment in (**a**) visible light and (**b**) ultraviolet light (370 nm). Photo Credit: Yale University Art Gallery

classically garbed women floating upwards into a sky filled with silhouetted oil derricks. As one of four lunettes celebrating Pennsylvania history and industry, *The Spirit of Light* conveys the power and wonder of electricity to create light [18]. In this preparatory figure study, Abbey focuses on capturing the graceful positioning and soaring upward movement of the two figures at the far right of the composition.

Severe efflorescence was present on dark passages of the painting, obscuring the composition and drastically altering the tonality and sheen of the work. The formation of efflorescent material corresponds to the location of selective, artist-applied coatings that visibly fluoresce under ultraviolet excitation (Fig. 23.1b). In preparation for treatment, research was conducted to identify the composition of the efflorescence and to examine its relationship to the locally applied varnishes and the materials within the stratigraphy of the painting. The findings yielded discoveries about Abbey's painting technique and significantly informed the subsequent cleaning of the painting.

23.2 Painting Materials and Technique

The Spirit of Light was painted on a linen canvas that was commercially primed with two off-white ground layers. The first ground layer is chalk-based and the second contains lithopone, or coprecipitated barium sulfate and zinc sulfide.[1] Like many of Abbey's preparatory studies, the oil paint layers were applied broadly and freely with diverse handling that ranges from thicker impasto to sweeping, dripping washes of thinned paint.[2] Abbey worked both wet-into-wet and wet-into-dry, scraping or scumbling in areas of compositional change. Paint application indicates that this study was worked and modified over a period of at least several months.

In addition to using both additive and subtractive painting techniques to achieve his aesthetic goals, Abbey appears to have mixed resins and oil-based varnishes into his paint, as well as applying them in discrete layers. This seemingly spontaneous and highly variable mixing is visible in the different intensities of fluorescence seen within and on top of the paint layers when viewed under ultraviolet light. The cross-sectional sample reveals Abbey's prevalent use of earth pigments in the background and shows the application of an oil-based coating on top of still wet paint layers (Fig. 23.2a, b). Lead sulfate was identified in this coating layer (Fig. 23.2c).[3]

Abbey's use of retouching varnishes, both soft and hard resins, and driers is well-documented in the Roberson Archives at the Hamilton Kerr Institute. Abbey's account from 1884–1904 records his frequent purchase of Vibert's Vernis à Peindre and Vernis à Retoucher in addition to the driers Siccatif de Courtrai and Siccatif de Haarlem [1].[4] Based upon observations of his painting technique, Abbey may have

[1] Characterization of the ground layers was conducted with Fourier transform infrared spectroscopy (FTIR) and field emission-scanning electron microscopy/energy dispersive spectroscopy (FE-SEM/EDS). Dispersed samples of the ground layers were analyzed with FTIR on a single diamond window using a Thermo Continuum infrared microscope coupled to a Nicolet 6700 bench-top spectrometer in transmission mode with the MCT-B detector. SEM-EDS was performed on a cross-sectional sample embedded in Bio Plastic™ resin and polished with Micro-Mesh® abrasive sheets. The sample was carbon coated and analyzed with a Hitachi SU8230 UHR Cold Field Emission SEM with a Bruker XFlash 5060FQ Annular EDS detector in back-scattered electron mode.

[2] The binding medium and coating material were analyzed using the RAdICAL + ESCAPE approach by pyrolysis/gas chromatography/mass spectrometry (Py/GC/MS) with thermally assisted hydrolysis methylation using tetramethylammonium hydroxide after [17].

[3] Lead was detected in the coating layer in the cross-sectional sample with FE-SEM/EDS. Lead sulfate was identified in a dispersed sample of the coating analyzed with FTIR. The coating was determined to be composed of a drying oil using Py/GC/MS.

[4] Siccatif de Courtrai and Siccatif de Haarlem are reported to be composed of copal dissolved in a solvent such as oil of spike or turpentine. Siccatif de Courtrai was noted to contain lead compounds to accelerate drying [4].

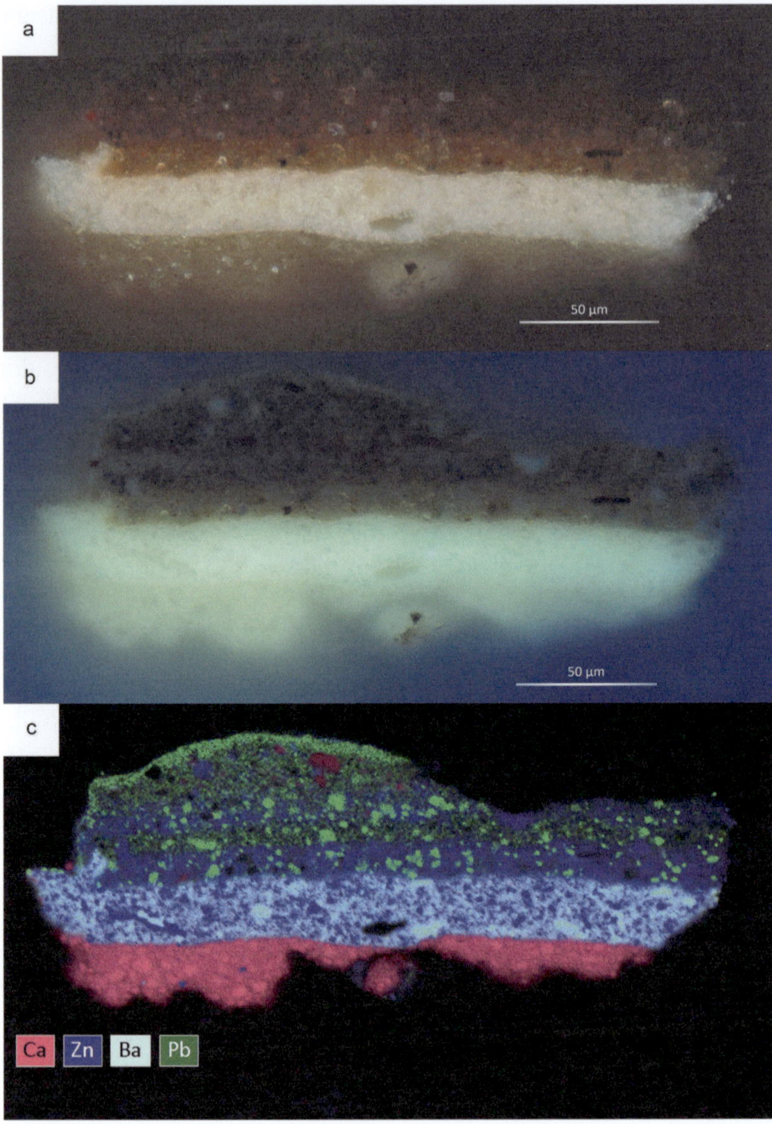

Fig. 23.2 Cross-section from brown background (**a**) epi visible darkfield mode (**b**) UV-induced visible autofluorescence mode (**c**) False color elemental map, 20 kV, 20 mA: calcium = pink, zinc = blue, barium = white, lead = green

layered and added these resins and driers freely to his paint and coating layers as he worked.

23.3 Visual Characteristics, Morphology, and Identification of Efflorescent Material

The efflorescence on *The Spirit of Light* closely corresponds to the distribution of a selectively applied oil-based coating, forming in drips and splatters. Under magnification, the efflorescent material appears white or colorless and crystalline, and has the greatest thickness in the valleys of the canvas weave (Fig. 23.3a). Although it appears powdery, it is tenaciously held to the surface and cannot be brushed away.

In order to observe the morphology of the efflorescent material, the top surface of a paint flake from the varnished brown background was imaged with SEM.[5] Imaging revealed rounded plates emerging from the varnished surface. The plates are homogenous in size, approximately 1–3 μm wide and 75–100 nm thick (Fig. 23.3b). There is a lower layer of plates parallel to the surface and an upper layer of lifted plates (Fig. 23.3c). Diffuse reflection of light by the irregular surface of the upper layer of plates causes the white appearance of the efflorescence. EDS determined that the plates contain carbon and zinc (Fig. 23.3d).

FTIR analysis of scrapings of efflorescent material revealed the presence of crystalline zinc carboxylates, confirmed by comparison with a reference sample of zinc stearate (Fig. 23.3e).[6] The crystalline species are characterized by a sharp COO^- band at 1538 cm^{-1} [3]. The sharp band observed at 1536 cm^{-1} in the spectrum of efflorescent material from *The Spirit of Light* indicates the presence of crystalline zinc carboxylates. Baij et al. propose a process in which free fatty acids in the binding medium combine with metal ions to form amorphous complexes, which eventually crystallize as metal-soap aggregates [3]. None of the carboxylate band characteristic of amorphous zinc soaps (1585 cm^{-1}) is discernable in the spectra collected from efflorescent material from *The Spirit of Light* [3].

The organic components of the efflorescence were identified with GC/MS as primarily stearic and palmitic acids, with some azelaic acid, present as metal carboxylates; trace levels of these same fatty acids were detected in their free form.[7] Phosphate was also identified in the efflorescent material.

---

[5]A paint flake sample was coated with carbon and analyzed with FE-SEM/EDS in secondary electron mode.

[6]The zinc stearate reference sample (purum, 10–12% Zn basis) was purchased from Sigma Aldrich, St. Louis, MO.

[7]Interpretation of the GC/MS results for efflorescence scrapings supports the proposed composition of metal carboxylates of stearic acid, palmitic acid, and azelaic acid, inorganic phosphate, as well as small amounts of free fatty acids. The analysis began with a two-step extraction and derivatization procedure and was created to fit the sample size limitations using ideas in [9, 16], and [14].

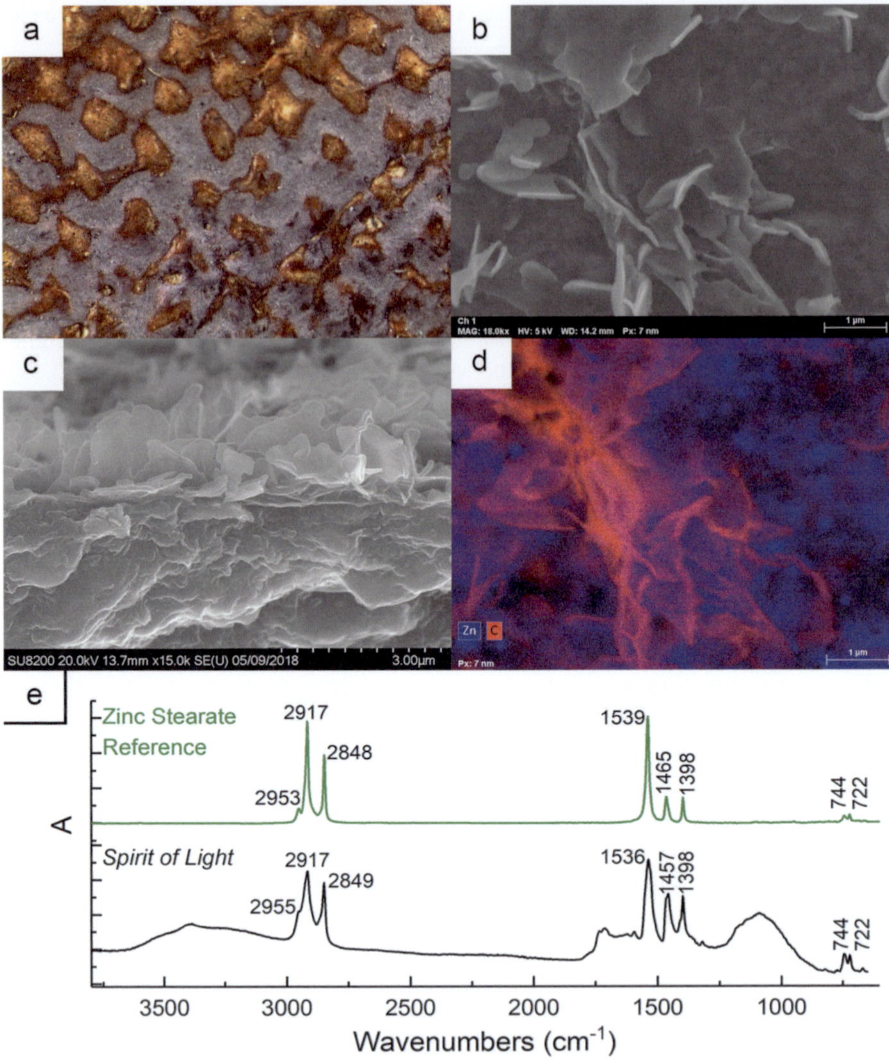

Fig. 23.3 (**a**) The efflorescence photographed at 16x (**b**) Secondary Electron Image (SEI) of efflorescence on top surface of paint chip, 15 kV, 18,000x (**c**) SEI of efflorescence in cross-sectional orientation, 20 kV, 15,000x (**d**) EDS false color elemental map of efflorescence plates, 5 kV, 18,000x: zinc = blue, carbon = red (**e**) FTIR spectra of zinc stearate reference sample (top) and efflorescence sample from *The Spirit of Light* (bottom)

Efflorescence samples were dissolved in a minimum volume of pyridine and analyzed for free fatty acid content. The remainder was then derivatized using a minimum volume of *N, O*-bis(trimethylsilyl) trifluoroacetamide for maximum conversion, and analyzed for trimethylsilylated carboxylates. The derivatized sample results included both the original free fatty acids and fatty acids present as metal carboxylates. An Agilent 7890A/5975C GC/MS system with a DB-5MS-UI column (30 m × 0.25 mm × 0.25 µm) and a CombiPAL autosampler was used for these analyses. The procedure and results will be described in more detail in a forthcoming publication.

23.4 Cleaning Approaches

The thin and underbound paint layers of *The Spirit of Light* are highly soluble and easily burnished, presenting a challenging surface from which to clean the disfiguring efflorescence. The paint layers are soluble in aromatic and aliphatic solvents, polar solvents, and aqueous solutions (pH between 6 and 8 with or without the inclusion of chelators). It is possible that the underbound nature of the paint layers contributes to *The Spirit of Light's* high solubility. This property may be exacerbated by the hydrolysis of the oil film as evidenced by the formation of metal soap and fatty acid efflorescence on the surface. The efflorescence appeared soluble in the aromatic solvents xylene and Shellsol A100, with both neat and gelled applications of these solvents greatly reducing the appearance of the efflorescence. Gelling Shellsol A100 and xylene with Carbopol increased the cleaning efficacy by achieving longer dwell times, enabling the removal of thicker efflorescence in the valleys of the canvas weave. However, this system was not safe for much of the delicate surface.

The silicone emulsifiers Momentive Velvesil Plus and Shin-Etsu KSG-350Z and the silicone solvents decamethylcyclopentasiloxane (D5) and octamethylcyclotetrasiloxane (D4) successfully reduced the appearance of the efflorescence without solubilizing the sensitive paint layers [5, 10]. Shin-Etsu KSG-350Z was preferred because of its wider commercial availability. The emulsifier was applied to the surface and agitated for at least 1 minute with a brush. This agitation step was crucial in evenly thinning the efflorescence in the valleys of the canvas weave (Fig. 23.4). The gel was cleared from the surface with D5 solvent. The addition of small percentages of aqueous chelating solutions and non-miscible aromatic solvents such as benzyl alcohol to form emulsions did not increase the efficacy of the cleaning.

This cleaning system is promising for paintings with sensitive surfaces and presents several advantages. Cleaning with the KSG-350Z emulsifier and silicone solvents allowed the original coatings to remain intact and restored subtle variations in gloss and saturation to the surface. The insolubility of zinc carboxylates in D5 solvent suggests that the cleaning will not mobilize soaps in the stratigraphy of the painting, although the mobility of metal carboxylates in silicone solvents has not yet been evaluated. This system also presents some challenges, such as the long evaporation time of the D5 solvent and the ease with which one can create tidelines during cleaning. When applied to the surface, the D5 saturates the efflorescence, making it difficult to differentiate between cleaned and uncleaned areas. The potential for silicone emulsions to leave residues, especially in porous paint surfaces, is yet to be fully explored.

The cleaning mechanism of the emulsifier during the removal of crystalline zinc soap efflorescence is uncertain. Crystalline zinc soaps have low solubility in D4 and D5; therefore, the effectiveness of the KSG-350Z emulsifier is surprising.[8] It is

[8] The solubility of crystalline zinc stearate and stearic acid in D4, D5, and the aromatic hydrocarbon Shellsol A100 were evaluated. Both compounds were determined to be minimally soluble in

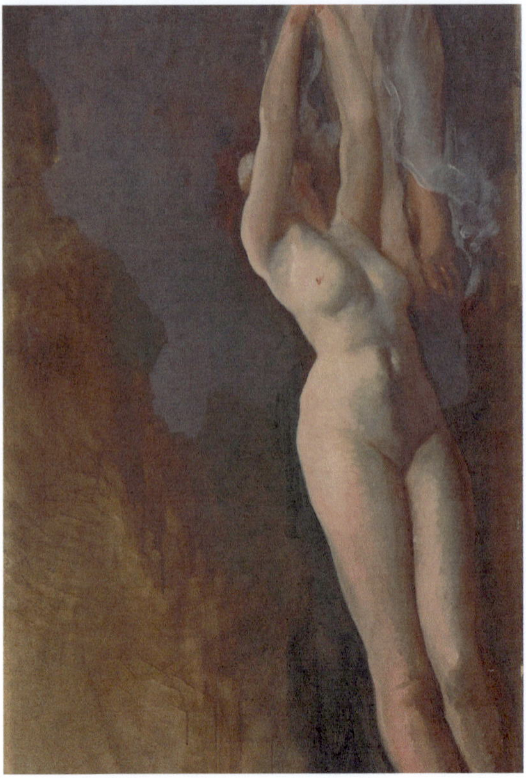

Fig. 23.4 Detail of *The Spirit of Light* during cleaning with Shin-Etsu KSG-350Z

possible that the emulsifier works on a primarily mechanical level to loosen, lift, and suspend the efflorescence flakes from the surface. It is also possible that the emulsifier flattens and redistributes the flakes across the surface.

23.5 Discussion and Conclusions

The selective distribution of zinc carboxylate efflorescence on *The Spirit of Light* suggests that efflorescence formation is closely related to the painting materials and the ways in which they were applied by the artist. The lithopone-containing upper ground layer is a potential source of zinc carboxylates, aggregates of which are visible in the cross-sectional sample and on the surface of the paint layers. Zinc carboxylates in ground layers have been previously linked to the formation of inorganic surface crusts, migrating aggregates, and free fatty acid efflorescence, but

siloxane-based solvents, suggesting that solubility was not a main factor in the efficacy of the Shin-Etsu KSG-350Z cleaning. Both compounds were soluble in Shellsol A100.

few studies have discussed the presence of efflorescence composed primarily of crystalline zinc carboxylates [7, 15].[9]

The potential presence of lead and zinc driers in the ground, paint, and coating layers may also be significant factors in efflorescence formation on *The Spirit of Light.* The mobile and reactive nature of these compounds has been noted in studies of nineteenth-century oil sketches executed on commercially primed canvases [19]. Archival research, visual observation, and analysis suggest that Abbey used commercially primed canvases and freely interlayered and mixed resins and driers into his paints. Mrs. Abbey noted that her husband "never went at it wet" and quoted him stating, "Sargent['s] work [will] never stand" [11]. This may suggest that Abbey readily used driers to paint works that he believed would remain stable with age. His desire to use stable painting techniques, perhaps compensating for his lack of formal training, may have led Abbey to use unsound painting practices.

In contemplating the unique relationship between the coating and the efflorescence, it is possible that zinc soaps formed within and on top of the varnish layer as a result of the reaction of lead and zinc driers with each other and with migrating free fatty acids: a phenomenon that has been noted in primary sources and could explain the presence of lead sulfate in the coating layer.[10] The formation of the stable compound lead sulfate within the coating layer, thereby providing few available lead ions, could explain the surprising absence of lead soaps. It is also possible that zinc carboxylates identified in the ground layer were mobilized by the varnish solvent, resulting in their migration to the surface. The identification of phosphate, possibly from bone black pigments, in samples of efflorescent material suggest that the composition of the paint layers may also contribute to efflorescence formation [12]. The impact of potentially high moisture environmental conditions in Abbey's studio and in museum storage could also be a contributing factor in efflorescence formation [2].

Imaging of the efflorescence contributed to a greater understanding of the sensitive paint surface, and characterization of the efflorescence explained its solubility characteristics, thereby informing cleaning methodology and testing. The use of Shin-Etsu KSG-350Z and the silicone solvent D5 for the cleaning of efflorescent material has not been previously noted; their effectiveness at reducing the appearance of the efflorescence without solubilizing sensitive oil paint layers presents a promising avenue for future investigation. Future study will concentrate on imaging the surface of *The Spirit of Light* after cleaning and analyzing the composition of the emulsifier after cleaning. Further research is ongoing to establish the mechanism for the formation of crystalline zinc carboxylates on the surface of Abbey's paintings.

---

[9] A similar morphology was imaged by Jimenez and Chua, who identified hydroxychlorides, sulfates, and/or oxalates in efflorescent material on an oil painting containing zinc white [8].

[10] Field notes a phenomenon in which both lead acetate and zinc sulfate driers are incorporated into a coating and react to form lead sulfate and zinc acetate, with efflorescence eventually forming on the surface [6].

Acknowledgements FE-SEM/EDS was carried out with assistance from Dr. Min Li at the Yale West Campus Materials Characterization Core. The authors would like to acknowledge the support and contributions of Ian McClure, Irma Passeri, Dr. Mark Mitchell, Lisa Hodermarsky, Sally Woodcock, Jason Wilson, Richard Wolbers, Matthew Cushman, Dr. Rosie Grayburn, Dr. Chris Petersen, Richard House, and Andrés Garcés.

References

1. Abbey, EA (1884–1904) Personal account. Roberson Archive, Hamilton Kerr Institute, Cambridge
2. Akerlund L (2013) Efflorescence in paintings and the role of moisture. http://www.collectiewijzer.nl/2013/01/18/efflorescence-in-paintings-and-the-role-of-moisture-by-luciana-akerlund/ Accessed 2 December 2017
3. Baij L, Hermans JJ, Keune K et al (2018) Time-Dependent ATR-FTIR Spectroscopic Studies on Fatty Acid Diffusion and the Formation of Metal Soaps in Oil Paint Model Systems. Angew Chem 130(25):7473–7476
4. Carlyle L (2001) Driers. In: Carlyle L (ed) The Artist's Assistant: Oil Painting Instruction Manuals and Handbooks in Britain 1800–1900 With Reference to Selected Eighteenth-century Sources. Archetype, London, p 41–54
5. Chung JY, Ormsby B, Lee J et al (2017) An investigation of options for surface cleaning unvarnished water-sensitive oil paints based on recent developments for acrylic paints. In: ICOM-CC 18th Triennial Conference Preprints, Copenhagen, 4–8 September 2017
6. Field G, Tilt C, Valpy AJ et al (1835) Chromatography: Or, A Treatise on Colours and Pigments, and of their Powers in Painting, &C. Tilt, London
7. Helwig K, Poulin J, Corbeil MC, et al (2014) Conservation Issues in Several Twentieth-Century Canadian Oil Paintings: The Role of Zinc Carboxylate Reaction Products. In: van den Berg K et al Issues in Contemporary Oil Paint, Springer, p 167–184
8. Jimenez ID, Chua L (2018) Micro-characterization of haze and degradation on zinc white oil-based painting – portrait of a Peranakan gentleman Mr. Tan Beng Wan, Singapore. Poster presented at the Conference on Modern Oil Paints, Rijksmuseum, Amsterdam, 23–25 May 2018
9. La Nasa J, Modugno F, Aloisi M et al (2017) Development of a GC/MS method for the qualitative and quantitative analysis of mixtures of free fatty acids and metal soaps in paint samples. Analytica Chimica Acta 1001:51–58
10. Lagalante A, Wolbers RC (2017) Particle-based silicone cleaning emulsions: studies in model paint systems for the cleaning of water-sensitive artworks. In: Angelova L et al (ed) Gels in the Conservation of Art: 193–199
11. Lockman, DWMC (1926) Mrs. Abbey at hotel. DeWitt M. Lockman interviews with artists, archival material, New York Historical Society
12. Loon A, Boon JJ (2005) The whitening of oil paint films containing bone black. In: 14th Triennial Meeting, the Hague, 12–16 September 2005 Preprints 1: 511–518
13. Lucas, EV, Rogers B (1921) Edwin Austin Abbey, royal academician: the record of his life and work. Methuen and Co, London
14. Mazurek J (2014) Paint Media Analysis by GC/MS: From Ancient to Modern. Poster presented at the 40th International Symposium on Archaeometry, Los Angeles, California, 19–23 May 2014
15. Noble P, Boon JJ (2007) Metal soap degradation of oil painting: aggregates, increased transparency, and efflorescence. In: Parkin HM (ed) AIC paintings specialty group postprints 19:1–15

16. Plater MJ, De Silva B, Gelbrich T et al (2003) The characterization of lead fatty acid soaps in 'protrusions' in aged traditional oil paint. Polyhedron 22: 3171–3179. doi:https://doi.org/10.1016/S0277-5387(03)00461-3
17. Schilling M, Heginbotham A, van Keulen H et al (2016) Beyond the basics: A systematic approach for comprehensive analysis of organic materials in Asian lacquers. Studies in Conservation 61: 3–27. doi: https://doi.org/10.1080/00393630.2016.1230978
18. Steffensen I (2002) Toward an Iconography of a State Capitol: The Art and Architecture of the Pennsylvania State Capitol in Harrisburg. The Pennsylvania Magazine of History and Biography 126(2):185–216
19. Zucker J (2007) Opaque to Transparent: Paint Film Defects in the Work of Frederick Church and the Hudson River School. In: Parkin HM (ed) AIC paintings specialty group postprints 19:33–41

Chapter 24
A Study of Softening and Liquefying Oil Paint on *Womenizer* by Alex Janvier

Kate Helwig, Jennifer Poulin, and Patrick Gauthier

24.1 Introduction

In November 2016, the National Gallery of Canada (NGC) opened a major retrospective exhibition that highlighted the career of the prolific painter Alex Janvier (1935–). The exhibition, entitled *Alex Janvier: Modern Indigenous Master*, featured over 150 paintings and works on paper spanning more than 60 years of creation. Alex Janvier, who lives and works in Cold Lake, Alberta, played a key role in the foundation of contemporary Indigenous art in Canada. He has devoted himself to the development of Indigenous art not only as an artist, but also as an activist, a teacher and a government consultant. Janvier was appointed into the Order of Canada in 2007, as a national recognition for his outstanding contributions to his field [1].

While Janvier's works are generally in very good condition, the large-scale, oil on canvas painting *Womenizer* (Fig. 24.1) is an exception; although the painting initially appeared to dry normally, areas of the work are now tacky, and some paint has become almost liquid. This article describes how the painting's history affected its materials and their conservation issues through a detailed analysis of the paint. The results provide insight into the phenomenon of softening and liquefying modern oil paint.

K. Helwig (✉) · J. Poulin
Canadian Conservation Institute, Ottawa, ON, Canada
e-mail: kate.helwig@canada.ca; jennifer.poulin@canada.ca

P. Gauthier
National Gallery of Canada, Ottawa, ON, Canada
e-mail: pgauthier@gallery.ca

K. J. van den Berg et al. (eds.), *Conservation of Modern Oil Paintings*,
https://doi.org/10.1007/978-3-030-19254-9_24

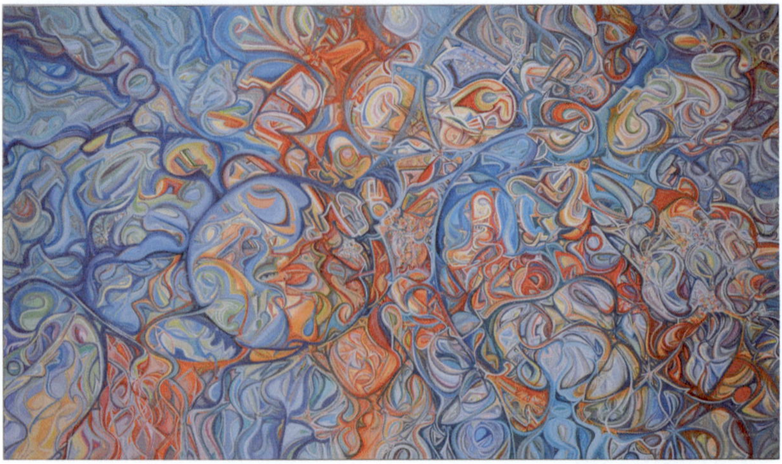

Fig. 24.1 *Womenizer,* 2004 by Alex Janvier; oil on canvas, 347.5 cm wide by 203.0 cm high, artist's collection. (Photographed on display at the National Gallery of Canada during the *Alex Janvier: Modern Indigenous Master* exhibition, photograph: Spencer Montcalm, National Gallery of Canada, 2017, with permission from the artist)

24.2 History and Condition of the Painting

Painted in 2004, *Womenizer* was one of the main pillar pieces of the 2016 Janvier retrospective exhibition at the NGC. Its presence in the show was of prime importance as it marked Janvier's return to his intricate and detailed designs after a short illness in the 90s that limited his abilities and forced him to adapt his practice. Like many of his large format works of that period, Janvier painted *Womenizer* with the canvas stapled onto a board and painted the image right to the edges. The canvas has a very open weave and, although both sides were prepared, the brown, oil-based ground did not fill the interstices between the threads. As a result, the thicker paint applications from the image layer easily travelled through the open weave and accumulated at the back in large sections, compressed between the board and canvas (Fig. 24.2). The work was painted in relatively thin layers, without significant impasto.

The painting was placed on a stretcher and hung in the artist's studio shortly after its completion. At that time, the paint was dry to the touch; it did not exhibit any signs of problematic drying or tacky paint [2]. According to the artist's records, *Womenizer* was taken down from display sometime between 2007 and 2012. It was removed from its stretcher and rolled from the right, with the image side facing in and without any interleaf, onto a small cardboard tube with a 12 cm diameter. The tube was shorter than the height of the canvas, leaving unsupported canvas on both ends of the roll. The rolled painting was sealed in a plastic bag and stored in a small room in which it remained until 2015, when it was taken out of the bag to be

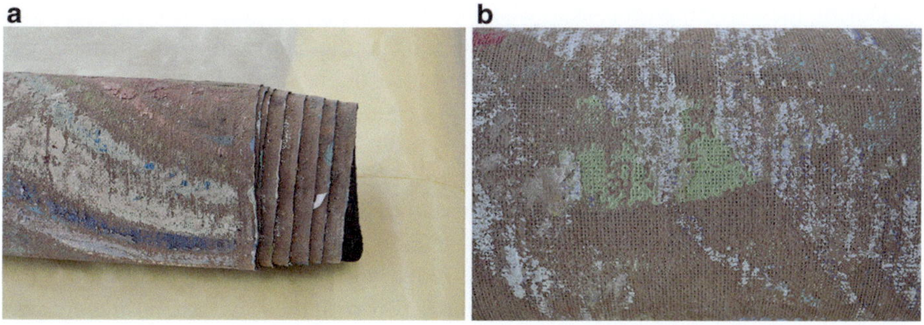

Fig. 24.2 (**a** and **b**) Details of the verso of *Womenizer* prior to unrolling. (**a**) Detail showing the tightly rolled painting, image side facing in; the brown ground and paint on the verso are visible. (**b**) Detail showing open weave canvas with paint from the recto going through to the verso. (Photographs: Spencer Montcalm, National Gallery of Canada, 2016, with permission from the artist)

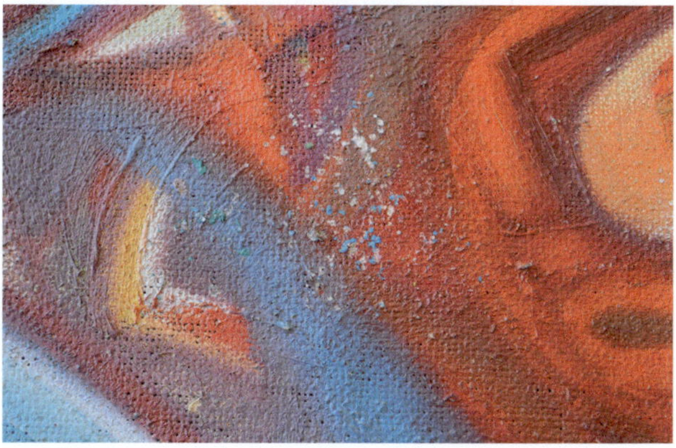

Fig. 24.3 Detail of the recto of *Womenizer*, showing an area with spots of wet paint transferred from the verso. (Photograph: Spencer Montcalm, National Gallery of Canada, 2016, with permission from the artist)

examined for the retrospective exhibition. At that point, the painting was judged too tacky to unroll safely in the artist's studio and was shipped to the NGC rolled.

At the NGC, it was noted that a strong, rancid smell emanated from the work and that areas of paint were so soft and sticky that the recto of the rolled painting was stuck to the verso. Unrolling the painting to prepare it for display proved challenging as small paint strands formed between the verso and the image layer during unrolling attempts. Testing revealed that the paint layer was sensitive to all the solvents tested, water included. Ultimately, the best way to avoid paint transfer during unrolling was to direct a gentle, narrow stream of warm air from a Leister welding pen at the meeting point between recto and verso while carefully unrolling the work. The warm stream broke the strands of paint without contact, and, more importantly, before they could pull on the paint layers. Even with this careful method, minor paint transfers and losses were inevitable. However, these damages are fairly small and barely visible at normal viewing distance (Fig. 24.3).

During the unrolling process, it was observed that the bottom end of the painting was rolled noticeably tighter than the top, and at the same time, the paint in the bottom third of the work was much softer and stickier than the top portion. The tackiest, softest paint in the lower portion was found to be closest to the inside of the roll along the right edge. The paint in the bottom right corner was so soft that the edge of the cardboard tube smeared the paint layer over a length of about two meters. The brown ground appeared relatively dry in most areas, even in the lower portion of the painting where much of the paint was tacky.

While the bottom and right edges exhibited soft, tacky paint, the left and top edges were almost completely dry. In comparison with the lower section, there was little smudging near the top right corner from the other edge of the cardboard tube. The top left corner, which was rolled less tightly and was on the outside of the roll, still exhibited sharp cracks along the fold over edges from the 2005 stretching. The paint layer at the corner folds did not show any signs of paint transfer from the 2005 stretching and subsequent removal from the stretcher. Drying cracks were also observed in certain colours throughout the image. These are all indications that the painting dried normally prior to being rolled.

Once the painting had been unrolled, analysis of materials was undertaken to determine why the paint became sticky and almost liquid in places during storage.

24.3 Methods of Analysis

Three softened or liquefied paint samples were taken from the lower right area of the work, where wet paint from another area of the composition had transferred to the surface as a result of rolling. Figure 24.3 shows an area with transfer of wet turquoise paint from the verso to the recto. Four samples of well-dried paint from the upper left were examined for comparison to the wet paint. A sample of the brown ground from the lower part of the painting was also analyzed. The micrographs of typical paint samples in Fig. 24.4 show the clear differences in rheology between the wet and well-dried paints.

The Appendix provides experimental details for all the analytical techniques employed. Analysis of the pigments and fillers was undertaken using a combination of Fourier transform infrared (FTIR) spectroscopy, scanning electron microscopy/energy dispersive spectrometry (SEM/EDX), x-ray diffraction (XRD), Raman spectroscopy and polarized light microscopy (PLM). To characterize the binding medium, the samples were first analysed using FTIR spectroscopy, followed by pyrolysis-gas chromatography-mass spectrometry (Py-GCMS) after methylation to determine the overall fatty acid profile of the oil. Finally, analysis by gas chromatography-mass spectrometry (GCMS) was performed after extraction and trimethylsilylation to determine the extractable, non-crosslinked portion of the oil. This method allows the determination of free fatty acids and their salts, but leaves ester bonds intact. The samples were pre-weighed to allow a comparison of the amount of extractable material among the samples.

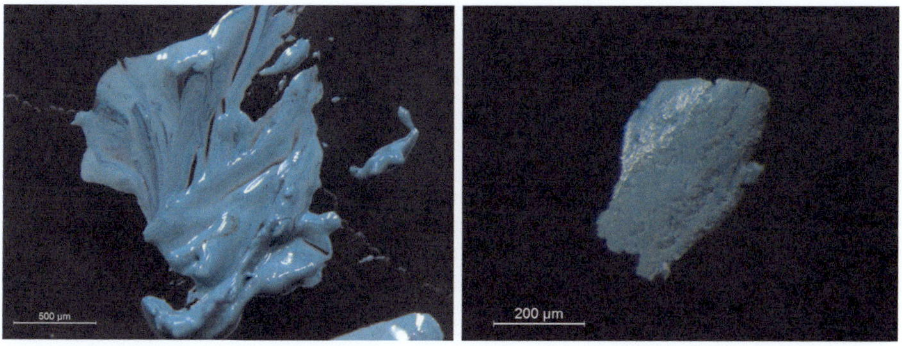

Fig. 24.4 Micrographs of samples of wet turquoise paint (left) and dry turquoise paint (right)

24.4 Results

24.4.1 Overall Composition of the Paint and Ground

The pigments and fillers identified in the paint and ground samples are listed in Table 24.1. All samples were found to contain a high concentration of titanium white (rutile form). Trace to minor amounts of zinc were detected in the paint samples using SEM/EDX, suggesting that small amounts of a zinc-based pigment such as zinc white or zinc sulfide are present. The orange paint differed from the other samples as it was found to contain barium sulfate and hydromagnesite in addition to titanium white. This sample also contained a higher concentration of zinc than the other paint samples. In addition to the white pigments and fillers, the paints all contain small amounts of various coloured pigments.

Like the paint samples, the ground contains primarily rutile titanium white. Calcium magnesium carbonate, red iron oxide, and traces of other pigments are also present. Although the elemental analysis indicated that there is a significant amount of zinc in the brown ground, no zinc-containing pigment was identified using XRD. Several white agglomerates in the ground were found to contain primarily zinc and oxygen, indicating that, although it was below the detection limits for XRD, some zinc white is present in the ground.

The medium of the paint and ground is drying oil. In the three wet paint samples, the FTIR spectra show, in addition to the ester band at about 1736 cm^{-1}, a strong band at about 1712 cm^{-1} indicating a high proportion of carboxylic acid groups (upper spectrum, Fig. 24.5). The four well-dried paints only show an ester band, indicating a low proportion of free carboxylic acid (middle spectrum, Fig. 24.5). While several of the paint samples show a weak, sharp band at 1541 cm^{-1} that could indicate the presence of small amounts of zinc fatty acid salts (zinc soaps), the only sample with a significant amount of zinc soaps is the dry, orange paint (bottom spectrum, Fig. 24.5, zinc fatty acid salt bands labelled in red). This is consistent with the higher concentration of zinc in the orange paint determined by SEM/EDX.

Table 24.1 Overall Paint Composition

Sample	Elements[a]	Pigments and fillers[b]	Medium (FTIR only)
Wet brown	C, O, Cd, Ti, Al, Si, S, Co, Se, (Zn, Cl, Fe, Mn, P, Ca)	Titanium white, cadmium red, cobalt blue, Pigment Yellow 97, Pigment Green 7	Drying oil, possible trace zinc soaps
Wet turquoise	O, Ti, C, Al, Si, Cd, (Zn, S, Co, Mg, P, Ca, Cl)	Titanium white, Pigment Blue 15, red iron oxide, probably ultramarine blue and cadmium red	Drying oil, possible trace zinc soaps
Sticky blue/purple	O, Ti, C, Al, Co, Fe, (Zn, Si, P, Mn, Cd, Mg, S)	Titanium white, cobalt blue, Pigment Blue 15, red iron oxide, manganese violet	Drying oil, possible trace zinc soaps
Dry turquoise	O, Ti, C, Cd, Al, (Si, S, Zn, Mg, Ca)	Titanium white, Pigment Blue 15, probably cadmium red	Drying oil
Dry purple/blue	Ti, O, C, Si, Al, (Cd, Na, Ca, Mg, Cl, S, P)	Titanium white, Pigment Violet 19	Drying oil
Dry blue	Ti, O, C, (Zn, Si, Al, Ca, Mg, Cl)	Titanium white, Pigment Blue 15	Drying oil
Dry orange	C, O, Ba, Ti, Zn, S, Cd, Al, Mg	Titanium white, barium sulfate, hydromagnesite, red iron oxide, organic yellow, possibly cadmium red	Drying oil, zinc soaps (major)
Brown ground	C, O, Ti, Zn, Fe, Ca, Mg, (Cd, Al, Si, S, P); white particle: Zn, C, O, Al, Ti, (Fe, Si, Ca)	Titanium white, calcium magnesium carbonate, red iron oxide, traces of brown-black pigment, ultramarine blue and Pigment Green 7	Drying oil, zinc formate dihydrate and zinc soaps (major)

[a]Chemical elements are classified qualitatively as major, minor or (trace). The categories are based on normalized weight percentages calculated by the Oxford AZtec x-ray microanalysis software
[b]Chemical formulae for the pigments and fillers identified are the following: titanium white = TiO_2, rutile form; cadmium red = CdS_{1-x} Se_x; cobalt blue = $CoAl_2O_4$; Pigment Yellow 97 = arylide (Hansa) yellow class ($C_{26}H_{27}ClN_4O_8S$); Pigment Green 7 = phthalocyanine green ($C_{32}Cl_{16}CuN_8$); Pigment Blue 15 = phthalocyanine blue ($C_{32}H_{16}CuN_8$); Pigment Violet 19 = quinacridone violet ($C_{20}H_{12}N_2O_2$); red iron oxide, = Fe_2O_3, hematite form; calcium magnesium carbonate = $CaMg(CO_3)_2$, dolomite form; ultramarine blue = $Na_{8-10}Al_6Si_6O_{24}S_{2-4}$; manganese violet = $(NH_4)_2Mn_2(P_2O_7)_2$; hydromagnesite = $Mg_5(CO_3)_4(OH)_2{\bullet}4H_2O$

The FTIR spectrum of the brown ground (dry but with a rubbery texture) shows an ester band from the oil at 1742 cm^{-1}; like the dry paints, no absorption band due to free carboxylic acid was visible. Several sub-fragments of the sample were analyzed; in some fragments high concentrations of zinc soaps were present, while in others zinc formate dihydrate (Fig. 24.6) was identified. The source of the formic acid could be internal or external to the paint. Within the paint, low molecular weight carboxylic acids, including formic acid, are formed during the curing and aging of the oil medium [3, 4]. Possible external sources of formic acid include the board to which the canvas was stapled while Janvier was working or the cardboard onto which the painting was later rolled.

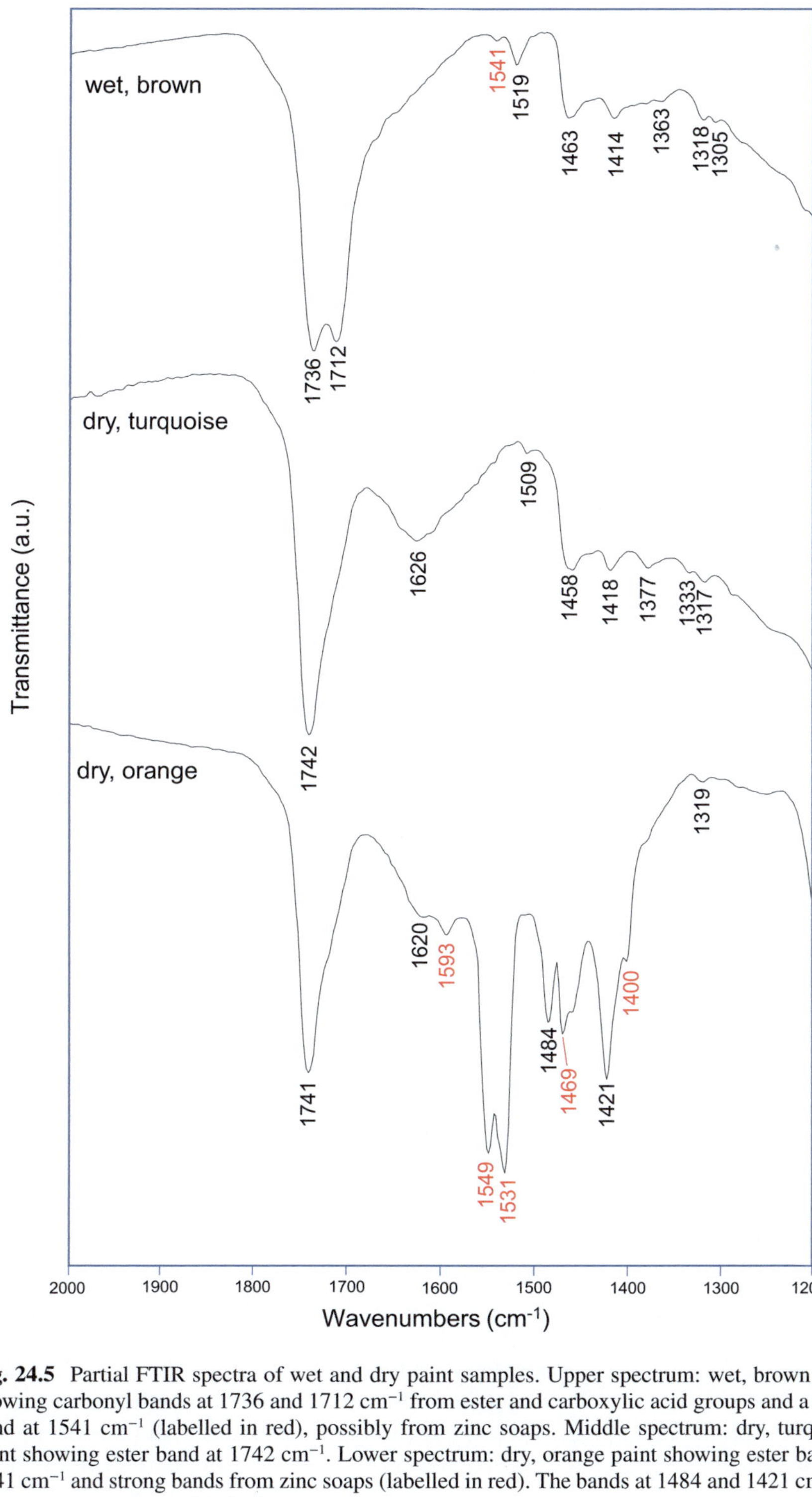

Fig. 24.5 Partial FTIR spectra of wet and dry paint samples. Upper spectrum: wet, brown paint showing carbonyl bands at 1736 and 1712 cm^{-1} from ester and carboxylic acid groups and a weak band at 1541 cm^{-1} (labelled in red), possibly from zinc soaps. Middle spectrum: dry, turquoise paint showing ester band at 1742 cm^{-1}. Lower spectrum: dry, orange paint showing ester band at 1741 cm^{-1} and strong bands from zinc soaps (labelled in red). The bands at 1484 and 1421 cm^{-1} in this spectrum correspond to hydromagnesite

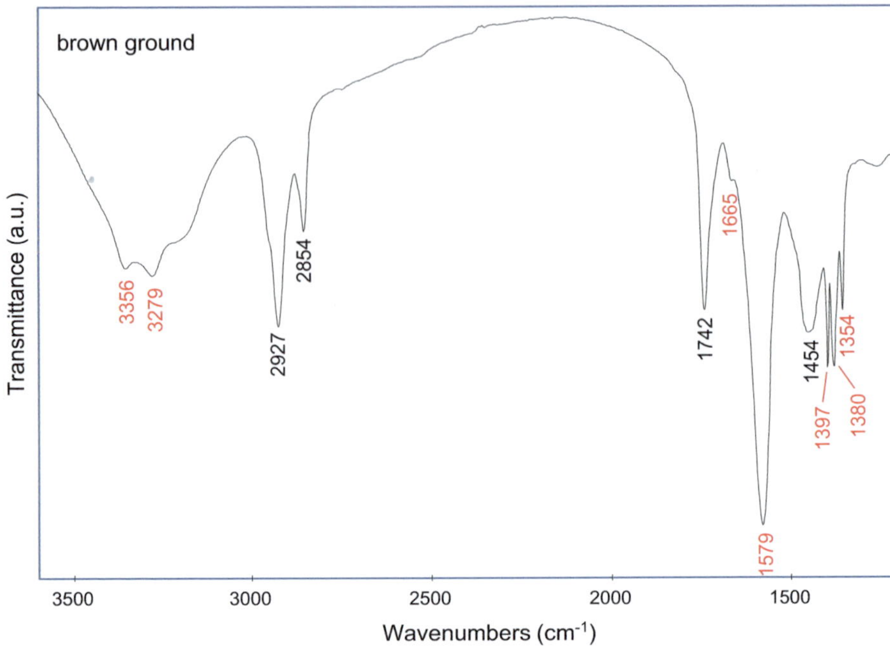

Fig. 24.6 FTIR spectrum of one area of the sample of brown ground, showing zinc formate dihydrate (labelled in red)

24.4.2 Overall Fatty Acid Profile of the Oil

To determine the overall fatty acid profile of the oil (including fatty acids present as glycerol esters as well as free fatty acids and fatty acid salts), the samples were analyzed by Py-GCMS after methylation. The results are listed in Table 24.2. The palmitate to stearate (P/S) ratio is between 1.4 and 2.1 for all the samples, consistent with linseed oil. Characteristic cyclic markers (predominantly methyl 9-(2-propylphenyl)nonanoate and methyl 8-(2-butylphenyl)octanoate) in the chromatograms indicate that the linseed oil has been heated [5, 6]. The samples all contain some oleic acid, a mono-unsaturated fatty acid. A small proportion of mid-chain oxidized fatty acids (hydroxy and epoxy stearates) was also identified. These compounds are formed as intermediate products during the oxidative curing and aging of the oil. The dry turquoise paint showed a higher abundance of these mid-chain oxidized fatty acids than the other samples. There are several possible reasons for this; for example, the paint may have been thicker in this area or sampling may have included paint originating deeper beneath the surface.

There is nothing in the overall fatty acid profiles that allows the wet paints to be differentiated from the well-dried paints. The ratio of diacids to saturated fatty acids (the A/P ratio in Table 24.2) is not significantly elevated and is similar for the wet and dry paints. Furthermore, the abundances of oleic acid and mid-chain oxidized

Table 24.2 Results for Py-GCMS after methylation[a]

	Wet brown	Wet blue/ purple	Wet turquoise	Dry turquoise	Dry purple/ blue	Dry blue	Dry orange	Brown ground
P/S	2.1	1.9	1.8	2.0	2.0	1.4	1.6	2.0
A/P	1.5	1.6	0.8	1.2	0.7	0.7	1.1	1.2
Su/A	0.2	0.2	0.2	0.2	0.2	0.2	0.2	0.2
O/S	0.0	0.1	0.5	0.4	0.5	0.4	0.1	0.2
[O] FA/ (P + S)	0.3	0.2	0.3	1.2	0.2	0.1	0.1	0.2

[a]*P/S* palmitate to stearate ratio, *A/P* azelate to palmitate ratio, *Su/A* suberate to azelate ratio, *O/S* oleate to stearate ratio, *mid-chain [O] FA/(P + S)* ratio of all mid-chain oxidized fatty acids to the sum of palmitate and stearate

fatty acids do not show a correlation with the dryness of the paint. The presence of these unsaturated and mid-chain oxidized fatty acids is not unusual in a young paint film [7, 8]. The fact that there are no polyunsaturated fatty acids remaining indicates that the paint has undergone the first stage of oxidative curing.

As well as components from the drying oil medium, the chromatograms of all the paint and ground samples show small amounts of beeswax and pentaerythritol. While beeswax is a common additive in modern oil paints [9], the presence of pentaerythritol is more unusual. It does not appear to be related to an alkyd resin component in the paint, since no corresponding phthalates were identified. The pentaerythritol may be associated with the titanium white; this polyol is sometimes used as an organic surface coating on titanium white to alter the surface properties of the pigment [10].

24.4.3 Extractable Fraction of the Oil

Because no significant differences between the wet and dry paints were observed using Py-GCMS, GCMS after trimethylsilylation was undertaken on a second set of fragments of the same samples. This method leaves ester bonds intact but allows for the determination of free fatty acids and salts of fatty acids. The extracted compounds correspond to the lower molecular weight, mobile fraction of the paint. The main compounds identified by GCMS after trimethylsilylation are listed below. A very similar series of compounds was identified by van den Berg in his study of extractable material in linseed oil paints [8]. It is important to note that while this method extracts much of the non-crosslinked material, glycerol di- and tri-esters as well as larger oligomeric fractions are not detected.

- glycerol
- short chain monoacids: caprylic (C8:0) and pelargonic (C9:0)
- diacids: suberic (diC8:0) and azelaic (diC9:0)

- glycerol monoesters of diacids: glycerol monosuberate (diC8:0-glyc) and glycerol monoazelate (diC9:0-glyc)
- hydroxy fatty acids: 9,10-dihydroxystearic (diOH-C18:0)
- long chain fatty monoacids: palmitic (C16:0), stearic (C18:0) and oleic (C18:1)
- glycerol monoesters of monoacids: glycerol monopalmitate (C16:0-glyc) and glycerol monostearate (C18:0-glyc)

The relative abundances of the compounds listed above were measured by integrating their peak areas in the chromatograms. For each sample, the sum of the peak areas for these 13 major compounds, corrected for sample weight, provided a measure of the total amount of extractable material per 100 micrograms of paint. Because the paints have similar compositions, this allows a qualitative comparison of the amount of extractable material to be made. While this approach is valid for the comparison of paints with similar compositions, the methodology could be refined in future work by comparing the quantity of the bound fraction of the oil medium to the extractable fraction.

The results for the total abundance of extractable material are illustrated in the gray bar chart in Fig. 24.7a. The wet and sticky paints (the first three bars) contain a high abundance of extractable material. Based on the FTIR results, none of these paints have a significant quantity of metal fatty acid salts, indicating that the majority of the extractable material is in the form of free fatty acids. The dry paints contain a much lower abundance of extractable material compared to the wet paints. In the case of the orange paint and the brown ground (the last two bars in Fig. 24.7a), strong bands from zinc fatty acid salts were observed in the FTIR spectra, indicating the retention of carboxylic acids in these paints as their zinc salts.

The coloured, stacked bar chart in Fig. 24.7b provides normalized data illustrating the relative amounts of the extracted compounds. Free glycerol, diacids, mid-chain hydroxy acids and monoglycerol esters of diacids are shown in green and blue. The less polar monoacids (and their glycerol monoesters) are shown in brown and yellow.

The relative amounts of the extracted compounds are very similar for the three wet paint samples. They all contain a high proportion of polar diacids; the diacids and their glycerol monoesters account for approximately 60–70% of the extractable compounds. The well-dried paints, on the other hand, show a lower proportion of diacids and a higher proportion of monoacids. The brown ground has a particularly high proportion of the short chain C8:0 and C9:0 monoacids (shown in yellow). The relative abundance of glycerol (shown in green) does not appear to correlate with the wetness or dryness of the paint.

24.5 Discussion

There have been several previous studies of modern oil paints that have liquefied and begun to drip over time. The liquefaction is generally associated with a specific type of paint having a problematic formulation and often applied in a thick layer.

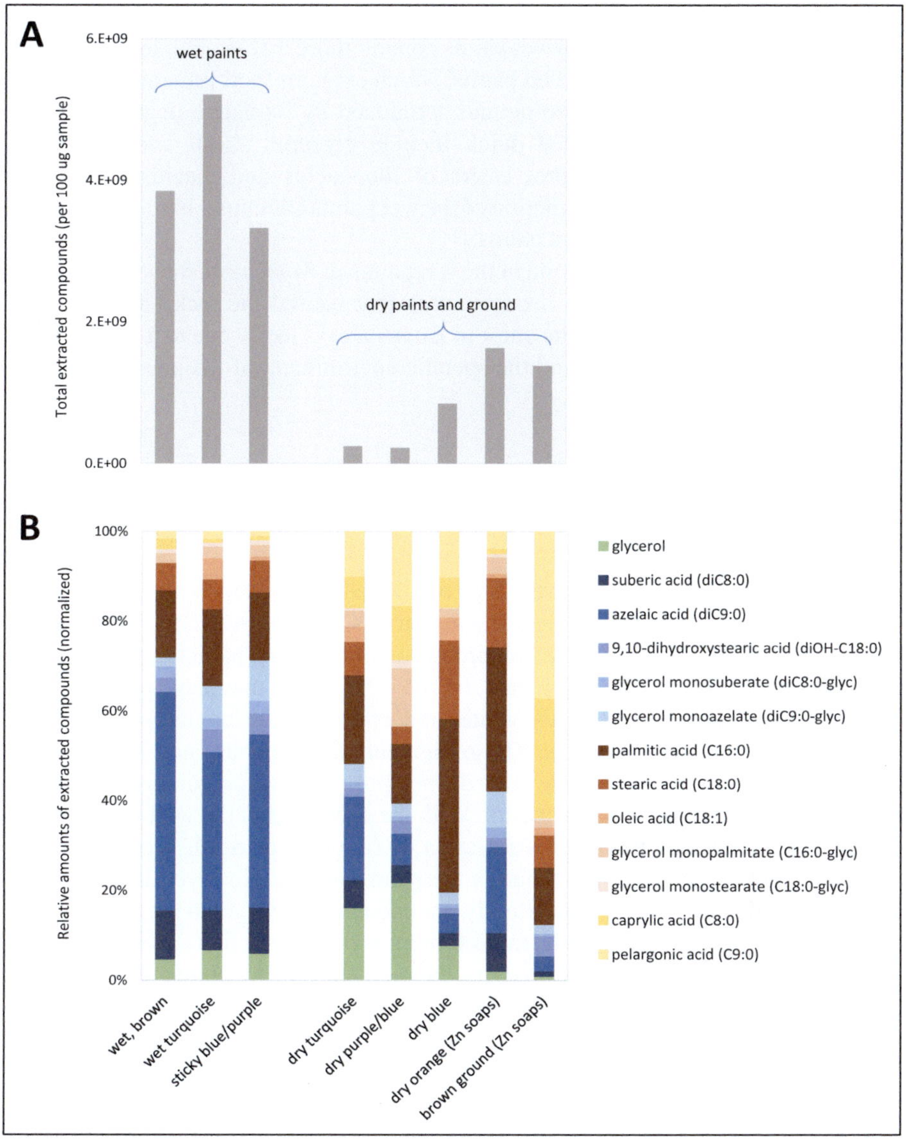

Fig. 24.7 (**a**) Total abundance of extracted compounds, measured as the sum of the peak areas per 100 micrograms of sample. Compounds extracted using BSTFA in pyridine. (**b**) Relative quantities of extracted compounds, normalized to 100%

Chemical analysis of these liquefied paints has identified the presence of slow drying oils (such as safflower oil) and pigments with lack of anchor points. It has been hypothesized that hydrolysis and oxidation reactions led to an abundance of polar compounds which, over time, separated from the bulk, leading to liquid or soft areas of paint [11–14].

The chemical composition of the wet and sticky paints in *Womenizer* is consistent with these previous studies. The wet paints contain more extractable, low molecular weight material than the well-dried paints, which explains their liquefied and sticky consistency. The extractable compounds, produced by oxidative degradation and hydrolysis of the initially cured paint, include glycerol, short- and long-chain monoacids, diacids, and glycerol esters of monoacids and diacids. Similar to previous work, the extractable fraction of the wet paints contains a higher proportion of polar compounds than the dry paints.

While the chemical composition of the wet paints in *Womenizer* shows similarities to previous studies, the set of circumstances that caused the tacky and liquefied paint differs. The softening of the paint in *Womenizer* is likely due to a combination of the original materials used and the specific environment of the painting while it was in rolled storage.

24.5.1 The Effect of the Original Materials

Neither the artist's technique nor the type of oil medium used in *Womenizer* appears to be a factor in the softening paint. The medium is heated linseed oil, which would be expected to dry well under typical circumstances, and the paint has been applied in relatively thin layers. The pigments, on the other hand, may have played a role. The paints are based on titanium white, and contain small amounts of coloured pigments, many of them organic. These pigments are generally non-reactive with the oil medium, are slow to dry, and do not provide anchor points to bind free carboxylic acids within the oil network. With the exception of the well-dried orange paint (which contained a high concentration of zinc fatty acid salts), only minor to trace amounts of zinc were identified in the paint samples. However, it is clear that compositional factors alone are certainly not enough to have led to the conservation issues seen on *Womenizer*. Firstly, the softening is not related to a specific colour or type of paint and the analysis showed that the pigments and medium are similar over the whole painting. Secondly, the liquefying paint appears to be an isolated case; no other paintings by the artist have been reported to exhibit similar problems, even works painted around the same time as *Womenizer*.

24.5.2 Environmental Factors

The local environment during the rolled storage appears to have played an important role in the softening of the paint in *Womenizer*. The fact that the painting is most tacky in the bottom third of the work, particularly near the right edge, which was on the interior of the most tightly rolled section, supports this. The painting was rolled some 3 to 8 years after its completion, after the first stage of oxidative curing had taken place, but during a critical period of the maturation process. It was stored in a relatively closed system, rolled and wrapped in plastic, for at least 4 years and possibly longer. As hydrolysis and oxidation reactions continued during this time, low molecular weight compounds were retained in the local environment (confirmed by the strong odour upon unrolling). The retention of volatile compounds softened and plasticized the paint and, as well, altered the atmosphere inside the enclosure. This altered atmosphere may, in turn, have had an effect on the ongoing curing reactions of the oil [15]. While this type of rolled storage is not typical for an oil painting, the results could be applicable to other situations where relatively fresh paint is placed in a closed environment, for example, works that are framed and glazed relatively soon after completion.

24.5.3 Implications for Conservation Treatment

The results show that the paint has undergone the first stage of oxidative curing. There is no evidence of unreacted polyunsaturated fatty acids, consistent with the observation that the painting was dry when it was first exhibited. In its current state, then, the liquefied paint is unlikely to polymerize further, and conservation treatments aimed at promoting crosslinking would not be effective.

Some of the extractable compounds identified are liquid at room temperature and act to soften and plasticize the paint. Analysis showed significant amounts of C8:0 (caprylic) and C9:0 (pelargonic) short chain monoacids in the extractable portion of the paint. These compounds are oily liquids at room temperature and have an unpleasant, rancid odour [16], explaining the strong smell when the painting was unrolled. Although they have been previously identified in a 5 year old oil film [4, 8], short chain monoacids evaporate over time and are not often found in aged oil paints.

These results suggest that natural evaporation could improve the consistency of the paint by removing low molecular weight volatile compounds. A previous study of heat treatments on test samples of liquefied paints [13] showed that some improvement in hardness can be achieved through evaporation. In light of this, two extraction units were placed over the unrolled work for several days to enhance evaporation of volatile components, followed by several weeks of natural evaporation. However, there was little change to the texture of the paint. Furthermore, additional tests with warm air in selected areas did not lead to improvements in hardness.

For the retrospective exhibition at the NGC, the painting was stretched, which also gave the back of the work a chance to off-gas. The work was placed on a self-standing wall for exhibition, with a low barrier surrounding it to prevent visitor contact. After the close of the exhibition, the painting was secured in a handling-travel-storage frame and covered with a polyethylene sheet to prevent dust from accumulating on the image. The back of the frame is open and allows for good air circulation. At the time of writing, the work is stored vertically in an environmentally controlled storage location allowing regular monitoring. Large areas of the paint still remain tacky; it is likely that this painting will remain a conservation challenge for cleaning as well as for storage and exhibition.

24.6 Conclusions

The softened and liquefied paints on *Womenizer* contain a high proportion of extractable, low molecular weight material compared to the well-dried paints. The high abundance of extractable components explains the liquefied and sticky consistency. The extractable portion of the wet paints contained more polar material than the dry paints, consistent with previous studies of liquefying paint. Since the oil medium has undergone the first stage of oxidative curing, it is unlikely to polymerize further and conservation treatments aimed at promoting crosslinking would not be effective. While multiple factors are likely involved in the softening of the paint, the local environment of the most tightly rolled section of the work while it was in storage appears to have played an important role.

Acknowledgements The authors would like to acknowledge the contributions of Susan Ashley Walker, Spencer Montcalm and Stephen Gritt (National Gallery of Canada), as well as Maeve Moriarty and Eric Henderson (Canadian Conservation Institute). Deepest thanks are due to Alex and Jacqueline Janvier for supporting the research and providing valuable background information.

Appendix: Experimental Details

For FTIR spectroscopy, samples were mounted on a low pressure diamond anvil microsample cell and analyzed using a Bruker Hyperion 2000 microscope interfaced to a Tensor 27 spectrometer. The samples were analyzed in transmission mode by co-adding 200 scans and using a 4 cm^{-1} resolution. Spectra were collected from 4000 to 430 cm^{-1} using a wide band MCT detector.

SEM/EDX was performed using an Hitachi SU3500 SEM integrated with a lithium-drifted silicon, light element X-ray detector and an Oxford Inca X-act analytical Oxford AZtec x-ray microanalysis system. The SEM was operated at an accelerating voltage of 20 kV in high vacuum mode. With this technique, elemental analysis of volumes down to a few cubic micrometers can be obtained for elements from boron (B) to uranium (U) in the periodic table at a level of approximately 0.1–1% or greater.

The XRD patterns were obtained with a Rigaku Rapid II equipped with a rotating anode and cobalt target. The patterns were measured at 40 kV and 15 mA, using a 0.3 mm collimator.

Samples were prepared as dispersions in Cargille Meltmount mounting medium ($n = 1.66$) and examined by PLM using a Leica DMRX polarizing light microscope.

Raman spectra were collected with a Bruker Senterra dispersive Raman microscope. An excitation wavelength of 785 nm was employed. The laser power at the sample was less than 1 microwatt. Either a 50× or 100× objective lens was used to produce analysis areas of approximately 2 or 1 micrometers in diameter respectively. The spectra were recorded using a grating and CCD detector combination that produced a spectral resolution of approximately 3–5 cm^{-1}.

Analysis by Py-GCMS was performed at 450 °C using an Agilent thermal separation probe (TSP) assembly with an Agilent 7890 gas chromatograph interfaced to an Agilent 5975 triple axis quadrupole mass spectrometer. Methyl derivatives of the samples were formed directly in the TSP using tetramethyl ammonium hydroxide (TMAH, 2.5% in methanol, Sigma-Aldrich). The TSP was ramped from 50 to 450 °C at a rate of 900 °C/min and held at that temperature for 3 min before cooling to 250 °C for the remainder of the run.

Analysis by GCMS was performed after extraction and derivatization with *N,O-bis*(trimethylsilyl) trifluoroacetamide (BSTFA, in pyridine; Pierce) for 60 min at 60 °C to form trimethylsilyl (TMS) derivatives. The silylated samples were analyzed with an Agilent 7890 gas chromatograph interfaced to an Agilent 5975 triple axis quadrupole mass spectrometer. The samples were weighed prior to analysis so that the relative amount of extracted material could be compared among the samples. Sample weights varied from approximately 25–75 micrograms.

References

1. Hill G (2016) Alex Janvier. National Gallery of Canada, Ottawa, p 12–13, https://doi.org/10.1016/j.jhin.2015.12.020
2. Janvier A, Janvier J (2017) Personal correspondence. Alex and Jaqueline Janvier with conservators at the National Gallery of Canada (Susan Walker Ashley, Patrick Gauthier and Spencer Montcalm).
3. Juita, Dlugogorski BZ, Kennedy EM, Mackie JC (2012) Low temperature oxidation of linseed oil: a review. Fire Sci Rev 1(3):1–36
4. van den Berg JDJ, van den Berg KJ (1999) Chemical changes in curing and aging of oil paints. In: Grattan D (ed) Preprints, 12th triennial meeting, ICOM committee for conservation, vol 1, Lyon, France, 1999. James and James, London, p 248–253
5. Perkins EG, Iwaoka WT (1973) Purification of cyclic fatty acid esters: a GC-MS study. J Am Oil Chem Soc 50:44–9, DOI: https://doi.org/10.1007/BF02886866
6. van den Berg JDJ, van den Berg KJ, Boon JJ (2002) Identification of non-crosslinked compounds in methanolic extracts of cured and aged linseed oil-based paint films using gas chromatography-mass spectrometry. J Chromatogr A, 950:195–211, https://doi.org/10.1016/S0021-9673(02)00049-3
7. Lazzari M, Chiantore O (1999) Drying and oxidative degradation of linseed oil. Polym Degrad Stab 65:303–313, https://doi.org/10.1016/S0141-3910(99)00020-8
8. van den Berg JDJ (2002) Analytical chemical studies on traditional linseed oil paints. AMOLF, Amsterdam, p 9-52, 131–152
9. Mayer R (1991) The artist's handbook of materials and techniques, 4th edn. Viking, New York, p 138–140
10. Laver M (1997) Titanium dioxide whites. In: West Fitzhugh E (ed) Artists' pigments: a handbook of their history and characteristics, vol 3. National Gallery of Art, Washington, p 295–339

11. Bronken IAT, Boon JJ (2014) Hard dry paint, softening tacky paint, and exuding drips on Composition (1952) by Jean Paul Riopelle. In: van den Berg KJ, Burnstock A, de Keijzer M, Krueger J, Learner T, de Tagle A, Heydenreich G (eds) Issues in contemporary oil paint. Springer, Heidelberg, p 247–262
12. Boon JJ, Hoogland FG (2014) Investigating fluidizing dripping pink commercial paint on Van Hemert's Seven-series works from 1990-1995. In: van den Berg KJ, Burnstock A, de Keijzer M, Krueger J, Learner T, de Tagle A, Heydenreich G (eds) Issues in contemporary oil paint. Springer, Heidelberg, p 227–245
13. Franken V, Heydenreich G, Jägers E, Müller W, Schulz J, Zumbuhl S (2014) Set back the race: treatment strategies for running oil paint. In: van den Berg KJ, Burnstock A, de Keijzer M, Krueger J, Learner T, de Tagle A, Heydenreich G (eds) Issues in contemporary oil paint. Springer, Heidelberg, p 333–349
14. Schultz J (2011) Liquefying oil paint: contemporary paintings in danger? In: Bridgeland J (ed) Preprints, 16th Triennial Meeting, ICOM Committee for Conservation, Lisbon, Portugal, 2011. James and James, London, paper 1320
15. López-Aparicio S, Grøntoft T, Odlyha M, Dahlin E, Mottner P, Thickett D, Ryhl-Svendsen M, Schmidbauer N, Scharff M (2010) Measurement of organic and inorganic pollutants in microclimate frames for paintings E-PS 7: 59–70
16. Bagby MO, Johnson RW, Daniels RW, Contrell RR, Sauer ET, Keenan MJ, Krevalis MA (2004) Carboxylic acids. In: Seidel A (ed) Kirk-Othmer encyclopedia of chemical technology, 5th edn, vol 5. John Wiley & Sons, Hoboken/New Jersey, p 27–78

Chapter 25
The Deterioration and Treatment of Some Late Paintings by Paul-Émile Borduas

Michael O'Malley

25.1 Introduction

Paul-Émile Borduas (1905–1960) was a seminal figure in the development of abstract painting in Québec in the 1940s. As a teacher, author, and mentor, he influenced a generation of younger artists and intellectuals in Montréal who were seeking to break with the rigid social mores of the day and push the boundaries of contemporary art. Inspired in part by stream-of-consciousness writing and Surrealism in France, Borduas's circle of artists, known as the *"Automatistes,"* developed a practice of spontaneous, non-figurative painting that boldly explored the creative possibilities of the subconscious mind (Fig. 25.1).

In 1948, Borduas was the principal author of a revolutionary text, *Rufus global*, which was not well received by his employer [3]. This ultimately led to the loss of his teaching position. He left Montréal in the early 1950s for Provincetown and New York. In 1955, he moved to Paris where he worked for 5 years until his premature death in 1960. It was during his Parisian period that he produced a series of iconic, black and white paintings that are the focus of this paper—stark, gestural compositions that feature thickly applied contrasting paints (Fig. 25.2).

The degree of preservation of the paintings from this period varies considerably. Many remain in surprisingly good condition, while others have deteriorated at an alarming rate, presenting wide cracks in the black paints and dramatic paint cleavage. While the results of an analytical study of Borduas's materials from this period have been published elsewhere, there are no publications that discuss the conservation treatment of these paintings [5, 6] . This paper presents an overview of the current knowledge gained from over two decades of conservation treatment of Borduas's black and white paintings at the Centre de conservation du Québec (CCQ). By necessity, conservators had to devise practical solutions to address the aesthetic

M. O'Malley (✉)
Centre de conservation du Québec, Quebec City, Canada
e-mail: michael.omalley@mcc.gouv.qc.ca

K. J. van den Berg et al. (eds.), *Conservation of Modern Oil Paintings*,
https://doi.org/10.1007/978-3-030-19254-9_25

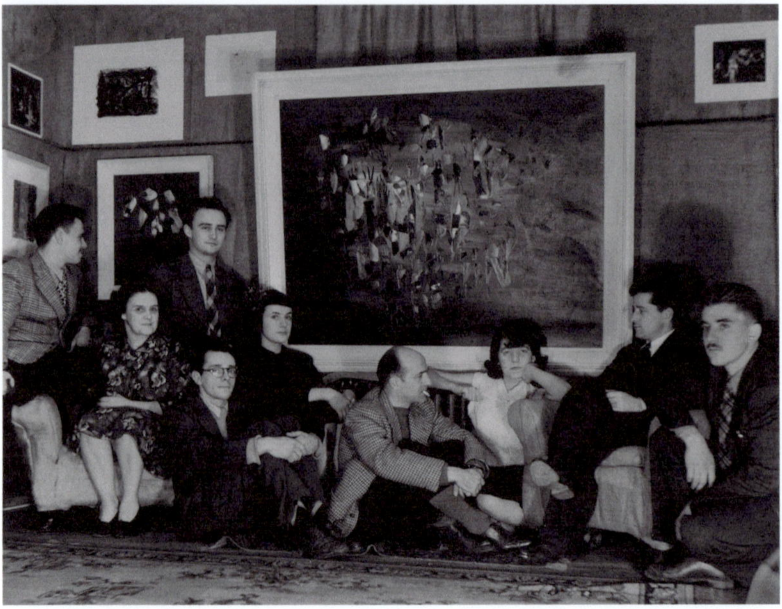

Fig. 25.1 Paul-Émile Borduas seated, centre, among a group of friends and artists gathered for a vernissage in Montréal. *"Seconde exposition des automatistes au 75, rue Sherbrooke Ouest, chez les Gauvreau (février 1947)"*. (Photo: © Maurice Perron, Musée national des beaux-arts du Québec (with the kind permission of Line-Sylvie Perron))

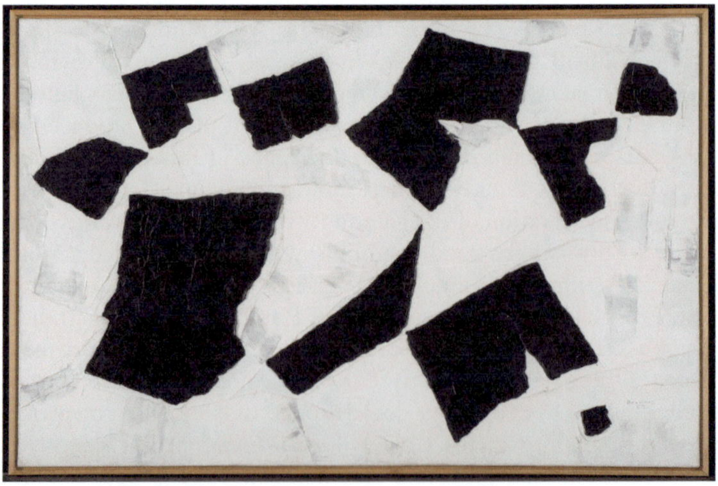

Fig. 25.2 Paul Émile Borduas, *Sans titre no. 34, Pierre angulaire,* 1957. Oil on canvas, 129 × 195 cm, Gift from National Museums Canada. Musée d'art contemporain de Montréal. (Photo: © Michel Élie, CCQ, 1992 © Estate of Paul-Émile Borduas/SOCAN 2019)

issues and the paint instability posed by these works. This was made possible in part by the groundwork laid by an unpublished thesis that first examined these condition issues in depth [2]. The analytical results from a study of eight paintings published by the Canadian Conservation Institute (CCI) also brought further information to light [5]. By revisiting these case studies of Borduas' Parisian paintings, the understanding of the changes and treatment options of these works can be enlarged.

The methodology used here has arisen organically over many years during the observation, study and treatment of paintings largely from the collection of the *Musée d'art contemporain de Montréal*. These paintings have required various interventions, including surface cleaning, consolidation, and in two cases, structural reinforcement. Some testing of modern consolidation materials was carried out to gain a better understanding of the performance of different adhesives. Mock ups were also made in an attempt to understand the possible effects of adding different adjuvants to thickly applied black paint samples. Crack maps drawn on Mylar sheets to document the alterations have provided valuable information about the evolution of these paintings.

25.2 Materials and Technique

Although Borduas was a prolific writer, he authored no texts describing his materials and techniques from this period. Through canvas stamps and sales receipts, it is known that Borduas purchased many supplies from (although not exclusively) the highly regarded Parisian firm Lefebvre-Foinet. He bought canvasses pre-mounted and pre-primed with a thin, lean lead white ground. He often top-coated the commercial ground with a second white paint layer consisting of lead white in either linseed or poppyseed oil [2].

Although the paintings from his Parisian period may look similar in appearance, they defy simple categorization with regard to their materials and technique of execution. Borduas applied the paint with knives or spatulas, leaving areas varying from thin scrapings to large shapes of raised impasto up to 1 cm thick. He claimed he used his paints straight from the tube, with no admixtures, although substantial quantities of linseed oil and turpentine were found in his studio [2]. In some instances, he applied the white paint first, leaving wells of scraped paint reserved for the black paints. Alternatively, black shapes were applied on top of the white layers. He did not apply overall surface coatings to these paintings, but occasionally he locally saturated some black passages, presumably sunken in, with brushed linseed oil.

25.3 Paint Instability and Disfiguring Alterations

The appearance of several paintings made by Borduas in his Paris studio has changed significantly over the past 60 years. These alterations include overall yellowing of white passages, localised yellowing as a result of binder exudates, canvas staining, and crack formation in some black and white paints that often leads to raised or active cleavage in the black passages.

A largely aesthetic condition issue is the problem of yellowing of some white paint passages. This appears to be the only alteration Borduas ever saw during his lifetime, as evidenced by correspondence on the subject with the Martha Jackson Gallery of New York [2]. The white paint sometimes darkens to various shades of ivory, as observed again most recently by the author in 2018 on a group of paintings stored together on a rack in the vault of the *Musée d'art contemporain de Montréal.* This phenomenon seems to occur after periods of prolonged storage in the dark or in dim light. In one case, *Sans Titre 38* (1958), the pronounced yellowing of the white paint may possibly be exacerbated by the presence of safflower oil in the paint binder (unpublished CCI analysis report LRA 4420, 2006). This yellowing can be reversed by exposing the painting to several hours of window-filtered sunlight, as witnessed during a treatment of this painting at the CCQ in 2007.

A different kind of yellowing phenomenon also occurs, localised surface yellowing, created by oily exudates that collect in the pores of the white impasto, or by drips from areas of dark paint, a feature found on many twentieth century modern paintings. Not only does the oil binder sometimes migrate to the surface, it also sinks down through the paint layers, from the oil-rich blacks to the white underlayers. Localised staining is seen on the reverse of the canvas in many of the paintings. This occurred during or shortly after the creation process, as seen in contemporary photographs of Borduas's studio, where canvasses are turned facing a wall. Analytical results show that the staining material is always raw linseed oil, which corresponds to the binding medium of the oil-rich bone black paint [5].

The problems encountered in some of the black paint passages are of another order of magnitude altogether. Approximately 25 of the black and white paintings from this period in the collection of the *Musée d'art contemporain de Montréal* show significant cracking in the black passages. The types of cracks fall broadly into two categories. The first are the relatively small drying cracks that often occur in the thicker outer ridges of the paint impasto. Smaller hairline cracks can also appear in transition areas between the thickly and thinly applied areas of paint that are a result of Borduas's spatula gestures [2].

The second, more problematic cracks are deeper, found within the thickly impasted black shapes. After an initial period of relative stability that lasted from 10 to 15 years, certain passages deteriorated rapidly and lead to paint delamination. Once this cracking process is initiated, the cracks tend to widen and lengthen until cleavage ensues. The result is that flat shapes of black paint are broken up into islands that can slide over the white underlayers and move out of plane. When the underlayers begin to show through the open cracks, this disrupts the continuity of the black paint and distorts the intended interplay of light, sheen, and texture.

The paint delamination appears to originate from within the white underlayers at the interface of the black impasto. A layer of artist-applied white paint is almost always seen adhered to the underside of the lifted black paint (Fig. 25.3). No exposed canvas is ever seen and the commercial white ground is left intact. However, fragments of the artist-applied white layer may sometimes be seen adhered to the white ground.

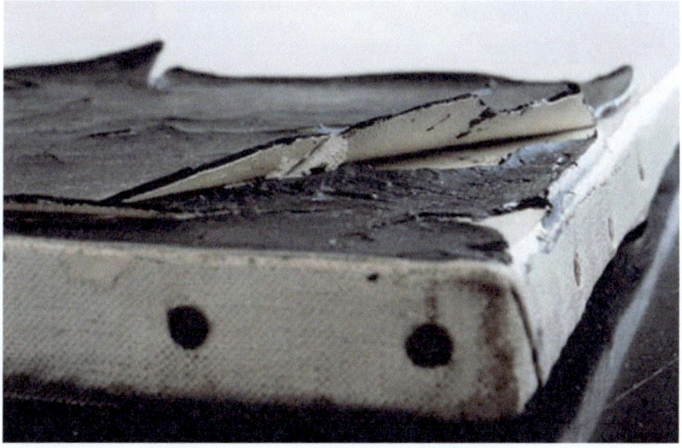

Fig. 25.3 Detail, *Sans titre no. 39,* 1958. Lifted black paint with a torn white underlayer and BEVA adhesive residue from a previous intervention. (Photo: © Michael O'Malley, CCQ, 1998)

Several possible reasons have been put forward to explain this cracking. However, the cause of this seemingly spontaneous deterioration that appears after some decades of normal ageing is not well understood and has yet to be fully explained. Paint cracking can be caused by the materials and techniques used, as well as several external factors, including environmental changes. Since these alterations appear in paintings in several different institutions across Canada, the source of the problems does not seem related to particular storage or presentation factors, although the precise conditions for each painting are not known. The technique of applying oil paint thickly with spatulas could possibly be a factor, but this does not explain why some Borduas paintings with thickly applied impasto from the same period show only minor changes in condition.

Drying and shrinking stresses most certainly play a part in the paint cracking, since the oil-rich bone black dries very slowly in comparison to the lead white layers below. In theory this respects the fat over lean painting principle. However, traces of white paint are sometimes found marbled into the edges of the black impasto, suggesting that an appropriate drying time for the white underlayers may not have been respected in some instances. The migration of the paint medium from the black paint into the underlying white layers, as seen by the canvas staining, may also be responsible in part for the cracking. It was initially supposed that this loss of paint medium could increase the pigment/volume concentration (PVC), which in turn could exacerbate embrittlement and drying stresses in the back paint. However, no correlation has been found between the amount of canvas staining and the condition of the black paint, and some paintings from this period with prominent canvas stains remain in very good condition.

The case of *Sans titre no. 34*, *Pierre angulaire* (1957) is of special interest since it is a large and well-documented painting. By studying photographs taken over time, a narrative of the crack formation in the black passages can be reconstructed. The two images below, taken in the 1970s and in the 1990s, show a detail of the largest shape in this painting (Fig. 25.4). Comparison suggests that the painting initially appears to have aged normally without much incident for 10 to 15 years after its creation. However, in the ensuing 20-year period, roughly half of the black shapes underwent a significant deterioration process.

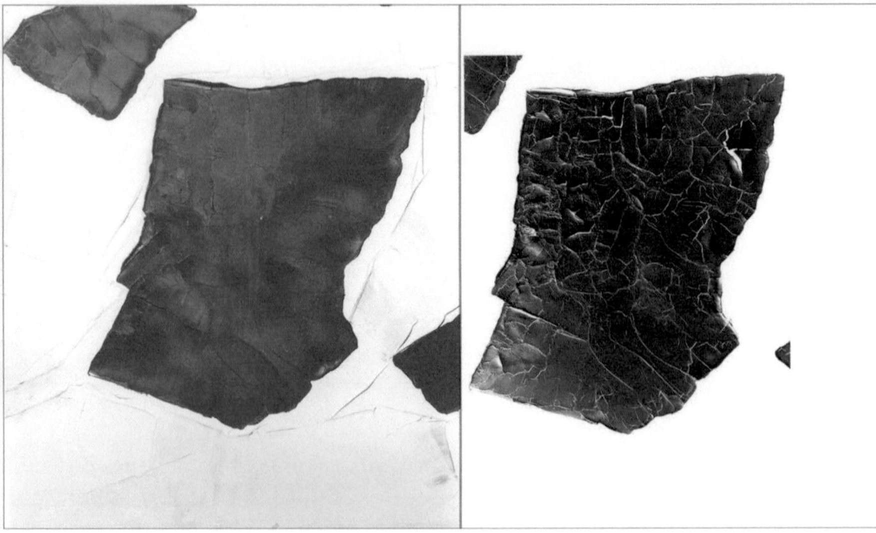

Fig. 25.4 This montage detail shows the same large black shape from *Sans titre no. 34, Pierre angulaire* (1957) taken in 1973 on the left, and in 1992 on the right. Crack maps reveal there has been no significant crack development in this black shape since the early 1990s. (Photo: © Michel Élie, CCQ, 1992)

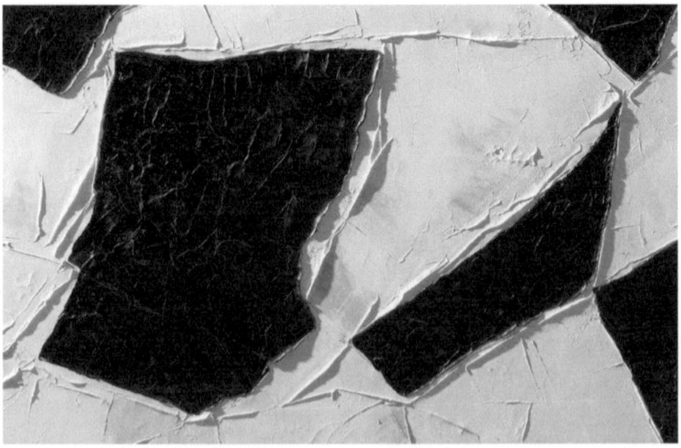

Fig. 25.5 Raking light detail, *Sans titre no. 34, Pierre angulaire,* 1957. The large black shape on the left is in very poor condition, while the adjacent black shape on the right remains in comparatively good condition. (Photo: © Jacques Beardsell, CCQ, 2009)

Not only does the amount of cracking in the blacks vary considerably from one painting to another, it even varies among the different black shapes on a given painting. This is also graphically illustrated in *Sans titre no. 34, Pierre angulaire* (Fig. 25.5), where a large damaged shape (left) lies adjacent to a black shape in comparatively good condition. In fact, the blacks in the entire lower right portion of this painting remain relatively intact, as opposed to those in the upper left portion, which display significant cracking and delamination. When observed under UV radiation, a similar degree of fluorescence emanates from all the black shapes, irrespective of their condition. The only difference is that the thicker, oil-rich areas of the paint fluoresce more strongly.

According to Challan-Belval, the location and configuration of the cracks may be related to the direction of the spatula strokes and to the paint thickness, yet this is not borne out here, where the black shapes are of apparent equal thickness, but can present distinctly different states of preservation and crack evolution [2]. However, the black shape in this painting with the greatest surface area (Fig. 25.4) is also the shape that displays the greatest amount of cracking.

A second example of the contrasting condition of black shapes within the same painting can be found in *Composition 24* (1959), in the collection of the National Gallery of Canada. Under UV radiation, the cracked black paint fluoresces much more strongly than the bands of stable black paint [2]. However, this marked difference in the fluorescence is exceptional here and not commonly found in other paintings. It was this difference in fluorescence in this particular case that first gave rise to the hypothesis that some of the black paints used by Borduas may have been different in nature.

25.4 Past Analytical Studies

These issues became the focus of research for a master's degree thesis at Queen's University [2]. Subsequently, the analytical results from paint samples taken from a total of eight paintings were published by the Canadian Conservation Institute [5].

As early as 1980, CCI microscopist Ian Wainwright noted the presence of two distinct black paints, as revealed by the variable pigment morphology seen in microtome thin sections [2]. Attempts were made to establish differences between the physical appearance and the chemical composition of unstable and stable black paints. For example, the more stable black often contained black and brown pigment particles of variable size, with a fairly equal distribution of smaller brown and larger black particles. By contrast, nine out of ten samples of unstable black paint from five different paintings were more homogenous, showing a much greater percentage of larger, more highly charred black particles. However, the possible effect of this variable pigment distribution and morphology on the condition of the paint has never been fully explored. Very subtle differences in some FTIR absorption peaks from black paint samples also suggest that Borduas may have used different paints of somewhat different composition. A majority of the stable blacks analyzed show small peaks at 2013 cm^{-1} that are absent from the majority of the stable blacks. However, the constituent material responsible for this peak has not been identified [5].

Three of the eight damaged paintings studied by Moffat and Miller had soft, waxy white underlayers beneath the black shapes. These underlayers contained high levels of lead soaps (lead stearate and lead palmitate, as determined by FTIR), with greater concentrations at the interface of the black and white paint [5]. During treatments, conservators have observed that these soft underlayers are heat sensitive, and are surprisingly soluble in ethanol and xylene, which is uncharacteristic of a normally cured lead white paint layer. However, high levels of lead soaps were not found in the underlayers of all these paintings, and they appear to be completely absent from the pre-primed grounds and the white impasto. When present, a high concentration of lead soaps seemed to affect the cohesion and mechanical properties of the white underlayers, which in turn could exacerbate the delamination of the overlying black paint. Questions remain, however, concerning the origin, progression, distribution, and effect of the metal soaps in these paintings.

Subsequent analyses undertaken at the Canadian Conservation Institute since 1993 on a few other Borduas paintings from this period (unpublished CCI reports ARS 3660, 1997; ARS 3715, 1998; LRA 4038, 2002; LRA 4420, 2006) have not revealed any findings that shed new light on the causes of these problems.

The limited number of samples analyzed from each painting and the inconsistent results have not allowed for easy explanations or extrapolations to account for the failure of some of the black paints. Several questions remain. Might the presence of undetected contaminants be a factor? Were there inadequate conditions in the manufacturing process of either the pigment or the paint? Does the cracking stem from the properties of the drying oil used as a binder or had several of these factors in combination - including painting techniques and environmental conditions - influenced the drying process differently? These are all further avenues for investigation in future analytical work.

Perhaps not by coincidence, art materials and thickly applied paint layers seem to be at the source of many condition problems observed in the paintings of several artists who worked in Paris in the 1950s. These include Jean-Paul Riopelle, Karel Appel, Asger Jorn, and Georges Mathieu, among others [1]. Similar random deterioration phenomena in black paints have also been documented in the work of Pierre Soulages, a contemporary of Borduas, who also obtained many of his supplies from the same Parisian firm, Lefebvre-Foinet [4]. However, analytical findings from Soulages's paintings revealed the presence of materials that differ somewhat from those used by Borduas. Unlike Borduas, Soulages preferred canvasses primed with thick oil-rich grounds, and his custom-made black paints were often formulated with a Flemish siccative containing copal or other resins. By contrast, no resins have been detected in Borduas's blacks [5].

25.5 Ivory Black Mock-Ups

If many of Borduas's black and white paintings from his Parisian period have survived in relatively good condition, showing only minor or insignificant crack development in the blacks, this seems to indicate that the artist's painting technique is not necessarily the cause of the problems observed. However, the application of oil paint in thick layers, with and without sufficient drying times of the underlayers bore further investigation. In an attempt to confirm or disprove that the process of paint application contributes to crack formation, an attempt was made to replicate the artist's layering technique. To this end, two sets of large oil paint mock-ups were made in the year 2000 at the Centre de conservation du Québec.

Linen canvasses pre-primed with a lead white ground were used. Standard amounts of Winsor & Newton ivory black paint were gradually modified by the addition of various admixtures to determine whether any factors that cause paint cracking or delamination could be identified. Samples of paint weighing 15 g were squeezed out for each black shape. The paint was then applied by spatula in thick and thinner passages in several columns and rows over previously-laid lead white paint layers. In half of each test canvass, the lead white underlayer was left to dry for 3 weeks, while the other half was left to dry for only 1 week before applying the black shapes. In one set, the PVC of the black paint was manipulated by gradually adding excess amounts of cold-pressed raw linseed oil, while turpentine diluent was added to the second set. Courtrai drier (Lefranc & Bourgeois, containing oxides of lead and manganese) was also added in incremental and excess amounts in both sets. The mock-ups were displayed in the lab and left to age naturally. While modern commercial materials were employed to make these mock-ups, the results are nonetheless of interest. At the time of this publication, no significant canvas staining has occurred, and no major cracking has occurred in any of the black shapes. However, as expected, some minor, hairline cracks have formed at some of the thick/thin transition areas of some shapes. The most frequent alteration produced by the addition of either excess oil, diluent, or drier is the same: a dense network of depressions and wrinkles as pictured below (Fig. 25.6), a condition not seen in Borduas's black paints. This suggests that Borduas did not add excessive amounts of diluent, oil or drier to his paint.

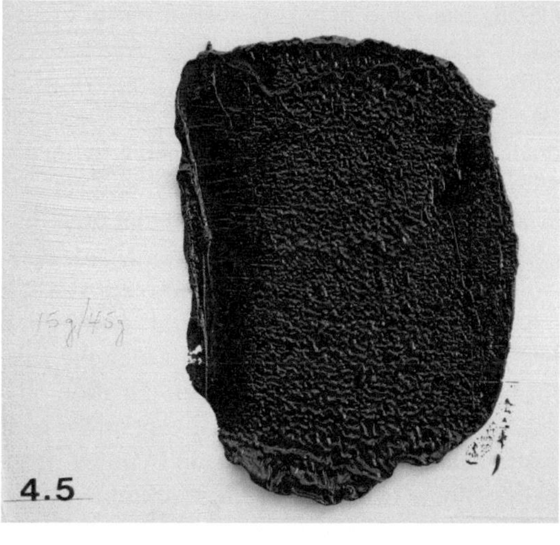

Fig. 25.6 Paint mock-up detail. In this detail, 4.5 g of raw linseed oil was added to 15 g of W&N ivory black tube paint. The most prominent alteration caused by the addition of excess oil is a dense network of paint wrinkles, a condition not seen in Borduas's paintings. No deep cracks have appeared after 18 years. (Photo: © Michael O'Malley, CCQ, 2018)

25.6 Past Structural Treatments

Several paintings were treated at the Centre de conservation du Québec for a major retrospective of Borduas's work in Montréal in 1998. While many works responded well to local consolidation, two of the paintings were deemed to be aesthetically unacceptable following repeated consolidation attempts. What should have read as smooth areas of monochrome black paint read instead as out-of-plane shapes disrupted by craquelure. It was thus decided to undertake two substantial structural treatments, i.e. a marouflage onto an aluminum honeycomb panel, and a stiff lining to a rigid synthetic fabric laminate. The goal of these structural reinforcements was threefold: (1) bring disfiguring deformations back into plane to restore the legibility of the paintings; (2) see whether stiffening the flexible canvas support would reduce the rate of deterioration of the black paint; and (3) compare the efficacy of a marouflage versus a stiff lining in holding deformations in plane.

Decisions to undertake such invasive treatments must be thoroughly justified and were weighed carefully. Concern was expressed about the potential stresses that might be induced by the application of humidity, heat, and pressure to the distorted black impasto. Both treatments proved initially to be successful. The paintings were stabilized, and their overall appearance and legibility improved.

It later became apparent, however, that the structural treatments did not halt or forestall crack development. Crack mapping over a 20-year period has revealed that following these two treatments, there was a significant evolution of fine craquelure in the black paints in both these paintings. The painting that underwent the stiff lining treatment, *Sans titre no. 41* (1959), required further local consolidation as little as 3 years later. This painting was again treated locally in 2008 for minor cleavage. In 2018, an examination revealed that there had only been a minor evolution of the craquelure, but the passages in the poorest condition once again require local consolidation and flattening.

The painting that was marouflaged to aluminum honeycomb, *Sans titre no. 39* (1958) has fared better overall in terms of its appearance and stability, although minor consolidation in one critical area was required in 2007. Figure 25.7 graphically illustrates the evolution of craquelure documented in this painting over a 20-year time period. The blue lines in the crack map show the size and location of cracks in 1998 immediately following the marouflage. Ten years later, an extended network of small hairline cracks had developed, shown by green lines. In 2018, however, an examination revealed that there has only been a slight evolution of the craquelure, and the painting has remained relatively stable and presentable.

The tendency observed here through crack mapping of the black paint over the past two decades shows an overall trend of subsiding crack development. What begins to emerge is a scenario in which the paintings initially appeared to be stable for possibly 15 years after their creation. By the early 1970s, some paintings required treatment for comparatively minor paint instability. By the late 1980s, dramatic deterioration had occurred in many of the paintings that had unstable black paints. Some evolution of the cracking was seen on the treated paintings after the

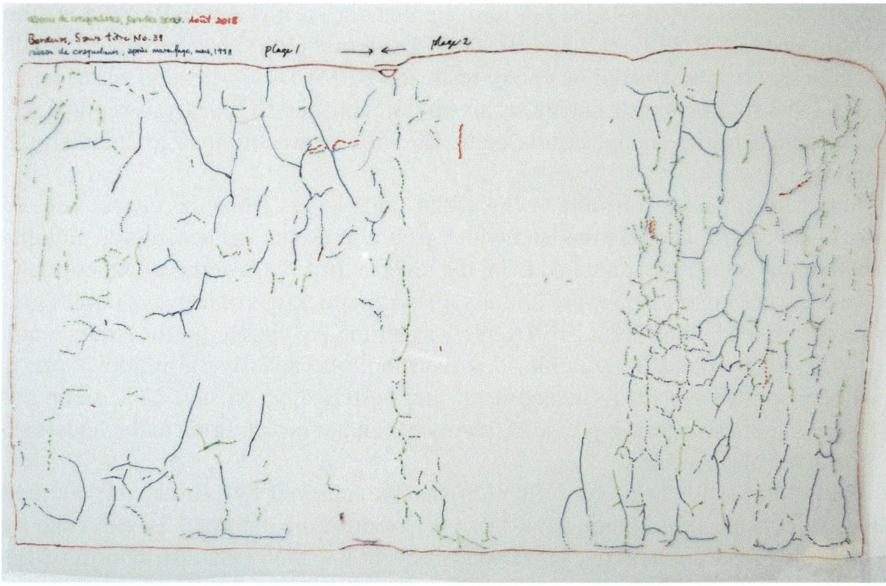

Fig. 25.7 Mylar crack map detail showing two of the black shapes from *Sans titre no. 39*. The blue lines show the size and location of cracks in 1998, post-marouflage. The green lines show a significant evolution of the craquelure by 2008. They appear to be extensions of existing cracks, or in some cases, new cracks that occur in areas between existing cracks. Dotted lines refer to the presence of small, hairline cracks, in many cases incompletely developed. By 2018, the red lines indicate that only relatively minor crack evolution has occurred. (Photo: © Michael O'Malley, CCQ, 2018)

1990s, but now the rate of crack development appears to have reduced significantly, almost nil, as typified by the red lines shown in Fig. 25.7. This trend has also been observed in four other works by Borduas in poor condition from this period belonging to the Musée d'art comtemporain de Montréal, namely: *Sans titre no. 34* (1957); *Sans titre no. 38* (1958); *Sans titre no. 41* (1959) and *Verticalité lyrique* (1960).

25.7 Local Consolidation

The ongoing curing, shrinking, and cracking of black paints have necessitated repeated conservation treatments for works in poor condition. Existing treatment reports from various Canadian institutions were first collected and compared to establish a repository of information, since no systematic study of the condition and the effect of various past treatments on these paintings had been undertaken, which had different treatment and storage histories. Institutional visits were also made to examine similar Borduas paintings.

Past adhesive choices were not always successful in holding the paint down. On *Pierre Angulaire* (1957) for example, several different adhesives had been tried.

Treatment reports from 1982–1998 indicate that wax-resin (1982), BEVA 371 film (1986), Acryloid B-72 (1992) and BEVA 371 solution (1998) had been used, while two adhesive residues based on epoxy resin and BEVA D-8 dispersion were identified in 1998 by FTIR instrumental analysis (unpublished CCI analysis report LRA 3715), which testify to the efforts made by various practitioners to stabilize the painting.

Initial attempts to consolidate the white underlayers from the canvas reverse were unsuccessful because the lead white ground prevents the penetration and diffusion of any consolidant applied from the reverse. In 1998, a series of aqueous and solvent-borne consolidants were initially tested on mock-ups of canvas samples preprimed with lead white paint. BEVA 371 was ultimately chosen for the strength and flexibility of its bond. In addition, this thermoplastic adhesive also allows previously consolidated paint to detach itself and be retreated. In Fig. 25.3, some old BEVA 371 adhesive from a previous treatment can be seen clinging to the underside of the lifted black shape.

Fairly successful local consolidation can be achieved by introducing warmed adhesive beneath lifted black paint from the front of the painting. This is done by syringe or brush application. However, an even and thorough distribution of the adhesive is never ensured, since it disappears beneath the black paint. Repeated applications may be necessary. Care must be taken when testing a solvent or a solvent mixture for the removal of any residual adhesive from the surface, since this may alter the surface gloss of the black paint. After complete solvent evaporation, which takes several days, gentle heat from a hot air tool and hand pressure are applied to lay down and re-adhere lifted paint. To address aesthetic considerations and permit easy access for retreatment, the cracks are not filled; they are simply inpainted when necessary with reversible media.

A fundamental concern was expressed at the outset of these consolidation treatments: Do such interventions provoke the lengthening of existing cracks or the formation of new cracks adjacent to treated areas? A study of the crack mapping indicates that repeated consolidation treatments over the years have not provoked or exacerbated the cracking in any significant way, and most paintings with areas of lifted paint treated previously with BEVA 371 have remained fairly stable. It appears, however, that such paintings will likely continue to require ongoing local consolidation treatments every decade or so until they reach an equilibrium state and all stresses are released. An unanswered question remains for the long-term conservation of these paintings: What is the effect of introducing heat and pressure to a painting that has a soft, waxy white underlayer, especially when a heat-activated adhesive that contains aromatic hydrocarbon solvents, synthetic resins, and wax has been applied?

Preventive measures, such as affixing polyester batting to backing boards and limiting the amount of travel for severely altered and vulnerable paintings are also now standard procedures for the conservation of these paintings.

25.8 Conclusions and Future Work

The variations found in analytical results and in the condition of the paintings point to an artist who may have experimented to some extent with his materials and painting technique, and who sometimes used, perhaps unknowingly, paints of inferior quality with varying composition. Many questions remain about the constituent materials of these paintings and more analysis is needed to better understand the causes of the yellowing and the severe cracking found in some paintings.

Local consolidation is often successful in stabilising these paintings. While additional structural treatments may initially improve the appearance of badly altered paintings, they do not forestall crack evolution. In fact, the opposite is true, and such structural treatments may even provoke an initial spike in crack development. Following two decades of monitoring these paintings, it now appears that crack development in the black paint is subsiding overall, and the worst period of the cracking seems to have occurred between the late 1970s and the early 1990s.

As digital technologies evolve and become more accessible, the surface mapping and documentation of these paintings will become more precise. Finally, the use of effective adhesives containing solvents that have less impact on the altered white underlayers should be more fully explored in the future.

Acknowledgments The contributions of current and former conservation scientists from the Canadian Conservation Institute are acknowledged here: David Miller (retired), Elizabeth Moffat (retired), Ian Wainwright (retired), Marie-Claude Corbeil, Kate Helwig, Jennifer Poulin, and Geneviève Sansoucy.

Marie-Noel Challan-Belval, former Chief Conservator at the Musée d'art contemporain de Montréal first documented these problems for her master's degree at Queen's University in 1991.

Gunnar Heydenreich, Gene Bourgeau, and Rachel Benjamin provided useful comments on this manuscript.

References

1. Boon J (2018) Microscopic Aspects of Softening, Exuding, Dripping and Delaminating Paints in Paintings Post 1950. Paper presented at the Conference on Modern Oil Paints, Rijksmuseum, Amsterdam, May 23–25, 2018
2. Challan-Belval M-N (1991) Materials, techniques and conservation problems related to the paintings of Paul-Émile Borduas from 1955–1960. Unpublished master's degree thesis, Queen's University
3. Gagnon F-M (1978) Paul-Émile Borduas – Biographie critique et analyse de l'œuvre. Fides, Montréal
4. Hélou-de la Grandière P, Le Hô A-S, Mirambet F (2008) Delaminating paint films at the end of 1950s: a case study on Pierre Soulages. In: Preparation for Painting. The artist's choice and its Consequences. Archetype publications, London, pp 156–162

5. Moffat E, Miller D (1994) Spectroscopic and chromatographic analysis of selected paintings from the Parisian period of Paul-Émile Borduas. J. IIC-CG 19:3–13
6. O'Malley M, Moffat E (2001) Paul-Émile Borduas, the blacks and whites of the later paintings (1955–1966), in Deterioration of artists' paints: effects and analysis. Extended abstracts of the presentations. ICOM-CC Working Groups Paintings 1 & 2 and The Paintings Section, UKIC. British Museum, London. 10–11 Sept 2001. Alan Phenix (ed), 2001, ICOM and UKIC, pp 84–87

Chapter 26
Colors Before Zero: Commercial Alkyd-Oil Enamel Paints in Early Reliefs by Jan J. Schoonhoven

Lydia Beerkens, Chiara Manfriani, Stijn Lenaerts, and Gwendoline R. Fife

26.1 Introduction

Jan Schoonhoven (Delft, 1914 – Delft, 1994) is widely known for his formal white reliefs, made out of cardboard, newspaper, glue and covered with matte white house-paint. Rather than producing a 'painted Zero-relief', Schoonhoven wanted his constructed surfaces to catch light and cast shadows over the squares, the verticals and horizontals, over the circles and diagonal shapes, whereby different images of the relief would be created depending on the available light and the hour of the day.

Before beginning his famous white serial reliefs in 1960, Schoonhoven experimented with materials including cardboard, newspaper, wallpaper glue and commercial colorful paints for his first so called 'informal' reliefs. Inspired by the Jackson Pollock (1912–1956) exhibition in 1958 at the Stedelijk Museum Amsterdam, he applied different colors of enamel paints by splashing and dripping them onto his abstract reliefs. During the entire process of making, the reliefs stayed in a horizontal position on the working table, also during the application of the paint layers, as can be seen from the direction of the drippings running down over the sides of the construction. In this way Schoonhoven avoided any planned or designed painted image, while working with colors [1]. Originally educated at the art academy to become a painter, he moved away from traditional painting materials altogether, making the radical choice to leave artist oil-paints and employ only commercial house-paints for the rest of his career. In his earliest works he used the alkyd paints and enamel paints produced for house-painters to cover and protect doors and windows etc. Enamel is a term loosely used to describe the industrial paints and house-paints of a very high

L. Beerkens (✉) · S. Lenaerts · G. R. Fife
SRAL Maastricht, Maastricht, The Netherlands
e-mail: lydia.beerkens@planet.nl

C. Manfriani
Maastricht University/SRAL Maastricht, Maastricht, The Netherlands

K. J. van den Berg et al. (eds.), *Conservation of Modern Oil Paintings*,
https://doi.org/10.1007/978-3-030-19254-9_26

gloss which bear a superficial resemblance to vitreous enamels. When unpigmented or transparent they are often termed 'lacquer' [2].

Schoonhoven purchased glossy house-paints, at the nearby house-painter in Delft, and used them from 1957 until 1960. From 1960 onwards, during the so called "NUL/ZERO period", he refrained from colors in favour of white as a 'non-color', and continued with matte white wall paint for the serial reliefs. Some of the first white reliefs show a glossy white enamel paint, either as the final paint layer or a preparation layer. The early preference of Schoonhoven for colorful enamel paint (alkyd-oil paint) and the transient period towards the white NUL reliefs was investigated. A better understanding of the original appearance is of particular significance since the flaking paint layers render the colored reliefs difficult to treat and surface dirt hides the colors and surface gloss Schoonhoven intended.

26.2 Research Questions and Approach

The goal of the research was to investigate via examinations and analysis of the paint-layers whether Schoonhoven had made a clear break in his working technique around 1960, when changing from colorful alkyd paints to matte white house-paints. With this study the authors wanted to gain a clearer insight into the potential step-by-step changes undertaken between producing the early works and the white reliefs-the core of the *oeuvre* by Schoonhoven. This was to be achieved through investigating reliefs from before, around, and after 1960. The research focused on:

- Comparing the paint layers of Schoonhoven's several reliefs from different years
- Investigating the paint layers medium in the early reliefs by Schoonhoven
- Studying the working technique up from the ground layer to determine how this developed over the years

To identify pigment materials and organic components, and obtain supporting data regarding the techniques of execution and the conservation problems, a multi-analytical approach – including Optical Microscopy (OM), Scanning Electron Microscopy (SEM) and Fourier Transform Infrared (FTIR) spectroscopy – was applied. This combination of techniques permits a parallel investigation of organic and inorganic materials, and the characterization of the artistic technique [4, 5].

The artworks from the early color period included in the research were:

- "Postoffice Relief" (1958), 115 × 265 × ca. 20 cm (KPN Art collection, permanent loan to the Rijksmuseum Amsterdam)
- "Ippolito" (1959) 80 × 100 × 11 cm (Centraal Museum Utrecht)
- "Vochanan" (1959) 49.5 × 74 × ca. 11 cm (private collection)

In addition a comparison was made with other works from around 1960 and later, where Schoonhoven made solely white reliefs (reliefs from the Rabo Art Collection, Museum Van Bommel Van Dam and a private collection).

26.3 Analytical Methods

Already detached paint material was collected from 5 reliefs dated between 1958 and 1960, and sampled for cross-section preparation. In total 21 samples were available for examination, and 18 were embedded. Before embedding, the paint flakes were photographed with a RoHS USB Digital Microscope to determine the possible origin in the reliefs and to study the layer build-up. The 18 selected samples were then embedded in a transparent polyester resin (Poly-pol PS 230 + hardener MEKP, Polyservice), cut on the surface by a DAP-7 polishing machine (Struers) on SiC papers and then polished (Micromesh polishing cloth) [6]. The first examination of the cross-sections was done with optical microscopy under different magnification (Leitz Aristomet research microscope, with both vis and UV sources, coupled with a Leica C3 digital camera). After an initial assessment for relevance a smaller selection of 6 samples was made for the next step in research. Further analysis for pigment identification included SEM- EDX (Jeol JSM-5800 Scanning Electron Microscope provided with an Oxford Instruments X-Act 10 mm2 SDD Detector) to obtain electron images and elemental maps. Test on the applicability of ATR-FTIR imaging (400 Perkin Elmer FTIR/NIR Microscope, controlled by a Spectrum IMAGE software) with previous modification of the resin blocks allowed a possible distinction between media in different layers (results not shown in the present paper as identification of the different binding media, dyes and pigments observed would benefit from further investigations) [7]. This research was correlated with the oral history and previously collected information regarding Schoonhovens' earliest relief painting technique [1].

26.4 Results Summary

The cross-sections showed a spectacular build-up of colored paint layers, and also indicated differences in paint medium and potential time intervals in their original application. Despite unknown locations many samples contained either card support residues, or the upmost layer visible on the art work surface, whereby the ordering of paint layers could be identified. Due to the considerable variability in layer build-up observed within the samples from individual works, the results presented focus on those layers potentially linking the works.

26.4.1 Postoffice

The paint samples collected from the "Postoffice" relief (1958) (Fig. 26.1) show a multi-layer stratigraphy, composed of different colors. The first layer applied by the artist is white. Then the sequence is highly variable, and includes orange, grey,

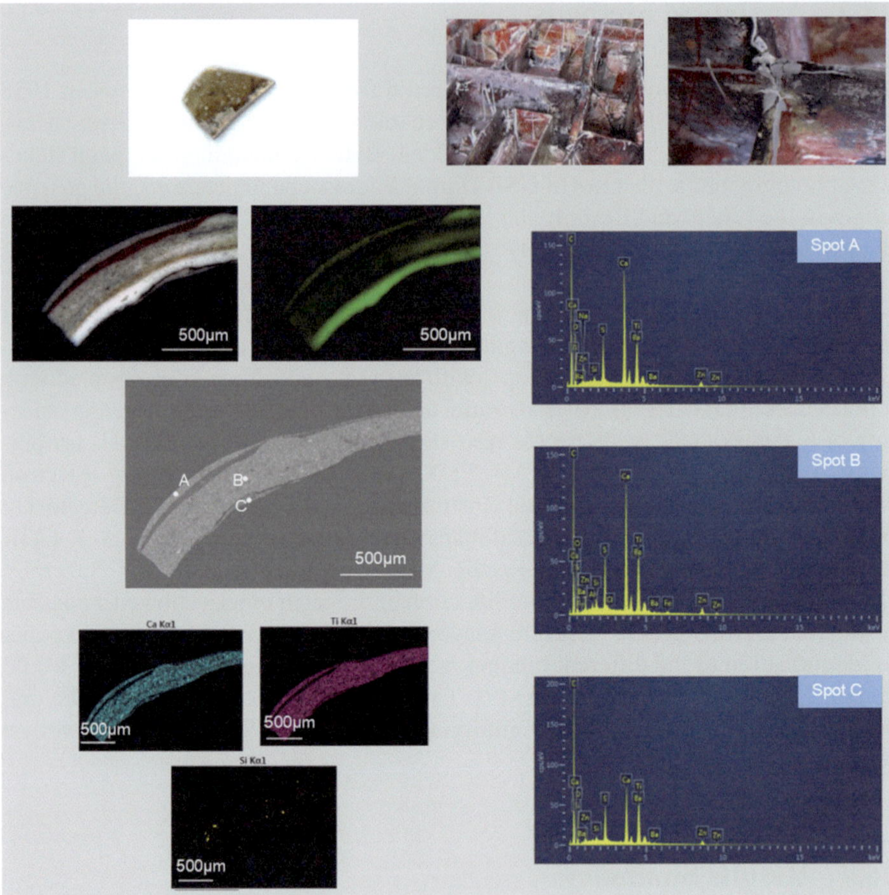

Fig. 26.1 Images of one sample from the Postoffice-relief. On the top right: two details of the relief showing the enamel paint applied with brush and in drippings on the surface; top left: sample #4 in Digital Microscope image, cross-section OM image in visible light and SEM electron image, spectrum and elemental maps

brown, and red, with apparently different mediums and potentially alkyd. However, prominent peaks for titanium (Ti), calcium (Ca) and barium (Ba), with related location association for oxygen (O), carbon (C), and sulphur (S) are relatively consistent between the layers and indicate the presence of the pigments titanium white, calcium carbonate, and barium sulphate in the paints used.

26.4.2 Ippolito

The paint flakes collected from the relief "Ippolito" (1959) (Fig. 26.2) show a multi-layer stratigraphy as well, with highly variable coloration, and the use of different house-paints is indicated. The lowest paint layer in these samples has a notably different pigment composition to that observed in the "Post Office" relief. The samples also show separation between the silver and grey layers, with no sign of dirt deposition.

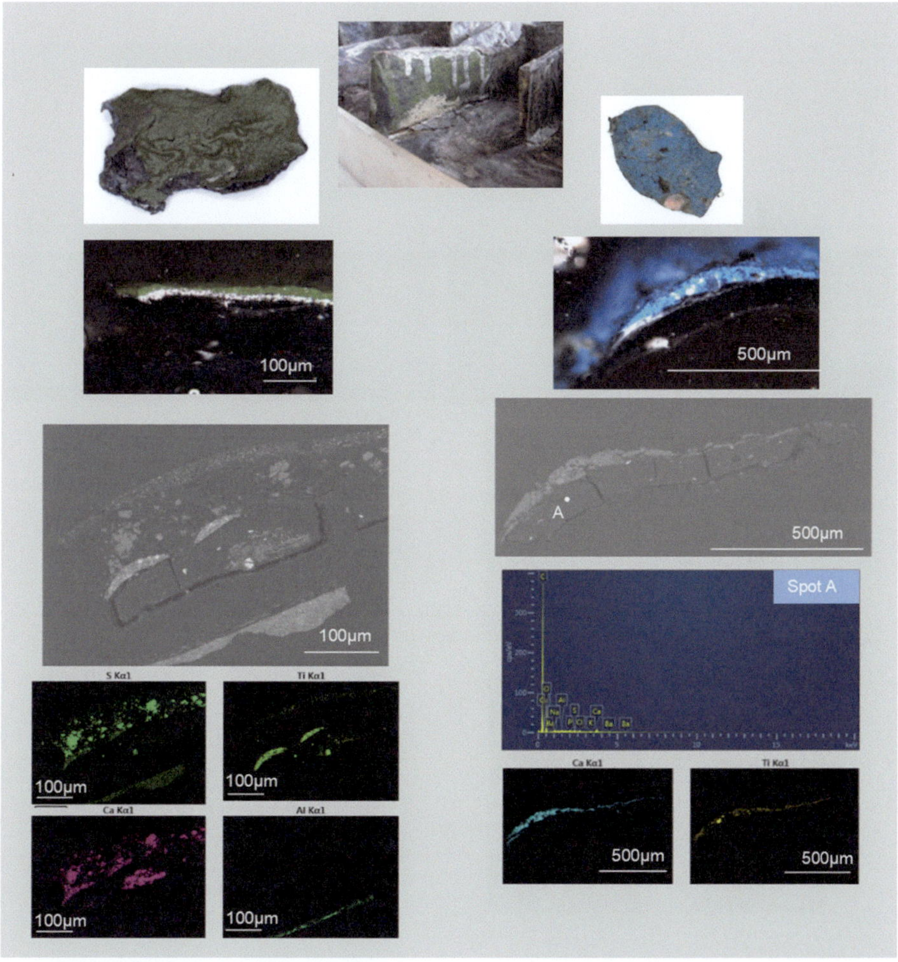

Fig. 26.2 Images of samples from Ippolito. From the top: detail of the artwork; samples #5 and #7 in Digital Microscope image, cross-section OM image in visible light and SEM electron image, spectrum and elemental maps

26.4.3 Vochanan

The sampled paint flakes collected from the relief "Vochanan" (1959) (Fig. 26.3) also show a multi-layer stratigraphy. In this work the variety of coloration in the layers is considerably more minimal than in the two other works but there is nonetheless a notable variety in pigment composition indicated between the layers. Time lapses between the application of subsequent layers is further indicated by the observation of dirt deposits between them.

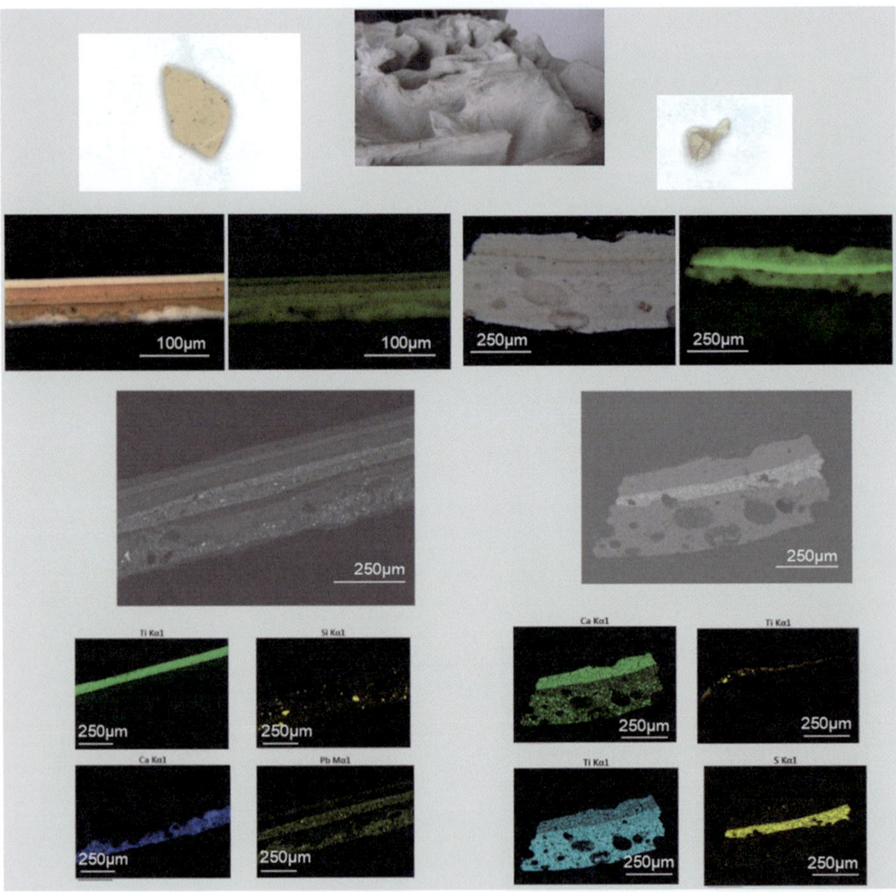

Fig. 26.3 Images of samples from Vochanan. From the left top: detail of the relief and; samples #1 and #2a in Digital Microscope image, cross-section OM image in visible light and SEM electron image and elemental maps

26.5 Conclusions

The analytical approach applied in this study optimized information-gathering from 'lost' materials which could not be re-adhered to the art works, and provided insights into Schoonhoven's processes.

As previously discussed, in the Postoffice relief, before applying multiple subsequent colors, Schoonhoven initially applied a glossy white paint. The combined results (Fig. 26.4) indicated this was identical to the white paint applied as an early layer in the Vochanan relief, above which Schoonhoven considered several beige layers, (with time intervals between) before selecting matte white paints that would appear in this, and his later white works (such as the Relief 74-4, Rabo Art collection). The Vochanan can thereby be regarded as a key work from the short transition period towards NUL.

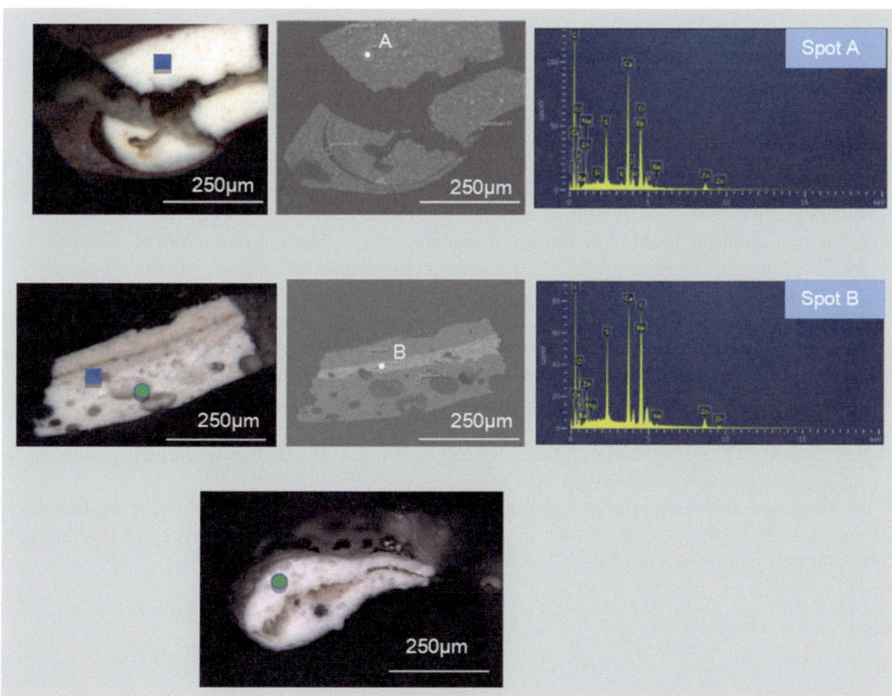

Fig. 26.4 Comparison between Postoffice-relief, Vochanan and the Relief 74-4, Rabo Art collection; from the top left: sample Postoffice #8 in OM visible light; SEM electron image and spectrum of the intial glossy white layer; sample Vochanan #2 in OM visible light; SEM electron image and spectrum of the intermediary glossy white layer; OM in visible light of sample Relief 74-4 #1. The initial paint layer in the Postoffice relief was determined identical to an early layer in the Vochanan relief (denoted by blue squares). In the Vochanan relief the uppermost matt white paint layer could be related to those seen in his later white works (denoted by green circles)

This research showed that Schoonhoven's shift in 1960 from color to white was not as abrupt as generally assumed in retrospective literature. It adds up to recent art historical research into various unknown reliefs from the period 1958–1960. Schoonhoven already applied a monochrome white on the relief 'Barbara' in 1958, still used a glossy enamel on his 'Servaas" relief from 1959, matte white on relief R59-1 (1959) and on two informal small reliefs R60-20 and R60-21 from 1960 [3]. Taking into account the regular 'freshing-up' of the reliefs by overpainting them in white, either by Schoonhoven himself or NUL colleagues and later assistants, it will be difficult to prove either his specific choice for (glossy) enamel or (matte) wall paint for the monochrome white reliefs from the transition period, or a more indifferent approach in this.

Jan Schoonhoven's use of enamel paint in the early colored reliefs can render an original (partial) glossy surface, whilst for his white serial reliefs Schoonhoven definitively preferred a matte surface. In cleaning treatments these different surface qualities of the paint should be considered and taken into account.

References

1. Beerkens, L., P. Noble, and S. Weerdenburg. 2017. From Pollock drips to perfect white in Jan Schoonhoven's Zero reliefs. In ICOM-CC 18th Triennial Conference Preprints, Copenhagen, 4–8 September 2017, ed. J. Bridgland, art. 0903. Paris: International Council of Museums.
2. Learner, T., and J. Crook. "The impact of modern paints." *Tate, London* (2000).
3. Melissen, A. "Jan Schoonhoven" Nai publishers, Rotterdam (2015), pp. 30–53
4. M. T. Doménech-Carbò, M. J. Casas-Catalàn, A. Doménech-Carbò, R. MateoCastro, J. V. Gimeno-Adelantado, F. Bosch-Reig, Analytical study of canvas painting collection from the Basilica de la Virgen de los Desamparados using SEM-EDX, FT-IR, GC and electrochemical techniques, in Fresenius Journal Analythical Chemestry 369, 2001, pp. 571–575
5. I. Joosten, "Applications of microanalysis in the cultural heritage field", in Microchimica Acta 161:3, 2008, pp. 295–299
6. J.H. Townsend and K. Keune, Microscopical techniques applied to traditional paintings, in Infocus Magazine 41:1, 2006, pp. 54
7. A. Van Loon, Improving the surface quality of paint cross-sections for analytical imaging studies with SR-FTIR ad Static-SIMS, in Color changes and chemical reactivity in seventeenth-century oil paintings, PhD thesis, University of Amsterdam, 2008

Chapter 27
Some Considerations When Cleaning Robert Ryman's Oil Paint(ings)

Lise Steyn, Klaas Jan van den Berg, Maartje Stols-Witlox, Ella Hendriks, and Louise Wijnberg

> *If someone is only seeing white, then they're not really experiencing my paintings.*
>
> – Robert Ryman [2]

27.1 Introduction

The case study described in this article makes up part of the *Cleaning Modern Oil Paints* research project, in which it is acknowledged that research is needed not only into the challenges modern *paints* offer conservators, but also the challenges that modern *paintings* pose as complex cultural objects with multiple values. The subjective and sometimes unclear or conflicting values inherent to modern paintings complicate conservation and cleaning treatments beyond that of the materials themselves and need to be acknowledged. This article presents the outcomes of studying one such complex artwork; a modern unvarnished oil painting by the American artist Robert Ryman.

L. Steyn (✉)
Cultural Agency of the Netherlands, University of Amsterdam, Amsterdam, The Netherlands
e-mail: l.steyn@dkh.nl

K. J. van den Berg
Cultural Heritage Agency of the Netherlands, Amersfoort, The Netherlands
e-mail: K.van.den.Berg@cultureelerfgoed.nl

M. Stols-Witlox · E. Hendriks
University of Amsterdam, Amsterdam, The Netherlands

L. Wijnberg
Stedelijk Museum Amsterdam, Amsterdam, The Netherlands

K. J. van den Berg et al. (eds.), *Conservation of Modern Oil Paintings*,
https://doi.org/10.1007/978-3-030-19254-9_27

27.2 About the Artist

Robert Ryman was born in Nashville Tennessee in 1930.[1] He moved to New York in June 1952 to pursue his dream of becoming a professional jazz saxophonist [16, 31]. In order to practice and attend auditions he needed a flexible job. In June of 1953 he found a position as a security guard at the Museum of Modern Art (MoMA), remaining for 7 years and coming into contact with various artists and artistic movements [14, 31, 38].[2] He encountered the artist Mark Rothko around 1957 or 1958. Ryman was greatly inspired by Rothko, whose work, according to Ryman had "… no reference to any representational influence. There was color, there was form, there was structure, the surface, the light – the nakedness of it, just there. There weren't any paintings like that" [31].

27.3 Ryman's Paintings

Ryman's paintings are generally characterized by his signature use of white paint and his methodical painterly methods and precise treatment of the paint surface (Fig. 27.1). Colaizzi notes: "No single method of application monopolized Ryman's attention." Ryman stated in 1969 that there "is never a question of what to paint, but only how to paint. The how of painting has always been the image – the end product" [7]. For this reason the artist was very explicit about the display conditions of his works, stipulating that they be exhibited against a neutral wall and that they should not be framed or glazed, unless directed by the artist. Ryman preferred daylight to a static artificial environment and stated that everything that is visible during display is integral to the work and cannot be altered, down to every bolt, piece of tape or staple as well as the distance of the work to the wall or surrounding works [29]. Ryman stated that he spent time "looking at the paintings under different lighting conditions, as my paintings work differently with changes of light" [16, 25].[3] It is therefore especially interesting that the artist's studio in New York, where he worked for most of his career, has no natural light in the area where he painted them against three white walls (Fig. 27.2). This area is lit artificially.

---

[1] Ryman passed away in New York in February 2019.

[2] Among other lesser-known artists, Ryman met and befriended Sol LeWitt and Dan Flavin, both whom were then employed at MoMA. During Ryman's time there, the MoMA bought works by- or held exhibitions of works by Robert Rauschenberg, Henri Matisse, Kazimir Malevich, Franz Kline, Jackson Pollock, Willem de Kooning, Robert Motherwell and Mark Rothko, to name a few [31].

[3] The transformative effect of light on Ryman's work was illustrated in an article by Jessica Morgan ([16], pp. 10–18) using photographs taken of one painting at different times of the day. In the photographs one can see the appearance of the painting change throughout the day due to the shifting of light. These effects are even more pronounced when the artworks are viewed in person.

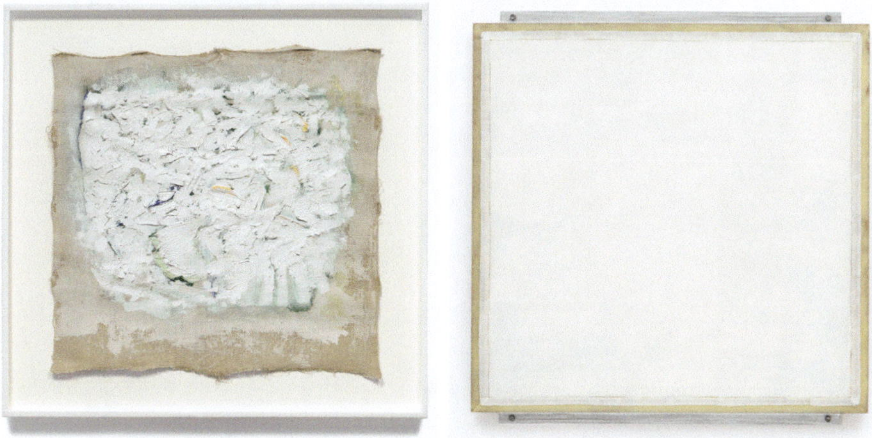

Fig. 27.1 (**a**) **Left**: Robert Ryman, *Untitled #1003*, 1960–61. Oil and gesso on unstretched linen canvas. 42 cm × 42 cm. The Greenwich Collection, Ltd. (**b**) **Right:** Robert Ryman, *Bond*, 1982. Enamelac on fiberglass and aluminum honeycomb panel with redwood edge with two aluminum fasteners and four hexagonal bolts panel: 76 cm × 76 cm.; overall: 81 cm × 76 in. The Greenwich Collection, Ltd. (Photographs by Bill Jacobson Studio, courtesy The Greenwich Collection, Ltd. All works by Robert Ryman © Robert Ryman/Artists Rights Society (ARS) New York)

Due to their interplay with reflected and absorbed light, the full effect of Ryman's paintings is difficult to document using photography. It is only when sharing a physical space with his artworks that one observes this full variety of surface appearances. The lack of figurative elements and colour in Ryman's paintings mean that this relationship of matte and gloss surfaces is extremely important to the experience of his work. Ryman states that his paintings:

> [...] are different when viewed from the left or the right side obliquely or from the front face. When they are photographed, for a record, there are three photos for each painting – left, right, and front. During the process of painting I will view them from both sides and move them from one wall to another wall in order to see the paintings under different conditions. [25]

Furthermore, wall labels should not be displayed alongside works as this could imply that the titles have a conceptual grounding, which they do not. Originally Ryman did not title his artworks, but it became necessary in order to differentiate between them for recordkeeping. The paintings are often accompanied by detailed installation and display instructions.

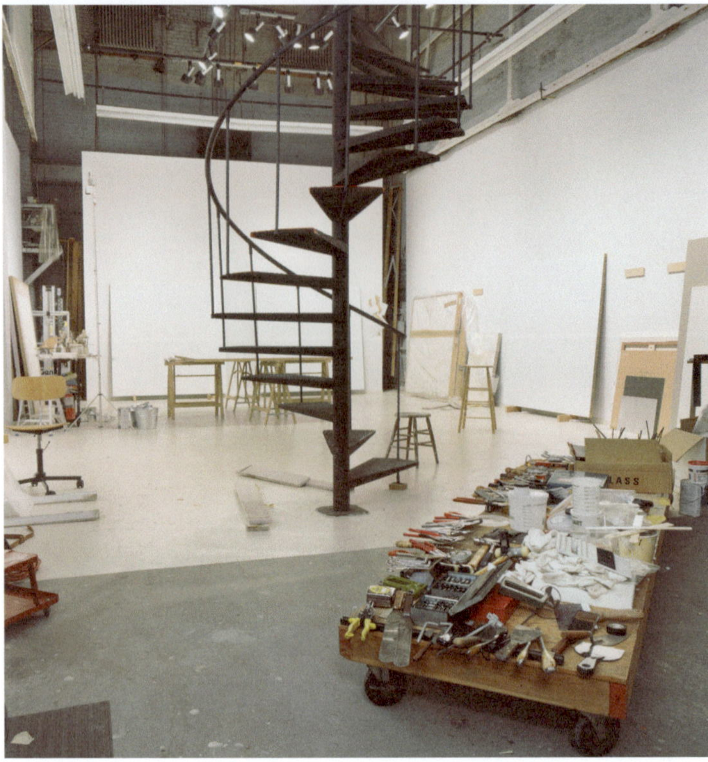

Fig. 27.2 Robert Ryman's Greenwich Studio in 1998. (Photograph by Bill Jacobson Studio, New York)

27.4 Ryman's Intent

The planned appearance of an artwork, its conceptual meaning and context as expressed by an artist, which is commonly referred to as the 'artistic intent', is an important and complex topic in conservation [11, 18, 24, 35, 37]. Conservators endeavour to preserve multiple values of an artwork [1, 18, 22]. Often however, these values are in conflict with one another and decisions must be made about which values to prioritize. Several decision-making models have been developed to assist conservators in dealing with these complex choices [1, 12, 15, 18, 28, 30, 35, 36]. In these models, and in the conservation of modern and contemporary art in general, artistic intent is a factor of major importance. Conservators often interview living artists about their materials, techniques, thought-processes, motivation and artistic concepts [3]. Ryman gave frequent interviews throughout his career to journalists, gallerists, art historians and conservation professionals. In one such videotaped interview titled: *Four Artists: Robert Ryman, Eva Hesse, Bruce Nauman, Susan Rothenberg (1987),* he describes three "procedures" in art, as he calls them;

representational, *abstract* and *realist* [4]. For Ryman, "representational" art signifies a narrative world confined to a frame that the viewer is drawn into, such as a landscape or naval battle scene. He describes "abstract" art as still utilising the rules of representational art by means of colours, narrative, association and illusion to elicit a response in the viewer; this illusion is often still isolated from the wall by a frame. Ryman's third procedure is "realist", because what it represents is real. There is no illusion, no picture, no narrative; the aesthetic experience is real and unconnected to the external world. He describes this "realist procedure" as the intentional drive behind his art, and what conservators would call 'artistic intent'.

27.5 White Paintings

Though Ryman's earliest works show some exploration of colour, from the mid-1960s onward his surfaces are predominantly white [8].[4] His paintings show no distinguishable narrative elements and, unsurprisingly, are often associated with monochrome art, minimalism and the Zero art movement. However, Ryman rejects these comparisons [17]. He is adamant that he does not paint white paintings. Rather, the lack of emotive colour and identifiable shapes is a means by which the viewer is forced to concentrate on the paint, its surface, texture, gloss, as well as the size and shape of the picture plane, and the work's proximity to the wall or other material elements of his technique [19]. Questioned in a 1971 interview with Phyllis Tuchman in *ArtForum* if he paints "white paintings", Ryman responded:

> No. It may seem that way superficially, but there are a lot of nuances and there's colour involved. Always the surface is used. [...] It's really not mono-chrome painting at all. The white just happened because it's paint and it does not interfere. I could use green, red, yellow, but why? It's a challenge for me to use paint and make something happen with it, without having to be involved in reds, greens, and everything which would confuse things. But I work with color all the time. I don't think of myself as making white paintings. I make paintings; I'm a painter. White paint is my medium [32].

In an article for *Art News* in 1986, Nancy Grimes quoted the artist as saying that white is "a neutral color that allows for a clarification of nuances in painting. It makes other aspects of painting visible that would not be so clear with the use of other colors" [13]. Thirty-one years after the Tuchman interview the artist was still unwavering about his use of white paint, as could be found in an exchange with Amy Baker Sandback in June 2002:

> Amy Baker Sandback: "The first question in the ArtForum interview is whether or not you make white paintings. Thirty years later the same question is still asked."
>
> Robert Ryman: "If someone is only seeing white, then they're not really experiencing my paintings. I think people are so used to looking into paintings as they would pictures and so

[4] With the exception of his series entitled *White Paintings* (2003-2005), which ironically incorporates the colours blue and balck.

> when they see white paint, they think that's the picture image. It is not, the white is just part of the structure of the painting in order to make things visible" [2].

Ryman has been clear about the lack of conceptual meaning behind his use of white paint throughout his career. The colour is not relevant other than to bring focus to the surface and to the "act of painting itself" [32].[5] The only exception occurred in a series of paintings from 2003-2005 in which the artist expressly stated that he had been asked about it so often that he had decided to give painting "white paintings" a try [25]. It is clear then that the artist did not endeavour to paint "pictures" [32], but rather to confront the viewer with the *materiality of paint*. The material is what demands consideration. Ryman was not the first artist to explore 'art for art's sake' or the first to paint works that bring focus on the materials rather than on a depicted scene. In fact, when asked in 1971 whether he was trying to "get away from the notion of object-hood", Ryman replies that during the 1950s and 1960s, the point "had really been gotten across strongly that painting was an object. So, I wanted to make a painting getting the paint across" [32]. Yet his works are still often compared to other artists who explored the theme of the 'white painting' such as Kazimir Malevich, Piero Manzoni, Ellsworth Kelly, Robert Rauschenberg and Josef Albers, rather than to an artist like Mark Rothko whose artworks are intrinsically closer to Ryman's *realist procedure* than any 'white painting'. Colaizzi [7] discusses this at length, stating:

> Rothko's own famous denial of the primacy of color in the interest of avoiding what he saw as trivial and decadent sensuality caused him to quip, 'Since there is no line, what is there left to paint with?' The missing term of this equation, between color and line, is *paint*: the substance whose form is always visibly conditioned by the hand. Unlike Ryman, Rothko never divorced paint from color.

What sets Ryman's work apart from other 'white paintings' is therefore his dedication to confronting the viewer with the materiality of painting, *paint* rather than a picture or conceptual construct.

27.6 Patina, Changes and Signs of Age

According to Ryman, the colour of the painting is not central to the understanding and enjoyment of his work, thus one might conclude that any treatment aimed at retaining artistic intent by retaining 'whiteness' would appear unjustified. But 'white' and 'clean' are different concepts. Did the artist intend for his paintings, even if 'whiteness' is not the focus, to appear clean and new? Did Ryman encourage surface dirt removal for this reason? Again, the prolific amount of detailed information from interviews with the artist proved invaluable. In a 2009 interview by Julie Barten, Ryman was asked about patina:

---

[5] In answer to the question, 'Does your use of white have any symbolic or mystical significance' Ryman simply responds "No."

> Julie Barten: "You've said over and over again that you like the effect of patination and [that] a little subtle layer of grime sometimes enhances the aesthetic. So, I think that could end up confusing conservators a little bit because the effect of patination can really cause subtle shifts in what are already subtle relationships among your paints."

> Robert Ryman: "Well yes that's true, yes. Well I just think things shouldn't be over-cleaned. It's all right to have, I mean things get dirty after a while, it's just natural. And as long as it doesn't really interfere with the aesthetic, it's all right. If it gets overlaid, dirty, or really starts to interfere, then it has to be cleaned. [...] Yeah, it doesn't have to look fresh and new all the time like it was just done. [...] But yeah, it's hard to say when something should be cleaned and when you just leave it. I guess some things through the years are going to develop a little crack maybe, or a little blemish. But I think that's OK. It's just a natural thing" [27].

Interestingly, when asked about a certain developing area of discolouration on a paper work Ryman stated: "I think I wouldn't do anything to it now, I mean, maybe if it got... even if it got darker, it would still be probably alright, as long as there weren't really strange things developing in the paper." When pressed further about the point at which discolouration would become 'unacceptable', Ryman responded that when discolouration became too noticeable, it might interfere because a discernible *shape* would appear. This exchange indicates that Ryman was mostly concerned with soiling once it started to create recognisable shapes and distracted the viewer from the paint. Though it remains difficult to establish at what point the artist considered a work "dirty" or when dirt started "to interfere", it appears that some restraint should be used when deciding if a Ryman is "dirty" enough to warrant treatment – or not. In the following sections we will explore how the above information relates to various values encountered in a case study painting by Robert Ryman.

27.7 Case Study

Untitled (1960) is a modern largely monochrome oil painting from the collection of the *Stedelijk Museum Amsterdam* (Fig. 27.3). The painting measures 1.65 m × 1.65 m and consists of two tabby weave cotton canvases stretched onto separate wooden stretchers that have been securely connected to each other with three metal bolts at the back.[6, 7] This diptych is an early example of Ryman's use of multiple elements to form one whole [5, 29]. The vertically oriented painting is divided in two-third and a one-third parts. The left stretched canvas combines a large area of impastoed swatches of green, blue, yellow, red and purple paint that has been partially covered in more articulated white paint with a small vertical area of exposed bare ground on the right (Fig. 27.4).[8] The right canvas consists of an even layer of white paint

[6] Samples of the support were identified as cotton by Bas van Velzen on 22 August 2017 using his *Modified Herzog Test* method [34].

[7] Not identified. Consists of a light-coloured wood type.

[8] 164,7 cm × 109 cm.

Fig. 27.3 Robert Ryman. *Untitled*, 1960, Collectie Stedelijk Museum Amsterdam. (Photograph by René Gerritsen 2017)

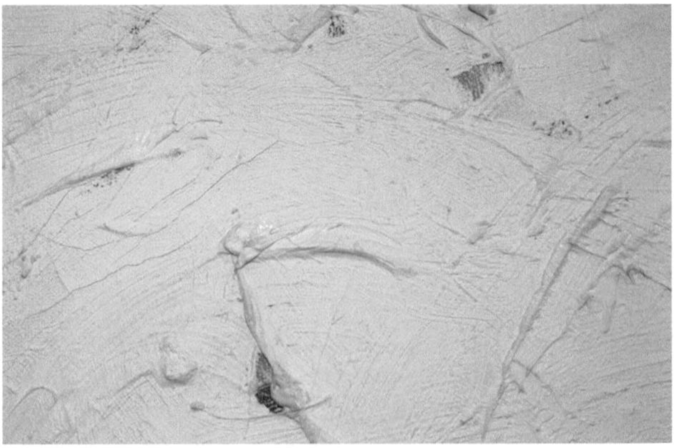

Fig. 27.4 Detail of coloured swatches under the white paint

applied with a brush.[9] The artist played with the delicate organisation of the tripartite composition in relationship to the two canvasses, which together form a square. Both canvasses have a white-grey coloured ground. This is left visible in a prominent area of the left canvas and is therefore given the same visual status as the painted areas, together forming the final composition. It is not known if Ryman

[9] 164,7 cm × 56,2 cm.

applied the ground himself. It is more thinly applied on the tacking margins than on the face of the painting and there are areas where thin paint appears to have been applied to the tacking margins. The overall condition of the painting appears to be stable and no urgent treatment issues were observed. There are no signs of delamination or flaking of either the paint or ground layers.

27.8 Concerns

The museum acquired *Untitled* (1960) in 1972. In 1973 surface dirt was removed using distilled water.[10] In 1990 surface dirt was removed using a weak alkaline solution and a small brush.

White museum walls make the spectator aware of the ageing of the work. This is especially evident on the borders of Ryman's work where it is in close contact with the wall and the surface has become noticeably and unevenly soiled over time.[11] In 2012 surface dirt was removed from the tacking margins using three parts water, one-part citric acid and one-part surfactant. Although conservators at the *Stedelijk Museum Amsterdam* adhere to have a policy of minimal intervention, they grew concerned that the painting had become noticeably unevenly soiled and that an overall cleaning treatment may be justified. A deeper understanding of the effects of cleaning is of particular importance for Ryman's paintings given his clear instructions that his paintings must not be framed (unless directed by the artist), as "the wall is necessary for the paintings to be complete" [25]. This means that the surface of the painting will remain exposed to atmospheric dirt while on display and therefore will continue to become dirty in the future.[12] This raised concerns about the possible long-term effects of repeated cleaning treatments.

27.9 Technical Analyses – The Fluorescing Paint Surface

A condition check of the painting was conducted using daylight, raking light, ultraviolet induced luminescence and microscopy. Some small aqueous cleaning tests were conducted on the edge of the painting in order to establish the solubility of the dirt and the appearance of the paint layer after its removal. The dirt could easily be

[10] From the condition record, *Stedelijk Museum Amsterdam*. Reference is made to cleaning the impastos, but it is not clear from the condition report whether only the left or both canvases were cleaned at this time.

[11] With unframed paintings from the 1960s and 1970s by artists such as Barnett Newman or Brice Marden, the tacking margins gained a particular importance. The thickness of the stretcher and colour is part of the painting because it is what causes the work to leave the wall plane and make it possible to view it from different areas.

[12] When in storage, measures can be taken to minimize the work's exposure to atmospheric dirt by placing it in a closed storage frame.

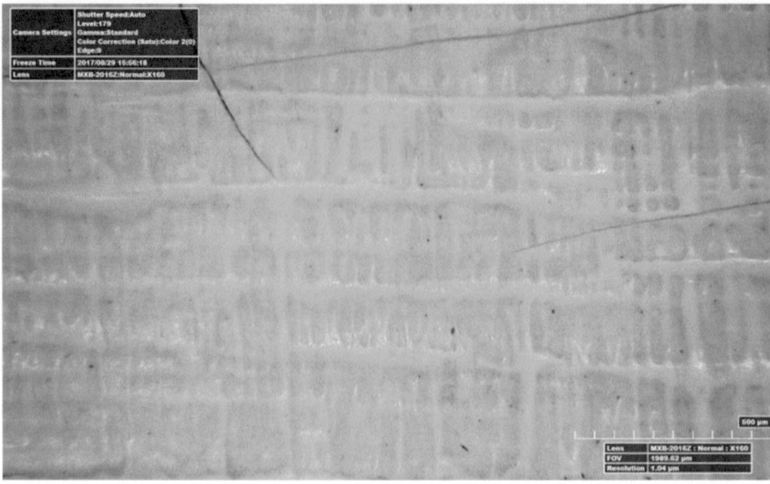

Fig. 27.5 Evaluation of the cleaning test (middle) using a HIROX RH-2000 microscope at a magnification of ×160. The surface gloss did not appear to have been affected by the treatment

removed on the right (flat) canvas using demineralised water and cotton swabs (Fig. 27.5). The paint surface is not visibly sensitive to the treatment and appears to remain intact.

The build-up and nature of the paint was investigated using embedded paint samples and an optical microscope (Fig. 27.6). The distribution of paint particles appeared uniform throughout the paint and no layer build-up could be identified. It appears that the artist applied the paint and did not work it or add different layers of white paint on top of one another. This is congruent with Ryman's statements that: "Usually, I never 'work' paint as I prefer to put paint down and leave it as it is" [25]. In order to establish the composition of the paint, samples were analysed using various methods.[13] Gas Chromatography Mass Spectrometry (GCMS) established that the binding medium of the paint was a drying oil.[14] Analyses using X-ray Fluorescence (XRF) and a Scanning Electron Microscope with Energy Dispersive X-Ray Spectroscopy (SEM-EDX) confirmed that the ground layers of both canvasses were the same, and contained calcium, sulphur, silicon and titanium. XRF

[13] XRF, XRD (and XPS) analyses and interpretation of the results were performed by Birgit van Driel, The results were documented in an unpublished report entitled: *X-ray analyses of Robert Ryman's Untitled 1960*. [10]

[14] The GCMS spectrum (not shown) gives a P/S ratio of 1.2, which would be indicative of linseed oil in traditional paints. However, it is known that for zinc white and other white paints non-yellowing oils are used. Ryman is known to have used Winsor White from Winsor&Newton often in his work ([27], p.26). Winsor&Newton used poppy oils until mid-1966 (Ian Garrett, formerly Winsor&Newton, verbal communication, to Klaas-Jan van den Berg 18 January 2019). Since non-yellowing drying oils, especially poppy, have a higher P/S ratio than linseed, the present value can only be explained by the admixture of a metal stearate to the paint by the manufacturer, which could lower the overall P/S ratio [33].

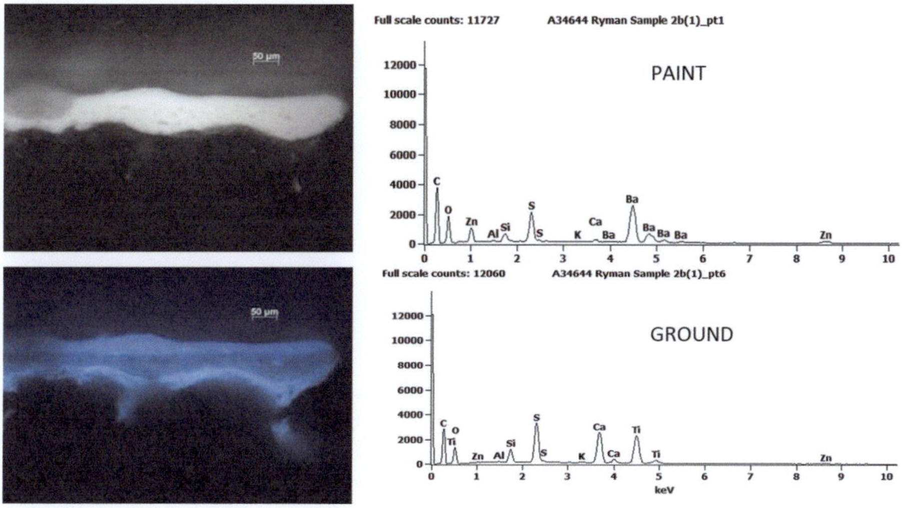

Fig. 27.6 Sample analyses of *Untitled* (1960). Top left: cross-section shows the uniformity of the paint layer as seen in bright field, and below the same sample under ultraviolet radiation. On the right: SEM-EDX spectra show the composition of the paint and ground layers

was used to identify the coloured paint swatches as Chrome yellow, Chrome green and Prussian blue.[15]

SEM-EDX analyses of several samples confirmed that one type of white paint had been used on both canvases. This consisted of titanium dioxide, zinc oxide, barium sulphate and calcium carbonate with a drying oil binder, likely linseed oil. Further analyses using X-Ray Diffraction (XRD) showed that the type of titanium dioxide used throughout the painting was of the anatase variant. As anatase titanium white can be vulnerable to photocatalytic degradation when exposed to ultraviolet radiation, [9] exposure should be avoided. However, ultraviolet radiation induced fluorescence had already been used as part of a condition check process prior to having this insight. This revealed crucial information that improved our understanding of the condition of the work and influenced the choice of safe methods of treatment.

The painting exhibits the distinct greenish fluorescence of zinc oxide white when exposed to ultraviolet radiation [6, 26]. It was observed that two separate areas did not show this. These areas were not apparent in daylight and appear to be different due to external causes. One area, starting at the top tacking margin of the left canvas, seems to be the result of a liquid that had dripped onto the painting (Fig. 27.7a). The dripping material left a tide line of a faintly fluorescent material, which was identified as a carbohydrate using GCMS analysis. The other dark area (or area of 'quenched' fluorescence) consisted of a palm-sized shape located on the right

[15] Unfortunately, the composition of the purple and red paints could not be identified with analysis of inorganic components; they probably consist of organic pigments.

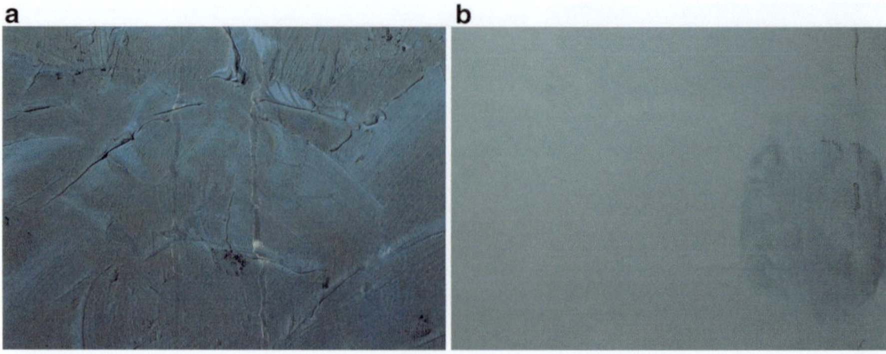

Fig. 27.7 (**a**) Detail of the 'drip area' observed under ultraviolet light, showing tidelines caused by the fluid. (**b**) Detail of what appears to be an old cleaning test under ultraviolet radiation

canvas (Fig. 27.7b). Within this shape, circular motions could be observed that are reminiscent of the motions used when removing dirt or varnish from a painting using a cotton swab. This may be the result of a previous cleaning treatment, demonstrating that cleaning may indeed change this aspect of the paint, even if it this is not visible under normal lighting conditions. Several experiments were conducted on samples from *Untitled* (1960) to investigate the nature of the phenomenon.[16] The preliminary results indicate that the paint binder isolates the bulk of the pigment particles from interaction with atmospheric moisture and molecular oxygen. It is presumed that aqueous treatments allow for the migration of water further into the bulk of the paint. Water molecules are thought to react with the oxygen vacancies on the zinc white surface, leading to a decrease in their concentration. As oxygen vacancies are related to the zinc white's green fluorescence, introducing water into a zinc white paint film is therefore hypothesized to reduce the paint's green fluorescence [20, 21]. A further indication that such an effect is possible is the response of cross-sections of this paint to air-exposure. When cross-sections of paint are dry-polished, their bright green fluorescence is observed to decrease in intensity within an hour after polishing. This is limited to the surface, because when re-polished, the initially observed fluorescence returns.[17]

[16] Experiments were conducted by Selwin Hageraats, who is currently conducting PhD research into the fluorescent properties of zinc oxide pigments at the University of Amsterdam.

[17] Research into this topic is ongoing and will be explored in greater detail in Steyn's PhD dissertation.

27.10 Discussion and Conclusions

Conservators must take into account a multitude of values held by different stakeholders when deciding whether or not to treat paintings to improve their aesthetic appearance [1]. One such value, known as artistic intent, is often prioritised and incorporated into decision-making by conservators in an attempt to retain the original meaning of an artwork, rather than just its appearance. That being said, no treatment can truly retain the original meaning of an artwork as it moves through time, increasingly further away from its original context and audience, necessarily resulting in compromise. One can infer from interviews with Ryman that restraint should be exercised when deciding whether or not to clean his paintings. He stated: "I just think things shouldn't be over cleaned. [...] things get dirty after a while, it's just natural. And as long as it doesn't really interfere with the aesthetic, it's all right." At the same time though, Ryman acknowledges that the point when dirt starts to interfere with the aesthetic is subjective and difficult to pinpoint, stating: "If it gets overlaid, dirty, or really starts to interfere, then it has to be cleaned" [27].

Ryman enlisted trusted conservators to clean his paintings when he found that they had reached this tipping point and had become too dirty [16, 27]. It follows that for *Untitled* (1960) too, when an uneven distribution of dirt distracts viewers to the extent that they are drawn out of their interaction with the artwork, this could justify a decision to clean. This study has shown that even though surface dirt removal on stable oil paint surfaces is generally considered to comply with a wish for minimal intervention, in fact this may not always be the case. For *Untitled* (1960), surface treatment using aqueous methods does not result in changes visible in daylight, however the changing UV fluorescence indicates that chemical changes may be occurring, the long-term effects of which are not yet known.

Glazing Ryman's paintings contradicts the artist's intent; therefore the paint surface needs to remain unprotected while the work is on display. This means that recurrent cleaning treatments are to be expected. Alternatively selective cleaning could be considered as opposed to overall cleaning. However, selective cleaning may result in surface dirt of differing composition and thicknesses, which may age and interact with treatments differently. This could possibly lead to dirt becoming unevenly ingrained and it becoming more difficult to achieve an overall even appearance with cleaning in the future.

In order to investigate the effects of repeated surface cleaning further, we are currently experimenting with the cleaning cycle of oil paints by conducting multiple cleaning, soiling and ageing tests on reconstructions and evaluating different levels of dirt adhesion. If results indicate that dirt removal becomes more difficult the longer dirt remains on the surface, this could be an argument in favour of frequent cleaning of Ryman's paintings. However, as discussed above, Ryman's attitude towards cleaning may provide equally valid arguments for extending the time between surface cleanings. This would additionally lower the possible risk of abrasion of the paint layer and mechanical damage to impastos. These risks are currently being investigated further.

Case studies such as this are instrumental in determining our attitudes towards cleaning. These decisions require conservators to examine their own beliefs, and deal with the complex relationship between the material and immaterial aspects within the same work that must be examined and negotiated anew for each treatment. Ryman's works are made with great craftsmanship, precision and attention to detail, resulting in a very pleasing aesthetic. His works initially leave his studio in pristine white condition, which makes it seemingly contradictory that he would not want this immaculate appearance to be preserved. It is therefore understandable that conservators who are unaware of Ryman's statements about the irrelevance of colour to his work and his views on patination and ageing could easily conclude that the maintenance of the colour and cleanliness of his paintings would be integral to the preservation of the meaning of the painting. There are multiple stake-holders when it comes to art, and artistic intent is only one of the issues that conservators consider when making treatment decisions. We hope that by discussing cleaning issues as they pertain to this case study by Robert Ryman we may contribute to this complex dialogue and make others aware of our findings. We found what was so eloquently once stated by Phaedrus: "things are not always what they seem" [23].

Acknowledgements This research was undertaken as part of the Cleaning Modern Oil Paints project (2015–2018) supported by Heritage Plus, a Joint Programming In-itiative of the European Commission. We would like to thank our Cleaning Modern Oil Paint project partners and research group for their valuable contributions to this research, in particular the Stedelijk Museum Amsterdam who supplied the case study painting discussed in this article. We are grateful to the following persons (in alphabetical order): Sandra Amann, Julie Barten, Madeleine Bischoff, Vera Blok, Meta Chavannes, Dana Cranmer, Paula de Cristofaro, Suzan de Groot, Michael Duffy, Isabelle Duvernois, Elizabeth Estabrook, Emilie Froment, David Gray, Mary Gridley, Selwin Hageraats, Ruth Hoppe, IJsbrand Hummelen, Ineke Joosten, Cecil Krarup-Andersen, Katrien Keune, Tom Learner, Margje Leeuwenstein, Frank Ligterink, Carol Mancusi-Ungaro, Luc Megens, Alan Phenix, Maja Rinck, Rachel Rivenc, Laura Rivers, Kirstin Robinson, Mikkel Scharff, Tatja Scholte, Robert Storr, Philip Tan, Birgit van Driel, Henk van Keulen and Sandra Weerdenburg,

References

1. Appelbaum, B., 2007. *Conservation Treatment Methodology* 1st ed., Oxford: Butterworth-Heinemann.
2. Baker Sandback, A., Tuchman, P. & Ryman, R., 2002. *Robert Ryman* G. Frei, ed., Zurich: Thomas Amann Fine Art Zurich.
3. Beerkens, L. & Abraham, L., 2012. *The Artist Interview : For Conservation and Presentation of Contemporary Art, Guidelines and Practice*, Amsterdam: Jap Sam Books.
4. Blackwood, M., 1987. *Four Artists: Robert Ryman, Eva Hesse, Bruce Nauman, Susan Rothenberg*, United States: Michael Blackwood Productions.
5. Clément, C., 1990. *Polyptyques: le tableau multiple du Moyen Âge au vingtième siècle : Musée du Louvre, Paris, 27 mars-23 juillet 1990.*, Paris: Ministère de la culture, de la communications, des grands travaux et du bicentenaire, Réunion des musées nationaux.
6. Clementi, C. et al., 2012. Photoluminescence Properties of zinc oxide in paints: A study of the effect of self-absorption and passivation. *Applied Spectroscopy*, 66(10), pp.1233–1241.

7. Colaizzi, V., 2017. *Robert Ryman*, New York, N.Y.: Phaidon. Available at: http://de.phaidon.com/store/art/robert-ryman-9780714849348/ [Accessed January 8, 2018].
8. Colaizzi, V. & Schubert, K. eds., 2009. *Robert Ryman: Critical Texts Since 1967*, London: Ridinghouse.
9. Driel, B.A. van, 2018. *Titanium White, Friend or Foe?: Understanding and Predicting Photocatalytic Degradation of Modern Oil Paintings*. Technical University of Delft.
10. Driel, B.A. van, 2016. *X-ray Analyses of Robert Ryman's Untitled 1960.*, Amsterdam.
11. Dykstra, S.W., 1996. The Artist's Intentions and the Intentional Fallacy in Fine Arts Conservation. *Journal of the American Institute for Conservation*, 35(3), pp.197–218.
12. Ferretti, V., Bottero, M. & Mondini, G., 2014. Decision Making and Cultural Heritage: An Application of the Multi-Attribute Value Theory for the Reuse of Historical Buildings. *Journal of Cultural Heritage*, 15(6), pp.644–655. Available at: https://doi.org/10.1016/j.culher.2013.12.007.
13. Grimes, N., 1986. Robert Ryman's "White Magic." *Art News*, p.89.
14. Harvey, M., 2010. MoMA I "ART WORK": Famous Former Staff. *Inside / Out*. Available at: https://www.moma.org/explore/inside_out/2010/07/01/art-work-famous-former-staff/ [Accessed October 11, 2018].
15. Henderson, J., 2011. Reflections on Decision-making in Conservation. *ICOM-CC 16th triennial conference Lisbon 19-23 September 2011: preprints.*
16. Hoban, S. et al., 2017. *Robert Ryman* S. Hoban & C. J. Martin, eds., New York: Dia Art Foundation.
17. Hudson, S.P., 2009. *Robert Ryman : Used Paint*, MIT Press.
18. Hummelen, I., Sillé, D. & Howland, A., 1999. *Modern Art : Who Cares ?*, Amsterdam: Foundation for the Conservation of Contemporary Art (SBMK) / Netherlands Institute for Cultural Heritage (ICN).
19. Kappeler, F., Lauchli, M. & Rausmüller, U., 1993. *Talk with the Artist: Robert Ryman*, Hallen für neue Kunst.
20. Lv, J. & Li, C., 2013. Evidences of VO, VZn, and Oi Defects as the Green Luminescence Origins in ZnO. *Applied Physics Letters*, 103(23).
21. Meng, X. et al., 2010. Temperature Behavior of Electron-acceptor Transitions and Oxygen Vacancy Recombinations in ZnO Thin Films. *Journal of Applied Physics*, 107(2).
22. Muñoz Viñas, S., 2005. *Contemporary Theory of Conservation*, Elsevier Butterworth-Heinemann.
23. Plato & Riley, H.T., The Fables of Phaedrus: Book IV The Weasel and the Mice. Available at: https://www.gutenberg.org/files/25512/25512-h/25512-h.htm [Accessed November 2, 2018].
24. Preissler, M.A. & Bloom, P., 2008. Two-year-olds Use Artist Intention to Understand Drawings. *Cognition*, 106(1), pp.512–518.
25. Robert Ryman, 2004. *Pace Gallery: Robert Ryman* Exhibition., New York: PaceWildenstein. Available at: https://www.pacegallery.com/exhibitions/11902/robert-ryman [Accessed June 14, 2018].
26. Romani, A. et al., 2010. Fluorescence Spectroscopy: A Powerful Technique for the Noninvasive Characterization of Artwork. *Accounts of Chemical Research*, 43(6), pp.837–846.
27. Ryman, R. et al., 2009. *Interview with Robert Ryman in His Studio 10 November 2009. Researchers from Getty Conservation Institute, Solomon R. Guggenheim Museum New York and Dia:Beacon (unpublished report)*, New York.
28. SBMK, 1999. Model voor Besluitvorming bij de Conservering en Restauratie van Moderne Kunst . Available at: https://www.sbmk.nl/uploads/model_voor_besluitvorming.pdf
29. Spector, N., De Wilde, E. & Van Beijeren, G., 1974. *Robert Ryman*, Amsterdam: Stedelijk Museum Amsterdam.
30. Stigter, S., 2016. *Between Concept and Material Working with Conceptual Art: A Conservator's Testimony*, Amsterdam: University of Amsterdam.
31. Storr, R., 1993. *Robert Ryman*, London: Tate Gallery London.
32. Tuchman, P., 1971. An Interview with Robert Ryman. *Artforum.*

33. Van den Berg, K.J., Burnstock, A. and Schilling, M. 2019. Notes on metal soap extenders in modern oil paints - history, use, degradation and analysis. in "Metal Soaps in Art – Conservation & Research", Eds F. Casadio et al. Springer. pp. 329–342.
34. Velzen, B. van, 2018. *The Universe Between Felt and Wire: A New Look into the Typology of Western Made Paper*, Amsterdam: University of Amsterdam. Unpublished.
35. Wetering, E. van de, 1978. Model for Taking Decisions in Conservation-Restoration: Theoretical Considerations with Respect to the Completion of Works of Art. In *Actes du cours international pour les restaurateurs*. Veszprém, pp. 15–27.
36. Wharton, G., 2015. Artist Intention and the Conservation of Contemporary Art. *AIC Objects Specialty Group Postprints*, 22, pp.1–12.
37. Wimsatt, W.K. & Beardsley, M.C., 1946. The Intentional Fallacy. *The Sewanee Review*, 54(3), pp.468–488.
38. Wolfs, R.C., 1988. *Robert Ryman: Verf Schilderen*. University of Amsterdam.

Chapter 28
Challenges of Surface Cleaning Paintings by Asger Jorn (1914–1973): An Inventory of Existing Practice

Juul Bierings, Lise Steyn, Maartje Stols-Witlox, and Klaas Jan van den Berg

28.1 Introduction

Asger Jorn (1914–1973) is a Danish artist who painted from the late 1930s until his death in 1973, with workshops in Silkeborg, Paris, Munich and Albissola [6]. Conservators frequently encounter difficulties when surface-cleaning his works, as Manya Lee-Jensen established in her dissertation [8]. However, information related to treatment options is scarce. This paper approaches these challenges from a phenomenological perspective by surveying different degradation phenomena that characterize Jorn's paintings, and by comparing the approaches to and methods chosen for surface cleaning. The outcomes may be applicable to modern oil paintings in a broader sense.[1]

[1] This research was part of the Cleaning Modern Oil Paints (CMOP) project.

J. Bierings
Freelance Paintings Conservator, Eindhoven, The Netherlands
e-mail: info@juulbierings.com

L. Steyn · K. J. van den Berg (✉)
Cultural Heritage Agency of the Netherlands, Amsterdam, The Netherlands

University of Amsterdam, Amsterdam, The Netherlands
e-mail: k.van.den.berg@cultureelerfgoed.nl

M. Stols-Witlox
University of Amsterdam, Amsterdam, The Netherlands

K. J. van den Berg et al. (eds.), *Conservation of Modern Oil Paintings*,
https://doi.org/10.1007/978-3-030-19254-9_28

28.2 Methodology

Nine interviews were conducted with experienced conservators working throughout Europe (Table 28.1). Conservators were asked about decision making, materials, results and challenges encountered during surface cleaning treatments of Jorn's

Table 28.1 Interviews

Interviewee	About	Method	When/where
Andersen CK (Assistant Professor, Pictorial art, The Royal Danish Academy of Fine Arts, Schools of Architecture, Design and Conservation, School of Conservation, Copenhagen, Denmark)	Experience through multiple treatments and visual investigation mainly in the service of Statens Museum for Kunst	Interview in person, with restoration reports, partly with objects	6-12-2016, Copenhagen
Baadsgaard E (Conservator of modern and contemporary paintings, head of department, Kunstkonserveringen, Denmark)	Experience through multiple treatments and visual investigation	Interview in person, with objects	8-12-2016, Århus
Bosc M (Freelance paintings conservator, France)	Treatment 2 paintings and experience gained during self-employment and work for Cobra Museum for Modern Art, Amstelveen	Interview via Skype, with conservation reports	16-12-2016
Bracht E (Retired painting conservator Stedelijk Museum Amsterdam)	Supervision multiple treatments, before 2009	Interview in person, with conservation reports	21-10-2016, Amsterdam
Bronken IAT (Project leader main exhibitions, collection management department, Nasjonalmuseet for kunst, arkitektur og design, Oslo, Norway)	Experience through multiple treatments and visual investigation	Interview in person	25-11-2016, Amsterdam
Froment E (Lecturer Conservation and restoration of paintings, program coordinator, University of Amsterdam)	Treatment 1 painting, during internship at the Stedelijk Museum Amsterdam in 2004	Interview in person, with personal documentation and conservation report	19-10-2016, Amsterdam
King A (Conservator of modern and contemporary paintings, Tate. London, England)	Treatment 1 painting in 2012	Written questionnaire, with restoration report	16-12-2016
Kuvvetli F (Conservator of modern and contemporary paintings, Kunstkonserveringen, Denmark)	Experience through multiple treatments and visual investigation.	Written questionnaire	18-12-2016
Speldekamp L (Conservator Gemeentemuseum Den Haag)	Treatment 1 painting in 2004	Interview in person, with object	28-10-2016, The Hague

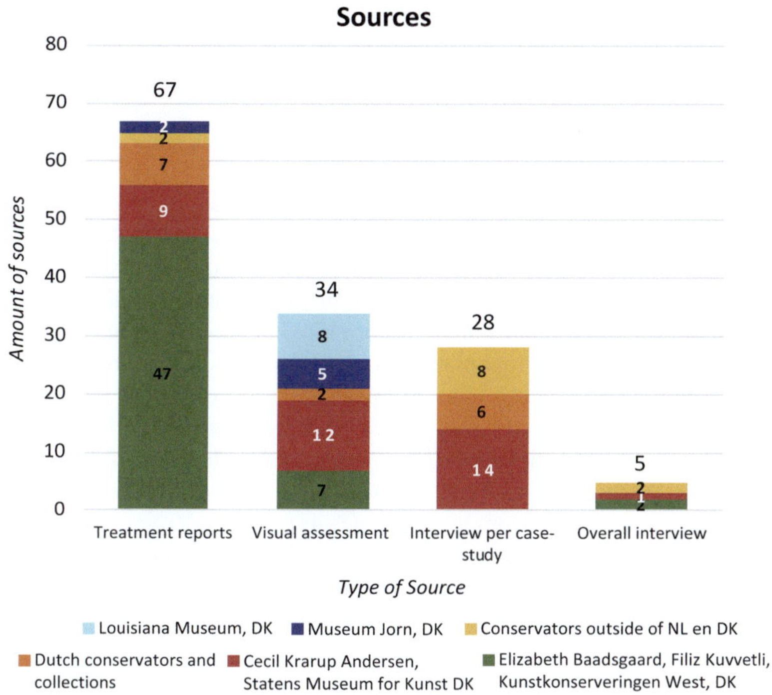

Fig. 28.1 Information sources

work. Half of the interviews were about specific conservations, the other half about general experience with cleaning Jorn's works. In addition, data was collected on 95 paintings (1939–1972), 61 of these paintings had been cleaned in the past. Figure 28.1 shows the sources provided by various institutions and individuals. Conservation reports and interviews with conservators are used to describe the restoration history of Jorn's paintings from 1999 to 2016. Thirty-four paintings were visually assessed with the naked eye, some were also studied under magnification, raking light and ultraviolet radiation.

28.3 Jorn's Materials and Working Methods

Jorn mostly applied pure colours, wet-in-wet in clearly visible brush strokes and impastos, opposed to his works prior to 1939 in which the paint was often thinly applied [2–5]. According to Guy Atkins, compiler of Jorn's *catalogue raisonné* and friend, Jorn experimented with different materials. Although mainly using oil, he also worked in media such as gouache, watercolour and acrylic, sometimes combining these with oil paint. Jorn bought both primed and unprimed canvases by the metre, and made use of other supports like masonite and cardboard [1]. According to Marie José van de Loo, daughter of Jorn's friend and Munich gallery owner, her father supplied Jorn with materials by Schmincke or Fritze Lacke when he was working in Munich, including paints, canvas, stretchers and frames.[2] Jacqueline de Jong, Jorn's partner from 1959 to 1970, said that in Paris he bought his paint at LeFebvre Foinet [13]. A paint tube on a preserved palette in Museum Jorn confirms this.[3]

De Jong mentioned Jorn's preference for working fast. To promote drying, he sometimes added siccative (by Lefranc Bourgeois) to his paint. She recalls Jorn added dammar and Venetian turpentine in his late work when he was able to afford them. Van de Loo remembers that Jorn diluted his paint with turpentine. Jorn did not varnish and often reworked his paintings. Jorn liked matte-gloss contrast, according to De Jong, so he added different materials to enhance this effect.[4] He created matte paint by adding sand or using gum Arabic, like often observed in yellow paint areas,[5] for instance in Jorn's 1962's 'Abstract minded figure' [10]. Although nowadays his thinly applied black paints often appear matte, Jorn aimed for an intense, saturated black, sometimes using furniture- and bicycle-lacquers and other synthetic materials in paintings from the 1960s [3, 4].[6]

Information from photographs further confirms Jorn's frequent use of commercial paint. Jorn's use of additives in oil paint leads to the question, whether these play a role in the pronounced degradation phenomena seen in his paintings.

28.4 Surface Phenomena

Jorn's paintings generally have complex surfaces with varying structures in specific colours. The different surface effects and degradation phenomena complicate cleaning: *'Texture has a lot to do with what it's like to clean a Jorn painting, flaking and texture is important, perhaps even more important than water-sensitivity because you cannot really get to the cleaning.'*[7]

The sensitivity of Jorn's paintings to mechanical action particularly complicates surface cleaning, its fragile impasto's requiring meticulous work in order to avoid breakage.[8] Age cracks were observed in 30 case-studies, leading to paint losses in 23 of them. This phenomenon is common in ultramarine blue, chromium oxide green and cadmium yellow, and manifests itself mainly in impastos. Sometimes the collapsing top layer reveals powdery underlying paint (Fig. 28.2a), which can be

[2] Wijnberg Froment [14].

[3] Palette from Jorn's atelier in Paris, 1972, collection Museum Jorn.

[4] Wijnberg Froment [14].

[5] Interview Andersen, Bracht, observations JB.

[6] Wijnberg Froment [14], interviews Andersen, Baadsgaard, Bosc, Froment, Speldekamp, observations JB, case-studies.

[7] Interview Andersen.

[8] All interviews, observations JB.

a b
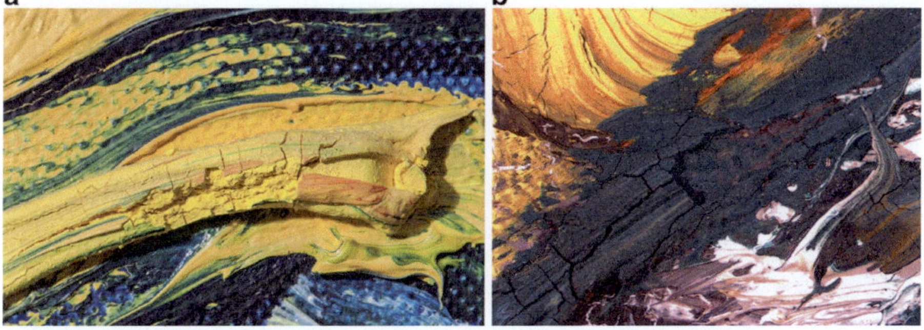

Fig. 28.2 (**a**) Paint loss in brittle impasto. Detail *Euphorism*, Asger Jorn, collection Museum Jorn, 1971/0274. (Photo: Cecil Andersen). (**b**) Detailed cracks and paint loss. *La Rose Féroce*, Asger Jorn, 1961, collection Gemeentemuseum Den Haag, 0333653. (Photo: JB)

misleading if the surface is still intact, because there is no visual indication that the underlying paint is so fragile.[9] With other cracks, the impastos with matt powdery surfaces have broken up into pronounced, solid flakes that easily detach from underlying paint layers, resulting in paint loss (Fig. 28.2b). Both types show the paint's vulnerability to mechanical action.[10]

Sensitivity of the paint to water, saliva and organic solvents is mentioned in one third of the painting treatments studied and in all interviews, sometimes further specified as polar solvents or alcohol. In particular matt, porous surfaces of impastos are mentioned as problem areas, mostly the dark blue, dark green and yellow paint areas.[11]

Almost all paints used by Jorn seem to be water-sensitive to a certain degree. Conservators do distinguish between more sensitive dark and less sensitive light paints. Some mention that white paints and paints with mixed-in white are generally not water-sensitive. Usually such paints have the closed, shiny surface that one would expect of oil paint.[12]

Most interviewees suspect a correlation between the pigment-binder ratio and water-sensitivity because of the matt appearance of the paint layers in question. Often when pigment-pickup occurs on lean paint, the influence of mechanical action is considered.[13] The aforementioned solvent-sensitive areas largely correspond to the areas of brittle paint. This raises the question if there is a connection between both phenomena. Filiz Kuvvetli hypothesizes that Jorn may have used *emulsion paints*, which results in a dry, velvet-like appearance that sometimes shows micro-

[9] Interviews Andersen, Bosc, Bronken, Froment, observations JB, case-studies.

[10] All interviews, case-studies.

[11] All interviews, observations JB, case-studies, Lee-Jensen [8].

[12] Interviews Andersen, Baadsgaard, Bosc, Bracht, Bronken, Froment, Kuvvetli, Speldekamp, observations JB, case-studies.

[13] Interviews Andersen, Speldekamp, observations JB.

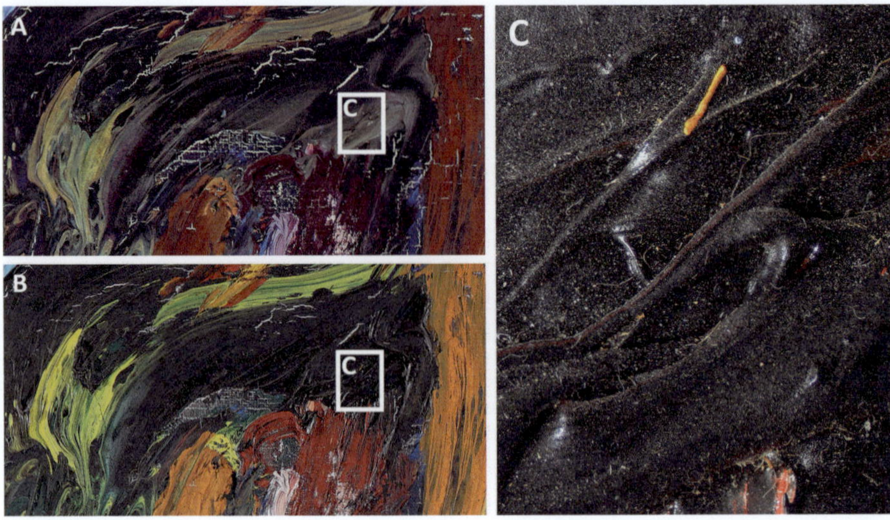

Fig. 28.3 (**a** and **b**) Detailed black paint stroke in daylight (**a**) and UV-induced fluorescence (**b**) of the glossy paint. *La Rose Féroce*, Asger Jorn, 1961, collection Gemeentemuseum Den Haag, 0333653. (Photo: Gemeentemuseum). (**c**) Detail indicated paint stroke with imbibed dirt. (Photo: LS)

bubbles on the surface due to mixing the paint.[14] In Kuvvetli's experience, such areas are difficult to clean due to brittleness, porosity and water-sensitivity.

In 20 case-studies, conservators observed large matt-gloss-contrasts within one colour, mainly in black paint.[15] While thinly applied areas are matt, the impastos are often glossy, as if the binder is unevenly distributed throughout the paint. A distinct UV-induced fluorescence can be observed where medium has accumulated (Fig. 28.3a, b). Furthermore, various white paints show yellow medium skins.[16] Several interviewees mention the risk of changing the degree of gloss by cleaning.[17]

Conservators observe soft and medium-rich paint with imbibed dirt, mostly on glossy impastos (Fig. 28.3c). This dirt is usually difficult to remove because it is attached to or even has become embedded in the paint surface.[18]

Bronken warns that soft paint layers as found in Jorn's works can in some cases conform to whatever is placed on them, even a sheet of melinex, making surface deformation another risk during cleaning. Marion Bosc has observed changes in the

[14] Research is being carried out by Kuvvetli to investigate this hypothesis.

[15] Interviews Andersen, Baadsgaard, Bosc, Froment, Speldekamp, observations JB, case-studies.

[16] Observations JB.

[17] Interviews Andersen, Kuvvetli, case-studies.

[18] Interviews Andersen, Bosc, Bronken, Froment, King, Speldekamp, observations JB, case-studies, Steyn [9].

shape of the paint layers caused by gravity, for example wrinkling of paint where the surface of the paint seems dry but the paint layer is still flexible.[19]

Some conservators believe that medium exudates observed on Jorn's paintings are related to an excessive use of medium. According to most who have observed this phenomenon, exudation can be an active process, a fact that was confirmed by Bosc after monitoring paintings over a period of 10 years.[20]

Finally, surface phenomena such as crust-formation, efflorescence, blooming and protrusions occur, mainly on black areas.[21]

28.5 Surface Cleaning Treatments

Jorn's paintings require a different approach than traditional oil paintings because of the aforementioned risks. Andersen states for Jorn's paintings the need for cleaning should be more crucial than for other paintings, due to the high risk of removing original material, possible unwanted optical or chemical changes and the time required. Especially soft paint with imbibed dirt is a risk, as the surface easily deforms during cleaning.[22] Not cleaning may however cause the dirt to become more imbibed over time. In Jorn's paintings colour and surface structure are very important and surface dirt can cause the image to lose its optical depth and saturated colours.

After the decision has been made to attempt cleaning a painting by Jorn, conservators take a relatively long time for preliminary studies and treatment. Conservators need to reserve sufficient time to conduct extensive tests because *'it is like exploring a landscape: things that could happen and go wrong once you start'*.[23] It is important to have realistic expectations of the outcomes, as compromises are often unavoidable when dealing with these complex artworks.[24]

Visual investigation is the first step to identifying vulnerabilities. UV-induced fluorescence is an important indicator because excessive medium, exudates and the soft paint usually fluoresce.[25] Most conservators interviewed advise cleaning tests for every colour area in order to map sensitivities and other characteristics, such as imbibed dirt or gloss changes.[26] This way, the cleaning level can be determined before choosing a cleaning system(s).

---

[19] This phenomenon has been observed while monitoring the Cobra Museum of Art collection for 10 years and changing from vertical to horizontal storage.

[20] Interviews Andersen, Baadsgaard, Bosc, Bronken.

[21] Interview Bronken, observations JB.

[22] Interviews Andersen, Bosc, Bronken, King.

[23] Interview Andersen.

[24] Interviews Andersen, Bronken, King.

[25] Interviews Andersen, Bronken, observations JB.

[26] Interviews Andersen, Bosc, Bronken, Kuvvetli, Speldekamp, case-studies.

Two approaches are employed to clean Jorn's painting. Firstly, a conservator may choose one method that can be applied to all or most areas of the painting. While this gives clarity about the material and treatment, it may lead to the skipping of sensitive areas and to different levels of cleaning (sensitive areas are cleaned less thoroughly). The conservator may reduce the overall level of cleaning, to prevent an imbalanced result.[27]

The second approach is to use tailored cleaning methods for different paints. This stems from the idea that each colour area has different chemistry and physical characteristics, and therefore requires a different treatment. This approach often makes it possible to clean the painting more thoroughly without skipping areas.[28]

Study of 61 treatments, including reports and interviews, shows that 48 treatments were executed with water-based systems, using cotton swabs[29] or soft sponges (PVA Blitzfix®).[30] Andersen notes that the fibers of cotton swabs remain on the paint surface and that a soft sponge follows the surface geography more easily. With 15 of the water-based treatments the conservators could only clean certain areas due to water-sensitivity. Other areas were cleaned with different systems, usually dry-cleaning with soft brushes or soft sponges, Wishab (Akapad soft) and make-up sponges.[31] Yet sometimes the paint turned out to be too fragile, soft or brittle even for these systems and was left uncleaned.[32]

A few conservators used non-polar solvents and although they noted little or no interaction with the paint they also did not consider such methods particularly effective in removing dirt.[33]

Sixteen of the 61 case-studies were (partially) dry-cleaned only. Tweezers were used in six treatments. In paintings with imbibed dirt all conservators used tweezers.[34] A fine pair of scissors or a scalpel was used in order to cut off remaining protruding fibres.

Most interviewees avoid systems that require clearance, such as gels and water with buffers/acids/alkalis.[35] This prevents the painting from exposure to moisture a second time. However, conservators are curious about the possibilities of rigid gels and other alternatives, like those used for water-sensitive paint on Karel Appel's paintings [11, 12].[36] Bosc nowadays uses Velvesil silicone gel for CoBrA paintings because of the safe and effective removal of dirt, but has not tried this on a painting by Jorn.

---

[27] Interviews Bosc, Bracht, Froment, Speldekamp, case-studies.

[28] Interviews Andersen, Bronken, King, Kuvvetli, case-studies.

[29] Interviews Andersen, Bosc, Bracht, Froment, Kuvvetli, Speldekamp, case-studies.

[30] Interview Andersen, case-studies.

[31] Interviews Andersen, Bosc, Bracht, Bronken, King, Kuvvetli, Speldekamp, case-studies.

[32] Interviews Andersen, Kuvvetli.

[33] Interview Andersen, observations JB, case-studies.

[34] Interviews Andersen, Bosc, Bronken, King, case-studies.

[35] Interviews Andersen, Bosc, Froment.

[36] Interviews Andersen, Bosc, Kuvvetli.

As it is not always possible to clean the entire painting to the same extent without damage, subtle differences, invisible to the museum visitor, are considered acceptable.[37] Conservators prefer waiting for more suitable options in the future.

Completed treatments are described as having resulted in clearly noticeable aesthetic improvement, particularly the re-appearance of matt-gloss differences and increased colour-saturation.[38]

The interviewees emphasise the importance of preventive measures to protect Jorn's paintings. They feel the paint surface needs to be protected from new dirt, especially when sticky paints are present.[39] The majority of Jorn's paintings are therefore placed behind glass. Conservators regard this a compromise, because glass creates a barrier between the viewer and immediacy of the paint. Some conservators advise monitoring Jorn's paintings in order to identify active processes and follow their development.[40]

28.6 Conclusion

Jorn is known to have experimented with painting materials in order to create various surface textures and effects. Although sometimes for financial reasons, they seem to have been a source of inspiration for Jorn and thus became an integral part of his technique, taking on more and more extreme forms.

This means that conservators often have to be creative during surface cleaning and combine different treatments in order to minimise risk and conduct treatments that fall within acceptable conservation norms and ethics.

Future research should focus on the material-technical reasons behind the degradation phenomena by chemical analysis – linking case-studies to knowledge on the influence of certain pigments and the composition of oil paints from suppliers like LeFebvre and LeFranc [7].

More insight into the causes of degradation phenomena will allow for a more informed use of current cleaning systems and will hopefully be a first step towards the development of new cleaning systems.

Acknowledgements The authors would like to express their gratitude to all interviewees (Table 28.1). They feel honoured that these conservators shared information, including uncertainties and concerns, for the purpose of this research. Openness about problems encountered in conservation practice is of paramount importance for the advancement of knowledge and the development of safer cleaning methods in conservation. Further thanks to: Gemeentemuseum Den Haag (Ruth Hoppe), Kunstkonserveringen West, Museum Jorn (Lucas Haberkorn), Statens Museum for Kunst Copenhagen, ARoS Århus, Stedelijk Museum Amsterdam, Sofia Johansson, Sanneke Stigter, Jan Hunger. LS was funded by the Netherlands Organisation for Scientific

[37] Interviews Bracht, Froment, Speldekamp, case-studies.

[38] Interviews Andersen, Froment, King, Kuvvetli, case-studies.

[39] Interviews Andersen, Bosc, Bronken, Bracht, Speldekamp.

[40] Interviews Andersen, Bosc, Bronken.

Research (NWO) through Heritage Plus, a Joint Programming Initiative of the European Commission.

References

1. Aggerholm D (2000) An investigation of conservation methods used on the COBRA-group paintings. Dissertation, School of Conservation, Copenhagen
2. Atkins G (1968) *Jorn in Scandinavia 1930–1953*. Copenhagen: Borgen
3. Atkins G (1977) *Asger Jorn: The Crucial Years 1954–1964*. Copenhagen: Lund Humphries
4. Atkins G (1980) *Asger Jorn. The final years 1965–1973*. Copenhagen: Lund Humphries
5. Atkins G (1986) *Asger Jorn. Supplement to the œurvre catalogue of his paintings from 1930 to 1973*. London: Asger Jorn Foundation, Lund Humphries
6. Baumeister R (2014) *Asger Jorn in images, words, and forms*. Zurich: Scheidegger 6 Spiess
7. Bronken IAT and Boon JJ (2015) Conservation challenges and physical changes in Composition (1952) by Jean-Paul Riopelle: a case study. In *Public Paintings by Edvards Munch and his Contemporaries: Change and Conservation Challenges*, ed. Frøysaker et al.: 334–348. London: Archetype publications
8. Lee-Jensen M (2016) Pigment transfer by the cleaning of oil paintings by Asger Jorn. A survey of blue and yellow paint layers in nine selected paintings. Dissertation School of conservation, Copenhagen, Denmark
9. Steyn L et al. (2017) *Cleaning Modern Oil Paints: The removal of imbibed dirt.* Poster presented at ICOM CC Copenhagen, 4–8 September 2017
10. Stols-Witlox M (2003) Asger Jorn's *Abstract Minded Figure* (1962). *Interdisciplinair tijdschrift voor Conservering en Restauratie* 4/1: 26–32
11. Volk A et al. (2016) A Partially Water-Sensitive Oil Painting by Karel Appel: Technical Investigation and a Tailored Approach to Surface Cleaning. *Zeitschrift für Kunsttechnologie und Konservierung* 30/1: 150–166
12. Wijnberg L et al. (2008) Watergevoelige olieverfschilderijen: werkwijze van Jasper Johns en Karel Appel. *kM* 66: 23–26
13. Wijnberg L (2014) Do we see what we know or do we know what we see? Conservation of oil paintings in the Stedelijk Museum Amsterdam. In *Issues in contemporary oil paint,* eds. KJ van den Berg et al.: 21–32. Switzerland: Springer
14. Wijnberg L and Froment E (2004) Interviews with Jacqueline de Jong and Marie José van de Loo in the context of the restoration of 'Im flügelschlag der schwäne'. Unpublished: Stedelijk Museum Amsterdam

Chapter 29
Decision-Making Processes Regarding the Treatment of Modern Oil Paintings (1950s–Present) Exhibiting Paint Dripping and Oil Exudates

Jazzy de Groot, Lise Steyn, Maartje Stols-Witlox, and Klaas Jan van den Berg

29.1 Introduction

Oil paintings (1950s–present) can exhibit oil exudates years after completion. Exudates have different appearances, varying from liquid, actively running drips that call for immediate remedial conservation action, to less dramatic transparent or opaque material that is secreted from the oil paint. While the first type actively drips, exudates of the second type often appear inactive. Not much attention is paid to solid and less conspicuous exudates and they are usually not the main subject of a publication. However, they are degradation phenomena that require attention. In order to obtain more insight into the treatment of older and often less conspicuous exudates, 28 practicing conservators from around the world with experience in treating modern paintings were contacted by phone or email. They were asked if, how and why they treated this kind of phenomena. In phase 2, more information was obtained through interviews (7) and questionnaires (3).

J. de Groot (✉)
De Groot Schilderijenrestauratie, Alkmaar, The Netherlands
e-mail: jazzy@degrootschilderijenrestauratie.nl

L. Steyn · M. Stols-Witlox
University of Amsterdam, Amsterdam, The Netherlands

K. J. van den Berg
University of Amsterdam, Amsterdam, The Netherlands

Cultural Heritage Agency of the Netherlands, Amsterdam, The Netherlands

K. J. van den Berg et al. (eds.), *Conservation of Modern Oil Paintings*,
https://doi.org/10.1007/978-3-030-19254-9_29

29.2 Appearances and Characteristics

Exudates have been described in conservation reports, the literature and in interviews as being solid to soft, sticky and/or liquid, stable to actively running and looking like either transparent brown or pigmented oil binder (Figs. 29.1, 29.2, 29.3, and 29.4). They are mostly associated with thick paint layers but can also manifest on thinner paints [1, 5, 6]. When soft, exudates can be thixotropic ([15], interview Bronken 2016, questionnaire Lister 2016). They can be sensitive to polar solvents and can also appear in relation to water sensitive and soft paint layers ([2, 7, 9], questionnaire Lister 2016). Under the influence of water or high relative humidity, the exudation processes can accelerate ([3, 12, 15], interview Bronken 2016, correspondence Boon 2016). Impasto can lose its shape, and sometimes the paint layer appears to be melting. Exudates alone may fluoresce and the paint layer from which they originate may show an uneven fluorescence [2, 5–7, 17]. In summary, with regard to the physical appearance of exudates, one can distinguish two extremes: type 1 where the paint layer appears to be melting, sometimes resulting in paint drips onto gallery floors; and type 2 which is a less obvious type of exudate. Type 2 can be transparent or opaque, and may appear to have formed slowly, eventually becoming solid. Variations occur between these two extremes.

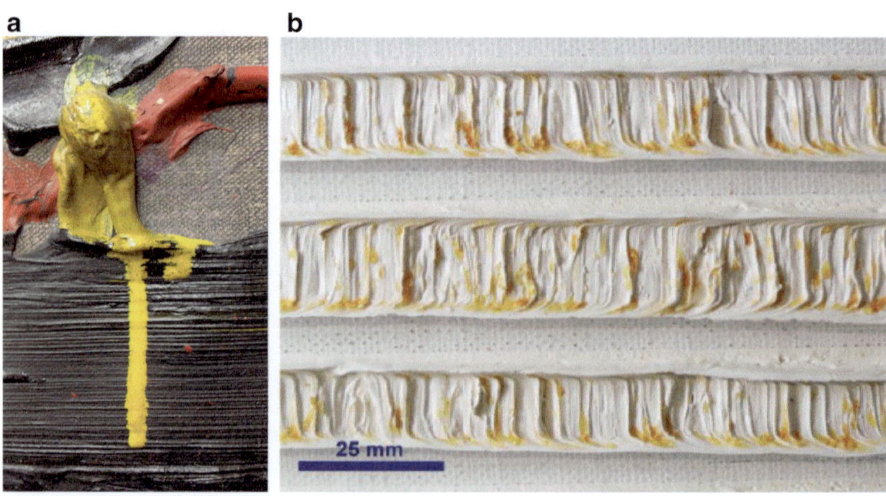

Fig. 29.1 Left: yellow and opaque exudate on a painting by Georges Mathieu (c. 1950, Musée National d'Art Moderne, Centre Pompidou). Relatively solid when touched, the drip grows very slowly (Questionnaire Alexia Soldano 2016). Right: transparent, oily exudate as a more even layer on a painting by Erik Oldenhof ([1], private collection)

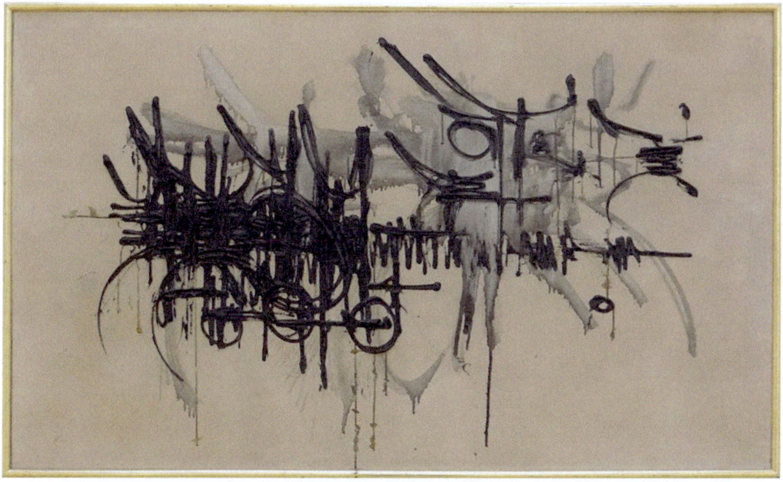

Fig. 29.2 Georges Mathieu, Théorème de Gödel, 1957, Musée des Beaux Arts de Lille. [16]

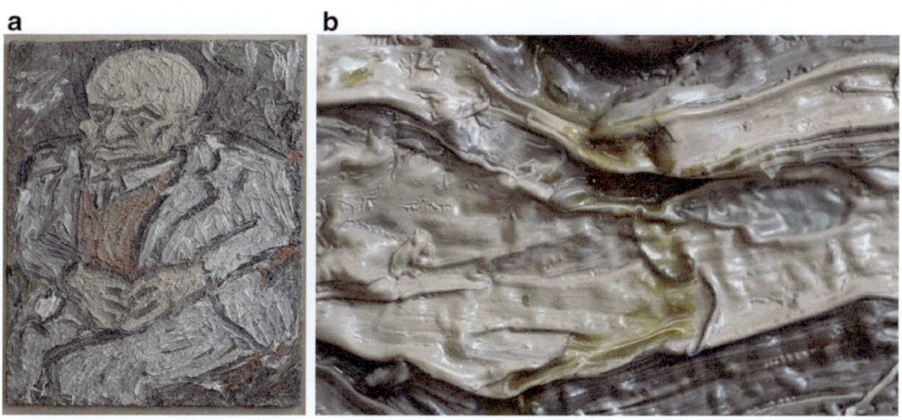

Fig. 29.3 Left: Portrait of Father (1978) by Leon Kossoff. Right: detail of exudate on *Portrait of Father*. (National Galleries Scotland, https://www.nationalgalleries.org/research/projects-263/conservation-of-leon-kossoffs-portrait-of-father-1978)

29.3 Causes, Consequences and Treatments

In the current literature, there is general agreement that a physical separation of fractions from the solid but not fully stable paint layer lies at the root of the exudation. Over time, the paint layer becomes more non-polar and starts repelling polar fractions [1, 3–6, 8, 11, 13, 15].

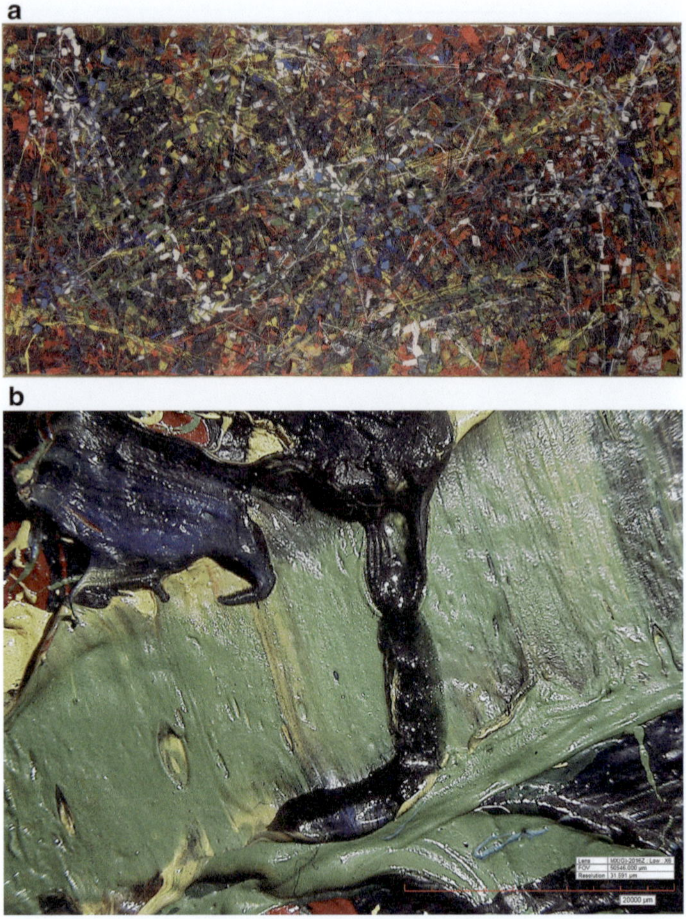

Fig. 29.4 Top: Jean-Paul Riopelle, *Composition*, 1952, Henie Onstad Art Centre's collection, Norway. Bottom: detail of exudate. [5, 6]

Secretions can affect other materials that they come into contact with. On the *Seven*-series by Frank van Hemert (1990–1995), dripping paint dissolved the paint layers underneath the drips [3]. Also, on a series by Tal R (2005) underlying layers of different composition were damaged [2]. The physical characteristics of exudates can also be influenced by underlying paint layers. For example, on *Composition* (1952) by Riopelle, the liquidity of exudates depends on the paint layers they crossed, suggesting a chemical interaction between the exudate and other paints (interview Bronken 2016).

Current research into possible treatment methods focuses both on exudate removal and on the possibility of stopping the exudation process. Interesting results have been achieved by exposing exudate and compromised paint layers to gamma radiation and heat or by the addition of metal ions to the paint layer [7, 12, 15]. No overall solutions have been found to date.

29.4 Removal of Actively Dripping Exudates

When it comes to recent, opaque and actively dripping exudates, several case studies have been published [2, 3, 7, 10, 12, 13, 15, 18].

Actively running exudates have been removed in different ways and to different extents, often in consultation with the artist. For example, when artist Otto Piene was asked for his opinion on the white drips of paint running over an orange paint layer in his painting *Harvest* (1993), the artist wished to stop the traces of the drips from getting any longer. The exudate was partly removed; afterwards, while the drips did continue to form, the previous traces remained unchanged [15].

Another approach was used on the paintings *Made in Heaven, Gate to Pink* and *Name, Rank and Departure* (2005) by Tal R, where drips of paint were running into other colour fields and were also found on the gallery floor. The artist found them disastrous. Exudates were partially removed, but only from contrasting colour fields. Conservators repeated the removal several times as the drips continued to form [2].

Depending on the artist intention and the extent to which exudates are considered disturbing, there is a general tendency towards removing or at least attempting to stop the dripping process.

29.5 Removal of Older Exudates

While multiple case studies have been published on actively dripping and liquefying paint, publications dedicated to (nearly) solid exudates are more rare. In many cases, the presence of solid exudates is mentioned as a side note.

The few publications that do deal with solid exudates demonstrate a variable attitude towards their removal.

In a publication concerning a consolidation treatment on *Théorème de Godel* (1957) by Georges Mathieu, the conservator mentioned that the light background was covered by transparent exudates (Fig. 29.2, [16]). These exudates had darkened and were solid. Based on old pictures of the painting it could be established that the exudates had manifested somewhere between 1970 and 1998. The conservator decided not to remove them because they did not appear to interfere with the perceived artist intent. The fact that the cause of the exudation process was considered to be inherent in the painting materials themselves appears to have made the exudates more acceptable to the author, as they could be considered an original part of the painting.

In the case of *Portrait of Father* (1978) by Leon Kossoff it was believed that the exudates were caused mainly by an external factor, namely heat (Fig. 29.3, [14], correspondence Maule 20-16). What may make this painting different in comparison with many other case studies, is that the painting is a figurative portrait and not an abstract and/or expressionist painting. The placement of the drips within a portrait were considered visually distracting. This observation, together with the belief that external factors were the main cause of the exudates led to conservators deciding to remove them.

In the conservation treatment of *Composition* (1952) by Jean-Paul Riopelle exudates were identified (Fig. 29.4). The conservator wished to remove them after she gained sufficient understanding of the phenomenon, but priority was given to a time-consuming consolidation treatment. Therefore, the exudates were not treated due to time limitations and because they were not considered visually distracting. While not obvious at first sight due to the artist's expressionist painting technique, the presence of the exudates has undeniably changed the painting (interview Bronken 2016). As demonstrated in these examples there is no one way to approach the treatment of exudates. It seems that the older the exudate, the more it is accepted as part of a work of art. The interpreted artist intention plays a major role, although the interpretation of intent can be quite challenging and subjective choices are difficult to prevent.

29.6 A Broader Perspective

To obtain a broader perspective on the approach to exudate treatment, conservators were asked their opinion regarding various treatment options and principles. During the first phase, 28 conservators were asked to look at pictures of different kinds of exudates (e.g. Fig. 29.1 bottom). At that point one conservator did not recognize exudates. Some conservators immediately explained how they would deal with them: 5 stated that they never considered removal and 11 mentioned explicitly that they did not treat them because they were not certain whether drips are the result of the artist's application techniques which may have included throwing or dripping paint onto the canvas.

During the interviews and in the questionnaires, it became clear that exudates can be hard to distinguish from drips due to the original application of the paint. Furthermore, as mentioned by 3 conservators, the possible presence of irregular application of varnish can confuse the reading of the surface. Under UV, uneven fluorescent paint layers and drips appear similar to passages with locally applied varnish. Since local varnishes may have been applied by the artist, removal would be considered controversial.

Uncertainty about the origin of older exudates leads to reluctance regarding their removal. Ten conservators (7 after first contact, 3 during interviews) out of the total of 28 suggested that even when they are certain that an exudate is not due to an artist's technique or intent, they still consider them as part of the original material and do not wish to remove them. On the other hand, when the an underlying paint layer is threatened by the presence of exudates on top, 2 conservators considered that removal is an option (when they saw pictures of exudates) and 3 more conservators concurred during interviews.

29.7 Conclusion

Opaque exudates may appear to be integral to the painting process and their presence does not always attract attention. With the naked eye it can be hard to distinguish between several possible scenarios including the possibility that binder separation may have occurred during the painting process (or even before application in the paint tube). In addition, the timing for the exudates to appear introduces a grey area: is an exudate still part of the painting process when it arises a day after completion, or after a week, a month or a year?

This study illustrates that conservators are uncertain how to deal with exudates and more discussion and understanding of these phenomena is necessary to inform decision making regarding their removal. As is often the case in conservation, context influences the decision-making process. On the basis of published treatments and interviews it was found that arguments for removal mainly revolve around the perception of the artist intention. As very little research has been performed regarding the chemical and physical consequences of not removing old exudates in particular, there is little technical information to form the basis for conservators to make their decisions. In the absence of such information, conservators are reluctant to remove old exudates, accepting them as a logical consequence of the artists' choice and original materials. More research on the identification of exudates and the mechanisms of their formation is needed in order to gain a deeper understanding of this phenomenon and to inform future treatment decisions.

Acknowledgements Conservators interviewed: Ida Bronken (private conservator, Norway), Marjan de Visser (private conservator, Netherlands), Zeph Benders and Laura Ledwina (Cultural Heritage Agency of the Netherlands,), Sanneke Stigter (University of Amsterdam, Netherlands), Vera Blok (private conservator, Netherlands) and Arielle Veerman (private conservator, Netherlands). Anne Baxter (private conservator, France), Kristin Lister (Art Institute of Chicago, USA) and Alexia Soldano (private conservator, France) filled in the questionnaire. Others who contributed to this research by email, phone or personal conversations: Pauline de la Grandière (private conservator, France), Jaap Boon (paint research chemist JAAP Enterprise, Netherlands), Gwendolyn Boevé-Jones (private conservator, Netherlands), Hélène de Ségogne (private conservator, France), Kate Seymour (Stichting Restauratie Atelier Limburg, Netherlands), Aviva Burnstock (Head of the Department of Conservation and Technology, The Courtauld Institute of Art, England), Claartje van Haaften (Hamburger Kunsthalle/private conservator, Germany), Ruth Hoppe & Leonoor Speldekamp (Gemeentemuseum Den Haag, Netherlands), Annegret Volk (private conservator, Germany), Michael Duffy (Museum of Modern Art, USA), Mirta Pavic (Muzej Suvremene Umjetnosti, Croatia), Femke van der Knaap, Nico Boeijink & Rob Boekel (private conservators, Netherlands), Mary Bustin (private conservator, England), Marion Bosc (private conservator, France), Lorraine Maule (National Galleries of Scotland, Scotland) and Jenny Schulz (private conservator, Germany). Leslie Carlyle (New University of Lisbon, Lisbon) reviewed the paper. LS was funded by the Netherlands Organisation for Scientific Research (NWO) through Heritage Plus, a Joint Programming Initiative of the European Commission.

References

1. Bayliss S, Van den Berg KJ, Burnstock A et al (2016) An investigation into the separation and migration of oil in paintings by Erik Oldenhof. Microchem J 124:974–982
2. Blomsterberg L, Havermark K, Segel K (2015) Cleaning off liquefying paint. In: Barros D'Sa A, Bone L, Clarricoates R et al (ed) Current technical challenges in the conservation of paintings. Archetype, London, p 14–27
3. Boon JJ, Hoogland FG (2014) Investigating fluidizing dripping pink commercial paint on Van Hemert's *Seven*-series works from 1990–1995. In: Van den Berg KJ, Burnstock A, De Keijzer M et al (ed) Issues in Contemporary Oil Paint. Springer, p 227–246
4. Boon JJ, Lister K (2014) Investigating aged oil-medium phase separating and exuding as drips from a 1960s painting by Pierre Soulages. In: Bridgland J (ed) ICOM-CC 17th Triennial Conference Preprints, 15–19 September 2014, Melbourne, Australia. Parijs, p 1317
5. Bronken IAT, Boon, JJ (2014a) Softening paint and drip formation in paintings by Jean-Paul Riopelle (1923–2002): Improving their condition with metal coordinating ions. In: Bridgland, J (ed) ICOM-CC 17th Triennial Conference Preprints, Melbourne 15–19 September 2014, art. 1303. Paris
6. Bronken IAT, Boon JJ (2014b) Hard dry paint, softening tacky paint, and exuding drips on *Composition* (1952) by Jean-Paul Riopelle. In: Van den Berg KJ, Burnstock A, De Keijzer M et al (ed) Issues in Contemporary Oil Paint. Springer, p 247–262
7. Bronken I, Boon J, Vila A (2015) In search of a solution for softening and weeping paints in contemporary paintings by Tal R and Frank van Hemert. In: Barros D'Sa A, Bone L, Clarricoates R et al (ed) Current technical challenges in the conservation of paintings. Archetype, London, p 28–41
8. Bronken IAT, Boon JJ, Corkery RW et al (2016) On the changing appearance of, and potential treatment options for, softening and dripping paints in CoBrA oil paintings. Stu In Cons 61:269–270
9. Burnstock A, Van den Berg KJ (2014) Twentieth century oil paint. The interface between science and conservation and the challenges for modern oil paint research. In: Van den Berg KJ, Burnstock A, De Keijzer M et al (ed) Issues in Contemporary Oil Paint. Springer, p 1–20
10. Corbeil MC, Helwig K, Poulin J (2004) Analysis of the painted oeuvre of Jean Paul Riopelle: from oil to mixed media. In: Roy A, Smith P (ed) Modern Art, New Museums Contributions to the Bilbao Congress 13–17 September 2004. Londen, p 170–174
11. Duffy M, McGlinchey C (2001) Weeping cadmium paint: a case study. Paper presented at Deterioration of artists' paints: effects and analysis. A joint meeting of ICOM-CC Working Groups Paintings 1 & 2 and The Paintings Section, UKIC, British Museum, London, 10–11 September 2001. Extended abstracts of presentations, p 81–83
12. Franken V, Heydenreich G (2014) Set back the race: treatment strategies for running oil paint. In: Van den Berg KJ, Burnstock A, De Keijzer M et al (ed) Issues in Contemporary Oil Paint. Springer, p 333–349
13. Hoogland FG, Van der Horst J, Boon JJ (2007) Liquefying oil paint in some late-Twentieth-Century paintings. In: Learner JS, Smithen P, Krueger JW et al (ed) Modern paints uncovered. Proceedings from the Modern Paint Uncovered Symposium. Tate Modern, London, May 16–19, 2006. Getty Publications, Los Angeles, p 282–283

14. National Galleries Scotland (2016) https://www.nationalgalleries.org/research/projects-263/conservation-of-leon-kossoffs-portrait-of-father-1978/condition-471. Accessed 12 Dec 2016
15. Schulz J (2011) Liquefying oil paint: contemporary paintings in danger? In: Preprints van ICOM CC Lisbon, ICOM committee for conservation 16th triennial meeting, Lisbon, 2011. ICOM CC, Paris, paper 1320
16. De Ségogne H (2014) *Théorème de Gödel* by Georges Mathieu, 1957. Study and restoration: consolidation through cohesive regeneration using a solvent. In: Van den Berg KJ, Burnstock A, De Keijzer M et al (ed) Issues in Contemporary Oil Paint. Springer, p 149–157
17. Volk A, Van den Berg KJ (2016) A partially water-sensitive oil painting by Karel Appel: technical investigation and a tailored approach to surface cleaning. Zeitschrift Für Kunsttechnologie Und Konservierung 30:150–166
18. Wijnberg L (2014) Do we see what we know or do we know what we see? Conservation of oil paintings in the Stedelijk Museum Amsterdam. In: Van den Berg KJ, Burnstock A, De Keijzer M et al (ed) Issues in Contemporary Oil Paint. Springer, pp 21–32

Chapter 30
Between Rejection, Ignorance, and Preference: The History and Dilemma of Modern Double-Sided (Recto/Verso) Paintings with Regard to Issues of Conservation and Presentation

Ulrik Runeberg

30.1 The Artist's Point of View – Various Reasons for the Rejection of Paintings

The re-use of the versos of canvases, cardboards, or wooden supports has generally been assumed to be a cost-saving action that is often attributed to financial problems and a lack of artist's materials. Although this certainly is one significant aspect, the reasons leading to the rejection of a painting can be many-fold, ranging from moral and religious aspects to political ones [3]. In addition, a decline of traditional academic approaches in the making of the arts and the rise of avant-garde ideas may have caused artists to reconsider their own oeuvres in a rather pragmatic and light-minded manner.

The re-use of a rigid and plain painting support is achieved by simply turning it. Paintings on canvas usually require additional measures, they need to be demounted, turned and re-stretched. In addition, the prior image now on the verso would often partially be blotted out, scraped off, covered through a layer of newspaper or overpainted with whitewash. Such layers consist in many cases of chalk ground [6] or left over painting materials that can range from transparent to solid and may contain traditional ingredients as well as modern components of that time [10] (Fig. 30.1).

Some hidden works on the verso, however, turn out to be linked to the painted subject on the main side. From the artist's point of view, they often refer to personal experiences and are not necessarily meant to be seen by a wider audience. Such considerations can contribute profoundly to the understanding of an artist's life as for example in the oeuvre of Otto Mueller: After overcoming a serious crisis with

U. Runeberg (✉)
Universität Münster (WWU)/Restaurierungszentrum der Landeshauptstadt Düsseldorf, Münster, Germany

K. J. van den Berg et al. (eds.), *Conservation of Modern Oil Paintings*,
https://doi.org/10.1007/978-3-030-19254-9_30

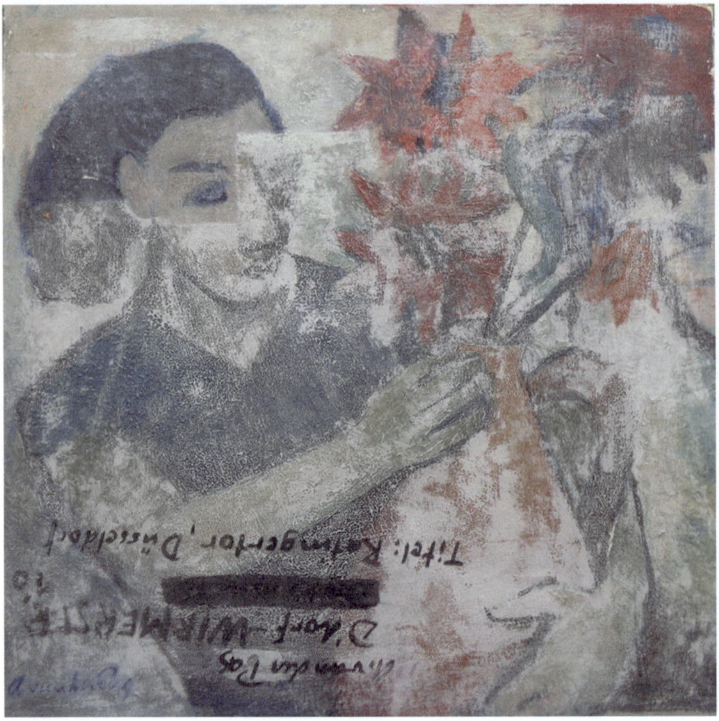

Fig. 30.1 Antonius van der Pas. Verso of *Ratinger Tor in Düsseldorf* (early 1960s). Ca. 77 cm × 77 cm. Oil on textile, mounted on plywood. Private collection. Image by the author

his fiancée, the artist turned and flipped the painting "The murder" (1919/1920), marked it with a "2" for verso and created on the new recto side "Double portrait with Maschka Mueller" from 1924/1925. [17].

Occasionally, the two sides of the same support were used by two individual artists, either with the intention to create works of a similar standard (Jessewitsch and, Schneider 2008, p. 136: Martel Schwichtenberg, "Frauenkopf",1920; and p. 492: Willy Robert Huth, "Hafenstadt", 1920), or in an imbalanced way while determining one side as superior to the other one, as in the case of the artist couple Heinrich Nauen and Marie von Malachowski ([25], p. 37).[1] Regardless of an artists` subjective evaluation, however, the later art historical assessment regarding the two sides of a work depends rather on scholar and economic considerations than on the artist's personal point of view. At times, disregard of an artist's own work may even lead to the destruction of significant pieces [3]. A rather creative approach within this process, though, has been observed in the recycling of disassembled paintings by the artists themselves, e.g. the use of painting fragments on hardboard as stabilizers for frame corners ([5], p. 155).

[1] Recto: Heinrich Nauen: Weiße Tulpen in violetter Vase (1911). Verso: Marie von Malachowski; Herbstblumen in violetter Vase (1911). P. 37.

30.2 History of the Presentation of Modern Recto/Verso Paintings

During the first half of the twentieth century, it was uncommon to put a painted verso side of a contemporary work on display. The verso generally was linked to negative associations and signified a complete taboo. In 1931, for instance, artists turned their paintings towards the wall of the Kunstverein in Cologne. This act of protest was a comment on exaggerated criticism on many avant-garde paintings (Fig. 30.2); [2].

In postwar time it became common practice to split painting supports bearing depictions on both sides, rather than exposing the two sides as recto and verso to the public. The splitting was executed especially by institutions with severely reduced collections of early twentieth century art, which seemed to compensate in this way the destructions of the war. First scattered cases of presenting double-sided paintings reach back to the late 1950s. Only about one decade later the presentation of twentieth century recto verso paintings became more common, a notable example being the recto/verso-exposition of an entire painting collection at the Museo de Arte in Sao Paulo in 1968, with several works carrying verso paintings.[2] By mounting the works in glass easels, visual access was provided to both sides. A similar milestone event in this context was set up a few years later in 1977 for a Tatlin-exhibition in Moscow, for which several recto-verso paintings were restored and deliberately exhibited double-sided [24]. This spirit of transparency also found its

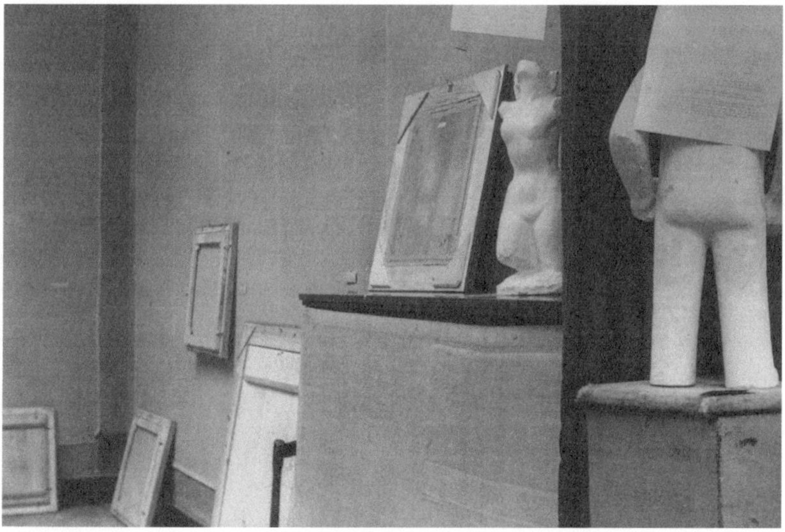

Fig. 30.2 Picture of an artists' exhibition in Cologne, Kunstverein (1931) to protest against overstated criticism on avant-garde art (see text)

[2] https://www.jcms-journal.com/articles/10.5334/jcms.1021221/ [10.02.2019].

way into the field of conservation. The application of translucent fiber-glass and other synthetic materials for relining recto/verso canvases allows for lasting visibility of both painting sides, despite its being relined [1].

During the following decades, elaborate and sophisticated systems such as prismatic elements and rotatable frames were developed, to present an "appropriate", visually correct format on both sides of a painting support [9]. A major presentation of double sided paintings in Germany was set up in Freiburg in 1991, called "Rücksichten" [16]. Around the turn of the millennium, the aim for transparency and sensation increased to such a degree that even museums in the hinterland internationally began to present verso-paintings without any visual barriers. Around the same time, a range of institutions started to provide open access to storage and deposit rooms which occasionally led the visitor to detect paint on the verso of art works – usually however without providing any kind of information [11].

In recent years, exhibitions seem to comprise a gamut of presentations [12], ranging from displays without accompanying information to rather extraordinary exhibitions on established artists and their oeuvre of rejected verso paintings, for example on Andrzej Wróblewski [4], Ernst Ludwig Kirchner [14], Franz Marc [15] and Lyonel Feininger [22]. Coincidentally and at an increasing pace, art historical literature in general appears to refer to the topic of verso paintings [e.g. 23]. Coherent guidelines regarding an appropriate and stringent presentation to date do not exist, which has an affect not only regarding art- historical issues but also with view to framing, restoration treatments and ethical issues involved [12].

30.3 Different Approaches Regarding the Exhibition of Discarded Verso Paintings

In this context, useful examples are provided by three municipal museums in Dortmund [U/Zentrum für Kunst und Kreativität], Münster [LWL Museum für Kunst und Kultur], and Düsseldorf [Kunstpalast] which present similar recto-verso-paintings by the artist Conrad Felixmüller (1897–1977) [catalogue raisonnée, 19] in profoundly different ways:

1. Image *"Beauty and youth"* (Fig. 30.3a, recto as chosen by artist), and *"depiction of a dancer"* (Fig. 30.3b, originally recto, rejected by the artist). The visitor has no visible access to, nor knowledge of the partially overpainted painting on the earlier painted reverse side (a dancer, ca. 1915), only the front side (c. 1931) is presented – this being the explicit wish of the artist in 1931.

Fig. 30.3 (continued) Reg.-Nr. GM93.3453. VG Bild-Kunst. Image by the author. (**d**) *Musizierende Knaben* ("Young musicians"). 100 cm × 70 cm. Oil on canvas. Signed and dated; Recto, as chosen by the artist in 1936. VG Bild-Kunst. Image by the author. (**e**) *Ruhrrevier II* (1920). Originally recto, later rejected and whitewashed by the artist [7]: The picture shows the verso after restoration in 1981. Reg.-Nr.: 1681 LM/83-59. LWL Landesmuseum Münster. VG Bild-Kunst. Image by the author. (**f**) *Der Geburtstagsbrief* ("The birthday-letter"). 95 cm × 75 cm. Oil on canvas. Signed and dated. Recto, as chosen by the artist in 1934. VG bild-Kunst. Image by the author

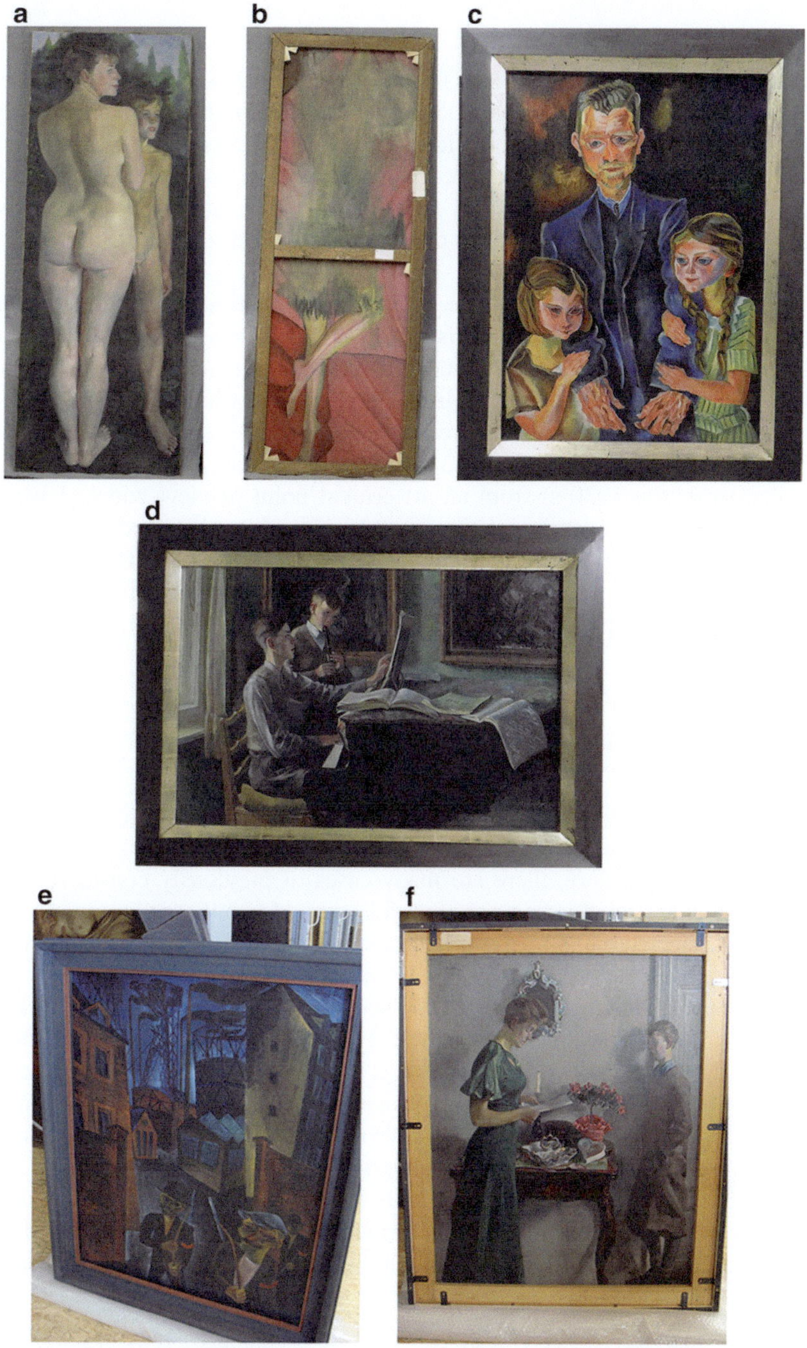

Fig. 30.3 Conrad Felixmüller: (**a**) *Schönheit und Jugend* (1932). ["Beauty and youth"]. 172 cm × 62 cm. Oil on canvas. Signed and dated. Image: Courtesy of Museum am Ostwall, Dortmund. VG Bild-Kunst. (**b**) Verso depicting a dancer/Ballerina; partially overpainted. (Described in Catalogue of 1984 incorrectly as: "Verso unfinished painting of a lying woman"). Courtesy of Museum am Ostwall, Dortmund. VG Bild-Kunst. (**c**) *Dr. Heinrich Becker and his daughters* (1924). Originally recto, later rejected by the artist: Not signed. Museum Kunstpalast,

2. Image *"Dr. Heinrich Becker with his daughters"*, (1924), (Fig. 30.3c, turned to the verso before 1934 and painted on the new recto side with "Boys playing instruments", Fig. 30.3d). Both sides are equally accessible, in a double-sided frame which is equipped with appropriate suspension rings to be fixed on a ceiling if required ([13]/2006). Based on art historical evaluation and contemporary taste, the expressionist painting, though, is regarded as the predominant and important painting.
3. Image *"Ruhrrevier II"* (Fig. 30.3e, 1920, originally recto. Rejected, overpainted and turned to the verso by the artist before 1934. Canvas reused in 1934 on the new and unpainted recto side, to depict *"The birthday letter"*, Fig. 30.3f). The ochre over-paint of the discarded image on the verso was removed in 1981, and the canvas was turned around to present '*Ruhrrevier II*' as the main picture again. The artist´s decision made in 1932 was overruled, since Felixmüller´s paintings of that time (the 1930´s) from an art critical point of view widely have been regarded as relatively uninteresting in style and subject [8]. According to statements made by Felixmüller, the removal of overpainting and the retrospective exhibition of earlier recto paintings in several cases clearly had been appreciated by the artist ([21], p. 44–45).[3]

30.4 Conclusions

This paper demonstrates that a rejected but fully signed painting on a verso in most cases requires and deserves a better attention than a half finished sketch or a failed composition. Furthermore, there are many cases of previously considered inconspicuous sketches which later were regarded to be of significant importance [as in the case of a rejected but almost finished sketch for an awarded and later lost painting "Portrait of Liz Bertram" (1928) by Heinrich Ehmsen, 20].[4]

Specific attention is particularly necessary to paintings on the verso that are not considered to be of the first rank, since they are more prone to neglect and thus subject to decay, grime deposits and flaking. The application of appropriate conservation interventions even on seemingly irrelevant verso-paintings is crucial for their protection and survival for their prospective research (e.g. temporal and removable vibration absorbers, consolidation treatments and visual documentation).

In this context, it is of utmost importance to shed light on the reasons that led artists to reject paintings. Relevant information of this kind would need to be included into adequate documentation systems to profoundly understand a painting

[3] In 1940, Felixmüller is stating [Translation]: »In course of the revision of my surviving expressionist paintings I realized that the color tones of my palette became duller and duller – ending up in dark paintings … there are only a few paintings which I did not destroy, and now I am glad that the painting "Workers on a night train" has been rescued«. ([21], p. 44–45).

[4] E.g. Heinrich Ehmsen/Portrait of Lis Bertram-Ehmsen. Sketch of lost painting, 1928, on verso of "Cathedrals" (1930's). (Osborn, 1928).

with a verso side, to help discuss approaches of presentation and to make accessible specialized knowledge to professional fellows as well as to the interested public.

The increasingly large compilation of case samples may stimulate further research in this field, merge into the provision of appropriate guidelines and evoke positive effects regarding the care and condition of recto verso paintings in general [18].

References

1. Althöfer H (1968) Restauriert und neu entdeckt. Aus der Arbeit der Restaurierung. Bildhefte des Kunstmuseums Düsseldorf, fig. 15.
2. Anon. (1931) Kölner Stadtanzeiger, evening edition # 634, „drittes Blatt". Schluß mit dieser Kunstkritik oder mit dieser Kunst? December 16.
3. Bünsche B (1999) Distanzierung und Zerstörung. Zum Frühwerk von Conrad felixmüller. In: Beiträge zur Erhaltung von Kunstwerken. Berlin, pp. 63–71.
4. De Chassey É, Dziewanska M (eds) (2015) Andrzej Wróblewski: Recto/Verso. Books No. 10, Museum of Modern Art in Warsaw.
5. De Vries A et al eds (2018) De Ploeg. Avant-Garde in Groningen 1918–1928. WBooks – Groninger Museum – Stichting 100 jaar De Ploeg.
6. Dietemann P et al. (2012) Ernst Ludwig Kirchners Bindemittel – Material und optische Bildwirkung. Conference contribution, Aufbruch in die Farbe, Staatliche Akademie der Bildenden Künste Stuttgart, 15–16. November.
7. Dittmar P (2012) Das Geheimnis der zweiten Leinwand – von erstaunlichen Funden und verborgenen Bildern. In: Die Welt, 07.12.2010.
8. Eggenstein G (2016) Conrad Felixmüller im Spannungsfeld der politischen Systeme in Deutschland. Conrad Felixmüller – »Kunst ist eine historische Angelegenheit«. Hengstenberg (editor), Kettler-Verlag Bönen, pp. 47–58.
9. Foulke R (2008) Stretchers for double-sided paintings. And: Prins, Steven; A modular design for double-sided stretchers. In: Stretchers and Strainers. Painting Conservation Catalog, Vol. 2, American Institute for Conservation (AIC). Pp. 193 and pp. 244.
10. Goetz E; van den Berg KJ (2005) Verkennend analytisch onderzoek naar de was verven van de Ploeg. In: Ploeg Jaarboek 2005, Stichting De Ploeg, Groninger Museum, p. 43–53.
11. Griesser-Stermscheg M (2013) Tabu Depot. Das Museumsdepot in Geschichte und Gegenwart. Böhlau, Wien, Köln, Weimar.
12. Grisebach L (2015) Werkintegrität. In: Der doppelte Kirchner. Die zwei Seiten der Leinwand. Wienand-Verlag, Köln, pp. 24–37.
13. Heydenreich G (1994) Restaurierungsdokumentation Gemälde, 09.03.2006, archive Restaurierungszentrum, Landeshauptstadt Düsseldorf, File "Felixmüller". Correspondence with Titus Felixmüller regarding the framing of the double-sided painting and conservation issues related to the presentation of both sides (25.04.1994).
14. Herold I (2015) Lorenz, Ulrike; Sadowsky, Thorsten (editors); Der doppelte Kirchner. Kunsthalle Mannheim / Kirchner Museum Davos, Wienand, Köln.
15. Klingsöhr-Leroy K (ed) (2016) Franz Marc. Zwischen Utopie und Apokalypse. Sieveking, München, 2016.
16. Ludwig J (1991) Rücksichten. Zeitung zur Ausstellung »Bilder von Hinten«, Freiburg im Breisgau, 25. October.
17. Posselt DW (2012) Otto Mueller – Einfach. Zweifach. Der Blick hinter die Fassaden als Auftrag für die Forschung. P. 131–133. In: Otto Mueller 1874–1930. Einfach-Eigen-Einzig. Lehmbruck-Museum. ARTeFACT, Apolda.
18. Runeberg U (2011) Übertüncht und abgedeckt – ein verborgener Heckel? In: Restauro, Callwey (8/2011), p. 8.

19. Spielmann H (1996) Conrad Felixmüller. Monographien und Werkverzeichnis der Gemälde. Wienand.
20. Osborn M Catalogue, "Wander-Ausstellung des Reichsverbandes Bildender Künstler. Georg Schicht Preis für das schönste Deutsche Frauen-Porträt 1928".
21. Penndorf J (1996) C. Felixmüller und seine Sammler in den Dresdner Jahren (1917–34), in: Conrad Felixmüller – Monografie und WVZ der Gemälde/ Wienand.
22. Philipsen C et al (eds) (2016) Lyonel Feininger: Paris 1912. Die Rückkehr eines verlorenen Gemäldes. Sandstein.
23. Springer Pr (2004); Das verkehrte Bild. Inversion als bildnerische Strategie. Delmenhorst.
24. Strigalev A and Harten J (eds) (1994) Vladimir Tatlin Retrospektive [Düsseldorf, Baden-Baden, Moscow and St. Petersburg], Du Mont, Cologne, p. 196.
25. Verein August Macke Haus e.V., Bonn (2011) Treffpunkt und Topos: Schloss Dilborn 1911–1931. Das Künstlerehepaar Heinrich Nauen und Marie von Malachowski und seine Gäste. Kettler, Bönen.

Part III
Model Studies

Chapter 31
Time-Gated Photoluminescence Microscopy as an Emerging Method for the Study of Paint Cross-Sections

Alessia Artesani, Roberta De Angelis, Austin Nevin, Gianluca Valentini, and Daniela Comelli

31.1 Introduction

In painting conservation, the study of the paint stratigraphy is fundamental and the most straightforward way of examining stratigraphic features is through the preparation of cross-sections. When cross-sections of paint samples are prepared they are typically examined with an epifluorescence microscope equipped with a UV light source, which is particularly advantageous for the examination of organic layers, for the detection of luminescent pigments including many lakes, and for distinguishing varnish layers from paint layers. Indeed many materials used in painting emit photons when excited by UV light. The process is generally known as photoluminescence (PL).

In this work, we present an improvement of basic fluorescence equipment and describe the benefits of time-resolved and spectrally-resolved PL microscopy for the study of paint cross-sections. In comparing the standard UV micro-images of samples with time resolved PL images, material heterogeneities become more apparent, and information related to the luminescence, such as spectrum and lifetime, can be measured. These parameters can be useful to understand the nature of the emitting species which we highlight in this work through a case study of the PL emission of two paint cross-sections from two eighteenth C. wall paintings, showing a complex stratigraphy and degradation products.

A. Artesani (✉) · G. Valentini · D. Comelli
Physics Department, Politecnico di Milano, Milano, Italy
e-mail: alessia.artesani@polimi.it

R. De Angelis
Department of Conservation and Built Heritage, University of Malta, Msida, Malta

A. Nevin
Department of Conservation, University of Gothenburg, Gothenburg, Sweden

K. J. van den Berg et al. (eds.), *Conservation of Modern Oil Paintings*,
https://doi.org/10.1007/978-3-030-19254-9_31

In order to better appreciate the benefits of a time-resolved approach, the basic principles of PL phenomenon are summarised [11, 13]. In particular, we refer to the Jablonski diagram which is a simplified scheme for describing the luminescence phenomena for simple molecules, but that can be extended to more complex luminescent species and crystalline solids (Fig. 31.1a).

Following absorption of light, molecules are usually excited from the ground state to a higher electronic level ($S_0 \rightarrow S_n$, n = 1,2,3,…) (Fig. 31.1b). The system reaches quickly (10^{-12} s or less) the thermal equilibrium ($S_{n>1} \rightarrow S_1$), and following this, it returns to the ground state. The transition can occur in a radiative or non-radiative way and in the former case the process produces the emission of photon and is generally referred to as *fluorescence*. The lifetime represents the average time that the molecule spends in the excited state before returning to the ground state, and for fluorescence, it is of order of 10^{-12}–10^{-9} s. It might happen that additional processes take place and molecules in state S_1 undergo a less likely spin conversion (called intersystem crossing) from the singlet state level to the triplet state ($S_1 \rightarrow T_1$). In this case the relaxation to the ground state ($T_1 \rightarrow S_0$, Fig. 31.1b) is forbidden by selection rules and the result is that the transition is less efficient with a longer lifetime, with typical values ranging from 10^{-6} to 10^0 s [11]. The process is generally referred to as *phosphorescence*.

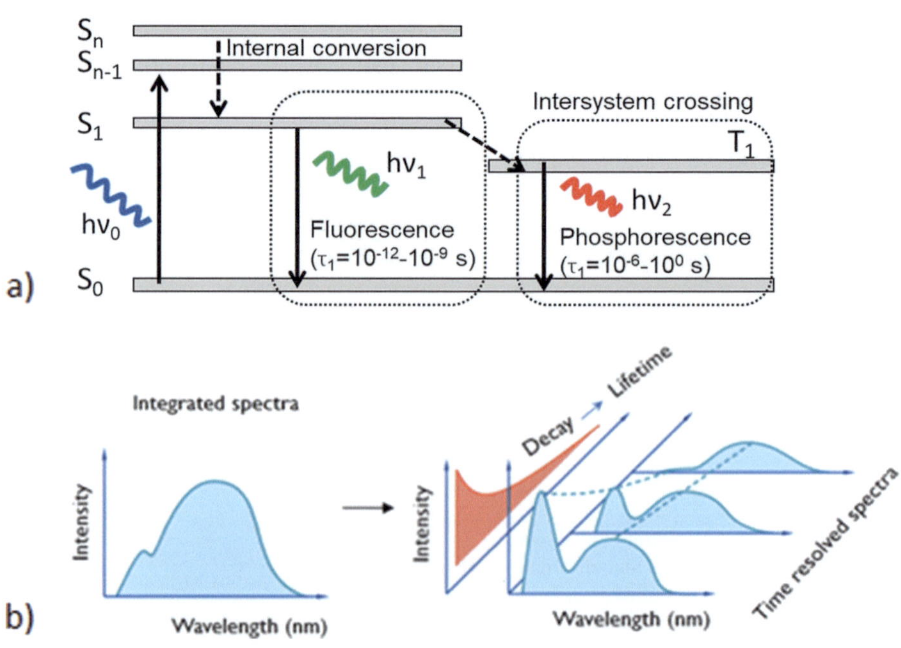

Fig. 31.1 (**a**) Simplified Jablonski diagram of the radiative transitions in luminescent compounds, in which are highlighted the fluorescence emission ($S_1 \rightarrow S_0$) and the *phosphorescence* emission ($T_1 \rightarrow S_0$). (**b**) Evolution of luminescence as integrated spectra and as a function of time, following pulsed excitation

When dealing with crystalline solids, we cannot more refer to molecular excited states, but more properly to electronic energetic levels and to luminescent centres associated with a crystal lattice. Like molecules, solids can exhibit luminescence. The radiative emission in this case is generated by electron-hole annihilation and governed by optical transition selection rules. The most frequent phenomena are (i) the exciton recombination with lifetime of order of hundreds of picoseconds and (ii) the radiative recombination via trap states levels associated with defects of the crystal lattice and by lifetime values ranging from few nanoseconds to tens of microseconds.

Whatever the emitting species, a radiative transition can be described in terms of the *emission spectrum*, where the emission intensity is measured as a function of the emitted energy (or wavelength). The time-resolved approach is based on the detection of the emission from materials at different delays with respect to excitation, and is helpful for discriminating amongst different luminescent phenomena in organic and inorganic compounds. In fact, when performing PL measurements in a steady state approach the emission spectrum is actually the integrated spectrum of the emissions that occur from different transitions. Instead, the use of a time-resolved approach allows the discrimination between spectral emissions occurring at different time-scale (Fig. 31.1b). In particular, emissions from less probable and long-lived transitions (i.e. from triplet states or trap state levels) can be highlighted with a time-gated approach. For the sake of completeness, it is stressed that PL techniques are not strictly analytical, since they cannot provide data for the chemical identification of compounds. Nonetheless, thanks to their high-sensitivity, time-resolved PL spectroscopy and imaging are widely used for the advanced characterization of materials in many sectors, including the study and fabrication of nano-materials. Hence a TRPL approach is particularly convenient for the study of cultural heritage, both for highlighting heterogeneity and for the detection of trace amounts of luminescent materials.

31.2 Materials and Methods

Here, we schematically describe the set up for microscopic analysis and highlight its main features (Fig. 31.2a). The whole set-up has been described in detail elsewhere [5]. We use a UV pulsed laser source ($\lambda = 355$ nm, third harmonic Nd:YAG laser) with a laser pulse width around 1 ns and a repetition rate of 100 Hz, implying excitation cycles composed of an almost instantaneous excitation at $t = 0$ followed by 10 ms of dark-mode, where excited fluorophores can relax back to the ground state. In this way, it is possible to probe emission decay kinetics with lifetimes ranging from 1 ns to 1 ms (Fig. 31.2b). The laser light is focused through an optical path onto the flat surface of the sample. In the present set-up, the magnification is made with the aid of a 15× reflective objective (NA = 0.55) and the whole system has a field of view of 1 mm of lateral size with a lateral resolution of around 1 μm [5]. Finally, the luminescence of the sample is collected and recorded via a time-gated

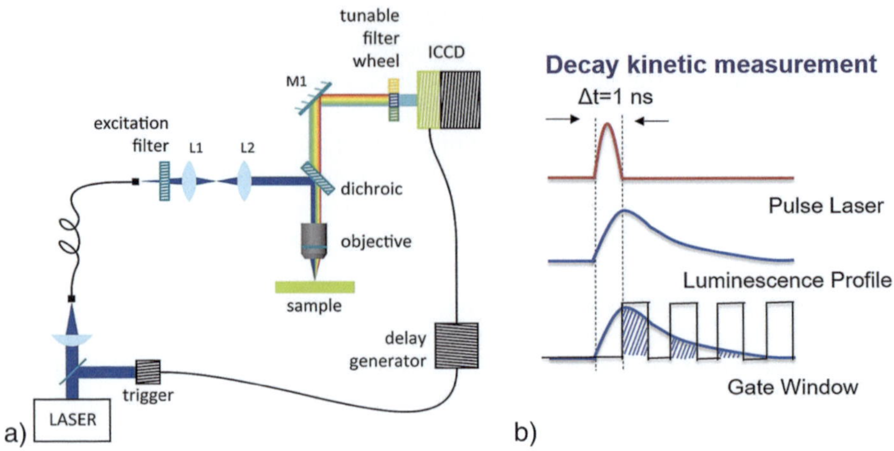

Fig. 31.2 (**a**) the scheme of the Time-Resolved PL microscope set-up (**b**) Time-resolved detection mode that illustrate the correspondence between the laser pulse, luminescence decay kinetic and the gate window of the ICCD camera

camera equipped with a filter wheel. The latter component, mounting ten band-pass filters with 40 nm of Full Width Half Maximum (FKB-VIS-40, Thorlabs Inc.), allows the multi-spectral detection in the 370–870 nm spectral region. The time-gated ICCD camera is synchronized with laser pulses through an electronic circuit and a delay generator. The time-gated camera is able to switch on and off in temporal windows ranging from 5 ns to the continuous mode. In this way, the preferred time delay and gate window is chosen and different decay kinetics can be properly sampled (Fig. 31.2b).

In this research, PL emission from samples has been studied by employing two different gate windows: (1) a *short* gate window that starts synchronously with the laser pulse and lasts 100 ns, (2) a *long* gate window, opening with a delay of 0.5 μs from the laser pulse and lasting for 10 μs. First the emission characterized by short lifetime (fluorescence) is detected. In this temporal regime, when considering paint layers, we expect to detect PL signal ascribed to the exciton recombination in inorganic pigments as well as the binder $S_1 \rightarrow S_0$ transition with a lifetime in the order of a few ns.

Second, in the *long* temporal regime, the *short* emissions have negligible intensity. Instead, we can detect long living emissions such as those from trap states in inorganic pigments with a crystalline structure. The long temporal window is thus more selective and the contribution from organic materials can be neglected, since oils and binding media do not contribute to phosphorescence.

In both *short* and *long* temporal regimes, we perform multi-spectral imaging and generate time-gated, multi-spectral imaging datasets. Within each dataset, every image corresponds to the emission in a specific temporal regime and in a specific spectral band centred at the value reported in the label. For example: BP400/40 indicates the use of a bandpass filter, centred at 400 nm with FWHM of 40 nm. Finally, when selecting a small area in a single layer, the time-gated spectrum of the corresponding material can be reconstructed. In a specific spectral band, it is also possible to reconstruct the emission decay kinetic of the luminescence process by varying the delay between the pulsed excitation and the acquisition gate window.

In order to illustrate the capabilities of the system for real case studies, we present the analyses of two micro-samples (C1 and C2) from two different wall paintings. Both wall paintings are located in Malta and are painted on a porous limestone (Globigerina Limestone). The chemical composition of the two micro-samples was studied previously [1]. Micro-sample C1 was taken from a wall painting that was painted using an egg-based binding medium in the eighteenth C and covered by at least four subsequent oil-based repainting schemes: two layers of whitewashing schemes (one probably dated to the nineteenth C. and the other to 1901), one redecoration Scheme (1903) and a later restoration Scheme (1963) [1]. Micro-sample C2 is from an oil painting dated to the 1750s.

Both wall paintings have been exposed to drastic environmental conditions (high temperature and humidity levels) and rainwater infiltration; they are affected by inherent oil paint degradation processes (soap formation) and deterioration due to soluble salt activity. Since the sampling was performed at different times, the C1 and C2 samples have been embedded in two different resins: Buehler EpoPlast resin and Tiranti Clear Casting polyester resin, respectively. Both samples were dry polished using Micromesh cloths with the aid of a MOPAS XS polisher.

31.3 Results

In Fig. 31.3, an image of the micro-sample C1 under white light illumination is shown together with a detail taken with a conventional UV microscope (LEICA DMRE, Mercury lamp excitation centred at 365 nm). This sample has a large number of layers consisting of (from bottom): the oldest paint layer (possibly eighteenth C.) (Layer 1), highly altered by the presence of an organic material; several white layers dated to 1901 (from Layer 2 to 5); repainting layers (red ochre and umber, layer number 6 and 7) associated to a restoration from 1963. The UV image highlights the stratigraphy and diversity in the optical properties of the two white paint layers: the first from layers 5 and 4 has a blue-green emission with sparkling fluorescent particles. The other white in layer 3 has a reddish fluorescence.

In this case, the time-gated PL imaging provides more information that can be useful for determining the identification of luminescent pigments on the basis of the spectral features of the emission at different time regimes. Here, we show the detail of a single image (BP380/20) with the *short* gated window (Fig. 31.3b, single image at short gate) and the entire spectral dataset achieved with the *long* gated window (Fig. 31.3b, four images at long gate).

In order to plot the distribution of material with similar spectral emission and for a better comprehension of the complex stratigraphy, we treated the data through the Spectral Angle Mapping (SAM) method that creates clusters of similar emitting areas [10]. This algorithm determines the spectral similarity between two sub-areas

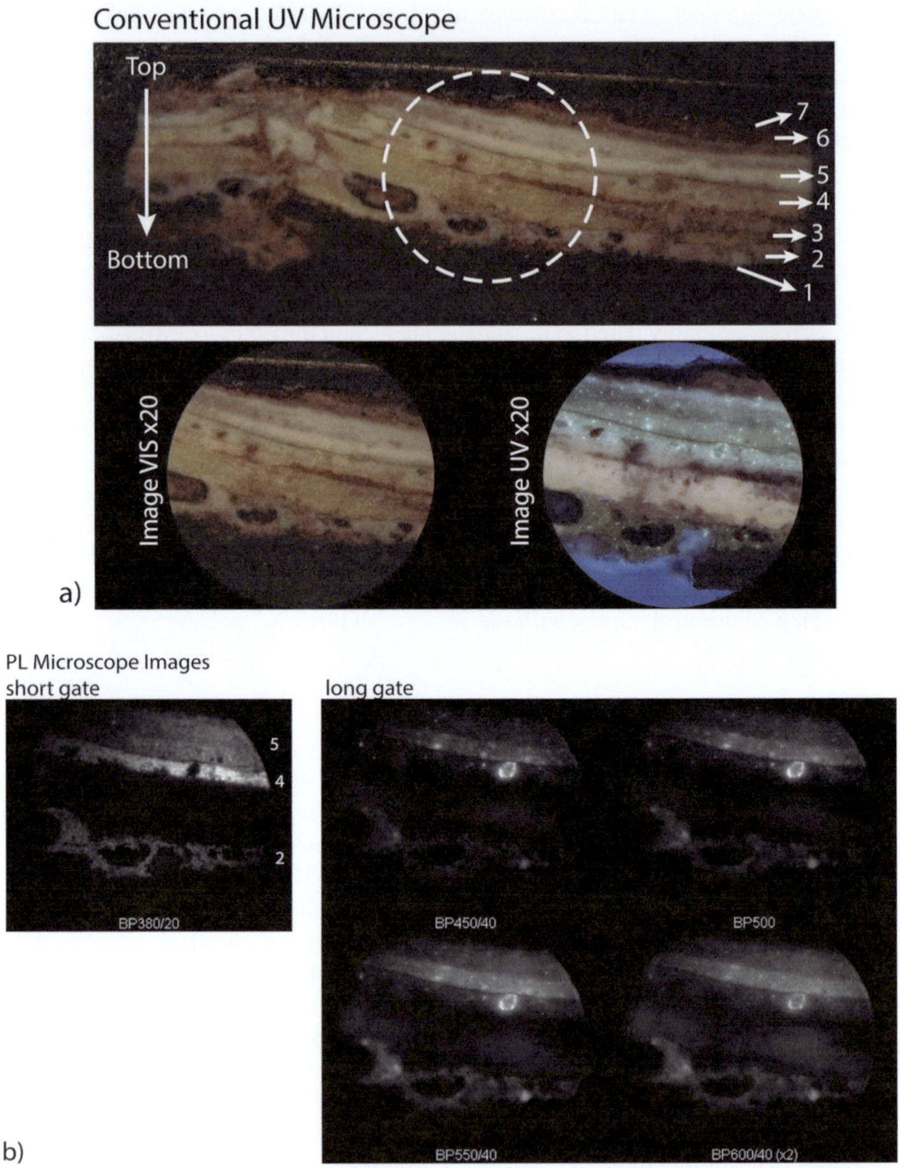

Fig. 31.3 (**a**) On the left, images of the paint stratigraphy of Sample C1 with white light (Image VIS) and under UV excitation (Image UV) obtained with a conventional UV microscope; (**b**) the PL single image (BP380/20) with a *short* gated window and (**c**) PL data set achieve with *long* gated window in four different spectral regions

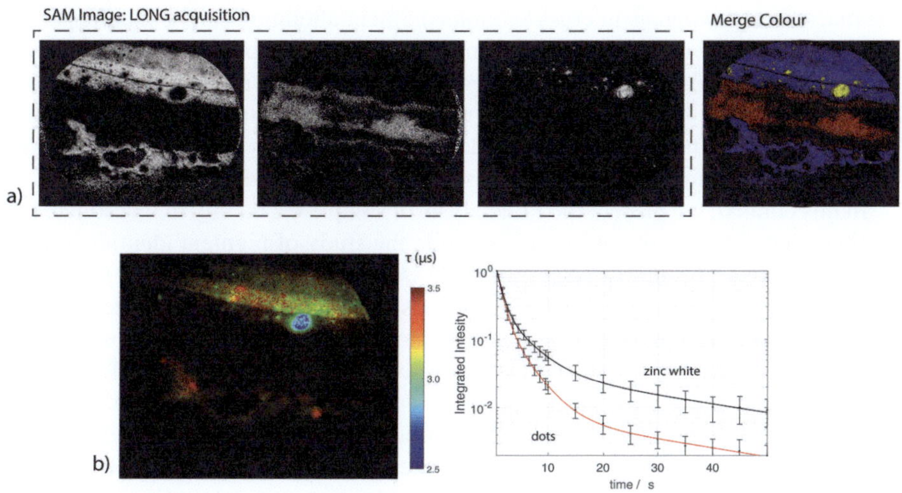

Fig. 31.4 (**a**) Spectral Angle Mapping (SAM) analysis on layers of the cross-section of sample C1 with similar emission indicate three main components highlighted in the Merge colour image. (**b**) Lifetime map (red colour longer emission, blue colour shorter emission) and the comparison of the decay kinetic between the zinc white and luminescent particles in layer 4

of interest by calculating the angle between the two spectra, treating them as vectors in a space with dimensionality equal to the number of bands [10].

Despite differences in the luminescence intensity, it is clear that layers 2, 4 and 5 all contain pigment with same emission (Fig. 31.4a). In particular, they are characterized by a sharp emission at 380 nm and a broad green emission with a peak at 550 nm, typical of zinc white (ZnO) [2, 3]. The spectral emission of layer 3 in the *short* and *long* regimes, and its emission lifetime (about 10 µs, data not shown), suggest that this thick layer consists of lead white ($2PbCO_3{\cdot}Pb(OH)_2$)[9].

We finally conclude the analysis of this sample focusing our attention on the distributed luminescent particles, detectable within the zinc white layers employing conventional UV-fluorescence microscopy (Fig. 31.3a). The time-gated microscopy analysis allows a better understanding of the emission from the particles. In the *short* regime, the particles do not emit. In contrast, at *long* acquisition times the same areas have a bright luminescence, which is more intense than that of zinc white, and particularly evident in the 400–500 nm band. The lifetime characterization reveals that the particles can be differentiated because of their decay kinetic, as displayed in Fig. 31.4b. In this figure, the calculated lifetime map is seen in which the red colour corresponds to longer lifetimes and blue to shorter lifetimes. The luminescent particles have a shorter lifetime ($<\tau>$ ~2.5 µs) with respect to the surrounding zinc white ($<\tau>$ ~5 µs). The fact that these luminescent particles displays a microsecond decay kinetic indicates that they could be related to crystalline compounds with a high density of crystal defects.

Besides revealing features of the original painting stratigrahy, the examination of cross-sections can be helpful in the detection of degradation products. Deterioration phenomena may be visually recognized as crusts, protrusions and efflorescence on the macro-scale, whereas in cross-sections pigment alteration can be manifested as salt formation and metal soap aggregation [12]. Metal soap aggregation has been documented both visually and through Fourier Transform infrared imaging of cross-sections [6]. While the luminescence of metal-based pigments has been investigated extensively, the luminescence of degradation products (salts and metal soap aggregates) is not well understood and is the subject of a recent research project [7].

In this context, we present the PL analysis of sample C2 mainly to investigate the luminescence of degradation products, while the study of pigment identification is briefly reported. The stratigraphy of C2 is less complex than sample C1. As visible from Fig. 31.5, the sample includes the porous substrate, a thin organic-rich brown-red preparatory layer and a red paint layer on top. The materials of this cross-section have been identified through Energy-Dispersive X-Ray spectroscopy (SEM-EDX) analyses. The white layer at the bottom of the sample is a lime plaster layer ($CaCO_3$). On top, a more transparent layer is made of an unidentified organic binder. In the main red layer an orange matrix in which other red and white pigment particles are distributed, is composed of an iron based red pigment (red ochre, Fe_2O_3) and calcium carbonate. The white particles are a lead-based pigments (lead white), whereas the red particles are vermillion (HgS). Finally a varnish layer is the thin transparent layer is present.

Included in the cross-section of C3 and mainly located at the surface of the stratigraphy, there are several agglomerates that are primarily composed of lead white (identified with SEM-EDX). These agglomerates present different optical properties with respect to the surrounding red paint, as they appear to be translucent in the visible image, while they are whitish under UV excitation with conventional microscopy (Fig. 31.5). The PL properties of these agglomerates can be studied through the use of the SAM algorithm applied to the multispectral dataset at the *long* microsecond timescale. The result is synthetically shown as a false colour image in Fig. 31.6.

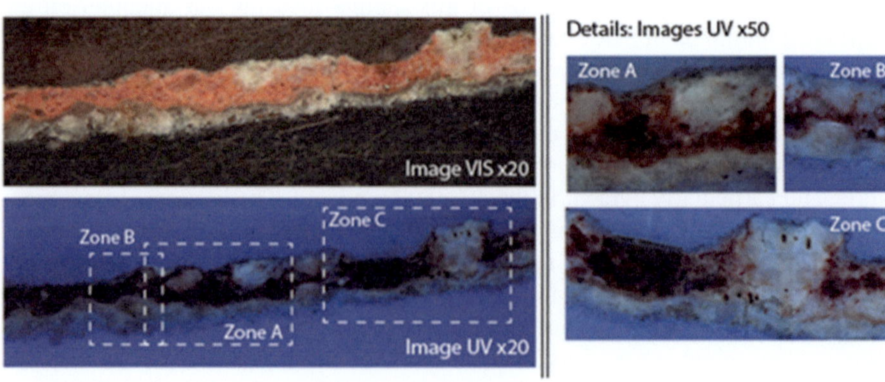

Fig. 31.5 On the left, images of the paint stratigraphy of the cross-section of sample C3 with white light (Image VIS) and under UV excitation (Image UV) obtained with a conventional UV microscope; on the right a detail of the UV images from different areas containing aggregate and protrusions

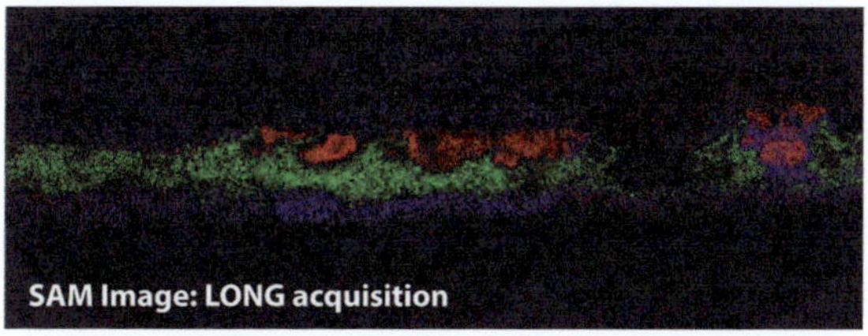

Fig. 31.6 Spectral Angle Mapping (SAM) analysis on the layers of sample C3 with similar emission indicate three main components highlighted in the false colour image: the red channel characterizes the white agglomerates, the green channel the distribution of vermillion and the blue channel the calcium carbonate of the ground

The SAM analysis clearly separates the main material components, such as the calcium carbonate substrate (Blue Channel) and the main red layer (Green Channel). Despite their different visual appearance in UV, the white and translucent agglomerates on top of the red layer have similar PL features, and belong to the same SAM cluster (Red Channel in Fig. 31.6). In particular, the emission of these agglomerates at *short* time-scale (data not shown) is centred at 450 nm, while the emission at *long* time-scale occurs at 550 nm. While it is not possible to propose material identification on the basis of PL features, a plausible suggestion is to relate these disruptions of the layer to the formation of degradation products. It is worth stressing that in this case the application of complementary analytical techniques, like FTIR mapping sensitive to chemical compounds for example, would complement luminescence data.

As a final consideration, we observe that PL suggests that calcium carbonate is present in the white agglomerate in zone C, as previously reported by SEM-EDX analysis. This indicates a possible eruption of the ground on top of the stratigraphy. The white protrusions present on top of the stratigraphy have similar luminescence properties under UV excitation and this may indicate that they share similar composition, such as the presence of lead white.

31.4 Conclusion

In this work we have shown the benefits of time-resolved PL microscopy applied to investigate painting cross-sections. In the past, the approach has been effectively employed to detect the presence of trace impurities and crystal defects in historical pigment powders [4, 8]. Here, the strategy is presented for multi-spectral microscopy for the identification of heterogeneous layers and mapping the presence of luminescent materials even in low concentrations. In this work, we have shown how the temporal resolution can be beneficially applied for investigating on the emission properties of paint layers, and recognizing degradation products within complex layer stratigraphy. In conclusion, while the data from luminescence imaging alone cannot be used as an analytical method, data improves the understanding of chemical differences in samples and the results from luminescence can inform the interpretation of stratigraphy and of reaction products. Of critical importance for a complete characterization is the comparison of photoluminescence data with other chemical methods which may provide molecular and elemental information.

References

1. De Angelis, R. *et al.* (2004) 'Oil Painting of Stone: A Case Study on Original Technique and Deterioration of an early 20th Century Painting by Giuseppe Cali in Malta', in *Proceedings of the 10th International Congress on Deterioration and Conservation of Stone: Stockholm June 27–July 2, 2004*.
2. Artesani, A. *et al.* (2017) 'A photoluminescence study of the changes induced in the zinc white pigment by formation of zinc complexes', *Materials*, 10(4). doi: https://doi.org/10.3390/ma10040340.
3. Artesani, A. (2017) 'Photoluminescence properties of zinc white', *Nuovo Cimento della Societa Italiana di Fisica C*, 40(2). doi: https://doi.org/10.1393/ncc/i2017-17100-0.
4. Bellei, S. *et al.* (2015) 'Multianalytical Study of Historical Luminescent Lithopone for the Detection of Impurities and Trace Metal Ions', *Analytical Chemistry*, 87(12), pp. 6049–6056. doi: https://doi.org/10.1021/acs.analchem.5b00560.
5. Comelli, D. *et al.* (2017) 'Time-resolved photoluminescence microscopy for the analysis of semiconductor-based paint layers', *Materials*, 10(11). doi: https://doi.org/10.3390/ma10111335.
6. Cotte, M. *et al.* (2017) 'Lead soaps in paintings: Friends or foes?', *Studies in Conservation*, 62(1), pp. 2–23. doi: https://doi.org/10.1080/00393630.2016.1232529.
7. *European research platform on ancient materials IPANEMA (CNRS, MCC, plateau de Saclay) at synchrotron SOLEIL - On-going PhD projects* (no date). Available at: http://ipanema.cnrs.fr/research/on-going-phd-projects/article/on-going-phd-projects?lang=en (Accessed: 11 March 2019).
8. Ghirardello, M. *et al.* (2018) 'Time-Resolved Photoluminescence Microscopy Combined with X-ray Analyses and Raman Spectroscopy Sheds Light on the Imperfect Synthesis of Historical Cadmium Pigments', *Analytical Chemistry*, 90(18), pp. 10771–10779. doi: https://doi.org/10.1021/acs.analchem.8b01666.
9. Gonzalez, V. *et al.* (2017) 'Revealing the Origin and History of Lead-White Pigments by Their Photoluminescence Properties', *Analytical Chemistry*, 89(5), pp. 2909–2918. doi: https://doi.org/10.1021/acs.analchem.6b04195.
10. Kruse, F. A. *et al.* (1993) 'The spectral image processing system (SIPS)-interactive visualization and analysis of imaging spectrometer data', *Remote Sensing of Environment*. doi: https://doi.org/10.1016/0034-4257(93)90013-N.
11. Lakowicz, J. R. (2006) *Principles of fluorescence spectroscopy, Principles of Fluorescence Spectroscopy*. doi: https://doi.org/10.1007/978-0-387-46312-4.
12. Osmond, G. *et al.* (2012) 'Metal stearate distributions in modern artists' oil paints: Surface and cross-sectional investigation of reference paint films using conventional and synchrotron infrared microspectroscopy', *Applied Spectroscopy*, 66(10), pp. 1136–1144. doi: https://doi.org/10.1366/12-06659.
13. Pelant, I. and Valenta, J. (2012) *Luminescence Spectroscopy of Semiconductors, Luminescence Spectroscopy of Semiconductors*. doi: https://doi.org/10.1093/acprof:oso/9780199588336.001.0001.

Chapter 32
Mechanical and Moisture Sorption Properties of Commercial Artists' Oil Paint by Dynamic Mechanical Thermal Analysis (DMA), Nanoindentation, and Dynamic Vapour Sorption (DVS)

Cecil Krarup Andersen, Ashley Freeman, Martin Nordvig Mortensen, Vincent Beltran, Michał Łukomski, and Alan Phenix

32.1 Introduction

It is well known that artists' oil paints—effectively filled polymer systems [20]—may differ considerably in their material properties depending on a variety of intrinsic factors including: binder composition; the nature of pigments/extenders; driers; and other additives. Similarly, oil paints may be differently affected by extrinsic factors such as ageing, changes in the environmental conditions, or exposure to conservation treatments like cleaning. Knowledge of the mechanical properties of paints and the way those properties change with variations in moisture content and temperature can inform active conservation treatments and passive preventive conservation measures.

The mechanical properties of artists' oil paints have been studied extensively by Mecklenburg and his co-workers using tensile testing to generate load/extension curves from which values for (elastic) modulus and tensile strength could be determined (eg. [10]). It is recognised, however, that oil paints, when cured, may be more or less viscoelastic depending on some of the intrinsic factors mentioned above. They may behave more glassy (elastic) or more viscous (plastic) depending

C. K. Andersen (✉)
Institute of Conservation, The Royal Danish Academy of Fine Arts, Schools of Architecture, Design and Conservation, Copenhagen, Denmark
e-mail: cka@kadk.dk

A. Freeman · V. Beltran · M. Łukomski · A. Phenix
Getty Conservation Institute, Los Angeles, CA, USA
e-mail: VBeltran@getty.edu; MLukomski@getty.edu

M. N. Mortensen
National Museum of Denmark, Kgs. Lyngby, Denmark
e-mail: martin.n.mortensen@natmus.dk

K. J. van den Berg et al. (eds.), *Conservation of Modern Oil Paintings*,
https://doi.org/10.1007/978-3-030-19254-9_32

on temperature and/or the rate of loading. If an oil paint shows a distinct softening (reduction in elastic modulus) when subjected to a temperature ramp, as it changes from the glassy state to the leathery/rubbery state, a glass transition temperature (T_g) may be determinable. The value of the T_g indicates the state in which the material likely exists at any operating temperature.

Within art conservation, research into the viscoelastic behavior of oil paints commenced with the influential study by Hedley, Odlyha and co-workers that was set in the context of cleaning. A specific point of interest was the long-term impact on paint properties of exposure to solvents and water [8]. As the first study in this subject area to use dynamic methods of thermal mechanical analysis, Hedley's group demonstrated the potential for significant changes to the mechanical properties of some oil paints caused by extraction (leaching) of plasticizing soluble components of the binder phase. Using the method of Dynamic Load Thermal Mechanical Analysis (DLTMA), this work documented the temperature-dependent softening transitions of a sample of commercial oil paint[1]. Three transitions were observed, the *onset* of the lowest-temperature one occurring at −20 °C. This transition was identified as the major change from glassy to viscoelastic state. It was attributed to "the glass transition temperature (T_g) of linseed oil itself".

The question of the temperature range over which artists' oil paints underwent the transition from glassy to leathery behaviour remained effectively unstudied experimentally for some time after Hedley and Odlyha's initial explorations; the main contribution on the topic was the review by Michalski [14]. Despite the "spotty collection of data" reported in the literature, Michalski concluded that there was sufficient evidence "to piece together the glassy/rubbery transition of oil paint" and that "the fundamental glass/rubber transition for oil paint starts near −30 °C (−22 °F) and ends before 0 °C (32 °F) for linseed, walnut and poppy oil, with or without pigments, with or without aging." On this evidence, one would expect most oil paints to be in the rubbery/leathery (viscoelastic), not glassy, state at room temperature. The topic of the T_g of artists' oil paint was revisited by Phenix [18], using DMA (Dynamic Mechanical Analysis) introduced by [8] two decades earlier.[2] Depending primarily on their pigment content and condition, the paints tested varied considerably in their thermo-mechanical properties, some showing effectively no distinct glass transition in the near-ambient temperature range, while others

[1] Test paint film of Grumbacher Burnt Sienna (linseed oil binder) prepared in 1978 by M. Mecklenburg.

[2] See Sect. 32.2.2 for instrumental details of DMA. Theoretical and experimental treatments of viscoelasticity have derived vector-based descriptors of mechanical properties in terms of complex modulus (E∗), storage or elastic modulus (E'), and viscous or loss modulus (E"), where E∗ = E' + iE". Usually this relationship is expressed as G∗ = G' + iG" where the G terms represent the corresponding shear moduli of the material. A further useful parameter, tan δ, is the ratio G"/G' [5]. The most common DMA test is a thermal scan during which changes in the parameters of G', G" and tan δ are measured as the temperature is increased at a set rate. The temperature at which any thermal transition occurs will depend on the frequency of the applied periodic stress. Conventionally, the temperature at the peak in the tan δ trace at 1 Hz is taken as the T_g of polymers ([4, 6]: 509). A comprehensive critical overview of the determination of T_g using DMA is provided by [2].

showed strong glass transitions with a narrow T_g peak and a considerable change of modulus that could be either above or below room temperature. However, only a very few paints had T_g values—as indicated by the temperature of the maximum value of the 1 Hz tan δ trace—that were below zero. The lowest value measured was ca. −4 °C for alizarin crimson in cold-pressed linseed oil.

Mechanical properties of paints, determined by tensile testing, have also been shown to vary with the degree of drying/curing and subsequent ageing, as well as pigmentation [5, 6, 10]. These very different properties are essential to conservation decisions such as flattening, consolidation and cleaning. Conservators know from experience that some pigments produce soft and flexible paint films and, therefore, are easier to deform when flattening deformations than for example paints that contain lead or zinc pigments. Earth pigments generally have a tendency to hydrolyse, which makes them sensitive to cleaning treatments [11]. In addition, modern commercial paints have shown to be sometimes problematic due to the addition of organic modifiers such as metal soaps, fatty acids, or oils that are only semi-drying [1, 9], not to mention inorganic extenders that are prone to chemical alteration by exposure to airborne pollutants to produce water-soluble salts [3]. Even without the presence of any hygroscopic, water-soluble salts, the mechanical properties of oil paints are known to be influenced by moisture content, water essentially having a plasticizing effect. This is known from practical experience of treating oil paintings, but it has also been demonstrated experimentally, albeit to a limited extent [9].

All of the topics considered above—the magnitudes of elastic (storage) and viscous (loss) moduli of oil paints of different formulation and condition; the temperature- and moisture content-dependency of those moduli— warrant further investigation. New instrumentation, particularly nanoindentation conducted under controlled environmental conditions, allows those properties to be explored at the sub-microscopic level [19, 22].

Conducted as part of the EU Heritage Plus project *Cleaning Modern Oil Paintings* (CMOP), the aim of the present study was to examine the mechanical properties of selected commercial Winsor & Newton oil paints. Conventional tensile testing was not a viable approach given the nature of the paint samples available to the CMOP group which were constrained by their size, their surface conformation and their mounting onto polyester support from which they could not be removed in many instances. The challenge for the study, therefore, was to find alternative approaches to probe the samples for parameters that characterized the respective physical/mechanical properties of each paint type. It was also desired that the test methods could provide information on changes in surface (versus bulk) properties that may be caused by cleaning liquids, or by shifts in temperature or relative humidity (RH). Given these constraints and aspirations, it was therefore decided to explore the use of nanoindentation paired with dynamic vapour sorption to analyse mechanical properties of selected paints, alongside conventional time-temperature testing by DMA.

32.2 Materials and Methods

32.2.1 Test Samples

The oil paint samples tested here were prepared at the Cultural Heritage Agency of the Netherlands (RCE) in 2006, and have been described previously in Mills et al. [16]. The pigments were ultramarine (UL), cobalt blue (CB), cadmium red (CR), cadmium yellow (CY), raw sienna (RS) and zinc white (ZW). A number of modifications had been made to the tube paints before they were painted out: 2% of aluminum stearate (-AS), 2% of zinc stearate (-ZS), 2% of fatty acids (-FA), 10% of linseed oil (-LN), 10% of alumina hydrate (-AH) were added to the different pure paints; oil was extracted from pure paints by use of filter paper (-EX). For more on sample preparation see Mills et al. [16]. These modifications introduced variations in physical properties that would challenge the instrumental testing. The paint films were applied onto, and usually inseparable from, a 100 μm polyester support (Melinex) providing a technical constraint for the mechanical testing. They were relatively uneven in surface. Since 2006 they had been stored in daylight and indoor conditions climatized for human comfort until February 2012, when they were unmounted and kept in darkness. The samples chosen for this study have not been artificially aged.

As most of the samples are prepared as brushed paint-outs, the brush strokes result in striations on the surface which can hinder the indentation testing. To lessen this effect, the smooth underside of the material that had been in contact with the plastic substrate was examined. Furthermore, yellow ochre (YO) which was prepared using a drawdown method by Mecklenburg in 1999 was also analyzed using indentation. The paints studied vary in their intrinsic physical properties, ranging from soft and pliable to stiff and brittle (see Table 32.1).

32.2.2 Dynamic Mechanical Analysis (DMA)

The Dynamic Mechanical Thermal Analysis was done using a similar method as reported by [18], which allowed determination of the glass transition temperatures of oil paints supported on polyester film. Testing was performed using a Triton Technology DMA 2000 instrument[3] in single cantilever bending mode, with samples ca. 9 mm wide and a nominal sample gauge length of 5 mm. The DMA tests reported here are time/temperature scans in air: the sample was subjected to a sinusoidal oscillating stress whilst the temperature was progressively increased. Initial cooling from ambient was effected by circulating liquid nitrogen through the DMA measuring head. The starting temperature for each experiment were in the region of

[3] http://www.perkinelmer.com/product/dma-8000-analyzer-qtz-window-ssti-clamp-n5330101t

Table 32.1 Details of paint samples and the tests applied.

Winsor and Newton "Artists' Oil Colour"[a]	DMA	DVS	UNHT	Comments on sample properties
Ultramarine (UL) and $MgCO_3$. Cast in 2006	Tested with three repetitions	Tested	Tested at different RH levels with seven repetitions	Semi-hard and easy to handle
Cadmium yellow (CY). Cadmium zinc sulphide, PY35 and $MgCO_3$, Cast in 2006	Pure and modified (AS, ZS, FA, EX, LN, AH)[b]. Tested with three repetitions	Tested	Before and after immersion in N-heptane,. Seven repetitions.	Semi-soft and easy to handle
Cobalt blue (CB). Cobalt zinc silicate, PY74 and $MgCO_3$. Cast in 2006	Too brittle to be tested	Tested		Very brittle and difficult to cut
Cadmium red (CR). $BaSO_4$ and zinc stearate. Cast in 2006	Pure and modified (AS, ZS, FA, EX, LN, AH)[b]. Tested with three repetitions	Tested		Semisoft and easy to handle
Raw sienna (RS). aluminium stearate. Cast in 2006		Tested		Soft and tacky surface
Zinc white (ZW). Zinc oxide, PW4 (ZnO). Cast in 2006		Tested		Hard and brittle
Yellow ochre (YO) prepared in 1999[c] by Marion Mecklenburg			Before and after immersion in *n*-heptane. Seven repetitions.	This sample was thicker and had a more even surface than the other samples

[a]Details on composition derived from Mills et al. [16]
[b]*AS* aluminium stearate added to the pure paint., *ZS* zinc stearate added, *FA* fatty acids added, *EX* Oil extracted, *LN* linseed oil was added, *AH* alumina hydrate added
[c]This paint was cast in 1999 whereas the others had been cast in 2006. All were tested in 2017

−20 °C, (well below the T_g of the paints under consideration) with the endpoint of the ramp at 50 °C. The heating rate was 2 °C per minute. The amplitude/displacement of the vibration was 30 μm. The thermal scans were carried out using vibration frequencies of 1 Hz and 10 Hz, though only the 1 Hz data are reported here.[4] Two or three repeat tests were done for each sample type. The samples were conditioned to the prevailing room environment (approximately 50% RH, 24 °C) prior to testing.

In these time/temperature scans, three main parameters are determined in relation to temperature: storage (elastic) modulus (E′), loss (viscous) modulus (E″) and tan δ, where: $\tan \delta = E''/E'$.

---

[4]In all the time-temperature DMA scans reported in this study, the 10Hz δ maxima were at higher temperatures than the corresponding 1Hz data, which confirms this is the glass transition.

32.2.3 Dynamic Vapour Sorption (DVS)

Water sorption isotherms were acquired using an Aquadyne DVS from Quantachrome Instruments fed with 99.999% pure nitrogen gas at a flow rate between 30 mL per minute and 200 mL per minute. The samples tested were approximately 1 cm by 1 cm dried Winsor & Newton paints on the polyester support. The instrument was programmed to record sample mass every 5 min at discrete RH steps; (30, 50, 70, 90, 70, 50, 30% RH) at 21 °C for the pure paint samples and the polyester support alone. For samples which were modified, a less time consuming scan was conducted to achieve an overview of their response. In this case, the instrument was programmed to record the sample mass once every 5 min at the following RH-steps; 50, 75, and 90% RH at 21 °C. As this overview showed no changes compared to the pure paints, the full RH scan was not done. Since ZW did not show significant change of mass with changed RH, a broader scan was made and hysteresis was not tested. Thus for ZW the programme was 10, 30, 50, 60, 70, 80, and 90% RH at 21 °C. Equilibrium was attained in most RH-steps except at the higher values. Sorption isotherms were calculated with the equilibrium mass at 50%RH as a reference point for the other equilibrium masses. For most RH conditions, the criteria for moving onto the next RH step was a mass change lower than 0.004% per minute with minimum and maximum measuring times of 600 and 650 min, respectively; for extreme RH levels (10, 30 and 90%), the minimum and maximum measuring times were increased to 800 and 850 min, respectively.

32.2.4 Ultra Nanoindentation Tester (UNHT)

Typical Instrumented Indentation Testing (IIT) consists of a probe, with a well-defined geometry, pressed into the surface of a sample producing a load-penetration curve that serves as the "mechanical fingerprint" of the material [7, 23]. Depending on the response of material, be it viscoelastic or otherwise, various iterations of an indentation scheme can be employed. However, there are three common stages of the indentation process: (1) loading, (2) pause, and (3) unloading. When a material exhibits viscoelastic behaviour, a sinusoidal load variation at a selected frequency can be superimposed upon the quasi-static loading process [17]. By analysing the response of a material to such load variations, its stiffness, which is represented by the dynamic reduced storage modulus (E_r'), can be defined. In the limit of linear viscoelasticity, E_r' can be calculated by using Sneddon's stiffness equation to relate stiffness and damping of material [21].[5]

For these tests, three pure samples (UL, CY, and YO) were characterized using an Ultra Nanoindentation Tester (UNHT) (Anton Paar, USA) equipped with a Berkovich (three-sided pyramidal) diamond indenter contained within a climatic

[5] E_r' can be calculated by:

chamber (Climats, France). A predetermined indentation depth of at least 6 μm was achieved by applying a predefined load at a constant strain rate of 0.4 s^{-1} with a sinusoidal load (amplitude of 10% of the maximum load and a frequency of 10 Hz) superimposed on the primary loading scheme. Upon reaching a depth of 6 μm, the load was held constant for 120 s, followed by unloading of the sample at a constant speed of 300 mN/min. The stiffness of the sample is directly proportional to the maximum applied load required to establish an appropriate indentation depth. The experimental procedure consisted of a series of indentations with a range of maximum loads to obtain the predetermined indentation depth. At least seven indents were carried out for each test condition.

To better understand how conservation procedures and treatments affect paint properties, two indentation studies were carried out to examine the effects of water sorption and immersion. Though immersion is not considered a conservation treatment, it can assist in revealing the most extreme response of a material to a solvent. The softening of UL by water sorption was examined by sequentially exposing the material to a range of relative humidity levels (35, 50, 60, 70, and 83% RH). For each level of RH, the sorption data from the DVS tests were used in order to establish the time to equilibrium prior to testing. The effects of immersion in *n*-heptane on the mechanical properties of CY and YO were also investigated. The test samples (10 × 10 mm) were prepared by separating the paint-outs from the polyester support and adhering the topside of each sample to a microscope slide using cyanoacrylate adhesive, exposing the flat underside for indentation. The reduced storage modulus was used to evaluate the effects of sorption and immersion on the samples.

32.3 Results and Discussion

32.3.1 Dynamic Mechanical Analysis (DMA)

The DMA results are focused on CR, CY and UL (Fig. 32.1) as well as the variants of CR and CY. The maxima of the tan δ peaks, signifying the respective glass transitions of each paint type, are listed in Table 32.2. Additionally, the effects of adding or extracting binder in these two paints is shown in Fig. 32.2 (CR, CRLN, CRLN) and Fig. 32.3 (CY, CYLN, CYEX).

$$E_r^{'} = \frac{\sqrt{\pi}}{2}\frac{1}{\beta}\frac{1}{\sqrt{A_p}}\frac{P_0}{h_0}\cos\delta$$

where β is a geometric factor of the indenter tip, A_p is the projected contact area, δ is the phase shift, P_0 is the amplitude of harmonic load, and h_0 is the amplitude of harmonic displacement.

Figure 32.1 shows that the different pure paints have quite different mechanical responses to temperature. CR is clearly the softest, most plastic of the three, with a T_g of 2 °C and the largest tan δ peak. The UL and CY paints transition from glassy to rubbery/leathery at higher temperatures (9 and 14 °C, respectively), but still below ambient temperature.

Figure 32.2 illustrates the influence of adding or extracting oil from the CR oil paints. Adding 10% oil shifts T_g to lower temperature and increases the magnitude of the softening transition. Extracting oil shifts T_g to higher temperature. T_g values listed in Table 32.2 for all the CR paints show that adding metal soaps effected minimal change to the softening transition temperature, whereas the addition of alumina hydrate made it slightly higher. The addition of oil and fatty acids made the transition temperature lower, as might be expected with a decrease in the relative pigment volume concentration (PVC).

Figure 32.3 illustrates the tan δ traces for similar variants of the CY paints. The positions of the tan δ maximum is effectively the same (15 °C) for both the pure CY and that paint with 10% oil added (CYLN), though the magnitude of the softening transition is somewhat larger for the latter paint, presumably on account of the larger binder proportion. The CYEX (oil extracted) is appreciably more glassy than the other two paints, with a tan δ maximum (T_g) of 34 °C, well above ambient temperature.

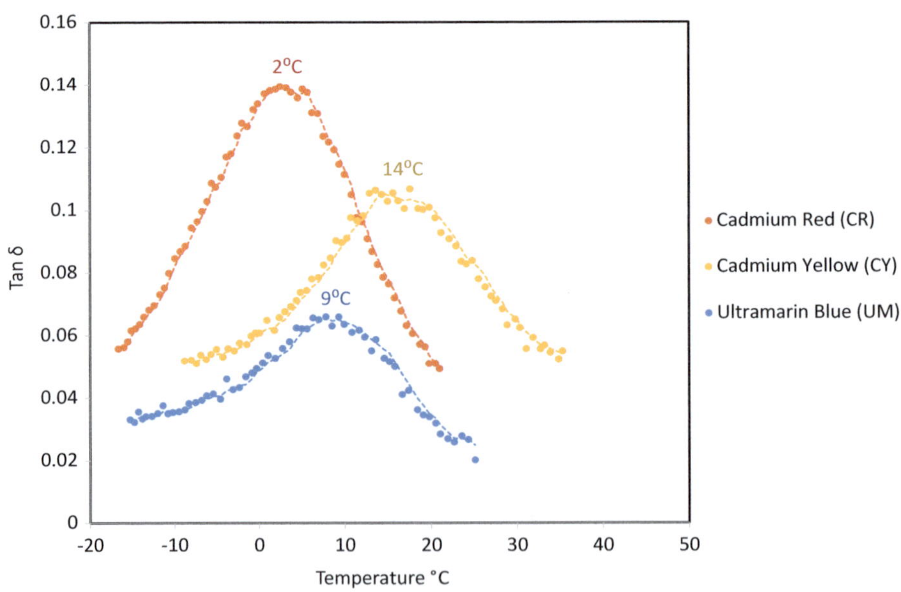

Fig. 32.1 DMA of pure Winsor & Newton paints conditioned to ambient climate: variation in 1 Hz tan δ with temperature; temperature range −20–50 °C, ramp rate 2°/min. The T_g is taken as the temperature corresponding to the maximum value of tan δ at 1 Hz, here: Cadmium red, CR (red trace): 2 °C, Ultramarine, UL (blue trace): 9 °C, Cadmium yellow, CY (yellow trace): 14 °C

Table 32.2 Average Tanδ peaks for cadmium red (CR) and cadmium yellow (CY) and their modifications

		Average 1 Hz Tanδ Peak °C (min/max values)	
Sample name	Modifications	Cadmium red	Cadmium yellow
CR/CY	Pure	2 (0/5)	15 (13/18)
-AS	Aluminium stearate added	−1 (−2/0)	33 (32/34)
-ZS	Zinc stearate added	2 (1/2)	19 (16/21)
-EX	Binder extracted	6 (6/6)	33 (30/35)
-LN	Cold pressed linseed oil added	−4 (−5/−2)	16 (15/17)
-FA	Fatty acid added	−4 (−4/−4)	33 (23/33)
-AH	Alumina hydrate added	7 (6/7)	36 (33/37)

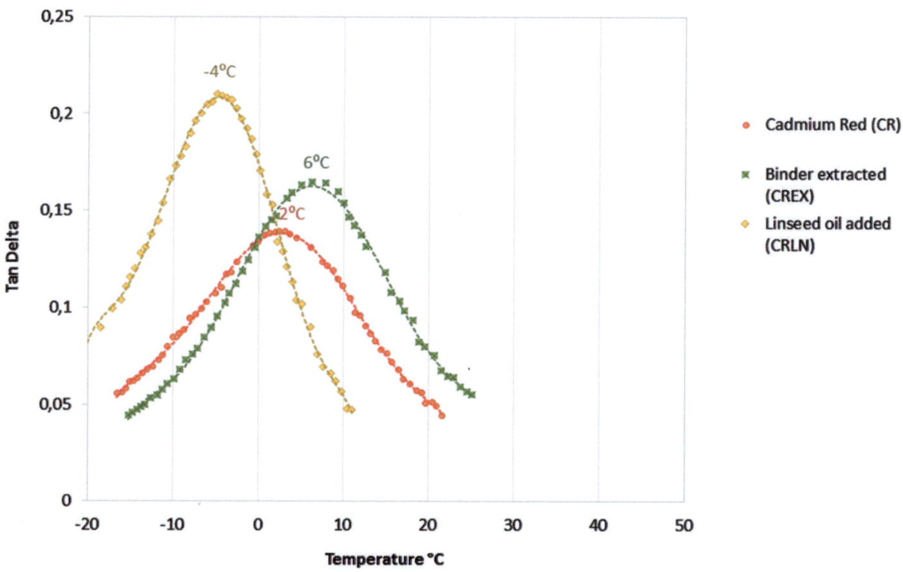

Fig. 32.2 DMA of medium-modified Winsor & Newton cadmium red oil paints: variation in 1 Hz tan δ with temperature; temperature range −20 °C–50 °C, ramp rate 2°/min. Cadmium red, CR, (red trace): 2 °C; medium extracted from cadmium red, CREX (green trace): 6 °C; and linseed oil added to cadmium red, CRLN (Yellow trace): −4 °C

Table 32.2 shows that adding metal soaps to CY increased the softening transition temperature just like the addition of alumina hydrate for both CY and CR. The addition of oil lowered the transition temperature for CR, whereas fatty acids made the transition temperature higher for CY and lower for CR. For both paints, the extraction of binder made the transition temperature higher.

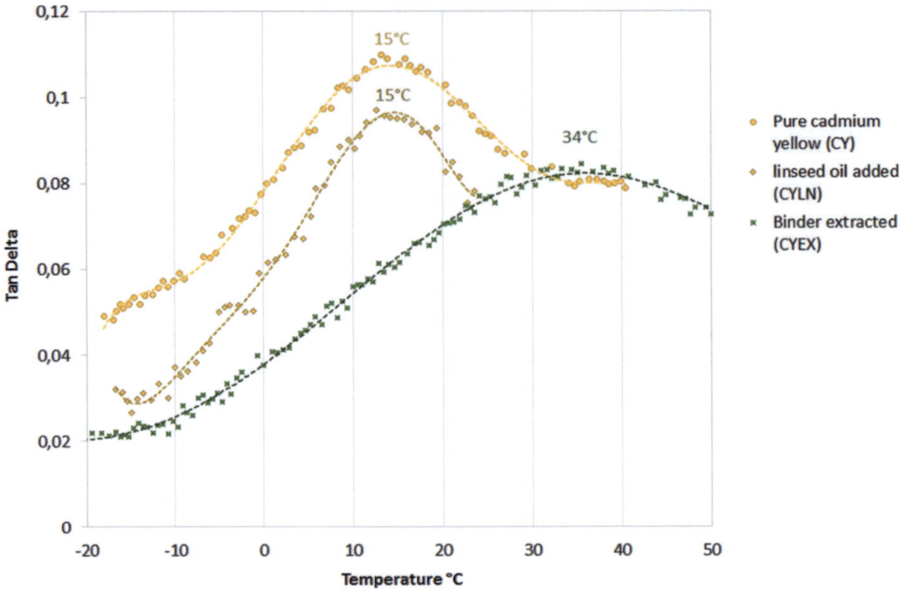

Fig. 32.3 DMA of medium-modified Winsor & Newton cadmium yellow oil paints: variation in 1 Hz tan δ with temperature; temperature range −20–50 °C, ramp rate 2°/min. Cadmium yellow, CY (yellow trace): 15 °C; Binder extracted from cadmium yellow, CYEX (green trace): 34 °C; and linseed oil added to cadmium yellow, CYLN (brown trace) 15 °C

For the various test paints studied, measured T_g values lay in the significant range of −5 and 36 °C. Dynamic mechanical analyses showed that T_g is affected by both pigments and additives. At temperatures common for art collections, the differently-formulated paints will therefore vary in stiffness and be elastic or viscous. At room and storage temperatures, some oil paints may experience state change between glassy and rubbery/leathery, and each paint type within a painting can be more or less viscous depending on its formulation. The addition or extraction of binder means a change in PVC. This parameter has implications for properties at different temperatures as the extraction of binder clearly shifted the transition temperature to a higher level. The addition of oil changed the properties of the CR resulting in a more pronounced state change at a lower temperature. The addition of metal soaps increased the softening temperature for CY but not for CR. Exactly how the metal soaps cause such a difference in T_g for one paint and not for the other is uncertain and warrants further study.

32.3.2 *Dynamic Vapour Sorption (DVS) and Nano-indentation (UNHT)*

Moisture sorption measurements were carried out at a range of RH values for the pure paint samples described in the experimental section (see Table 32.1), except for YO which was added to the study later. Furthermore, as done for DMA, modifications of cadmium red (CR, CRAS, CRZS, CREX, CRLN, CRFA, CRAH) were tested. The polyester support was tested alone and showed no moisture absorption.

In Fig. 32.4, it is shown that the paint samples responded very differently to changes in RH. While ZW was unaffected by a humidity of 90% RH, UL adsorbed as much as 12% moisture (excluding the mass of the polyester supports they were cast on). The fact that dried oil paint can take up moisture to such a high degree is somewhat surprising since oil, per se, is hydrophobic. However, a film of dried oil is a product of oil autoxidation during curing and ageing [15], which can change the properties of the material. Because of the large sorption of moisture in UL, it was decided to also test the changes in stiffness corresponding to the uptake of water vapour.

No hysteresis was observed in the sorption isotherms in Fig. 32.4. The paint samples in Fig. 32.4 (as well as the samples not shown) were almost unaffected by changes in humidity under 60% RH, indicating no significant moisture sorption. Above 60 or 70% RH, most samples showed increased moisture sorption, with the

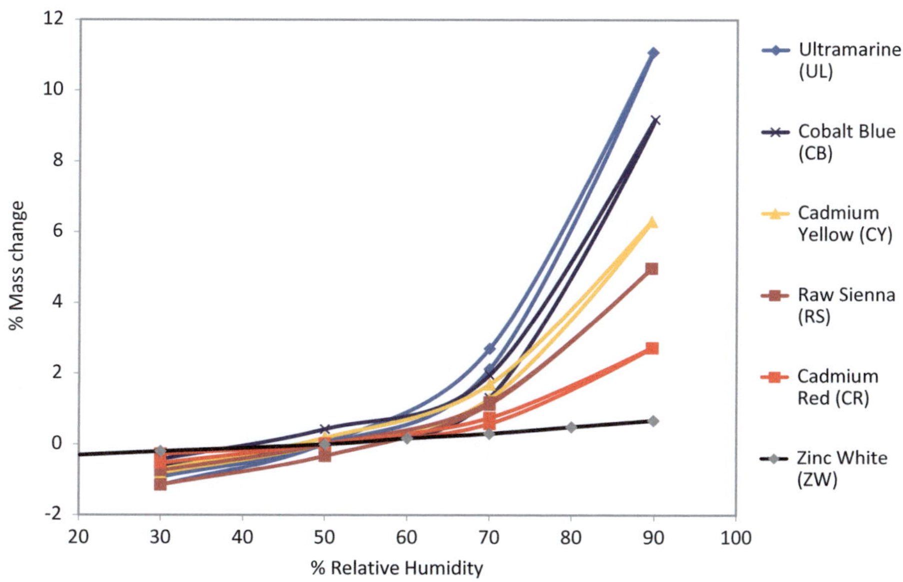

Fig. 32.4 Sorption isotherms for UL (blue line), CB (dark blue line), CY (yellow line), RS (brown line) CR (red line) and ZW (dark grey line) Winsor & Newton paint samples measured by DVS at 21 °C. The % mass change values are calculated with 50% RH as reference point

exception of ZW. The modifications to CR had no effect on the sorption characteristics of the paint.

The indentation results shown in Figs. 32.5 and 32.6 examined the effects of water sorption and immersion in *n*-heptane on selected samples. Measurements were carried out in a climatic chamber. Figures 32.5 shows the effect of water sorption on the stiffness of UL; the stiffness of the UL sample (which is conveyed by the applied load and E_r') is indirectly proportional to the change in mass caused by the water sorption measured with DVS. Figure 32.5a depicts the measured load and subsequent penetration depth into the UL sample during exposure to discrete RH conditions. To ensure that the tests reached the predetermined depth (6 μm), the applied load was adjusted for each RH step. Indentations performed at 35% RH required a load which was 4 times greater than the load needed at 50% RH to reach the desired testing depth; similarly, much smaller loads were required when the sample was exposed to high RH conditions. This necessary adjustment in the testing parameters signals a change in the mechanical properties of UL and, more specifically, its sensitivity to water. Figure 32.5b summarized the E_r' of UL determined at each RH with comparison to its corresponding percent mass change determined by DVS. Note that nano-indentation tests were not performed above 83% RH. At this humidity UL becomes deliquescent and difficult to characterize in a quantitative way. Known to be water sensitive, UL becomes softer, which is indicated here by a low value of moduli when exposed to higher humidity conditions.

Indentation tests were also performed to monitor the change in mechanical properties after immersion. CY and YO samples were chosen for this test because they have a similar stiffness. The samples were immersed in *n*-heptane for 24 h and allowed to dry for two weeks. Figure 32.6 depicts the average of 7 measurements obtained before and after immersion of CY and YO, while also listing the maximum indentation load (mN). As shown in Fig. 32.6, the mechanical properties of both CY

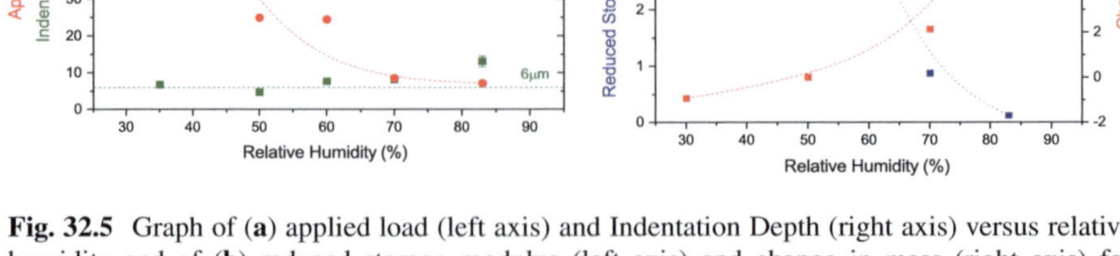

Fig. 32.5 Graph of (**a**) applied load (left axis) and Indentation Depth (right axis) versus relative humidity and of (**b**) reduced storage modulus (left axis) and change in mass (right axis) for ultramarine. Each data point represents an average of seven measurements, with error bars representing the standard deviation. Prior to indentation testing, samples were conditioned at each RH, with equilibration times determined from the sorption data

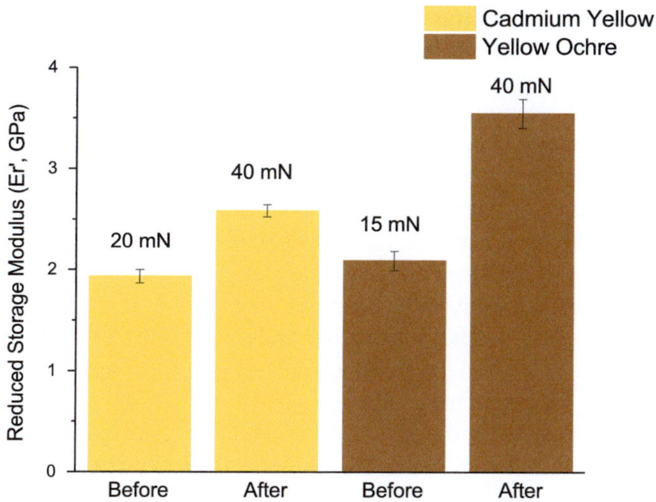

Fig. 32.6 Graph of reduced modulus measured before and after immersion in *n*-heptane followed by a minimum of two weeks of drying for cadmium yellow (yellow) and yellow ochre (brown). Each bar represents an average of seven measurements, with error bars representing the standard deviation. Listed above each bar is the maximum applied load used during indentation

and YO display a higher stiffness (larger E_r') after immersion. This behaviour may be attributed to extraction of plasticizing, non-polar low molecular weight components of the paint binder as a result of the solvent immersion process. However, additional testing is needed to clarify the physical/chemical mechanisms by which the samples are changing.

32.4 Conclusions

This study explored different ways of monitoring the physical properties of Winsor & Newton oil paints. The paints had dried for 11 years, except for the yellow ochre sample which dried for 18 years. Their behaviour was studied while the samples were on a polyester substrate. Glass transition, moisture sorption and stiffness at different relative humidity levels and before and after immersion in *n*-heptane were examined.

The assumption that oil paints generally have a T_g below 0 °C (32 °F) was challenged by the present DMA results which confirmed previous findings by Phenix [18]. The DMA results showed that addition of metal soaps and pigment volume concentration can affect the glassy-to-viscoelastic state transition in modern oil paints, and transitions as high as 36 °C were measured in cadmium yellow with a high pigment volume concentration. This indicates that modern oil paints can change properties when exposed to changes in temperature within the typical temperature range of museums or historic houses.

The amount and effect of moisture sorption on the stiffness of the test paints was also examined by DVS and nano-indentation. DVS measurements showed that the oil paints adsorbed moisture to different extents above 60% RH. Winsor & Newton ultramarine (UL) showed the highest moisture sorption which resulted in dramatic softening as seen by indentation. The stiffening of cadmium yellow (CY) and yellow ochre (YO) after immersion in *n*-heptane was furthermore demonstrated.

Nano-indentation is a relatively new technique within the conservation field and the presented results demonstrated that it is sufficiently sensitive to trace alterations in the mechanical properties of oil-based paints resulting from water sorption and immersion. Nano-indentation allows for the quantitative analysis of samples too small for other mechanical characterization techniques. Due to its high sensitivity, and spatial resolution, it can provide unique information useful for practitioners managing the care of artistic materials.

The results shown here indicate that for each paint a change in temperature or humidity can result in very different effects on mechanical properties. This complex situation means that for each work of art different areas of the paint layer can be either soft and viscous or stiff and brittle at a given temperature and relative humidity. It seems that 60% RH marks a transition above which the paints can change properties rapidly.

Acknowledgements This project received support from Heritage Plus, a Joint Programming Initiative of the European Commission and the Getty Conservation Institute.

The authors also wish to thank Lise Steyn, Bronwyn Ormsby, Klaas Jan van den Berg, Mikkel Scharff, Jørgen Wadum, Tom Learner, Ilaria Bonaduce, Aviva Burnstock, and Marion F. Mecklenburg.

References

1. BOON, J.J., and HOOGLAND, F.G. 2014. Investigating Fluidizing Dripping Pink Commercial Paint on Van Hemert's *Seven*-Series Works from 1990–1995. In *Issues in Contemporary Oil Paint* VAN DEN BERG, K.J, et al. (Eds). Cham: Springer. 227–246.
2. CHARTOFF, R. P., WEISSMAN, P. T. and SIRCAR, A. 1994. The application of dynamic mechanical methods to Tg determination in polymers: an overview. Assignment of the Glass Transition. ASTM International.
3. COOPER, A., BURNSTOCK, A., van den BERG, K. J., & ORNSBY, B. 2014. Water sensitive oil paints in the twentieth century: a study of the distribution of water-soluble degradation products in modern oil paint films. Issues in contemporary oil paint. Springer. 295–310.
4. DUNCAN, J. 2008. Principles and applications of mechanical thermal analysis. *Principles and Applications of Thermal Analysis, 2*. In GABBOTT, P., Ed. Oxford: Blackwell. 119–163.
5. ERHARDT, D., TUMOSA, C. S. & MECKLENBURG, M. F. 2005. Long-term chemical and physical processes in oil paint films. Studies in conservation, 50, 143–150.
6. GEARING, J. 1999. Dynamic mechanical (thermal) analysis. In BROWN, R. 1999. *Handbook of polymer testing: physical methods*, CRC press. New York: Marcel Dekker. 501–531.
7. HAY, J. L., and PHARR, G., M. 2000. Instrumented Indentation Testing. In KUHN, H. and MEDLIN, P., Eds., ASM Handbook Volume 8, Mechanical Testing and Evaluation. Materials Park, Ohio: ASM International. 232–243.

8. HAGAN, E., CHARALAMBIDES, M., LEARNER, T. J., MURRAY, A. & YOUNG, C. Factors Affecting the Mechanical Properties. Modern Paints Uncovered: Proceedings from the Modern Paints Uncovered Symposium, 2007. Getty Publications, 227-.
9. HEDLEY, G. & ODLYHA, M. The moisture softening of paint films and its implications for the treatment of fabric supported paintings. First published in: Traitement des supports. Travaux interdisciplinaires. Paris, 2, 3 et 4 novembre 1989, 1989. ARAAFU, 157–162. Reprinted in: VILLERS, C. (ed.); HEDLEY, G. Measured Opinions: Collected papers on the conservation of paintings. London: UKIC 1993: 99–102.
10. HEDLEY, G; ODLYHA, M; BURNSTOCK, A; TILLINGHAST, J; and HUSBAND, C A study of the mechanical and surface properties of oil paint films treated with organic solvents and water. In Book. *Cleaning, retouching and coatings: technology and practice for easel paintings and polychrome sculpture: preprints of the contributions to the Brussels Congress, 3–7 September 1990*. Mills, John S.; and Smith, Perry (Editors). International Institute for Conservation of Historic and Artistic Works, London, United Kingdom (1990) pp. 98–105
11. LEE, J., BONADUCE, I.., MODUGNO, F., LA NASA, J., ORMSBY, B., VAN DEN BERG, K.J. 2018. Scientific investigation into the water sensitivity of twentieth century oil paints. *Microchemical Journal*, 138: 282–95.
12. MECKLENBURG, M. F., TUMOSA, C. S., ERHARDT, D., VANDIVER, P. B., MASS, J. L. and MURRAY, A. 2005. The changing mechanical properties of aging oil paints. In *Materials Issues in Art and Archaeology VII: Symposium held November 30-December 3, 2004, Boston, Massachusetts, U.S.A.* Materials Research Society Symposium Proceedings, V. 852. VANDIVER, P. B., MASS, J. L. and MURRAY, A., Eds. Warrendale, PA: Materials Research Society. 13–24.
13. MECKLENBURG, M. F., TUMOSA, C. S. & VICENZI, E. P. 2013. The influence of pigments and ion migration on the durability of drying oil and alkyd paints. In *New Insights into the Cleaning of Paintings (Cleaning 2010), Proceedings from the Cleaning 2010 International Conference, Universidad Politécnica de Valencia.* MECKLENBURG, M. F., CHAROLA, A. E. and KOESTLER, R. J., Eds. Washington DC: Smithsonian Institution Scholarly Publications. 59–67.
14. MICHALSKI, S. 1991. Paintings: Their response to temperature, relative humidity, shock, and vibration. Art in Transit: Studies in the Transport of Paintings, National Gallery of Art, Washington DC, 223–248.
15. MILLS, J. S. and WHITE, R. 1999. The Organic Chemistry of Museum Objects, 2nd ed. Oxford: Butterworth.
16. MILLS, L., BURNSTOCK, A., DE GROOT, S., MEGENS, L., BISSCHOFF, M., VAN KEULEN, H., DUARTE, F., and VAN DEN BERG, K. 2008. Water sensitivity of modern artists' oil paints. In *ICOM Committee for Conservation 15th Triennial Conference, 22–26 September 2008,* New Delhi: Preprints, Bridgland, J., Ed. Paris: ICOM Committee for Conservation. 651–659.
17. OLIVER, Warren Carl, and George Mathews PHARR. 2004. "Measurement of Hardness and Elastic Modulus by Instrumented Indentation: Advances in Understanding and Refinements to Methodology." Journal of Materials Research 19 (1): 3–20. https://doi.org/10.1557/jmr.2004.19.1.3.
18. PHENIX, A. 2011. Thermal mechanical transitions in artists' oil paints and selected conservation materials: a study by Dynamic Mechanical Analysis (DMA). *Postprints of the Paintings Specialty Group, American Institute for Conservation meeting, Los Angeles, 2009.* BUCKLEY, B., Ed. Washington DC: American Institute for Conservation. 72–89.
19. SALVANT, J., BARTHEL, E. and MENU, M. 2011. Nanoindentation and the micromechanics of Van Gogh oil paints. *Applied Physics A: Materials Science and Processing,* 104, 509–515.
20. SATO, K. 1976. The mechanical properties of filled polymers. *Progress in Organic Coatings,* 4, 271–302.

21. SNEDDON, I.N. 1965. The relation between load and penetration in the axisymmetric Boussinesq problem for a punch of arbitrary profile. *International Journal of Engineering Science 3* (1): 47–57.
22. VANLANDINGHAM, M. R., CHANG, N. K., DRZAL, P., WHITE, C. and CHANG, S. H. 2005. Viscoelastic characterization of polymers using instrumented indentation. I. Quasi-static testing. *Journal of Polymer Science Part B: Polymer Physics*, 43: 1794–1811.
23. VANLANDINGHAM, Mark R. 2003. Review of Instrumented Indentation. *Journal of Research of the National Institute of Standards and Technology*, 108 (4): 249–265.

Materials:

24. Artists' oil paints, Winsor & Newton Artiststs' Oil Colour, Harrow, UK
25. Hydrated alumina, *purum*, Brocades-Stheeman & Pharmacia, Amsterdam, before 1993
26. Aluminium stearate; technical grade, British Drug Houses Ltd, Poole UK, *c.*1993
27. Zinc stearate; technical grade, Aldrich 30, 756-4, Steinheim Germany
28. Fatty acids; heptadecanoic acid 98^{n}/n, Aldrich [506-12-7], Milwaukee WI, USA
29. Linseed oil; *koudgeslagen*, De Bonte Hen, Zaandam, Netherlands *c.* 1992

Chapter 33
A Preliminary Investigation into the Behavior of Modern Artists' Oil Paints in a Hot and Humid Climate

Claire Grech, Nicole Tse, Augustine Doronila, Xiaofei Duan, Amerrudin Ahmad, and Musrizal Mat Isa

33.1 Introduction

A painting's environment is understood to have a crucial role in determining its condition, from film formation through to longer term ageing. Despite the global use of modern oil paints, the majority of research into their response to their environment has been conducted within the temperate climatic conditions of North America and Europe [13]. Hence there is a misalignment of this research to paintings produced, exhibited, and stored in regions experiencing more diverse environments.

This study is a preliminary investigation of the behavior of reference paint films on canvas aged in Kuala Lumpur and Melbourne. This research was initiated during the Australian Research Council Linkage project, 'The behaviour of Western artists' materials in tropical environments'. It is a collaboration between Balai Seni Negara, the National Art Gallery of Malaysia, located in Kuala Lumpur, and the Grimwade Centre for Cultural Materials Conservation at the University of Melbourne in Australia, with expertise from the Trace Analysis for Chemical, Earth and Environmental Sciences Platform at the University of Melbourne. This paper expands on previous research [6], examining links between the current condition of

C. Grech (✉) · N. Tse
The Grimwade Centre for Cultural Materials Conservation, School of Historical and Philosophical Studies, Faculty of Arts, The University of Melbourne, Parkville, Australia
e-mail: cgrech@student.unimelb.edu.au; nicoleat@unimelb.edu.au

A. Doronila · X. Duan
Trace Analysis for Chemical, Earth and Environmental Sciences Platform, The University of Melbourne, Parkville, Australia
e-mail: adoro@unimelb.edu.au; duanx@unimelb.edu.au

A. Ahmad · M. Mat Isa
Balai Seni Negara, Kuala Lumpur, Malaysia

K. J. van den Berg et al. (eds.), *Conservation of Modern Oil Paintings*,
https://doi.org/10.1007/978-3-030-19254-9_33

differently formulated oil paints and differences in their history of environmental exposure.

33.2 Background

Tropical climates are likely to promote oxidation and hydrolysis of oil paint at an accelerated rate compared to temperate conditions. Broadly they act synergistically with other modes of deterioration, supporting mould growth, insect activity, acidic deterioration by pollutants, and photochemical degradation [13, 22, 28]. More specifically Kuala Lumpur is classified as having a tropical rainforest climate with outdoor annual historical weather records measuring between 78 and 90% RH and temperatures from 23.2 to 26.2 °C [4, 19], and Melbourne is classified as an oceanic climate – a subtype of the temperate group [21].

Current research indicates that the more complex formulations of modern oil paints contributes to their higher incidence of stability issues and unusual degradation, such as the formation of efflorescence and metal soap protrusions, phase separation, softening and liquefaction of paint, as well as sensitivities to aqueous cleaning [1, 7, 9, 10, 18]. Amongst this research there is a small but expanding group of studies examining the effects of in-situ ageing in hot and humid climates. Significant degradation in relatively young paintings and reference paint films have been identified, particularly in those that contain zinc oxide, and have suggested that the lifetime of paintings in these environments may be reduced [11, 13, 16, 19]. If significant chemical differences are present in paint films constructed from the same materials, but aged in different climates, they may respond differently to conservation treatments. Given the wider discussions on museum environments, and that few paintings enter collecting institutes immediately after completion and are therefore exposed to non-standard climates, it is timely that studies examine the ageing of a wide variety of paint films in-situ in tropical climates to inform their care.

33.3 Experimental

33.3.1 Artists' Oil Paints

The two proprietary ranges investigated were Winsor & Newton (W&N) Artists' Oil Colours and Marie's Oil Colours. W&N is an English fine art materials company founded in 1832. Their Artists' Oil Colours are well researched within the conservation literature, and their products are marketed and available worldwide [3]. Marie's is a Chinese artists' materials company, established in 1919, with products stocked in over sixty countries [2]. Relatively little is known about the formulation and performance of their Oil Colours range, as limited product information is available in

English from the manufacturer and no known conservation studies exist on their behavior. These ranges were selected as they represent modern-day, globally available artists' colourmen paints produced in established and emerging industrialized economies respectively. Artists' paints labelled 'titanium white', 'cadmium red' and 'burnt sienna' were selected for study from each brand. The W&N paints were sourced from Nanyang Art Supply Company in Kuala Lumpur; the company was founded in 1943 and have imported W&N products into Malaysia intermittently during their history, and were the exclusive distributor between 1996 to at least 2004 [25]. Marie's paints and commercially available pre-primed cotton canvas were sourced from Balai Seni Negara in the early 2000s.

33.3.2 Project Design and Methods

The parameters of this research have been framed by the geographic location of the ageing samples, and access to scientific instruments and resources. These factors have determined when and how frequently sampling has occurred, and the type and number of analyses undertaken at different times. In 2003 the importance of having naturally aged oil paint samples in known tropical climate conditions was recognized, and paint films were therefore prepared with the aim to undertake chemical analysis at a later date. This set of paint films aged long-term at Balai Seni Negara in tropical conditions will henceforth be referred to as the 'tropical' set and were prepared in 2003. The set aged at the Grimwade Centre in temperate conditions will be referred to as the 'temperate' set and were prepared in 2010 (Table 33.1). To achieve uniform paint thickness, five layers of masking tape was applied in two strips 10 cm apart to primed cotton canvas. The oil paint was applied directly from the tube along the top edge and then drawn down using a straight edge of Perspex. Both the 2003 tropical and 2010 temperate sets were prepared at Balai Seni Negara and underwent the first three months of curing in the ambient indoor conditions of the conservation laboratory. Environmental readings in the laboratory were collected from 8th to 24th July 2014, and while incomplete for the total period of natural ageing, the ranges are comparable to other environmental data recorded in the gallery and provide a general indication of the environment in which the films have been ageing, ranging 24–26 °C and 56–74% RH.

Three months after application and initial curing, the temperate set was relocated to the Grimwade Centre and hung in the climate-controlled conservation laboratory with environmental parameters within 20±2 °C and 50±5% RH. The tropical set was hung vertically on an interior wall in the Balai Seni Negara conservation laboratory near a window which was intermittently opened as needed for human comfort. This exposed the paint films to outside levels of temperature, relative humidity, any potential external pollutants, and is likely to have influenced photochemical oxidation during the ageing process. Each set was exposed to the artificial lighting of the laboratory spaces during work hours.

Table 33.1 Summary of paint films – preparation, sampling dates, and analysis

Set	Ageing location	Brand and range	Brand color	Date prepared	Years sampled	Age of samples (years)	Analysis			
							OM	SEM-EDS	ATR-FTIR	XPS
Tropical	Balai Seni Negara, Kuala Lumpur, Malaysia	W&N Artists' Oil Colour	Titanium White*	Aug. 2003	2011	8	X		X[b]	
					2018	15	X[a]	X[a]	X[a]	X
			Cadmium Red	Aug. 2003	2011	8	X		X[b]	
					2018	15	X[a]	X[a]	X[a]	
			Burnt Sienna	Aug. 2003	2013	10	X		X[b]	
					2018	15	X[a]	X[a]	X[a]	X
		Marie's Oil Colours	Zinc-Titanium White	Aug. 2003	2011	8	X		X[b]	
					2018	15	X[a]	X[a]	X[a]	
			Cadmium Red	Aug. 2003	2011	8	X		X[b]	
					2018	15	X[a]	X[b]	X[a]	
			Burnt Sienna	Aug. 2003	2013	10	X		X[b]	
					2018	15	X[a]	X[b]	X[a]	
Temperate	Grimwade Centre, Melbourne, Australia	W&N Artists' Oil Colour	Cadmium Red	Nov. 2010	2013	3	X		X[b]	
					2018	8	X[a]	X[a]	X[a]	
			Burnt Sienna	Nov. 2010	2013	3	X		X[b]	
					2018	8	X[a]	X[a]	X[a]	X
		Marie's Oil Colours	Zinc-Titanium White	Nov. 2010	2013	3	X		X[b]	
					2018	8	X[a]	X[a]	X[a]	
			Cadmium Red	Nov. 2010	2013	3	X		X[b]	
					2018	8	X[a]	X[b]	X[a]	
			Burnt Sienna	Nov. 2010	2013	3	X		X[b]	
					2018	8	X[a]	X[b]	X[a]	

Set	Ageing location	Brand and range	Brand color	Date prepared	Years sampled	Age of samples (years)	Analysis			
Baseline	Grimwade Centre, Melbourne, Australia	W&N Artists' Oil Colour	Cadmium Red	Nov. 2012	Dec. 2012				X[b]	
			Burnt Sienna	Nov. 2012	Dec. 2012				X[b]	
		Marie's Oil Colours	Zinc-Titanium White	Nov. 2012	Dec. 2012				X[b]	
			Cadmium Red	Nov. 2012	Dec. 2012				X[b]	
			Burnt Sienna	Nov. 2012	Dec. 2012				X[b]	

*The W&N Titanium White sample from the temperate and baseline sets have been excluded here due to unresolved data. The sample from the tropical set has been retained as it provides an interesting preliminary understanding of a widely used paint

[a]Paint film analyzed in triplicate – three different positions or cross-sections

[b]Paint film analyzed in duplicate – two different positions or cross-sections

Following the acquisition of an ATR-FTIR in 2011, chemical analysis of the paints began and a third set, of 'baseline' samples – were painted out from the original tubes in 2012. This baseline set was intended to provide an understanding of the young paint films following curing at the Grimwade Centre. As other researchers joined this long-term study, additional analytical techniques were utilized and further data was collected (Table 33.1). Sampling sites for chemical analysis were randomly selected and undertaken in homogenous areas of the paint film away from the edges of the draw downs and previous sampling sites. The original composition of each paint for the temperate and tropical sets (Table 33.2) has been inferred from the chemical analysis of the cured baseline paint films where possible. This may represent a limitation to the study; however, care was taken to differentiate between the fresh composition and the ageing behaviors.

The time difference between the preparation of the 2003 tropical set and the 2010 temperate set is also recognized as a limitation of this preliminary research, as is the lack of chemical data on these paints during their early ageing. However, the possibility of evaluating eight years of real time ageing in an under-researched climate was considered worthwhile in order to identify any possible areas of concern to guide future research.

33.3.3 Optical Microscopy

Paint film surfaces and polished cross-sections were examined with an Olympus BX51 Microscope using incident polarized light and incident ultraviolet (UV) light (U-LH100-3 100W halogen lamp and a U-LH100HG 100W mercury lamp respectively). Examination under UV-light used the following filters: Olympus U-MWU2 (excitation BP 330-385, BA420, DM400), Olympus UMWBV2 (excitation BP 400-440, BA475, DM455). The imaging system comprised an Olympus DP70 camera and a DELL Optiplex GX280 computer.

33.3.4 Scanning Electron Microscopy

A Phillips (FEI) XL30 ESEM-TMP was operated via an FEI Graphical User Interface, and was coupled to an OXFORD INCA energy dispersive spectrometer. A tungsten filament was used as the electron source providing a beam of up to 30 kV at spot sizes (beam current control) of 1–9. For EDS analysis a beam of 15 kV and spot size of 5 was used, with back scattered electron (BSE) and secondary electron (SE) detectors for data collection. The EDS used a liquid-nitrogen cooled Si-Li detector with an area of 10 mm^2 and an ATW2 thin detector window which allowed collection of x-rays between B and U. High vacuum mode was used, with gold coating on paint films for surface analysis and carbon coating on mounted cross-sections to render the samples conductive. Spot analysis of particles within

Table 33.2 Proposed composition of the selected oil paints, including information available from manufacturers

Brand and range	Brand color (pigment used)	Oil listed on label	EDS	ATR-FTIR	GC-MS*	Conclusions
W&N Artists' Oil Colour	Titanium White (PW6)	Safflower	C, O, Ti, Zn, Ba, Mg, Al, S	Drying oil, $BaSO_4$, $MgCO_3$, TiO_2	Drying oil, P/S = 1.9, 12-hydroxyoctadecanoic acid	Titanium dioxide and zinc oxide in a drying oil, with hydrogenated castor oil, barium sulfate, magnesium carbonate, aluminum-based additive(?)
	Cadmium Red (PR108)	Linseed/ Safflower	C, O, Cd, Se, S, Ba, Mg	Drying oil, $BaSO_4$, $MgCO_3$	Drying oil, P/S = 2.4	Cadmium sulfo-selenide, - possibly co-precipitated - with barium sulfate, in drying oil, with magnesium carbonate extender
	Burnt Sienna (PR101)	Safflower	C, O, Fe, S, Al, Mg	Drying oil, iron oxide, $MgCO_3$	Drying oil, P/S = 1.4	Iron oxide pigment in a drying oil, with magnesium carbonate, quartz, aluminum-based additive
Marie's Oil Colours	Zinc-Titanium White	Refined botanic oil	C, O, Zn, Ti, Si, Al, Mg?	Drying oil, metal stearates, TiO_2	Drying oil, P/S = 1.1	Zinc oxide with titanium dioxide in a drying oil, with metal stearates
	Cadmium Red	Refined botanic oil	C, O, Ba, S, Ca, Zn, Si, Al, Cl?	Drying oil, synthetic organic red, $BaSO_4$	Drying oil, P/S = 1.3	Synthetic organic red colorant in drying oil, with barium sulfate extender
	Burnt Sienna	Refined botanic oil	C, O, Fe, Ba, S, Si, Al, K?, Ti?	Drying oil, $BaSO_4$	Drying oil, P/S = 1.9	Iron oxide pigment in drying oil, with aluminum silicates, quartz, barium sulfate extender

*Grech et al. [6] – Paint films were analyzed in triplicate, samples (0.5–1.0 mg) were derivatized with Meth Prep II for total fatty acid content

cross-sections to identify their elemental composition was carried out on visually similarly particles (≥3 per cross-section to confirm composition).

33.3.5 Fourier Transform Infrared Spectroscopy

Fourier transform infrared spectroscopy was undertaken using a Bruker Alpha-P Fourier Transform Infrared spectrometer with a diamond attenuated total reflection window (Bruker Optik GmbH, Germany) (ATR-FTIR). For each absorbance spectrum 32 scans were compiled at a spectral resolution of 4 cm^{-1} in the range of 4000–400 cm^{-1}, with 32 background scans. Data acquisition and processing were performed with OPUS version 7.5. 'Surface' analyses were undertaken directly on the paint films without sampling – the penetration depth under these conditions was in the range 0.3–3 μm. To understand the distribution of certain moieties within the paint, samples were excised with a scalpel and the underside analyzed; this method was limited by inconsistent depth of detection, however could identify differences in composition in the lower regions relative to the surface.

33.3.6 X-Ray Photoelectron Spectroscopy

Data was acquired using a Kratos Axis ULTRA X-ray Photoelectron Spectrometer (Thermo Scientific, UK) incorporating a 165 mm hemispherical electron energy analyzer. The incident radiation was monochromatic Al Kα X-rays (1486.6 eV) at 150 W (15 kV, 15 mA). Survey (wide) scans were taken at an analyzer pass energy of 160 eV and multiplex (narrow) high resolution scans at 20 eV. Scanned area was ~0.8 × 0.3 mm and the depth was <10 nm (volume ~2400 μm^3). Survey scans were carried out over 1200-0 eV binding energy range with 1.0 eV steps with a dwell time of 100 ms. Narrow high-resolution scans were run with 0.05 eV steps and 250 ms dwell time. Base pressure in the analysis chamber was 1.0×10^{-9} torr and during sample analysis 1.0×10^{-8} torr.

33.4 Results and Discussion

33.4.1 Formulation

Table 33.2 reports the inferred composition of the selected oil paints from previous and current analysis with SEM-EDS, ATR-FTIR and GC-MS [6].

The presence of an oil binder was confirmed in all paints with similarities in pigments observed across corresponding colors. Paints labelled 'Titanium White' contained high proportions of titanium and some zinc, likely present as titanium dioxide (TiO_2) and zinc oxide (ZnO); paints labelled 'Burnt Sienna' contained iron, likely present as synthetic iron oxides; and the W&N 'Cadmium Red' contained cadmium, selenium, sulfur and barium, likely cadmium sulfo-selenide co-precipitated with barium sulfate ($BaSO_4$). Most of the other paints were also identified to contain $BaSO_4$ through SEM-EDS and ATR-FTIR; characteristic peaks ca. 1181, 1060 and 983 cm^{-1} – were identified through comparison to the spectrum of an analytical grade reference sample. The Marie's Cadmium Red paint – although it was not labelled 'hue' – was determined to be an organic red, rather than a cadmium or other inorganic red, from the elemental data and the ATR-FTIR spectra.

A range of other compounds that can modify the properties and reactivity of these paints were suggested and, in some cases, identified. Magnesium carbonate ($MgCO_3$) was detected in all three W&N paints using both ATR-FTIR and SEM-EDS at multiple points. W&N are reported to have replaced magnesium carbonate with dolomite, a calcium-magnesium carbonate in 1975 (Garrett 2011a, b, cited in [8], p. 303), however multiple peaks in these ATR-FTIR spectra aligned with known peaks for magnesium carbonate (specifically those at 1481 and 1417 cm^{-1}) rather than the slightly shifted main peak of dolomite (ca. 1431 cm^{-1}) ([23], p. 103). Furthermore, SEM-EDS spot analysis across multiple particles in all W&N cross-sections suggested a magnesium carbonate, rather than a calcium-magnesium carbonate.

Aluminum and silicon were detected in most of the paints, possibly present as surface coatings on the pigment particles [20], or as additives such as quartz (SiO_2), kaolin ($Al_2O_3.2SiO_2.2H_2O$), aluminum stearate, or aluminum hydroxide ($Al(OH)_3$) ([17], pp 25–26; [14]). Quartz particles were detected using SEM-EDS in both the W&N and Marie's burnt sienna samples.

Characteristic peaks for aluminum stearate at 1588 and 1469 cm^{-1} were not identified in any of the paints, however may have been present below the detection limit of the instrument or have been obscured by the magnesium carbonate peaks in the W&N paints [10]. Hydrogenated castor oil (HCO) was only identified in the W&N Titanium White paint [6, 10, 17, 18]. A defined peak in the baseline spectra of the Marie's Zinc-Titanium White was suggestive of the inclusion of metal stearates by the manufacturer [15, 26].

33.4.2 Preliminary Observations of Change

Significant visual changes have been observed in paints in the tropical set. The burnt sienna films from both brands appear to be particularly susceptible to change, likely due to their increased hygroscopicity compared to the cadmium and titanium/zinc pigments. Early within the first year the Marie's paint film exhibited signs of biodeterioration and changes in surface texture; the formation of white crystalline entities on the W&N paint film surface was first documented in 2010 (7 years), although likely occurred earlier. By early 2018 (15 years) they were quite extensive across the film causing significant visual disruption. In 2017 (14 years), a noticeable change in color of the Marie's Cadmium Red paint film was observed relative to the temperate-aged film - it appeared much browner, suggesting poor stability of the organic colorant.

All paints in the tropical set, regardless of color or brand, were observed in 2014 and again in 2018 (11 and 15 years respectively) to have a matte appearance, exhibited poor adhesion at the paint/priming-canvas interface, were extremely friable during sampling, and have developed a network of micro-cracks across the surface. In comparison, paints in the temperate set are still relatively soft and glossy, with good film cohesion.

33.4.3 Reactivity of Magnesium Carbonate Additive

As the most visually disruptive change observed during this study, the formation of white crystalline aggregates across the surface of the tropical-aged W&N Burnt Sienna film was highlighted as a potential concern for conservators, as it appeared to indicate a particular susceptibility of this specific formulation to its environment that would require treatment if identified on an artwork. This behavior was not observed on any of the other paint films in either environment. When imaged using SEM, the aggregates appeared as needle-like crystals protruding from the paint's surface (Fig. 33.1), and were identified to contain magnesium, carbon, and oxygen

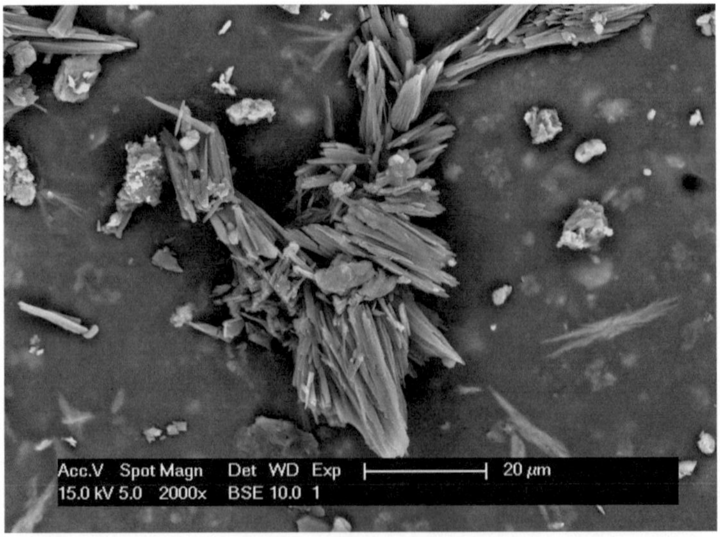

Fig. 33.1 SEM BSE image of the white crystalline entities on the surface of the tropical W&N Burnt Sienna film, aged in Kuala Lumpur for 15 years, at ×2000 magnification, 15 kV

using EDS. They bore a resemblance to crystals previously reported in the literature as hydrated magnesium sulfate complexes, such as epsomite. Their formation has been strongly linked to W&N's inclusion of magnesium carbonate extenders in their paints and exposure to humidity and sulfur dioxide [8, 12].

Sulfur was detected only in the bulk paint film by EDS, however XPS was able to detect trace amounts of sulfur at the surface of both the tropical and temperate W&N Burnt Sienna films. A slightly higher amount was detected at the surface of the tropical-aged film compared to the temperate-aged film. The presence of sulfur on the latter may indicate either that this reaction is also occurring, albeit at a decreased rate, in the conservation laboratory in Melbourne, or it may be a remnant from the first three months of ageing in Kuala Lumpur. Both the binding energies of the magnesium (50.9 eV) and sulfur (169.6 eV) supported the presence of a magnesium sulfate compound on the surface, rather than magnesium carbonate or elemental sulfur (163–164 eV) [27].

Changes in the characteristic absorbance peaks for magnesium carbonate – a doublet ca. 1417 and 1481 cm^{-1} and a small sharp peak around 3648 cm^{-1} – were identified in all W&N paints at multiple points across the surfaces of all paint films compared to the available baseline spectra (Fig. 33.2) ([23], p. 103). Minor decreases in the intensity of the peak at 1481 cm^{-1} were observed relative to the peak at 1417 cm^{-1} for both the tropical-aged and temperate-aged Cadmium Red and Titanium White films and the temperate-aged Burnt Sienna film. This peak was no longer observable in spectra taken at the surface of the tropical-aged Burnt Sienna film, however measurements of the undersides of excised paint samples did not exhibit this loss at all, appearing similar to the presentation of the magnesium carbonate in the baseline-paint spectra of the W&N Burnt Sienna. The increased reactivity of the Burnt Sienna film relative to the Cadmium Red and Titanium White films may be linked to the increased binder content of the paint and the relative hygroscopic behavior of iron oxide pigments. Although monitoring for airborne pollutants was not carried out in the vicinity of these samples during their ageing, the localization of this change to the paint/air interface does suggest that the environment of the paint film has been crucial in driving the formation of the magnesium sulfate crystals at the surface. This observation correlates with previous research into the formation of magnesium sulfate compounds which have identified it to be predominantly a surface phenomenon [8, 12]. While the specific relationship between the observed decrease of the peak at 1481 cm^{-1} and the formation of these crystals in this environment is unknown, further research monitoring for this change in artworks may provide a way to identify at-risk paint films in similar environments.

33.4.4 Reactivity of the Paint Binder

Overall, after eight years of ageing, the spectra of the temperate-aged films have not diverged greatly from their respective baseline-paints (Figs. 33.2 and 33.3). Although there has been a slight broadening and shifting towards slightly lower

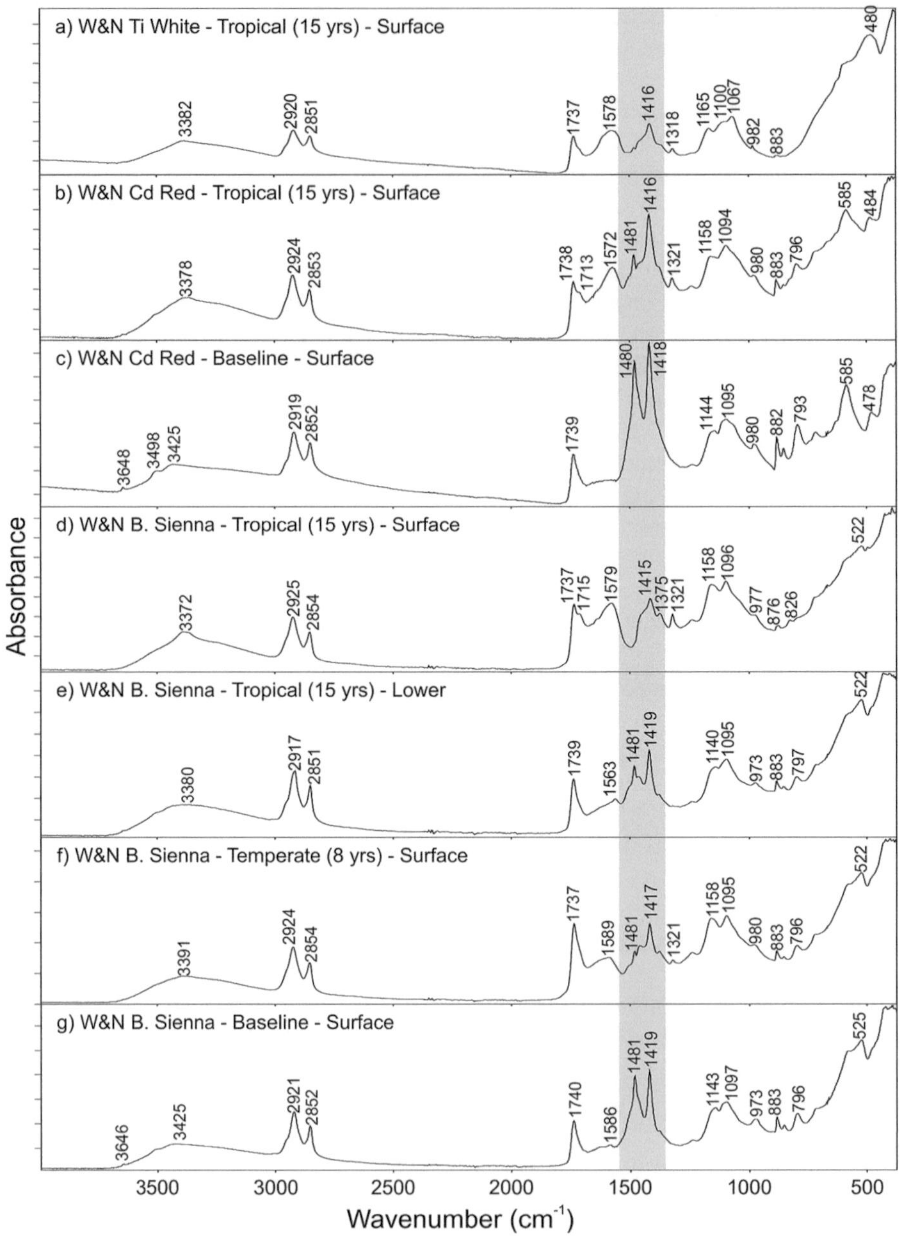

Fig. 33.2 Comparison of ATR-FTIR spectra of W&N films related to the changes in the characteristic magnesium carbonate peak at 1481 cm^{-1} (highlighted in box). Titanium White tropical sample, 15 years, surface **(a)**; Cadmium Red set: tropical sample, 15 years, surface **(b)**, baseline sample, surface **(c)**; Burnt Sienna set: tropical sample, 15 years, surface **(d)** and lower region **(e)**, temperate sample, 8 years, surface **(f)**, baseline sample, surface **(g)**

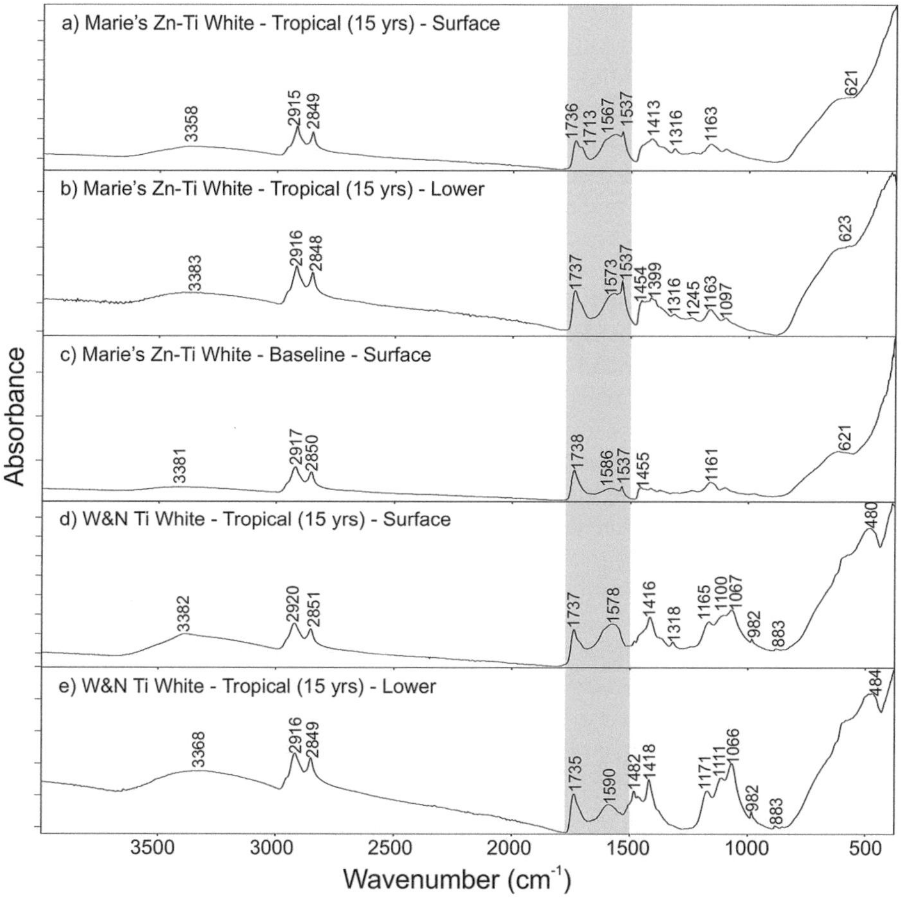

Fig. 33.3 Comparison of ATR-FTIR spectra of Marie's Zinc-Titanium White with the W&N Titanium White, highlighting differences in behavior of each film following ageing. Marie's Zinc-Titanium White set: tropical sample, 15 years, surface **(a)** and lower region **(b)**, baseline sample, surface **(c)**; W&N Titanium White tropical sample, 15 years, surface **(d)** and lower region **(e)**

frequencies of the carbonyl band ν(C=O), the maxima have still remained centered around ~1738 cm^{-1}, signifying a gradual, and not unusual, rate of formation of carboxylic acids through the hydrolysis of acylglyceride esters [24]. Contrastingly, by eight years, films from the tropical set had begun exhibiting splitting of the carbonyl peak - most spectra contain a defined shoulder peak, if not a shift in the peak maxima value to around ~1710 cm^{-1}.

These observations indicate that the formation of carboxylic acids through hydrolysis of the paint was occurring at a much greater rate in the tropical set, and that significant quantities of these uncoordinated fatty acid moieties were also still present in these films despite the concurrent broadening and strengthening in intensity of a broad peak between 1650–1500 cm^{-1} related to the formation of a variety of amorphous metal carboxylates [10, 24, 26]. This correlates to previous studies which have identified significant amounts of metal carboxylates in relatively young paintings produced in Southeast Asia [11, 13, 16].

Comparing differences in the spectra acquired at the surface and lower down in the paint films has revealed similar trends to those discussed earlier regarding the localized degradation of magnesium carbonate within the paint films. All spectra collected at the surface have exhibited both a more defined carbonyl shoulder or split, and a greater absorption of the broad metal carboxylate peak relative to the carbonyl peak when compared to the spectra of samples excised from deeper within the paint films (Fig. 33.3). The W&N surface spectra have also tended to exhibit a small, but sharp peak around 1321 cm^{-1}, within the context of the other changes observed in these paint films, these are likely related to the formation of oxalates at the surface [1].

These observed differences suggested that there was a differential distribution of certain moieties within the films, and that degradation products were more prevalent at the air/paint interface than deeper down in the paint films. This may be due to the air/paint interface serving as a reaction site due to the diffusion of ambient water vapor and possibly airborne pollutants into the paint, and/or may also be from the migration of highly polar, low-molecular weight organic species within the film to the surface [1, 10, 18]. This is significant within the context of artworks as the changing condition of the surface of the paint films will affect their visual appearance and their response to treatment.

Evidence of significant phase separation of organic species was observed for the tropical-aged W&N Titanium White and Cadmium Red paint films (Fig. 33.4d–i). It was identified in multiple cross-sections from each film through examination under UV light and SEM. Comparable behavior has been observed for modern oil paint films containing both zinc oxide and aluminum stearate over similar time scales, most notably in studies by Osmond et al. [15] and Bayliss et al. [5]. HCO has also been proposed to contribute to this behavior [10]. Despite also containing zinc oxide this type of phase separation was not observed in any of the samples from the Marie's Zinc-Titanium tropical-aged film (Fig. 33.4a–c) although this film did exhibit differences between the ATR-FTIR spectra of its surface and lower down in the paint film (Fig. 33.3). These discrepancies in behavior suggest that under these elevated environmental conditions these behaviors are strongly influenced by paint formulation.

33.5 Conclusions

This preliminary study, contributing to a broader shift towards place-based conservation, has highlighted the reactivity of the investigated modern oil paints to conditions of elevated temperature and humidity, postulating links between formulation, degradation and ageing environment. Significant chemical differences were observed between paint films of a similar age that were aged in different

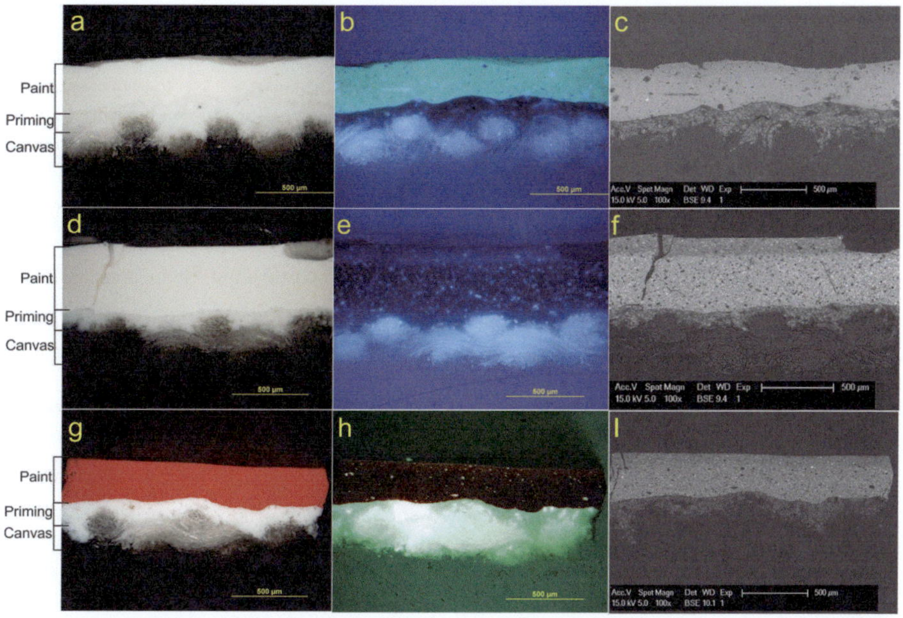

Fig. 33.4 Comparison of tropical-aged paint films at ×100 magnification, highlighting the phase separation in the W&N samples, but not the Marie's, following ageing. Marie's Zinc-Titanium White reflected light image **(a)**, UV fluorescence (UVF) image (Olympus U-MWU2 filter) **(b)**, BSE image **(c)**; W&N Titanium White reflected light image **(d)**, UVF image (Olympus U-MWU2) **(e)**, BSE image **(f)**; W&N Cadmium Red reflected light image **(g)**, UVF image (Olympus UMWBV2) **(h)**, BSE image **(i)**

environments. Acidic deterioration of paint films has been suggested to be a significant and present risk in Southeast Asia, particularly for collections located in urban centers, and highlights the need for further in-situ studies within museum environments. Paint films containing magnesium carbonate, particularly in combination with earth pigments, and aluminum additives in combination with zinc oxide are proposed to have greater vulnerability in these environmental conditions; these results will inform the direction of future research.

Acknowledgments The authors gratefully acknowledge Selina Halim who did initial research around the topic in 2007. Special thanks go to Graham Hutchinson, TrACEES Platform, University of Melbourne, who performed the SEM analysis, and Dr Barry Wood at the Centre for Microscopy and Microanalysis, University of Queensland, for the XPS analysis. This research has been supported by the Australian Research Council, an Australian Government Research Training Program Scholarship, and a Melbourne Research Scholarship and PhD Fieldwork Grant from the University of Melbourne.

References

1. Banti D, La Nasa J, Lluveras Tenorio A, Modugno F, van den Berg KJ, Lee J, Ormsby B, Burnstock A, Bonaduce I (2018) A molecular study of modern oil paintings: investigating the role of dicarboxylic acids in the water sensitivity of modern oil paints. RSC Adv 8:6001-6012
2. Shanghai SIIC Marie Painting Materials Co., Ltd. (2018) Introduction. http://www.chinamaries.com/En/About/Default_144396663052566528.aspx Accessed 12 April 2018
3. ColArt Contract Manufacturing Ltd. (2018) Store locator. http:www.winsornewton.com/au/shop/store-locator-retailers Accessed 23 December 2018
4. Tse N, Labrador A, Scott M, Balarbar R (2018) Preventive conservation: People, objects, place and time in the Philippines. Stud Conserv 63(sup1):274-281
5. Bayliss S, van den Berg KJ, Burnstock A, de Groot S, van Keulen H, Sawicka A (2016) An investigation into the separation and migration of oil in paintings by Erik Oldenhof. Microchem J 124:974-982
6. Grech C, Tse N, Doronila A, Duan X (2015) A preliminary study on the effects of a tropical climate on the chemical ageing of proprietary artists' oil paints. Paper presented at the 4th Meeting of the Asia Pacific Tropical Climate Conservation Art Research Network, Cheng Shiu University, Kaohsiung, 25-27 November 2015
7. Bronken IAT, Boon JJ (2014) Hard dry paint, softening tacky paint, and exuding drips on Composition (1952) by Jean-Paul Riopelle. In: A Burnstock, M de Keijzer, J Krueger, T Learner, A de Tagle, G Heydenreich, KJ van den Berg (eds) Issues in contemporary oil paint. Springer International Publishing, pp 247-262
8. Cooper A, Burnstock A, van den Berg KJ, Ormsby B (2014) Water sensitive oil paints in the twentieth century: A study of the distribution of water-soluble degradation products in modern oil paint films. In: A Burnstock, M de Keijzer, J Krueger, T Learner, A de Tagle, G Heydenreich, KJ van den Berg (eds) Issues in contemporary oil paint. Springer International Publishing, pp 295-310
9. Izzo FC, Van den Berg KJ, Van Keulen H, Ferriani B, Zendri E (2014) Modern oil paints – formulations, organic additives and degradation: some case studies. In: A Burnstock, M de Keijzer, J Krueger, T Learner, A de Tagle, G Heydenreich, KJ van den Berg (eds) Issues in contemporary oil paint. Springer International Publishing, pp 75-104
10. Osmond G (2014) Zinc white and the influence of paint composition for stability in oil based media. In: A Burnstock, M de Keijzer, J Krueger, T Learner, A de Tagle, G Heydenreich, KJ van den Berg (eds) Issues in contemporary oil paint. Springer International Publishing, pp 263-282
11. Osmond G, Ebert B, Drennan J (2014) Zinc oxide-centred deterioration in 20th century Vietnamese paintings by Nguyễn Trọng Kiệm (1933-1991). AICCM Bulletin 34(1):4-14
12. Silvester G, Burnstock A, Megens L, Learner T, Chiari G, van den Berg KJ (2014) A cause of water-sensitivity in modern oil paint films: the formation of magnesium sulphate. Stud Conserv 59(1):38–51
13. Tse N, Chayabutra S, Kamolchote K, Panmanee S, Khlungwisarn T, Puskar L, Best S (2014) Artist oil paints in Thailand. In: J Bridgland (ed) ICOM-CC 17th Triennial Conference Preprints, Melbourne, 15-19 September, 2014. International Council of Museums, Paris, art. 1314, 9pp
14. Standeven HAL (2013) Oil-based house paints from 1900 to 1960: an examination of their history and development, with particular reference to Ripolin enamels. J Am Inst Conservat 52(3):127-139
15. Osmond G, Boon JJ, Puskar L, Drennan J (2012) Metal stearate distributions in modern artists' oil paints: surface and cross-sectional investigation of reference paint films using conventional and synchrotron infrared microspectroscopy. Appl Spectrosc 66(10):1136-1144
16. Ebert B, MacMillan Armstrong S, Singer B, Grimaldi N (2011) Analysis and conservation treatment of Vietnamese paintings. In: J Bridgland (ed) ICOM-CC 16th Triennial Conference

Preprints, Lisbon, 19-23 September, 2011. Critério Produção Gráfica, Lda, Almada, art. 1305, 8pp
17. Izzo FC (2011) 20th century artists' oil paints: a chemical-physical survey. PhD Thesis, Universita Ca'Foscari Venezia, Venice
18. Mills L, Burnstock A, Duarte F, de Groot S, Megens L, Bisschoff M, van Keulen H, van den Berg KJ (2008) Water sensitivity of modern artists' oil paints. In: J Bridgland (ed) ICOM-CC 15th Triennial Conference Preprints, New Dehli, 22–26 September, 2008. Allied Publishers, New Delhi, pp 651-659
19. Tse N, Sloggett R, Roberts A (2008) A preliminary understanding of oil paintings in tropical Southeast Asia. In: J Bridgland (ed) ICOM-CC 15th Triennial Conference Preprints, New Dehli, 22–26 September, 2008. Allied Publishers, New Delhi, pp 641–650
20. Croll S (2007) Overview of developments in the paint industry since 1930. In: TJS Learner, P Smithen, JW Kreuger, MR Schilling (eds) Modern Paints Uncovered: Proceedings from the Modern Paints Uncovered Symposium. Getty Conservation Institute, Los Angeles, pp 17-29
21. Peel MC, Finlayson BL, McMahon TA (2007) Updated world map of the Köppen-Geiger climate classification. Hydrol Earth Syst Sci 11:1633-1644
22. Erhardt D, Tumosa CS, Mecklenburg MF (2005) Long term chemical and physical processes in oil paint films. Stud Conserv 50:143-150
23. Learner TJS (2005) Analysis of modern paints. Getty Conservation Institute, Los Angeles
24. van der Weerd J, van Loon A, Boon JJ (2005) FTIR studies of the effects of pigments on the aging of oil. Stud Conserv 50(1): 3-22
25. Lian WJ (2004) 21st December. [Interview with N. Tse]. Nanyang Art Supply Company, Kuala Lumpur
26. Robinet L, Corbeil MC (2003) The characterization of metal soaps. Stud Conserv 48(1):23-40
27. Moulder JF, Chastain J (1992) Handbook of x-ray photoelectron spectroscopy: a reference book of standard spectra for identification and interpretation of XPS data. Physical Electronics Division, Perkin-Elmer Corporation, Eden Prairie
28. Thomson G (1986) The museum environment 2nd ed, Butterworths & Co Ltd, London

Chapter 34
Inside the Forger's Oven: Identification of Drying Products in Oil Paints During and After Accelerated Drying with Increased Temperatures

Diana Blumenroth, Stephanie Dietz, Wolfgang Müller, Stefan Zumbühl, Walter Caseri, and Gunnar Heydenreich

34.1 Introduction

In 2010, the international art market was unsettled by the Beltracchi forging scandal. During his trial in 2011, Wolfgang Beltracchi admitted to having forged 14 paintings by modernist artists like Heinrich Campendonk, Max Ernst, Fernand Leger and Kees van Dongen [12]. Beltracchi's use of pigments that were not yet in production for the alleged Campendonk painting was just one of the many pieces of evidence which finally exposed his criminal activity [7, 16]. During and after his time in jail, Beltracchi and his wife published two books; appeared in a

D. Blumenroth (✉) · S. Dietz · G. Heydenreich
TH Köln, Cologne Institute of Conservation Sciences, Cologne University of Applied Sciences, Köln, Germany
e-mail: diana.blumenroth@th-koeln.de; stephanie.dietz@th-koeln.de; gunnar.heydenreich@th-koeln.de

W. Müller
H. Schmincke & Co.-GmbH & Co.KG, Erkrath, Germany
e-mail: mueller@schmincke.de

S. Zumbühl
Bern University of Applied Sciences, Bern, Switzerland
e-mail: stefan.zumbuehl@hkb.bfh.ch

W. Caseri
Swiss Federal Institute of Technology ETH, ETH Zürich Laboratory for Multifunctional Materials, Zürich, Switzerland
e-mail: walter.caseri@mat.ethz.ch

K. J. van den Berg et al. (eds.), *Conservation of Modern Oil Paintings*,
https://doi.org/10.1007/978-3-030-19254-9_34

documentary and hosted a television show.[1] In total, Beltracchi claims to have forged over 300 artworks, most of them still unidentified [1]. In his autobiography "Selfportrait" and the documentary "Beltracchi: The Art of Forgery" he explained and boasted about his skills and the tricks of his trade. He was also aware of the possible scientific investigation of his paintings and therefore reused old stretched canvases, avoided modern glues and considered examination by X-radiography, where he included underlying paint into his compositions. He even stated that he had analysed paints in advance to avoid detection of modern pigments in his works ([2], p. 529). Evidence for simulating ageing, such as an artificial patina or drilled exit holes from 'wood pests' were also found on Beltracchi's paintings [4]. Among other aspects, he also explained his method for the rapid drying of oil paints into hard paint films, which was a prerequisite for selling his artworks successfully. However, no analytical evidence has yet been reported that confirms the accelerated drying of his paints, despite his constant reinforcement of myths to this end.

Art forgers apply several methods to speed up the drying and hardening of oil paints. The (in)famous Van Meegeren added phenol formaldehyde resin to his paint medium to produce a hard and cracked paint surface ([5], p. 188, [20]). Subsequently he 'baked' the painting at circa 100 °C for about two hours ([10], p. 163). However, additions such as phenolic resin can now be detected through instrumental analysis [5]. Interestingly, Beltracchi also stated that he "baked" his paintings – which raised several questions such as: How did he 'treat' his paintings? Can we detect differences between his 'baked' and naturally cured oil paints? Can we find evidence that Beltracchi 'baked' his paintings through analysing paint samples? Is it feasible to determine the presence of heat aged oil paint through analysis and can this provide enough evidence to substantiate his intent to defraud?

34.2 Thermal Treatment of Oil Paints

Wolfgang Beltracchi described his method to promote accelerated drying of oil paints in his book and the documentary [2, 3], where his intention was to produce a hard paint surface. He reported how "art dealers" would put the tip of a needle onto the surface of paintings in order to check firmness ([2] pp. 95–96). In the book and documentary, Beltracchi described and showed his 'oven', which opened to one side and appeared to be a wooden structure similar to a chest of drawers with metal bars on which the paintings were horizontally placed. In his book he mentioned that one oven was as large as "150 × 200 × 100 centimetres" ([2], p. 243) and that another "should accommodate paintings as big as 140 centimetres" ([2], p. 368).

For the regulation of temperature he used an electric heating panel, a sauna thermostat and a timer clock ([2], p. 234). The stretched painting was placed face up on

[1]Beltracci, W., Beltracci, H.: Selbstporträt. Rowohlt Verlag GmbH (2014)
Beltracci, W., Beltracci, H.: Einschluss mit Engeln: Gefängnisbriefe vom 31.8.2010 bis 27.10.2011. Rowohlt Verlag GmbH (2014)
Arne Birkenstock: Beltracchi: Die Kunst der Fälschung. Senator (2014)
Der Meisterfälscher. 3 Sat (2014-2016)

small bars, some distance from the heating mat. In the documentary Beltracchi wore gloves while handling a painting and claimed that the mat temperature was 400 °C at the surface. On the other hand, in his book ([2], p. 234–235) he stated with regard to temperature that: "In the first phase of drying, the oven was set to a lower temperature" and furthermore he described "the designated 50 °C…". However, he also mentioned that he had problems with the temperature control setup within the oven, which once even caused a fire.

Previous studies have indicated that the thermal treatment of oil paint may have a significant impact on the drying process. In 2011, Zumbühl and his colleagues reported that the accelerated ageing conditions often influences the drying mechanisms and that products from these processes are detectable using FTIR analysis. Under specific ageing conditions (e.g. T/RH) different chemical reactions take place and/or become dominant [8]. Furthermore, the chemical composition of the paint may also have a significant influence on reaction pathways. Zumbühl et al. found that none of the paint samples exposed to environmental accelerated conditions matched the composition of ~100-year old naturally aged paint samples and pointed out the importance of relative humidity for carboxylate formation [21]. Relative humidity (RH) is a relevant aspect in the context of thermal ageing, since the water saturation concentration at 60 °C is around 10-times higher, relative to 20 °C. This means that in an open system, without efficient humidification equipment, the relative humidity is far below "normal" conditions. In a different context, Franken et al. used heat for the curing of oil paints. Here the authors stated that in order to cure or harden liquefied oil paints on modern and contemporary paintings, a minimum of 60 °C was necessary and that the hardening of defective liquefied oil paints was mainly a physical process attributed to the evaporation of the low molecular weight ageing products from the binder [9]. These aspects are considered more thoroughly in the present study.

34.3 Experimental

34.3.1 Sample Preparation

Commercially prepared oil paint (Kremer) was used, with lead white[2] (basic lead carbonate) and zinc white[3] (zinc oxide) pigments, within a linseed oil binding medium. Among other reaction products, these paints were used to investigate soap (metal carboxylate) formation during ageing. Paint films (100 μm wet film) were applied with a film applicator[4] onto silicone-coated polyester foil. For pre-drying, the films were stored for one month under ambient environmental conditions (T:

[2] Kremer Pigmente: Kremserweiß in Leinöl (lead white in linseed oil) Lead white is most commonly used to imitate historical paintings.

[3] Kremer Pigmente: Zinkweiß in Leinöl (zinc white in linseed oil) Zinc white is known to be very reactive in oil binders.

[4] Filmapplicator BYK No.:2041/1197347 on HOSTAPHAN® foil (The oil paint was thinned before the application: 50 g zinc white + 5 ml Shellsol® T (Kremer Pigmente); 30 g lead white + 1 ml Shellsol® T (Kremer Pigmente))

20–22 °C, RH: 40–55%, indirect daylight). Thereafter the samples were aged in a laboratory oven.[5] Since the precise conditions within Beltracchi's oven and his exact procedure were not known, one batch of samples were exposed to 40 °C, one to 60 °C and another to 80 °C each for a period of eight weeks maximum, with the system RH at between 6% and 12%[6]. The period of eight weeks was chosen as Beltracchi does not state a precise length and this corresponds with earlier studies [21]. Conditions in the oven were stable, as the door was opened only once a week to remove one sample from the batch. The oxygen concentration was kept at atmospheric level. After thermal ageing, one series of samples were stored in a dark room (T: 20–22 °C, RH: 40–55 and the other set in indirect daylight exposure(T: 20–22 °C, RH: 40–55 %, indirect daylight), for analysis at a later stage. Besides the heat treated samples, two sets of untreated oil paint films were stored, one in dark storage (T: 20–22 °C, RH: 40–55%) and another in normal indoor conditions (T: 20–22 °C, RH: 40–55 %, indirect daylight), positioned close to a window.

Transmission FTIR measurements were performed before and after sample derivatisation using sulfur tetrafluoride SF_4. The selective formation of acyl fluoride C(O)F from carboxylic acids C(O)OH and carboxylates C(O)OM allows the detection of acids formed as intermediate or final reaction products. Micro-samples (∅ ca. 20 μm) were rolled on to a CVD diamond window prior to being placed in a gas reaction chamber. The derivatisation treatment was performed in an all-Teflon® chamber (volume 10 cm^3). Prior to SF_4 exposure the chamber was purged with gaseous nitrogen N_2 for 10 min, then the water-free chamber was flushed with gaseous sulfur tetrafluoride SF_4. The samples were then exposed to the reactive gas for 12 h. Following treatment, the chamber was purged with N_2 and the toxic gas was adsorbed onto activated charcoal and analysis was carried out immediately after derivatisation.

34.3.2 Instrumentation

FTIR measurements were executed on a Bruker® Hyperion 3000/Tensor 27 infrared spectrometer and the micro-samples were analysed in transmission mode at 4 cm^{-1} spectral resolution and 32 scans. The detection area of a measurement was 100 μm in diameter (spot size).

X-ray photoelectron spectra were recorded on a Sigma 2 spectrometer (Thermo Scientific) using a polychromatic Al Kα X-ray source. The binding energy was calibrated with regard to the C(1s) main line, was set to 284.8 eV.

Electrospray Ionization Mass Spectroscopy (ESI) was conducted in positive ion mode using a Bruker maxis 4G instrument. Samples were prepared by dissolving 10 mg of the paint in a mixture of 1 ml chloroform and 0.5 mls methanol. After centrifugation, 1 ml of the solution was diluted with methanol to achieve the final concentration.

[5] Oven manufactured by Memmert

[6] 40 °C RH: 8–12 %; 60 °C RH: 6–8%

Colour measurement was performed using a Datacolor CHECK 3 (Light source and observation angle: D65 10 Deg), which included an integrating sphere.

34.4 Results and Discussion

34.4.1 Changes in optical and mechanical properties

As a first step, changes in optical appearance were evaluated. The lead white samples had turned brownish at 80 °C within only one week. One reason for the significant colour change noted was probably the lack of light, which would usually destroy the chromophores formed in the oil-paint. Another possibility may include reactions between the lead white pigment and the oil, which have been reported elsewhere [6, 17, 18]. Exposure of the lead white samples to 60 °C and 40 °C respectively caused smaller colour shifts to a slightly yellowish shade. Colour change was also observed for the zinc white samples, however to a lesser extent and at a slower rate than the lead white group. The least change in colour was observed for the zinc white samples aged at 40 °C (Fig. 34.1). From this, it appears that 80 °C was most likely higher than the temperature used by Beltracchi, as

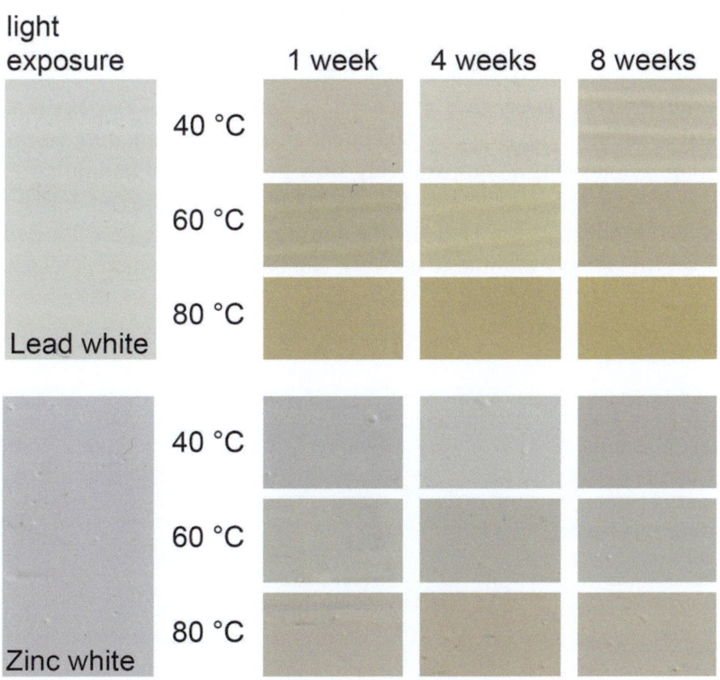

Fig. 34.1 In contrast to the oil paint samples that were exposed to light, the changes in colour of the thermally aged samples are dependent on temperature and time

similar brownish discolouration has not been observed in his paintings containing lead and zinc white.

Related changes were noted regarding the mechanical properties of the samples. The exposure of the lead white samples to 80 °C caused severe brittleness of the films, whereas the samples exposed to 40 °C and 60 °C remained more flexible. At 40 °C, the paints had set, but did not harden within the first four weeks, whereas the samples exposed to 60 °C appeared harder. After 8 weeks all of the samples appeared firm, where most would have likely passed the 'art dealer needle test' mentioned by Beltracchi. These changes in mechanical properties were significant when compared to the samples stored under ambient environmental conditions. For the thermally-aged samples, pronounced brittleness had developed without added physical stress. Even though the IR spectra showed only minor chemical alterations (see next section) there was a strong correlation between the thermal ageing conditions and changes in mechanical properties, as has been previously reported by Erhardt et al. [8]. (Fig. 34.2)

34.4.2 Chemical changes during thermal ageing

Previous studies have shown there is a complex relationship between environmental conditions such as temperature, relative humidity, light exposure and the resulting chemical reactions and products found in drying oils ([11, 14, 15, 21]). Of particular relevance are: oxidative cross-linking reactions, oxidative fragmentation, hydrolysis of triglycerides and the saponification of acidic ageing products. Two key reactions are of particular interest in this context. Oxidative changes are generally fast processes, whereas soap formation is a comparatively slow process under natural ageing conditions and, as a secondary reaction, is dependent on the acids formed during oxidative fragmentation and hydrolysis [19]. In the presence of multivalent ions (pigments, additives), an irreversible ionomeric network can be formed, which is responsible for

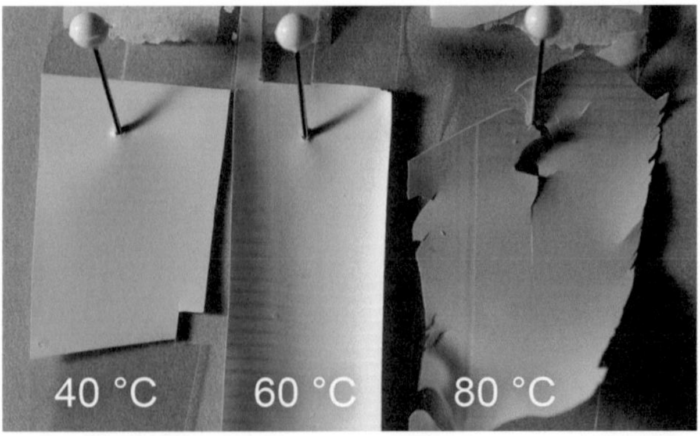

Fig. 34.2 Lead white oil paint samples aged in the oven for 8 weeks. Changes in the mechanical properties and a loss of flexibility were found to be dependent on temperature

the increased hardness and reduced elasticity of very old oil paints. Thus, carboxylate formation under accelerated ageing conditions is of specific interest.

Since these reaction pathways lead to different functional groups, FTIR analysis can be employed to characterise the ageing state of paint samples. However, the interpretation of the IR-spectra from oil-based paint samples often suffers from interfering bands of the different compounds, namely binder, oxidative ageing products, carboxylates, as well as pigments and fillers. Thus, a derivatisation technique using sulfur tetrafluoride SF_4 was used to discriminate overlapping signals [22]. Following derivatisation, primary carboxylic acids C(O)OH and carboxylates C(O)OM are transformed into an acyl fluoride C(O)F with a characteristic and a well-separated absorption band at ~1840 cm^{-1}. This single IR-band enables the characterisation of all acids formed as intermediates of the hydrolysis and oxidation reactions of the oil binder. Following the ratio between the relative signal intensity of the acyl fluoride signal at ~1840 cm^{-1} relative to the triglyceride ester band ~1730 cm^{-1} imparts information on the ageing state. Furthermore, strongly IR-absorbing pigments such as basic lead carbonate, are transformed into a non-IR active component, which again makes it possible to characterise previously overlapping signals.

As a first step, soap formation during ageing was investigated, and the results focus on the thermally aged samples (40 °C and 60 °C) containing zinc white pigment (Fig. 34.3). Generally, old zinc white paints show a very high degree of saponification (Fig. 34.3). This dominant ageing reaction during the later ageing period leads to a characteristic increase in the broad carboxylate bands $\nu_a C(O)O^-$ $\approx$1580 cm^{-1} and a corresponding decrease of the $\nu C{=}O$ ester band $\approx$1735 cm^{-1} in the IR-spectrum. The results of this study show that the pre-dried zinc white paints contain only small quantities of zinc soaps, similar to those expected in a young, flexible oil paint. Although the mechanical properties changed considerably during thermal ageing, the saponification of the oil appears to have been limited under all ageing conditions.

As mentioned, after derivatisation the fluorinated acid band (acyl fluoride C(O) F) shifts to $\approx$1840 cm^{-1} and is therefore separated from other carbonyl bands arising from esters C(O)OR and ketones $R_2C(O)$. This makes it possible to measure the relative amounts of carboxylic anion $C(O)O^-$ compared to the carboxylic acids C(O) OH and corresponding carboxylates C(O)OM within an oil paint sample. The data (Fig. 34.3) shows that carboxylic acids were formed to a lesser extent during thermal ageing, which was confirmed using XPS, where no significant increase in oxygen-containing functional groups were detected. Hence the chemical alteration in these samples was very limited in relation to "real" photooxidative ageing (Fig. 34.3), therefore an increase in acidic functionality is not responsible for the increased brittleness seen in these samples. This was also verified by the determination of acidity, as after FTIR-derivatisation, no significant change of the acyl fluoride absorption band was observed when compared to the pre-dried (not thermally aged) samples. This limited change in the concentration of carboxylic acids suggests that hydrolysis was not a relevant degradation reaction in the thermally-aged group, which is supported by the very dry conditions in the oven during ageing (Fig. 34.4).

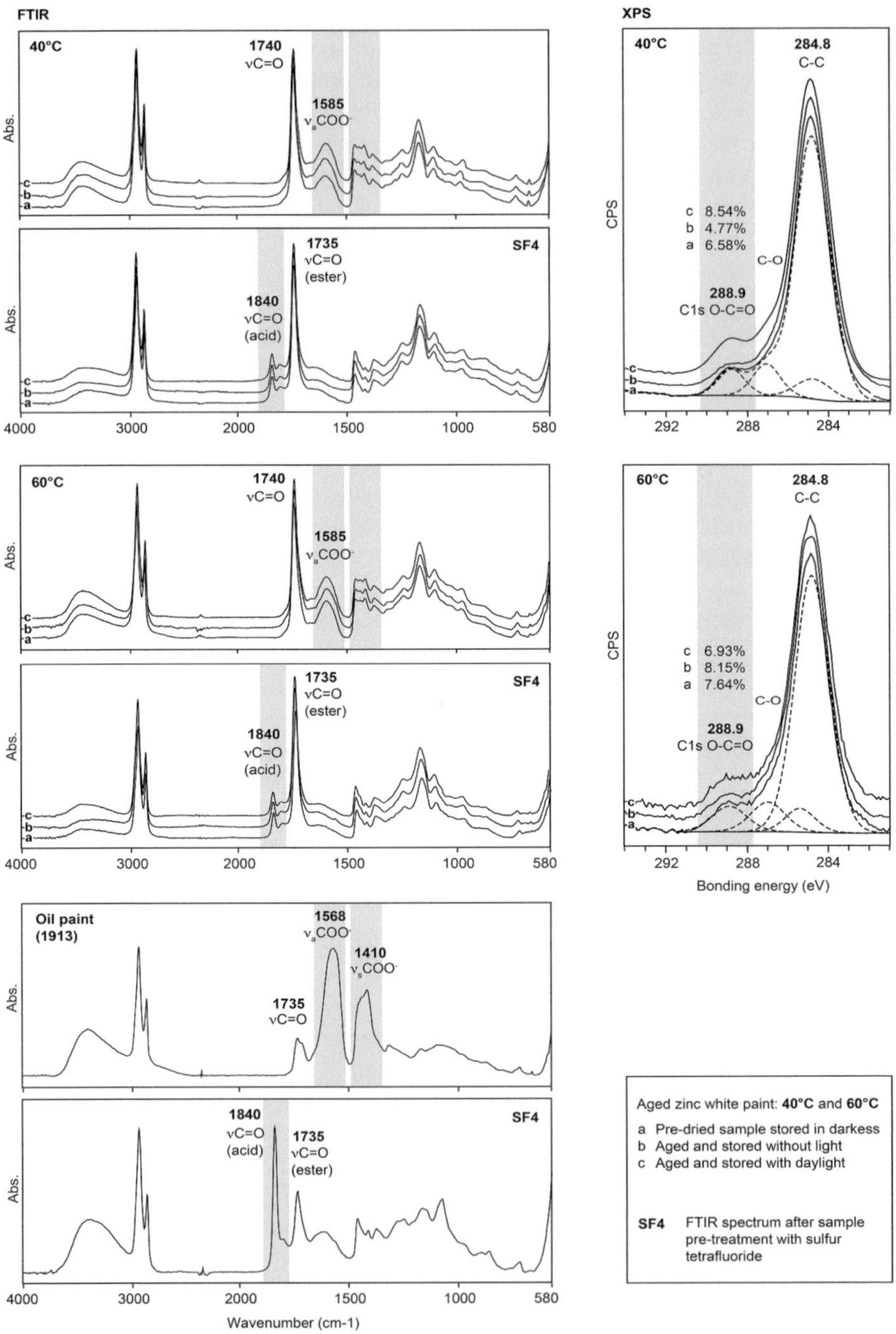

Fig. 34.3 FTIR analysis shows the difference between naturally aged and thermally treated samples. νC=O(ester) ≈1740 cm^{-1}; νC=O (acids) ≈1715 cm^{-1}; $\nu_a CO_2^-$ ≈1590 cm^{-1} and $\nu_s CO_2^-$ ≈1410 cm^{-1} (carboxylates)

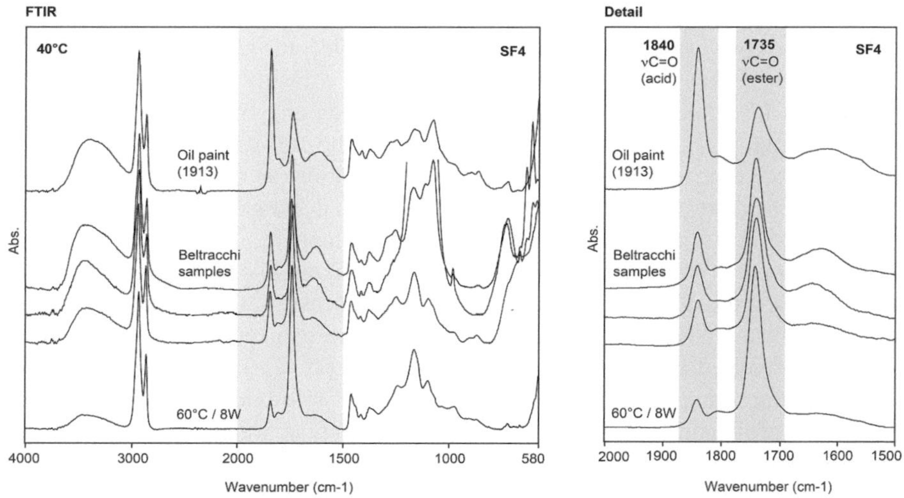

Fig. 34.4 After SF_4 treatment the absence of hydrolysis in the thermally treated samples contrasts with the naturally aged oil paint sample. The samples of Beltracchi's paintings show a less distinctive pattern. νC=O ≈1840 cm^{-1}(acyl fluorides: fluorinated acids and carboxylates); νC=O ≈1740 cm^{-1} (ester)

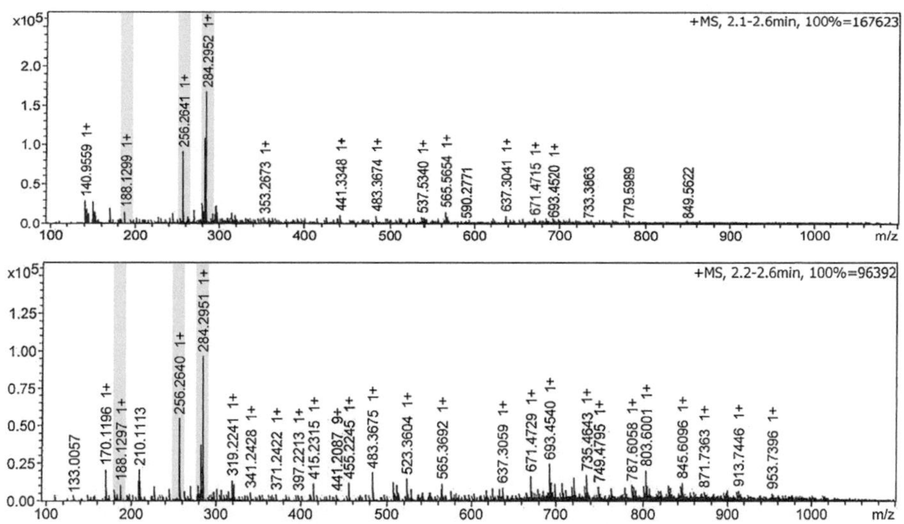

Fig. 34.5 MS-Analysis Electrospray ionization (ESI) above: lead white 105 years old, ESI pos. mode, thinning (1:3); below: lead white 1 week 60 °C, ESI pos. mode, thinning (1:100); azelaic acid 188 m/z; palmitic acid 256 m/z; stearic acid 284 m/z

In Fig. 34.5 the positive ESI-MS spectra of an approximately 100-year old lead white paint sample and a lead white paint sample aged for 1 week at 60 °C are shown. Both ESI spectra look similar at first glance, especially the main signals, with azelaic, palmitic and stearic acids highlighted in the chart. However, due to the different weights and dilutions of the two ESI samples, the abundance of the heat-treated lead white sample had to be multiplied by a factor of ~ 25× in order to compare. Thus, the concentration of free fatty acids and fragments in the heat-treated sample is significantly higher compared to the 100-year old sample, which is presumably due to different degrees of saponification, as only free acids can be detected. This difference in fatty acid concentration is the largest difference recorded thus far; with further evaluation of the ESI spectra planned.

Extrapolating from these combined results, is it likely that Beltracchi's artificial ageing method did not simulate the natural ageing processes of oxidation, hydrolysis and saponification, but mainly aimed to mimic the mechanical properties of paint. It has been observed that at temperatures of around 60 °C, low molecular weight ageing products can evaporate easily [9]. This is of relevance for the mechanical properties of a paint film since these low molecular weight products act as plasticisers. These products are described by Van den Berg et al. [19] among others as low molecular weight (un)saturated aldehydes, ketones, alcohols, acids and hydrocarbons. Furthermore, the relative binder volume influences paint mechanics, as verified by Franken et al. [9] through weighing and thermogravimetric analysis, where the reduction of a significant amount of the binder matrix components lead to an increase of the pigment volume concentration and consequently a decrease in elasticity. This can be traced by the relative decrease of IR-signal from the binder in relation to the invariable signals of the IR-active pigments and fillers (i.e. a relative solids:binder ratio).

34.4.3 Comparing Accelerated Aged Samples to Beltracchi

To explore these findings with respect to fakes, the same group of analytical techniques were applied to selected Beltracchi paint samples. Results revealed a rather low acidity and a moderate degree of saponification in zinc white containing paint samples removed from Beltracchi's "Pechstein" imitation (Figs. 34.4 and 34.6). The results were not as distinctive as seen in the thermally aged series, nonetheless the ageing state remains distinct from "naturally aged" samples. Differences may be ascribed to the that fact that Beltracchi's methods are not fully known and that his paintings have now experienced 30–40 years of natural ageing; hence, these paint samples contain degradation products from post-manufacture (natural) ageing. Furthermore, we do not know whether Beltracchi treated all his paintings in the oven in the same manner.

Nonetheless, the results demonstrate that the ageing state of Beltracchi's paint sample are not comparable to a more than 100-year old sample from a painting of the "classical modernism" period. The data shows that hydrolysis (and carboxylate formation) were not the dominant alteration reactions in any of the thermally aged paint samples tested. This confirms that the aim of hardening the fresh oil paints via

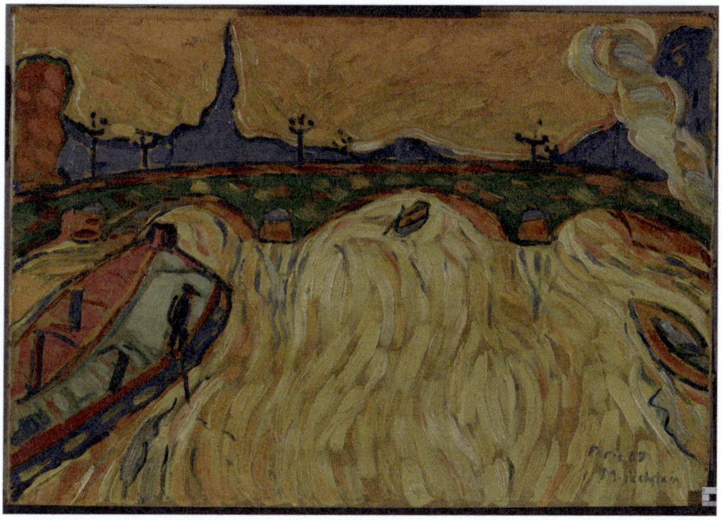

Fig. 34.6 Wolfgang Beltracchi in the style of Max Pechstein *Seine-Brücke mit Frachtkrähnen*

heat treatment was predominantly about achieving a physical resemblance to aged oil paints rather than chemical.

34.5 Reversibility of Colour Changes

Even though the 80 °C aged paints changed colour in the most dramatic way, the 40 and 60 °C batches also had a more yellowish tone in comparison with the naturally aged paints. This might help us to understand another statement by Beltracchi: "In order to get back their natural shade, they needed exposure to air and sunlight; so I leaned them against the balcony trellis of a south facing window for a few hours." ([2], p. 501). It is likely that he repeated this over of days or weeks. To evaluate his procedure, the thermally aged samples were subsequently exposed to indirect daylight under ambient indoor conditions. After a few weeks, changes were observed in that the yellow colour had visibly faded. The samples were measured with a spectrophotometer before exposure to light, and again after 6 and 17 weeks. Five spots were measured on every sample and averaged in order to determine the colour difference Delta E of the CIELAB colour space.[7] A $\Delta E*$ above 5 is considered to be an obvious change in colour ([13], p.131). As seen in (Fig. 34.7) ΔE values over 10 were detected. The greatest changes were measured after the first six weeks, after that, the fading continued more slowly, however this was still visible to the naked eye. The most pronounced colour changes were noted for the samples heated to

[7] CIEDE1976 was used to describe colour difference.

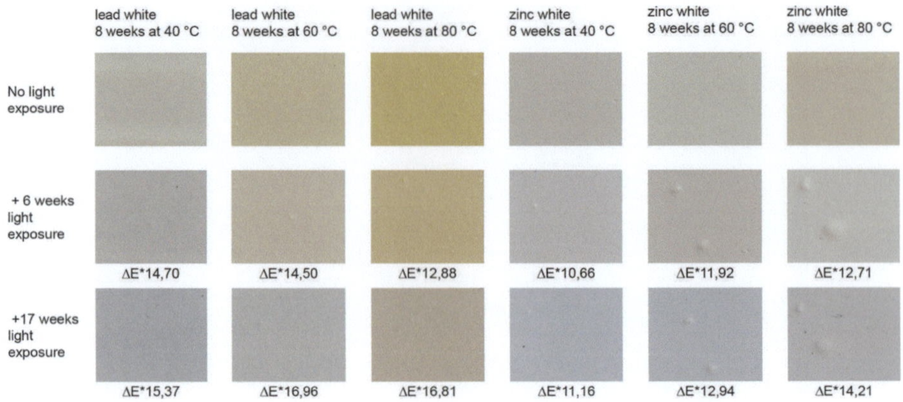

Fig. 34.7 After being aged in the oven for 8 weeks the samples were exposed to light in indoor conditions. ΔE∗ was measured 5 times and averaged. The most apparent changes developed in the first 6 weeks; after this, the changes were slower

80 °C where the strong brownish shade turned to an ivory tone, which was still however, far removed from the original white (Fig. 34.7).

34.6 Conclusions

The study demonstrated distinct differences between naturally cured and heat-treated (white pigmented) oil paints. Severe changes in paint properties such as colour and flexibility were observed after thermal ageing. Even though the paints appeared firm, similarly to naturally aged materials (which was Beltracchi's aim), they remain chemically different. One of the key reasons for this was the very low relative humidity in the thermal ageing oven, which hinders hydrolysis and saponification reactions. Thus, the chemical alterations noted were not comparable to those present in paint samples from a naturally aged painting from the early twentieth century. Minor differences were also detected between our artificially aged samples and samples from Beltracchi's paintings by means of FTIR. The fact that the differences were minor are not surprising since Beltracchi's paintings have additionally naturally aged over the last 30–40 years. Nevertheless, the curing state of these forgeries is not comparable to real paintings of the "classical modernism" period that they were intended to mimic. Thus, analysis of the paint media from queried works may assist the identification of forged and artificially aged paintings. However, it would be useful to explore whether the baked oil paints develop chemical profiles that are closer to naturally aged oil paints over time, and to explore the effect of other pigments on the resulting paint physical and chemical properties.

References

1. Baurmann J, Tönnesmann J (2018) Wolfgang Beltracchi: "Es war ein Schock für unsere Kinder". ZEIT Nr. 19/2018, 3. Mai https://www.zeit.de/2018/19/wolfgang-beltracchi-kunstfaelscher-interview/seite-2 (18.09.2018)
2. Beltracchi W and H (2014) Selbstportrait. Rowohlt, Reinbek
3. Birkenstock A (2014) Beltracchi: Die Kunst der Fälschung. Senator: Berlin
4. Blumenroth D; Nieberle J, Heydenreich G (2014) Möglichkeiten und Grenzen kunsttechnologischer Untersuchungen: Heinrich Campendonks Gemälde im Kontext des Beltracchi Fälschungsskandals. In: Schmuhl B, Bauer-Friedrich T (eds) Original bis ... Fälschungen zwischen Faszination und Betrug: 260–275 https://www.th-koeln.de/mam/downloads/deutsch/hochschule/fakultaeten/ kulturwissenschaften/campendonk_blumenroth-nieberle-heydenreich 2014.pdf (17.10.2018)
5. Breek R, Froentjes W (1975) Application of pyrolysis gas chromatography on some of van Meegeren's faked Vermeers and Pieter de Hooghs. Studies in Conservation 20 (Issue 4): 183–189
6. Carlyle L, Binnie N, Kaminska E, Ruggles A (2002) The Yellowing /bleaching of oil paintings and oil paint samples, including the effect of oil processing, driers and mediums on the colour of lead white paint. In: Vontobel R (edt) ICOM-CC 13th Triennial Meeting Preprints, Rio de Janeiro 22–27- September 2002. James and James: London: 328–337
7. Eastaugh N (2014) Lessons from the Beltracchi case: Conceptual aspects, methodology and protocols around the scientific view on the material of art. In: Keazor, Henry; Öcal,Tina (eds) Der Fall Beltracchi und die Folgen. Interdisziplinäre Fälschungenforschung heute. De Gruyter, Berlin:79–91
8. Erhardt D, Tumosa C, Mecklenburg M (2000) Natural and Accelerated Thermal Aging of Oil Paint Films. In: Roy A, Smith P (eds) Tradition and Innovation: Advances in Conservation, Contributions to the IIC Melbourne Congress, 65–69. IIC Preprints, Melbourne Congress. International Institute for Conservation of Historic and Artistic Works: London: 65–69
9. Franken V, Heydenreich G, Jägers E, Müller W, Schulz J, Zumbühl S (2014) Set Back the Race: Treatment Strategies for Running Oil Paint. In: van den Berg K J, Burnstock A, de Keijzer M, Krueger Jay, Learner T, de Tagle A, Heydenreich G (eds): Issues in Contemporary Oil Paint. Springer International Publishing: Switzerland: 333–249
10. Hebborn E (1997/1999) Der Kunstfälscher. DuMont: Köln
11. Lazzari M, Chinatore O (1999) Drying and oxidative degradation of linseed oil. Polymer degradation and stability. 65 (2): 303–313
12. Landgericht Köln 27.10.2011 Aktenzeichen:110 KLs 17/11 http://www.justiz.nrw.de/nrwe/lgs/koeln/lg_koeln/j2011/110_KLs_17_11_Urteil_20111027.html (18.09.2018)
13. Lübbe E (2013) Farbempfindung, Farbbeschreibung und Farbmessung – Eine Formel für die Farbsättigung. Springer Vieweg: Wiesbaden.
14. Mallégol J, Lemaire J, Gardette J-L (2000a) Drier Influence on the Curing of Linseed Oil. Progress in Organic Coatings 3(2–4): 107–113
15. Mallégol J, Gardette J-L, Lemaire J (2000b) Long-Term Behavior of Oil-Based Varnishes and Paints. Fate of Hydroperoxides in Drying Oils. Journal of the American Oil Chemists' Society77 (3): 249–255
16. Nadolny J, Eastaugh N (2014) The analytical results of a group of Beltracchi forgeries and some historical context to their reception. In: Keazor H, Öcal T (eds) Der Fall Beltracchi und die Folgen. Interdisziplinäre Fälschungenforschung heute. De Gruyter: Berlin: 59–77
17. Stols-Witlox M, Megens L, Carlyle L (2012) 'To prepare white excellent...': reconstructions investigating the influence of washing, grinding and decanting of stack-process lead white on pigment composition and particle size. In: Eyb-Green, S, Townsend J, Clarke M, Nadolny J, Kroustallis S (eds) The artist's process: technology and interpretation. Archetype: London: 112–129

18. Townsend J, Carlyle L, Ja-Hyn C, Campos M F (2011) The Yellowing/ bleaching behaviour of oil paint: further investigations into significant colour change in response to dark storage followed by light exposure.In: Bridgland J (edt) ICOM -CC 16th Triennial Conference: 1–10
19. Van den Berg JDJ, Van den Berg KJ, Boon JJ (1999) Chemical changes in curing and ageing oil paints. In: Bridgeland J, Brown J (eds), ICOM-CC, 12th Triennial Meeting. James & James, London: 248–253
20. Wallert A, van de Laar M (2011) Working Methods, tricks, materials and techniques: Top quality forgeries. In: Lammertse F, van Es J (eds) Van Meegeren's Vermeer The Connoisseur's Eye and the Forger's Art. Museum Bojimans Van Beuningen:Rotterdam: 66–99
21. Zumbühl S, Scherrer N C, Ferreira E, Hons S, Müller M, Kühnen R, Navi P (2011) Accelerated aging of drying oil paint - an FTIR study on the chemical alteration. Zeitschrift für Kunsttechnologie und Konservierung 25 (2):339–351
22. Zumbühl S, Scherrer N C, Eggenberger U (2014) Derivatization technique to increase the spectral selectivity of two-dimensional Fourier Transform Infrared Focal Plane Array imaging: Analysis of binder composition in aged oil and tempera paint. Applied Spectroscopy 68 (4): 458–465

Chapter 35
The Influence of Metal Stearates on the Water Sensitivity of Modern Oil Paints

Donatella Banti, Aviva Burnstock, Ilaria Bonaduce, Jacopo La Nasa, Judith Lee, Anna Lluveras-Tenorio, Francesca Modugno, Bronwyn Ormsby, and Klaas Jan van den Berg

35.1 Introduction

The sensitivity of unvarnished modern oil paintings to surface cleaning using water or saliva is well known to conservators. The causes of water sensitivity in well bound oil paint surfaces have been linked to paint formulation, artists' technique, the condition of the paint and environmental factors. Paint additives may influence the ageing and the degradation processes thus rendering the paint water sensitive [5]. Additives to modern oil paints include metal soaps (MS – e.g. metal stearates or palmitates, but all commonly called stearates[1]) and free fatty acids (FFAs) added as dispersing agents to improve the wetting properties of the pigments [4, 9]. Metal stearates can also be present in the paint as a result of interaction of paint degradation products and metals in pigment [17, 20].

[1]The term metal stearates is used in conservation research to either indicate generically metal soaps or technical stearates, which usually define a mixture of stearates and palmitates of different proportions depending on the source.

D. Banti (✉) · A. Burnstock
Department of Conservation and Technology, Courtauld Institute of Art, London, UK
e-mail: aviva.burnstock@courtauld.ac.uk

I. Bonaduce · J. La Nasa · A. Lluveras-Tenorio · F. Modugno
Department of Chemistry and Industrial Chemistry, University of Pisa, Pisa, Italy
e-mail: ilaria.bonaduce@unipi.it; jacopo.lanasa@for.unipi.it; anna.lluveras@for.unipi.it; francesca.modugno@unipi.it

J. Lee · B. Ormsby
Tate, London, UK
e-mail: Judith.Lee@tate.org.uk; Bronwyn.Ormsby@tate.org.uk

K. J. van den Berg
RCE, Cultural Heritage Agency of the Netherlands, Amsterdam, The Netherlands
e-mail: K.van.den.Berg@cultureelerfgoed.nl

K. J. van den Berg et al. (eds.), *Conservation of Modern Oil Paintings*,
https://doi.org/10.1007/978-3-030-19254-9_35

It has been hypothesised that metal stearates may have a role in water sensitivity [4, 5], and studies have shown that added stearates may increase [18] or decrease [21] water sensitivity of oil paint films. Investigation and analysis of water sensitive paints has also found an increased presence of acids at or near the surface of water sensitive paints [6, 9, 14]. This has led to the hypothesis that that the sensitivity of some oil paints may be due to the formation of free polar species [15].[2]

This study investigated the role of metal stearates and other metal carboxylates and their related free fatty acids and diacids on the water sensitivity of modern oil paints. The paints in this study were produced from tubed oil paints of three different manufacturers, which were used both unmodified (i.e. directly from the commercial tubes) and modified with additives. These paints were prepared as swatches in 2006 and first tested for water sensitivity in 2008 [16]. The swatches were re-examined in the present study that included an assessment of change in water sensitivity and the appearance of the films since 2008.

Chemical characterisation of selected paints was undertaken using a range of analytical methods including Gas Chromatography coupled with Mass Spectrometry (GC/MS), High Performance Liquid Chromatography with Electrospray Ionisation-Quadrupole-Time Of Flight Mass Spectrometry (HPLC-ESI-Q-TOF), Fourier Transform Infrared (FTIR) spectroscopy, and Scanning Electron Microscopy with Energy-Dispersive X-Ray spectroscopy (SEM-EDX). GC/MS was used to identify and quantify only the organic components in the paint sample that after ageing and curing are not bound to the polymeric network and are thus free and mobile species — i.e. free fatty acids (FFA), free diacids (DCA), metal carboxylates of free fatty acids and diacids (MC) [23, 24]. The results provided a molecular characterisation of the free species in these paints, and a means of examining the possible correlation between water sensitivity and the degree of hydrolysis and saponification of the selected oil paints.

35.2 Methodology

The swatches investigated are listed in Table 35.1, which includes the corresponding acronyms used during the studies. The paint swatches were prepared in 2006 from tubed artists' oil paints made by three different manufacturers (Winsor & Newton, Talens and Old Holland – Fig. 35.1a). For each brand the following tube colours were included: Ultramarine, Cobalt Blue, Chrome Green, Cadmium Yellow, Cadmium Red, Zinc White and Raw Sienna (Fig. 35.1a). The swatches were prepared with unmodified paint directly from the commercial tube and with 6 modifications: leached oil, and addition of one of the following: 2% Aluminium Stearates, 2% Zinc Stearates, 10% Linseed Oil, 2% Fatty Acids, 10% Aluminium hydroxide.

The paint swatches produced were split in two and a half was naturally aged (NA), (Fig. 35.1b) while the other half was artificially light aged (LA)[3] for an equiv-

[2] With the term 'free' in this article the authors refer to species not bound to the polymeric network of the cured and aged oil.

[3] Illumination was provided by 36W Philips colour 96.5 fluorescent lamps, with UV filtering (transmission 15 μWatt/lumen), rendering a measured output of 10,000 lux at the sample surface.

Table 35.1 Manufacturers, tube colours and modifications used for producing the paint swatches and related acronyms used in the study

Manufacturer		Tube Colour		Modification	
Name	Acronym	Name	Acronym	Name	Acronym
Winsor and Newton	WN	Ultramarine	UL	Unmodified	–
		Cobalt Blue	CB	2% Aluminium stearates	AS
Talens	TA	Chrome Green	CG	2% Zinc stearates	ZS
		Cadmium Yellow	CY	Leached oil	EX
Old Holland	OH	Cadmium Red	CR	10% Linseed oil	LN
		Zinc White	ZW	2% Fatty acids	FA
		Raw Sienna	RS	10% Aluminium hydroxide	AH

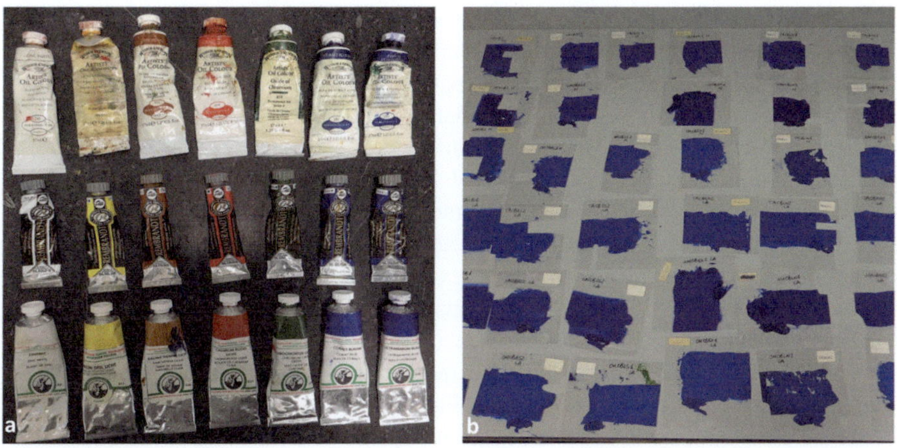

Fig. 35.1 (**a**) Original tubes used for the preparation of the swatches. (**b**) examples of paint swatches used in the study

alent of 24 years exposure in museum conditions, resulting in a total of 304 paint swatches. Until February 2012 the swatches were stored in daylight indoors and since then they have been stored in the dark.

Water sensitivity tests and preliminary assessments were undertaken in 2016 with the main purposes of recording changes in sensitivity and of guiding the selection of paint swatches for further analysis. Water sensitive and non-water sensitive paints and/or paints that had shown a surface change or a change in water sensitivity since 2008 were selected for further study. The surfaces of the swatches were examined using Hirox and SEM microscopy.[4]

Temperature was 25 °C and RH 60% on average. The total ageing time of 1390 h is calculated to be equivalent to twenty-four years of exhibition in recommended museum conditions, 200 lux, eight hours a day, assuming reciprocity.

[4]Two different Hirox microscopes were used at RCE Amsterdam.

– HIROX RX-2000 with high-performance zoom lens MXB 2016z (magnification range: 20–160x).

Water sensitivity tests were carried out based on a standardised method of swab rolling used in the previous assessment in 2008 [18]. Where a visible trace had been left by the 2008 tests, the new tests were carried out in a different area of the sample. The water sensitivity tests were conducted using saliva and deionised water and results were recorded using the criteria used by Mills et al. [16]. Swatches were ranked for sensitivity as follows:

1. No paint removed after 30 swab rolls: not sensitive.
2. Paint removed within 21–30 swab rolls: moderately sensitive.
3. Paint removed within 11–20 swab rolls: sensitive
4. Paint removed within 10 swab rolls: very sensitive.

Tests with saliva were carried out three times by two different conservators and tests with deionised water twice. The results where averaged and are discussed in Sect. 3.1.

Samples from the selected swatches were subject to GC/MS analysis.

The method used in the gas chromatographic analyses was developed at the Department of Chemistry of the University of Pisa [12, 13]. Samples from each swatch were subjected to two independent quantitative GC/MS analyses: one to determine free fatty (FFA) and free dicarboxylic acids (DCA) that will herein be indicated generally as free acids (FA = FFA + DCA) and the other to determine the related free metal carboxylates (MC) – metal soaps of fatty acids (MS) and metal carboxylates of dicarboxylic acids. The determination is based on a two-step analytical approach entailing the subsequent use of two silylating agents, 1,1,1,3,3,3-hexamethyldisilazane (HMDS) for the analysis of free acids (FA) only and N,O-bis(trimethylsilyl)trifluoroacetamide (BSTFA) for the analyses of free acids plus metal carboxylates (FA + MC). The quantitative data on the metal carboxylates only (MC) was calculated by difference with the previous analyses. Full experimental details on the methodology and data treatment are reported elsewhere [2].

The parameters investigated via GC/MS were:

A/P: ratio of azelaic to palmitic acid content, which is an indication of the level of oxidation of the paint and in this case, it is referred to the free species only (i.e. either the free acids (FA) or the sum of free acids and metal carboxylates (FA + MC) of these species).

Weight %: amount of free acids (FA) or of related free metal carboxylates (MC). These amounts, expressed as the percentage on the total weight of the sample, can be used as an indication of the degree of hydrolysis and of the metal carboxylate formation in the paint.

- HIROX RX-2000 with triple objective turret MXB-2500REZ (Magnifications in the Low-Range: 35–250x; Mid range: 140–1000x; High-Range 350-2500x). Images were processed with a KH8700 3D viewer.

The SEM was a Hitachi S4000 with a Field Emission Gun (FEG) used at 25 kV accelerating voltage. The EDX system is an Oxford Instruments INCA system used at 25 kV voltage. This voltage gives enough energy to both image and produce sufficient K, L and M lines for most of the elements within the samples.

Σ dicarboxylic acids (weight %): amount of free dicarboxylic acids (DCA) or of related free metal carboxylate. It is the percentage weight of the sum of all the free diacids or of related metal carboxylates over the total weight of the sample. This is can be used as an indication of the level of oxidation of the free fraction of the paint.

HPLC-ESI-Q-TOF analysis of the triglyceridic profile completed the analysis of the cobalt blue paints achieving the identification of the oil type [11]. The presence of characteristic polyunsaturated triacylglycerols enabled us to distinguish paints containing linseed oil, characterised by high concentrations of linolenic acid and linoleic acid, and paints containing safflower oil, containing high amounts of linoleic acid and other specific triacylglycerols.

35.3 Results and Discussions

35.3.1 Water Sensitivity Tests

The tests displayed low reproducibility in the number of rolls, due to individual approaches to testing, including the size, compactness, angle and pressure of the swab, saliva and judgement about when paint was removed. These are variables one would expect between conservators in practical cleaning of paintings. The paint surfaces were not homogenous and may therefore present different sensitivity on testing. The variation in results was more significant for saliva tests that may reflect also individual saliva variation.

Standardisation of cleaning tests has been used in other studies [22] and the possibility of robotic cleaning tests might improve reproducibility in analytical studies [10], however the results reflect real variance in conservation practice.

The results obtained with deionised water on light-aged swatches are shown in Fig. 35.2 that compares the results obtained in 2008 (2a) with those of 2016 (2b).

In general, the paint swatches were more sensitive to saliva than to deionized water. However, no significant difference was found between the naturally aged and the light-aged swatches. This is consistent with what observed in the 2008 tests [16].

According to 2008 results, the paint layers containing three pigments showed some water sensitivity: ultramarine (UL), cadmium yellow (CY) and cobalt blue (CB) along with two modified paint swatches in the Talens cadmium red (TACR). In 2016, the water sensitivity tended to increase.

Some Talens and Old Holland raw sienna (TARS and OHRS) and Old Holland cobalt blue (OHCB) swatches that showed no sensitivity in 2008 were sensitive in 2016. Cobalt blue swatches showed the most significant increase in sensitivity and were thus selected for further analysis. Swatches containing raw sienna from Winsor and Newton (WNRS) were also selected as representative of non-water sensitive paint. Moreover, these swatches had developed visible fatty acid efflorescence on the surface since 2008.

a

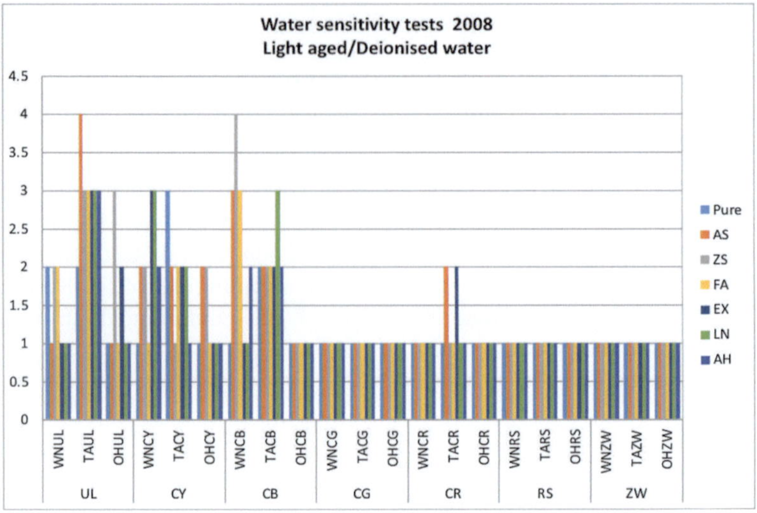

b

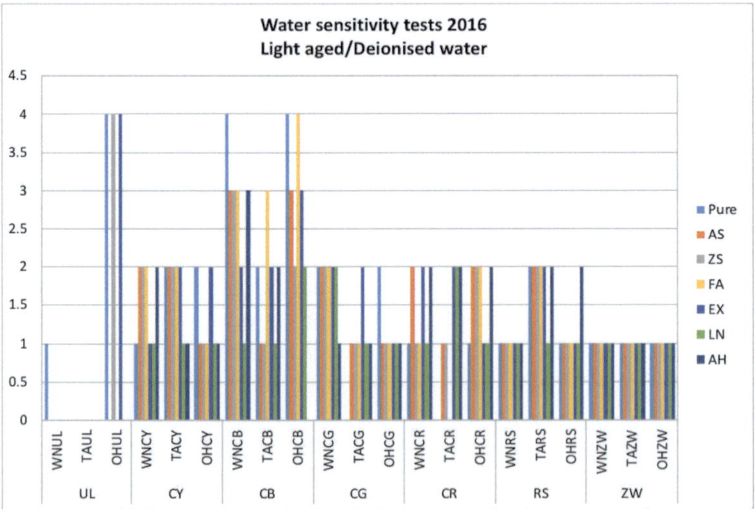

Fig. 35.2 Results of the 2008 (**a**) and 2016 (**b**) water sensitivity tests using deionised water on light aged swatches. In the y axis is indicated the sensitivity of the paint swatches as follows: *1*. No paint removed after 30 swab rolls: not sensitive; *2*. Paint removed within 21–30 swab rolls: moderately sensitive; *3*. Paint removed within 11–20 swab rolls: sensitive; *4*. Paint removed within 10 swab rolls: very sensitive

The effect of additives to the manufactured tube paints on sensitivity showed no clear trends. At the concentrations used in these paint swatches, the additives did not significantly affect the sensitivity of the surface to water or saliva. The addition of aluminium or zinc stearates had an inconsistent effect enhancing or lessening the water sensitivity of some swatches.

35.3.2 GC/MS Analyses

Analyses were performed in triplicate and confidence intervals were calculated at 95%. Results of the analyses are shown as graphs in Fig. 35.3. It presents the most significant results of the GC/MS analysis of samples from the prepared swatches. The results show that water sensitivity does not correlate with the relative amount of free fatty and dicarboxylic acids (FFA and DCA) or of related metal carboxylates (MC), as both water sensitive and non-sensitive paints contain similar amount of these species [1, 2]. The Winsor & Newton and Talens Cobalt Blue paints (WNCB and TACB) were found to have different water sensitivities as well as different molecular profiles of the mobile phase (Fig. 35.3). Research on paint formulation was supplemented by analyses using SEM-EDX and HPLC-ESI-Q-TOF ([1, 2, 19], Appendix 2.14) that revealed that different pigments and oils were used for the Winsor & Newton and the Talens cobalt blue paints.

Co-Zn Silicate (PB74) bound in safflower oil and linseed oil was used for the Winsor & Newton Cobalt Blue paint (WNCB). Cobalt Aluminate (PB28) bound in safflower oil was used instead in the Talens Cobalt Blue paint (TACB, Fig. 35.4).

The type of pigments and the oil composition have a significant effect on the curing and ageing of the oil paint swatches and thus on their chemical profiles. Therefore, the different chemical profiles of the two blue paints found via GC/MS analysis could be due to their different paint formulations.

Through handling of the paint swatches it was observed that the paints showed also differences in mechanical properties, in that the Winsor & Newton cobalt blue swatches were brittle and lifted very easily from the Melinex support, while the Talens cobalt blue were slightly more flexible. The results for the analysis of Winsor & Newton cobalt blue swatches are in agreement with observations made in a study of paintings by Jean-Paul Riopelle [7]. In the study it was found that most passages of cobalt blue paint that were unstable and easily lifting had a value of A/P above 2. As a comparison, Talens Cobalt Blue paints that showed a much lower A/P ratio (in the range 0.0–0.1) were soft and pliable upon handling.

Swatches of Raw Sienna from Winsor & Newton were found to be all non-sensitive to deionised water (albeit moderately sensitive to saliva), despite all having a high A/P ratio in the metal carboxylates fraction (Fig. 35.3). This highlights that a high A/P in the non-network bound components of the paint (FA and FA+MC) is not sufficient to result in water sensitivity of the paint film.

Quantitative GC/MS results showed that the relative content of FFA, DCA and non-network bound metal carboxylates (MC) was below 10% for all paints examined (Fig. 35.3). However, FTIR absorptions at 1710 cm^{-1} in Winsor & Newton Raw Sienna and at 1589 cm^{-1} in Winsor & Newton Cobalt Blue and 1592 cm^{-1} in Talens Cobalt Blue (Fig. 35.5a–c) are ascribed to acidic moieties (1710 cm^{-1}) and carboxylate groups (1589/1592 cm^{-1}) that are in higher amount than 10%.[5] This indicates that

[5] FTIR spectra of fresh paint samples produced in the lab with either 10% w/w stearic acid or Zn Stearate showed a CO stretching band (at 1710 cm^{-1} and 1540 cm^{-1}, respectively) whose relative intensity with respect to the CO stretching of ester bonds as 1743 cm^{-1} was significantly, lower

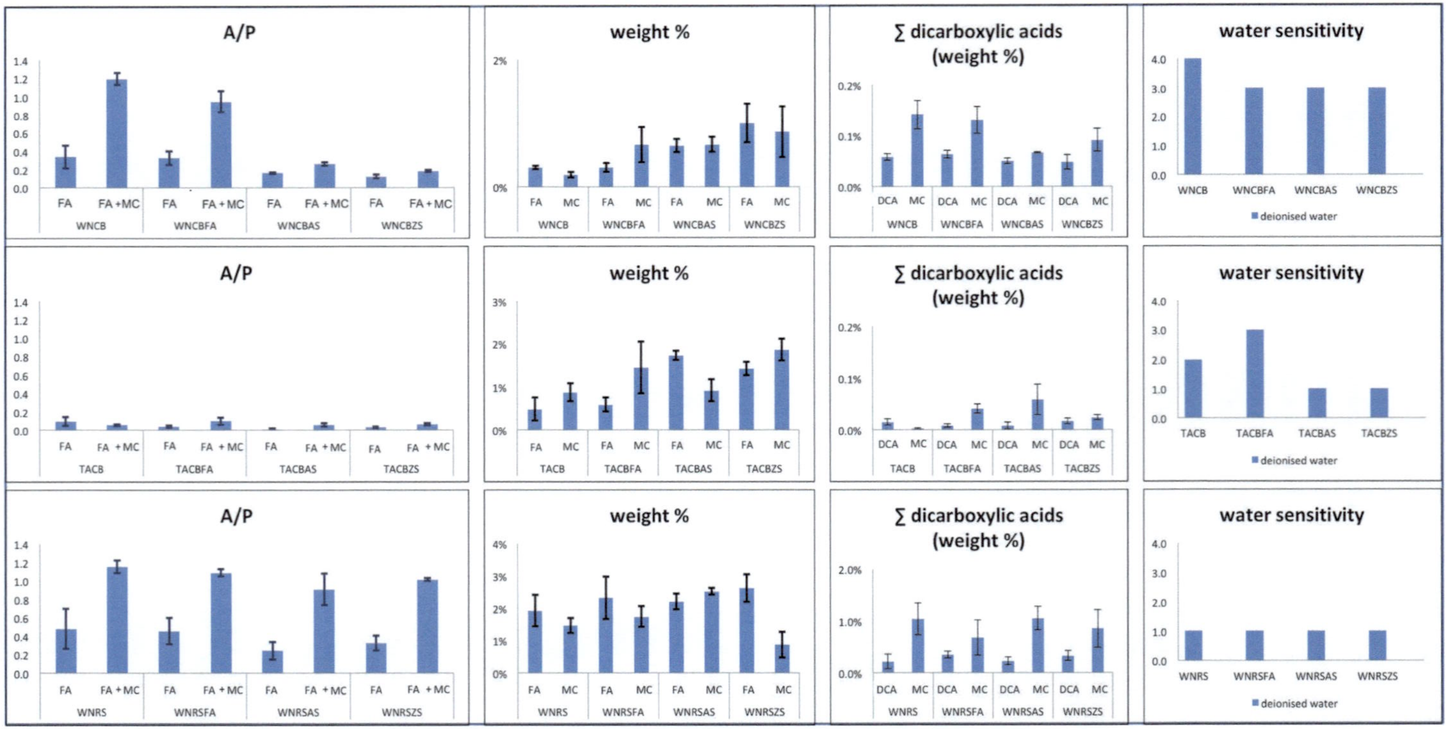

Fig. 35.3 Chemical parameters obtained by GC/MS analysis and corresponding water sensitivity tests results for the Winsor & Newton Cobalt Blue (WNCB – upper row), Talens Cobalt Blue (TACB – middle row) and Winsor & Newton Raw Sienna (WNRS – lower row) paint series. FA indicates the results obtained for the free (i.e. the non-network bound) acidic species; MC indicates the results obtained for the free (i.e. the non-network bound) metal carboxylates; DCA indicates the results obtained for the non-network bound dicarboxylic acid species (modified from Banti et al. 2018 – Published by The Royal Society of Chemistry)

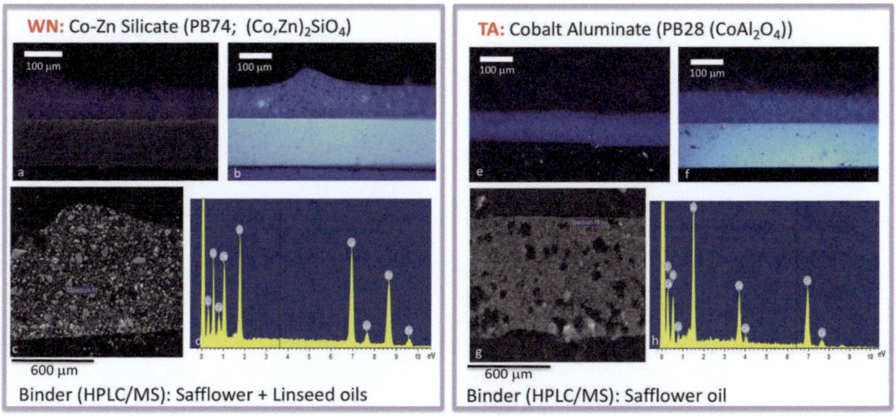

Fig. 35.4 Cross sections micrographs (**a**; **b**; **e** and **f**), SEM images (**c** and **g**) and EDX analysis (**d** and **h**) of Winsor & Newton (**a–d**) and Talens (**e–h**) cobalt blue paint swatches

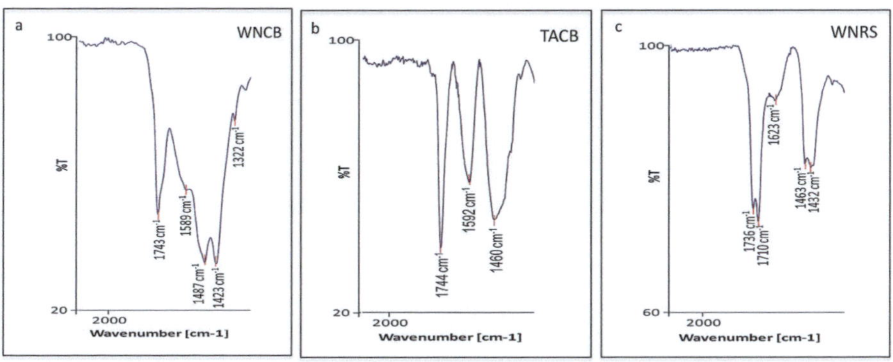

Fig. 35.5 Detail of FTIR spectra (range 1750–1300 cm^{-1}) of Winsor & Newton Cobalt Blue (**a**), Talens Cobalt Blue (**b**) and Winsor & Newton Raw Sienna (**c**) swatches

the majority of the acidic and metal carboxylates groups in the paint are bound to the polymeric/ionomeric network formed upon aging of the oil binder [2, 3, 8].

The quantitative GC/MS results also highlighted that the development of fatty acid efflorescence on the paint surface is linked to a high level of free fatty and dicarboxylic acids present in the paint. An increase in value in the weight % of free fatty and dicarboxylic acids found via GC/MS parallels the increase in fatty acid florets density and dimension on the surface (Table 35.2). This confirms again that the

than that observed in WNCB, TACB and WNRS swatches in Fig. 35.5 [2]. This pointed to the fact that the peaks of the acid at 1710 cm^{-1} and the carboxylate salts at 1589 and 1592 cm^{-1} in the swatches were due to species present in quantities higher than the amount found via GC/MS (approx. 2%) for the non-network bound species.

Table 35.2 Correlation between the increase in fatty acids efflorescence on the paint swatches surface (visible in the micrographs in the first and fourth columns) and the amount of free acids present in the paint swatches and found through GC/MS analysis (third and sixth columns). The higher the amount of free acids, the higher the density and dimension of the fatty acids' florets on the surface

Micrographs	Sample	Weight %	Micrographs	Sample	Weight %
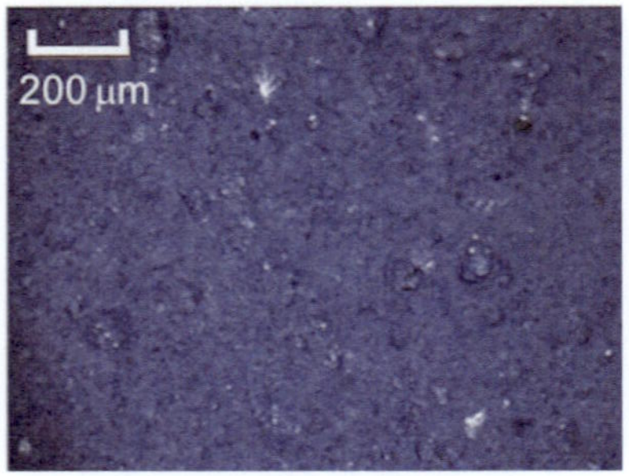	Winsor &Newton Cobalt blue 2% Zinc Stearate	1.0 ± 0.3	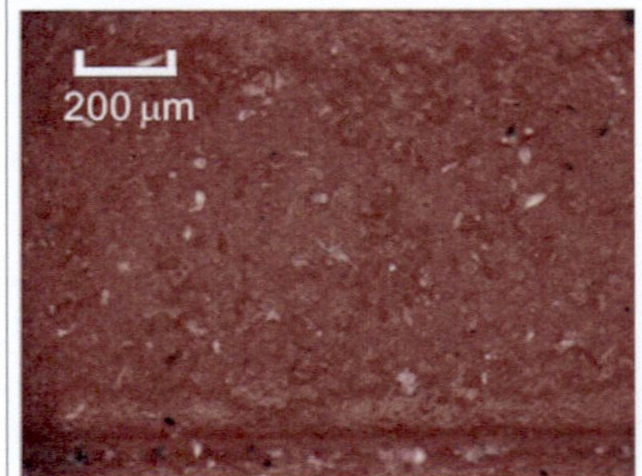	Winsor &Newton Cadmium Red	1.4 ± 0.2
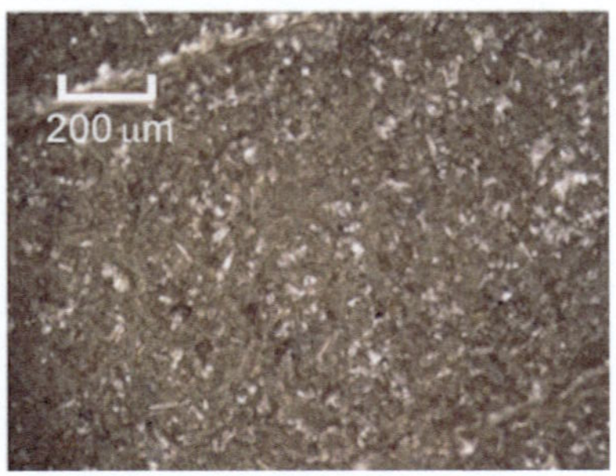	Winsor &Newton Raw Sienna	1.9 ± 0.5	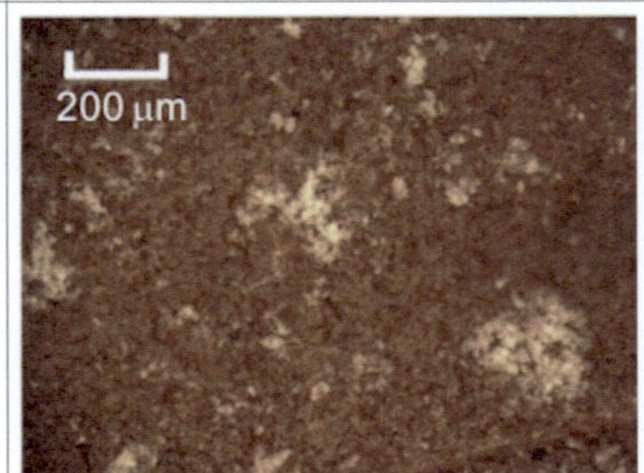	Winsor &Newton Raw Sienna 2% Fatty Acids	2.3 ± 0.7

degree of hydrolysis in the paint is not proportional to the water sensitivity found in the paint, but to the amount of efflorescence that may develop on the surface.

Talens Cobalt blue samples (TACB) also showed similarly high relative amounts of free fatty acids, but no fatty acid efflorescence. The amount of dicarboxylic acids in these swatches is significantly lower, however a full understanding of these results would require further analysis into the degree of oxidation of the paint and determination of the pigment/binder ratio.

35.4 Conclusion

This study examined the relationship between empirical swab tests of water sensitivity and the quantitative analysis of fatty acids, dicarboxylic acids and related metal carboxylates in selected oil paint films. The selected swatches from Winsor & Newton and Talens had developed increased water sensitivity or fatty acid efflorescence during 8 years of ageing in ambient indoor conditions. Results indicated that, although water sensitivity is influenced by pigment, oil type and additives, there was no correlation between sensitivity and the proportions of free acids and metal carboxylates, including metal soaps in the paints. This suggests that condition of the polymeric/ionomeric network in oil paint is likely to be a significant factor in the development of water sensitivity. Research is still necessary to investigate the complex paint system further.

Acknowledgements This work was performed within the context of the CMOP project (Heritage Plus Joint Call project 2015–2018) and partly funded by EU/H2020, partly by NWO (The Netherlands), AHRC (UK) and MIUR (Italy).

The authors would like to thank William Luckhurst, King's College London for granting access to the SEM-EDX instrument and the Netherlands Institute for Conservation, Art and Science (NICAS) (Conference grant) and the Anna Plowden Trust (Anna Plowden/Clothworkers CPD Grants) for awarding generous grants to Donatella Banti to attend the CMOP Conference in May 2018.

References

1. Banti D (2017) The influence of metal stearates on the water sensitivity of modern oil paints. Dissertation, Courtauld Institute of Art
2. Banti D, La Nasa J, Lluveras Tenorio A, Modugno F, van den Berg KJ, Lee J, Ormsby B, Burnstock A, Bonaduce I (2018) A molecular study of modern oil paintings: investigating the role of dicarboxylic acids in the water sensitivity of modern oil paints. RSC Advances 8: 6001–6012
3. Boon JJ (2006) Processes inside paintings that affect the picture: chemical changes at, near and underneath the paint surface. In: Boon JJ, Ferreira E (eds) Reporting Highlights of the De Mayerne Program MOLART Report 13: 21–32
4. Burnstock A, van den Berg KJ, de Groot S, Wijnberg L (2008) An Investigation of Water-Sensitive Oil Paints in Twentieth-Century Paintings. In: Learner TJS, Smithen P, Krueger JW,

Schilling MR (eds) Modern paints uncovered: proceedings from the Modern Paints Uncovered Symposium, Tate Modern, London, 16–19 May 2006: 177–188
5. Burnstock A, van den Berg KJ (2014) Twentieth Century Oil Paint. The Interface Between Science and Conservation and the Challenges for Modern Oil Paint Research. In: van den Berg KJ, Burnstock A, de Keijzer M, Krueger J, Learner T, de Tagle A, Heydenreich G (eds) Issues in Contemporary Oil Paint; Springer, Heidelberg: 1–20
6. Burnstock A, Lee J, van den Berg KJ, Ormsby B (2016) Water sensitivity of modern oil paint films in Dall'olio all'acrilico, dall'impressionismo all'arte contemporanea: Studi, ricerche, indagini scientifiche ed interventi conservativi: proceedings from the VII International Congress on Colour and Conservation, Milano 13–14 November, 2015, ed. Cesmar7 and Il Prato publishing house srl: 66–76
7. Corbeil MC, Helwig K, Poulin J (2011) Jean Paul Riopelle: The Artist's Materials. Getty Conservation Institute, Los Angeles: 67–72
8. Hermans JJ, Keune K, van Loon A, Iedema D (2015) 'An infrared spectroscopic study of the nature of zinc carboxylates in oil paintings' *J. Anal. At. Spectrom.* 30: 1600–1608
9. Izzo FC, van den Berg KJ, van Keulen H, Ferriani B, Zendri E (2014) 'Modern Oil Paints – Formulations, Organic Additives and Degradation: some case studies', in, Issues in Contemporary Oil Paint; ed. van den Berg KJ, Burnstock A, de Keijzer M, Krueger J, Learner T, de Tagle A, Heydenreich G, Springer: 75–104
10. Keefe M, Ormsby B, Soldano A, Phenix A, Learner T, Tucker C, Behr A, Meyers G, Rheinhardt C, Boomgaard T, Peitsch C (2011) Art and industry: novel approaches to the evaluation and development of cleaning systems for artists' acrylic latex paints. Coatings Tech, American Coatings Association, September 2011: 26–39
11. La Nasa J, Ghelardi E, Degano I, Modugno F, Colombini MP (2013) 'Core shell stationary phases for a novel separation of triglycerides in plant oils by high performance liquid chromatography with electrospray-quadrupole-time of flight mass spectrometer', Journal of Chromatography A, 1308: 114–124
12. La Nasa J, Modugno F, Aloisi M, Lluveras-Tenorio A, Bonaduce I (2018) Development of a GC/MS method for the qualitative and quantitative analysis of mixtures of free fatty acids and metal soaps in paint swatches. Analytica Chimica Acta 1001: 51–58
13. La Nasa J, Lluveras-Tenorio A, Modugno F, Bonaduce I (2018) 'Two-step analytical procedure for the characterization and quantification of metal soaps and resinates in paint samples', Heritage Science, 6: 57
14. Lee J (2017) Investigating water sensitivity of Modern oil paints made by British manufactured paints by Winsor & Newton, PhD Thesis, Courtauld Institute of Art
15. Lee J, Bonaduce I, Modugno F, La Nasa J, Ormsby B, van den Berg KJ (2018) 'Scientific investigation into the water sensitivity of twentieth century oil paints' Microchemical Journal, 138: 282–295
16. Mills L, Burnstock A, van Keulen H, Duarte F, van den Berg KJ (2008) Water sensitivity of modern artists' oil paints, ICOM Committee for Conservation (ICOM-CC) 15th Triennial Meeting, 22–26 Sept 2008, New Delhi: 651–659
17. Mills JS, White R (1999) 'The Organic Chemistry of Museum Objects', Butterworths, 2nd Ed,
18. Mills L (2008) Water Sensitive Oil paints: an experimental investigation characterizing the causes of the phenomena and analysis of a case study 'Les Animaux', 1961 by Karel Appel. Dissertation, Courtauld Institute of Art
19. Modugno F, Di Gianvincenzo F, Degano I, van der Werf ID, Ilaria Bonaduce I, van den Berg KJ (2019) On the influence of relative humidity on the oxidation and hydrolysis of fresh and aged oil paints. Scientific Reports, 9(1):5533
20. Sawicka A, Burnstock A, Izzo FC, Keune K, Boon JJ, Kirsch K, van den Berg, KJ (2014) 'An Investigation into the Viability of Removal of Lead Soap Efflorescence from Contemporary Oil Paintings' in Issues in Contemporary Oil Paint; ed. van den Berg, KJ, Burnstock A, de Keijzer M, Krueger J, Learner T, de Tagle A, Heydenreich G, Springer: 311–332

21. Tempest H, Burnstock A, Saltmarsh P, van den Berg KJ (2013) Sensitivity of Oil Paint Surfaces to Aqueous and Other Solvents in New Insights into the Cleaning of Paintings Proceedings from the Cleaning 2010 International Conference Universidad Politécnica de Valencia and Museum Conservation Institute Ed.: Mecklenburg, MF, Charola, AE, Koestler, RJ: 107–114
22. Tiedemann LM, Kunov-Kruse AJ (2016) Surface cleaning acrylic emulsion paints: evaluating conductivity adjusted aqueous solutions. In: Selva Bonino, V. (eds) Dall'olio all'acrilico, dall'impressionismo all'arte contemporanea, VII International Congress on Colour and Conservation, Milano 13–14 November, 2015 Il Prato Publishing House Srl: 169–175
23. van den Berg JDJ (2002) Analytical chemical studies on traditional linseed oil paints, PhD Thesis, Universiteit van Amsterdam
24. van den Berg JDJ, van den Berg KJ, Boon JJ (1999) 'Chemical changes in curing and ageing oil paints' In: 12th Triennial Meeting Lyon 29 August – 3 September 1999; Vol. 1, ICOM-CC ICOM Committee for Conservation. – Lyon: 248–253

Chapter 36
The Rate of Solvent Action on Modern Oil Paint

Stefan Zumbühl

36.1 Introduction

Solvents are often indispensable in art conservation. One fundamental issue however is that empirical experience of solvent use on classical oil paints cannot be directly transferred to modern paint systems, which are distinct from traditional oils in several ways. Aside from the use of different raw materials, the advent of tube paints greatly influenced material properties via the use of additives to facilitate industrial processing and to extend shelf life. Of particular relevance is also the introduction of synthetic organic pigments (SOP) that lead to a successive replacement of almost all heavy metal containing pigments over the twentieth century. Although only a few of those newer colorants are directly soluble in organic solvents, their small particle size has proven to render modern paints very sensitive to solvent interaction [1]. In addition, the different ageing behaviour of modern oil paint is a key factor in influencing paint properties with respect to swelling and solubility. This paper focuses on specific solvent properties and solvent-related parameters to describe the rate of action of solvents on modern oil paints within the first few seconds of solvent interaction, which is arguably the most critical period in any solvent-based cleaning procedure.

36.1.1 Influence of Oil Paint Properties on Solvent Action

Oil paint follows a complex ageing process, which effectively influences interactions with solvents. Of particular relevance are oxidative cross-linking reactions, oxidative fragmentation, hydrolysis, and carboxylate formation. These four reaction

S. Zumbühl (✉)
Bern University of Applied Sciences, Bern, Switzerland
e-mail: stefan.zumbuehl@hkb.bfh.ch

K. J. van den Berg et al. (eds.), *Conservation of Modern Oil Paintings*,
https://doi.org/10.1007/978-3-030-19254-9_36

pathways form the focus of the following discussion, as they influence material properties and solvent interactions. The solvent chemical structure, on the one hand, influences solvation behaviour, including the potential to interact at the intermolecular level with solvent molecules; the polymeric matrix consisting of cross-linked glycerides, on the other hand, largely defines the swelling behaviour, while low molecular weight decomposition compounds are primarily associated with leaching behaviour. Of course, these pathways can vary over time and with pigment type, amongst other factors.

36.1.2 Ageing of Oil Paint

In order to model solvent action on modern oil paints it is relevant to understand ageing properties. Primarily, oil-based binders dry or cure via the formation of a chemical network promoted by radical chain reactions [16]. These cross-linking and fragmentation chemical pathways have been thoroughly investigated [2, 5]. After drying, polymer network decomposition becomes increasingly relevant over time, whereas the material properties of the paint film depend on the equilibrium conditions established between cross-linking and fragmentation reactions.

Two fragmentation reactions are of specific relevance in this context (Fig. 36.1a): 1). Peroxide homolysis, which can be induced by light or metal catalysts [6, 9]. The degradation of oil paint is primarily driven by the β-elimination of the alkyl radicals ("oxidative cleavage") leading to an irreversible fragmentation of the fatty acid chain and the formation of a carboxylic acid group as the final reaction product [2, 9]. In particular the increased acidity has a significant influence on the water activity of a material and the "hydrolytic cleavage" of the triglycerides. As a consequence, different dicarboxylic acids such as azelaic acid are formed [16]. Thus, it is evident that the soap formation process and the formation of "ionic cross-links" between two carboxylate groups and a multivalent metal ion are essential in stabilising aged oil paints (Fig. 36.1b). The main differences between classical and modern oil paints are therefore not the degradation reactions per se, but stabilisation via the formation of an ionomeric network by free cations, which are associated with differing solvent sensitivity.

36.1.3 Influence of Solvent Properties on Solvent Action

It is known that the variable interactive ability of solvent molecules influences their effect on an oil paint. However, solvent properties do not only control the polarity effects of solvent action; they also have a decisive influence on the rate of action. While Phenix explored the swelling behaviour of oil paints [11] and Sutherland investigated the potential extractability of low molecular compounds during long-solvent immersions [14], the focus of this paper is the rate/speed of these processes as explored across the first 30 s of solvent interaction with a paint film.

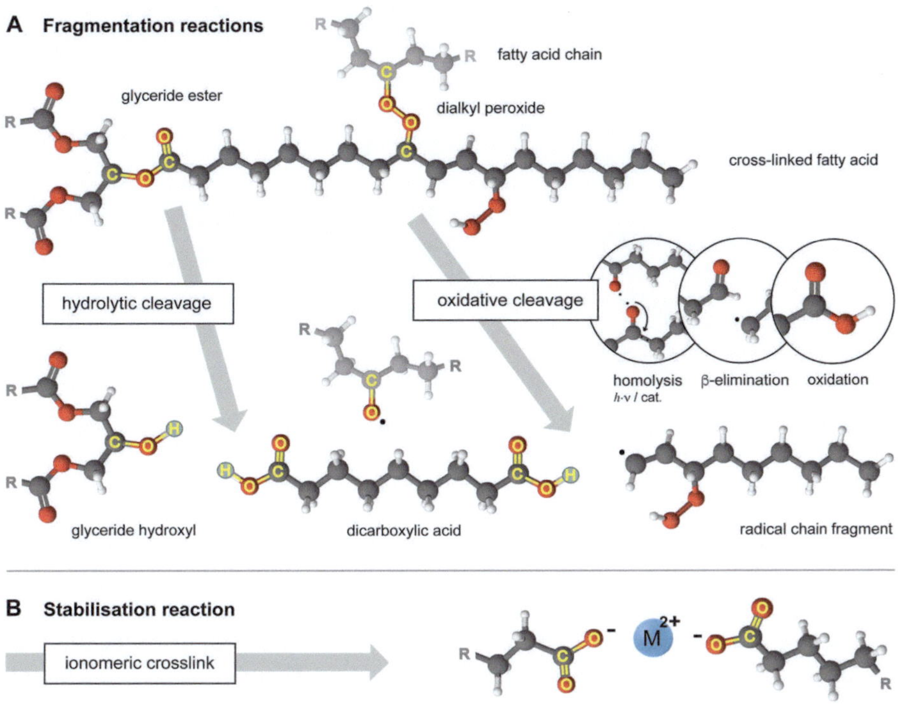

Fig. 36.1 (**a**) Fragmentation reactions of fatty acids in oil paint. (**b**) Stabilisation of acidic ageing products by ionomeric crosslinks. R = glyceride rest

36.1.4 Principles of the Process of Dissolution

In contrast to solubility charts (by Teas, Hansen and others) used in conservation [4, 15], thermodynamics form the basis of all modern dissolution concepts, where the process of dissolution is described via a change in enthalpy and entropy [12]. To characterise the different properties of dissolution of materials, a chart was developed using a rating based on values for enthalpy and entropy (Fig. 36.2), as well as diffusion rate. Looking at the thermodynamic process of dissolution in energetic terms, the following equation must be taken into account: $\Delta G\text{m} = \Delta H\text{m} - \text{T}\Delta S\text{m}$. The requirement for a spontaneous dissolution (mixing) process is that the Gibbs free energy of the solution ΔHm is lowered relative to the undissolved state. The enthalpy of mixing ΔHm follows the commonly known rule of "like dissolves like" and requires similar intermolecular solvent-solvent and solvent-solute forces for successful mixing. Since readily mixing solvents often exhibit a positive and small ΔHm-value, the change in entropy ΔSm at a given temperature T is the key driving force behind the processes of swelling and/or dissolution. This forms the background of the proposed solubility chart, which was developed to describe the rate of solvent action. This is based on two fundamental concepts of solution chemistry: the

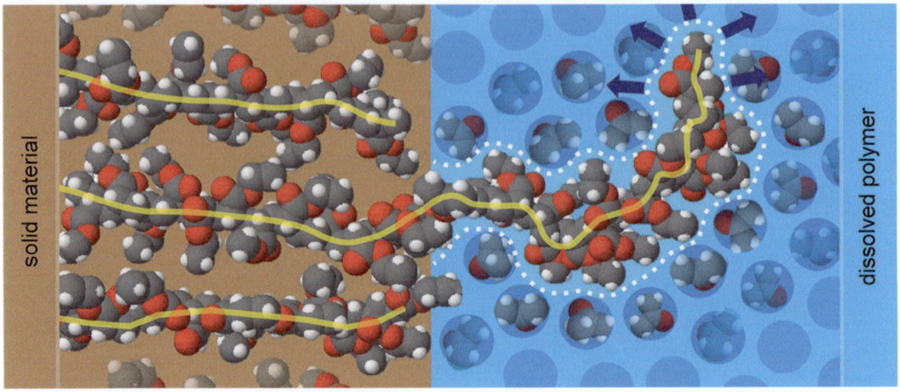

Fig. 36.2 Graphic visualisation of the freedom of conformation of the polymer chain (ethyl acrylate) and of the cavity formation (dashed line) during the mixing process. This change in entropy requires the overcoming of the cohesive energy density of the liquid (blue arrows)

"Flory Huggins-Theory", which describes the thermodynamic properties of polymer solution [3] and, the widely used "free-volume theory" that was developed by Vrentas and Duda to describe diffusion in polymeric systems [13].

The rate of solvent action is primarily driven by the distribution of the material molecules into the liquid solvent. The Flory Huggins-Theory describes the mobility of a single macromolecule as a process of self-diffusion, which is influenced by the properties of the surrounding solid or solvent molecules. According to this theory, the major source of entropy associated with the mixing of a polymer should be translational, since the whole polymer chain can access more configurations by being moved from a sample of a pure polymer into a solvent (Fig. 36.2). On the one hand, the chain dynamics are strongly influenced by the segment mobility of the polymer chain (T_g) and the molecular mass, as well as the cross-linking or crystallisation of the polymer system. On the other hand, the liquid phase (solvent) has a significant influence on the process of mixing as well. The easier a solvent molecule can be replaced to incorporate the solute molecule into the liquid (freedom of movement), the higher is the entropy of mixing. The key factor is the so-called "cavitation energy" which corresponds to the cohesive energy density of the liquid (Fig. 36.2).

36.1.5 Solubility System to Visualise the Rate of Solvent Action

To present an overview on the swelling rate of a material, a new solubility chart has been used (Fig. 36.3). Therein, the solvation properties define the range of polarity wherein a specific material can be dissolved. In order to visualise this complex interaction, 50 selected solvents were divided into 7 subgroups (I-VII) along the

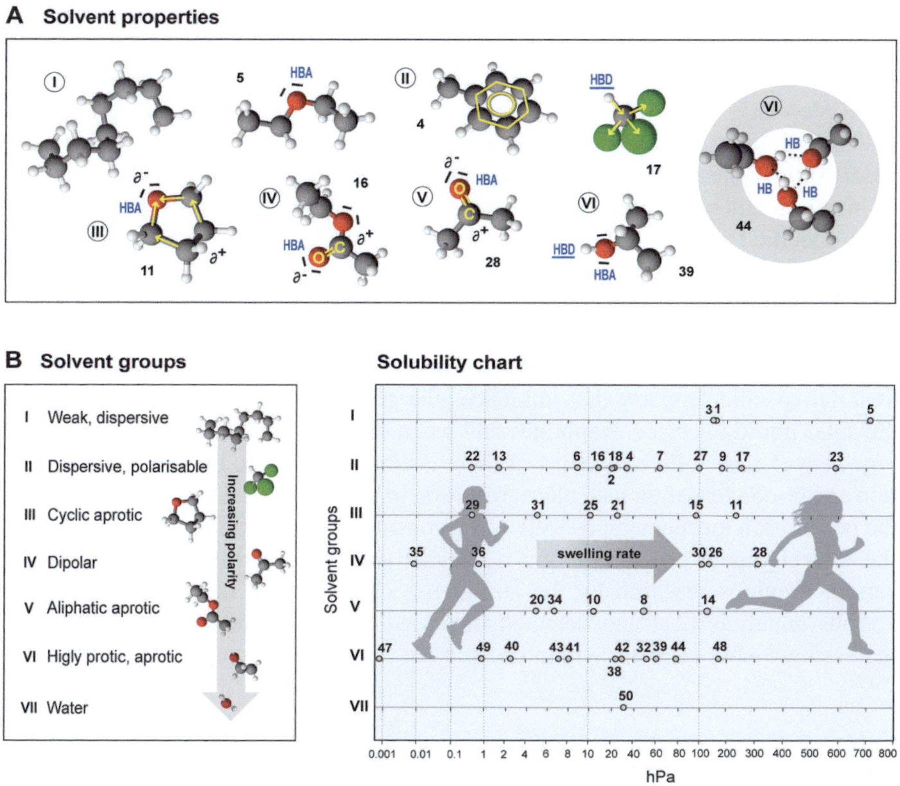

Solvents:

1	n-Hexane	**13**	o-Dichlorobenzene	**25**	Morpholine	**37**	Acetonitrile
2	Tetrachloroethene	**14**	Ethyl acetate	**26**	Methyl ethyl ketone	**38**	2-Butanol
3	Tetrachloromethane	**15**	1,2-Dimethoxyethane	**27**	1,2-Dichloroethane	**39**	2-Propanol
4	Toluene	**16**	Butyl acetate	**28**	Acetone	**40**	1-Pentanol
5	Diethyl ether	**17**	Chloroform	**29**	N-Methylpyrrolidinone	**41**	1-Butanol
6	o-Xylene	**18**	Methyl-i-butyl ketone	**30**	1,3-Dioxolane	**42**	1-Propanol
7	Trichloroethylene	**19**	1,1-Dichloroethane	**31**	N,N-Dimethylformamide	**43**	Cellosolve
8	1,4-Dioxane	**20**	Cyclohexanone	**32**	t-Butanol	**44**	Ethanol
9	1,1,1-Trichloroethane	**21**	Pyridine	**33**	Propionitrile	**47**	Glycerol
10	Diethyl carbonate	**22**	Acetophenone	**34**	Acetic anhydride	**48**	Methanol
11	Tetrahydrofuran	**23**	Dichloromethane	**35**	Sulfolane	**49**	Formamide
12	Butylamine	**24**	Ethyl formate	**36**	Dimethyl sulfoxide	**50**	Water

Fig. 36.3 (**a**) Solvents classified into different solvent groups. The characteristic structural elements are highlighted. Yellow arrows: electron shifts by inductive effects. Yellow ring: polarisable element; Dipolar groups: ∂^+/∂^-; Aprotic groups: HBA; Protic groups: HBD. No. 44 shows the cluster formation of ethanol molecules. (**b**) Solubility chart with 7 solvent groups. The polarity increases from group I to VII

y-axis, based on their interaction and structural properties [18]. Each group comprises solvents with similar solvation properties, which are linked to the enthalpy change. In the solubility chart (Fig. 36.2b) the solvent vapour pressure (hPa 25 °C) was used as an entropy-relevant value and applied to the x-axis (data: [7]).

In theory, the rate of solvent action within each group of solvents increases from left to right, whereby the increasing intermolecular interaction within the solvent slows down the process of mixing. Kamlet and Taft, for example, used the square of the Hildebrand-Parameter ∂_H^2 as an entropy relevant factor to describe the process of dissolution [12]. There are two reasons why the vapour pressure was used here to describe the cohesive property of the solvent: (1) molecules of similar size exhibit an inverse relationship between the cohesive energy density and the vapour pressure (both specified using the SI derived unit of pressure [Pa]), and thus the vapour pressure is a good indicator to describe the ease of entropy change; (2) in a practical context, the solvent transport after application is relevant, since the rate of solvent action is also influenced by the diffusion rate of the solvent into the polymeric material. In principle, this aspect is not part of the swelling process itself, yet it is a material-specific property that influences the general reaction rate. The solvent at the solid/liquid interface evaporates and gas molecules penetrate into the free volume between the solid molecules. If there is a high affinity to the polymer, adsorption and re-condensation take place, resulting in swelling of the material. Since both the penetration mechanism, as well as the entropy change of dissolution is linked to the solvent vapour pressure, this is a relevant value to describe the rate of solvent action. Previous research by the author has characterised the solubility rate of various materials [21] and detected increased leachate concentrations [20] with increasing vapour pressure within one solvent class/group.

36.2 Experimental

To explore this relationship, the solvent-specific swelling and leaching rate of 42 solvents (Fig. 36.2) was investigated within the first few seconds of solvent immersion. The solvent sensitivity of a nine-year old oil paint (Schmincke® Norma™) with different organic pigments (PV19:PR189) was compared to the behaviour of a zinc white paint (PW4), which is known to form an ionomeric polymer network. The two paints were aged under normal indoor light conditions (behind window glass) under ambient environmental conditions. The immersion-swelling test was performed on self-supporting paint films. The standardised 300 μm thick films (wet paint thickness) were prepared using an Erichsen GmbH film applicator. A segment of 1 mm width was fixed in a circular glass holder with an inner diameter of 4 mm (free length) described in [19]. Extracted components were quantified gravimetrically and characterized with Fourier transform infrared spectroscopy (FTIR) [17].

36.3 Results and Discussion

36.3.1 The Swelling Rate of Modern Oil Paint

Regarding the swelling of these paint samples, the results showed a general consistency with data reported in the literature [10]. Oil paint contains extensive hydrocarbon backbones and thus shows the strongest effects within the solvent group II, which is characterised by strong dispersive force interaction. Similarly, high swelling was observed in group III, with cyclic aprotic solvents. These solvents are gently polarised due to structural induction effects and thus likewise exhibit relatively strong dispersive force interactions (Fig. 36.2a). The solvation properties appeared to change marginally during drying, since oxidative cross-linking primarily produces apolar and aprotic groups, which are often sterically well-protected. Even in very old oil paint, however, only a marginal widening of the solubility range has been observed [10].

The rate of swelling, however, strongly depended on the ease of displacing chain segments of fatty acids into the solvent, which can only partially be derived from the degree of cross-linking of the system. Analysis of the collected data suggests that the degree of saponification, that is the ionic network, is the dominant dimension affecting swelling. Within the zinc white based paint, the rate of swelling was strongly limited by zinc carboxylates with only a few solvents showing significant swelling after 30 s of immersion (Fig. 36.4a). Modern oil paints with chemically inert pigments (e.g. synthetic organic pigments SOP), on the contrary, had extremely high rates of swelling. After only 10 s of immersion, volume increases in excess of 70% were reached in any chlorinated solvent. Beyond this, the paints evaluated proved to be sensitive across a broad range of solvents along the polarity spectrum (Fig. 36.4b). These findings have highly relevant implications for conservation practice. They concur with swab tests carried out on modern oil paints containing synthetic organic pigments, which showed that these paints can be extremely sensitive to a wide range of fast swelling solvents in combination with mechanical stress [1]. The low wipe resistance, as seen with swabbing action, is further enhanced by the typically small particle dimension of the SOPs, which adds to the swelling induced destabilisation of the binder.

36.4 Conclusions

In general, the solvent sensitivity of modern oil paint is not the same as that of classical oil paint. The distinct chemical composition of the binder in modern oils compared to classical oil paint is significant. Nevertheless, the differences in solvation properties are minimal. Of particular interest are modern oil paints containing non-soluble synthetic organic pigments, which exhibit a very fast swelling rate that is distinctly different to the behaviour of classical oil paints containing zinc

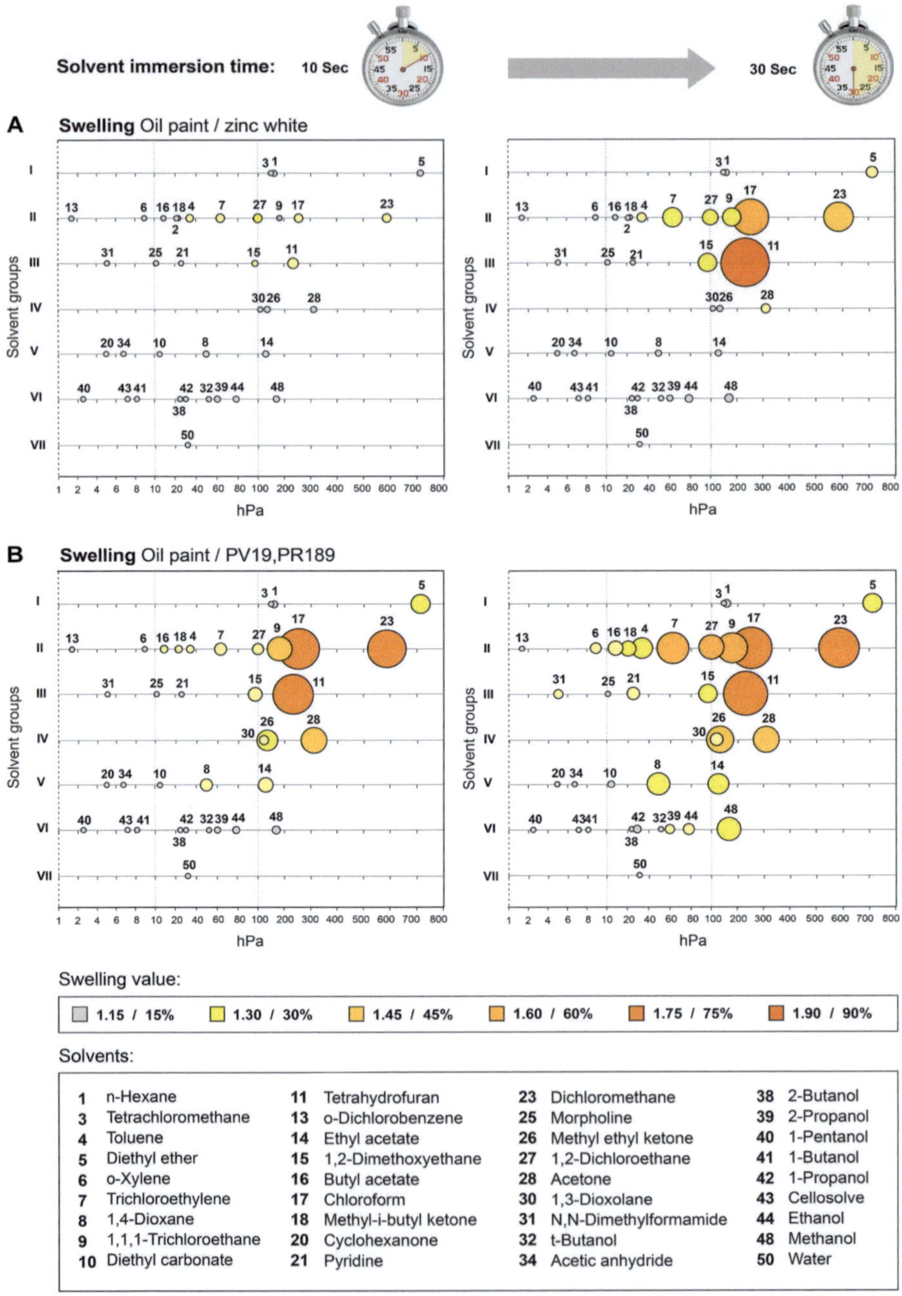

Fig. 36.4 Rate of swelling of modern oil paint after 10 s (left) and 30 s immersion time (right) (**a**) Saponified paint containing zinc white, (**b**) aged oil paint containing synthetic organic pigment only

or lead white. The main difference is the formation of an ionomeric network by free cations within the paints containing the lead and zinc ions. The lack of multivalent ions in modern oil paints (such as those containing organic colorants only) leads to different states of ageing, with different solvent sensitivities. The significant swelling observed is accompanied by a high rate of component leaching [20], which, when combined, have a detrimental impact on material behaviour. These findings strongly emphasise the need for a comprehensive, tailored approach to cleaning modern oil paints.

References

1. Blumenroth D, Zumbühl S, Scherrer N, Müller W. (2014). Sensitivity of modern oil paints to solvents – Effects on synthetic organic pigments. In: Van den Berg KJ, Burnstock A, de Keijzer M, Krueger J, Learner T, de Tagle A, Heydenreich G (eds) Issues in contemporary oil paint, Springer, Cham: 351–362.
2. Chan HW-S. (1987). Autooxidation of unsaturated lipids. Academic Press, London
3. Flory PJ. (1941). Thermodynamics of high polymer solutions. Journal of Chemical Physics 10:51–61.
4. Hansen M. (2007). CRC Handbook of polymer-liquid interaction parameters and solubility parameters. CRC Press, Taylor & Francis Group, Boca Raton
5. Kamal-Eldin A. (2003). Lipid oxidation pathway. AOCS Press, Champaign
6. Mallégol J, Gardette J-L, Lemaire J. (2000). Drier influence on the curing of linseed oil, Progress in Organic Coatings 39:107–113.
7. Marcus Y. (1998). The properties of solvents. Wiley series in solution chemistry 4, Wiley Chichester
8. Mecklenburg MF, Tumosa CS, Vicenzi EP. (2010). The influence of pigments and ion migration on the durability of drying oil and alkyd paints. In: Fuster-Lopez L, Charola EA, Mecklenburg MF, Domenech-Carbo M-T (eds) Cleaning 2010 – New insight into the cleaning of paintings, Universidad Politecnica de Valencia, Valencia: 27–28.
9. Mallégol J, Gardette J-L, Lemaire J. (2000). Long-term behavior of oil-based varnishes and paints – Photo- and thermooxidation of cured linseed oil. Journal of the American Oil Chemist's Society 77:257–263.
10. Phenix A, and Sutherland K. (2001). The cleaning of paintings – Effect of organic solvents on oil paint films. Reviews in Conservation 2:47–60.
11. Phenix A. (2002). The swelling of artists' paints in organic solvents – Part 2: Comparative swelling powers of selected organic solvents and solvent mixtures. Journal of the American Institute for Conservation 41:61–90.
12. Reichardt C. (1990). Solvents and Solvent Effects in Organic Chemistry. VCH Verlagsgesellschaft, Weinheim.
13. Sharma J, Tewari K, Arya RK. (2017). Diffusion in polymeric systems–A review on free volume theory, Progress in Organic Coatings 111:83–92.
14. Sutherland K. (2003). Solvent-extractable components of linseed oil paint films. Studies in conservation 48:111–135.
15. Torraca G. (1984). Solubility and solvents for conservation problem, ICCROM International Centre for the Study of the Preservation and the Restoration of Cultural Property, Rome
16. Van den Berg JDJ, Van den Berg KJ, Boon JJ. (1999). Chemical changes in curing and ageing oil paints. In: Bridgeland J, Brown J (eds), ICOM-CC, 12th Triennial Meeting. James & James, London: 248–253

17. Zumbühl S, Ferreira ESB, Scherrer NC, Schaible V. (2013) The Non-ideal action of binary solvent mixtures on oil and alkyd paint – Influence of selective solvation and cavitation energy. In: New insights into the cleaning of paintings: Proceedings from the Cleaning 2010 International Conference, Smithsonian Contributions to Museum Conservation 3, Smithsonian Institution Scholarly Press, Washington DC: 97–105.
18. Zumbühl S. (2014). Parametrization of the solvent action on modern artists' paint systems. Studies in Conservation 59:24–37.
19. Zumbühl S, Attanasio F, Scherrer N, Müller W, Fenners N, Caseri W, „Solvent Action on Dispersion Paint Systems and the Influence on the Morphology – Changes and Destruction of the Latex Microstructure", Postprints of the Symposium; Modern Paint Uncovered, Tata Modern London, Tom Learner (Ed.), 17–19. May 2006, Tate Modern (2008) 255–266.
20. Zumbühl S, Scherrer, NC, Engel N, Müller W. (2014) The Kinetics of Dissolution of Varnishes – The Influence of Vapour Pressure on the Rate of Solvent Action, ICOM-CC 17th Triennial Conference Preprints, Melbourne 15–19 September, J. Bridgland (Ed.), Art. 1611, 11pages.
21. Zumbühl S. (2019) Solvents, solvation, solubilization and solution: The solubility of materials – An introduction for conservators, including solubility data of selected conservation materials. HDW Publications, Bern.

Part IV
Conservation Treatment

Chapter 37
Water-Sensitivity in Modern Oil Paintings: Trends in Phenomena and Treatment Options

Miriam Gillman, Judith Lee, Bronwyn Ormsby, and Aviva Burnstock

37.1 Introduction

Water-sensitivity in unvarnished twentieth century oil paintings is a well-known phenomenon influenced by paint formulation, curing, and degradation, environmental factors, and artists' modification of painting materials. When displayed unglazed, an accumulation of grime may significantly affect the appearance of an unvarnished work. Removal of surface dirt from water-sensitive paintings presents challenges for conservators, as swab rolling with traditional aqueous methods can result in pigment removal or other surface changes. Recent research has focused on new systems and application methods for aqueous solutions that give the conservator greater control over moisture exposure to vulnerable paint surfaces, while retaining the cleaning efficacy of water [9, 21]. The present study builds on research by Chung et al. [6] that investigated a range of surface cleaning methods designed for acrylic emulsion paint films, on swatches of water-sensitive Winsor & Newton (W&N) Artists' Oil Colour paint.

M. Gillman
The Barnes Foundation, Philadelphia, USA
e-mail: mgillman@barnesfoundation.org

J. Lee · B. Ormsby
Tate, London, UK
e-mail: Judith.lee@tate.org.uk; Bronwyn.ormsby@tate.org.uk

A. Burnstock (✉)
Courtauld Institute of Art, London, UK
e-mail: Aviva.Burnstock@courtauld.ac.uk

K. J. van den Berg et al. (eds.), *Conservation of Modern Oil Paintings*,
https://doi.org/10.1007/978-3-030-19254-9_37

37.1.1 Characteristics and Causes of Water-Sensitivity in Modern Oil Paintings

Previous studies have investigated surface characteristics associated with water-sensitivity including fatty acid efflorescence, polar surface degradation products, tacky paint, and soft, sticky medium skins or organic exudates [3]. These characteristics both increase the vulnerability of unvarnished surfaces to dirt and demand different approaches to surface cleaning. Subsequent research has characterised the chemical and morphological features concomitant with water-sensitivity in oil paintings, highlighting causal factors [2].

In well-bound oil paint, water-sensitivity manifests from very high (pigment transfer within one or two swab rolls) to slight, with light staining after fifty rolls. Trends have been documented in water-sensitive pigments across brands, including ultramarine, cadmium, cobalt, and chrome-based colours [2, 4, 19]. White pigments containing zinc, lead and titanium are generally non-water-sensitive. This may relate to their tendency to react with fatty acids and form metal soaps, or carboxylates, which impart stability to the paint film [14].

One documented cause of surface water-sensitivity occurs in some paints containing magnesium carbonate, an additive indicative of W&N oil paint formulations. Upon exposure to a sulphurous chemical environment, the hygroscopic degradation product magnesium sulphate heptahydrate ($MgSO_4 \cdot 7H_2O$), or epsomite, is formed. [4, 16]. Water-soluble epsomite crystals are often partially embedded in medium skins on the surfaces of particularly sensitive passages of paint.

Chemical analysis of twentieth century commercial oil paints has identified components that may influence water-sensitivity, including metal stearates, free fatty acids, and both semi-drying and non-drying oils [11]. Addition of metal stearates may lead to elevated levels of free fatty acids (Banti et al., this volume), and it has also been hypothesised that metal soaps may promote hydrolysis [10]. Inhibition of the curing of the oil medium has further been proposed as a cause of sensitivity (Bonaduce, this volume).

Recent examination of water-sensitive W&N oil paint swatches found high levels of diacids at some paint surfaces, potentially causing vulnerability to polar solvents [11]. In an earlier study of water-sensitive W&N swatches, increased concentrations of carboxylic acid and hydroxyl functional group compounds at paint surfaces were correlated to skins of organic material [5]. Fourier transform infrared spectroscopy (FTIR) analysis showed split carbonyl bands characterised by absorptions at 1740/1710 cm^{-1} in many of the swatches, indicating that ester bonds at the surface of oil paint films may hydrolyse, catalysing oxidative degradation processes and producing diacids [5]. Split carbonyl absorptions at 1740/1710 cm^{-1} have also been observed in water-sensitive paintings ([2]; Lee et al. this volume), suggesting a relationship between oxidation and/or hydrolysis and water-sensitivity.

37.1.2 *Approaches to Cleaning*

This project evaluated various approaches to cleaning, selected based on promise in research and treatment case studies related to water-sensitive oil paintings. [6] explored the effect of adjusting pH and conductivity of aqueous solutions, finding that lowering the pH to 5 resulted in less pigment pickup and improved cleaning efficacy.

Differing application methods for aqueous reagents have been assessed with the aim of reducing moisture exposure and mechanical action on unvarnished paintings, including brushes, isoprene makeup sponges and tissue or solvent barriers [9, 18]. The polysaccharide gels agar and gellan have also been investigated in relation to vulnerable painted surfaces and provide a method for controlled application of aqueous cleaning solutions [21].

37.1.2.1 Water-in-Oil Microemulsions

Mineral spirits-based water-in-oil (WiO) microemulsions were developed for surface dirt removal from acrylic painted surfaces [13]. Microemulsions optimise aqueous methods for cleaning water-sensitive surfaces by dispersal of the aqueous phase in a nonpolar solvent system, harnessing the cleaning efficacy of water while minimising the risk of surface swelling. Chung et al. [6] investigated their potential for soil removal from water-sensitive oil paint samples, selecting those with the lowest proportions of water and surfactant and high proportions of solvent. These included Series 3b WiO microemulsions, in which the nonpolar component (aliphatic hydrocarbon solvent Shellsol D38 or D40) is the continuous 'oil' phase [13].[1] The surfactant component is anionic sodium dioctylsulfosuccinate (NaDOSS) at ~65% w/w dissolved in Shellsol D38 or D40 [13]. The activity of the microemulsion is controlled through modifying the proportions of these three phases and the properties of the aqueous phase.

37.1.2.2 Silicone Systems

Silicone-based materials can be used in free solvent form, emulsions, microemulsions, or as surface barriers [18]. Despite their low polarity, silicone materials have hydrogen bond acceptor sites, which can facilitate their interaction with polar paint surfaces. It has been posited that they may protect the paint surface during the application of more aggressive polar cleaning solutions [18]. Silicones are odourless, with low surface tension and polarity. They were previously considered non-toxic,

[1] Four series of microemulsions *(1, 2, 3a, 3b, 4)* were developed following collaborative research between the Dow Chemical Company, Tate, and the Getty Conservation Institute (GCI), varying in formulation and surfactant components [12].

but the use of cyclic siloxanes D4 and D5 was recently restricted by the European Union due to environmental hazards associated with their bioaccumulation (Commission Regulation [EU] [7]/35).

37.1.3 Aims

The aim of this study was to explore the potential of selected free aqueous systems, polysaccharide gels, WiO microemulsions and silicone emulsifier systems, and the impact of varying methods of application for removal of surface dirt from five naturally soiled modern oil paintings that included water-sensitive passages with a variety of surface characteristics. The materials and techniques of the paintings were examined to identify probable causes of water-sensitivity in each case. Comparative evaluations of the safety and efficacy of the cleaning materials and application methods trialled in this study were carried out with a view to providing guidelines for the treatment of other water-sensitive oil paintings.

37.2 Methodology

37.2.1 Selection and Examination of Test Paintings

The five paintings in this study were selected from a wider collection of water-sensitive works dating from the mid-late twentieth century, primarily purchased from flea markets in London as research case studies.[2] Elemental analysis and FTIR spectroscopy of samples from each painting identified magnesium carbonate, suggesting the use of W&N paints. The paintings all included reasonably homogenous water-sensitive passages large enough for comparative cleaning tests. The degree of water-sensitivity was assessed systematically using deionised water and pre-rolled cotton swabs. One roll was defined as a full backward and forward roll of the swab, and sensitivity was determined based on up to 50 swab roll counts.[3] If no pigment was removed after 50 counts, the paint surface was considered non-water-sensitive (Table 37.1).

[2] Paintings sourced by Aviva Burnstock.

[3] This method was based on sensitivity testing in previous studies: [6, 13].

Table 37.1 Water-sensitivity of test paintings

Title	Date	Artist	Dimensions (cm)	Support	Water-sensitivity
Clown and Goat	1975	Unknown	64 × 76	Canvas	High
Sick Child	1976	Student	57 × 42	Paper	Moderate
Vase of Flowers	1977	Mollie Rhodes	46 × 36	Canvas board	Moderate
The Pond	1970s	Unknown	51 × 41	Hardboard	High
Portrait of a Woman	1991	Jonathan Flatford	46 × 56	Hardboard	Slight

37.2.2 Analytical Methods and Materials

The paintings were photographed in visible and ultraviolet (UV) light with a Canon EOS 5D digital SLR camera. Photomicrographs were taken using a Leica M165C microscope fitted with a Leica DFC 500 camera.

Samples were prepared as cross-sections embedded in polyester resin, wet-ground and dry-polished. Samples were examined with a Leica DM4000 M LED microscope with a Leica Kubler Codix UV light source and photographed with a Leica DFC450C camera and LAS V4.6 software.

The inorganic components of the paint and ground layers were identified using SEM-EDX. Cross-sections were carbon-coated and analysed with a JEOL JSM-T100 scanning microscope and INCA microanalysis software suite.

Paint surfaces were characterised with back-scattered and secondary electron imaging using a FEI Quanta 650 field emission gun scanning electron microscope. Elemental mapping of paint samples was carried out using a Brucker Flat Quad 5060 F detector.

FTIR analysis was conducted using a Thermo Scientific Nicolet iN10 MX microscope using a single diamond cell. Sixty four scans were collected at a resolution of 4 cm^{-1} across a wavenumber range of 4000–600 cm^{-1}. Data was processed using Omnic 8 software.

A Rhopoint Novo-Gloss Glossmeter was used to measure surface changes, together with HIROX 3D high-resolution digital microscope, mounted on a photo stand with a ring-light diffuser attachment and microscope zoom lens (MXB-2500REZ).

37.2.3 Cleaning Materials

The systems tested in this study were selected from the most recent list of cleaning materials used at the Cleaning of Acrylic Painted Surfaces (CAPS) workshops.[4] Appendix 2 lists the component proportions of the microemulsions and silicone emulsifier gels.

37.2.3.1 Free Aqueous Systems

Pilot testing explored free aqueous systems, assessing the effects of adjusted pH and conductivity, surfactants, and chelators upon the test paintings. Four variations of water adjusted for pH and conductivity were prepared at pH 4.5 and pH 5, 1 mS/cm and 5 mS/cm. These adjusted waters were then used to prepare surfactant solutions with ECOSURF™ EH-6 and EH-9 at concentrations of 0.5% and 1%. Solutions of triammonium citrate (TAC) and ethylenediaminetetraacetic acid (EDTA) chelating agents were each prepared at a concentration of 1% (w/v) at pH 5 and pH 6.5.

Varied application methods for the free aqueous systems and microemulsions were compared, including hand-rolled and pre-rolled cotton swabs, isoprene makeup sponges, a wet-strength tissue barrier, and a D5 cyclomethicone barrier.

37.2.3.2 Rigid Polysaccharide Gels

The most successful aqueous cleaning agents were loaded into polysaccharide gels. Agar gels at 3% (w/v) and 4% (w/v), and gellan at 2% (w/v) and 3% (w/v) were prepared with deionised water. These hydrogels were tested for their inherent cleaning efficacy, and also used for clearance of gels containing chelating agents and surfactants. These included agar and gellan gels at the above percentages made with citric acid and EDTA solutions at concentrations of 0.5% (w/v) and 1% (w/v), pH 5. Surfactant gels were prepared with 0.5% ECOSURF™ EH-9 concentrate.[5]

37.2.3.3 WiO Microemulsions

Four microemulsions from *Series 3b* (4, 14, 24, 25) were prepared with deionised water, selected with the aim of minimising residues and pigment removal by keeping the proportions of surfactant minimal and the aqueous content low (10–40%). Me3b-14 and Me3b-24 (20 and 40% aqueous content, respectively) were selected to

[4]The Cleaning of Acrylic Painted Surfaces (CAPS) project is part of the Research into Practice Initiative at the GCI.

[5]It is necessary to add the surfactant to the gel in concentrate form in order to avoid heating the EH-9 above its cloud point of 61°C.

further investigate modification of the aqueous phase. One of each microemulsion was prepared with adjusted water at pH 4.5 and 1 mS/cm and one with 0.5% (w/v) TAC. All microemulsion tests were cleared with Shellsol D40.

37.2.3.4 Silicone Systems

Silicone emulsifier gels were prepared using Shin-Etsu KSG-350Z at 20% and 40% aqueous content. The impact of different aqueous phases on their cleaning efficacy was explored, using water at pH 4.5 and 1 mS/cm, 1% EDTA at pH 5, and 1% citric acid at pH 5. The KSG-350Z gel was combined with D5 cyclomethicone and the aqueous phase dispersed through this mixture by weight. The gels were cleared with D5.

37.2.4 Assessment of Cleaning Tests

Cleaning materials were assessed based on criteria developed in the CAPS workshops, also used on water-sensitive oil paints by Chung et al. [6]. Each system was rated on a scale of one (least satisfactory) to five (most satisfactory) for: soil removal efficacy; swelling; pigment/coating removal; changes in light reflection; visible residues; ease of application; practicability; and health and safety. The higher the overall rating, the safer the system was deemed. For each system, a star diagram was constructed comparing these variables to assess overall performance. Surfaces were evaluated before and after testing using light and HIROX microscopy, UV fluorescence, SEM imaging and gloss measurements.

37.3 Results and Discussion

37.3.1 Technical Examination of Test Paintings

The most water-sensitive works were *Clown and Goat* and *The Pond*, where paint of all colours (except those mixed with white pigment) was removed immediately on contact with a damp swab. *Vase of Flowers* was very sensitive in certain passages, while *Sick Child* was sensitive in the areas without an oiling out layer. The most highly sensitive passages in all five paintings included cadmium-based pigments, following a trend noted in previous studies [2, 19].

37.3.2 *Materials and Techniques of Test Paintings*

37.3.2.1 Clown and Goat

Clown and Goat is painted on a finely woven canvas from Roberson & Co., with a commercial priming composed of lead and titanium white extended with alumino-silicates and kaolin. The most sensitive areas were the large, thickly painted passages of cadmium red paint and the dark blue background, composed of a mixture of Prussian blue, ultramarine, ochre and bone black. The water-sensitivity of the paler areas of green, blue, and pink paint was dependent on the proportion of titanium white pigment.

The green paint contained mixtures of chrome yellow, Prussian blue, titanium white and iron oxide pigments, and crystalline zinc stearate identified with FTIR. Magnesium carbonate was found in all samples of paint analysed, suggesting the use of oil paint manufactured by W&N. Also found throughout the painting were barium sulphate and aluminium hydroxide, previously identified in W&N paints [11]. In passages of cadmium red-containing paint, superficial water-soluble crystals associated with elemental magnesium and sulphur indicated the presence of epsomite, visible in the SEM secondary electron image in Fig. 37.1.

The painting was moderately soiled with imbibed dirt, displaying craquelure that was potentially vulnerable to capillary penetration of cleaning reagents. The surface appeared matte with subtle variations in gloss, and several areas of medium exudate

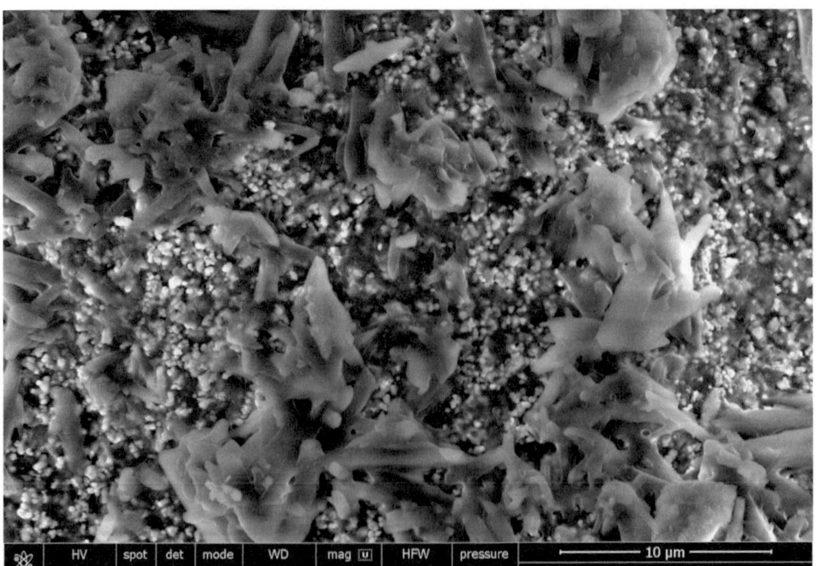

Fig. 37.1 Secondary electron image of crystalline structures on the surface of *Clown and Goat*

were noted in the dark blue background paint. Cleaning tests were focused on areas of cadmium red and the blue background

37.3.2.2 Sick Child

Sick Child is painted directly on paper sized with gelatine, mounted on acidic card. The paint is cracking and flaking, exacerbated by undulation of the aged paper support. Areas containing titanium white were water-resistant while the predominantly iron oxide containing paints used for the remainder of the composition were moderately water-sensitive, as well as passages of green containing viridian and ultramarine. Cleaning tests were carried out throughout areas of darker, matte paint.

The artist recalled using W&N paints (consistent with the barium sulphate found throughout) and applying a linseed oil layer to the painting in 1978. This coating is uneven, with matte water-sensitive passages adjacent to non-sensitive glossy areas. All over the surface, patches of a water-soluble whitish haze indicate free fatty acid efflorescence. No epsomite was found, and the cause of water-sensitivity may be associated with the formation of polar degradation products at the paint surface.

37.3.2.3 Vase of Flowers

Vase of Flowers was painted on Daler Artists' Board commercially primed with titanium white in a PVA binder. The background is mainly composed of titanium white and barium sulphate, with slight water-sensitivity in areas with higher proportions of coloured pigment.

Paints used for the table, vase, and flowers were highly sensitive and prone to blanching on application of water. The brown paint of the table and vase contains red and brown iron oxide pigments, umber, and green earth. FTIR analysis of this paint detected a carbonyl absorption at 1709 cm^{-1}, observed in other studies of water-sensitive oil paintings [1, 5, 11]. This indicates oxidation and/or hydroxylation of the paint, potentially with an abundance of free azelaic acid species influencing overall water-sensitivity.

The primary pigments used for the flowers are cadmium orange and yellow. Magnesium carbonate and barium sulphate were identified in these paints, signifying the use of W&N products. FTIR analysis of the cadmium paint showed a weak signal for amorphous metal soaps, suggesting that a lack of metal soap stabilisation may contribute to the water-sensitivity of these passages. The painting is matte overall, with imbibed soil in areas of impasto and a uniform layer of surface dirt. Cleaning tests were carried out on the brown, green, orange and yellow areas of paint.

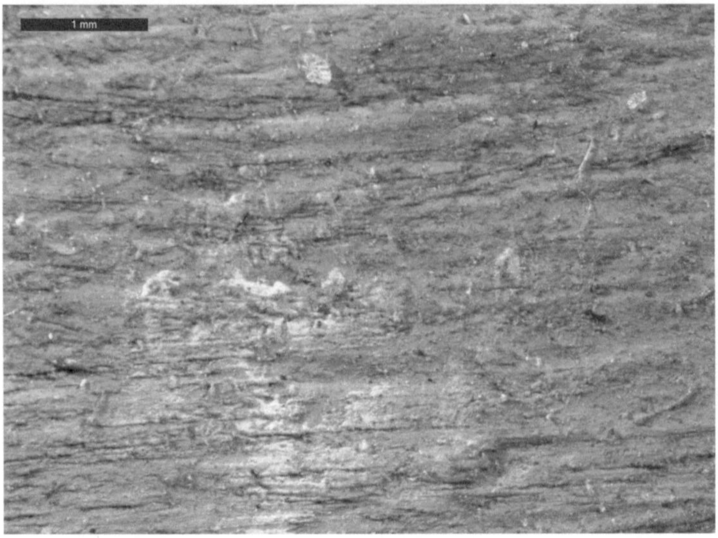

Fig. 37.2 Photomicrograph of water-sensitive blue paint of *The Pond* with particulate and imbibed dirt

37.3.2.4 The Pond

The Pond is painted on a textured hardboard panel with two ground layers. The lower commercial white priming is composed of titanium white and kaolin with alumino-silicates. The upper peach-coloured ground is artist-applied, containing titanium white, barium sulphate, chalk and an iron oxide pigment. Apart from a strip of cerulean blue and titanium white-containing paint at the top of the composition, the entirety is highly water-sensitive. This appeared particularly related to the green passages, mixtures of viridian, chrome yellow, ultramarine and green earth. Magnesium carbonate was found throughout the composition, suggesting W&N paints, but no epsomite was detected.

The paint was friable, highly impastoed and vulnerable to mechanical action. The surface, coated with yellowish grime, exhibited variable gloss related to the proportion of binder to pigment and presence of medium skins. Cleaning tests were carried out on the most homogenous passages of water-sensitive green and blue paint, illustrated by the light microscope details of the surface in Fig. 37.2.

37.3.2.5 Portrait of a Woman

Portrait of a Woman was painted in thin layers on artists' hardboard with a titanium white acrylic ground. The flesh tones and the sitter's blue garment were non-water-sensitive, while yellow ochre brushstrokes in the hair showed slight sensitivity. The most consistently water-sensitive element was the red background, where cleaning tests were primarily focused.

The background paint contained cadmium red, iron oxide red, and titanium white. There were variations in thickness and sensitivity all over, with pigment removed from the heavier areas after ten swab rolls and the thinner areas after thirty. SEM-EDX and FTIR analysis of the red paint identified alumina and barium sulphate. FTIR also found crystalline zinc stearate and associated amorphous metal soaps throughout the paint layers. Magnesium was identified in blue and beige passages of paint with EDX analysis, but FTIR did not confirm the presence of magnesium carbonate.

37.3.3 Evaluation of Cleaning Materials

37.3.3.1 Free Aqueous Systems

Testing free aqueous systems informed the selection of the most effective and low risk cleaning systems. pH 4.5 waters removed slightly less pigment than pH 5.

There were observable differences between the effect of water at 1 and 5 m/S, consistent with previous research [17]. Therefore, the adjusted water chosen for loading into selected application methods was pH 4.5 and 1 mS/cm. The safety of cleaning systems was determined by the numerical rating of the CAPS cleaning criteria in tandem with visual evaluation of pigment removal or optical changes to the surface.

Surfactants and Chelator Solutions at Set pH and Conductivity

ECOSURF™ EH-9 surfactant solutions were safer and more effective at soil removal than EH-6. Concentrations of surfactant >0.5% w/v were too active for water-sensitive paint surfaces, leading to spreading and foaming, residue rings and blanching or gloss changes. The surfactant solutions made with pH 5 water were slightly more active than those that contained pH 4.5 water.

Solutions of chelating agents TAC and EDTA acid (1% w/v) were prepared at pH 5 and 6.5, with conductivities ranging from 7.7 to 10.7 mS/cm. Both mixtures were very effective in removing dirt, but resulted in pigment loss and left visible residues. Subsequent tests aimed to limit the interaction of the aqueous phase with the paint by loading these reagents into alternate application methods.

37.3.3.2 Application Methods for Aqueous Systems

Agar

Agar gels were prepared ~5 mm thick and applied to the surface of the test paintings for one minute, then cleared with the corresponding hydrogel. Massaging or weighting the gel improved surface contact and cleaning efficacy.

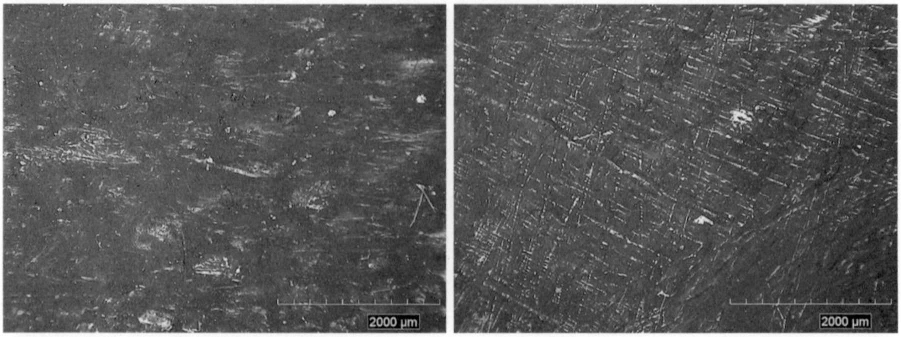

Fig. 37.3 HIROX images of paint from *Clown and Goat* before (left) and after (right) cleaning with 3% agar gel with 1% citric acid

4% w/v agar forms a rigid gel that conformed poorly to textured paint surfaces and did not release enough water for sufficient soil removal. Of the agar gels, 3% w/v gels with 1% w/v added chelating agent were the most effective across the test paintings. A 3% w/v agar gel with 0.5% w/v ECOSURF™ EH-9 was also successful - loading the surfactant into a gel eliminated foaming and spreading associated with surfactant-containing free aqueous systems. In one instance application of citric acid 1% in agar removed the medium skin from an area of dark blue background paint from *Clown and Goat* (Fig. 37.3).

Gellan

The gellan gels were looser, wetter and more transparent than agar. 3% w/v gellan gel conformed poorly to textured paint surfaces and had low cleaning efficacy. Thinner pieces of gel weighted down on the paint surface improved the surface contact of 3% w/v gellan. This may be more effective for paintings with a higher degree of water-sensitivity, or a water-sensitive substrate prone to swelling, as in *Sick Child*. The most useful gellan gels were 2% w/v with 1% w/v chelating solution. However, 2% w/v gels tended to disintegrate when applied to highly textured paint surfaces, with visible gellan residues after cleaning.

Overall, polysaccharide gels were most useful for all the test paintings in terms of cleaning efficacy with the least unwanted surface change. The star diagram in Fig. 37.4 shows the highly rated results of 2% gellan with added chelating solution on *Clown and Goat*. The transparency of gellan is an advantage, rendering any pigment transfer visible, while yellowish agar gels disguise pigment loss until after removal. A more transparent agar gel may be made using a finer grade of agar powder and/or repeating the heating and cooling cycle during preparation [15].[6]

[6] Agar agar E406 was used in this study.

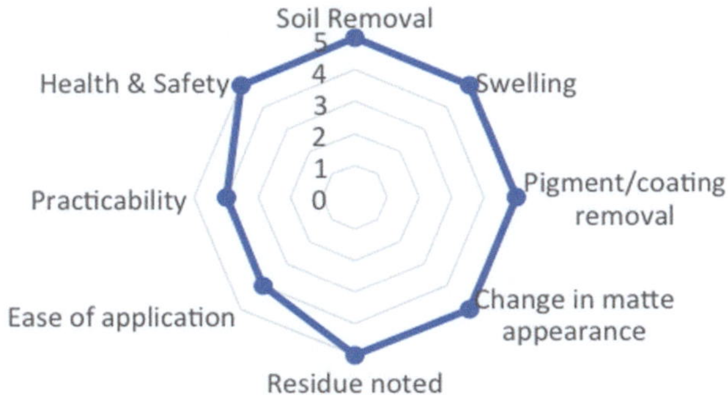

Fig. 37.4 Star diagram for 2% Gellan with 1% EDTA pH 5 on blue passage of *Clown and Goat*

Gellan was more effective than agar in conforming to textured paint surfaces. However, other studies have improved the surface contact of agar through brush application while the agar is warm and semi-solid state, letting it gel on the surface preceding removal [8, 20]. Agar gels that retain water were effective for application of reagents to the flatter, less sensitive passages of paint on *Clown and Goat* and *Portrait of a Woman.* Gellan gels were more useful for removing dirt from textured surfaces with moderate to high water-sensitivity, such as *Vase of Flowers* or *The Pond*, but posed a residue risk.

Silicone Systems

Free aqueous solutions and microemulsions were tested following the application of hydrophobic D5 silicone solvent. In this study the D5 neither prevented pigment pickup, nor provided additional cleaning efficacy. Slow-evaporating D5 left surfaces slippery, making it harder to control the subsequent action of the cleaning reagent.

Silicone emulsifier gels made with Shin-Etsu KSG-350Z presented similar handling problems, with the swab unrolling and spreading the cleaning reagent. Potential residue issues may relate to the slow evaporation of D5, further explored by Ormsby et al. in the present volume. The length of time the silicone solvent remains on the surface compromises visual evaluation of the success of the clearance step, which may be mitigated by the use of faster-evaporating hexamethyldisiloxane as the clearance solvent.

The gels made with 20% aqueous content were harder to control, while 40% gels were thicker and easier to manipulate in a limited cleaning area. The 20% gels had minimal cleaning efficacy with no perceptible change to the paint surfaces. 40% gels proved too active, resulting in immediate pigment loss and visible residues. Test areas of two paintings cleaned using the gels were examined several days after the solvent had evaporated, and showed a glossy sheen compared to adjacent control areas (Fig. 37.5).

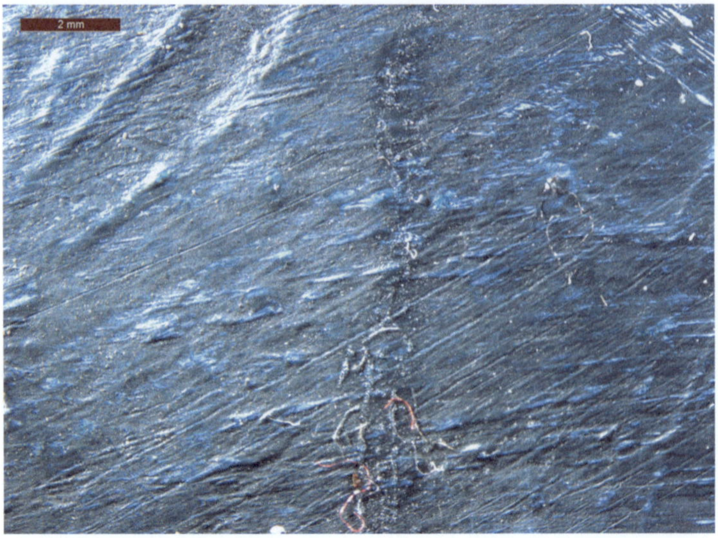

Fig. 37.5 Photomicrograph before cleaning *Clown and Goat* (left), and residues after cleaning (right) with Shin-Etsu KSG-350Z emulsifier (40% aqueous content)

WiO Microemulsions

Iso-octane was more compatible with the paint surfaces in this study than Shellsol D40, suggesting a preference for this solvent used in microemulsion formulations for cleaning water-sensitive oil paints. Pilot studies ascertained that most of the water-sensitive surfaces could tolerate up to 20% aqueous solutions, but Me3b-4 and Me3b-14 (40% and 20% water content respectively) removed paint and blanched the surface.

Addition of 0.5% w/w. TAC to selected microemulsions with lowered aqueous content proved successful for surface cleaning. Me3b-24 (10% aqueous phase, 10% surfactant phase) with TAC had consistently favourable results across the test paintings, with satisfactory soil removal, minimal or no pigment loss or gloss changes, and minimal discernible residues (Fig. 37.6). Most of the microemulsions containing a higher proportion of surfactant left visible skins or residue rings.

Application with isoprene makeup sponges reduced the risk of pigment pickup and removed soil more evenly than swabs. However, the sponges tended to disintegrate after about ten passes. The most successful application method for microemulsions was swabbing through wet-strength tissue. This was particularly useful for paintings with cracked or friable paint, as it reduced mechanical action. Me3b-25 with deionised water was too active for most surfaces alone, but when swabbed through wet-strength tissue soil was evenly removed with minimal pigment loss or surface changes (Fig. 37.7).

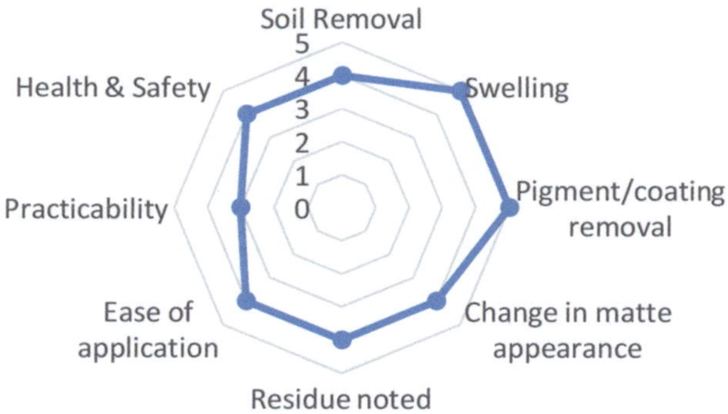

Fig. 37.6 Star diagram for Me3b-24 in 0.5% TAC solution through wet-strength tissue on red passage of *Portrait of a Woman*

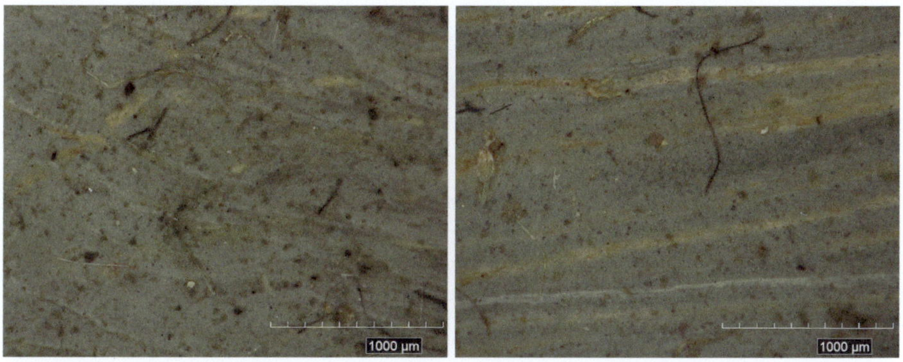

Fig. 37.7 HIROX images taken before cleaning *The Pond* (left) and after cleaning with Me3b-25 through wet-strength tissue (right)

37.4 Conclusions and Further Work

The paintings chosen for this study exhibited characteristics common to unvarnished water-sensitive oil paintings, including embedded surface dirt, efflorescence, medium exudates and variable gloss. Technical examination and elemental analysis of the paintings identified pigments, additives, and degradation products linked to the development of water-sensitivity.

Consistency with published research was shown in the potential of tailoring aqueous systems to the needs of water-sensitive oil paintings. Investigation of surface dirt removal exhibited some clear trends. Lowering the pH of aqueous systems proved effective, together with the use of gels loaded with reagents, sponges and wet-strength tissue. Alternative methods of using gels might include brushing agar on warm, varying and/or repeating the application times, and experimenting with other grades of powder. WiO microemulsions were promising in terms of cleaning efficacy, with some qualifications on the control of their effect on the paint surface. The silicone emulsifiers showed promise for dirt removal, but the percentage of the aqueous phase requires further adjustment.

This study relates to the importance of understanding the nature and possible causes of sensitivity in developing treatments for water-sensitive oil paintings. Overall, the significant variation shown in the results reflects the need for further studies in the field and careful testing in each case prior to treatment.

Acknowledgments Thanks to Tomasz Goral, William Luckhurst, Ben Blackburn, Joyce H. Townsend, Angelina Bartoletti, Jae Youn Chung, and colleagues at the Courtauld Institute of Art.

The purchase of the HIROX KH-8700 digital microscope used in this study was supported in part by the Horizon2020 funded NANORESTART project (agreement 6466063).

Appendices

Appendix 1: Materials and Suppliers

Supplier	Materials
Kremer Pigmente, Germany	D5 cyclomethicone
	ECOSURF™ EH-6
	Shellsol D40
Sigma-Aldrich, UK	Acetic acid
	Citric acid
	ECOSURF™ EH-9
	EDTA (ethylenediaminetetraacetic acid)
	Hydrochloric acid ~37%
	Iso-octane (2,2,4-trimethylpentane)
	NaDOSS ~97% (dioctylsulfosuccinate sodium salt)
	Sodium hydroxide
	TAC (triammonium citrate)
Shin-Etsu Silicones of America, USA.	Shin-Etsu KSG-350Z
Special Ingredients Ltd.	Agar agar powder (E406)
	Gellan powder (low acyl)
Ungfu Mall (Amazon)	Cotton swabs
VWR International	Ammonium hydroxide 32%

Appendix 2: Components of WiO Microemulsions and Silicone Emulsifier Gels

2.1 Microemulsions

System	Aqueous phase	% Aqueous	% Shellsol D40	% NaDOSS (65% in D40)	pH
Me3b-4	Deionized water	40	20	40	6
Me3b-19		10	20	70	
Me3b-24		10	70	20	
Me3b-25		10	80	10	
Me3b-14	pH 4.5, 1 mS/cm water	20	40	40	4.5
Me3b-24		10	70	20	
Me3b-14	0.5% TAC solution	20	40	40	5
Me3b-24		10	70	20	

2.2 Silicone Emulsifier Gels

	Aqueous phase	% Aqueous	% Shin-Etsu KSG-350Z	% D5 cyclomethicone	pH
1	pH 4.5, 1 mS/cm water	20	50	30	4.5
2		40		10	
3	1% citric acid solution	20		30	5
4		40		10	
5	1% EDTA solution	20		30	
6		40		10	

References

1. Bay, L (2017) *Water sensitive oil paintings and investigation of phenomenology from works at Tate*, final year project, Department of Conservation & Technology, Courtauld Institute of Art
2. Bay, L et al (2017) Water sensitivity of modern oil paintings. In: Bridgland J, (ed) ICOM-CC 18th Triennial Conferences Preprints. Copenhagen, September 2017
3. Burnstock A, van den Berg KJ (2014) Twentieth century oil paint: The interface between science and conservation and the challenges for modern oil paint research. In: Burnstock A et al (eds) Issues in Contemporary Oil Paint. Springer, Heidelberg, pp 1–19
4. A. Burnstock, K.J. van den Berg, S. de Groot and L. Wijnberg, An Investigation of Water-Sensitive Oil Paints in 20th Century Paintings. In Learner, T. (ed.) Reprints of the Modern Paints Uncovered conference, London 2006. Los Angeles: Getty, 2008. 177–188

5. Burnstock A et al (2015) Water sensitivity of modern oil paint films. In: Bonino VES (ed) Dall'olio all'acrilico, dall'Impressionismo all'arte contemporanea: Studi, ricerche, indagini scientifiche ed interventi conservativi. CESMAR7/il Prato, Padova, pp 66–76
6. Chung JY et al (2017) An investigation of options for surface cleaning unvarnished water-sensitive oil paints based on recent developments for acrylic paints. In: Bridgland J, (ed) ICOM-CC 18th Triennial Conferences Preprints. Copenhagen, September 2017
7. Council of European Union (2018) Commission regulation (EU) no 2018/35 of 10 January 2018 amending Annex XVII to Regulation (EC) No 1907/2006 of the European Parliament and of the Council concerning the Registration, Evaluation, Authorisation and Restriction of Chemicals (REACH) as regards octamethylcyclotetrasiloxane ('D4') and decamethylcyclopentasiloxane ('D5'). In: OJ L 6 (11.1.2018): 45–47
8. Cremonesi P, Casoli A (2017) Thermo-reversible rigid agar hydrogels: their properties and action in cleaning. In: Angelova LV et al (eds) Gels in the Conservation of Art, London, 2017. Archetype, London, p 19
9. Daudin-Schotte M, van Keulen H (2014) Dry cleaning: Research and practice. In: Burnstock A et al (eds) Issues in Contemporary Oil Paint. Springer, Heidelberg, pp 363–73
10. La Nasa J et al (2018) Development of a GC/MS method for the qualitative and quantitative analysis of mixtures of free fatty acids and metal soaps in paint samples. Analytica Chimica Acta. doi:https://doi.org/10.1016/j.aca.2017.11.017
11. Lee J et al (2017) The chemical characterization of water-sensitive modern oil paint swatches by Winsor & Newton. In: Bridgland J, (ed) ICOM-CC 18th Triennial Conference Preprints. Copenhagen, September 2017
12. Ormsby B et al (2015) Cleaning acrylic paints: mineral spirits-based microemulsions and the question of residues. In: Bonino VES (ed) Dall'olio all'acrilico, dall'Impressionismo all'arte contemporanea: Studi, ricerche, indagini scientifiche ed interventi conservativi. CESMAR7/il Prato, Padova, pp 97–106
13. Ormsby B et al (2016) Mineral spirits-based microemulsions: A novel cleaning system for painted surfaces. J American Institute for Conservation 55(1): 13–31
14. Osmond G (2014) Zinc white and the influence of paint composition for stability in oil-based media. In: Burnstock A et al (eds) Issues in Contemporary Oil Paint. Springer, Heidelberg, pp 263–81
15. Sawicka A et al (2017) Challenges of a virgin water-sensitive surface: designing a cast agar gel cleaning system for Patrick Heron's Still-life against the Sea:1949. In: Angelova LV et al (eds) Gels in the Conservation of Art, London, 2017. Archetype, London, p 87
16. Silvester G et al (2014) A cause of water-sensitivity in modern oil paint films: The formation of magnesium sulphate. Studies in Conservation 59(1):38–51
17. Soldano A, van den Berg KJ (2014) Investigation into the surface conductivity of water-sensitive modern oil paints. In: Burnstock A et al (eds) Issues in Contemporary Oil Paint. Springer, Heidelberg, pp 407–16
18. Stavroudis C (2015) Silicone-based solvents in conservation. As free solvents and components of gel systems and microemulsions. In: Bonino VES (ed) Dall'olio all'acrilico, dall'Impressionismo all'arte contemporanea: Studi, ricerche, indagini scientifiche ed interventi conservativi. CESMAR7/il Prato, Padova, pp 176–83
19. Tempest H et al (2010) Sensitivity of oil paint surfaces to aqueous and other solvents. In: Mecklenburg M et al (eds) New Insights into the Cleaning of Paintings: Proceedings from the Cleaning 2010 International Conference. Smithsonian: Washington DC, p 107
20. Volk (2016) A partially water sensitive painting by Karel Appel: Technical investigation and a tailored approach to surface cleaning, Zietschrift fur Kunsttechnologie un Konservierung 1, 150–166.
21. Volk, A, van den Berg KJ (2014) Agar – A New Tool for the Surface Cleaning of Water

Chapter 38
Modern Oil Paintings in Tate's Collection: A Review of Analytical Findings and Reflections on Water-Sensitivity

Judith Lee, Bronwyn Ormsby, Aviva Burnstock, and Klaas Jan van den Berg

38.1 Introduction and Aims

In the context of paintings conservation, water-sensitivity of modern oil paintings has been defined as the removal of pigment and/or binding medium as a result of gentle rolling of a cotton swab dampened with water, over the paint surface [2, 27, 39]. Recent research has sought to expand the range of cleaning options for water-sensitive modern oil paintings and other modern painted surfaces, and to develop an understanding of the causes of water-sensitivity [2, 4, 11, 14, 17]; some surveys of modern oil paintings that link analytical findings, paint surface characteristics and solvent sensitivity have been published [4, 17, 25].

One known cause of water-sensitivity in modern oil paints is the formation of water-soluble salts, present as degradation products in aged oil paints. The British artists' paint manufacturer Winsor & Newton (W&N) frequently use magnesium carbonate as an extender, which can form water-soluble magnesium sulphate heptahydrate (i.e. epsomite) upon exposure to sulphur dioxide [36]. However, there are many examples of water-sensitive paintings where epsomite is not detected suggesting additional or alternative causal factors.

J. Lee (✉) · B. Ormsby
Conservation Department, Tate, London, UK
e-mail: Judith.Lee@tate.org.uk; Bronwyn.Ormsby@tate.org.uk

A. Burnstock
Courtauld Institute of Art, London, UK
e-mail: Aviva.Burnstock@courtauld.ac.uk

K. J. van den Berg
Cultural Heritage Agency of the Netherlands, University of Amsterdam, Amsterdam, The Netherlands
e-mail: K.van.den.Berg@cultureelerfgoed.nl

K. J. van den Berg et al. (eds.), *Conservation of Modern Oil Paintings*,
https://doi.org/10.1007/978-3-030-19254-9_38

As numerous factors may influence the development of water-sensitivity, such as environmental conditions (i.e. UV/visible light and relative humidity) [22]; artists' modification of commercially prepared tube paints; paint formulation e.g. the binding medium [24], pigment type [24, 25, 39] metal soaps and driers [2, 6, 27] as well as any preparatory and/or underlying paint layers, it is difficult to unravel what is naturally a complex and dynamic set of interdependent reactions and outcomes.

Future studies of simplified model oil paint samples, with the application of computational modelling will help to develop an advanced understanding of the complex reaction pathways of modern oil paint which influence the development of degradation and ageing phenomena. However in order to inform further study, it is important to have a body of scientific and technical information pertaining to the materials, technique, physical and chemical characteristics of case-study paintings that present the degradation phenomena of interest. This information also helps to contextualize the results of studies that rely solely upon model samples.

This paper therefore aims to provide an overview of technical and analytical studies of 24 water-sensitive modern oil paintings in Tate's collection. The relationship between obtained compositional information and water-sensitivity is explored, whilst also presenting additional technical information that may inform future studies. Whilst some of the findings are published [4, 17, 25] in this paper, data derived from the group of selected paintings is reviewed in the context of recent developments in our understanding of the chemistry of modern oil paints [6, 7]. Through taking a collective view across 24 paintings, this study has been able to capture the variation in appearance and chemical characteristics of water-sensitive modern oil paintings likely to be encountered by conservators, supported by analytical findings that further our understanding of the condition of specific paint passages.

38.2 Methodology

38.2.1 Case Study Paintings

To date, 24 water-sensitive oil paintings in Tate's collection (see Table 38.1) have been examined to investigate the possible causes and characteristics of water-sensitivity. This includes works by a wide range of artists, and varying paint colours and surfaces. The group of paintings date from 1925 to 2011, although most were painted in the 1950s and 1960s. All were listed as unvarnished oil paintings, most were painted on commercially prepared canvas, with either oil, acrylic or polyvinyl acetate priming layers, and three (Auerbach, T00418; Brett, T00285; Davie, T01748) were painted on hardboard supports.

The environmental exposure history of the paintings was varied. Of the 24 paintings, Tate acquired ten within three years after completion (see Table 38.1), hence these works have been largely displayed or stored under museum conditions.

Table 38.1 24 selected water-sensitive oil and oil-containing paintings from Tate's collection

Artist, title, year, accession number	Year acquired by Tate	Known support and priming	Binding medium (with any organic additives detected using GCMS)	Known surface cleaning history	Pigment of water-sensitive passages (* = where pigment not identified)	Surface appearance of water-sensitive passages	Pigment of non-water-sensitive passages (* = where pigment not identified)	Degradation phenomena (other than water-sensitivity) noted
Karel Appel *People, Birds and Sun*, 1954 (T04163)	1986	Coarse, plain weave hemp canvas, with thin white ground	Oil	Surface-cleaned using damp swabs in 1985 (aside from water-sensitive areas); dry surface cleaned in 2000.	French ultramarine, Prussian blue, green*, organic red (PR83), chrome yellow, unidentified orange*, various ochres, purple (mixture of ultramarine and organic red), bone black	Glossy and medium-rich passages with wrinkling, pigment-medium separation, some matte paint passages	Zinc white and passages mixed with zinc white	Fatty acid efflorescence; weeping paint
Frank Auerbach, *Oxford Street Building Site I*, 1959–60 (T00418)	1961	Unprimed hardboard	Oil	Dry surface-cleaned in 2000.	Mars red, ochre, chrome yellow and green*	Generally medium-rich surfaces, some passages with the most medium rich passages also presenting distinct UV fluorescence.	N/A	N/A

(continued)

Table 38.1 (continued)

Artist, title, year, accession number	Year acquired by Tate	Known support and priming	Binding medium (with any organic additives detected using GCMS)	Known surface cleaning history	Pigment of water-sensitive passages (* = where pigment not identified)	Surface appearance of water-sensitive passages	Pigment of non-water-sensitive passages (* = where pigment not identified)	Degradation phenomena (other than water-sensitivity) noted
Francis Bacon, *Figure in a Landscape*, 1945 (N05941)	1950	Commercially prepared canvas with double ground layer (chalk / lead white lower ground, lead and zinc white upper ground)	Oil (with some beeswax given C30-OH; carnauba wax given C32-OH)	N/A	Organic red, unidentified orange*, yellow, unidentified purple*, Prussian blue, unidentified green*, yellow ochre	Matte appearance with no visible medium skin.	Black*, pale blue*, lead and zinc white passages and mixtures thereof	N/A
Derek Boshier, *Shy Cowboy*, 1980 (T07475)	1998	Plain weave linen canvas with acrylic (pEA/MMA) priming	Oil (with some castor wax given 12-hydroxystearic acid)	Surface dry-cleaned and with lightly moistened swabs in 2002.	Cadmium sulphoselenide, ultramarine, mixed green (Hansa yellow + bone black), umber and ultramarine mixture	Well-bound surfaces of moderate impasto paint with variable gloss. No UV fluorescence.	Bone black, umber, mixed pink passage, lead and zinc white	N/A

Dorothy Brett, *Massacre in the Canyon of Death: vision of the sun God*, 1958 (T00285)	1959	Hardboard with white priming	Oil	N/A	Umber, ochre, iron oxide mixed with cadmium sulphide	Matte, and diluted passages; some impasto paints that appear medium rich. No UV fluorescence.	Light green (copper arsenate/ titanium white mixture)	N/A
Avinash Chandra, *Hills of Gold*, 1956 (T00724)	1965	Medium weight plain weave canvas, commercially prepared	Oil	N/A	Cadmium orange, organic red, mixed black; organic red and carbon black	Variable gloss; some matte impasto, some medium-rich impasto. Distinct UV fluorescence likely relating to zinc white.	Chrome yellow (lead chromate)	Fatty acid efflorescence
Alan Davie, *Image of the Fish God*, 1956, (T01748)	1973	Hardboard with artist-applied oil bound zinc white priming layer	Oil	N/A	Prussian blue, iron oxide, chrome yellow and the underbound barium sulphate + zinc white in oil priming layer.	Variable gloss with both matte impasto and medium-rich surfaces. Some localized areas of pigment-medium separation. No UV fluorescence.	Discoloured unpigmented oil (an artist-applied coating applied in some areas).	N/A
Adrian Heath, *Painting in Brown and Black*, 1960 (T00396)	1960	Commercially primed canvas	Oil (with some castor oil present given detection of ricinoleic acid in the vermilion passage)	N/A	Vermilion, cadmium red	Matte impasto paint; no UV-fluorescence characteristics.	Bone black, light brown (iron oxide)	N/A

(continued)

Table 38.1 (continued)

Artist, title, year, accession number	Year acquired by Tate	Known support and priming	Binding medium (with any organic additives detected using GCMS)	Known surface cleaning history	Pigment of water-sensitive passages (* = where pigment not identified)	Surface appearance of water-sensitive passages	Pigment of non-water-sensitive passages (* = where pigment not identified)	Degradation phenomena (other than water-sensitivity) noted
Patrick Heron, *The Long Table with Fruit*, 1949 (T06850)	1994	C. Roberson plain weave commercially prepared canvas	Oil	Surface-cleaned with damp swabs	Unidentified orange*, vermilion, lemon yellow (mixture of PY31, barium chromate and traces of lead white)	Well-bound paint of moderate gloss.	Unidentified dark blue*, light green*, dark green*, pale blue*	N/A
Per Kirkeby, *The Siege of Constantinople*, 1995 (T07460)	1998	Plain weave canvas with artist-applied white ground	Oil	N/A	Bright yellow*, brick red (iron oxide)	Matte surfaces and diluted paints, with much layering.	N/A	Localised areas of lifting paint
Ferdinand Kulmer, *Brown Picture*, 1960 (T00429)	1961	Unknown: canvas not accessible during examination.	Oil and oil-modified alkyd passages given dimethyl phthalate (with some castor wax given 12-hydroxystearic acid; and some natural resin given tetradehydroabietic acid and dehydroabietic acids)	N/A	Iron oxide, raw sienna, ultramarine and Prussian blue, ochre	Well-bound glossy surfaces, with some wrinkling; diluted and matte paint. No distinct UV-fluorescence.	N/A	N/A

Joan Mitchell, *Chord II*, 1986 (T12921)	2009	Commercially prepared Lefebvre Foinet (19 Rue Vavin 2, Rue Bréa) extra fine canvas	Oil (with some castor wax given 12-hydroxystearic acid; and some natural resin given tetradehydroabietic acid and dehydroabietic acids detected in the phthalocyanine green passage)	Dry surface-cleaned in 2010	Ultramarine, dark green (phthalocyanine green), cadmium sulphide yellow and zinc chromate green, cadmium sulphoselenide	French ultramarine paints diluted and generally matte in appearance. Other colours tend to appear well-bound and medium-rich. Dark phthalocyanine green paints have drying cracks. No UV-fluorescence.	Lead white, cadmium sulphoselenide	Pigment-medium separation of French ultramarine passages.
Piet Mondrian, *Composition with Yellow Blue and Red*, 1937/9–42 (T00648)	1964	Lefebvre-Foinet canvas, with commercial priming	Oil	N/A	Cadmium yellow	Somewhat variable surface gloss (moderate to matte), paint appears well-bound, some age craquelure.	N/A	Localised areas of lifting paint; cadmium yellow degradation
Emile Nolde, *Sea B*, 1930, (T00865)	1966	Coarse open weave canvas, with artist-applied Mars orange ground layer.	Oil	Some areas cleaned with deionized water/ white spirits in 1990, dry-cleaned in 2003.	Prussian blue, cadmium sulphide (orange)	Stiff paint with strongly retained brush strokes. Paint appears well-bound, medium skins visible in some areas under magnification (~50×) [3].	Cadmium zinc sulphide	Localised areas of lifting paint

(continued)

Table 38.1 (continued)

Artist, title, year, accession number	Year acquired by Tate	Known support and priming	Binding medium (with any organic additives detected using GCMS)	Known surface cleaning history	Pigment of water-sensitive passages (∗ = where pigment not identified)	Surface appearance of water-sensitive passages	Pigment of non-water-sensitive passages (∗ = where pigment not identified)	Degradation phenomena (other than water-sensitivity) noted
Ray Parker, *Untitled*, 1959, (T00441)	1961	Commercially prepared fine weave canvas	Oil and oil-modified alkyd given presence of dimethyl phthalate.	N/A	Organic red (azo) + cadmium sulphide mixture; cadmium sulphide + mercury sulphide + titanium dioxide mixture	Variable gloss evident in raking view, flat and fluid paint, little to no impasto. No UV-fluorescence.	Mixed pink (PR81 + titanium white)	N/A
Pablo Picasso, *Nude Woman in a Red Armchair*, 1959 (N06205)	1953	Lead white in oil commercially primed canvas [20]	Oil	dry surface cleaned	Cadmium sulphoselenide, chromium oxide	Uniform, matte/lean water-sensitive cadmium red passages, chromium oxide applied with moderate impasto, with glossy surface visible under magnification (~50×)	Zinc white (and some lead carbonate); zinc white only	Fatty acid efflorescence
Pablo Picasso, *The Three Dancers*, 1925 (T00729)	1965	Fine, plain weave linen canvas with thinly applied grey paint (lead white and bone black) in oil ground [21].	Oil	Surface cleaned using moistened swabs	Localised passages of mixed green (Prussian blue, iron oxide, cadmium sulphide) and iron oxide containing paints.	Glossy, medium rich brown passages with some fine wrinkling in places. Water-sensitive mixed green paint appeared highly diluted/lean, reticulated, and matte.	Black, zinc white, Prussian blue-zinc white mixture, Mars Yellow, Mars Orange	N/A

Jean-Paul Riopelle, *Perspectives*, 1956 (T00123)	1957	Lefebvre Foinet commercially prepared canvas, with lead white in oil ground.	Oil	Swab-cleaned in the past	Prussian blue mixed with cerulean blue, cadmium sulphoselenide, chrome yellow, ochre, cerulean blue + titanium white	Generally matte surfaces, red paint passages associated with medium skins and subtle UV fluorescence [3].	Lead carbonate, bone black	Brittle and cracked impasto; recurrent delamination. Fatty acid efflorescence
Wiliam Sasnal, *Gaddafi I*, 2011 (T14241)	2014	Extremely fine- weave cotton canvas, sized and has PVAc white priming	Oil (with some castor oil present in cadmium yellow paint given detection of ricinoleic acid)	N/A	None[a]	N/A	Bone black, cadmium zinc sulphide, cadmium sulphide, unknown lime green colourant, ultramarine, iron oxide	Impasto remains soft
William Scott, *Reclining Nude (Red Nude)*, 1956, (T00811)	1965	Commercially prepared canvas, with lead white priming	Oil (with some castor wax given detection of 12-hydroxystearic acid)	N/A	Cadmium sulphide	Highly uniform and matte surface, no significant age cracks; no UV-fluorescence	Cadmium sulphoselendide	N/A
Matthew Smith, *The Young Actress*, 1943 (N05440)	1943	Plain weave canvas, with thin white priming layer	Oil	Dry surface cleaning	French ultramarine, cadmium sulphide.	Highly uniform and glossy surface, with some green UV-fluorescence noted in French ultramarine passages	Cadmium sulphoselenide	Localised areas of lifting paint; fatty acid efflorescence

(continued)

Table 38.1 (continued)

Artist, title, year, accession number	Year acquired by Tate	Known support and priming	Binding medium (with any organic additives detected using GCMS)	Known surface cleaning history	Pigment of water-sensitive passages (∗ = where pigment not identified)	Surface appearance of water-sensitive passages	Pigment of non-water-sensitive passages (∗ = where pigment not identified)	Degradation phenomena (other than water-sensitivity) noted
Richard Smith, *Painting*, 1958, (T01588)	1972	Plain weave canvas, with thin white priming layer	Oil (with some castor wax given 12-hydroxystearic acid)	N/A	Cadmium sulphide, and cadmium sulphoselenide,	Matte and lean, heavily diluted	Lead carbonate	Localised areas of lifting paint
Clyfford Still, *1953*, 1953 (T01498)	1971	Plain weave cotton canvas with animal glue size layer and no ground layer	Oil	Dry surface cleaned in 1979, 2004 and 2011.	Pure areas of French ultramarine, red, yellow (cadmium yellow), black	Variable gloss, some matte some medium rich with localized areas of wrinkling. Probable cadmium and ultramarine degradation.	Lead carbonate	Localised areas of lifting paint Cadmium yellow degradation
Ethel Walker, *The Zone of Love: Decoration*, 1930–2 (N05668)	1946	Medium weight linen plain weave canvas, commercially prepared with white priming.	Oil	N/A	Cadmium zinc sulphide, French ultramarine and cobalt blue (cobalt aluminate).	Glossy and medium rich, pure colours, applied in discrete areas only	All passages mixed with white.	N/A

[a]This painting was included in the study as surprisingly all passages were non-water-sensitive despite the presence of areas painted using colours (pigments) that are frequently water-sensitive

The remainder were acquired between 14 and 45 years after completion (see Table 38.1), and thus exposed to more varied environmental conditions. The selected paintings had minimal previous conservation treatment, apart from dry surface cleaning. Table 38.1 summarises the known conservation treatment history and relative water-sensitivity of different passages of paint together with the results of the analysis of the pigments present.

38.2.2 Assessment of Water-Sensitivity

Discrete water-sensitivity tests (described elsewhere) were carried out using cotton swabs dampened with deionized water to identify water-sensitive and non-water-sensitive paint passages [27, 39].

38.2.3 Sampling Methodology

Where appropriate, samples were taken from accessible areas that were representative of major water-sensitive and non-water-sensitive paint passages. Surface scrapings were taken for analysis, and in a more limited number of cases cross sections were prepared, and samples for Scanning Electron Microscopy Energy dispersive X-ray spectroscopy (SEM-EDX) were obtained for imaging and analysis.

38.2.4 Analytical Techniques

38.2.4.1 Light Microscopy and Cross Section Analysis

A Dino-Lite digital USB microscope was used to document paint surfaces. Light microscopy of embedded cross sections was carried out using a Leica light microscope (DM4000) and Leica Application Suite (LAS software). Cross section samples were prepared using a polyester resin (Tiranti LTD, 'Clear Casting AM' resin) and cured using a liquid hardener (Butanox M-50; methyl ethyl ketone peroxide solution in dimethyl phthalate, supplied by Tiranti LTD).

38.2.4.2 Scanning Electron Microscopy Energy Dispersive X-Ray Spectroscopy (SEM-EDX)

Cross sections were carbon coated for SEM-EDX analysis using an Edwards 306 carbon coatings system that provides a coating of carbon of depth ~10–20 nm carbon. An SEM Hitachi S4000 scanning electron microscope, with an accelerating voltage of 25 keV was used with an Oxford instruments INCA EDS system for the analysis of embedded cross sections. Data processing was carried out using INCA microanalysis suite 4.02.

The surfaces of paint samples were examined using a LEO 1455 variable-pressure scanning electron microscope (VP-SEM) with INCA microanalysis suite. Environmental Scanning Electron Microscopy with Energy-Dispersive X-ray Spectroscopy (ESEM-EDX) was carried out using a Philips XL30 ESED-FEG instrument fitted with an Oxford INCA EDS analysis system. EDX analysis was done under standard ESEM conditions: H20 mode, 10 mm, 20 kV accelerating voltage, 0.7 torr water vapour pressure.

38.2.4.3 Fourier Transform Infrared Spectroscopy (FTIR)

Transmission FTIR analysis was carried out on a Thermo scientific Nicolet iN10 MX microscope using a diamond cell; 64 scans were collected at a resolution of 4 cm^{-1}, the collection time was 22 s. Data was obtained and processed using OMNIC 8 software.

38.2.4.4 Gas Chromatography Mass Spectrometry (GC-MS)

Samples were derivatised using Meth Prep II (Grace™ Alltech™). GC was carried out on a Varian CP-3900 GC coupled with a 1200 L MS detector. Oven program: 80 °C ramped to 320 °C at 10 °C/min then held for 5 min at 320 °C. Total run time was 29 min. Split (20:1) injection volume was 1 μL, and the helium flow was 1.0 mL/min. MS conditions: source temperature: 220 °C; transfer line temperature: 270 °C, injection port set at 300 °C. Column: Phenomenex Zebron ZB-5 column (30 m length; 0.25 mm i.d.; 0.25 μm film thickness). EI mode (70 eV); scan group 1: 45–300 amu; Group 2: 45–700 amu at 16 min, every 1 s.

38.2.4.5 Electrospray Ionisation Mass Spectrometry (ESI-MS)

Direct injection ESI-MS analyses were carried out on paint samples of 0.1–0.5 mg after solvent extraction using ethanol [42, 44]. The extract was mixed 1:1 with 20 mM ammonium acetate (NH_4Ac) in ethanol. Samples were delivered to the MS system with a Micromass CapLC system: MS analysis was carried out with a Micromass Q-tof-2, equipped with a nano probe and ESI source. MS data were collected and interpreted using MassLynx 4.0 software (Waters). All spectra were collected in high mass resolution mode.

38.2.4.6 Statistical Analysis

An unsupervised multivariate pattern analysis method, principal component analysis (PCA) was used to investigate the relationship between selected organic components detected using GC-MS and ESI-MS analysis and water-sensitivity. For GC-MS data, chromatographic peak areas of palmitic, stearic and oleic acid, and C6-C10 diacids were obtained and normalized to their sum. The data matrix of 8 × 75 variables was analysed using PCA of the correlation matrix. ESI-MS negative mode data provided the most interesting information about the polar extractable fraction; the intensity of 29 significant [M-H]- ions, *m/z* (including short chain monoacids, diacids, hydroxy acids, and glycerides containing diacids) were normalized to their sum. The data matrix of 29 × 84 variables was analysed by PCA of the correlation matrix, using XLSTAT 2015.3.01 software (Addinsoft). Further details, including the data matrixes used for the PCA and associated lists of assigned ions identified in the ESI-MS mass spectra are available elsewhere [25].

38.3 Results and Discussion

38.3.1 Aspects that Could Be Correlated to Water-Sensitivity

The visual characteristics that appeared to be related to water-sensitivity across the group of selected paintings are described below. Key analytical results pertaining to the chemical composition of the binding medium that appeared relevant to water-sensitivity are also presented, offering associations with visual aspects where possible.

38.3.1.1 Colour and Pigment Type

Previous studies have highlighted a strong pigment dependence with regards to water-sensitivity [11, 24, 27]. The visible surface characteristics of water-sensitive and non-water-sensitive paint passages across the group of paintings were compared. Based on empirical observation, the most reliable visual indicator for water-sensitivity was the presence of paint passages made using pigments typically associated with water-sensitivity. This included passages made using the following inorganic pigments: cadmium yellow (PY 35), cadmium orange (PO23), chrome yellow (PY34:1), iron oxide (PB 6), French Ultramarine (PB 29), or Prussian blue (PB 27) pigments, used unmixed with zinc white (PW 4) and/or lead white (PW 1) (Fig. 38.1 presents the commonly encountered water-sensitive pigments for the group of paintings). Other visible surface characteristics, discussed in Sect. 38.3.2, proved to be less reliable indicators for water-sensitivity.

The water-sensitivity of some paint passages could also be ascribed to the use of organic pigments, that are inherently vulnerable to solvents [5]. As examples,

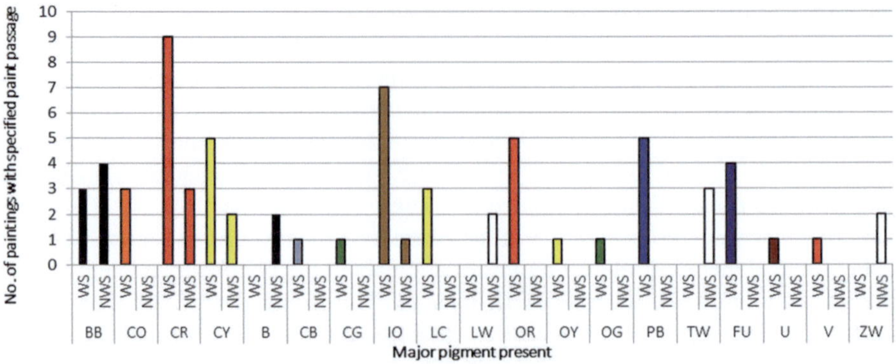

Fig. 38.1 Illustrates the number of paintings with a paint passage containing a specified major pigment, characterised as either water-sensitive (WS) or non-water-sensitive (NWS). Colours are specified as follows: *BB* bone black, *CO* cadmium orange, *CR* cadmium red, *CY* cadmium yellow, *B* carbon black, *CB* cerulean blue, *CG* cobalt green, *IO* iron oxide, *LC* lead chromate, *LW* lead white, *OR* organic red, *OY* organic yellow, *OG* organic green, *PB* Prussian blue, *FU* French ultramarine, *U* umber, *V* vermilion, *TW* titanium white, *ZW* zinc white

organic pigments were found in water-sensitive passages of the Appel (red; PR 3) Chandra (azo red), Parker (unidentified organic red), Mitchell (phthalocyanine green, PG7), and Still (organic reds; PR3 and PR1) paintings. Across this particular group of paintings there seemed to be a higher incidence of water-sensitivity associated with the use of inorganic pigments, rather than organic pigments. However this also reflects the material choices of the artists and the fact that amongst this group of paintings paints based on inorganic pigments were more commonly used than those based on organic pigments.

38.3.1.2 Lean Paints

The paintings by Chandra, Mitchell and R. Smith had water-sensitive paint passages that appeared (based on visual inspection), matte and heavily diluted, presumably to achieve fluid brush strokes, characteristic drip marks, or a staining effect. Highly diluted paints are generally underbound which is likely to contribute towards the observed water-sensitivity and a general vulnerability to mechanical action during cleaning [12].

38.3.1.3 Epsomite

Magnesium sulphate heptahydrate or epsomite can form as a degradation product of magnesium carbonate, used as an extender in some modern oil paint formulations [16, 17]. As a water-soluble salt, epsomite is one possible cause of water-sensitivity in magnesium carbonate-containing paints. Magnesium carbonate (the presence of which is considered a prerequisite for the formation of epsomite) was identified using FTIR in twelve water-sensitive samples and one non-water-sensitive sample[1], taken from seven of the 24 case-study paintings (Chandra, Davie, Heron, Mitchel, R. Smith, Scott, and Walker). Of the 13 samples where FTIR confirmed magnesium carbonate, 11 contained cadmium pigments (yellow, orange or red) or chrome yellow as the sole or as a significant pigment. Epsomite was confirmed in water-sensitive passages (in a total of 6 samples, 3 of which were cadmium or chrome colours) taken from four of the twenty-four paintings studied, using a combination of FTIR, SEM-EDX (elemental mapping of intact surface samples) and X-ray Photoelectron Spectroscopy (XPS) (see Table 38.2). For all samples where epsomite was identified, the presence of magnesium was confirmed using SEM-EDX and/or the presence of magnesium carbonate confirmed using FTIR. Since XPS (the most sensitive technique used in this study for detecting epsomite) was not used for all samples, it is possible that epsomite may have been present in a greater number of samples than those listed in Table 38.2. Epsomite may have been present below the detection limit for FTIR, or present without sufficiently developed crystals for detection using SEM-EDX. However it is emphasized that epsomite was considered a likely or possible cause for observed water-sensitivity only where its presence was either confirmed as described above, or where magnesium carbonate was present in sufficient amounts to be detected using FTIR. This was the case for only specific water-sensitive paint passages of 8 of the 24 paintings. This suggests that amongst this group, the majority of the cases of water-sensitivity could not be attributed to the formation of epsomite.

38.3.1.4 The Molecular Composition of the Binding Medium

As discussed above (see Sect. 38.3.1.3), the possibility that epsomite was the cause of water-sensitivity was considered unlikely in most cases. This provided the opportunity to investigate possible relationships between the molecular composition of the binding medium and water-sensitivity.

Previous studies have noted relatively high levels of azelaic acid in modern oil paints based on titanium white, burnt sienna, raw umber, green earth, ultramarine, cobalt blue, cobalt violet and Prussian blue pigments; a possible relationship to water-sensitivity was highlighted [19]. Azelaic acid is a polar dicarboxylic acid (i.e. a diacid) that derives from oxidative scission (also referred to as 'oxidative degradation') of unsaturated fatty acids. A paint film with a higher azelaic acid /diacid content will be more polar. In addition, because oxidative degradation (which leads to fragmentation) and oxidative polymerization (which leads to a crosslinked paint film) represent competing reaction pathways [12], an oil paint film containing a relatively high azelaic acid (i.e. diacid) content, may also be less crosslinked [12]. It has been suggested that water-sensitivity may relate to heightened relative polarity

[1]A total of 90 samples taken from the 24 case-study paintings were analysed using FTIR.

Table 38.2 Identification of epsomite in the selected case study paintings

Painting	Water-sensitive paint passage	EDX or XPS [∗ = major element(s)]	FTIR	Interpretation; paint composition	Evidence suggesting epsomite/additional comments	References
Francis Bacon, *Figure in a Landscape*, 1945 (N05941)	Carbon black	C, Ca, Cl, Mg, Zn, Al, Si, Pb, Ca, Fe, Zn, As	–	Carbon black paint, with possible magnesium carbonate and calcium carbonate extenders. Many metal ions present; highly mixed sample.	SEM images showing crystalline entities at the paint surface. Elemental mapping/further analysis was not carried out.	[17]
Avinash Chandra, *Hills of Gold*, 1956 (T00724)	Organic red	C, O, Mg, Al, Si, S, Cl, Fe	Oil, synthetic organic red (possibly PR2 or PR10?) magnesium carbonate.	Synthetic organic red (pigment contains Cl on alumina base) magnesium carbonate, silica, possible aluminium stearate, iron oxide	SEM-EDX imaging and elemental mapping showing rod shaped crystals, rich in magnesium and sulphur.	[3, 4]
	Cadmium red	O, C, S, Si, Cd, Mg, Zn, N, Pb, Ba, Ca	Oil, organic red (possible PR10) magnesium carbonate	Oil, cadmium-zinc sulphide, organic red (PR10 possible) magnesium carbonate, barium sulphate, silica.	XPS results consistent with hydrated magnesium sulphate.	[45]
	Chrome yellow	C, O, Al, S, Cr, Zn, Pb*, Ba*	Oil, magnesium carbonate, barium sulphate, zinc stearate.	Lead chromate (chrome yellow), magnesium carbonate, barium sulphate, zinc stearate, possible aluminium stearate.	SEM/EDX showing crystals rich in both magnesium and sulphur.	[3, 4]
Patrick Heron, *The Long Table with Fruit*, 1949 (T06850)	Pale lemon yellow	N/A	Oil, magnesium carbonate, organic yellow (possible PY31)	Oil, organic yellow (PY31?) with magnesium carbonate extender	Light scattering surface crystals visible using light microscopy and presence of magnesium carbonate.	[17]
Piet Mondrian, *Composition with Yellow Blue and Red*, 1937/9-42 (T00648)	Cadmium yellow	C, O, S*, Cd*, Zn, Ba, Mg	Oil, likely beeswax, possible magnesium sulphate heptahydrate (1331, 1658, 1101 cm^{-1})	Oil, beeswax, cadmium zinc sulphide, barium sulphate, possible epsomite	FTIR suggested possible epsomite, and SEM imaging showed crystalline entities (no elemental mapping carried out).	[16]

Epsomite as detected amongst the case study paintings has been documented before, as per the cited references coupled with a lower degree of crosslinking and/or metal soap formation (saponification) of modern oil paint films [2].

To investigate this hypothesis, and any possible relationship between water-sensitivity and the azelaic acid/diacid acid content, samples were taken from both water-sensitive and non-water-sensitive passages of selected paintings (see Table 38.1) and analysed using GC-MS and ESI-MS (see Sects. 38.2.4.4 and 38.2.4.5) [25]. The non-crosslinked, non-saponified polar fraction of the binding medium including glycerides, free fatty acids and free diacids were characterized using ESI-MS (negative mode). Free fatty acids and free diacids, including those originating from hydrolysed glycerides[2] [37], and also those bound as metal soaps, were analysed using GC-MS. Unlike the ESI-MS analysis, the GC-MS protocol detects fatty acids and diacids that are bound as metal soaps and/or are bound to a wider crosslinked network only via ester bonds[3]. The non-hydrolysable, crosslinked portion of the paint network was not measured at all in this study[4].

As stated above, the relative content of azelaic acid (a C9 diacid) is considered a measure of the degree of oxidative degradation of the oil binder [15]. In this study, the overall degree of oxidative degradation (referred to as 'degree of oxidation') of the oil binder is best inferred through considering the relative azelaic acid content derived from the GC-MS protocol, as opposed to the ESI-MS protocol, since the GC-MS protocol and sample preparation procedure employed (see Sect. 38.2.4.4) facilitates the detection of a greater proportion of the azelaic acid/diacid content present.[5]

Principal component analysis (PCA) of normalized peak area ratios of selected fatty and diacids deriving from the GC-MS data set was carried out to explore the role of the azelaic acid content [25] as a measure of the degree of oxidation [15]. The resulting biplot (Fig. 38.2) shows that many non-water-sensitive and water-sensitive samples cluster together (see blue highlighted region, Fig. 38.2). This indicates that there is little difference between water-sensitive and non-water-sensitive samples in terms of the selected variables, which suggests there is no clear relationship between the overall degree of oxidation and the incidence of water-sensitivity amongst this group of paintings. This is consistent with findings of other studies of model paint films [2].

However, a possible exception to this is the cluster of predominantly water-sensitive samples, identified using PCA that are associated with positive values of Principal Component 1 (PC1) (i.e. the samples within the region highlighted in grey, Fig. 38.2), which are correlated with the diacid content. This indicates these sam-

---

[2] Unlike ESI-MS analysis, GC-MS cannot measure intact glycerides. Glycerides are hydrolysed as part of the sample preparation method and the resulting free fatty acids/diacid species are then detected.

[3] The derivatization procedure used for the GCMS analysis involves hydrolysis of ester bonds. Neither ESI-MS nor GC-MS can detect species that are covalently bound to the wider crosslinked network via non-hydrolysable bonds. The crosslinked network was not studied here.

[4] For example, pyrolysis GC-MS is a technique that can be used to study the crosslinked fraction.

[5] Any diacids that are part of the crosslinked network were not detected in this study.

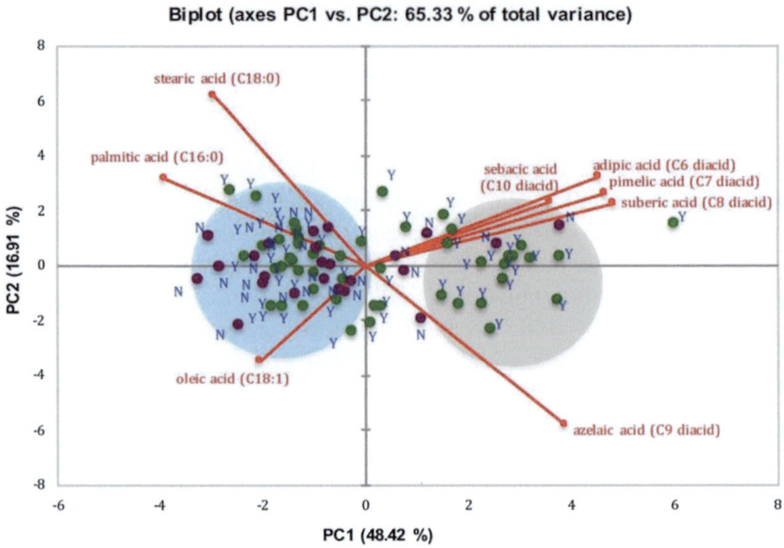

Fig. 38.2 PCA biplot of GC-MS data showing the relationship between the water-sensitive samples (illustrated in green and labeled Y), non-water-sensitive samples (illustrated in purple, and labeled N), and the content of selected fatty acids and diacids, which together are the measured variables, which are shown in red. The red lines represent how each of these variables relate to the principal components (*PC* principal component) which are the axis that represent the 'directions' that best separate out the samples

ples may be more oxidized relative to the others. This data is presented again in Fig. 38.3, coloured to indicate pigment type, and show that the majority of the samples that cluster toward positive values of PC1 are either iron oxide, French ultramarine or Prussian blue water-sensitive paints. These samples were taken from nine different paintings (Auerbach, Brett, Davie, Kulmer, Mitchell, Nolde, Picasso (*Three Dancers*), Riopelle, and M. Smith) indicating that the clustering observed is more correlated with pigment type rather than the source of the samples.

To compliment the results discussed above, PCA analysis of the ESI-MS data set pertaining to the polar non-crosslinked and non-saponified fraction was also carried out (see Sect. 38.2.4.6) [25]. The ESI-MS analysis did not, for example, detect azelaic acid species that were bound as metal soaps, but did give an indication of which samples had the greater relative amount of free azelaic acid (and azelaic acid containing mono, di and tri–glycerides). This non-crosslinked and non-saponified fraction thus represents the binding medium fraction that is considered most likely to be vulnerable during exposures to polar solvents[6], and is hereby referred to as the 'extractable' fraction.

[6] Metal soaps (excluding those of group I and II metal cations which are water-soluble) are not soluble in polar solvents.

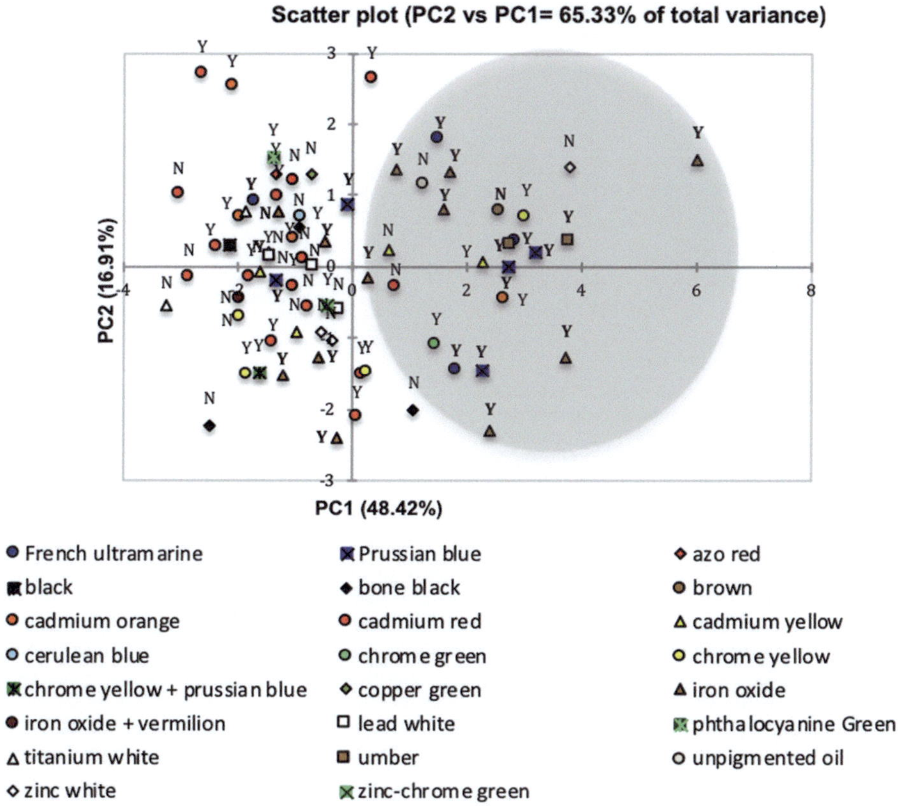

Fig. 38.3 PCA scatter plot of GC-MS data of samples from paintings, which corresponds to the biplot shown in Fig. 38.2. The points which represent paint samples, are now coloured to represent the main pigment present. Water-sensitive samples are indicated Y, non-water-sensitive samples, N. Samples highlighted inside the grey circle are associated with more positive values of PC1. These samples capture 18 out of a total of 21 samples taken from iron oxide, Prussian blue and French Ultramarine water sensitive paint passages that were present across nine different paintings

It is perhaps unsurprising given that iron oxide, Prussian blue and French Ultramarine paints appeared the most oxidized (based on the GCMS results), to find that they also contained a higher proportion of extractable diacid species than paints containing other pigments; as shown in Fig. 38.4, the iron oxide, Prussian blue, and French ultramarine paints cluster toward positive values of PC1 (grey region of Fig. 38.4a), which correlates with the diacids and with glycerides containing only diacids (grey region Fig. 38.4b). This is likely to relate to the fact that these pigments do not readily form metal soaps in oil, (and certainly not to the same extent as pigments such as zinc and/or lead white for example) [28, 30], hence non-cross-linked diacid species tend to remain unbound within the paint films.

As discussed above, a combination of heightened polarity (from the heightened azelaic acid content and/or relatively high degree of oxidation) coupled with a prob-

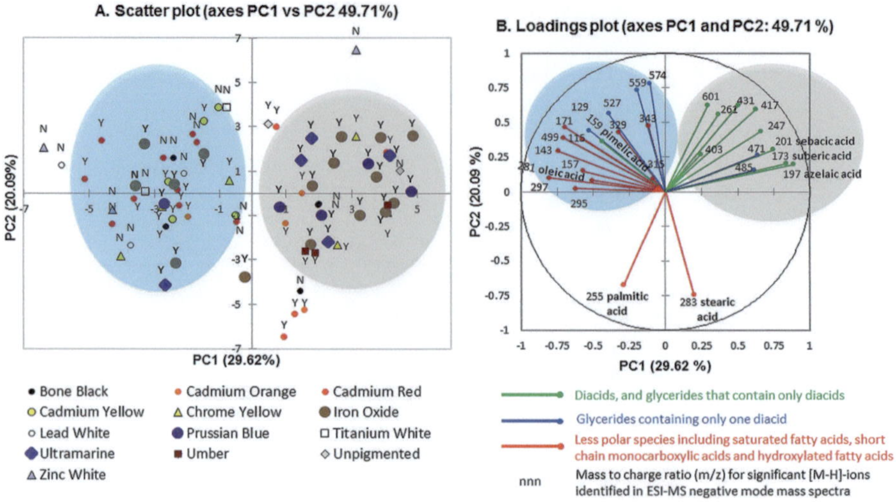

Fig. 38.4 The PCA scatter plot of ESI-MS negative mode data is shown in (**a**) and the associated loadings plot, shown in (**b**). Points on the scatter point (**a**) represent samples of various colours as indicated. Water-sensitive samples and non-water-sensitive samples are labelled Y and N respectively. The loadings plot (**b**) shows how the measured variables (i.e. the 29 [M-H]-ions *m/z* that were used for the PCA analysis) relate to the two principal components, PC1 and PC2. The measured variables are broadly divided into three families as indicated: diacids and glycerides containing only diacids; glycerides containing one diacid; other less polar species such as palmitic/stearic acids

able lower degree of polymerisation may help explain the frequent observation of water-sensitivity with iron oxide, Prussian blue, and French ultramarine paints. Whilst the mechanisms that result in pigment/binding media loss in response to water-exposure during cleaning are not yet fully understood, a swelling mechanism [31, 39], driven by polar-interactions seems a probable explanation. An oil paint that is both comparatively polar, possibly also less crosslinked and/or lacking in metal soap interactions that serve to stabilize the paint [8], may well be more vulnerable to swelling with the ingress of polar solvents such as water [31]. The presence of sufficient free and accessible carboxylic acid groups, or other polar functional groups in a paint film, may facilitate more rapid transport of water into the paint film, leading to swelling and pigment removal during cleaning [31, 32, 34]. It would be useful to carry out diffusion experiments [1] in order to assess the transport of water through water-sensitive oil paints.

Recent evaluations of tailored cleaning systems applied to oil paint mockups, including artificially aged water-sensitive iron oxide and French ultramarine paints (see Ormsby et al. [29], these proceedings), have suggested that modifying the pH of aqueous cleaning solutions may help to reduce pigment loss and swelling during cleaning [14]. It is possible that paints such as water-sensitive iron oxide, Prussian blue and French Ultramarine paints may be more sensitive toward pH effects during cleaning, at least partly, due to the fact these paints have a tendency to be more oxidized and contain more diacids species [25] (and therefore more accessible carboxylic acid groups). Iron oxide paints are also known to have a tendency to be more hydrolysed than paints of other pigments, which influences their mechanical properties and response to solvents [41].

However, it is also important to recognize that Fig. 38.4 also shows that many water-sensitive and non-water-sensitives samples of colours other than iron oxide, Prussian blue and French ultramarine, also cluster together (blue region, Fig. 38.4a). These are not correlated with the diacid content but rather less polar species (blue region, Fig. 38.4b), hence for many of the samples taken, the free azelaic acid content does not appear to be directly related to water-sensitivity. This indicates that whilst azelaic acid may be a contributing factor towards water-sensitivity, particularly for paints that include pigments associated with comparatively higher degrees of oxidation/azelaic acid content, there are clearly other factors to be explored. The nature of the crosslinked/ionomeric network (including relative size and polarity) is likely to be a critical factor in terms of the development of water-sensitivity and response to solvent exposure which requires further study [2, 6]. The specific influence of factors that influence the development of the crosslinked/ionomeric network such pigment-type, pigment-particle size, environmental conditions, paint additives, and possibly conservation treatment also requires investigation.

38.3.2 Aspects Not Directly Correlated with Water-Sensitivity

Observations and analytical results that could not be directly correlated with water-sensitivity are discussed below.

38.3.2.1 Other Ageing/Degradation Phenomena

Amongst this group of paintings were occasional examples of 'weeping paints' [10] fatty acid efflorescence [35], and localized flaking and delamination (see Table 38.1). Pigment-medium separation that related to the presence of excess binding medium and surface-wrinkling was also observed (see Fig. 38.5). These examples of ageing phenomena were not as common as the incidence of water-sensitivity, nor did they appear to be directly associated with water-sensitivity, amongst the paintings examined.

38.3.2.2 Medium Skins and UV-Fluorescence

Medium-rich surfaces or 'medium skins' that exhibit UV-fluorescence, have been cited as a feature of water-sensitive paints [11, 27]. This type of water-sensitive surface was observed in this group of paintings most frequently in French

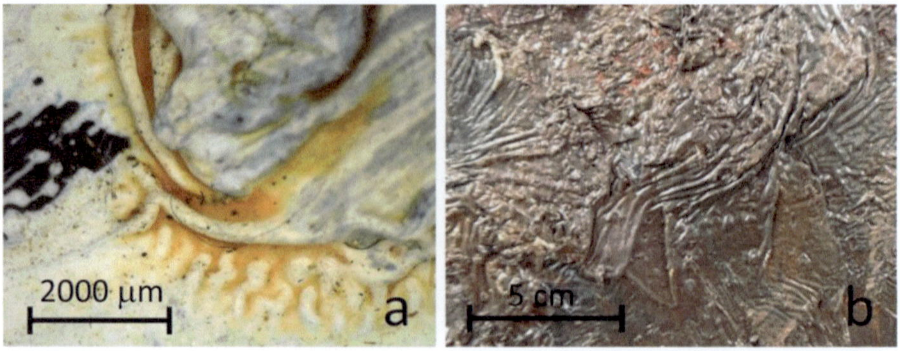

Fig. 38.5 (**a**) Photomicrograph (65× magnification) of non-water-sensitive mixed pale blue passage of Davie's, *Image of the Fish God*, 1956, showing pigment-medium separation caused by the presence of excess binding medium, with wrinkling and imbibed surface-dirt. Photo © Tate; artwork © Estate of Alan Davie (**b**) Detail of Kulmer's *Brown Picture*, 1960 showing glossy, medium rich water-sensitive brown paints (earth colours, and earths mixed with French ultramarine), and distinct surface wrinkling. Photo © Tate; artwork © reserved

ultramarine, Prussian blue or iron oxide passages,[7] which correlates well to the analysis of the organic fraction (see Sect. 38.3.1.4). Medium skins form as the oil paint initially dries to a solid film and can further develop on ageing. Pigment-medium separation may be influenced by the oil type, pigment-medium ratio, the hydrophilicity and oil-absorption of the pigments, the porosity of the ground or underlying paint layers, and may also suggest phase-separation behavior relating to mobile components of the binding medium [38]. However, since there were many examples of water-sensitive passages amongst the group of paintings that appeared well-bound, yet matte in quality and lacking any visible medium skin (or associated UV-fluorescence), it seems neither the presence nor absence of a medium skin, nor UV-fluorescence is a reliable visual indicator for all types of water-sensitivity. However, the presence of a medium skin may alert the conservator to water-sensitive passages that may be particularly vulnerable to gloss changes during surface-cleaning due to the potential for disruption of the sensitive medium-skin (see Ormsby et al. [29] in these proceedings).

38.3.2.3 Metal Soaps

Metal soaps are common oil paint additives which function as rheology modifiers or as driers [13, 40], and which influence the ageing pathways of oil paint, and possibly the development of water-sensitivity [2]. Metal soaps may also form in oil paints as a result of pigment-medium interactions [26], forming amorphous metal soaps that contribute toward the stabilization of aged oil paints [18, 43]. Although metal

[7] Possibly suggesting a link between medium-skin formation and degree of oxidation, given these paints seemed more oxidised than others.

soaps were widely identified in samples taken from all of the paintings using FTIR [25], none of the paintings presented metal soap protrusions or disfiguring metal soap crusts often seen in older works [9, 33].

The detection of metal soap additives including the identity of the metal cation present using FTIR is complicated by the fact that metal soaps are included as additives at levels below typical detection limits. In addition, where crystalline metal soaps were identified using FTIR (including the metal cation) it is possible that they had formed *in-situ* rather than included as an additive. Crystalline aluminium, copper, manganese, and zinc soaps were occasionally identified[8] in both water-sensitive and non-water-sensitive samples taken from the case study paintings. Water-soluble metal soaps such as sodium or potassium soaps were not detected using FTIR in any of the samples analysed and are generally not believed to be a cause for the water-sensitivity observed.

Amorphous metal soaps were frequently identified in paint samples taken from the case study paintings[9] and there was no clear relationship noted between the identification of amorphous metal soaps and the incidence of water-sensitivity. However, oil paints containing pigments that readily form metal soaps in oil e.g. lead or zinc white which were consistently non-water-sensitive paints, did contain higher quantities of amorphous metal soaps than paints of other colours [25]. Further studies may investigate the relationship between pigment type (or cation type) and the relative significance of the formation of stable covalent crosslinks as opposed to amorphous metal soap interactions, in relation to paint stability and the development of water-sensitivity. It has been suggested before that the nature of the ionomeric network is a major factor in development of oil paint degradation phenomena and is deserving of further research [8, 9].

38.3.2.4 Other Detected Organic Additives

The binding media of the group of paintings was confirmed as predominantly oil, using a combination of FTIR and GC-MS analysis. However, the precise seed origin of the oils present was not identified in this study,[10] with the exception of the detection of castor oil in two instances[11]. Semi-drying oils such as safflower oil are widely used in modern oil paint formulations [19] and may render an oil paint film less crosslinked and more vulnerable during cleaning [31]. The relationship between water-sensitivity to the type or blend of seed oil used, as well as associated oil-processing methods [13], also requires further investigation. In addition to the main

[8] Crystalline Al, Cu, Mn and Zn soaps were identified based on the presence of sharp infrared absorption bands at 1584, 1586, 1569 and 1540 cm^{-1} respectively, and the presence of associated metal ions confirmed using EDX analysis.

[9] On account of a broad infrared absorption band appearing in the 1500–1600 cm^{-1} region.

[10] This requires other analytical protocols. See van Dam et al. [42].

[11] GCMS identified of ricinoleic acid, a marker for castor oil, in a very soft non-water-sensitive cadmium yellow impasto passage, of the Sasnal painting, and in a water-sensitive cadmium red passage of the Adrian Heath painting (see Table 38.1).

seed oil binder, other organic additives can be present in modern oil paintings given that artists have access to a range of commercially manufactured mediums[12] at their disposal, to modify drying times, to create impasto, or to achieve other surface effects for example. Organic additives may also form part of the paint formulation e.g. castor wax has been used as a rheology modifier and/or suspension aid [19]. The presence of these types of organic additives could also impact on the response of oil paints to solvents, which is also an area warranting further investigation.

Across the group of paintings explored, in addition to oil and castor wax (identified in both water-sensitive and non-water-sensitive paints), natural resin (identified in water-sensitive paints) and oil-modified alkyd resin (identified in both water sensitive and non-water-sensitive paints) were occasionally identified using GC-MS analysis[13] (see Table 38.1). In this study it was not possible to make associations between water-sensitivity and the presence of castor wax, natural resin and oil-modified alkyd given the infrequency with which these elements were detected. However, it may be speculated that the addition of components that do not contribute significantly toward polymerisation may render the oil paint matrix more vulnerable toward solvents and mechanical action. Conversely the presence of added oil-modified alkyd media may contribute toward a more stable paint due to the enhanced crosslinking of the alkyd polymer [34].

38.4 Conclusions

Water-sensitive paintings present highly variable surfaces however it appears that pigment type/colour remains the most reliable visual indicator for water-sensitivity. In modern oil paintings, paint containing selected pigments unmixed with zinc white and/or lead white paints are more likely to be water-sensitive, including a range of inorganic pigments (the incidence of water-sensitivity was greatest for inorganic pigments in this study) and synthetic organic pigments.

Amongst the group of paintings examined, several complex interrelated factors appear to have contributed to the development of water-sensitivity. For a small number of paintings, epsomite was identified as a likely causal factor, for others, under-bound paints proved to be physically vulnerable, while in others, the presence of organic pigments made paint layers inherently vulnerable to solvents.

When considering the oil fraction the combined analytical results indicated that (with the possible exception of water-sensitive iron oxide, Prussian blue and French Ultramarine paints) water-sensitive paints did not prove to be generally more

[12] Winsor & Newton for example manufacture a range of mediums, such as Liquin (an oil-modified alkyd) and others: http://www.winsornewton.com/uk/shop/oils-solvents-mediums-and-varnishes/oil-colour/mediums (accessed 17.1.19).

[13] Castor wax was identified based on the presence of 12-hydroxystearic acid; natural resin was identified based on the presence of tetradehydroabietic acid and dehydroabietic acids; oil-modified alkyd paints were identified based on significant amounts of dimethyl phthalate.

oxidized than non-water-sensitive paints, nor did the relative proportion of the major components of the polar extractable fraction generally correlate with water-sensitivity. However, iron oxide, Prussian blue, and French Ultramarine oil paints tended to be water-sensitive, more oxidised, and contained a higher proportion of extractable diacid species than other colours. Hence for these colours, a combination of heightened polarity and insufficient polymerisation may both contribute to water-sensitivity. In general a paint film that is not sufficiently polymerised and/or saponified, will have a greater likelihood of being water and solvent sensitive, and vulnerable to mechanical action. The relative size and polarity of the polymeric network in water-sensitive and non-water-sensitive modern oil paints requires further investigation [2, 23, 25] as do factors which may influence the development of the polymeric/ionomeric network, including: environmental conditions during curing and ageing; the type and processing of the seed oil used; the presence of organic additives; added metal soaps; driers; and most obviously, pigment type.

The investigation of the causes of water-sensitivity based on technical and scientific studies of a group of arguably unrelated paintings is inherently challenging, and particularly without the use of modelling and big data tools. Nonetheless, the combined data has offered an expanded view of the similarities and differences across the works studied and has reinforced the influence of pigment type as a key indicator for caution when approaching the interpretation of paint surfaces and conservation treatment. Further explorations using defined model paint samples under controlled conditions will prove useful for investigating how specific variables influence the development of the polymeric/ionomeric network and eventually how they may influence water-sensitivity.

Acknowledgments The authors gratefully acknowledge the support of the Heritage Plus, a Joint Programming initiative of the European Commission, as part of the work was undertaken in the context of the project "CMOP – Cleaning Modern Oil Paints" (2015–2018). The authors would like to thank those who have contributed toward the investigation of the group of paintings: Lucia Bay, Anna Cooper, Annette King, Rachel Scott, Joyce Townsend, Francesca Modugno, Ilaria Bonaduce, Nelly von Aderkas, Art Néss Proaño Gaibor, William Luckhurst, Marc Vermeulen, Alan Phenix, Tom Learner, Tate Conservation and Research departments.

References

1. Baij, L., J. J. Hermans, K. Keune, and P. D. Iedema. 2018. "Time-Dependent ATR-FTIR Spectroscopic Studies on Solvent Diffusion and Film Swelling in Oil Paint Model Systems." *Macromolecules*. 51 (18), pp. 7134–7144. https://doi.org/10.1021/acs.macromol.8b00890.
2. Banti, D., J. La Nasa, A.L. Tenorio, F. Modugno, K. Jan Van Den Berg, J. Lee, B. Ormsby, A. Burnstock, and I. Bonaduce. 2018. "A Molecular Study of Modern Oil Paintings: Investigating the Role of Dicarboxylic Acids in the Water Sensitivity of Modern Oil Paints." *RSC Advances* 8 (11). https://doi.org/10.1039/c7ra13364b.
3. Bay, L. 2016. "Documenting Water Sensitive Oil Paintings in the Tate Gallery." Final Year Dissertation, Postgraduate Diploma, Courtauld Institute of Art, University of London.

4. Bay, L., A. Burnstock, J. Lee, B. Ormsby, and K.J. van den Berg. 2017. "Water Sensitivity of Modern Oil Paintings." Edited by J Bridgland. *ICOM-CC 18th Triennial Conference Preprints, Copenhagen, 4–8 September 2017*. Paris: International Council of Museums.
5. Blumenroth, D., S. Zumbühl, N. C. Scherrer, and W. Müller. 2014. "Sensitvity of Modern Oil Paints to Solvents. Effects on Synthetic Organic Pigments." In *Issues in Contemporary Oil Paint*, edited by K J Van den Berg, A Burnstock, M de Keijzer, Jay Kreuger, T Learner, Alberto de Tagle, and G Heydenreich, 351–362. Switzerland: Springer.
6. Bonaduce, I. 2018. "The Organic Chemistry of Modern Oil Paints." Paper presented at *Conference on Modern Oil Paints*. Amsterdam, 23–25 May 2018.
7. Bonaduce, I. et al. 2019 "Conservation issues of modern oil paintings: a molecular model on paint curing". *Accounts of Chemical Research*. In Press.
8. Boon, J. J., S. L. Peluvé, O. F. van den Brink, M. C. Duursma, and D. Rainford. 1997. "Molecular Aspects of Mobile and Stationary Phases in Ageing Tempera and Oil Paint Films." In *Early Italian Paintings – Techniques and Analysis*, edited by T Bakkenist, R Hoppenbrouwers, and H Dubois, 35–56. Limburg Conservation Institute, Maastricht.
9. Boon, J. J. 2006. "Processes inside Paintings That Affect the Picture: Chemical Changes at, near and underneath the Paint Surface." In *Reporting Highlights of the De Mayerne Programme: Research Programme on Molecular Sutdies in Conservation and Technical Studies in Art History*, edited by J J Boon and E S Ferreira, 21–31. The Hague: NWO.
10. Boon, J. J, and Hoogland. F.G., 2014. "Investigating Fluidizing Dripping Pink Commercial Paint on Van Hemert's Seven-Series Works from 1990–1995." In *Issues in Contemporary Oil Paint*, edited by K J Van den Berg, M de Keijzer, Alberto de Tagle, Jay Kreuger, and G Heydenreich, 227–246. Switzerland: Springer.
11. Burnstock, A., K J van den Berg, S. de Groot, and L. Wijnberg. 2006. "An Investigation into Water Sensitive Oil Paints in Twentieth-Century Paintings." Edited by T Learner, J W Smithen, and M R Schilling. *Modern Paints Uncovered: Proceedings from the Modern Paints Uncovered Symposium*. Los Angeles: Getty Conservation Institute.
12. Burnstock, A., and van den Berg., K.J. 2014. "Twentieth Century Oil Paint: The Interface Between Science and Conservation and the Challenges for Modern Oil Paint Research." In *Issues in Contemporary Oil Paint*, edited by K J van den Berg, A Burnstock, Matthijs de Keijzer, Jay Krueger, Tom Learner, Alberto de Tagle, and Gunnar Heydenreich, 1–19. Switzerland: Springer.
13. Carlyle, L., and Witlox. M., 2005. "Historically Accurate Reconstructions of Artists' Oil Painting Materials." In *Art of the Past – Sources and Reconstructions*.
14. Chung, J. Y., B. Ormsby, J. Lee, A. Burnstock, and K. J. Van den Berg. 2017. "An Investigation Options for Surface Cleaning Unvarnished Water-Sensitive Oil Paints Based on Recent Developments for Acrylic Paints." In *ICOM-CC 18th Triennial Conference Preprints, Copenhagen, 4–8 September 2017*, edited by J Bridgland. Paris: International Council of Museums.
15. Colombini, M., F. Modugno, R. Fuoco, and A. Tognazzi. 2002. "A GC-MS Study on the Deterioration of Lipidic Paint Binders." *Microchemical Journal* 73: 175–185.
16. Cooper, A. 2012. "Water-Sensitive Oil Paints in the 20th Century A Study of the Distribution of Water-Soluble Degradation Products in Winsor & Newton Artists' Oil Colour Paint Swatches, with Case Studies from Tate's Collecion." Final Year Dissertation, Postgraduate Diploma, Courtauld Institute of Art, University of London.
17. Cooper, A., A. Burnstock, K. J. van den Berg, and B. Ormsby. 2014. "Water Sensitive Oil Paints in the Twentieth Century: A Study of the Distribution of Water-Soluble Degradation Products in Modern Oil Paint Films." In *Issues in Contemporary Oil Paint*, edited by Klaas Jan van den Berg, Aviva Burnstock, Matthijs de Keijzer, Jay Kreuger, Tom Learner, Alberto de Tagle, and Gunnar Heydenreich. Switzerland: Springer.
18. Hermans, J. J., K. Keune, A. van Loon, R. W Corkery, and P.D. Iedema. 2016. "Ionomer-like Structure in Mature Oil Paint Binding Media." *RSC Advances* 6: 93363–93369.

19. Izzo, F.C., K.J. van den Berg, H. Van Keulen, B. Ferriani, and E. Zendri. 2014. "Modern Oil Paints – Formulations, Organic Additives and Degradation: Some Case Studies." In *Issues in Contemporary Oil Paint*, edited by K J van den Berg, A Burnstock, M de Keijzer, Jay Kreuger, T Learner, Alberto de Tagle, and G Heydenreich, 75–104. Switzerland: Springer.
20. King, A., Townsend, J., Ormsby, B., 2017(a) "Nude Woman in a Red Armchair 1932 by Pablo Picasso" in *Tate Papers* no.28 Autumn 2017, URL: https://www.tate.org.uk/research/publications/tate-papers/28/picasso-nude-woman-red-armchair (accessed 29.3.19)
21. King, A., Townsend, J., Ormsby, B., 2017(b) "The Three Dancers 1925 by Pablo Picasso" in *Tate Papers* no.28 Autumn 2017, URL: https://www.tate.org.uk/research/publications/tate-papers/28/picasso-three-dancers (accessed 29.3.19)
22. Labuza, T, and L R Dugan Jr. 1971. "Kinetics of Lipid Oxidation in Foods." *C R C Critical Reviews in Food Technology* 2 (3): 355–405. https://doi.org/10.1080/10408397109527127.
23. La Nasa, J., J. Lee, I. Degano, A. Burnstock, K.J. van den Berg, B. Ormsby, and I. Bonaduce. 2019. "The Role of the Polymeric Network in the Water Sensitivity of Modern Oil Paints." *Scientific Reports* 9 (1): 3467. https://doi.org/10.1038/s41598-019-39963-z.
24. Lee, J., B. Ormsby, A. Burnstock, M. R. Schilling, and K. J. van den Berg. 2017. "The Chemical Characterisation of Water Sensitive Modern Oil Paint Swatches by Winsor & Newton." In ICOM-CC 18th Triennial Conference Preprints, Copenhagen, 4–8 September 2017, ed. J. Bridgland, art. 1604. Paris: International Council of Museums.
25. Lee, J., I. Bonaduce, F. Modugno, J. La Nasa, B. Ormsby, and K.J. van den Berg. 2018. "Scientific Investigation into the Water Sensitivity of Twentieth Century Oil Paints." *Microchemical Journal* 138. https://doi.org/10.1016/j.microc.2018.01.017.
26. Mazzeo, R, S. Prati, M. Quaranta, and E. Joseph. 2008. "Attenuated Total Reflection Micro FTIR Characterisation of Pigment-Binder Interaction in Reconstructed Paint Films." *Anal Bioanal Chem* 392: 65–76.
27. Mills, L., A. Burnstock, F. Duarte, S. de Groot, L. Megens, and M. Bischoff. 2008. "Water Sensitivity of Modern Artists' Oil Paints." *ICOM-CC 15th Triennial Conference New Delhi*. New Delhi: Allied Publishers.
28. Noble, P., and J. J. Boon. 2007. "Metal Soap Degradation of Oil Paintings: Aggregates, Increased Transparency and Efflorescence." In *AIC Paintings Specialty Group Postprints : Papers Pres. at the 34th Annual Meeting of the AIC of Historic & Artistic Works Providence, Rhode Island, June 16–19* , edited by H M Parkin, 1–15. Washington: AIC.
29. Ormsby, B., J. Lee, I. Bonaduce, and A. Lluveras-Tenorio. 2019. "Chapter 2: Evaluating Cleaning Systems for Use on Water Sensitive Modern Oil Paints: A Comparative Study". In *Conservation of Modern Oil Paintings*, ed. K. J. van den Berg, I. Bonaduce, A. Burnstock, B. Ormsby, M. Scharff, L. Carlyle, G. Heydenreich and K. Keune, 11–35. Cham: Springer.
30. Osmond, G. 2012. "Zinc White: A Review of Zinc Oxide Pigment Properties and Implications for Stability in Oil-Based Paintings." *AICCM Bulletin* 33 (1): 20–29.
31. Phenix, A. 2002. "The Swelling of Artists' Paints in Organic Solvents. Part 2, Comparative Swelling Powers of Selected Organic Solvents and Solvent Mixtures." *Journal of the American Institute for Conservation*. https://doi.org/10.2307/3179897.
32. Phenix, A. 2013. "Effects of Organic Solvents on Artists' Oil Paint Films: Swelling." In *New Insights into the Cleaning of Paintings Proceedings from Th Cleaning 2010 International Conference Universidad Politécnica de Valencia and Museum Conservation Institute*, edited by M F Mecklenburg, A E Charola, and R J Koestler, 69–76. Washington D.C.: Smithsonian Institution Scholarly Press.
33. Plater, M John, B. De Silva, T. Gelbrich, M. B. Hursthouse, C. L Higgit, and D. R Saunders. 2003. "The Characterisation of Lead Fatty Acid Soaps in 'protrusions' in Aged Traditional Oil Paint." *Polyhedron* 22: 3171–3179.
34. Ploeger, R. 2008. "The Characterization and Stability of Artists' Alkyd Paints." PhD Thesis. *Department of I.P.M. Chemistry*. University of Torino.

35. Schilling, M. R., D. M. Carson, and H. P. Khanjian. 1998. "Evaporation of Fatty Acids and the Formation of Ghost Images by Framed Oil Paintings." *WAAC Newsletter*. http://cool.conservation-us.org/waac/wn/wn21/wn21-1/wn21-106.html.
36. Silvester, G., A. Burnstock, L. Megens, T. Learner, G. Chiari, and K. J. van den Berg. 2014. "A Cause of Water-Sensitivity in Modern Oil-Paint Films: The Formation of Magnesium Sulphate." *Studies in Conservation* 59 (1): 38–51.
37. Sutherland, K R. 2001. *Solvent Extractable Components of Oil Paint Films*. Amsterdam: FOM Institute for Atomic and Molecular Physics.
38. Bronken, I.A., and J. J Boon. 2014. "Softening Paint and Drip Formation in Paintings by Jean-Paul Riopelle (1923-2002): Improving Their Condition with Metal Coordinating Ions." *ICOM-CC 17th Triennial Conference Melbourne 15-19 September 2014: Preprints*.
39. Tempest, H., A. Burnstock, P. Saltmarsh, and K. J. van den Berg. 2013. "Sensitivity of Oil Paint Surfaces to Aqueous and Other Solvents." In *New Insights into the Cleaning of Paintings Proceedings from Th Cleaning 2010 International Conference Universidad Politécnica de Valencia and Museum Conservation Institute*, edited by M F Mecklenburg, A E Charola, and R J Koestler, 107–114. Washington D.C.: Smithsonian Institution Scholarly Press.
40. Tumosa, C. S. 2001. "A Brief History of Aluminum Stearate as a Component of Paint." *WAAC Newsletter* 23 (3). http://cool.conservation-us.org/waac/wn/wn23/wn23-3/wn23-304.html.
41. Tumosa, C. S., and M. F. Mecklenburg. 2010. "Oil Paints : The Chemistry of Drying Oils and the Potential for Solvent Disruption." *New Insights into the Cleaning of Paintings: Proceedings from the Cleaning 2010 International Conference*. https://doi.org/10.1016/j.sleep.2007.08.003.
42. van Dam, E., van den Berg, K. J., Ness Proaño Gaibor, A., van Bommel, M., 2015. "Analysis of Triglyceride Degradation Products in Drying Oils and Oil Paints Using LC-ESI-MS." *International Journal of Mass Spectrometry* Volume 413, February 2017, Pages 33–42 https://doi.org/10.1016/j.ijms.2016.09.004
43. van den Berg, J.D.J., K. J. van den Berg, and J. J. Boon. 1999. "Chemical Changes in Curing and Ageing Oil Paints." In *ICOM-CC 12th Triennial Meeting*, edited by J Bridgland, 248–253. Lyon: James and James.
44. van den Doel, A., 2015. "Understanding 20th Century Oil Paint Chemistry through ASCA Analysis of a Designed Experiment." Masters Thesis. Amsterdam: Radboud University.
45. Winnats Scientific Services, Winnats. 2018. "Report ID: 1D8UP02 XPS Analysis of Paints and Pigments ."

Chapter 39
Analytical Study to Monitor the Effectiveness of a Combined Liquid-Dispensing and Micro-aspiration System for the Cleaning of Modern Oil Paintings

Antonella Casoli, Paolo Cremonesi, Pierre-Antoine Héritier, and Stefano Volpin

39.1 Introduction

Aqueous systems frequently offer the most appropriate approach to the surface cleaning of painted artworks. Furthermore, adjustments to aqueous conditions such as pH and conductivity can enhance activity, facilitating tasks such as: the removal of film forming materials including varnishes and protective coatings, retouching and overpaint, as well as residual adhesives and consolidants. For this latter task, traditional approaches rely on organic solvents [10, 14]. Fortunately, aqueous based systems, which are intrinsically devoid of toxicity, are characterized by a higher number of 'control parameters' which facilitate the development of safer (lower-risk) treatments for artworks. Unfortunately, contemporary oil-based paint media often exhibit pronounced sensitivity to water, with several causal factors. For example, extensive medium hydrolysis, resulting in the formation of low-molecular weight, hydrophilic fragments and the use of additives in paint formulations such as magnesium carbonate, can enhance the hydrophilic character of the paint [1, 3, 16].

Devising an appropriate aqueous treatment therefore requires two distinct, sequential phases: firstly, tailoring the aqueous phase to achieve the most appropriate composition in terms of pH, the type and concentration of ionic components, and possible 'adjuvants' such as chelators and surfactants [19]; and secondly, and

A. Casoli
Department of Chemistry, Life Sciences and Environmental Sustainability, University of Parma, Parma, Italy

P. Cremonesi (✉)
Freelance Conservation Chemist, Lodi, Italy

P.-A. Héritier
Freelance Conservator, Atelier Héritier sàrl, Geneva, Switzerland

S. Volpin
Superintendency for Cultural Heritage, Trento, Italy

K. J. van den Berg et al. (eds.), *Conservation of Modern Oil Paintings*,
https://doi.org/10.1007/978-3-030-19254-9_39

perhaps even more crucially, devising a safe means of delivering the aqueous solution to the paint surface. We are fortunate that Wolbers, alongside others, have provided valuable guidance on the most relevant factors to help devise appropriate, low-risk aqueous solutions [19].

This paper contributes to the second aspect: a safer way of delivering the aqueous solution to the painted surface. This approach has been informed through research focused on contemporary paint media, including important water diffusion studies [11, 13, 18], which illustrate the need to minimise contact time between water and the paint surface. The proposed application method relies on the combination of delivering a precisely controlled amount of the tailored aqueous cleaning solution onto the paint surface via a power-driven syringe infusion pump, followed by immediately collecting it through a flexible tubing connected to a surgical micro-aspirator [5]. Micro-aspiration enables the quick and efficient removal of the aqueous phase from the surface, leaving it dry to the touch very soon after treatment [4].

The apparatus is composed of various tools used in the medical/personal health care sectors including: an infusion pump as the liquid-dispensing unit and a surgical micro-aspirator to remove the cleaning solution from the surface. These two instruments can be fitted with a variety of tools such as: extension tubing, simple brushes, luer-lock brushes, needles of various diameters, aspiration tips made with different materials, which may even be custom made via 3D-printing [5].

In a previously introduced apparatus [4], the liquid dispensing unit consisted of a simple, battery powered syringe driver. Although effective and convenient, this unit suffered drawbacks such the advancing of the plunger syringe in discrete steps, which meant that the operator had to adjust to a discontinuous delivery of liquid. To amend this, a larger AC powered syringe pump was later adopted, the Graseby 3100 (Smiths Medical), as shown in Fig. 39.1. This unit facilitates a continuous action on the syringe plunger, thus providing a more precisely controlled flow of liquid. As per the previous

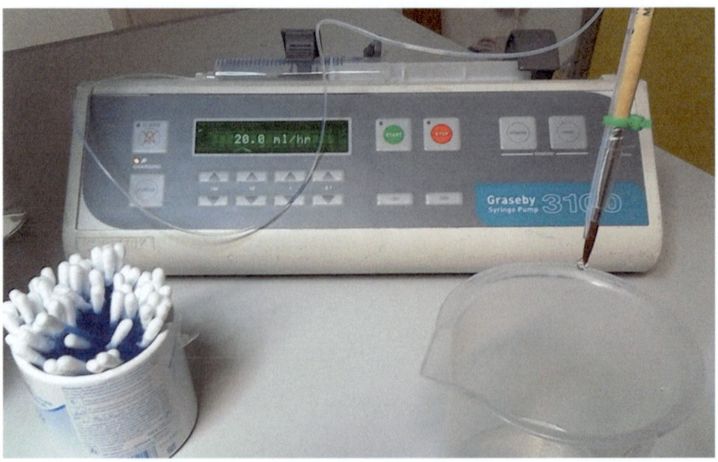

Fig. 39.1 The syringe driver used as controlled liquid-delivery unit

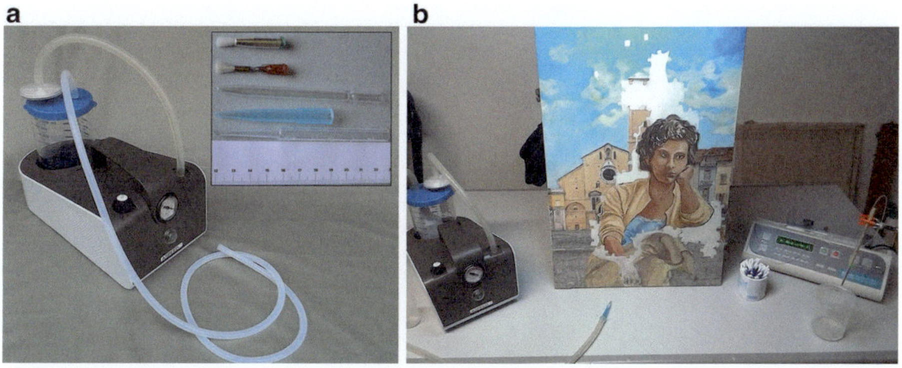

Fig. 39.2 The surgical micro-aspirator with different aspiration tips, the test painting and the working setup

setup, the syringe is connected to luer-lock fine-bore Teflon™ tubing, which delivers the aqueous solution onto the ferrule of the 'cleaning brush'; which can be any type and size of brush connected to the delivery unit. The single-pump surgical aspirator, Aspeed 2 (Alma Medical), is shown in Fig. 39.2. This instrument operates quietly (noise 55 dBA) and provides 15 l/min air supply. A 1 ml Eppendorf pipette tip inserted into the open end of the silicone tubing is then used to micro-aspirate the surface.

Analytical investigations were planned for two purposes: to characterise the binding medium and the pigments present in the test areas and to monitor interactions with the paint with respect to the aqueous leaching of pH-sensitive mobile components and pigment pickup. Investigations compared both traditional swab-rolling and micro-aspiration techniques, supported by visual and microscopic examination, on an ~40 year old unvarnished oil painting, and of an additional group of young, water-sensitive oil-based paint films which had been subjected to the same treatment.

Previous testing of this apparatus on sensitive paint layers, in comparison with traditional swab-rolling, demonstrated an effective soiling removal action [4]. Furthermore, unlike the swab cleaning tests, paint pickup and surface abrasion were not detected when the aspirated samples were examined under the microscope. Further analytical characterization was then planned for this study, to explore the perception of the surface being immediately dry after treatment, and any risks associated with the procedure.

39.1.1 Aims of the Present Study

For this study, an unvarnished oil painting on canvas created by one of the authors in the late 1970s was selected, which had been naturally aged and soiled in a domestic environment for almost 40 years (Fig. 39.3). Paint components were identified in selected areas via: non-invasive identification of pigments by X-Ray Fluorescence spectroscopy (XRF) and micro-invasive analysis of the layer structure via two

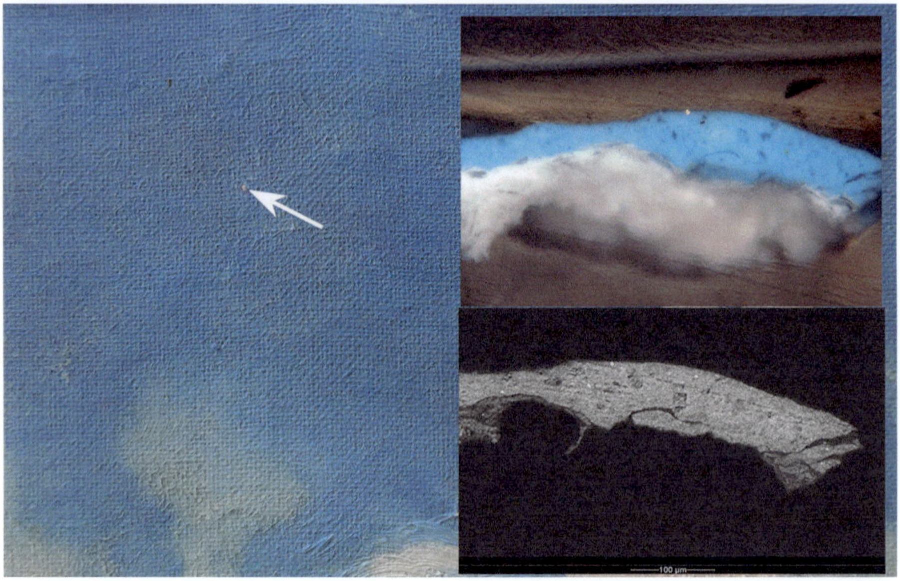

Fig. 39.3 Detail of the sampling point (left) and cross-section images by light microscopy (top) and by SEM (bottom)

cross-sections, and the paint medium via Gas Chromatography-Mass Spectrometry (GC/MS). This information was also used to verify the effectiveness of the cleaning method (as described later). Following the characterization of the binding medium, the painting was surface-cleaned with an aqueous solution, comparing two different techniques: traditional swab-rolling and the controlled delivery of liquid micro-aspiration. A simple 'trap' was devised for detecting any potential pigment loss in the latter where a hollow, plastic cotton swab was inserted into the aspiration tip. This cotton tip, together with one from the swab rolling procedure were analyzed by XRF. The areas cleaned using the two procedures were then sampled for GC/MS analysis, to ascertain whether the cleaning procedures had removed any original organic material. Only a few cleaned areas were sampled and analyzed for this work, therefore these results may not prove the 'absolute safety' of this technique. However, they suffice to demonstrate that it is 'safer' than the more traditional swab-rolling application with respect to the parameters evaluated.

39.2 Materials and Methods

39.2.1 Tailoring the Aqueous System

Surface pH and conductivity were also measured via an Agarose gel pellet, revealing a neutral, low conductivity surface. The stock solution was then prepared as a 12 mM potassium citrate aqueous solution, buffered to pH 8.5. The cleaning solution was loosely based on these surface parameters and obtained by diluting the high-conductivity solution to a theoretically hypertonic condition (about 15× the painting surface conductivity) ([19], p. 161). Although "*...the link between sensitivity and swelling of the paint film in the case of oil films is not as clear as it is with acrylics*" [17] hypertonic conditions were still adopted in this specific study, aiming to promote a potential greater interaction of the aqueous cleaning solution with a paint film of higher hydrophilic character caused by the high, ionizing pH deliberately chosen. In the expectation of the authors, these drastic conditions could enhance any differences between the two application methods in terms of relative degree of interaction with the paint layer. The cleaning solution was applied onto the surface through two cotton swabs, pre-dried on blotting paper from excess solution, and gently rolled for 10 s each; alternatively, it was delivered onto the 'cleaning brush' via the syringe pump set at a constant rate of 5 ml/h, with simultaneous micro-aspiration. The brush was gently stirred on the surface for 20 s.

39.2.2 Samples

For the preliminary investigation, two micro-samples were collected from the painting to prepare cross-sections [15], which were prepared by embedding the micro-samples in cold-cured polyester resin (Glass Resin – Colorchimica s.p.a.). Cross-section analysis was performed using visible and ultraviolet (UV) light microscopy, followed by Scanning Electron Microscopy (SEM) coupled with an electronic microprobe (EDS). Another 3 scraping micro-samples were taken from the painting surface, of different colours, for the analysis of the organic binding media, using Gas Chromatography – Mass Spectrometry (GC/MS).

39.2.3 Instrumentation

39.2.3.1 Optical Microscopy (OM)

A Nikon TK-1270E optical microscope was used in reflected light. Dark field observations were performed using fixed oculars of 10× and objectives with different magnifications (5, 10, 20 and 40×). Photomicrographs were recorded using an Olympus DP70 digital scanner camera directly connected to the microscope.

39.2.3.2 Energy Dispersive X-Ray Fluorescence (ED-XRF)

X-ray fluorescence analysis (ED-XRF) was carried out using XRF Niton XL6t GolD+ in the following modalities: silver anode; maximum RX tube voltage of 50 kV, with the possibility of differentiated spectra at low (6 kV), medium (20 kV) and high (50 kV) voltages to better differentiate light elements; measuring area:

3 mm in diameter; acquisition time: variable from 60 to 120 s; detector: Peltier-cooled silicon multichannel, beryllium window; detectable elements: Z > 12, i.e. from magnesium onwards.

39.2.3.3 Electronic Microscopy with Electronic Microprobe (SEM-EDS)

A SEM (Philips XL 20 SEM) equipped with an energy dispersive X-ray spectrometer (Phillips EDAX PV9900) with a light element detector type ECON was initially used. An environmental SEM (Philips Quanta FEI 200) equipped with an energy dispersive X-ray spectrometer (EDX) by Link Analytical Oxford (Link, UK), model 6103 was also used. The samples prepared for optical microscopy were analysed, some having been gold-coated. Several analyses were performed on each layer of the cross-sections and the elemental analysis was carried out at 30 kV acceleration voltage, with a 100 s lifetime and 30 mm working distance. The samples examined in low vacuum, included a chamber pressure set at 23 Pa.

39.2.3.4 Gas Chromatography – Mass Spectrometry (GC/MS)

The analytical procedure used was based on those described by Bersani et al. [2] for the analysis of fatty acids, involving derivatisation methyl esters. Paint samples of 0.8 mg each were selected, and an internal standard was used for quantitative results: 100 μl 2,3 mM tetracosane, tR = 21,2. The sample was treated with 4 N-HCl in methanol (1 ml) and n-hexane (1 ml) for 2 h at 50 °C, with addition of 100 μl 2,3 mM tetracosane (internal standard). The n-hexane phase, which contains fatty acid methyl-esters, was used for gas chromatographic analysis (1 μl). Separation of the methyl ester of fatty acids was done following this temperature program: isothermal at 80 °C for 2 min, with 20 °C/min heating up to 280 °C and isothermal at 280 °C for 6 min (total run time 18 min). The samples were analysed using a Focus GC (Thermo Scientific) equipped with a split-splitless injector, coupled to DSQ II (Thermo Scientific) with single quadrupole. The mass spectrometer was operated in the EI positive mode (70 eV). The carrier gas was used in constant flow mode (He, purity 99.995%) at 20 mL/min. The column was an RXI-5 (Restek), with a 0.25 μm (30 m × 0.25 mm × 0.25 μm) methyl-silicone (5% phenyl) coating and the injector used in splitless mode. The mass spectrometer was operated in EI positive mode (70 eV), over a mass range of 40–500 m/z. Chromatograms were acquired both in total ion chromatogram (TIC) mode and selected ion monitoring (SIM) mode.

39.3 Results and Discussion

In the area investigated, the combined analytical campaign revealed that the 40-year old painting structure is as follows (Fig. 39.4): in direct contact with the textile support is a white ground composed of silicates and titanium white, with traces of zinc white

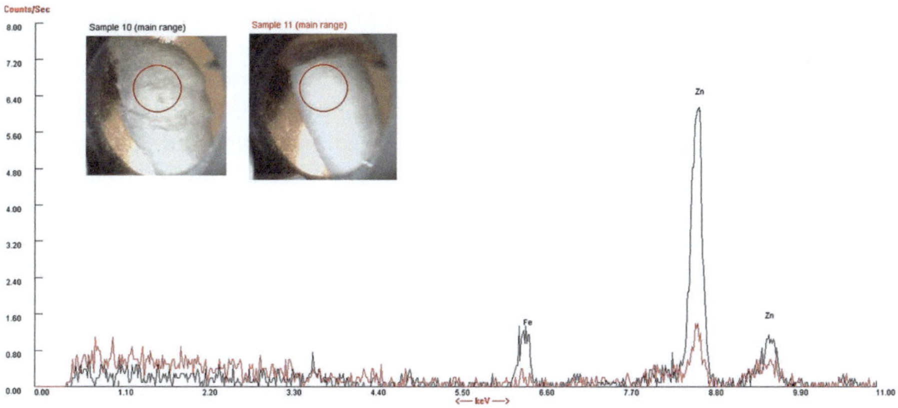

Fig. 39.4 Comparison of the XRF spectra from traditional swab-rolling (sample 10 – black line) and swab inserted into the aspiration tip (sample 11 – red line)

Table 39.1 Fatty acid composition of the original blue paint sample and of samples 12 and 13, calculated in relation to the tetracosane internal standard

Fatty acid	t_R	Original blue paint %	Sample 12 (swab cleaned) %	Sample 13 (brush applied and micro-aspirated) %
Azelaic acid	10.21	9.1	3.07	3.26
C14	10.88	4.0	0.04	0.14
C15	11.78	2.6	0.03	0.06
C16:1	12.69	2.9	0.01	0.02
C16	12.91	40.4	3.56	7.16
C18:1	16.19	5.7	0.25	0.51
C18	16.69	26.4	5.85	8.85
C20	20.18	3.5	0.25	0.33
C22	22.14	5.4	0.16	0.21

(XRF analysis identified silicon, titanium and zinc). This is followed by the light blue color, which is composed of a mixture of zinc white, an organic blue pigment on calcite, and a minor quantity of titanium white. The paint binder was confirmed as lipid through the presence of several fatty acids in the chromatographic profile.

Table 39.1 lists the fatty acids identified in the original blue paint scraping, including azelaic acid, however, specific characterization of the oil source was not possible due to the common occurrence of these fatty acids in most oils of interest. Nonetheless, two unsaturated fatty acids (palmitoleic and oleic acids) were also identified, together with several saturated fatty acids, including archidic acid (C20) and behenic acid (C22): the latter two possibly indicating the presence of safflower oil (*Carthamus tinctorius*) which is a commonly used oil binder for paints after the 1960s [1]. Based on these findings, the historical drying oils, such as linseed, poppy seed and walnut oil [12] could not be identified, however they may be present as a minor component.

Two paint fragments were also collected for GC/MS analysis from the cleaned areas of the painting: sample 12 s was taken from the swab-cleaned area and sample 13 s was taken from the micro-aspirated area. The chromatographic profiles of these two samples are similar to the profile of original paint sample. The relative amounts of fatty acids in sample 12 s, from the swab-cleaned area, compared to the tetracosane internal standard were lower than those in sample 13 s, from micro-aspirated area, as also shown in Table I. Though only one sample was taken from each area, the results suggest that the swab cleaning (sample 12) resulted in the reduction of lipids, which can be interpreted as the leaching of a larger amount of lipid material from the swab-cleaned areas when compared to the more controlled delivery via micro-aspiration (sample 13).

The XRF data (Fig. 39.4) obtained from the cleaning swabs (for both methods) concur with the GC/MS analysis in the sense that the XRF spectra taken from the swab-rolled cotton tip (sample 10) and of those inserted into the aspiration tip (sample 11) reflect a similar pattern. It is clear that the zinc signals, attributed to the white pigment pickup, are very different. Here the low-energy XRF spectra (which increases detection of light elements) confirms this trend: the amounts of zinc, calcium, iron, sulfur, barium and silicon are significantly higher (around 4×) in the rolled cotton tip as compared to the cotton inserted into the aspiration tip.

39.4 Further Developments

The use of this type of system may be advantageous for the aqueous cleaning of painted surfaces, and particularly works of art with more hydrophilic surfaces, whether contemporary, or more traditional paintings. In any case it is necessary to control the amount of water used in order to avoid or reduce risks such as the mechanical retraction of the support and ground, swelling, blanching, as well as medium hydrolysis and/or leaching. Additional risks with the use of excessive moisture include the possibility of opening cracks, which can encourage realignment forces [8]. It is also known [7] that dry paint films have residual internal tensions, which undergo interfacial phenomena due to the surface tension and capillarity action of water. These combined considerations prompted investigations aimed at developing a method for the controlled delivery of aqueous solutions (with/without adjusted pH and conductivity, chelating agents and other materials). This innovative micro-aspiration technique facilitates control over fluid dynamics, as well as contact with the support during application. This system combines ergonomics with significantly decrease risk to paint surfaces, particularly when compared to traditional techniques such as cotton swab rolling.

In an effort to improve the system further, a more precise and versatile instrument was evaluated and eventually selected; consisting of a manual pneumatic dispenser with a digital mechanism (Pneumatic Airpulse Dispenser), capable of working in a pre-set or manual cycle for the application of fluids or pastes. A digital readout between 0.1 and 9.9 s allows speed adjustment, facilitating control of the quantity of aqueous material delivered. Specific devices have been included in Fig. 39.5: models 8000D or FU-180. The dispenser is powered by compressed air, which is purified through an on-line water-oil filter. The aqueous solutions are contained in a syringe-tanks of various capacities, fitted with polyethylene pistons with double wiping edges.

The correct use of a micro-aspiration system combined with a simultaneous controlled delivery of an aqueous cleaning solution also requires the correct choice of accessories, such as tools-brushes, where choice is based on brush softness. For more precise control of the application time, a more advanced tool was engineered (and 3D printed), combining a set of aspirated brushes and blowers powered by compressed air purified through a triple air-filtration system, quality ISO 8573-1: 2010 (Fig. 39.5). Several prototypes made via 3D printing allowed the final design of this new tool: each exploiting the same principle of the double-chamber. The first chamber allows gentle aspiration through six paired luer-lock *needle-paintbrushes* (14G), made of soft nylon hair, where a thin space in the centre of each brush promotes easier aspiration. The second chamber, inserted inside the same unit, conveys a stream of purified air which can be regulated to speed up the drying of residually humid areas of the paint surface. This tool enables higher control and reduces the risks of tidelines or "coffee-ring effect" [6]: a capillary flow causing ringed stains from dried liquid drops combined with suspended dirt particles, conveyed towards the edges of the drop. This is a result of the high capillary flow, often inherent to the cleaning of matte, porous paint layers. The precise control over both the amount of applied water and its retention considerably reduce the risks of blooming of the paint layer and the swelling of deeper layers.

Several tests and verifications were successfully conducted either on paintings bought at flea markets or on painting fragments, which shared significant sensitivity

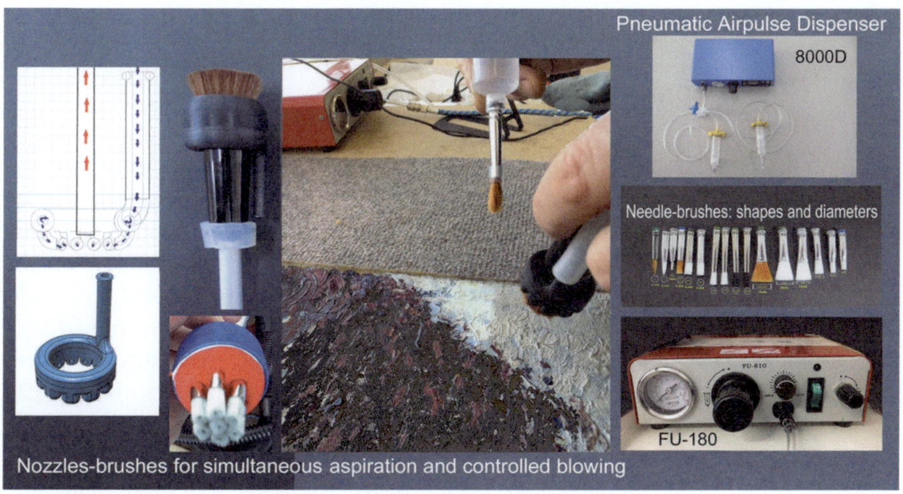

Fig. 39.5 The pneumatic dispenser with accessories

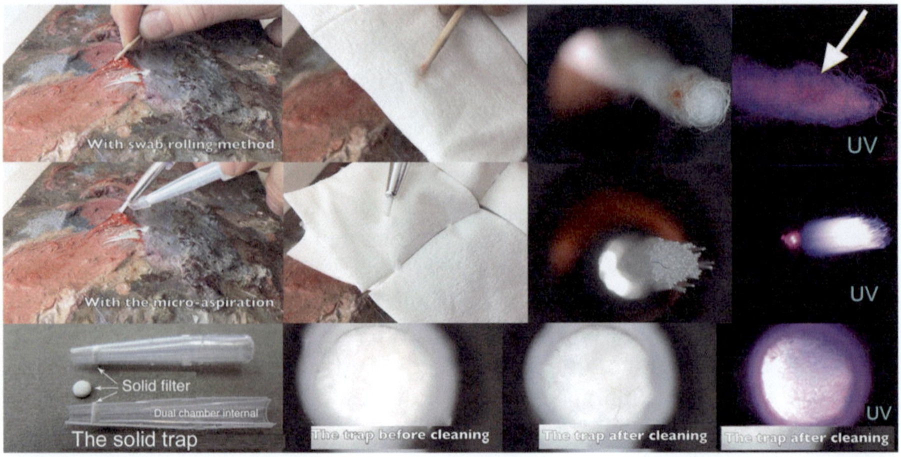

Fig. 39.6 Monitoring the mechanical impact on oil paint before and after cleaning by swab-rolling and combined liquid-dispensing and micro-aspiration system: naked-eye and microscope examination under visible and UV light

to water and to leaching. The comparative testing on the original oil-color palette of Bram Van Velde, ca. 1965 (results not shown here) is one example among the ten surfaces that we have submitted to both cleaning methods (Fig. 39.6).

The risk of colour extraction, as a result of swelling and hydrolysis of the painted surface, could be monitored and comparatively assessed with both techniques, swab rolling and micro-aspiration/controlled fluid delivery. Evidence was collected by simply comparing, under the microscope and with UV illumination, the residues trapped (or not) in a 'solid trap' specially devised for this study, and the visual appearance of the cotton swabs.

39.5 Conclusions

The combination of the controlled delivery of aqueous cleaning solutions and micro-aspiration offers a novel way of using aqueous systems on water sensitive surfaces that can minimize the risk of leaching and surface disruption. To explore interactions such as leaching and surface disruption, this study focused on the impact of mechanical action during cleaning, comparing the micro aspiration technique with traditional swab rolling. Among a range of methods that we applied after the current study to improve the examination of the treated paint surfaces, were microphotography, visible and UV examination, and microscopic examination [9]. Results from the evaluations discussed as part of this ongoing work suggest that the micro aspiration technique effectively reduced the amount of fatty acids leached out from the surface of the oil paint areas evaluated when compared to traditional swab cleaning. This leads to a lower levels of pigment pickup and surface change, thanks to the decreased contact time of the aqueous solution with the paint surface.

This system may also be of use with organic solvents for removing film forming materials such as varnishes, binding media, and adhesives through the modification of specific components, such as a Teflon™ coated membrane pump for the aspiration of organic solvents. Different tools could also be used for the delivery of the solvents to paint/coating surfaces including an all-plastic syringe, with no rubber ring at the end of the plunger which is compatible with organic solvents of different polarities. Alternatively, a pneumatic dispenser manufactured for the application of adhesives [5] could be used. Most recently, a prototype solvent dispenser has been custom made from an appropriately fitted steel safety can, connected to an argon cylinder as the solvent propelling gas; and analytical studies are in progress to monitor the performance of this controlled delivery/micro-aspiration system for use with organic solvents.

References

1. Banti D, Nasa JA, Lluveras Tenorio A, Modugno F, van den Berg KJ, Lee J, Ormsby B, Burnstock A, Bonaduce I (2018) Molecular study of modern oil paintings: investigating the role of dicarboxylic acids in the water sensitivity of modern oil paints. RSC Advances 8: 6001–6012
2. Bersani D, Berzioli M, Caglio S, Casoli A, Lottici PP, Medeghini L, Poldi G, Zannini P (2014) An integrated multi-analytical approach to the study of the dome wall paintings by Correggio in Parma Cathedral. Microchemical journal, 114: 80–88
3. Cooper A, Burnstock A, van den Berg KJ, Ormsby B (2013) Water sensitive oil paints in the 20th century. A study of the distribution of water-soluble degradation products in Winsor & Newton Artists' oil colour paint swatches, Wight case studies from Tate's collection. In: Issues in contemporary oil paint (ICOP). Book of abstracts. Amersfoort, The Netherlands, Cultural Heritage Agency, Ministry of Education, Culture and Science, p 95–97
4. Cremonesi P (2018) Combination of a liquid-dispensing and micro-aspiration device for the cleaning of sensitive painted surfaces. Studies in Conservation 63 (6): 315–325
5. Cremonesi P, Héritier PA (2017) Un approccio innovativo alla pulitura di superfici dipinte sensibili: la combinazione simultanea di erogazione controllata di liquido e micro-aspirazione. Il Prato, Padova
6. Deegan RD (1997) Capillary flow as the cause of ring stains from dried liquid drops. Nature 389: 827–829
7. Delcroix G, Havel M (1988) Phénomènes physiques et peinture artistique, Paris, EREC
8. Fuster Lopez L, Mecklenburg MF, Castell M, Guerola V (2008) Filling materials for easel paintings: when the ground reintegration becomes a structural concern, in: *Preparation for Painting: The Artist's Choice and its Consequences*, edited by Townsend, JH, Doherty, T, Heydenreich, G, and Ridge, J, 180–186. London: Archetype Books
9. Héritier PA (2018) A study on the impact of cleaning processes by swab rolling and combination of a Liquid-Dispensing and Micro-Aspiration Device for the Sensitive Painted Surfaces as Picasso's. Paper presented at the symposium *Around Picasso – An insight into the relationship between material choices and failure mechanisms*, November 29 2018 Museu Picasso, Barcelona, Spain
10. Masschelein-Kleiner L (1991) Les Solvants. Brussels: IRPA-KIK
11. Michalski S (1990) A physical model of the cleaning of oil paint, in: Cleaning, retouching and coatings. Technology and practice for easel paintings and polychrome sculpture edited by Mills J.S. and Smith P. 85–92. London: IIC

12. Mills JS (1966) The gas chromatographic examination of paint media. Part I. Fatty acid composition and identification of dried oil films. Studies in Conservation, 11: 92–108
13. Ormsby B, Learner T (2009) The effects of wet surface cleaning treatments on acrylic emulsion artists' paints – a review of recent scientific research. Reviews in Conservation 10: 29–41
14. Phenix A, Sutherland K (2001) The cleaning of paintings: effect of organic solvents on oil paint films. Reviews in Conservation 2: 47–60
15. Plester J (1956) Cross-sections and chemical analysis of paint samples. Studies in Conservation 2: 110–157
16. Silvester G, Burnstock A, Megens L, Learner T, Chiari G, van den Berg KJ (2014) A cause of water-sensitivity in modern oil paint films: the formation of Magnesium Sulphate. Studies in Conservation 59 (1): 38–51
17. van den Berg KJ, Soldano A (2014) Investigation into the Surface Conductivity of Water-Sensitive Modern Oil Paints, in *Issues in Contemporary Oil Paint*, edited by Burnstock A, Matthijs de Keijzer M, Krueger J, Learner T, de Tagle A, Heydenreich G, and van den Berg KJ. 407–17. Switzerland: Springer International Publishing
18. Withmore, PM, Morris, HR, Colaluca, VG (2007) Penetration of liquid water through waterborne acrylic coatings, in *Modern Paints Uncovered*, edited by. Learner TJS, Smithen P, Krueger J. and Schilling MR. 217–23. Los Angeles, CA: The Getty Conservation Institute
19. Wolbers R (2000) Cleaning Painted Surfaces. Aqueous Methods. Archetype Publications, London

Chapter 40
Ketone Resins Varnishes on Canvas Paintings from the Collection of the Munch Museum: An Interdisciplinary Transnational Research Project

Eva Storevik-Tveit, Irina Crina Anca Sandu, Charlotte Nora Stahmann, Klaas Jan van den Berg, Ineke Joosten, Joy Mazurek, Michael Schilling, Giancarlo Lanterna, Oriana Sartiani, and Sony George

40.1 Introduction

In the middle of the twentieth century, conservators sought more stable materials for varnishing paintings that had optical properties similar to traditionally used triterpenoid resins [22, 23, 28]. Natural resins display excellent optical and surface coating properties [10, 31, 38], but after long-term exposure to light, radiation, temperature fluctuation, oxygen and pollution they turn yellow, cross-link and show solubility in increasingly polar solvents [10, 11, 38].

In response to these issues, synthetic low molecular weight (lmw) resins such as ketone resins were introduced as a promising alternative to natural resins [5, 11].

In 1918 the Badische Anilin- & Soda-Fabrik in Ludwigshafen firstly patented the production of a cyclohexanone resin by the condensation of cyclohexanone or its derivatives with formaldehyde (Fig. 40.1a) (patent from [4]). Following this, AW2®

E. Storevik-Tveit · I. C. A. Sandu (✉)
Conservation Department, Munch Museum, Oslo, Norway
e-mail: irina.sandu@munchmuseet.no

C. N. Stahmann
University of Applied Science Cologne, Cologne, Germany

K. J. van den Berg · I. Joosten
RCE – Cultural Heritage Agency of the Netherlands, ZC, Amsterdam, The Netherlands

J. Mazurek · M. Schilling
Getty Conservation Institute – Laboratory of Material Characterization, Los Angeles, CA, USA

G. Lanterna · O. Sartiani
Opificio delle Pietre Dure, Fortezza da Basso, Florence, Italy

S. George
NTNU-The Norwegian Colour and Visual Computing Laboratory, Gjøvik, Norway

K. J. van den Berg et al. (eds.), *Conservation of Modern Oil Paintings*,
https://doi.org/10.1007/978-3-030-19254-9_40

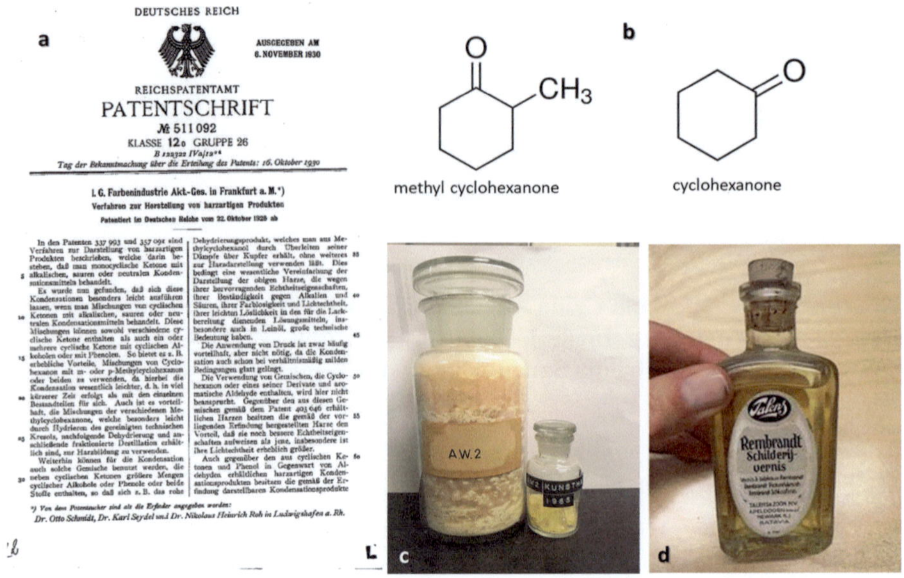

Fig. 40.1 (**a**) Image of the first page of the patent for ketone resins, no. 339107/1.7.1918; (**b**) chemical formula of the 2 monomers; (**c**) AW2® from the SMK, Denmark (obtained from the Archlab access in 2017); (**d**) varnish from the collection of Munch Museum (Talens brand)

resin was manufactured by the polymerisation of cyclohexanone and methyl cyclohexanone in the presence of an alkaline catalyst [1, 21].

Ketone resins [8, 15, 29, 40] consist of oligomeric compounds of 6–10 monomeric units of cyclic structure and an average molecular weight of 600–800 Da [23]. They are products of a self-condensation reaction of 2 monomers (Fig. 40.1b), cyclohexanone and methyl cyclohexanone (AW2, MS2®) or of cyclohexanones only (Ketone Resin N®, Laropal K8®) [12, 23].

The photosensitive carbonyl groups in cyclohexanones are prone to oxidation [8], therefore the manufacturers reduced them to hydroxyls for obtaining more stable materials (MS2A®, MS2B®) [34].

Besides the carbonyl groups, cyclohexanone resins have functional groups that are vulnerable to oxidation, including ether groups, the tertiary carbon atoms (alkyl radicals are produced and subsequently react with oxygen to originate peroxy radicals), hydrogen atoms in α-position to the carbonyl group (easily abstracted to produce alkyl radicals) and double bonds between carbon atoms. Hence, propagation reactions take place both in the unreduced and reduced cyclohexanone resin types [8].

The ageing of cyclohexanone may show similarities with that of natural resins. The literature reports the radical mechanisms (Norish I and Norish II reactions) at the basis of photochemical aging of ketone coatings (especially the one generated by UV radiation), as also for natural resins (i.e., Dammar) [8, 12]. Reactions of main-chain scission were considered as dominating (Norish II), leading to free unstable macro-radicals. More recent studies [14, 39] showed that Norrish type ageing products are not found in natural ageing of natural triterpenoid resins because natural ageing of varnishes generally does not involve UV radiation in controlled environments where UV filters are used. In fresh triterpenoid resins, large amounts of radicals develop within weeks after application as a varnish due to the greatly increased surface to volume ratio [14].

When UV-based ageing is followed by thermal aging, the number of unsaturated carbonyl compounds increases. Unstable aldols derived in the course of condensation reactions dehydrate, forming α,β-unsaturated aldehydes and ketones. Photochemical and thermal aging performed simultaneously, create conditions for the formation of similar chromophore derivatives. During further aging, isolated double bonds rearrange making a conjugated system of double bonds, which intensifies the yellowness of the coating [8, 12].

Dietemann et al. [14] and van Doelen et al. [39] showed that for triterpenoid varnishes new compounds responsible for yellowing are formed both in light and in darkness, but seem to be more pronounced in darkness since they are not photobleached in light. Upon ageing, the amount of analysable materials as measured by GC/MS decreases since such excessively oxidized compounds cannot be identified with this analytical method [14]. This behaviour can be seen as a similar in ketone varnishes due to similarities in the chemical composition with the natural resins as triterpenoids and hence the difficulty to characterize the aged coatings of these resins when UV radiation is not involved.

40.2 AW2 as Varnish Material for Paintings and Conservation

Initially AW2 was developed as a substitute for less stable natural resins, used in the printing ink and lacquer production, to increase gloss and hardness [13]. It was favoured in industry for properties including solubility in a wide range of solvents [17], compatibility with other film-forming materials, its pale hue and lack of yellowing, adhesion to metal surfaces, weathering properties and the fact that it is non-saponifiable [21]. In 1967 BASF stopped the production of AW2 resin. In Germany the remaining stock of AW2 resin was sold to conservators, at latest until 1971 [23].

In conservation field AW2 resin was used as a retouching varnish, final varnish, retouching medium [23] or as an addition to lining adhesives [40] mainly used between the 1950s and the 1970s. It was applied both as self-prepared resin in solvents and as a ready-made varnish containing AW2 which may have been Talens "Rembrandt Picture Varnish"[1] mentioned in the conservation records (see ***Collection Survey*** below). For improving the applicability, its filming and visual properties and

[1] Talens Picture Varnish consisted of AW2® at least from 1963 and Ketone resin N from 1969, according to the Talens' archival records (KJvdB, accessed 30 August 2016).

ageing as coating, different additives were used for AW2 resins, such as beeswax, castor oil, bleached stand oil [9, 27, 33].[2]

AW2 resin shares similar visual appearance and physical properties with triterpenoid resins [10] and fulfils many of the criteria required for conservation materials, being easily applied by brush or spray. Even in high concentrations (up to 50% solids content) the varnishes make solutions of low viscosity that can level evenly on the paintings' surface forming a glossy film [13].

As drawbacks, the material's high gloss, brittleness and consequently its sensitivity to mechanical stress were mentioned.

40.3 The AW2 Varnish Project at Munch Museum

Two questions were posed when the topic of ketone resins varnishes came out at Munch Museum: *a) Why did conservators varnish Munch's paintings with ketone resins? b) Were matte surfaces Munch's original intention?*

It was known quite early that Munch preferred to keep the surfaces matte [25, 26].[3] Later analyses and visual examinations indicate that Munch used a number of different materials and application techniques to achieve this effect [35].

Between the 1880s and the early 1900s Munch lived in Europe [16, 24, 41], where he became familiar with both his predecessors and contemporary artists' use of materials and artistic expressions, many preferring matte surfaces. There was an increasing interest during the late nineteenth century in imitating fresco paintings [7]. The Impressionists tried different techniques to produce lean oil colours to create these surfaces [7, 19, 20], while many contemporary European and American artists did not varnish their paintings [30]. Gauguin, as occasionally Munch, also painted on unprepared cotton canvases [19].

Munch also emphasised the dramatic effect of oil painting rather than the cool and dry style of late fresco painting [37]. He experimented, for example, with the use of resins, oils and mixtures, as for the *The Sick Child* (National Museum of Art, Architecture and Design in Oslo), where he probably added resin to the oil medium [2] that caused the surface to become glossy. Munch's experiments with materials and surface gloss were in the past mistakenly interpreted as later conservation treatments or the other way round. Examples include the addition of varnish layers in limited areas in connection with linings, to consolidate weak paint layers, or to protect specific passages of paint.

Recent treatment of Munch's paintings at the museum has been limited to stabilization of the works, while more invasive treatment was undertaken forty years ago.

---

[2] None of these additives were present in Talens Picture Varnish (KJvdB).

[3] «…my only aim is to get the paint to work its way [?] into the canvas I have never loved the oily glossy genre which Soot and others specialise in that is why it would be a pure misunderstanding to varnish my things.»

This included varnishing Munch's paintings while the artist still lived [3] to protect them from dirt. The first treatments of Munch's paintings in the Munch Museum between 1945 and 1960 reflected the trends of the time: 35% of the paintings were lined after cleaning with thinned Sunlight soap [18]. This treatment may have caused uneven gloss on some paintings, but on others, unevenness was the result of an intentional effect by the artist combined with deterioration of some materials. The reason for varnishing was to lessen the visual impact of the unevenness. Several paintings were varnished with AW2 mixed with wax to reduce the gloss. Some paintings, like *Kiss* (MM.M62), were varnished with AW2 because of uneven gloss, but later conservators decided to remove the varnish and cover the painting with a wax layer to reduce the gloss in line with the artist's intentions.

The present study aimed to identify paintings in the collection varnished with AW2 and other ketone resins; to evaluate the condition of varnish and paint layers; to analytically characterize AW2 and other ketone resins and to investigate the observed degradation phenomena of the varnish and its effect on paint layers. The wider aim was to provide the basis for a suitable method for the removal of ketone varnishes. Conservators' evaluation of the paint and varnish layers of all paintings reported as varnished with AW2 suggested that only some paintings should be considered for cleaning. The first and most important reason to remove the varnish from these paintings were cupping in the varnish that led to delamination between paint and ground. Secondly, the paintings appeared blanched, glossy or discoloured.

Analytical investigation of samples of resin beads from the reference collection and varnishes from paintings using FTIR-ATR, microFTIR, Py-GC/MS and DTMS was carried out in collaboration with laboratories from GCI (Los Angeles) and RCE (Amsterdam). The methods for characterizing these samples are given elsewhere [36].

40.3.1 Collection Survey Outcomes

A survey of the conservation treatment reports in the archives of the Munch Museum showed that 23 paintings (1882–1940) had been varnished with ketone resins between the 1950s and 1980s. Among these, 16 paintings are documented as being varnished with AW2 between the 1957 and 1970 (Fig. 40.2), 6 paintings with Ketone-N and 1 with MS2A.

Furthermore, 11 paintings are recorded as varnished with the commercial varnish Talens-Rembrandt Picture Varnish (matt), presumably containing AW2 or Ketone-N resin (*vide supra*). Most of the reviewed reports lack information regarding additives, solvents and description of the application method.

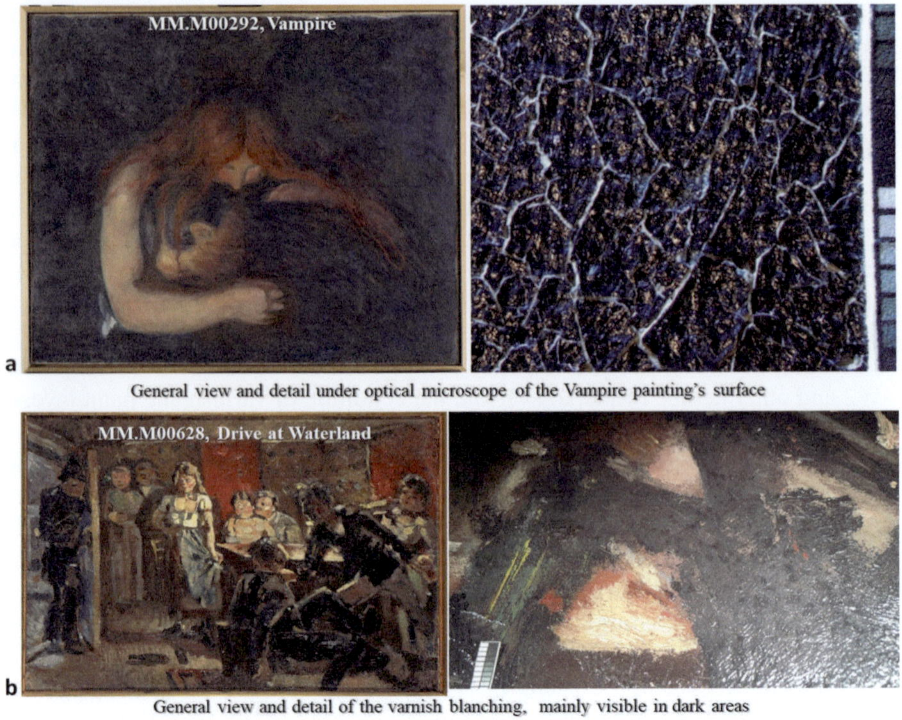

Fig. 40.2 General views and details of the surface appearance of 2 paintings varnished with AW2 in 1957: (**a**) *Vampire* (Woll M 335), 1893, oil on canvas; (**b**) *Drive at Vaterland* (Woll M 82), 1882–83, oil on cardboard (Photo credits: Munch Museum)

All 16 paintings varnished with AW2 underwent macroscopic and microscopic investigation and were photographed under visible light and UV radiation (an UV fluorescence induced photography standard was established using a TL-D Blacklight Blue low-pressure lamp, emitting wavelength ca. 340 nm – 405 nm).

40.3.2 Development of an In-House Database of Ketone Resins References

Several ketone varnish samples were collected from different European institutions and MID-ATR-FTIR spectra were acquired together with GC/MS and DTMS data. Small quantities of resins: AW2 (6 samples, brands), Laropal K8 (**2** samples), MS2A (3 samples), Ketone N (**2** samples), unspecified Cyclohexanoneharz (1 sample) and **7** samples from varnishes (e.g. Rembrandt from Royal Talens) constitute the reference collection of ketone resins at the MUM up to date.

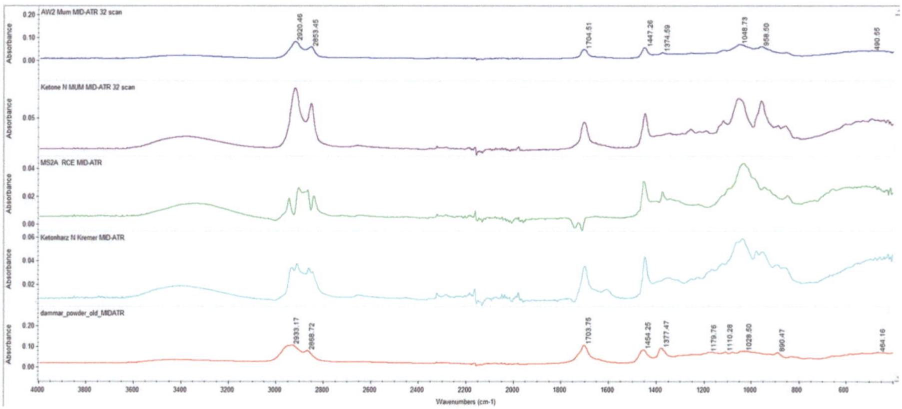

Fig. 40.3 MID-ATR FTIR spectra of several ketone resins (AW2, Keton-N and Ketonharz N, MS2A) and Dammar resin

As shown in a previously mentioned chapter [36] it is difficult to discriminate between the synthetic and natural resins using only FTIR spectroscopy (the only tool available at the Munch Museum), therefore further characterisation of these resins was undertaken using GC/MS and DTMS available in other laboratories. The more specific and detailed discussion around the effectiveness of chromatographic and mass spectrometric tools in identifying ketone resins is given in this chapter.

As an example, Fig. 40.3 shows the differences in ATR-FTIR spectra between batches of samples of the reduced ketone resin MS2A and unreduced Ketone Resin N and AW2 compared to natural resins, such as Dammar. The -C=O band absorption at around 1700 cm^{-1} which is present in the spectra of the unreduced ketone resins, is not identified in the MS2A spectra (Fig. 40.3b).

Infrared spectra of AW2 and MS2 show a band at around 1380 cm^{-1} corresponding to absorption by methyl groups, whereas this band is much lower in intensity in the spectra of Ketone Resin N and Laropal K8 [32]. Laropal K8, available from BASF, is identical to Ketone Resin N according to its manufacturer and some authors also showed that these two resins have identical spectroscopic features [12]. Laropal K8 is still widely used, for example, in Italian (OPD) restoration workshops as final varnish covering a coating of natural resin (Dammar or Mastic) [6].

40.3.3 Paint Mock Ups in White and Blue Colors – Preparation, Varnishing and Light Ageing

The paint mock-ups were produced in 2009 (master thesis of E. Tveit, 2011), using synthetic ultramarine and zinc white paint tubes from Old Holland, first artificially aged at SINTEF Institute in Trondheim, using an Atlas SC 600 MHG chamber and then stored for 9 years in laboratory environment.

Three types of paint formulations were obtained: **A** – 26 ml chalk was added to the paint from tube (40 ml), **B** – unmodified applied directly from the tube on the canvas, **C** – the paint from tube was diluted in turpentine oil (Fig. 40.4a).

A total of 18 mock-ups on canvas were obtained and used further for the varnish project. The surface of all mock-ups appear matte regardless of the observation angle, except for the B-paint with zinc with a half gloss, but all the paint surfaces, also the white B, exhibit chalking. Although the PVC (Pigment Volume Concentration) differs in the three formulations, the paint straight from the tube is also porous and chalking.

The paint mock-ups were varnished using 9 different recipes with AW2 and wax in different concentrations dissolved in White Spirit 009 (Rembrandt from Talens) with an aromatic content of 22 wt % (Fig. 40.4b).

The solutions were obtained from naturally aged AW2 beads (reference collection of the museum) and bleached beeswax. They were applied by brush in one layer on the mock-ups and on glass slides, twice each. One set of the varnished mock ups and glass slides was light aged (daylight fluorescent tubes, ordinary tubes of light mastec 36 Watt, cool white, CIT 50/08, Gamma, the Netherlands, at c. 8500 Lux) at RCE for 4 months (from July to November 2017) at constant RH (50%) and temperature (20 C°) using an ageing chamber consisting of a self-constructed with ventilation and light ceiling.

The ageing protocol was chosen to imitate the ageing conditions for the paintings varnished with AW2 and kept in the storage or other museum environment for 50–60 years after the application of the varnish. When not exhibited they were kept in a dark and acclimatized storage room. The most exhibited painting with AW2 is *Vampire* (Woll M 335 – Fig. 40.2a) exposed to a min. of 7.6Mlx∗h after varnishing.

40.3.4 Visual and Microscopic Characterization of Mock-Ups After Varnishing

For the mock-ups covered with the 9 recipes of varnish the gloss and the saturation on their surface are higher with the increasing amount of AW2 resin; in particular there is a huge change between 35% and 50% amount of resin. The mock-ups covered with 20% AW2 varnishes are matte, the samples covered with the 50% solution exhibit strong gloss and darkening. There are only slight changes in saturation when the amount of AW2 is 20% because of the variation in wax content. Less change in gloss and saturation is seen in the mock-ups with 20% AW2 and 10% wax. Gloss and saturation changes are largest in B and A paints. Some cracks are visible in samples varnished with the largest amount of AW2. The varnishes do not show any yellowing, nor blanching or cupping. Only in areas where the varnish has been subjected to mechanical stress (e.g. sampling areas), cuppings appear.

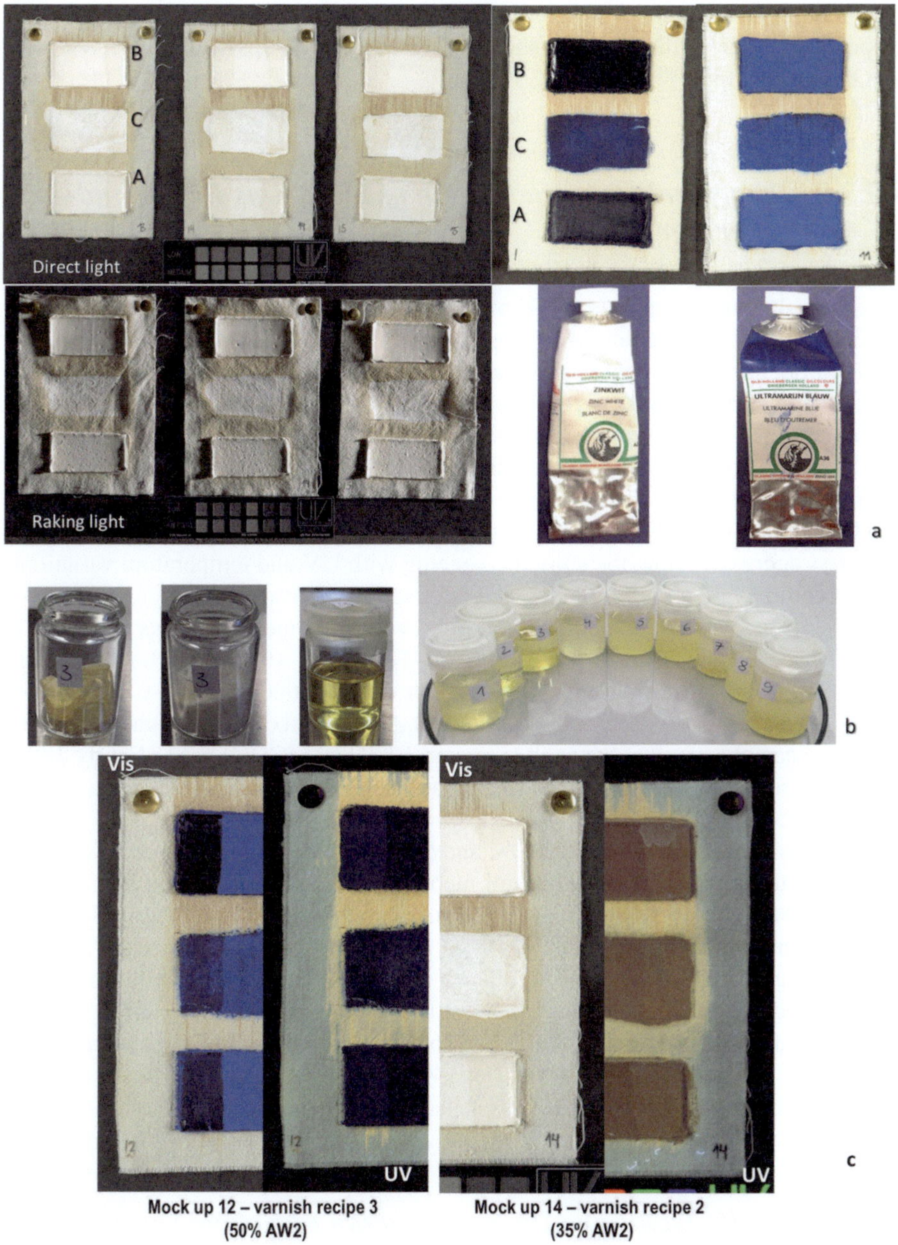

Fig. 40.4 (**a**) White and blue colors paint tubes and mock ups according three recipes (A, B, C) in direct and raking light; (**b**) The preparation steps for 1 (sol. 3) of the 9 solutions of AW2 varnish; (**c**) selection of mock ups for 2 recipes/2 mock-ups of varnish after varnishing

When comparing the Munch paintings varnished with AW2 with the mock-ups appearance after varnishing with the 9 recipes, the mock-ups with 50% AW2 present the most similar gloss.

40.3.5 Analytical Characterization of Mock-Ups Before and After Ageing

The varnished mock-ups and the 9 solutions applied on glass slides were characterized before and after ageing using Fourier Transform Infrared Spectroscopy (FTIR-ATR), optical microscopy and Scanning Electron Microscopy (SEM from RCE) imaging on surfaces and cross sections of the paint layers, and CIE L∗a∗b∗ colorimetry. As the data from colorimetric investigation do not give consistent patterns of total colour variation according the type of mock-up/composition of paint and recipes of AW2 resin varnishes, these will not be discussed here.

The ageing was only done using light sources without UV component and therefore less aggressive than the combined ageing with UV and temperature variations, which could explain the minimal changes observed after ageing.

When wax was added to the solution of varnish some new peaks appeared in FTIR spectra, such as: a shoulder at 1736–1741 cm^{-1} (for -COO bond), around 1171–1184 cm^{-1} (esters groups) and also in the 700–600 cm^{-1} (C-C str) region. When comparing two spectra of solutions of AW2 in mixture with wax (at 10 and 20%) we can observe that the peak at around 1730 cm^{-1} (COO) has a slightly different value and intensity, the same for the ones at 720–730 cm^{-1} (C-C str), as shown in Fig. 40.5a.

Figure 40.6 (a and b) shows the OM and SEM imaging performed before and after ageing on the surfaces of both types of mock ups. The optical imaging does not display any differences in the surface appearance or ultraviolet fluorescence of the varnished paints before and after ageing. SEM imaging of the paint surface at 150 x and 5000 x magnification showed no significant changes in the surface after ageing. Paint A presented a coarser surface than the paint B, probably due to differences in pigment volume concentration. Some small white spots were observed on some areas within the varnish solution mixed with wax (Fig. 40.6b), that could point to inhomogeneity of the varnish solution or a resin – wax separation, to be investigated further.

40.3.6 Multi-Technique Characterization of Ketone Varnishes on Munch's Paintings

Characterisation of varnish samples taken from paintings required a different procedure than for mock-ups and was far more challenging. Preliminary analysis identified ketone resins using FTIR-ATR in some samples from the 16 paintings recorded

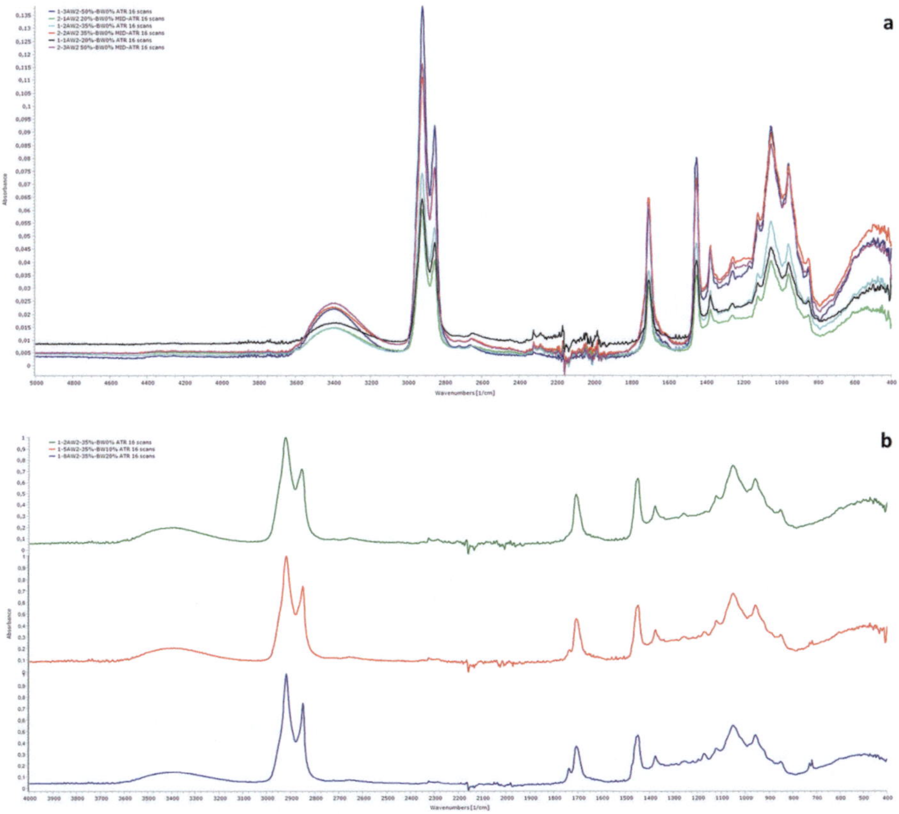

Fig. 40.5 (**a**) Different FTIR spectra of aged and not aged AW2 solutions (20%, 35 and 50% AW2 in WS); (**b**) FTIR spectra of artificially aged AW2 solutions (35% AW2 mixed with wax in different concentrations)

as varnished with AW2. From this group a selection of 10 paintings was made for sampling and analysis with GC/MS and DTMS.

It was also possible to identify other materials in the samples including glues, wax, natural or synthetic resins. As it was not possible to obtain cross-sections from original paint layers, it was not clear whether these materials were original additives to the paint, additives to the varnish or residues of consolidants. The analytical characterization of triterpenoid and ketone resins using only FTIR spectroscopy and Py-GC/MS was possible in few cases.

DTMS analysis of some varnish samples corroborated the presence of AW2 as hypothesized initially according the survey on the collection and preliminary screening with ATR-FTIR spectroscopy (e.g. in sample MM.M127). Py-GC/MS also confirmed the preliminary screening of a ketone resins in several samples [36].

Figure 40.7 shows the results of DTMS analysis of varnish samples from 3 Munch's paintings. Among the real samples analysed with this technique, in 7 cases the ketone resin content was identified but only in 2 (MM.M 927B and MM.M639)

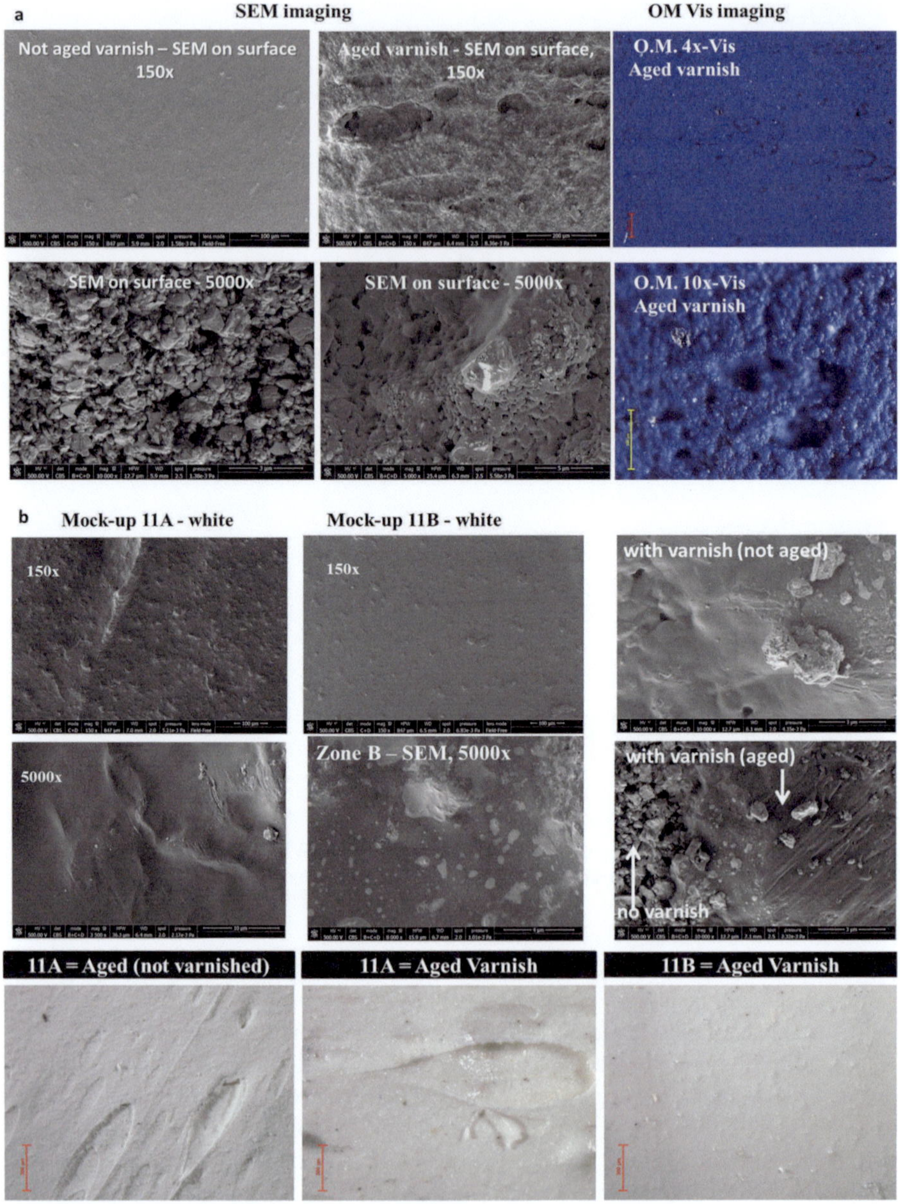

Fig. 40.6 (**a**) OM and SEM imaging on surfaces before and after ageing of blue mock-ups varnished with recipe 2 (AW2 35% in WS); (**b**) SEM and OM imaging on surfaces of white mock-ups varnished with recipe 5 (35% AW2 + 10% bees wax)

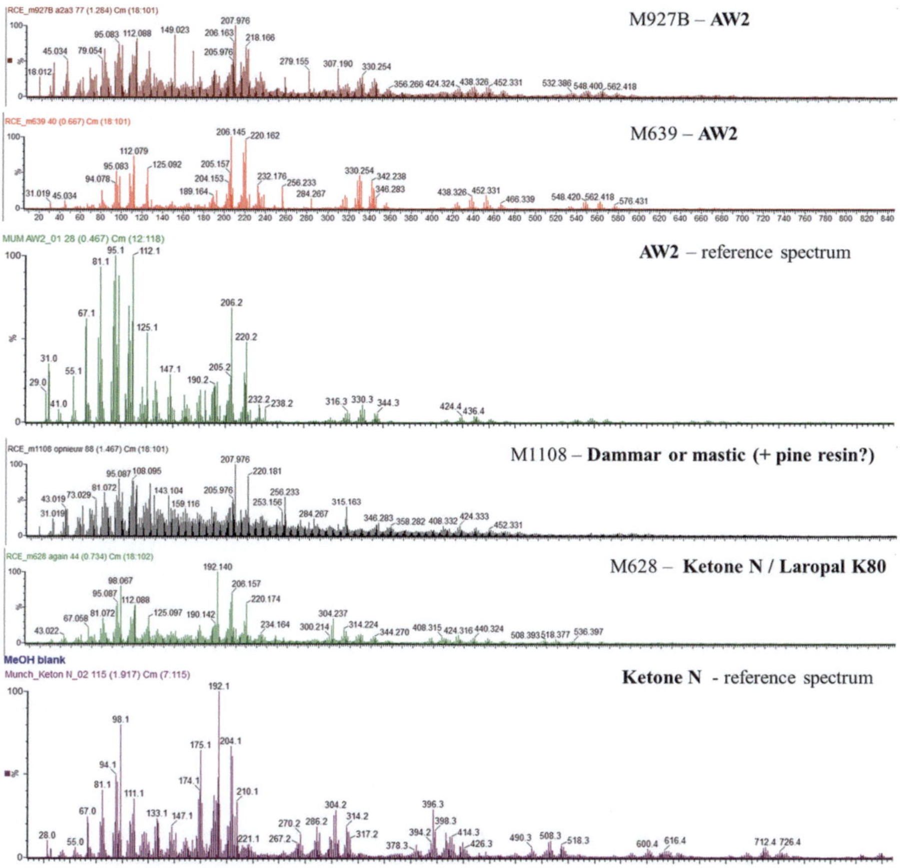

Fig. 40.7 DTMS spectra for the identification of ketone resins varnishes on three of Munch's paintings in comparison with 2 reference spectra (AW2 and Ketone-N)

it was possibly to unambiguously assess the presence of AW2, while for the other paintings other ketone varnishes (e.g. Ketone-N) or mixture of organic coatings were identified.

40.3.7 Short Discussion Around Cleaning Necessity and Options

The varnished works were painted in different periods and probably had different glosses before varnishing. It is difficult today to say how these surfaces were changed by the varnish, but some of them have had a much matter appearance with less saturated colours if we think to Munch as an artist preferring matt surfaces [18, 35]. This is possible to assume by comparing the paintings to others painted during the same period that have not been varnished. Today the surfaces of the paintings have different degrees of gloss, depending on cracking and blanching, the thickness of applied varnish or the addition of wax. Yellowing, blanching, cracking, cupping and partial loss of adhesion to the underneath paint layer were detected on some paintings, on other paintings only minor visual disturbances were observed. Blanching on several paintings varnished with AW2 (Fig. 40.2) has also been noted elsewhere [12, 23].

Removal of dirt and varnish is challenging due to the condition of the paint surface and layers beneath it, in Munch's works. These exhibit porous paint with weak internal cohesion and adhesion to the support, as a consequence of Munch's paint techniques, together with deterioration caused by handling and storage [35].

40.4 Conclusions

This study showed that of the 23 paintings included in the survey, 16 are documented as being varnished with AW2. However of the 10 paintings that were analysed only 7 included ketone resin, and of these only 2 were varnished with AW2. Results for other works were compromised due to sample size or contamination, though wax and natural resins were identified.

Several organic materials were identified in samples from the 7 paintings, such as AW2, Laropal K8 and MS2A, together with beeswax and/or paraffin, animal glue, drying oils and egg. This suggests that the method used for sampling may have included underlying materials. Traces of natural resin (pine resin) and gum or starch were identified in few cases. There is no evidence that the artist applied varnishes in the traditional sense, covering the entire paint surface, on the contrary, he criticized the varnishing practice in his letters. Therefore the most credible hypothesis is that the other materials (e.g. natural resins residues) identified in the varnish layers besides the synthetic resins were added with later conservation treatments.

Experimental mock-ups light aged for 4 months aimed to reproduce the conditions the paintings were exhibited after being varnished with ketone resins. The mock-ups showed no significant change in surface appearance after aging, on examination using optical microscopy and SEM imaging. MID-ATR-FTIR spectra showed no indications of chemical changes in samples of ketone resins after ageing. More analyses should be performed to understand the possible interaction between paint and varnish layers. This will provide a basis for the development of an improved methodology for varnish removal.

Acknowledgements The Archlab access provided through grant 654028 (H2020-INFRAIA-2014-2015; IPERION) to the Opificio delle Pietre Dure (OPD; Florence, Italy), Centre for Art Technological Studies (CATS; Copenhagen, Denmark) and the Cultural Heritage Agency of the Netherlands (RCE; Amsterdam, Netherlands) is acknowledged. Hilde LeClerq (IRPA) and Monica Galeotti (OPD, Florence) from the IPERION CH-Archlab are gratefully mentioned.

We also wish to thank to: Beatrice Kromp (BASF), Isabella Blank (BASF), Hans Portsteffen (TH Köln), Ester Ferreira (TH Köln), Stefanie Dietz (TH Köln), Line Berg Härström (Munch's Hus), Mille Stein (NIKU), Oda van Maanen (van Gogh Museum), Laura Kolkena (Stedelijk Museum), René de la Rie (UiA), Pauline Lehmann Banke (SMK), Troels Filtenborg (SMK), David Buti (CATS), Anita Schink (Staatsgalerie Stuttgart), Julian v.d. Berg (RCE), Henk v. Keulen (RCE), Wim Genuit and Shell Global Solutions International B.V for making available the DTMS instrumentation. Andrea Casini and Misra Kaya (Erasmus interns at the MUM) are also kindly acknowledged for their work on mock-ups and the cross-sections.

References

1. Allwinn R (1976) Harze; Sythetische; Cyclohexanon-Harze, Ullmanns Encyklopädie der technischen Chemie (Bd. 12). 4. Aufl. Weinheim/New York, p. 551–552;
2. Aslaksby TE, Ustvedt Ø (2009) Edvard Munch: Det syke barn: Historien om et mesterverk, The sick child: The story of a masterpiece, Nasjonalmuseet, Oslo, 64;
3. Avisinnlegg «Munch-bildene blir som de er!» av Bernhard Rostad, Aftenposten 29.11.1980;
4. Badische Anilin- & Soda-Fabrik in Ludwigshafen a. Rh.: Verfahren zur Darstellung von Kunstharz. filing date: 1.7.1918, publishing date: 13.07.1920. Patent document: DE0000003f 107A;
5. Borgioli L, Cremonesi P (2001) Le resine sintetiche usate nel trattamento di opera policrome, Collana I Talenti, Il Prato, p. 142–145;
6. Bracco P, Sartiani O, Toso C (2009) «Per te parte porto». Tecnica artistica, stato di conservazione, e intervento di restauro dei dipinti, *Andrea Mantegna. La Pala di San Zeno. Studio e conservazione,* M. Ciatti e P. Marini (eds.), Edizioni Edifir, Firenze, p. 125–174;
7. Callen A (2000) The art of impressionism: Painting technique & the making of modernity, Yale university press, New Haven og London, p. 101;
8. Ciabach J (1998) Modern varnishes. Ageing and possibility of stabilization. In: Firnis. Material – Ästhetik – Geschichte, Internationales Kolloquium. Braunschweig 15. bis 17. Juni 1998, p. 105–108;
9. de la Rie ER (1993) Polymer additives for synthetic low-molecular-weight varnishes, Preprints of 10th Triennial Meeting, ICOM-CC, Vol. 2, Washington D.C., August 1993, Lawrence, Kan: Allen Press, p. 566–579;
10. de la Rie ER (1992) Stability and Function of Coatings used in Conservation, Polymers in Conservation, N.S. Allen, M. Edge, C.V. Horie (eds.), RSC Publication, p. 75;
11. de la Rie ER, McGlinchey CW (1990) New synthetic resins for picture varnishes, Cleaning, Retouching, and Coatings, J.S. Mills and P. Smith (eds.) London, International Institute for Conservation of Historic and Artistic Works, p. 168–173;
12. de la Rie ER, Shedrinsky AM (1989) The Chemistry of Ketone Resins and the Synthesis of a Derivative with Increased Stability and Flexibility, Studies in Conservation, 34, London, p. 9–19;
13. de la Rie ER (1987) The Influence of Varnishes on the Application of Paintings, Studies in Conservation, 32, London, p. 1–13;
14. Dietemann P, Higgit C, Kalin M, Edelmann MJ, Knochenmuss R, Zenobi R (2009) Aging and yellowing of triterpenoid resin varnishes – Influence of ageing conditions and resin composition, Journal of Cultural Heritage, 10, p. 30–40;
15. Doerner M (1994) Malmaterial und seine Verwendung im Bilde, 18, Auflage, Stuttgart;
16. Eggum A (1995) Edvard Munch: Malerier-skisser og studier, J.M. Stenersen forlag as, Oslo, 50, 61;

17. Feller RL, Bailie CW (1971) Solubility of Aged Coatings Based on Dammar, Mastic, and Resin AW-2, Bulletin of the American Group of the International Institute for Conservation (12/2), p. 72–81;
18. Ferrer J, Storevik Tveit E (2016) Gode intensjoner og uante Resultater, Konsekvenser av tidligere konserveringsbehandlinger av Edvard Munchs malerier Norsk museumstidsskrift, årgang 2, no. 1, 30–43;
19. Jirat-Wasiutymski V, Newton jr. HT (1998) "Absorbent grounds and the matt aesthetic in post-impressionist painting", 235–239, in A. Roy og P. Smith (ed.), IIC: Contribution to the Dublin congress, 7–11 september, 1998, International institute for conservation of historic and artistic works (IIC), London, p. 235;
20. Hensick T, Olivier K, Pocobene G (1997) "Puvis de Chavannes's allegorical murals in the Boston public library: History, technique and conservation", Journal of the American institute for conservation (JAIC), vol. 3, no. 6, p. 1–26;
21. Hill A, Wilson WC, Hale DK, (1946), Manufacture and applications of AW2® resin – British intelligence objectives sub-committee;
22. Horie V (1986) Materials for Conservation. Organic consolidants, adhesives and coatings, London, p. 258;
23. Koller J, Baumer U (2001) Kunstharzfirnisse. Teil III: Die niedermolekularen (nichtpolymeren) Kunstharzfirnisse, Restauro, 107, p. 26–38;
24. Lucie-Smith E (1972) *Symbolist Art*, Thames and Hudson, London, p. 184;
25. Ludvik Ravensberg's diary, LR 536, entry 1910, Munch Museum archives;
26. Ludvik Ravensberg diary, MM LR 537, Date 22.4.1909–16.7.1909, Munch Museum archives;
27. Marra A (2011) Materiali di sintesi per la verniciatura, A scuola di restauro: Le migliori tesi degli allievi dell'Istituto Superiore per la Conservazione ed il Restauro e dell'Opificio delle Pietre Dure negli anni 2005–2007, Roma, p. 113–119;
28. Massa V, Scicolone G (1998) Le vernici per il restauro. I leganti, Nardini Editore, Firenze;
29. Masschelein-Kleiner L (1984) Teil III. Anwendungsbeispiele. Lösungsmittel in der Restaurierung. 5. Auflage. Wien, p. 141–143;
30. Mayer L, Myers G (2004) American Impressionism, varnishing and mattness, Journal of the American Institute for Conservation (JAIC), 43, 3 Fall/winter 2004, p. 237–254;
31. Menck-Schaa A, Schulte K, Reuber L, Portsteffen H (2008) Firnis von matt bis glänzend, München, p. 3–11;
32. Mills JS, White R (1987) The Organic Chemistry of Museum Objects, Butterworths, London, p. 118–120;
33. Raft K (1985) A Preliminary Report on the Possibility of Using Bleached Beeswax to Improve the Resolubility of Picture Varnishes Based on Polycyclohexanones, Studies in Conservation, 30, p. 143–144;
34. Routledge V, (2000) The Development of MS2A Reduced Ketone Resin in Painting Conservation, Waac Newsletter, vol. 22, no. 2;
35. Sandbakken EG, Tveit ES, "Edvard Munch's monumental sketches (1909–16) for the Aula of Oslo University, Norway: Conservation issues and treatments", Studies in Conservation, 57, International Institute for Conservation of Historic and Artistic Works (IIC), London, 2012, p. 258–267;
36. Sandu I.C.A, van den Berg K. J, Candeias A, Sandbakken E. G, Storevik Tveit E, van Keulen H, Chemical analysis in cultural heritage. Part II. Physicochemical approaches in art and archaeology investigation. Chapter 5 – Multi technique and multiscale approaches to the study of ancient and modern art objects on wooden and canvas support, I. van den Werft and L. Sabbatini (eds.), Physical Sciences Review, de Gruyter Publ, Published Online: 2019-01-10 | DOI: https://doi.org/10.1515/psr-2018-0016;
37. Tøjner PE (2000) Munch in his own words, Forlaget Press a.s. Oslo and Søren Fogtdal a.s. Copenhagen, p. 163;

38. van der Doelen GA (1999) Molecular studies of fresh and aged triterpenoid varnishes, Amsterdam;
39. van Doelen GA, van den Berg KJ, Boon JJ (1998) A comparison of wheatherometer aged dammar varnishes and aged varnishes from paintings, in J. Goupy, J.P. Mohen (eds.), Proceedings of the Conference "Art et Chemie"- La couleur, Paris, France, p. 146–149;
40. Wehlte K (2009) Werkstofe und Techniken der Malerei, Freiburg, p. 258;
41. Ydstie I, Guleng M (eds.) (2008), Munch blir 'Munch': Kunstneriske strategier: 1880–1892, Munch-museet, Oslo, p. 233.

Chapter 41
Designing Paint Mock-Ups for a Study of Novel Surface Cleaning Techniques for Munch's Unvarnished Aula Paintings

Lena Porsmo Stoveland, Bronwyn Ormsby, Maartje Stols-Witlox, Tine Frøysaker, and Francesco Caruso

41.1 Introduction

Unvarnished, modern oil paintings can pose serious challenges for cleaning [7, 18]. Burnstock and van den Berg [2] identified surface cleaning as one of the most critical conservation interventions. Edvard Munch's monumental unvarnished oil paintings in the Aula of the University of Oslo (1909–16) have a history of issues associated with cleaning (Fig. 41.1a). When first installed in 1916, these expressionist paintings were praised for their bright colours [1]. However, their appearance has been partly compromised by decades of exposure to dust and airborne particles from cigarette smoke, heaters, and their location in an urban environment [10]. Since the first cleaning campaign in 1925, several conservation interventions have employed

L. P. Stoveland (✉) · T. Frøysaker
Conservation Studies, Department of Archaeology, Conservation and History, University of Oslo, Oslo, Norway
e-mail: l.p.stoveland@iakh.uio.no; tine.froysaker@iakh.uio.no

B. Ormsby
Conservation Department, Tate, London, UK
e-mail: Bronwyn.Ormsby@tate.org.uk

M. Stols-Witlox
Department of Conservation and Restoration, University of Amsterdam, Amsterdam, The Netherlands
e-mail: M.J.N.Stols-Witlox@uva.nl

F. Caruso
Conservation Studies, Department of Archaeology, Conservation and History, University of Oslo, Oslo, Norway

Present Affiliation: Department of Art Technology, Swiss Institute for Art Research (SIK-ISEA), Zurich, Switzerland
e-mail: francesco.caruso@iakh.uio.no; francesco.caruso@sik-isea.ch

K. J. van den Berg et al. (eds.), *Conservation of Modern Oil Paintings*,
https://doi.org/10.1007/978-3-030-19254-9_41

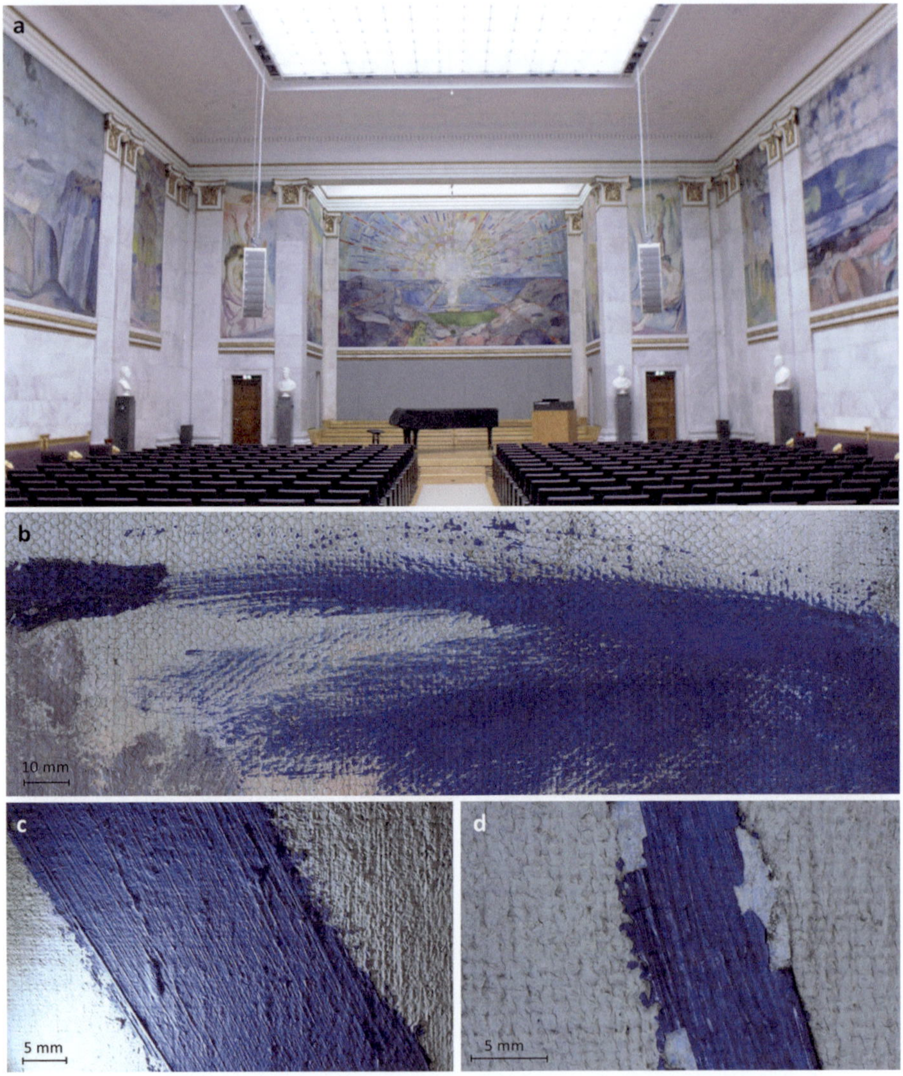

Fig. 41.1 Edvard Munch's paintings (1909–16) in the Aula of the University of Oslo (**a**) with detail of an area of diluted cobalt blue paint (**b**), an area with visible brush marks (captured in raking light) (**c**), and cobalt blue paint with poor adhesion to the ground (**d**)

traditional approaches for removal of dirt and soot from the paintings with limited success. The present study aims to address the efficacy and risks of a selection of novel cleaning methods on paint mock-ups of fragile areas of the Aula paintings.

This paper describes the methodology and decision processes employed for the preparation of experimental paint mock-ups destined to be used as testing substrates for the novel dirt and soot removal methods. This is the first stage in a comparative cleaning study using soft-particle blasting, CO_2-snow, and Nanorestore Gel® Peggy and Dry series (techniques discussed in [9, 14, 22]).

41.2 Background

The choice of method(s) for surface cleaning should be governed by the nature of the dirt, the type, condition and properties of the paint or substrate, as well as the sensitivity of the work of art toward the chosen system(s) [2]. In the case of the Aula paintings, previous research has shown that the removal of soot particles could not be achieved using commonly used dry methods, and that wet cleaning carried unacceptable risks, such as dissolution of the ground and pigment removal [13]. The last cleaning campaign carried out from 2009 to 2011 using dry polyurethane sponges, removed a significant amount of surface dirt. However, Frøysaker concluded that while the treatment made the paintings cleaner, a thin, dull veil of greyish soot and dust remained on the surface; with visual and physico-chemical consequences [11]. Evaluating and optimising methods for the safe removal of the remaining soiling is therefore the goal of a wider study.

41.3 Materials and Methods

Direct testing of novel cleaning methods on Munch's Aula paintings is not possible due to the uncertainty of the effects and results on the most fragile areas of ground and paint. Therefore, the production of mock-up paint samples is considered appropriate for the initial evaluation of any risks associated with cleaning methods. An important consideration - which precedes the choice to create mock-ups - pertains to their exact role(s) in the research. For this study, the aim was to produce test samples that are similarly vulnerable to mechanical action/friction, with a degree of water sensitivity that is roughly representative of the Aula paintings. The group of samples need to be intercomparable, providing a systematic, comparative set of representative samples, which are of enough likeness to the Aula paintings to provide useful results during the initial evaluations of the novel methods. While previous research into the composition of the Aula paintings has guided mock-up preparation, detailed copying would result in too many variables to reasonably assess and interpret outcomes of cleaning tests.

41.3.1 Preparation of Mock-Ups

In consultation with the conservators involved in the most recent conservation treatment, three areas of chalk-glue ground and cobalt blue paint were chosen as the basis for the mock-ups. The properties of these areas are considered representative of the cleaning issues encountered, such as sensitivity to aqueous and polar solvents, and mechanical action [4]. The porous chalk-glue ground retains soil and dissolves quickly in aqueous solvents. The cobalt blue oil paint is typically moderately water-sensitive (less than ultramarine and more than Cerulean blue) [20] and appears as both diluted (Fig. 41.1b) and more thickly applied impasto passages with clearly defined brush strokes (Fig. 41.1c). The areas of diluted cobalt blue paint are lean and powdery, while the impasto areas tend to have poor adhesion to the layers below (Fig. 41.1d). Based on those areas, three mock-ups were prepared and divided into 72 individual samples, including reference samples as controls.

The mock-ups were produced according to published information on Munch's technique [12, 21], subject to contemporary material availability, and were prepared as described below, according to the following stratigraphy (Figs. 41.2 and 41.3):

1. Stretched linen canvas, glue size, chalk-glue ground
2. 1 + undiluted cobalt blue oil paint
3. 1 + turpentine diluted cobalt blue oil paint

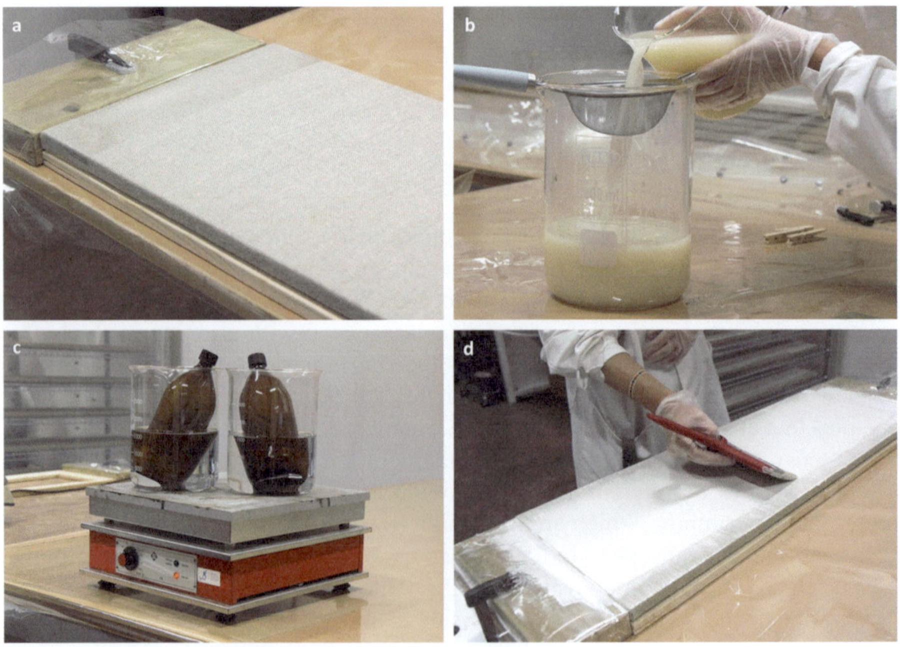

Fig. 41.2 Canvas mounted on stretcher (**a**), preparation of glue size (**b**), prepared glue and chalk-glue ground (**c**), and application of ground (**d**)

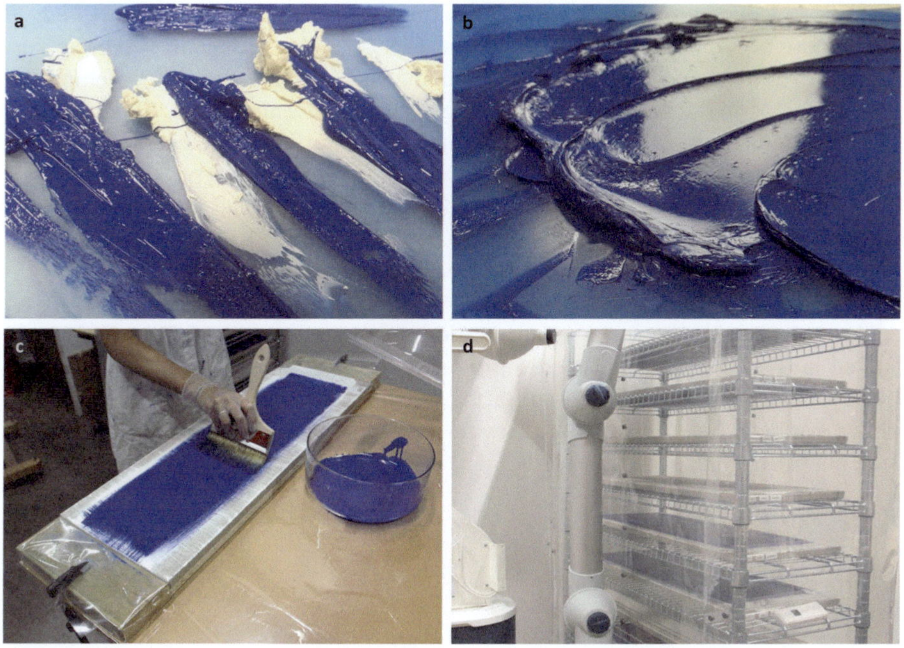

Fig. 41.3 Mixing of prepared cobalt blue and barium sulfate in poppyseed oil (**a**), consistency of the finished paint (**b**), application of the diluted paint (**c**), samples stored on the drying shelf (**d**)

41.3.2 Canvas

An unprimed, plain weave, double thread linen canvas (068 from Claessens, 395 g/m²) was chosen as the fabric substrate. The weave pattern and thread density resemble those of the canvas of *Alma Mater*, one of the Aula paintings. The rough canvas texture results in a visible topography across individual ground and paint layers, and as such, contributes to the Aula paintings' vulnerability towards mechanical action. The canvas was thoroughly soaked and rinsed seven times in warm (40 °C) and cold (19 °C) distilled water to remove most of the PVA-sizing used in production [5] and to reduce the dimensional response of the linen fibers to moisture. After semi-drying at room temperature, the canvas was steam-ironed and stretched onto wooden stretchers.

41.3.3 Glue Size

Although it is not clear (due to past treatments) to which extent size layers are present in the Aula paintings [12] it is possible that the canvases Munch used were either commercially pre-sized, or that he/his assistants sized them. It was common practice to isolate artist's canvases with sizing layers to stiffen the support, reduce the amount of ground required to cover the surface, and to protect the fibres against oil [19]. Animal glue was chosen as a sizing material based on a contemporary recipe by Doerner (1921) [8]. Doerner suggested a concentration of 70 parts (70 g) of glue to 1000 parts (1 L) of water, and that the proportion could be either reduced or increased to account for strength differences of the various types of glues. Tests demonstrated that a solution of 5 g of hide skin glue (Hasenleim aus Hasenfellen, 63025 Kremer) with 100 g of distilled water (5 g + 100 g) resulted in a thin and barely visible film that sealed the weave; whereas a 7 g concentration created a thicker layer that did not correspond with observations of the Aula paintings. After soaking overnight, the lower concentration was heated to just below 40 °C and stirred until homogenous. The resulting solution was then sieved into two glass bottles, one for the sizing and one for the chalk-glue ground. The size was applied with a synthetic brush, dipped half way in, as recommended by Doerner [8], making sure not to soak the canvas through.

41.3.4 Chalk-Glue Ground

A chalk-glue ground was chosen because it represents one of the most water-sensitive compositions of the grounds identified in the Aula paintings [12]. It was prepared using readily available chalk (Champagner Kreide, 58000 Kremer) with a good spectral match to grounds in the Aula paintings *The Sun* and *Alma Mater* (Fourier-transform infrared (FTIR) spectroscopy analysis, unpublished report) and the warm glue solution (described above). A 1.25:1 ratio by mass was chosen as it resulted in a ground layer with a similar visual appearance to the Aula paintings. Chalk was added gradually to the glue with constant manual stirring until a smooth consistency was achieved, then poured into a glass bottle to soak overnight. The bottle was re-heated to just below 40 °C and shaken carefully before application, to ensure homogeneity without creating air bubbles. The ground was applied to the sized canvases with a soft synthetic brush, wide enough to cover the desired area of application in one layer, with some working in of the ground to ensure evenness. Despite the thinness of the glue size, the ground did not penetrate through to the reverse side of the canvas. As Munch left large areas of ground exposed, the chalk-glue ground layer serves as the uppermost surface layer of sample type 1, i.e. without any additional paint layer. For sample types 2 and 3, the ground was left to dry for 48 h before applying subsequent paint.

41.3.5 Oil Paint

The oil paint for the mock-ups was manually prepared, as currently-available cobalt blue artist oil paints differ from traditional versions. Low vacuum scanning electron microscopy coupled with energy-dispersive X-ray (SEM-EDX) (FEI Quanta 450, X-Max^N Oxford 50 SDD) analysis of a sample of one of Munch's blue colours in *The Sun* (in poor structural condition and sensitive to water), showed the presence of aluminium, cobalt, sulfur and barium, with traces of nickel, lead, phosphorus and zinc. The presence of elemental aluminium, cobalt, sulfur and barium suggested a combination of a cobalt oxide-aluminium oxide (cobalt blue) and barium sulfate as filler. The relatively high counts for aluminium to cobalt may point to an aluminium-containing additive like hydrated alumina [4]. The elemental distribution maps for barium and sulfur overlap. Traces of nickel and phosphorus have been linked to the production of cobalt blue [16]. Analysis by FTIR spectroscopy (Thermo Fisher Scientific) in transmittance mode of the same blue colour indicated the presence of sulfate and a drying oil [20].

For this study, the paint composition was simplified to limit the number of variables. The oil paint was prepared from cobalt blue pigment ($CoO \cdot Al_2O_3$. Kobaltblau mittel, 45710 Kremer), commercially available barium sulfate (Amertek, UK) and refined poppyseed oil (Mohnöl 73600 Kremer). Kremer Pigmente is not aware of any surface coatings on the cobalt blue pigment [15]. Poppyseed oil was chosen as it is recommended for Kremer cobalt blue paints to reduce yellowing and to set off the fast drying action of the pigment. However, paints based on poppyseed oil have been associated with poor drying properties and cracking paint [8], and it has been suggested that they can be more affected by solvent diffusion than paints based on linseed oil [23].

The pigment was mixed in a ratio of 1 g to 0.58 g poppyseed oil and barium sulfate filler in a ratio of 1 g to 0.39 g oil using a glass muller on a frosted glass plate. This amount of oil added was just enough to disperse the pigments. The mixtures were stored in closed containers overnight to allow the pigment and filler particles to absorb the oil. By grinding and mixing the pigment and filler in oil separately, the differences in handling properties and textures could be observed, and the pastes easily mixed in different ratios through trials (Fig. 41.3a). The final paint was mixed in a 2:1 mass ratio of dry pigment to filler, with a content of 34% oil to dry ingredients by mass, and a pigment volume concentration (PVC) of 31%. The poppyseed oil and barium sulfate filler provided the paint with a short, buttery consistency (Fig. 41.3b), comparable to that of a commercially available cobalt blue tube paint (Winsor and Newton). The paint was applied to the ground layer using a hog hair bristle brush (Fig. 41.4c), where the bristle stiffness enabled the production of brush strokes similar to those observed on the Aula paintings (Fig. 41.1c).

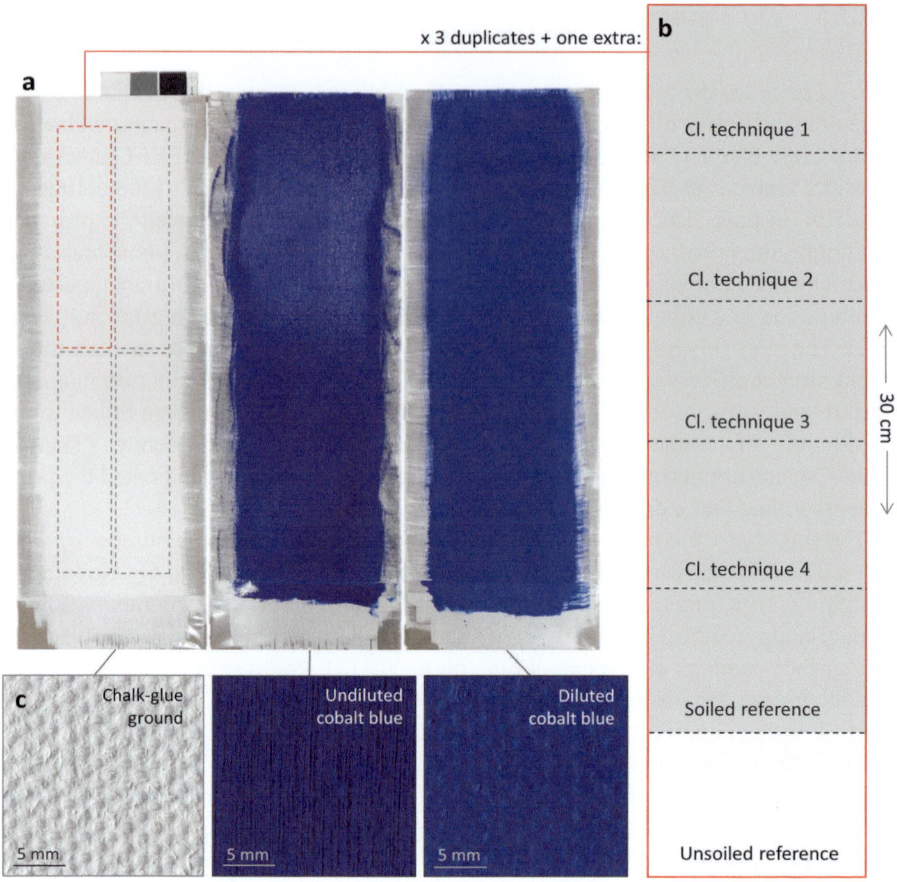

Fig. 41.4 Prepared mock-ups (**a**) with illustration of duplicates for each cleaning technique (**b**) with detail of each sample type (**c**)

41.3.6 Diluted Oil Paint

Munch painted several areas of the Aula paintings using diluted paint (thinned with turpentine or white spirit), which have been noted as sensitive to cleaning [12]. Therefore, a portion of the prepared cobalt blue oil paint was diluted with natural pine turpentine (Balsamterpentin, Ottossons Färgmakeri) to create a lean paint with a matte surface. Through trials, it was found that a paint/turpentine ratio of 1.5:1 created an appearance similar to areas of the Aula paintings (Fig. 41.1b). The diluted paint was then applied in a single layer using a hog hair bristle brush (Fig. 41.3c), resulting in a smooth surface with few visible brush marks, conforming to the structure of the ground and canvas (Fig. 41.4c).

41.3.7 Limitations

Recognizing inherent differences between deliberately simplified mock-ups and complex painting systems is crucial in experiment-driven cleaning studies. The preparation of the mock-ups clearly demonstrates that compromise is required to produce surfaces that are appropriate for the subsequent systematic, comparative cleaning studies. The mock-ups were created to represent challenging areas to clean, based on previous experience and initial studies into Munch's paintings, augmented by the analysis of one blue paint sample. The mock-ups were created using simple compositions and are composed of relatively easily available materials. Thus, the mock-ups are not historically accurate reconstructions and cannot be considered as such [6].

Differences in mechanical behaviour between modern materials and the products available during Munch's lifetime may also be expected due to different processing methods and pigment particle size and shape [6]. Nonetheless, the samples produced do share characteristics with Munch's paintings, such as visual similarity between the ground surface roughness and the varied appearance of the cobalt blue oil paints. Also important was the choice to produce simplified layered mock-ups. Using brushes (rather than drawdown bars) was also appropriate as the surface characteristics of the paints and the adhesion of the applied ground and paints to the substrate are relevant factors for the evaluation of novel cleaning methods. The process of using hand-prepared art materials and brush-applied paints instead of producing uniform single-layered samples results in surfaces with a greater degree of variation, but with enough consistency to facilitate the systematic comparison of cleaning procedures. This would not be possible if discrete tests were carried out on the Aula paintings *in-situ*. In addition, the mock-ups provide the opportunity to gain experience in manipulating and optimising the novel cleaning techniques on surfaces that are comparable, within reason, to the original painting surfaces, but which can be pushed further to help explore and characterise the potential risks associated with novel cleaning treatments.

41.4 Further Steps

The painting mock-ups were left to dry for 5 months under protection from dust. Duplicate samples (three of each sample type per cleaning technique) were cut out from the individual mock-ups (Fig. 41.4), and are subsequently undergoing a procedure of artificial soiling [17] and accelerated photo-chemical ageing at elevated temperatures and relative humidity. These steps in the sample preparation will be published elsewhere. An unsoiled and a soiled reference area have been left on the duplicates and an unaged, unsoiled reference set is stored in ambient, dust free conditions for comparison. The sample surfaces and subsequent cleaning tests will be characterised and documented using a selection of techniques: visual assessment, photography, colourimetry, glossimetry, attenuated total reflectance FTIR analysis, digital 3D microscopy and SEM-EDX analysis.

41.5 Conclusion

The limitations of using simplified mock-ups to represent paintings or naturally aged samples are well recognized and render comparison with in-situ cleaning effects challenging [3]. Nonetheless, the value of appropriate mock-ups to evaluate the effects of new cleaning methods in preliminary research is undeniable. The present cleaning study will lead to a greater understanding of the general effects and the practical application of the selected novel cleaning methods on unvarnished paint surfaces and to contribute to informed strategies for removing surface dirt from unvarnished oil paintings by Edvard Munch.

Acknowledgements The authors wish to thank Judith Lee (Tate Britain, UK) for advice on mock-ups, and Calin Constantin Steindal (Museum of Cultural History, University of Oslo) for the analysis of the paint sample from the Aula.

References

1. Berman PG (2011) Strategisk modernitet: Munch og Universitetet. In: Anker P, Berman PG (eds) Edvard Munchs Aulamalerier. Messel Forlag, Oslo, pp 47–70
2. Burnstock A, van den Berg KJ (2005) A pilot study of the effects of tri-ammonium citrate solutions used for surface cleaning paintings. In: AIC Paintings Speciality Group Postprints 17. AIC, pp 56–64
3. Burnstock A, van den Berg KJ (2014) Twentieth Century Oil Paint. The Interface Between Science and Conservation and the Challenges for Modern Oil Paint Research. In: van den Berg KJ, Burnstock A, de Keijzer M, Krueger J, Learner T, de Tagle A, Heydenreich G (eds) Issues in Contemporary Oil Paint. Springer International Publishing, Cham, pp 1–19. doi:https://doi.org/10.1007/978-3-319-10100-2_1
4. Burnstock A, van den Berg KJ, de Groot S, Wijnberg L (2007). An Investigation of Water-Sensitive Oil Paints in Twentieth-Century Paintings. In: Learner TJS, Smithen P, Krueger JW, Schilling MR (eds) Modern Paints Uncovered. Getty Conservation Institute, Los Angeles, pp 177–188
5. Carlyle L, Hendriks E (2009) Visiting Claessens, Artists' Canvas Manufacturers. News in Conservation, IIC 11 (April): 4–5
6. Carlyle L, Witlox M (2005) Historically accurate reconstructions of artists' oil painting materials. In: Clarke M, Townsend JH, Stijnman A (eds) Art of the Past: Sources and Reconstructions. Archetype Publications, London, pp 53–59
7. Daudin-Schotte M, Bisschoff M, Joosten I, van Keulen H, van den Berg KJ (2013) Dry Cleaning Approaches for Unvarnished Paint Surfaces. In: Mecklenburg M, Charola AE, Koestler RJ (eds) New Insights into the Cleaning of Paintings: Proceedings from the Cleaning 2010 International Conference Universidad Politécnica de Valencia and Museum Conservation Institute. Smithsonian Institition Scholary Press, Washington D.C., pp 209–219

8. Doerner M (1969 (1921)) The materials of the artist and their use in painting with notes on the techniques of the old masters. Rupert Hart-Davis Ltd., London
9. Domingues J, Bonelli N, Giorgi R, Fratini E, Gorel F, Baglioni P (2013) Innovative Hydrogels Based on Semi-Interpenetrating p(HEMA)/PVP Networks for the Cleaning of Water-Sensitive Cultural Artifacts. Langmuir 29 (8): 2746–2755
10. Frøysaker T (2007) The paintings of Edvard Munch in the Assembly Hall of Oslo University. Their treatment history and the Aula-project. Restauro Forum für Restauratoren, Konservatoren und Denkmalpfleger 113 (4): 246–257
11. Frøysaker T (2016) Munch's ambivalence to Alma Mater: making, changes and challenges. Zeitschrift für Kunsttechnologie und Konservierung 30 (2): 338–350
12. Frøysaker T, Liu M (2009) Four (of eleven) unvarnished oil paintings on canvas by Edvard Munch in the Aula of Oslo University. Preliminary notes on their materials, techniques and original appearances. Restauro Forum für Restauratoren, Konservatoren und Denkmalpfleger 115 (1): 44–62
13. Frøysaker T, Miliani C, Liu M (2011) Non-Invasive Evaluation of Cleaning Tests Performed on "Chemistry" (1909–1916). A Large Unvarnished Oil Painting on Canvas by Edvard Munch. Restauro Forum für Restauratoren, Konservatoren und Denkmalpfleger 117 (4): 53–63
14. Hubacek O (2016) Energy and Motion – Über die Reinigung von Jackson Pollocks Number 32. Restauro Forum für Restauratoren, Konservatoren und Denkmalpfleger 122 (4): 14–17
15. Kremer D. Personal communication 4.1.2019
16. Geldof M, Steyn L (2013) Van Gogh's Cobalt Blue. In: Jansen L (ed) Van Gogh's Studio Practice. Mercatorfonds, Amsterdam, pp 256–267
17. Ormsby B, Soldano A, Keefe M, Phenix A, Learner TJS (2013) An empirical evaluation of a range of cleaning agents for removing dirt from artist's acrylic emulsion paints. In: The AIC Painting Speciality Group Postprints, 38th Annual Meeting of the American Institute for Conservation and Historic Works, Milwaukee, Wisconsin, May 11–14, 2010, vol 23. AIC, Washington D.C., pp 77–87
18. Perry, R (1990) Problems of dirt accumulation and its removal from unvarnished paintings: A practical review. In: Hackney S, Townsend J, Eastaugh N (eds) Dirt and pictures separated: papers given at a conference held jointly by UKIC and the Tate Gallery, January 1990. The United Kingdom Institute for Conservation of Historic and Artistic Works, London, pp 3–7
19. Stols-Witlox M (2017) A Perfect Ground. Preparatory Layers for Oil Paintings 1550–1900. Archetype Publications. London, pp xi-xvi
20. Stoveland LP, Stols-Witlox M, Ormsby B, Caruso F, Frøysaker T (2019) Edvard Munch's monumental Aula paintings: a review of soiling and surface cleaning issues and the search for new solutions. In: Clarricoates R, Dowding H, Wright A (eds) Interactions of Water with Paintings. Archetype Publications. London, pp 85–99
21. Topalova-Casadiego B (2008) Tekniske aspekter ved Edvard Munchs maleri. In: Woll G (ed) Edvard Munch. Samlede malerier. Catalogue raisonné, Volume II. Cappelen Damm, Oslo, pp 424–458
22. Tsang J-S (2011) Soot removal from acrylic emulsion paint test panels: A study of dry and non-contact cleaning. In: Bridgland J (ed) ICOM-CC 16th Triennial Conference Preprints, Lisbon, 19–23. September 2011. Critério, Almada, pp 1–9
23. Tumosa CS, Mecklenburg MF (2013) Oil Paints: The Chemistry of Drying Oils and the Potential for Solvent Disruption. In: Mecklenburg MF, Charola AE, Koestler RJ (eds) New Insights into the Cleaning of Paintings: Proceedings from the Cleaning 2010 International Conference, Universidad Politécnica de Valencia and Museum Conservation Institute. Smithsonian Institution Scholary Press, Washington D.C., pp 51–58

Chapter 42
Analysis of Cleaning Efficacy and Clearance of Silicone-Based Pickering-Type Emulsions Used in the Cleaning of Water-Miscible Oil Paints

Jennifer M. Myers and Matthew Cushman

42.1 Introduction

In recent years, efforts to minimize the extraction of surfactants, low-molecular-weight, and water-soluble components during aqueous cleaning through careful control of solution pH, conductivity and viscosity, for example, have shown promise; however, many of these approaches utilize hydrogels of cellulose ethers, polysaccharides, or poly(acrylic acid), all of which deliver a substantial amount of water to the paint film. This experiment is a continuation of recent efforts at the Winterthur/University of Delaware Program in Art Conservation (WUDPAC) to investigate the use of silicone-based particle emulsions in the cleaning of water-sensitive paint films. Their use avoids the need for surfactants that could irreversibly and unpredictably mix with mobile surfactants within the paint film. The use of surfactants in a cleaning system may cause difficulties with clearance and incur undesirable and unpredictable additive properties of hydrophilic lipophilic balance (HLB) levels and critical micelle counts (CMC), which have yet to be explored in this context, but may contribute to unanticipated surface sensitivity or soil retention. Given the demonstrated sensitivity of the paint film, additional beneficial outcomes may include a reduction in surface disruption resulting from pigment loss and medium leaching.

Water-miscible oil paints present many challenges to the conservator owing to their varied formulations. Patent literature indicates the use of polyoxyethylene-based (POE) nonionic surfactants with unsaturated fatty acid chains, rendering the paint mixable with water, while also cross-linking with fatty acids in the drying oil

J. M. Myers (✉)
Winterthur/University of Delaware Program in Art Conservation, Newark, DE, USA
e-mail: myersjm@udel.edu

M. Cushman
Winterthur Museum, Garden & Library, Winterthur, DE, USA
e-mail: mcushm@wintherthur.org

K. J. van den Berg et al. (eds.), *Conservation of Modern Oil Paintings*,
https://doi.org/10.1007/978-3-030-19254-9_42

component. Other additives, such as stabilizers, are also noted[1] [15]; the varying formulations between manufacturers may contain varying components not previously identified. A study of one water-mixable oil line, Grumbacher Max, POE surfactant was identified in addition to calcium or zinc stearates, cobalt drier, castor wax, wetting agents, and a polymeric dispersant found in some samples [9]. Industry literature discusses the formulations of the water-miscible paints in terms of working properties, such as gloss, film hardness, and drying times [7]. Changes in surface gloss due to loss of surfactant, as well as the movement of surfactants and soils into and out of both acrylic and water-miscible oil paints with changes in relative humidity have been observed [5] [18]. To address these challenges, Richard Wolbers (WUDPAC) and Anthony Lagalante (Villanova University) identified KSG-350z[2] as a potential emulsifier for testing, while advocating the use of silicone solvents for saturating test samples in order to provide a hydrophobic barrier, aiming to limit the infiltration of water contained within the cleaning system [10].

Research into the cleaning efficacy and clearance of silicone-based, Pickering-type emulsions on acrylic paints was performed by WUDPAC graduate students [2]. X-ray fluorescence (XRF) analysis indicated that silicon (Si) was present in elevated levels on acrylic surfaces following standardized cleaning protocols. A related study was conducted on samples of water-miscible oil paints [8]. This experiment used similar cleaning systems to the acrylic study and included data from SEM imaging, colorimetry and glossimetry to evaluate the effects of cleaning water-miscible oil paints on paint surfaces. Residues from the cleaning system were noted on cleaned samples of the paint-outs and further investigation was suggested. Residues from silicone solvents using octacyclotetrasiloxane (D4) as a clearance agent [6][3], regardless of the aqueous phase delivered, were identified using ATR-FTIR [4]. In other studies, surfactant and silicone residues from microemulsions and particle-based emulsions used on water-sensitive surfaces (see Ormsby et al. in this volume) were identified using X-ray photoelectron spectroscopy (XPS) [14] [13].

42.2 Experiment

42.2.1 Soiling Protocol

Two sets of five colors of Winsor & Newton Artisan Water-Mixable Oil Colors were painted out onto individual ground glass borosilicate slides and naturally aged 1.5 years in the paintings studio at Winterthur, with recorded RH levels between

[1] The components include amine soaps of fatty acids and thixotropic stabilizers based on castor oil derivatives.

[2] Manufactured by Shin-Etsu Chemical Company, Ltd., Tokyo. Available from Kremer Pigmente, Germany.

[3] The use of octacyclotetrasiloxane (D4) has been banned in Europe.

32–60% and temperature ranging from 17.2–25 °C[4]. The colors chosen were titanium white, ultramarine blue, viridian, alizarin crimson, and yellow ochre. Both sets of paint-outs were soiled using HITT Latent Fingerprint Powder;[5] composed of sintered lycopodium spores with the addition of nylon particles, visible under ultraviolet illumination. The fingerprint powder was made into a slurry with hexamethyldisiloxane (HMDS)[6], then applied with a synthetic brush to one half of each slide.

One set of paint-outs was artificially aged through temperature and relative humidity (RH) cycling. Cycling repetitions were suggested to simulate sufficient aging of these paints and to allow migration of surfactant into and out of the paint films [11]. The cycling consisted of five 24-hour cycles with a range of 20–80% RH and 10–30 °C [17]. The paint-outs were tested for water sensitivity with pH 6 adjusted water at 6000 μS, with swabs rolled on the surface. The different colors displayed variable loss of gloss and pigment transfer with as few as four and as many as 60 swab rolls. The most sensitive colors were alizarin crimson and yellow ochre. These observations were consistent with previous studies of traditional Winsor & Newton oil paint water sensitivity relating to the composition of binding medium, including increased oxidation combined with decreased cross-linking or raised levels of water-soluble diacids observed in some colors (including iron oxides and non-zinc oxide containing pigments) [12]. The addition of an emulsifier, or surfactant, was found to cause disruption in the cross-linking process, as well as an increase in the mobility of the molecules within the water-miscible oil paint film [16]. Migrated surfactant may also be a contributing factor to this observation, however, the presence of surfactant was not explicitly confirmed in this study.

42.2.2 Cleaning Protocol

Approximately one half of the soiled portion of each cycled and uncycled paint sample was cleaned. First, the surface was saturated with a 1:4 mixture of silicone solvents, D4, and decamethyl-cyclopentasiloxane (D5), to aid in the exclusion of the aqueous phase within the prepared emulsion. A cleaning system consisting of an aqueous phase in a 5:2:1 mixture by weight was used:

- Shin-Etsu KSG 350z
- Phenyltrimethicone (PTM)
- 0.5% citrate buffered with phosphate at pH 7

This was applied with a synthetic brush and/or cotton swab, then agitated on the surface. Where soil reduction was more difficult, the cleaning system was adjusted

[4] Eclimate notebook datalogger recordings at Winterthur from February 28, 2016 through August 31, 2018.

[5] HITT Marking Devices, HITT Companies, Inc., Santa Ana, California

[6] Sigma Aldrich Corporation

to swell the surface and dislodge the soiling material [17]. The formula that was found most effective consisted of the following in a 5:1:1:1:1 mixture:

- KSG-350z
- PTM
- 0.5% w/w diethylenetriaminepentaacetic acid (DTPA) adjusted with NaOH to pH 8
- Benzyl alcohol
- Xylene

This second system was developed to swell the surface in an attempt to remove the tightly-bound artificial soiling. Bulk removal of the emulsion was performed with cotton swabs. Both emulsions were then cleared using a 9:1 mixture of D5: HMDS (HMDS was added to accelerate evaporation).

42.3 Results

42.3.1 Cleaning Efficacy and Clearance

Examination with fluorescence microscopy (see Fig. 42.1a–c), visual observation, and scanning electron microscopy (SEM) indicated that the silicone emulsions were effective in the removal of the artificial soiling. Qualitatively, there was a significant difference between the ease of cleaning between the cycled and uncycled soiled samples. The cycled samples proved difficult to clean, resulting in the adjustment of the first cleaning system to include a solvent phase, as well as an increase of the pH

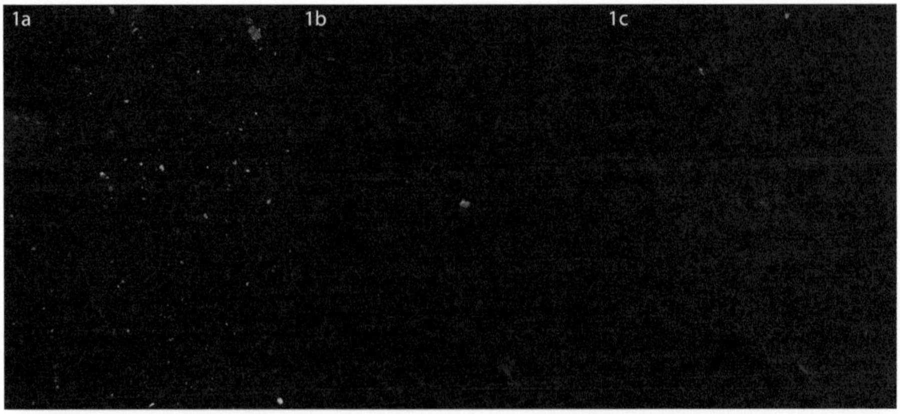

Fig. 42.1 a–**c** Alizarin crimson paint-outs. *Left to right* (**a**) Sample before cleaning with fluorescing, nylon particle-containing soiling (10x magnification, UV illumination) (**b**) Sample after cleaning with second silicone emulsion formulation (10x magnification, UV illumination) (**c**) Sample after cleaning with second silicon emulsion formulation (10x magnification, visible light)

of the aqueous phase. This was done to swell the surface in an effort to remove embedded soiling. Yellow ochre, alizarin crimson, and titanium white had soft surfaces that retained more of the fingerprint powder; additionally, these surfaces were easily abraded. Cleaning the uncycled samples was more straight-forward, utilizing the initial system to remove a large amount of soiling. The same three colors, however, required a pass with the second cleaning system.

One initial aspect of this research study was to use random sampling of cleaned areas to quantify the residues left behind by these emulsions to evaluate clearance. SEM imaging showed few occurrences of residual silicone globules, with fewer than two observed per cm^2 (see Fig. 42.2).

42.3.2 Surface Morphology

The surfaces of the alizarin crimson and yellow ochre have a cratered appearance at 10x, often with soiling residing in the recesses. Pigment was picked up on the cotton swabs following cleaning with the second formulation required to remove the soiling. These recesses may represent the loss of migrated surfactant or an oil component, creating voids and areas that harbor soiling material. It is also possible that fatty acid solubility in silicone solvents (namely HMDS, D4, and D5) is nonzero; if so, medium leeching may be a factor. Surface disruption could also result from the mechanical cleaning action and use of a cotton swab. Additionally, exposure to aqueous or solvent components held within the emulsion may contribute

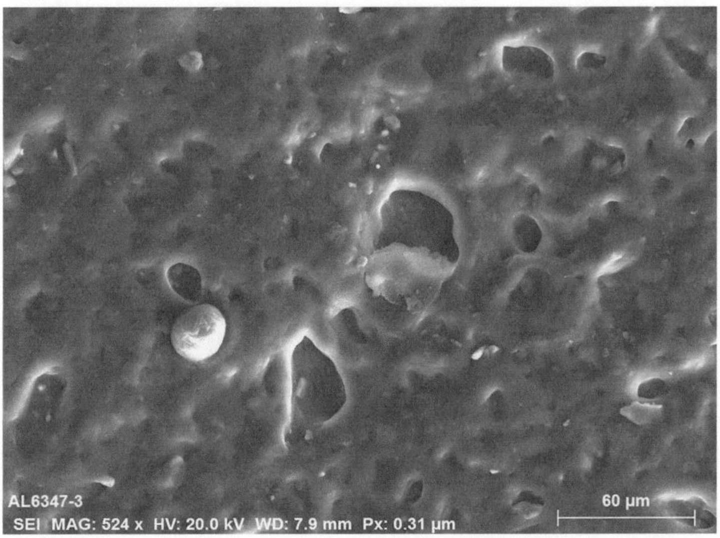

Fig. 42.2 Single, globular silicone particle identified on Alizarin crimson cycled and cleaned sample (rounded particle in lower left). SEM image courtesy of Dr. Judy Rudolph

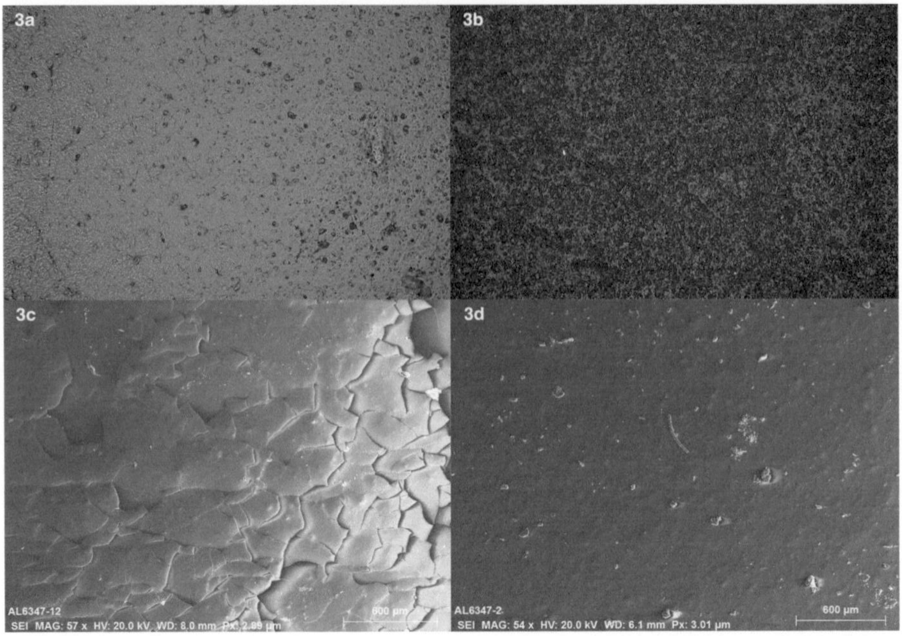

Fig. 42.3 a–d Alizarin crimson paint-outs. *Top* (**a**) Cycled and unsoiled on *left*. (**b**) Cycled and cleaned with the second silicone emulsion formulation on *right* (10x objective. Epi Brightfield filter cube) *Bottom* (**c**) Cycled and unsoiled on *left*. (**d**) Cycled and cleaned with the second silicone emulsion formulation on *right*. (SEM images courtesy of Dr. Judy Rudolph)

to the surface disruption, as the amounts of these materials reaching the surface are not eliminated. This disruption of the surface is readily apparent under epi-brightfield (see Fig. 42.3a). The surfaces of the yellow ochre, alizarin crimson, and titanium white samples were noticeably less reflective after cleaning. SEM imaging suggests that the appearance of cracking at the surface was reduced after cleaning, likely the result of swelling (see Fig. 42.3b). In some regions, the cleaned surfaces appear softened and more porous. Preliminary observations suggested that the surface gloss recovered somewhat over time.

42.3.3 Cross-Sections

Cross-section analysis was not part of the initial protocol, however it was performed after the discovery of distinct differences in the ease and efficacy of cleaning between the cycled and uncycled samples. Observations indicated that RH fluctuations and/or levels relate to this difference in soil retention. Cross-sectioned samples were taken from cycled and uncycled soiled paint-outs in order to characterize interaction between the artificial soiling and the paint film.

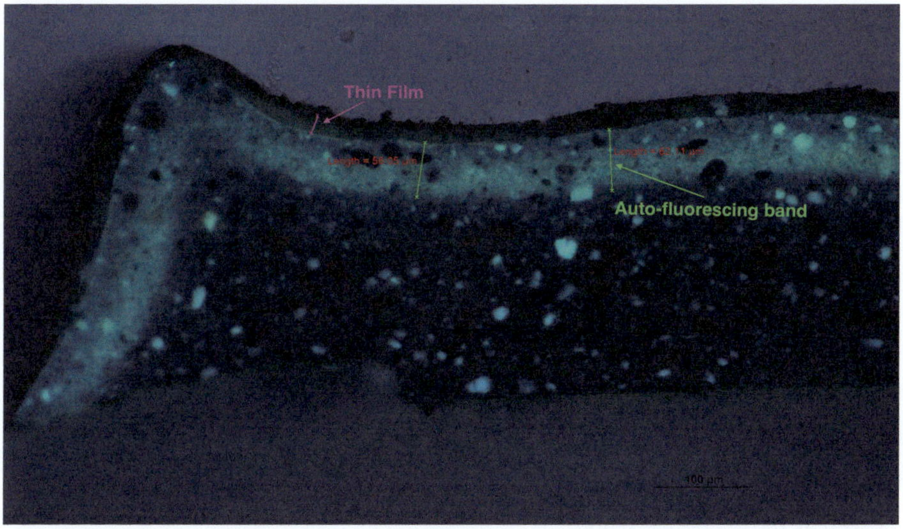

Fig. 42.4 Soiled and uncycled ultramarine blue cross-section viewed with UV illumination and 20x objective

The layer of soiling did not appear to have migrated into the surface of the paint film with the possible movement of surfactant. In all samples, the soiling accumulated in a uniform thickness, tightly conforming to the topography of the surface. However, uncycled samples exhibited soiling within a thin, fluorescing film above the paint surface, possibly the result of surfactant migration and/or medium skins found in other modern formulations of Winsor & Newton oil paints [3] [4]. The lack of such films visible in the cycled samples may help explain the difficulties encountered with cleaning the cycled samples. Additionally, all samples exhibited a UV-fluorescing band, ~40–60 microns in thickness, near the surface of the paint film (see Fig. 42.4). Similar phenomena have been noted in studies of both modern and traditional oil paints. In one study [3], the bands were analyzed in comparison to the bulk of the film and found to have higher concentrations of hydroxyl groups and carboxylic acids near the surface in the majority of the Winsor & Newton oil paints examined. The characterization of this auto-fluorescent band as also found in water-miscible oil paints will be considered in future investigations.

42.4 Conclusions

Silicone-based particle emulsions are a surfactant-free option for the cleaning of water-miscible oil paints, however consideration should be given to the changes in the paint film observed during this study. Following the cleaning protocol described resulted in the observation of minimal residues (<2 particles/cm^2), the impact of which have yet to be ascertained in the medium to longer-term. It is possible these residues may affect adhesion of future varnishes to the surface of cleaned paint films [1]. These results were noted in both sets of paint-outs and with both the original and adjusted cleaning systems.

The cycling and/or elevated RH and temperature protocol contributed to a more challenging cleaning process, as the soiling was more difficult to separate from the surface, particularly on the softer films. The soiling material conformed to, but did not migrate into, the paint film, contrary to observations for acrylic paint systems that underwent similar environmental conditions [11].

Changes were noted in the surface morphology, such as the softening of paint film, loss of fine surface cracks, and loss in gloss after cleaning. These changes may be related to the application method, the swelling of the surface due to interactions with paint components and/or the aqueous or solvent phases within the silicone-based emulsions. Auto-fluorescent layers were observed in cross-sections that may relate to effects seen in standard modern oil paints. These layers may represent phase or medium separation, and occurred in samples that were exposed to cycles, as well as those that were not.

42.5 Further Research

Water-miscible oil paints are increasingly being used and will likely come across the easels of painting conservators in the future. Investigation into the prevalence and availability of water-miscible oil paint products, their presence in collections, and their use by artists would help us understand these paints and their conservation challenges more fully. Further research may include the identification of possible brand variations in surfactant type, length of ethoxylate chains, and/or functional groups on the modified oil. One study examined the specific components of Grumbacher Max water-mixable oil colors, using FTIR and GC-MS [9]. Further studies could characterize these alterations within a singular product line and/or between manufacturers of similar products in the United States, Europe and further afield.

The characterization of the auto-fluorescent phase separation layers observed in cross-section analysis may assist in the identification and understanding of the behavior of mobile components present in these paints, with a particular emphasis on how they compare to traditional oil paints. This study suggests that phase separation occurs regardless of temperature and relative humidity levels or fluctuations the paints were subjected to. These mobile components may originate from additives, emulsifiers, or drying oil oxidation or hydrolysis products, which may vary across product lines. The period of time before these separations occur also remains to be studied. The presence of a thin film or medium skin at the surface of the uncycled sets could be monitored through studies of surface morphology over time, and after cleaning to determine the rate and extent which the surface gloss forms and returns.

The ongoing study of these relatively new paint products will prove invaluable to conservators working with contemporary collections. Given the transitionary status these paints are occupying in schools and studios, there may be misidentification due to these materials being described by manufacturers and users as being similar to/the same as traditional oil paints, in terms of handling and appearance. Understanding the components of these materials may be necessary as future works labeled "oil paint" may in fact behave completely differently. The identification of these materials in paintings may become critical in the treatment of surfaces with unique sensitivities. It is the case that water-miscible oil paints differ in composition from traditional oil paints, affecting possible aging and water or solvent-sensitivity, and contributing to the complex issues already faced by conservators of modern and contemporary paintings, such as zinc soap and epsomite salt formation.

Acknowledgements This research project would not have been possible without assistance and support from Winterthur/University of Delaware Program in Art Conservation faculty and staff, including Richard Wolbers, Dr. Rosie Grayburn, Dr. Judy Rudolph, Debra Hess Norris, and Dr. Joyce Hill Stoner. Funding for attending the Conference on Modern Oil Paints was generously provided by the Society of Winterthur Fellows and the Graduate Department at the University of Delaware.

References

1. Baade B (2018) Personal communication. University of Delaware, Newark, DE. March 13, 2019
2. Beller A, Summer J, Albertson G (2016) Liquitex® Acrylic Emulsion Cleaning Study: Testing siloxane-based cleaning materials. Poster presented at IIC Congress, Los Angeles, 12–16 September 2016
3. Burnstock A, Lee J, van den Berg KJ, Ormsby B (2016) Water sensitivity of modern oil paint films. In: Selva Bonino VA (ed) CESMAR7 – Colore y Conservazione 2015. 7th International Conference Color and Conservation 2015, 13-14 Nov 2015, Milan, Italy. Il Prato, p 66–76
4. Chung JY, Ormsby B, Lee J, Burnstock A, van den Berg KJ (2017) An Investigation of options for surface cleaning unvarnished water-sensitive oil paints based on recent developments for acrylic paints. In: Proceedings from Modern Materials and Contemporary Art, 18th Triennial Conference, International Council of Museums (ICOM) and the Committee for Conservation (ICOM-CC), Copenhagen, 2017
5. Digney-Peer S, Burnstock A, Learner T, Khanjian H, Hoogland F, Boon J (2004) The migration of surfactants in acrylic emulsion paint films. In: Roy A, Smith P (eds) Modern Art, New Museums: Contributions to the Bilbao Congress, International Institute for Conservation of Historic and Artistic Works, London, 13-17 September 2004, Stud in Conservation 49(sup2): 202–207. https://doi.org/10.1179/sic.2004.49.s2.044
6. ECHA (2018) Identification of octamethylcyclotretrasiloxane (D4) as a substance of very high concern because of its PBT and vPvB Properties (Article 57 D & E) Support Document. European Chemical Agency. Updated 13 June 2018. Available via https://echa.europa.eu/documents/10162/13638/msc_svhc_supdoc_d4_en.pdf/83af6327-5fbd-6218-60a2-6053eed4143f. Accessed 8 Mar, 2019
7. Foster A (2007) The Performance and Properties of Artisan Water Mixable Oil Colour Compared with Other Oil-Based Paints by Winsor & Newton. In: Learner TJS, Smithen P, Krueger JW, Schilling MR (eds) Modern Paints Uncovered: Proceedings from the Modern

Paints Uncovered Symposium, The Getty Institute, London, 16–19 May 2006, Getty Publications, p 53–57

8. Hartman D, Wingel K, Nigro E, Porell M, Teeter K (2017) New Applications of Pickering-type emulsions for the cleaning of water-dispersed oil paints. Poster presented at Modern Materials and Contemporary Art, 18th Triennial Conference, International Council of Museums (ICOM) and the Committee for Conservation (ICOM-CC), Copenhagen, 4–8 September 2017
9. Hodgkins RE, Berrie BH, Lomax SQ (2015) Multi-analytical approach to characterization and comparison of Grumbacher Pre-tested and Max water mixable oils. Poster presented at TECHNART International Conference, Catania, 27–30 April 2015
10. Lagalante AF, Wolbers RC (2017) Particle-based silicone cleaning emulsions: studies in model paint systems for the cleaning of water-sensitive artworks. In: Angelova LV, Ormsby B, Townsend J, Wolbers R (eds) Gels in the Conservation of Art, Archetype Publications, London, p 193–199
11. Lagalante AF, Wolbers RC (2016) The cleaning of acrylic paintings: new particle-based water-in-oil emulsifiers. In: Selva Bonino VE (ed) CESMAR7 – Colore y Conservazione 2015. 7th International Conference Color and Conservation 2015, 13-14 Nov 2015, Milan, Italy. Il Prato, p 163–167
12. Lee J, Bonaduce I, Modugno F, La Nasa J, Ormsby B, van den Berg KJ (2018) Scientific investigation into the water sensitivity of twentieth century oil paints. Microchemical J 138: 282–295
13. Ormsby B, Lee J, Bonaduce I, Lluveras-Tenorio A (2019) Evaluating cleaning systems for use on water sensitive modern oil paints: a comparative study. In: van den Berg KJ, Bonaduce I, Burnstock A, Ormsby B (eds) Conservation of Modern Oil Paints, Springer, New York. Submitted March 2019
14. Ormsby BA, Willneff EA, Clark MB Lopez J JR, Keefe MH, Phenix A (2016) Cleaning acrylic paints: mineral spirits-based microemulsions and the question of residues. In: Selva Bonino, VE, (ed) CESMAR7 – Colore y Conservazione 2015. 7th International Conference Color and Conservation 2015, 13–14 Nov 2015, Milan, Italy. Il Prato
15. Schilling MR, Mazurek J, Learner TJS (2007) Studies of modern oil-based artists' paint media by gas chromatography/mass spectrometry. In: Learner TJS, Smithen P, Krueger JW, Schilling MR (eds) Modern Paints Uncovered: Proceedings from the Modern Paints Uncovered Symposium, The Getty Institute, London, 16–19 May 2006, Getty Publications, p 129–139
16. Udell NA, Hodgkins, RE, Berrie BH, Meldrum T (2017) Physical and chemical properties of traditional and water-mixable oil paints assessed using single-sided NMR. Microchemical J 133: 31–36. https://doi:10.1016/J.MICROC.2017.03.013
17. Wolbers R (2018) Personal communication. Winterthur Museum, Library, & Garden, Winterthur, DE. April 11, 2018
18. Ziraldo I, Watts K, Luk A, Lagalante AF, Wolbers RC (2016) The influence of temperature and humidity on swelling and surfactant migration in acrylic emulsion paint films. Stud in Conservation 61(4): 209–221. https://doi.org/10.1179/2047058414Y.0000000156

Chapter 43
Improving the Surface Cleaning of Water Sensitive Oil Paint by Use of Alternative Application Methods

Mirjam Hintz, Klaas Jan van den Berg, Maartje Stols-Witlox, and Lise Steyn

43.1 Introduction

Water sensitivity in modern oil paints can pose a significant challenge during surface cleaning, as sensitivity can lead to the removal of pigment and undesired changes to the surface appearance of paints. Although, non-polar solvents or dry cleaning methods can sometimes be used for the cleaning of sensitive paints, they may not produce satisfactory cleaning results. Hence, when applying wet cleaning systems to sensitive surfaces either the mechanical action, specifically abrasion, or the amount of water, or both, must be reduced. Reducing the mechanical action during cleaning has not been the subject of extensive systematic research, although investigations have been carried out using rigid hydrogels specifically for this purpose [2, 14].

One known cause for the water sensitivity of oil paints is the formation of water-soluble magnesium heptahydrate sulphate (epsomite) crystals, which can protrude through the upper medium skin of paint films, rendering them water sensitive [6, 10]. Silvester et al. showed that magnesium sulphate and sulphite hydrate crystals could be produced on $MgCO_3$ containing test paint films, through exposing them to atmospheric sulphur dioxide (SO_2). This methodology was followed in the present study using contemporary W&N Artist's oil paints. Cleaning tests were performed, using aqueous solutions and one silicone emulsion, which were applied using an ultrasonic mister. Moisture and soil were then removed with a brush and/or tissue. This study also investigated the use of a non-polar cyclic silicone solvent (D5),

M. Hintz (✉)
Golden Artist Colors, Frankfurt, Germany

K. J. van den Berg
Cultural Heritage Agency of the Netherlands, Amsterdam, The Netherlands

M. Stols-Witlox · L. Steyn
University of Amsterdam, Amsterdam, The Netherlands

K. J. van den Berg et al. (eds.), *Conservation of Modern Oil Paintings*,
https://doi.org/10.1007/978-3-030-19254-9_43

proposed for the cleaning of sensitive paints, which can also be used for the temporary hydrophobization of sensitive surfaces, prior to the application of aqueous cleaning solutions/systems [13].

43.2 Materials and Methodology

43.2.1 Sample Preparation

To produce water sensitive paint samples for the cleaning tests, five contemporary W&N Artists' Oil Paints (Table 43.1) were painted out on 200 μm thick Melinex® sheets, then artificially aged and soiled.[1] Samples were then exposed to SO_2 vapours using the same methodology was used as described in Silvester et al. [10]. Prior to the SO_2 incubation, Scanning Electron Microscopy with Energy Dispersive X-ray spectroscopy (SEM/EDX) analysis was carried out on all paints. Results confirmed that all paint samples, except the cadmium red paint, contained the $MgCO_3$ filler known to be used in W&N paints.

To document the impact of soiling and the SO_2 incubation, the water sensitivity and the surface conductivity of the samples were recorded before and after SO_2 incubation, on both soiled and unsoiled surfaces (Table 43.1). Regarding surface conductivity, a correlation was observed between the values obtained and water sensitivity. The more water sensitive ultramarine and cobalt blue samples had significantly higher conductivity values after SO_2 incubation (Table 43.1), which suggests the formation of water-soluble salts [11].

The longer the samples were exposed to SO_2, the more water sensitive they became as evaluated using the swab-roll method described above. The incubation process produced samples with varying degrees of water sensitivity between colours and consistent sensitivity within the same colour. The cadmium red (non-$MgCO_3$ containing) samples did not develop water sensitivity and were excluded from further assessment. It is noted that the water-sensitivity of these samples differed to the

Table 43.1 Water sensitivity and conductivity of soiled samples before and after SO_2 incubation

	Before SO_2 incubation, soiled		After SO_2 incubation, soiled	
Samples	Swab rolls until pigment pick-up	Avg. conductivity in μS/cm	Swab rolls until pigment pick-up	Avg. conductivity in μS/cm
Cadmium lemon	<30	19 (16–23)	9	235 (250–300)
Oxide of chromium	20	34 (16–69)	14	183 (210–330)
Cobalt blue deep	7	457 (430–490)	3	3232 (2800–3700)
French ultramarine	14	60 (56–65)	3	1003 (870–1050)
Cadmium red	<30	–	<30	–

[1] For the composition of artificial soil, which was brush applied, see Ormsby et al. [9].

W&N paint swatches, dating from 1965, investigated by Cooper et al. [4],[2] where the samples with confirmed epsomite exhibited extreme sensitivity to water at the surface while the bulk paint film showed less sensitivity. However, in the current study, the cobalt blue, chrome green and ultramarine samples remained water sensitive to the same extent for over 300 swab rolls. Only the cadmium yellow samples gradually decreased in sensitivity after about 100 swab rolls. This difference may be explained by the fact that the paints used in the current study are relatively young, and thus not fully crosslinked.[3]

Differently shaped crystals were observed using digital microscopy (HIROX) on the cobalt blue and ultramarine samples. SEM-EDX analysis of sample surfaces showed that these crystalline areas contained very high amounts of sulphur and magnesium. As no XRD analysis was executed within this study, these crystals were not further identified.

43.2.2 The Ultrasonic Mister

The ultrasonic mister (ProMedix PR-800)[4] was used for the application of the eight aqueous solutions and the silicone emulsion as outlined in Table 43.2. This device was designed for medical purposes and facilitated the use of chemicals. The mouth piece was exchanged for a flexible latex tube, so that the mist could be applied locally and in a controlled manner. The ultrasonic action breaks up water into tiny droplets, thus wetting a larger surface area using less water. The misting action aimed to weaken the bond between the dirt and paint surface, so that dirt removal could then be completed using additional methods. For this study, the following combinations were evaluated:

(a) Swab rolling
(b) Mister (Fig. 43.1a), and Evolon tissue (Fig. 43.1c) dampened with clearing solvent (see Table 43.2), dabbed with a sponge for even pressure
(c) Mister, brush (Fig. 43.1b) and Evolon tissue dampened with clearing solvent, dabbed with a sponge
(d) Mister and brush

[2] The paint swatches were prepared by W&N for quality control, specifically to investigate the drying times of batches of paint [4].

[3] A reviewer suggested that this water sensitivity may also be explained by pronounced hydrolysis during the artificial ageing. Since the samples were exposed to high humidity levels during the SO_2 incubation, the authors believe that the contribution of hydrolysis to the water sensitivity could have been significant.

[4] Inspiration to use a mister for cleaning was derived from a previous case study, in which a non-sensitive, high impasto oil painting had been successfully cleaned with an aqueous solution which was misted onto the surface while the dirt was picked up with a brush, pulp tissue pads and cotton swabs [12].

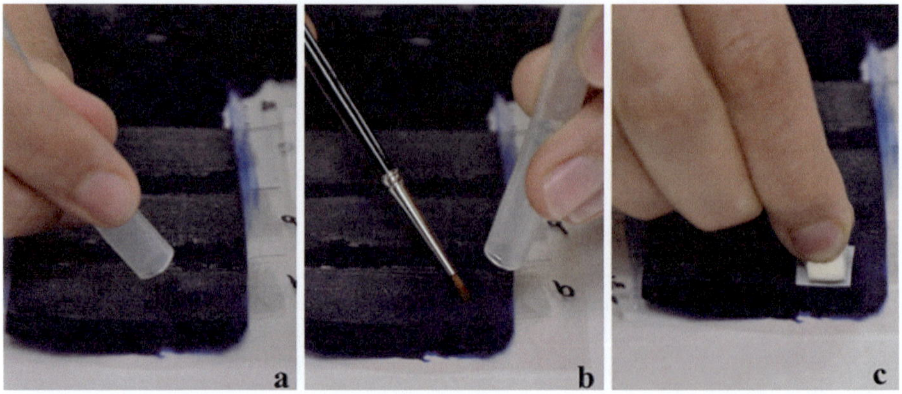

Fig. 43.1 Aqueous cleaning solution is applied with the mister (**a**). The soil is agitated and picked up with a sable brush (**b**). Damp Evolon tissue is dabbed with a sponge several times to slightly increase the mechanical action (**c**)

43.2.3 Aqueous Solutions

The water used for numerous aqueous solutions and gels (Table 43.2), was buffered to a pH of 5, to presumably minimize ionization, swelling or blanching of the paint surface [3]. Volatile buffering materials (acetic acid & ammonium hydroxide) were chosen to omit the necessity for clearance. The citric acid chelating solution was also buffered to pH 5 with sodium hydroxide. Low concentrations of triammonium citrate were also used with and without additions of non-ionic surfactants. Surfynol® 61 (3,5-Dimethyl-1-hexin-3-ol) is a non-ionic and volatile surfactant, which might not require rinsing and thus could be a valuable cleaning agent for the surface cleaning of water sensitive paints.[5]

43.2.4 Hydrogels and Cyclomethicone D5

Agar gel was prepared with deionised (DI) water at 8% w/w. concentration, in a double boiler with thermometer, left to cool, and then heated for a second time whereby the buffered solution (same as in test 2) was added to reduce the gel to 4%, then applied once at ~40 °C, and removed after 2 min. Two 4% w/w. xanthan gels were prepared with DI water and reduced to 2% with the buffered solution used in test 2. To one of the two xanthan gels 1% Shellsol T and 1% Pluronic® L64 (a non-ionic, water soluble polyethylene glycol - polypropylene glycol surfactant) were

[5] Balcar et al. [1] reported that Surfynol 61 left depostits on the surfaces of various plastics, when used in greater concentrations than 0.1% in distilled water. See also the study by K. Mürer: https://www.hornemann-institut.de/german/epubl_ha_ausgabe.php?haid=876&ha_sort=au&l=o&th=M&spra=deutsch&sort=ea_name&ref=/german/epubl_hs_arbeiten.php . Accessed March 2019.

Table 43.2 Water sensitivity determined by swab roll tests[a] and conductivity[b] before and after SO_2 incubation of soiled samples

No.	System type	System type	Clearance material	Average cleaning results Pigment pick-up	Cleaning efficacy	Gloss integrity
1	Deionized water (DI)	Aqueous solutions	n/a	3) faint stain	3) moderate	2) change visible in daylight
2	Buffered water at 1mS/cm, pH 5		n/a	3- faint stain	3) moderate	3) change visible in raking light
3	0.5% citric acid, pH 5 with NaOH		DI	3- faint stain	3) moderate	3) change visible in raking light
4	0.5% citric acid, pH 5 with NaOH & 0.5% Pluronic L64		DI	3- faint stain	3) moderate	3) change visible in raking light
5	0.1% Surfynol 61 in pH 5 buffered water		n/a	3- faint stain	3) moderate	3) change visible in raking light
6	1% TAC		DI	3- faint stain	3) moderate	3) change visible in raking light
7	1% TAC & 1% Pluronic L64		DI	3- faint stain	3) moderate	3) change visible in raking light
8	1% TAC & PEG 4000		DI	3- faint stain	3) moderate	2) change visible in daylight
Averaged cleaning results of all aqueous solutions (1–8) according to application method						
Swab			30 swab rolls with one swab	2) minor pigment pick-up	4) slightly effective	2) change visible in daylight
Mister and Evolon tissue			3 Evolon tissues moistened with DI	4) very faint stain	2) little effect	3) change visible in raking light
Mister, brush, Evolon tissue			3 Evolon tissues moistened with DI	4) very faint stain	3) moderate	3) change visible in raking light
Mister and brush			3 Evolon tissues moistened with DI	4) very faint stain	4) slightly effective	3) change visible in raking light
9	D5- based silicone emulsion (50% DI, 30% PEG, 20% D5)	Silicone emulsion	3 Evolon tissues moistened with D5	5) no pigment/ stain	2) little effect	4) slight shift
10	4% agar gel, pH 5 buffered water	Rigid gels	n/a	5) no pigment/ stain	3) moderate	3) change visible in raking light

(continued)

Table 43.2 (continued)

No.	System type	System type	Clearance material	Average cleaning results: Pigment pick-up	Cleaning efficacy	Gloss integrity
12	2% xanthan gum, pH 5 buffered water aqueous gels	Aqueous gels	2 swabs, each 20 swab rolls with DI[c]	3) faint stain	3) moderate	3) change visible in raking light
13	2% xanthan gum, pH 5 buffered water & D5 barrier layer		2 swabs, each 20 swab rolls with DI	4) very faint stain	3) moderate	4) slight shift
14	2% xanthan, buffered pH 5 + 1% Pluronic L64 + 1% Shellsol T	Oil-in-water emulsions	2 swabs, each 20 swab rolls with DI	3) faint stain	4) slightly effective	3) change visible in raking light
15	2% xanthan, buffered pH 5 + 1% Pluronic L64 + 1% Shellsol T & D5 barrier layer	Oil-in-water emulsions	2 swabs, each 20 swab rolls with DI	4) very faint stain	4) slightly effective	3) change visible in raking light

Pigment pick-up: 1) substantial pigment pick-up, 2) minor pigment pick-up, visible with stain, 3) faint stain, 4) very faint stain, 5) no pigment/stain
Cleaning efficacy: 1) no effect, 2) little effect, 3) moderate, 4) slightly effective, 5) complete removal
Gloss integrity: 1) significant change, 2) visible change, i.e. visible in normal light, 3) slight change, i.e. only visible in raking light, 4) slight shift, 5) no change visible
[a]Cotton swabs were dipped half-way into deionized water, slightly blotted prior to application and then rolled on a section of the paint surfaces until pigment was visible on the swabs. One swab roll is equal to a frontward roll and a backward roll
[b]The conductivity was measured with a drop of deionized water, placed on the samples for 1 min and then transferred to a Horiba B-173 Twin Cond Conductivity meter. Of each sample three readings were taken
[c]In addition to clearing with two swabs (each 20 swab rolls with DI), the xanthan gels were also cleared using five Evolon tissues moistened with DI. The pigment-pick up was not reduced using the Evolon tissues for clearing.

added for increased cleaning power. The gels were applied by brush, and removed after 2 min. The xanthan gels were tested with and without a prior application of cyclomethicone D5, a volatile and highly hydrophobic cyclic silicone solvent, used to create a temporary barrier between the gels and the paint surfaces.[6]

[6]The use of the silicone solvent D4, D5, and D6 is presently reviewed by the European Chemical Agency (ECHA) and restricting regulations will enter into force after 31. January 2020. Restrictions are derogated for their use in conservation and restoration, medical purposes and other niche markets. D4 is known to be hazardous for human health and is classified as a reprotoxic substance. www.echa.europa.eu/documents/10162/26cc3562-d0a0-1e09-b1eb-72f6bdf9309a. Accessed 9 Mar 2019, [5].

43.2.5 Silicone Emulsion

The D5 solvent-based silicone emulsion contained 50% DI water, 30% Polyethylene glycol 4000 (a hydrophilic and high molecular polymer ethylene glycol with low-foaming tendency), and 20% D5. The solvent-surfactant-water ratio was based on R. Wolbers [13].

43.2.6 Evaluation Methodology

Results were evaluated using a number of factors such as pigment pick-up, cleaning efficacy, changes in gloss, visible residues and handling properties as employed in similar studies [3]. Effects were rated directly after the samples were dry via visual observation and subsequently investigated with Micro Reflectance Transformation Imaging (Micro RTI), digital microscopy (HIROX), SEM-EDX and SEM imaging.

43.3 Results and Discussion

43.3.1 Aqueous Solutions

The most striking result in terms of pigment pick-up, was that no one cleaning solution proved to be more acceptable, and that significant changes were observed depending on the application method. Swab action caused consistently more pigment pick-up than combined applications using the mister. For the cobalt blue samples, all of the aqueous solutions caused substantial pigment pick-up, which was partly reduced by a process of applying the solution several times, always stopping before pigment was removed and letting the surface dry before continuing (Table 43.3).

Cleaning (soiling removal) efficacy also depended on the application method. Swab action always resulted in the cleanest surfaces, whereas the efficacy for the combination methods depended on how they were used. Humidifying the surface with the mister followed by dirt pickup using the Evolon tissue performed poorly as the tissue was too dense to imbibe the dirt.[7] If a brush was used in combination with the mister however, cleaning efficacy improved. All of the aqueous cleaning solutions had moderate cleaning efficacy. Ultimately, 0,1% (v/v.) Surfynol® 61 in pH 5 buffered water (test 5) produced the best cleaning results, followed by 0,5% (w/w) sodium citrate solution with an 0,5% Pluronic® L64 addition (test 4).

Concerning gloss integrity, the impact of cleaning tests depended largely on the paint colours (pigment type). On the chrome green and cadmium yellow samples, the reduced gloss was only visible in raking light, whereas on the ultramarine samples, most tests caused a gloss reduction that was visible in normal light. On the cobalt blue samples, most tests caused a significant reduction in gloss, where the structure of the Evolon tissue and cotton swabs left a visible pattern after drying,

[7] See also Vergeer et al., these Proceedings.

Table 43.3 Cleaning systems and summary of results

No.	System type	System type	Clearance material	Average cleaning results Pigment pick-up	Cleaning efficacy	Gloss integrity
1	Deionized water (DI)	Aqueous solutions	n/a	3) faint stain	3) moderate	2) change visible in daylight
2	Buffered water at 1mS/cm, pH 5		n/a	3- faint stain	3) moderate	3) change visible in raking light
3	0.5% citric acid, pH 5 with NaOH		DI	3- faint stain	3) moderate	3) change visible in raking light
4	0.5% citric acid, pH 5 with NaOH & 0.5% Pluronic L64		DI	3- faint stain	3) moderate	3) change visible in raking light
5	0.1% Surfynol 61 in pH 5 buffered water		n/a	3- faint stain	3) moderate	3) change visible in raking light
6	1% TAC		DI	3- faint stain	3) moderate	3) change visible in raking light
7	1% TAC & 1% Pluronic L64		DI	3- faint stain	3) moderate	3) change visible in raking light
8	1% TAC & PEG 4000		DI	3- faint stain	3) moderate	2) change visible in daylight
Averaged cleaning results of all aqueous solutions (1–8) according to application method						
Swab			30 swab rolls with one swab	2) minor pigment pick-up	4) slightly effective	2) change visible in daylight
Mister and Evolon tissue			3 Evolon tissues moistened with DI	4) very faint stain	2) little effect	3) change visible in raking light
Mister, brush, Evolon tissue			3 Evolon tissues moistened with DI	4) very faint stain	3) moderate	3) change visible in raking light
Mister and brush			3 Evolon tissues moistened with DI	4) very faint stain	4) slightly effective	3) change visible in raking light

(continued)

Table 43.3 (continued)

				Average cleaning results		
No.	System type	System type	Clearance material	Pigment pick-up	Cleaning efficacy	Gloss integrity
9	D5- based silicone emulsion (50% DI, 30% PEG, 20% D5)	Silicone emulsion	3 Evolon tissues moistened with D5	5) no pigment/ stain	2) little effect	4) slight shift
10	4% agar gel, pH 5 buffered water	Rigid gels	n/a	5) no pigment/ stain	3) moderate	3) change visible in raking light
12	2% xanthan gum, pH 5 buffered water aqueous gels	Aqueous gels	2 swabs, each 20 swab rolls with DI	3) faint stain	3) moderate	3) change visible in raking light
13	2% xanthan gum, pH 5 buffered water & D5 barrier layer		2 swabs, each 20 swab rolls with DI	4) very faint stain	3) moderate	4) slight shift
14	2% xanthan, buffered pH 5 + 1% Pluronic L64 + 1% Shellsol T	Oil-in-water emulsions	2 swabs, each 20 swab rolls with DI	3) faint stain	4) slightly effective	3) change visible in raking light
15	2% xanthan, buffered pH 5 + 1% Pluronic L64 + 1% Shellsol T & D5 barrier layer	Oil-in-water emulsions	2 swabs, each 20 swab rolls with DI	4) very faint stain	4) slightly effective	3) change visible in raking light

Pigment pick-up: 1) substantial pigment pick-up, 2) minor pigment pick-up, visible with stain, 3) faint stain, 4) very faint stain, 5) no pigment/stain
Cleaning efficacy: 1) no effect, 2) little effect, 3) moderate, 4) slightly effective, 5) complete removal
Gloss integrity: 1) significant change, 2) visible change, i.e. visible in normal light, 3) slight change, i.e. only visible in raking light, 4) slight shift, 5) no change visible

causing a more matte appearance (Fig. 43.2).[8] In addition, cotton fibres would frequently adhere to wet cobalt blue paint. RTI and SEM investigations of the range of samples revealed that the application of solutions with the mister, in combination with a brush created the smoothest surfaces at the micro-scale, however the imbibed dirt was not efficiently removed.

43.3.2 *Silicone Emulsion*

The silicone emulsion resulted in no pigment pickup regardless of the application method. However, the dirt was not fully solubilized, which resulted in poor soiling removal. The high percentage of surfactant (30%) included in the formulation also raised concerns about clearance and possible residues. Clearance was however easily carried out with D5 without any visible change.

[8] Reduction in gloss after cleaning was also noted by Myers, these Proceedings, Ch. xx, [7].

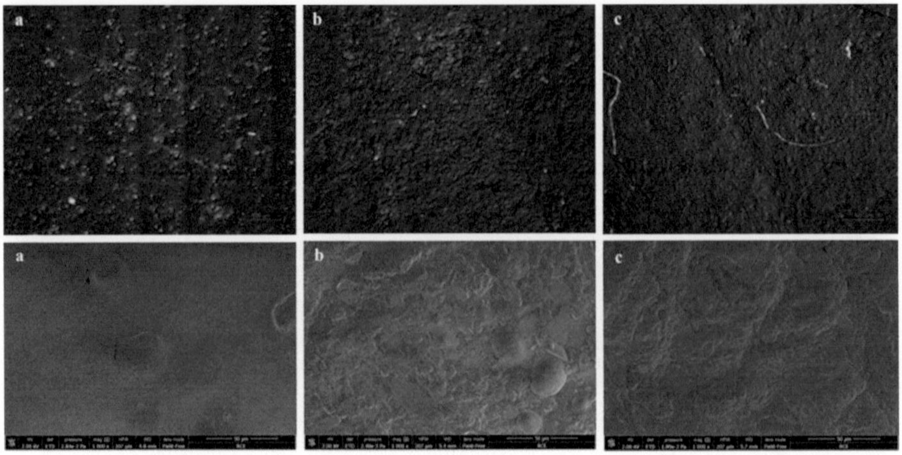

Fig. 43.2 Micro RTI images in default mode, x5 magn. (upper row) and LV SEM-BSE images (x1000 magn.) (bottom row) of cobalt blue, cleaned with pH 5 buffered DI and applied with (**a**) mister & brush, (**b**) mister & brush & Evolon, and (**c**) swab. It can be seen that the mister and brush application does not damage the surface structure of the paint (**a**), whereas the additional use of Evolon (**b**) and the swab (**c**) flatten/roughen the paint surface. In c cotton fibres are visible (top row). SEM images (**a**) and (**b**) show starch globules, which have been removed in (**c**). SEM-BSE: Ineke Joosten

In previous cleaning studies, some silicone-based emulsions and mineral-spirits-based [8] micro emulsions have been reported to result in pigment pickup when applied by swab [3]. For the silicone emulsions used in this study, foam was created inside the mister nebulizer and did not transfer onto paint surfaces.

43.3.3 Agar Gels

No pigment pick-up occurred with the agar gel and the dirt removal efficacy was moderate. Gloss integrity after treatment was dependent on the sample colours. The cadmium yellow and chrome samples green exhibited no changes in gloss, whereas the ultramarine and cobalt blue samples once again showed significant reductions in gloss. It was noted that during application, the agar gels resulted in swelling at the paint surface which has also been noted by others (Ormsby et al., this volume), resulting in blanching and a temporary lightening of the colour until the paint was fully dry.

43.3.4 Xanthan Gels

Protecting the paint surfaces with D5 improved the performance of the xanthan gels by reducing pigment pick-up and enhancing gloss integrity. Compared to the agar options, the xanthan gel resulted in more pigment pick-up but offered enhanced cleaning. The addition of surfactant and solvent for tests 14 & 15 unsurprisingly improved the cleaning efficacy by enhancing the cleaning power of the gel. The xanthan gel was however less transparent than the agar gel, which made the observation of subtle discolouration due to pigment pick-up more difficult. However, the xanthan gel was more controllable than the agar gel due to its more fluid consistency.

43.4 Conclusions

The phenomenon of increased water sensitivity due to the presence of magnesium sulphate or sulphite salts was successfully reproduced through SO_2 incubation of $MgCO_3$ containing oil paints; though others have recently reported the use of high humidity alone to induce similar effects (Ormsby et al., this volume). The combined cleaning assessments revealed clear tendencies, which may inform cleaning practice. The present study has reinforced the growing understanding that application methods are key for lowering the risks associate with surface cleaning sensitive paints and can be as, or even more important than the choice of cleaning solution. Assessment using an ultrasonic mister proved promising, where dirt layers were pre-softened selectively to facilitate the ensuing cleaning procedure.

Acknowledgements The authors would like to thank: Paolo Cremonesi, Ineke Joosten, Suzan de Groot, Laura Maria Ledwina, Julia van den Burg, Birgit Reissland (RCE); Martin Jürgens (Rijksmuseum); Katharine Waldron (Courtauld Institute of Art).

LS was funded by the Netherlands Organisation for Scientific Research (NWO) through Heritage Plus, a Joint Programming Initiative of the European Commission.

Materials and Suppliers

Agar (05040)	Sigma-Aldrich
Evolon® CR (2219000)	Deffner & Johann
Xanthan gum (63450)	Kremer Pigmente
Acetic acid (45741)	Sigma-Aldrich
Ammonia (64700)	Kremer Pigmente
Citric acid (27109)	Sigma-Aldrich
Sodium hydroxide (71690)	Sigma-Aldrich
Triammonium citrate (A1332)	Sigma-Aldrich
Pluronic® L64 (435449)	Sigma-Aldrich
Surfynol® 61 (78010)	Kremer Pigmente
Polyethylene glycol 4000	Sigma-Aldrich
Cyclomethicone D5 (87081)	Kremer Pigmente
Shellsol T (70460)	Kremer Pigmente

References

1. Balcar, N. et al. (2012) Studies in Cleaning Plastics. http://popart-highlights.mnhn.fr/active-conservation-of-plastic-artefacts/studies-in-cleaning-plastics/index.html. Accessed 9 Mar 2019.
2. Bartoletti et al. (2018) Reviving Whaam! The painting goes back on display. https://www.tate.org.uk/about-us/projects/nanorestart/reviving-whaam. Accessed 30 Mar 2019.
3. Chung JY, et al. (2017). An investigation of methods for surface cleaning unvarnished water-sensitive oil paints based on recent developments for acrylic paints. In ICOM-CC 18th Triennial Conference Preprints, Copenhagen, 4–8 September 2017, ed. J. Bridgland, art. 1309. Paris: International Council of Museums
4. Cooper A, Burnstock A, Van den Berg K, Ormsby B. (2014) Water Sensitive Oil Paints in the Twentieth Century: A Study of the Distribution of Water-Soluble Degradation Products in Modern Oil Paint Films. In: Van den Berg KJ et al (eds) Issues in contemporary oil paint. Springer, Dordrecht: 295–310
5. ECHA (2019) Annex XV Restriction Report. Proposal for a Restriction. www.echa.europa.eu/documents/10162/26cc3562-d0a0-1e09-b1eb-72f6bdf9309a. Accessed 9 Mar 2019.
6. Lee J, et al (2018). Scientific investigation into the water sensitivity of twentieth century oil paints, Microchemical Journal, 138: 282–295
7. Myers JM (2019) Analysis of cleaning efficacy and clearance of silicone-based Pickering-type emulsions used in the cleaning of water-miscible oil paints. These Proceedings.
8. Ormsby B, et al. (2019) Evaluating cleaning systems for use on water sensitive modern oil paints: a comparative study. These Proceedings, Chapter xx.
9. Ormsby B, et al. (2013). An empirical evaluation of a range of cleaning agents for removing dirt from artists acrylic emulsion paints. In: AIC specialty group postprints, 23rd edn: 77–87
10. Silvester G, et al. (2014) A cause of water-sensitivity in modern oil paint films: the formation of magnesium sulphate. Studes in Conservation 59(1):38–51
11. Soldano A and Van den Berg KJ (2014). Investigation into the surface conductivity of water-sensitive modern oil paints, In: Van den Berg KJ et al (eds) Issues in contemporary oil paint. Springer, Dordrecht: 407–417
12. Stangier, Susanne (2014). Fleurs Grises – A Delicate Surface. Difficult to Clean – Essential to Protect. In: Van den Berg KJ et al (eds). Issues in contemporary oil paint. Springer, Dordrecht: 205–212
13. Stravroudis C. (2012). More from CAPS3: Surfactants, silicone-based solvents, and micro-emulsions. In: WAAC Newsletter, 34(3): 24–27
14. Volk A and Van den Berg KJ (2014). Agar – A New Tool for the Surface Cleaning of Water Sensitive Oil Paint? In: Van den Berg KJ et al (eds). Issues in contemporary oil paint. Springer, Dordrecht: 389–406

Chapter 44
Evolon® CR Microfibre Cloth as a Tool for Varnish Removal

Michelle Vergeer, Klaas Jan van den Berg, Saskia van Oudheusden, and Maartje Stols-Witlox

44.1 Introduction

Varnish removal is one of the most frequently carried out conservation treatments on paintings, but can prove to be a challenging and high-risk activity, particularly when carried out on modern, solvent-sensitive paint surfaces. The solvent application technique can significantly affect the paint surface, through the choice of solvent itself and the associated mechanical action.

A relatively new addition to the conservator's tool box is Evolon® CR[1] microfibre cloth, which is a non-woven microfilament textile often used as a compress for varnish and/or overpaint removal with an appropriate solvent. Evolon® was first introduced in 2013 at the Institute for Conservation Science in Cologne, after it was investigated by the Staatliche Akademie der Bildenden Künste in 2006 [1, 10]. In the UK it was used for the removal of graffiti ink in the conservation treatment of a painting by Mark Rothko (2012–2014) [2]. It was introduced in the Netherlands soon after this, where its use spread quickly.[2] Presentations at the annual AIC meetings in 2017 and 2018 also demonstrated a growing interest in Evolon®, where both case-study treatments and one scientific study into its properties and behavior were presented [8, 9, 11].

The manufacturer describes Evolon® as a textile consisting of a mixture of polyamide and polyester fibres spun into segmented filaments. These filaments are uniformly spread out on a surface and then split and fixed into microfilaments using high pressure water jets. This creates a partially hydrophobic/hydrophilic material

[1] Hereafter called Evolon®. https://evolon.freudenberg-pm.com/en Accessed 19.3.19.

[2] One day conference 'Picture Meeting' held at the Dutch Cultural Heritage Agency in January 2016: 'Show and tell: practical paintings conservation issues'.

M. Vergeer (✉) · K. J. van den Berg · S. van Oudheusden · M. Stols-Witlox
University of Amsterdam, Amsterdam, The Netherlands
e-mail: info@vergeerschilderijenrestauratie.nl

K. J. van den Berg et al. (eds.), *Conservation of Modern Oil Paintings*,
https://doi.org/10.1007/978-3-030-19254-9_44

with strong absorbent characteristics [4], produced in thicknesses varying from 30 to 240 grams per m^2 [5]. The conservation materials supplier Deffner & Johann sells one type at 77,4 grams/m^2, which has been used in this research (Evolon® CR 2018) [3].

This study provides an overview of the current use of Evolon® by conservators and discusses underlying considerations.[3] The outcomes will help conservators to make informed choices regarding the use of Evolon® for varnish removal.

44.2 Methodology

To evaluate the use of Evolon®, nine interviews were conducted with conservators, in person or via Skype, which lead to a representative inventory of application methods. The 18 questions asked focused on experiences with Evolon® including application methods, advantages and disadvantages noted, treatment types and any circumstances where conservators would choose to either use or not use Evolon®. They were also asked their thoughts on this tissue for varnish removal compared to a cotton swab.

Based on the observations and concerns noted, experiments were then devised to assess the properties of Evolon® for varnish removal using the application methods described. Assessments focused on the efficacy of removal, effects on the paint surface and the amounts of solvent used in comparison to the amounts used with a cotton swab. Additionally, a test was developed to help visualize and compare the distribution of solvent for different application methods. The practical focus of this study, including visualization techniques, offers conservators useful information that is closely linked to conservation practice.

44.2.1 Case Study Assessments

The assessments were conducted on three oil paintings from the study collection of the Cultural Heritage Agency of the Netherlands (RCE) and the University of Amsterdam (UvA). Two of the paintings were dated to the twentieth century, with paint layers sensitive to polar solvents, making them particularly interesting for this study. The third was an unstretched nineteenth century canvas painting, which was not particularly solvent sensitive, but offered larger testing areas.

The first case study (Fig. 44.1) is a modern oil painting made by Ernst van de Wetering in 1962, which has been used for past research projects [6]. As part of this history, six different strips of varnish had been brush-applied in 1974, and part of

[3] Conducted by Michelle Vergeer, 3-month research project during a post-graduate traineeship in Conservation and Restoration of Paintings at the University of Amsterdam (UvA).

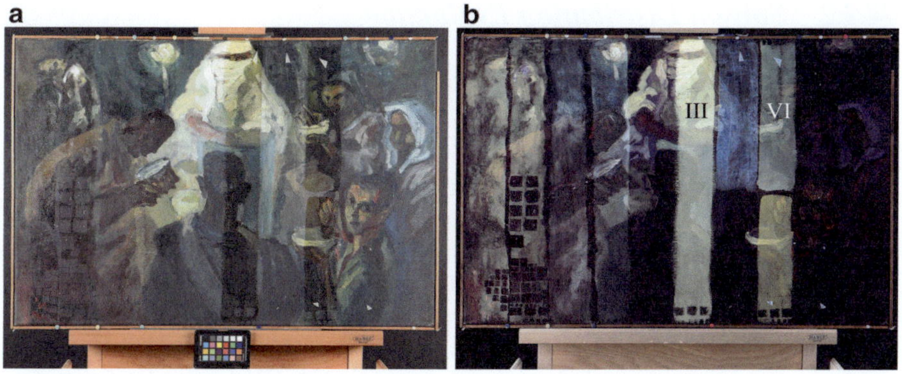

Fig. 44.1 Ernst van de Wetering, untitled, signed and dated 1962, canvas on a stretcher, 67 × 94.7 cm (**a**) daylight (**b**) UV induced fluorescence. Tests were performed on varnish strips III and VI

the painting had been left unvarnished [12]. From these six strips, two areas were selected for testing[4] based on their coherent gloss and strong UV fluorescence, which enabled the monitoring of the varnish removal process with both daylight fluorescent tubes and UV-induced fluorescence.

The second case study consisted of four remnants (strips of 3 × 25 cm) of a c ~40-year old[5] monochrome modern oil painting, which had been partially coated with dammar.[6] These monochrome surfaces, without any significant impasto, allowed easy comparison of the varnish removal test results with unvarnished areas.

For all case studies, the assessments were executed using ethanol after loose surface dirt was removed using a dry makeup sponge.[7] Ethanol was selected as it is commonly used by conservators, it successfully dissolved the varnishes on all case studies, and the results were more comparable when the same pure solvent was used. The removal test results were evaluated using the naked eye, microscopy (stereomicroscopy, Micro Reflectance Transformation Imaging (Micro-RTI), Hirox®3D Digital Microscopy) and photography in both UV induced fluorescence and with daylight fluorescent tubes.[8]

---

[4] [12]: Varnish strip III consists of 4 parts Venetian Turpentine, 2 parts sandarac, 8 parts oil of turpentine and 1-part nut oil. Varnish strip VI consists of 10 parts oil of turpentine and 5 parts mastic. Both recipes were made after recipes from the De Mayerne manuscript.

[5] Oral communication Klaas Jan van den Berg, October 2016.

[6] Varnished (April 2016) with multiple layers of dammar, then artificially aged 755 h in a dark room (50 °C, 30–80% RH, Vötsch VC0020) subsequently aged for 336 h in natural daylight next to a glass window.

[7] White polyurethane sponge, Deffner & Johann, Germany

[8] Single-Lens Reflex Camera (SLR): Nikon D750 with KINO-FLO 400 UV-lights and Broncolor Siros 800S RFS 2.1 flashlights. Stereo microscope: Leica MZ95, Hirox: a RH-2000 digital microscope with MXB-2500REZ dual illumination zoom lens and MXB-2016Z zoom lens (UV).

44.2.2 Solvent Distribution Evaluation

To relate the distribution of solvent to different application techniques, a visualization test was developed. Ethanol was stained with a 0.1% w/w fluorescein isothiocyanate (FITC) protein indicator dye which has a bright yellow/green UV-induced fluorescence, introduced for staining proteinaceous layers in paint cross-sections by Wolbers [13].

Large petri dishes were prepared with a substrate of gelatin (5% w/w) in a deionized water gel. The FITC-stained ethanol was applied to Evolon® strips using the chosen techniques (as described by conservators) and left for 30 s on the gelatin substrate. This was sufficient for the FITC to stain the gelatin substrate. In total, 60 tests were carried out in a dark room illuminated with UV lamps. Each test was immediately documented, filmed and photographed with a digital camera, both during and after removal of the Evolon®.

44.3 Results

44.3.1 Conservator Interviews

The interviews revealed that conservators use Evolon® in several different ways (Fig. 44.2). This Figure illustrates that the main variations center around the method of application of solvent to the Evolon®. After solvent application, some conservators also covered the Evolon® with Melinex® and/or applied additional mechanical action such as gentle rubbing over the surface. Conservators use Evolon® both on canvas and panel supports, however the majority of applications were carried out on relatively smooth paint surfaces; many considered Evolon® unsuitable for impasto surfaces as they considered it to be too rigid to follow the paint texture.

The conservators also noted that Evolon® has a strong absorbency and that less mechanical action was required with its use. They also remarked that they tended to use less solvent when compared to cotton swab cleaning and were able to remove the varnish more quickly. These aspects are all potentially advantageous when cleaning solvent sensitive paint surfaces. The application method can also be systematically repeated (depending on local solvent sensitivity variations) across an entire surface, using the same amount of solvent and application time; and multiple conservators can work on the same painting using this system without noticeable differences between hands. The used Evolon® can then be saved for future reference and serve as a record, for example, of the exact location of overpaint removed.

The conservators also described cutting Evolon® into different sizes and shapes; some preferred squares varying in size from ~2 × 2 cm up to 40 × 40 cm. Others cut pieces to irregular shapes, following the contours of the painted composition, which can limit the visibility of tidelines - one of the general disadvantages of using tissue. As an ideal scenario, one could avoid tidelines by using a piece of Evolon® as large

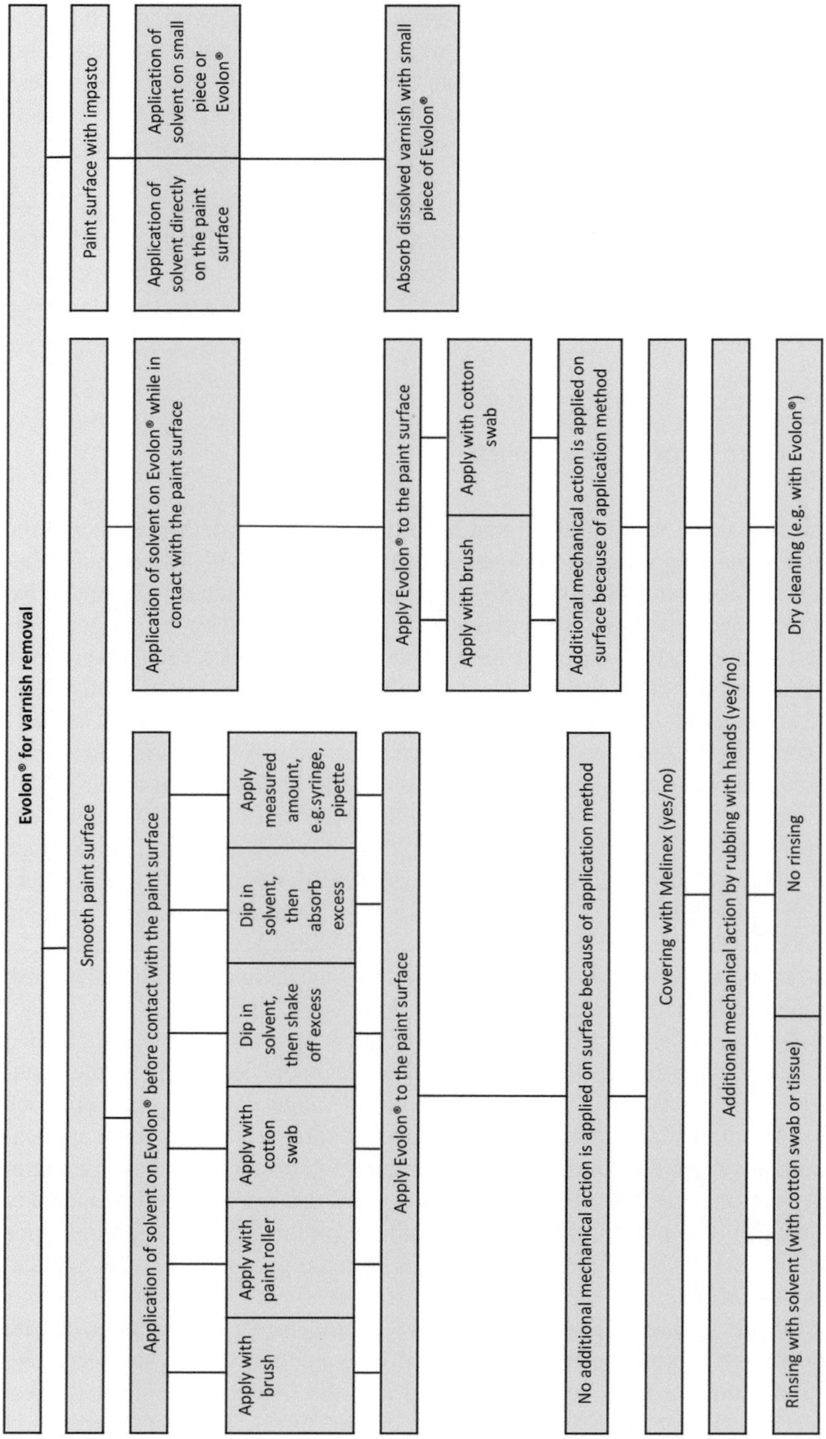

Fig. 44.2 Overview of application methods of Evolon® as described in 9 interviews

as the entire painted surface, however the opaqueness of the material, which is considered its main disadvantage, and variations in paint sensitivity across a work of art necessitates due caution. Some conservators mentioned that they noticed the pattern of the weave structure of Evolon® on the surface of the painting after use. This is thought to be caused by residual varnish/surface dirt and was also observed during tests on the case studies.

Even though the high absorbency of Evolon® is considered an advantage, it can also be a cause for concern, as components of the paint layer may be extracted through leaching whilst the paint film is swollen with solvent. This has yet to be explored, and remains a concern relating to the general use of solvents, as has been described by many (e.g. [7]).

44.3.2 Case Study Assessments

The varnish removal tests revealed that even though the Van de Wetering and the monochrome painting remnants showed sensitivity to polar solvents (such as ethanol), both surfaces could be successfully cleaned using Evolon® techniques. No pigment particles were picked up and no tissue fibers remained on the surface. This contrasted with the swab cleaning tests, where small particles of pigment were picked up after a few rolls, the surface was not fully cleaned, and both varnish residues and cotton swab fibres remained behind on the paint surface (Fig. 44.3a, b).

This confirms conservators' observations on the suitability of Evolon® for cleaning sensitive paint surfaces, however, the decreased mechanical action required with Evolon® may also result in less efficient soiling removal. In some cases, a non-fluorescent deposit (possibly residual surface dirt) was noted on the paint surface after varnish removal with Evolon®. In one test, the imprint of the Evolon® structure was visible in this deposit, however this was easily removed with a dry make-up sponge. As was described in the interviews, Evolon® turned out to be unsuitable for impasto-rich paint surfaces as it is too rigid to follow the texture, resulting in varnish residues pooled around impasto brushstrokes (Fig. 44.3c, d).

By weighing the amount of ethanol used for each varnish removal method it became clear that the amounts required varied considerably depending on the application technique. Tests performed on the nineteenth century painting (area of each test ~13 x 45 cm) revealed that for a similar level of varnish removal, some application techniques required almost double the amount of solvent as others. For example, with a cotton swab 2,87 grams of solvent was needed per 100 cm^2, compared to 1,55 grams per 100 cm^2 for an Evolon® technique (in this case, by applying twice the weight of Evolon® of ethanol on the tissue with a pipette before tissue application to the surface).

Even though the use of Evolon® requires less solvent, the area of contact with the paint surface is much larger than achieved with a cotton swab. One interesting observation is that the more solvent was used with the Evolon®, the more fluorescent material was transported to the reverse of the canvas. This was particularly

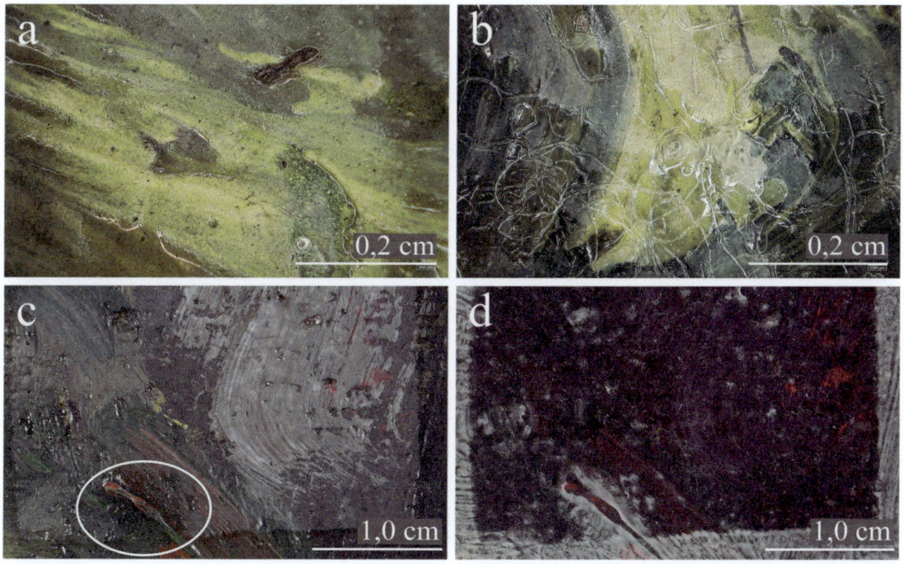

Fig. 44.3 Comparison between results of varnish removal with Evolon® and cotton swab on the Van de Wetering painting. (**a**) With Evolon®: few residues. (**b**) With cotton swab: many fibre/varnish residues. (**c** and **d**) With Evolon®: varnish residues surrounding the impastos (c daylight, d UV)

noticeable when the tissue was fully saturated with solvent (2,64 grams of solvent per 100 cm^2). In this case, the saturated Evolon® was unable to absorb more material, so the excess solvent was transported to the reverse of the canvas.

44.3.3 Solvent Disttibution Evaluation

FITC dye-based assessments of the application techniques lead to the general observation that when more solvent is used with Evolon®, the solvent spreads out over a larger surface area, extending beyond the contact zone. Spreading also occurs when mechanical action is applied (e.g. rubbing over the Evolon® on the surface), when the Evolon® is covered with Melinex® and when contact time is increased (Fig. 44.4 a–d). These tests also provide information about the evenness of solvent application and spreading. Particularly when the solvent was applied after the Evolon® was placed onto the gelatin surface, an uneven distribution of solvent was noted. In some tests an imprint of the pattern of the Evolon® tissue weave was also observed within the stain. This seems to be the consequence of the slightly uneven weave structure of Evolon®, which leads to some areas having a longer contact time with the tissue, resulting in stronger staining (Fig. 44.4c). Such factors need to be taken into consideration when selecting an appropriate application method and require further exploration.

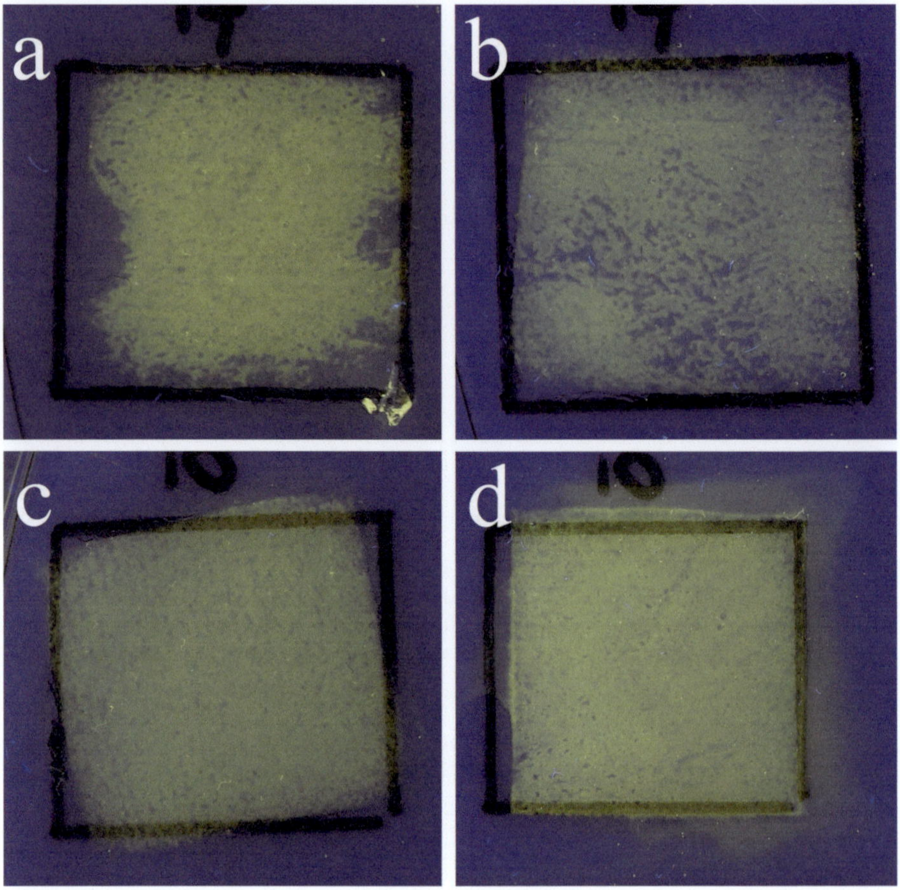

Fig. 44.4 Gelatine surface (UV photograph), after application of Evolon® tissue of 1.5 × 1.5 cm with two drops of ethanol
(**a**) No Melinex®
No mechanical action,
(**b**) With Melinex®
No mechanical action,
(**c**) No Melinex®
With mechanical action,
(**d**) With Melinex®
With mechanical action
The pictures show an increase in solvent spreading from a-d. In c, the pattern of the weave structure of Evolon® is visible as vertical lines

44.4 Conclusions

This research explored the practical use of Evolon® for varnish removal from a range of oil paintings. Interviews with conservators provided an overview of experiences and ways of using Evolon®. The interview results then guided the experimental set-up used to investigate some of the observations made by conservators in more depth. The range of assessments undertaken confirmed that Evolon® is potentially advantageous for varnish removal from modern solvent-sensitive painted surfaces with little impasto, as less mechanical action is required when compared to cotton swabs.

In general, the use of Evolon® required less solvent. However, tests on case studies investigating the influence of application method on the amount and distribution of solvent, showed considerable variations. Evaluations using FITC-stained ethanol proved to be a useful way to visualize solvent distribution and clearly demonstrated variations with different application techniques. Hence the choice of application method strongly influences the results regarding solvent distribution to the painting surface.

In addition to knowing whether Evolon® can absorb solvent extractable paint components, this research has indicated that the penetration of fluorescent material (likely solubilized varnish) to the reverse of the canvas requires attention, as this may have undesirable long-term consequences for the degradation of the canvas, aside from drawing varnish components through the paint structure.

Acknowledgments This research was conducted at the Cultural Heritage Agency of the Netherlands (RCE) and the University of Amsterdam (UvA) and is part of IPERION-CH www.iperionch.eu (7.1c). The authors would like to thank everyone involved: Maarten van Bommel (UvA), Katrien Keune (Rijksmuseum Amsterdam, RMA/UvA), Kate Seymour (Stichting Restauratieatelier Limburg, SRAL), Han Neevel (RCE), Sofia Johansson (former trainee at RMA) and all interviewed conservators: Gwendoline Fife (SRAL), Kathrin Kirsch (Restauratieatelier Amsterdam), Anna Krekeler (RMA), Mireille te Marvelde (Frans Hals Museum), Kari Rayner (former trainee at Hamilton Kerr Institute), Julie Ribits (former trainee at Mauritshuis), Marjan de Visser (Restauratieatelier Marjan de Visser), Lisette Vos (RMA) and Nienke Woltman (RMA).

References

1. Albrecht B (2013) Der Blick ins Freie: Konservierung und Restaurierung von fünf Gemälden des Bernhard Gottfried. Master thesis, Cologne Institute of Conservation Sciences
2. Barker R, Ormsby B (2015) Conserving Mark Rothko's *Black on Maroon* 1958: The construction of a representative sample and the removal of graffiti ink. In: Tate Papers, no. 23, Spring 2015. https://www.tate.org.uk/research/publications/tate-papers/23/conserving-mark-rothkos-black-on-maroon-1958-the-construction-of-a-representative-sample-and-the-removal-of-graffiti-ink. Accessed 5 Mar 2019

3. Evolon® CR – Deffner & Johann (2018) https://www.deffner-johann.de/evolon-cr-en.html. Accessed 5 Sept 2018
4. Evolon® technology: revolutionary, unique, invented by Freudenberg (2018) https://evolon.freudenberg-pm.com/evolon_technology/technology. Accessed 5 Sept 2018
5. Evolon® Technical Data (2016) Freudenberg Performance Materials, Colmar
6. Lamers M (2016) The contribution of selected analytical techniques in monitoring surface effects caused by varnish removal from a modern oil painting. Unpublished article, University of Amsterdam
7. Phenix R, Wolbers R (2012) Removal of varnish: organic solvents as cleaning agents. In: Hill S, Rushfield A (eds) Conservation of easel paintings. Routledge, New York, pp 524–554
8. Rayner K (2017) The Evolution of a Method: Optimizing the Use of Evolon® CR to Poultice Varnish on a Large Scale. In: Abstract book 45th Annual Meeting AIC, p 107
9. Ribits J (2017) Lifting the Microfiber Veil: Utilizing Evolon® fabric at the Mauritshuis to remove aged varnish from Hendrick Heerschop's A Visit to the Doctor. In: Abstract book 45th Annual Meeting AIC, p 114
10. Schink A (2006) Vergleichende Untersuchungen zur Verwendung von Microfasertüchern für die Oberflächenreinigung von Gemälden. Thesis, Staatliche Akademie der Bildende Künste
11. Smelt S (2018) Evolon®: Its Use from a Scientific and Practical Conservation Perspective. In: Abstract book 46th Annual Meeting AIC, p 77
12. Stols-Witlox M (1995) Slotvernis voor olieverfschilderijen. Recepten voor slotvernis in de Nederlandse bronnen gedurende de zeventiende tot en met de negentiende eeuw. Thesis, University of Leiden
13. Wolbers R (2000) Cleaning painted surfaces, aqueous methods. Archetype Publications, London

Chapter 45
Can Cellulose Beads Save *The Circling of the Planets*? Cellulose-Based Consolidating Filler to Stabilise Lifted Brittle Flakes on a Large Painting

Karolina Soppa, Anita Hoess, Matthias Läuchli, Silja Meyer, Thomas Geiger, Nadim C. Scherrer, Stefan Zumbühl, and Tobias Haupt

45.1 Introduction

Artists sometimes modify modern oil paint to achieve specific effects. Unfortunately, the outcome of such modifications is not always predictable. In the case of Augusto Giacometti's *The Circling of the Planets* („Das Kreisen der Planeten", 1907) at the Kunsthaus Zürich (Fig. 45.1), the artist himself was confronted with the problem of flaking paint on the painting soon after completion. In the following years, the lean tempera (see Fig. 45.1b, c) on non-absorbent oil-rich *tempera grassa* has gone through repeated interventions of consolidation with wax, polyvinylacetate, polyamide, BEVA 371™, and probably proteinaceous glues. Now, it is again in poor condition (Fig. 45.1b). The paint has reached a detrimental state of degradation: it is stiff, fragile, sometimes porous and thus highly absorbing (Fig. 45.1c). Moreover, in some areas, due to absorption of former high refractive index consolidants, the paint has darkened significantly, particularly in the blue paint areas. Furthermore, widespread lifting of paint flakes (~ 0.2–5 mm) is observed on a smooth, non-polar ground (Fig. 45.2a), again mainly restricted to the blue paint areas. Securing the detaching lifted paint flakes with common consolidation techniques was not

K. Soppa (✉) · S. Meyer · N. C. Scherrer · S. Zumbühl
Bern University of the Arts, Bern, Switzerland
e-mail: karolina.soppa@hkb.bfh.ch

A. Hoess · M. Läuchli
Läuchli Eysler Hoess – Ateliergemeinschaft Konservierung Restaurierung, Wallisellen, Switzerland

T. Geiger
Cellulose and Wood Materials, Swiss Federal Laboratories for Materials Science and Technology (Empa), Duebendorf, Switzerland

T. Haupt
Kunsthaus Zürich, Zürich, Switzerland

K. J. van den Berg et al. (eds.), *Conservation of Modern Oil Paintings*,
https://doi.org/10.1007/978-3-030-19254-9_45

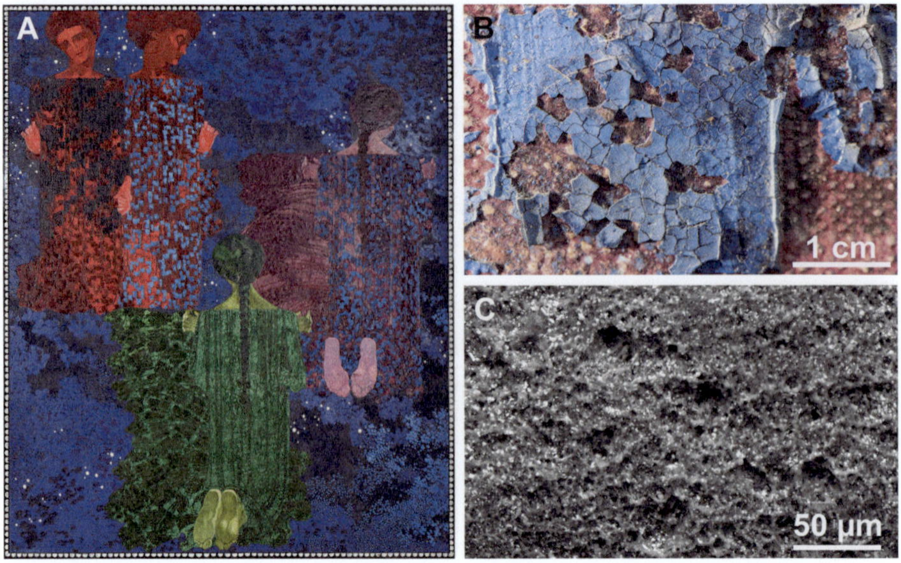

Fig. 45.1 (**a**) Augusto Giacometti, 1877–1947, *The Circling of the Planets* („Das Kreisen der Planeten"), 1907, 268.0 x 218.0 cm, lean tempera and tempera grassa on canvas, Kunsthaus Zürich, 1967, © Kunsthaus Zürich

(**b**) Before conservation: Substantial material losses, highly absorbent and lifted flaking paint of the cobalt blue lean tempera on partially hydrolyzed and saponified ground (FTIR analysis), © Kunsthaus Zürich

(**c**) Porous surface at the bottom of the cobalt blue lean tempera flake (SEM-SE), © Kunsthaus Zürich

feasible due to the rigidity of the material. In order to solve this technical problem, our approach was to stabilize the rigid paint flakes by filling the gaps between ground layer and flake using a custom water-based filler preparation having methyl cellulose (MC) as the binder. Methyl cellulose is the most non-polar type of commercial cellulose derivatives that is still water-soluble; it is available in different molecular weights. The method has proven suitable for porous, highly absorbing lean tempera paint, as well as for solid *tempera grassa* paint and ground. To fill larger volumes, as present in this object, however, the addition of one or more suitable bulking agents was essential. Furthermore, a colorant to mask the whitish appearance of the raw consolidating filler in the main colour area, was added to the aqueous methyl cellulose solution to avoid additional retouching.

The potential bulking agents must meet certain criteria. They should be lightweight, micron-sized, compatible with methyl cellulose, facilitate the fluidity of the filling, and exhibit adequate long-term stability, similar to the binder. The filled binder solution must be of a high viscosity with minimal shrinkage. Furthermore, the consolidating filler should be applicable using the smallest tip size of a micro-dispensing system (0.11 or 0.15 mm inner tip diameter) to guarantee precise placement of the high viscosity material under every detached and lifted flake. The smallest, lightweight spherical filler commonly used in conservation treatments,

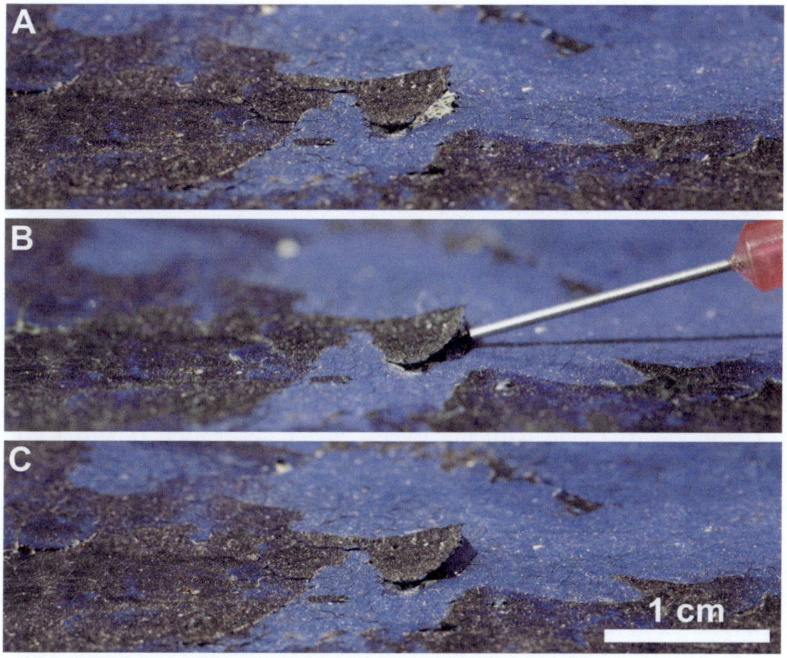

Fig. 45.2 (**a**) Flaking paint, (**b**) Application of the filling mixture with the micro-dispenser and (**c**) the coloured filling between flake and ground, © Kunsthaus Zürich

3M™ *Scotchlite*™ *S22* glass bubbles (30–35 μm in diameter), is too large for the small size micro-dispensing tips. Furthermore, they break easily under applied pressure. Cellulose beads made of regenerated cellulose, in turn, are lightweight, round, white coloured, hydrophilic, dimensionally stable and, available in diameters of 5, 10, 30, and 50 μm. Additionally, in order to increase the bond strength of the methyl cellulose/beads mixture, cellulose fibres should be admixed. The shortest-fibre cellulose product commonly used in conservation, *Arbocel® BWW40*, contains fibres that are too long (200 μm) for the smallest size tips (110–150 μm) of the micro-dispensing system. For this reason, we selected a shorter fibre micro-crystalline cellulose product *Vivapur® CS 4 FM* (hereafter described as VP4) with a stated fibre length of 1–7 μm. To gauge the shrinkage and the adhesion of the cellulose beads, we decided to compare them to the S22 glass bubbles commonly used in conservation.

45.2 Materials and Methods

Following a thorough characterisation of the ground, former conservation materials, pigments and the binding media of the original paint by Fourier-transform infrared spectroscopy (FTIR), we focused on investigating the absorbency and topography of the lean blue tempera paint using optical microscopy and scanning electron microscopy with secondary electron imaging (SEM-SE).

45.2.1 Materials and Application of the Consolidating Filler Mixture

Different cellulose fillers were tested to find the best bulking agent for the consolidating filler with methyl cellulose as the binding medium. Cellulose beads are rounded, white coloured and prepared via the dissolution, shaping and regeneration of cellulose [2]. They typically find commercial use in cosmetic products. We chose the *Cellulobeads D-10*, which are the smallest cellulose beads of the group in the *Cellulobeads* range with the lowest water absorption (Daito Kasei data sheet 2013). Their density is 0.71 g/cm^3, and according to the data sheet they are <15 μm in diameter (Fig. 45.4a, b). *Cellulobeads D-30* (average diameter 30 μm) were also tested for direct comparison with 3M™ *Scotchlite*™ *S22* glass bubbles. The other kind of organic material included in the test formulations of filler components was the fibrous microcrystalline cellulose, *Vivapur® CS 4 FM*, which is white coloured and has a density of 1.5 g/cm^3, with a particle size of 1–7 μm. The 3M™ *Scotchlite*™ *S22* glass bubbles were used for a second test paste. This material is transparent with a density of 0.19 g/cm^3. Two methyl cellulose types served as binding media: *Benecel*™ *A4C* (3% w/w in deionised water) and *Benecel*™ *A4M* (2% and 2.5% w/w in deionised water). The solutions were prepared with the hot-cold method and were stored refrigerated for >12 hours. Additionally, the consolidating filler needed to show a similar colour as the original blue, red, light and dark green, and black paint areas in order to mask the whitish cellulosic appearance of the raw formulation in order to avoid additional retouching of all fillings. Accordingly, the filling colour was adjusted variously using the pigments phthalocyanine blue 515,400, PB15:4, Irgazin Rubin DPP-TR, PR264, chromium oxide green, PG17, phthalocyanine green, PG7, and lamp black, PBk7. The filler mixtures were blended with the aid of a Hauschild SpeedMixer™. Finally, the filling paste was applied under the lifted flakes in a controlled and time-efficient way using a micro-dispenser.

45.2.2 Analytical and Test Methods

All filling materials used were thoroughly characterised. Particle size analysis was performed with a laser diffraction particle size analyser. The morphology of the filler materials was examined by scanning electron microscopy. The composition and crystallinity of the different cellulose fillers were investigated by FTIR spectroscopy. Dynamic Vapour Sorption (DVS) measurements were carried out in order to test the water vapour uptake of the different bulking agents (the commonly used *Arbocel® BWW40* in comparison to the new cellulosic fillers) as well as the final filling mixture.

45.2.3 Tests for Tensile Strength and Shrinkage of the Consolidating Filler

For the tensile strength test, we produced separate test flakes (1 cm^2) to simulate the lean tempera paint. Chalk ground showed an absorbency similar to the lean tempera paint. Therefore, chalk ground (chalk + 5% w/w gelatine type A, 180 Bloom; 8:5 w/w) was used in preliminary tests as a model for the lean tempera, while alkyd paint was chosen to simulate the non-polar (oily)ground and paint. The drying time for the filling was 1 week at ambient conditions of 50 ± 5% RH and 22 ± 2 °C. Using a tensile testing machine (described in [3]), the tensile strengths of *Benecel™ A4M*, 2% and three mixtures at different ratios of MC/cellulose beads/microcrystalline cellulose were investigated according to the method described in [1]. Ten samples were tested in each test series.

In order to investigate the shrinkage upon drying of the mixtures, the film thickness was determined optically using a calibrated Leitz DMRB microscope.

45.3 Results and Discussion

45.3.1 Filling Materials

The results of the particle size analysis showed a very narrow size distribution for the D-10 cellulose beads; they were only slightly larger than the D-5 grade. Most *Vivapur* fibre lengths were in the range of 2–20 μm (Fig. 45.4a). 3M™ *Scotchlite™ S22* glass bubbles and D-30 cellulose beads had similar size distributions (Fig. 45.4a). However, SEM-SE imaging revealed an irregular, elongated, oval-like form instead of a spherical shape for the D-30 cellulose beads (Fig. 45.4b). FTIR spectroscopy confirmed the presence of the cellulose spheres by detecting regenerated cellulose, as well as for *Arbocel® BWW40* and crystalline cellulose for *Vivapur*. No other compound could be detected. Regenerated cellulose has fewer crystalline regions (Cellulose II) and thus is more amorphous, exhibiting greater swelling than highly crystalline native cellulose (Cellulose I). DVS measurements showed lower water uptake for the D-10 cellulose beads than *Arbocel® BWW40* cellulose fibre, but the uptake progressed more rapidly, especially at levels of 40–70% RH (Fig. 45.4f).

45.3.2 Film Formation Properties and Adhesion

In a first step, the mechanical properties of binary systems of methyl cellulose binder and filler was tested (Fig. 45.4a–c). To characterise the film formation properties, the material was applied on Melinex and glass slides using a film application tool (120–140 μm wet film thickness). After drying, all films containing pure cellulose beads of different particle sizes cracked or curled up on the Melinex and exhibited high cohesion. This effect was also influenced by the type of methyl cellulose. In contrast to *Benecel™ A4M* (2%), the A4C (3%) showed lower shrinkage (pure MC-films). In contrast to these binder-only films, films containing S22 glass bubbles had no cracks, and shrinkage was not influenced by the binding medium. However, in comparison to all cellulosic fillers (beads and fibres), the cohesion was comparably weak (Fig. 45.4e). The third system containing microcrystalline cellulose (A4M 2%: VP4; 5: 1.5) was very viscous and resulted in a flexible film with hardly any cracks. Furthermore, these pre-tests showed, that it is important to calculate the binder content in a specific manner. Additionally, cellulose beads absorb water, whereas glass bubbles are inert. The observed reduced shrinkage of the A4M methyl cellulose combination with cellulose beads led to the decision to continue with this grade of methyl cellulose.

Pure *Benecel™ A4M* 2% and *Benecel™ A4M 2% + Cellulobeads D-10* or S22 glass bubbles achieved almost the same tensile strength (Fig. 45.3). Combinations of D-10 and VP4 with three different binding media ratios, however, exceeded these results by far and were able to pass the 0.11 mm or 0.15 mm tips of the microdispenser (Fig. 45.2). The mixture with phthalocyanine blue, PB15:4 (0.5 g) achieved also a very high adhesive strength as did other coloured mixtures.

The filler mixtures with D-10 cellulose beads as well as microcrystalline cellulose VP4 (always 1.5: 0.5) were tested with three different methyl cellulose A4M ratios. As expected, the greater the amount of binding media, the higher the shrinkage: (a) **4**: 1.5: 0.5, 111 ± 6.4 μm; (b) **5**: 1.5: 0.5, 98 ± 3.1 μm; (c) **6**: 1.5: 0.5, 77 ± 2.7 μm. Since mixture (b) was easily applicable with the smallest microdispensing tip and showed a low shrinkage, this material combination was regarded as the most compliant mixture with an addition of phthalocyanine blue, PB15:4 (film thickness 102 ± 2.5 μm, Fig. 45.4d). The cohesion of these films was very high. The films peeled off the Melinex and curled up, but had fewer cracks than films with pure cellulose beads. Thus, it could be concluded that, before using MC fillings with pure cellulose beads or in combination with microcrystalline cellulose, it is strongly recommended to run tests to judge the behaviour of larger volumes.

After evaluation of this mixture, the water sorption behaviour was tested. Dynamic Vapour Sorption measurements revealed that the combination of regenerated cellulose beads, microcrystalline cellulose and methyl cellulose resulted in less water uptake than for regenerated cellulose only. The water uptake of the filler formulation between 40–70% RH was similar to *Arbocel® BWW40* initially, but reduced further on at higher RH values (Fig. 45.4f).

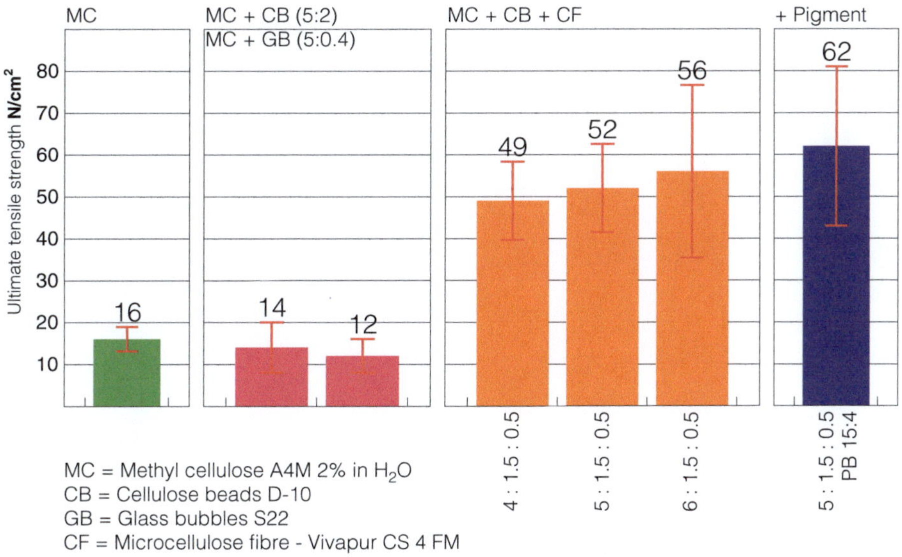

Fig. 45.3 Results of the ultimative tensile strength (Zwick, Z2.5, testing speed 50 mm/min, the results are presented above with the calculated averages and standard deviations) on a model: Alkyd paint flakes represented the ground and chalk ground flakes the absorbent paint layer

45.3.3 Final Mixture Preparation and Application

Based on the results of the individual test systems, a final consolidating filler mixture for the painting was produced. Preparation of a blue-coloured filling is outlined exemplarily: first, 0.4 g of phthalocyanine blue PB15:4 was pre-wetted for 2 minutes in approximately 10 drops of 2-propanol. Then, 3 g of deionised water was added and the paste was homogenised using the SpeedMixer™ (30 sec, 1200 rpm), and filtered 5 times through a mesh (120 μm). 1.87 g of the pigment suspension were then added dropwise to 3.33 g of 3% *Benecel*™ *A4M* and homogenised again with the SpeedMixer™ (60 sec, 3500 rpm). Now, the bulking agents, 1.5 g D-10 cellulose beads, and 0.5 g microcrystalline cellulose were added and stirred in with the SpeedMixer™ (60 sec, 3500 rpm). At the end, the resulting mixture was approximately 2% A4M; it was highly homogenised, and agglomerations of phthalocyanine blue were rare under microscopic observation. Once prepared, the mixtures were filled into a syringe and stored in a fridge. If kept under seal, it could be stored for several days. Open storage was more problematic. At room temperature some mixtures showed mould growth in the beaker after two weeks, but never in a syringe. After one week, the first mixture started to separate. We observed that more rigid fillings maintained a homogeneous state over a longer period.

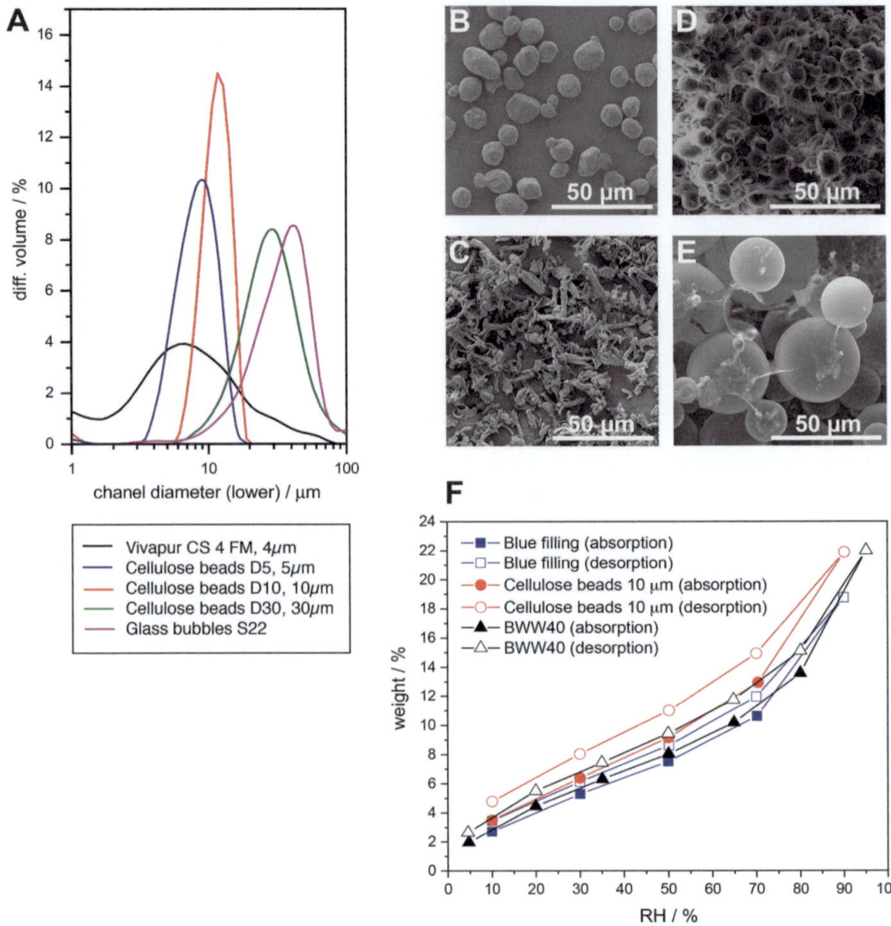

Fig. 45.4 (**a**) Particle size distribution of *Cellulobeads D-10*, *Cellulobeads D-5*, *Vivapur™ CS 4 FM*, and S22 glass bubbles; (**b**) *Cellulobeads D-10* (SEM-SE); and (**c**) *Vivapur™ CS 4 FM* (SEM-SE); (**d**) Final filling mixture after tensile strength test (SEM-SE); (**e**) S22 glass bubbles and MC A4M after tensile strength test (SEM-SE); (**f**) DVS of *Arbocel® BWW40*, D-10 cellulose beads, and the blue final filling mixture

The final mixture was applied to vulnerable areas of flaking paint using the above-mentioned micro-dispensing system with pedal control. The pressure used was 3–6 bar. The tips of the syringe were always immersed in water when not in use to prevent clogging. In most cases the fillings dried quickly and remained homogenous (Fig. 45.2b, c). The amount used on the whole painting was approximately 30 g.

45.4 Conclusion

For reasons of applicability with the micro-dispensing system, tensile strength and the lowest shrinkage, best performance was achieved with the binder system *Benecel*™ *A4M* methyl cellulose, *Cellulobeads D-10*, and *Vivapur*™ *CS 4 FM* microcrystalline cellulose at a ratio 5: 1.5: 0.5 plus pigments. The more microcrystalline cellulose fibres present in the mixture, the more flexible it was and fewer cracks were visible in the dry film. The results are very promising and offer a new strategy for treating delicate cases such as this highly fragile and large piece by Augusto Giacometti.

Acknowledgements With special thanks to Ars Rhenia (project finance). Further thanks to: Laura Conti (Empa), Marcel Giger (H. Sigrist & Partner AG), Anja Huch (Empa), Sarah Lämmlein (Empa), Hanspeter Marty (Kunsthaus Zürich), Kerstin Mürer (Kunsthaus Zürich), Brigitte Lienert-Pärli (Bern University of the Arts) and Jörg Scheller (Zurich University of the Arts).

Notes

1. For materials and application see also [3].

Materials

Cellulobeads, D-5, D-10, D-30 (Daito Kasei Kogyo Co.)
Vivapur™ *CS 4 FM* microcrystalline cellulose (J. Rettenmaier & Söhne, JRS) https://www.jrs.eu/jrs_en/jrs-the-company/product-overview/vivapur.php
Arbocel® BWW40 (Kremer); 3M™ *Scotchlite*™ *S22* glass bubbles (Kremer)
https://www.3m.com/3M/en_US/company-us/all-3m-products/~/3M-Glass-Bubbles-S22/?N=5002385+3292670797&rt=rud
Benecel™ *A4C* and *A4M* methyl cellulose (Ashland)
Pigments: phthalocyanine blue, PB15:4 (Heubach); Irgazin Rubin DPP-TR, PR264 (Kremer); chromium oxide green, PG17 (Gamblin); phthalocyanine green, PG7 (Kremer); and lamp black, PB7 (Kremer)

Equipment and Instrumentation

SpeedMixer™, DAC 150.1 FVZ (Hauschild & Co KG).
Micro-dispenser, JBE1113 and C100-4.0 (H. Sigrist & Partner AG). http://www.dosiersysteme.ch/documents/JBE_1113.pdf
Precision weighing device (Sartorius, Practum224-1S).
Micrometer, sensitivity 1 μm (Lorentzen & Wettre Kista).
Particle size analysis, LS 13320 (Beckman Coulter, Indianapolis, USA).

Scanning electron microscopy (Nova NanoSEM 230 instrument, FEI, Hillsboro, Oregon, USA) at an accelerating voltage of 5 keV. The cellulose beads were deposited on a holder and sputtered with 7 nm platinum (BAL-TEC MED 020 Modular High Vacuum Coating Systems, BAL-TEC AG).

FTIR (Perkin Elmer System 2000) with single element LN-MCT detector (settings: 16 scans/4 cm^{-1} resolution/range 4000-580 cm^{-1}).

Dynamic Vapor Sorption measurements were carried out with a VTI-SA+ dynamic vapour sorption analyser (TA-Instruments, New Castle, USA). A sample weighting ~10 mg was placed in the sample cup, and the precise weight of the sample compared to the reference was registered at a resolution of 0.1 μg, while the relative humidity (RH) surrounding the cups was controlled. The sample was first dried at 90 °C and 0% RH for eight hours, followed by equilibration at 23 °C. Then the humidity conditions were set to 10%, 30%, 50%, 70%, 90%, 70%, 50%, 30% and 10%, respectively.

Tensile testing machine, Zwick 1120 (Zwick GmbH &Co) – testing speed 50 mm/min.

References

1. Comiotto, A. 2009. 'Fine-tuning atmospheric plasma to improve the bondability and coatability of plastics in modern and contemporary art and design', in T. Bechthold (ed.), *Future Talks 009: The Conservation of Modern Materials in Applied Arts and Design*. Munich: Die Neue Sammlung, 33–39.
2. Gericke, M., Trygg, J. and Fardim, P. 2013. 'Functional Cellulose Beads: Preparation, Characterization, and Applications.' *Chemical Reviews* **113,** 4812–4836.
3. Soppa, K., Laaser, T., Krekel, C., Genton, M. and Seidel, T. 2014. 'Adhesion and penetration of sturgeon glue and gelatines with different Bloom grades', in J. Bridgland (ed.), *Preprints of the ICOM-CC 17th Triennial Conference, Melbourne, 15-19 September 2014*, art. 1313. Paris: International Council of Museums. [https://www.icom-cc-publications-online.org/publicationDetail.aspx?cid=2d3b3fbd-537a-47eb-b0dd-0c2062031706]

Correction to: Conservation of Modern Oil Paintings

Correction to:
K. J. van den Berg et al., *Conservation of Modern Oil Paintings*, https://doi.org/10.1007/978-3-030-19254-9

The book was inadvertently published with an incorrect editor affiliation for Prof. Heydenreich as:

Gunnar Heydenreich
FH Köln
Cologne Institution for Conservation Science
Köln, Nordrhein-Westfalen, Germany

It should read instead:

Gunnar Heydenreich
TH Köln
Cologne Institute of Conservation Sciences
Köln, Germany

The affiliation has been updated in the book.

The updated version of the book can be found at
https://doi.org/10.1007/978-3-030-19254-9

K. J. van den Berg et al. (eds.), *Conservation of Modern Oil Paintings*,
https://doi.org/10.1007/978-3-030-19254-9_46

MIX
Papier aus verantwortungsvollen Quellen
Paper from responsible sources
FSC® C105338

If you have any concerns about our products,
you can contact us on
ProductSafety@springernature.com

In case Publisher is established outside the EU,
the EU authorized representative is:
Springer Nature Customer Service Center GmbH
Europaplatz 3, 69115 Heidelberg, Germany

Printed by Libri Plureos GmbH
in Hamburg, Germany